# The Mycota

## A Comprehensive Treatise on Fungi as Experimental Systems for Basic and Applied Research

## Volume 13

The fungi represent a heterogenous assemblage of eukaryotic microorganisms and have become favored organisms for research at the cellular and molecular level. Such research involvement has been stimulated by interest in the biotechnological application of fungi in processes related to industry, agriculture and ecology. Considering both yeasts and mycelial fungi, THE MYCOTA highlights developments in both basic and applied research and presents an overview of fungal systematics and cell structure. Foremost authorities in research on mycology have been assembled to edit and contribute to the volumes.

All volumes/chapters are systematically reviewed by the series editor(s) and respective volume editor(s).

Minou Nowrousian • Jason E. Stajich
Editors

# Fungal Genomics

Third Edition

Springer

*Editors*
Minou Nowrousian
Dept. of Molecular & Cellular Botany
Ruhr-University Bochum
Bochum, Germany

Jason E. Stajich
Dept. of Plant Pathology &
Microbiology
University of California-Riverside
Riverside, CA, USA

ISSN 2945-8048 ISSN 2945-8056 (electronic)
The Mycota
ISBN 978-3-031-75665-8 ISBN 978-3-031-75666-5 (eBook)
https://doi.org/10.1007/978-3-031-75666-5



Editorial Contact: Sabine Schwarz
This Springer imprint is published by the registered company Springer Nature Switzerland AG
The registered company address is: Gewerbestrasse 11, 6330 Cham, Switzerland

# Volume Preface to the Third Edition

Since the second edition of Volume XIII of *The Mycota* in 2014, the already vibrant and fast-developing field of fungal genomics has further broadened its impact through a widening field of omics-based applications that resulted in, for example, the ability to answer questions that were previously technically challenging or impossible to address as well as the discovery of novel biological phenomena like RNA editing and giant transposons in ascomycetes.

This is reflected in this third edition that contains 11 articles covering genomics research with ascomycetes, basidiomycetes, and basal fungal lineages and topics ranging from basic research to pathogenicity and industrial applications.

Of course, this still does not mean that this volume can comprehensively cover all aspects of this fast-developing field. Similar to the first and second editions, it is rather meant to illustrate the increasing impact of genomics on fungal research by combining articles on a broad range of species and biological problems. (We apologize to those whose work or favorite topic is not covered).

The volume is divided into four sections, the first of which, *genome sequences and beyond*, contains chapters that illustrate how the availability of long-read sequencing and a willingness to collaborate resulted in the discovery of giant transposons (Chap. 1), how novel sequencing-based techniques can be used to analyze the three-dimensional chromatin architecture (Chap. 2), and how sequencing-based approaches can be used to study extremophilic fungi that are difficult to culture under laboratory conditions (Chap. 3). The second section, on *cell and developmental biology*, addresses questions that are important for fungal biology, including a chapter on high-throughput genetic screens that can improve on generation and mapping of genetic mutations (Chap. 4), a chapter on the recently discovered RNA editing that is essential during fruiting body development in filamentous ascomycetes (Chap. 5), and two chapters that describe recent advances in the analysis of the many roles of codon usage in fungi (Chap. 6) and the emerging field of analyzing epigenetic regulation in early-diverging fungi (Chap. 7). The third section, *genomics for biotechnology*, highlights the fungal diversity in plant biomass-converting enzymes (Chap. 8) and work with species from the biotechnologically important genus *Trichoderma* (Chap. 9). The fourth section focuses on *interactions—symbioses, mutualisms, and pathogens* and contains a chapter on the results from the emerging pathogen *Candida auris* (Chap. 10), as well as a chapter on genomic signatures of the ectomycorrhizal symbiosis in fungi (Chap. 11).

We are extremely grateful to all authors who have made the production of this third edition possible. All are experts in their fields, and we appreciate their willingness to contribute to this project which will be useful for all scientists interested in advances in fungal genomics.

Bochum, Germany
Riverside, CA, USA
July 2024

Minou Nowrousian
Jason Stajich

# Series Preface

## The Mycota Series: 2022

*The Mycota* is an encyclopedic book series published by Springer as a comprehensive treatise spanning the Fungal Kingdom through a focus on fungi as experimental systems for basic and applied research. Articles in this series present introductory information in a comprehensive manner to broaden our general knowledge, along with state-of-the-art information on selected topics in fungal biology. Thus, these articles serve as a timeless reference source for historically relevant discoveries in fungal biology to drive further advances in the field.

*The Mycota* was founded in 1994 by Karl Esser and Paul Lemke. Since then 15 volumes have been published, with several now in their third edition. The steadily growing interest in *The Mycota*, reflected in particular by the high number of e-book downloads, encouraged Springer to continue this publication with a new international Series Editorial board beginning in 2022.

Historically, the study of fungi originated as a sub-discipline of botany and was a largely descriptive discipline until the early 19$^{th}$ century. Subsequent experimental research has propelled the field of mycology, and achievements in the genetics and molecular biology of fungi have benefited studies in the related fields of fungal biochemistry and biophysics, plant pathology, medical mycology, and systematics. Ground-breaking research has been carried out using fungi as recognized by the receipt of nine Nobel Prizes, including the revolutionary 1945 Nobel Prize in Physiology and Medicine for the discovery of penicillin and the foundational 2006 Nobel Prize in Chemistry for establishing the molecular basis of eukaryotic transcription. All such studies expand our knowledge of fungi and of fungal processes and improve our ability to understand, utilize, and control fungi for the benefit of humankind.

Fungi have invaded every conceivable ecological niche. Saprobic forms abound, especially in the decay of organic debris. Pathogenic forms exist with both plant and animal hosts. Fungi even grow on other fungi. They are found in aquatic as well as soil environments, and their spores have been found in the air of extreme environments. Fungi can be variously associated with plants as symbionts in the form of lichens, mycorrhizae, and endophytes and also occur as overt pathogens. Association with animal systems varies; examples include the predaceous fungi that trap nematodes, the micro-fungi that grow in the anaerobic environment of the rumen, and medically important pathogens afflicting humans. Currently, taxonomists have identified about 150,000 species, but it is thought that over 90% of fungal species remain undescribed, with conservative estimates ranging from 2 to 5 million fungal species on earth.

From this perspective, there are new topics in fungal biology to explore, which will expand the current volumes and take them in new directions. Further, there will be new volumes in areas (e.g., Cryptomycota) that were not covered before.

Our understanding of the evolution of fungi is still incomplete and mainly based on species that can be grown in culture. But recent environmental DNA analyses have revealed a highly diverse form of eukaryotic life that branches with the Fungi, and thus the resulting and highly diverse clades were named the cryptomycota. The discovery of novel intermediate forms will redefine the fungal tree of life. Other topics will consider the adaptation of fungi to climate changes, the occurrence of fungal pathogens in the environment, and the dispersal of fungi during global pandemics.

Fungal constituents of the microbiome have received much less attention thus far, yet recent findings from clinical and animal studies clearly establish fungi as a significant component of the oral, gastrointestinal, pulmonary, and skin microbiomes. Finally, the relevance of fungal biology to society is reflected by the increasing number of fungal related human diseases.

In the history of pharmacology, fungi have always been sources of useful molecules for humans, but the diversity of molecules that can be obtained from fungi is still largely underexplored. However, efforts in fungal genomics provide resources for data-driven genome mining and large-scale comparisons to explore the molecular repertoires produced by fungi. The result will be new compounds with applications in the pharmaceutical and agricultural science industries.

Fungi can also help the entire planet and may, for example, be relevant in specific sectors, such as that linked to pollution from plastic. Fungi produce a wide range of enzymes that have the potential to break down the chemical bonds of plastic polymers, and in this context the potential role of marine fungi in plastic degradation may be of major relevance. Finally, new biomaterials from fungal species may open the door to alternatives to fossil-based materials, and thus reduce environmental pollution.

For consistency throughout this series of volumes, the names adopted for major groups of fungi should be followed according to the following paper, which gives an overview of all of the orders in the fungal kingdom: https://pubmed.ncbi.nlm.nih.gov/32660385/

We are grateful to Springer for continuing *The Mycota* series and are especially thankful to all of the Volume Editors in selecting topics and assembling experts from diverse fields of fungal biology.

Sydney, Australia — Dee Carter
Delhi, India — Anuradha Chowdhary
Durham, NC, USA — Joseph Heitman
Bochum, Germany — Ulrich Kück
July 15th, 2022

# Contents

**Part 1 Genome Sequences and Beyond**

1 ***Starship* Discovery: A Collaborative Approach to Uncover Massive Transposable Elements** . . . . . . . . . . . . . . . . . . . . . . . . . . 3
S. Lorena Ament-Velásquez, Emile Gluck-Thaler, Andrew S. Urquhart, Megan C. McDonald, and Aaron A. Vogan

2 **The Three-Dimensional Chromatin Architecture in Fungi**. . . . . 15
Satya Dev Polisetty and Kaustuv Sanyal

3 **Fungi from Extreme Environments: Genome Sequences and Beyond** . . . . . . . . . . . . . . . . . . . . . . . . . . . . . . . . . . . . . . . . . . . . 33
Cene Gostinčar, Claudia Coleine, Nina Gunde-Cimerman, and Jason E. Stajich

**Part 2 Cell and Developmental Biology**

4 **Recent Advances in High-Throughput Genetics in Fungi**. . . . . . 51
Joshua W. Lyon and R. Blake Billmyre

5 **Genome-Wide A-to-I RNA Editing During Sexual Reproduction in Filamentous Ascomycetes** . . . . . . . . . . . . . . . . . 67
Zhuyun Bian, Diwen Wang, and Jin-Rong Xu

6 **A Symphony of Roles for Codon Usage in Fungal Genomics** . . . . . . . . . . . . . . . . . . . . . . . . . . . . . . . . . . . . . . . . . . . . . . 75
Abigail Leavitt LaBella

7 **Epigenetic Regulation in Early-Diverging Fungi**. . . . . . . . . . . . . 101
Carlos Lax and Victoriano Garre

**Part 3 Biotechnology**

8 **Fungal Diversity Related to Plant Biomass Degradation** . . . . . . 119
Miia R. Mäkelä, Peicheng Sun, Mao Peng, and Ronald P. de Vries

9 **Genomes and Genomics of the Genus *Trichoderma***. . . . . . . . . . . 139
Monika Schmoll

**Part 4 Interactions: Symbioses, Mutualisms, and Pathogens**

**10 Genomic Innovation and Virulence Evolution in the Emerging Human Fungal Pathogen *Candida auris*** . . . . . . . . . . . 163
Hugh Gifford, Johanna Rhodes, Duncan Wilson, and Rhys Farrer

**11 General Trends of Genomic Signatures of the Ectomycorrhizal Symbiosis in Fungi: A Comparison Across Multiple Lineages** . . . . . . . . . . . . . . . . . . . 189
Marisol Sánchez-García, Shingo Miyauchi, Elodie Drula, Francis Martin, and Martin Ryberg

# About the Editors and Contributors

## About the Series Editors

**Dee Carter** is Professor of Microbiology at the University of Sydney. She has a degree in microbiology and biochemistry from Otago University, New Zealand, and a PhD from Imperial College, London, UK. After postdoctoral fellowships in Montpellier, France, and UC Berkeley, California, USA, she moved to Australia in 1995 to take up a lectureship at the University of Sydney. Dee's research has encompassed 1) the population genetics and ecology of medically important fungi, including yeast and mold pathogens; 2) responses of fungi to host and antifungal stress, including the production of variant morphological forms; and 3) the transcriptome and proteome response to antifungal therapy, particularly during synergistic interactions between antifungals and natural products. As well as research, Dee teaches microbiology including mycology, epidemiology, and molecular biology at Sydney University.

**Anuradha Chowdhary** is a Professor of Medical Mycology at the Vallabhbhai Patel Chest Institute, University of Delhi, India. She received her medical degree in 1992 and MD (Microbiology) degree in 1996 from the University of Delhi, India and PhD, from Faculty of Medical Sciences, Radboud University Medical Center, Netherlands. Dr Chowdhary's research interests include molecular ecology and population genetics of pathogenic fungi, antifungal drug resistance mechanisms, especially Aspergillus and Candida spp., and the epidemiology of systemic mycoses. She is currently working on molecular epidemiology of Candida auris, terbinafine-resistant dermatophytes, and azole-resistant Aspergillus fumigatus.

**Joseph Heitman** is James B. Duke Professor and Chair, Department of Molecular Genetics and Microbiology, Duke University, Durham, NC, USA. His research studies model and pathogenic fungi addressing fundamental questions of scientific and medical importance. Pioneering studies with Baker's yeast revealed how immunosuppressive natural products interdict signaling cascades via FKBP12-drug complexes and discovered TOR as a globally conserved nutrient sensor targeted by the immunosuppressive, antiproliferative drug rapamycin. His research discovered unisexual reproduction of pathogenic microbes, with implications for pathogen emergence, how sex generates diversity, and how sex evolved. Dr. Heitman's lab has further developed genetic and genomic approaches elucidating molecular principles of fungal virulence, identifying therapeutic targets, and illustrating convergent evolution of fungal mating-type loci with mammalian, insect, and plant sex chromosomes, defined the calcium-activated protein phosphatase calcineurin as a globally conserved fungal virulence factor, and elucidated functions of RNAi in microbial pathogen genome integrity, hypervirulent outbreak lineages, and drug resistance via epimutation.

**Ulrich Kück** is a Professor of General and Molecular Botany at the Ruhr-University in Bochum, Germany. He is being graduated in Biology and Chemistry and has a longstanding experience in the molecular biology of fungi, algae, and plants. His research with fungi has focused on two general aspects: 1st, Genetic engineering of the secondary metabolism of biotechnically relevant filamentous fungi, including functional genomics and proteomics. 2nd, Molecular genetic analysis of cellular growth in filamentous fungus, with the focus on sexual development. In particular, he is interested in the function of mating-type loci and the involvement of conserved signaling complexes, such as the striatin-interacting phosphatases and kinases (STRIPAK) complex, in the control of cellular and developmental processes.

## About the Volume Editors

**Minou Nowrousian** earned her diploma and Ph.D. in biology at the Ruhr-University Bochum in the laboratory of Dr. Ulrich Kück working in molecular biology of fungi. During her postdoc at Dartmouth Medical School in the laboratory of Dr. Jay Dunlap and Dr. Jennifer Loros, she worked on circadian rhythms in *Neurospora*. She is currently a research fellow at the Department of Molecular and Cellular Botany at the Ruhr-University Bochum. Her group combines molecular genetics and bioinformatics to study the molecular mechanisms and evolution of fungal development. A focus of her work is the analysis of the regulation of gene expression during development using genomics and transcriptomics.

**Jason Stajich** earned his Ph.D. Genomics and Genetics at Duke University in laboratory of Dr Fred Dietrich studying bioinformatics, evolution, and comparative genomics of Fungi focused on *Cryptococcus* and *Coprinoposis*. This was followed by postdoctoral as a Miller Research Fellow in laboratory of Dr John Taylor at the University of California, Berkeley using population and comparative genomics to study *Coccidioides*, *Neurospora*, and *Batrachochytrium*. He is currently a Professor in the Department of Microbiology and Plant Pathology at the University of California, Riverside. He develops computational tools for genome assembly, annotation, comparative genomics, phylogenomic analyses, and metagenomic studies of Fungi and Bacteria. His research team tests hypotheses about Fungi and their interactions with plants, animals, and other microbes with work on microevolution of *Aspergillus fumigatus* and *Clavispora lusitaniae* in cystic fibrosis lung environment, genomics of black yeast fungi in the Chaetothyriales and Dothideomycetes, the amphibian pathogen *Batrachochytrium*, role of Basidiobolus in gut health and microbiomes of amphibians and reptiles, and molecular ecology of fungi and bacteria in arid environments.

## Contributors

**S. Lorena Ament-Velásquez** Division of Population Genetics, Department of Zoology, Stockholm University, Stockholm, Sweden

**Zhuyun Bian** Department of Botany and Plant Pathology, Purdue University, West Lafayette, IN, USA

**R. Blake Billmyre** Department of Pharmaceutical and Biological Sciences, College of Pharmacy, University of Georgia, Athens, GA, USA

Department of Infectious Diseases, College of Veterinary Medicine, University of Georgia, Athens, GA, USA

**Claudia Coleine** Department of Ecological and Biological Sciences, University of Tuscia, Viterbo, Italy

**Ronald P. de Vries** Westerdijk Fungal Biodiversity Institute, CT, Utrecht, The Netherlands

**Polisetty V. S. Satya Dev** Molecular Mycology Laboratory, Molecular Biology and Genetics Unit, Jawaharlal Nehru Centre for Advanced Scientific Research, Bengaluru, India

**Elodie Drula** Architecture et Fonction des Macromolécules Biologiques (AFMB), CNRS, Aix-Marseille Université, Marseille, France

INRAE, UMR 1163, Biodiversité et Biotechnologie Fongiques, Marseille, France

**Rhys Farrer** Medical Research Council Centre for Medical Mycology at the University of Exeter, Department of Biosciences, Faculty of Health and Life Sciences, Exeter, UK

**Victoriano Garre** Departamento de Genética y Microbiología, Facultad de Biología, Universidad de Murcia, Murcia, Spain

**Hugh Gifford** Medical Research Council Centre for Medical Mycology at the University of Exeter, Department of Biosciences, Faculty of Health and Life Sciences, Exeter, UK

**Emile Gluck-Thaler** Department of Plant Pathology, University of Wisconsin-Madison, Madison, WI, USA

Wisconsin Institute for Discovery, Madison, WI, USA

**Cene Gostinčar** Biotechnical Faculty, Department of Biology, University of Ljubljana, Ljubljana, Slovenia

**Nina Gunde-Cimerman** Biotechnical Faculty, Department of Biology, University of Ljubljana, Ljubljana, Slovenia

**Abigail Leavitt LaBella** Department of Bioinformatics and Genomics, The North Carolina Research Campus, Kannapolis, NC, USA

**Carlos Lax** Departamento de Genética y Microbiología, Facultad de Biología, Universidad de Murcia, Murcia, Spain

**Joshua W. Lyon** Department of Pharmaceutical and Biological Sciences, College of Pharmacy, University of Georgia, Athens, GA, USA

Department of Infectious Diseases, College of Veterinary Medicine, University of Georgia, Athens, GA, USA

**Miia R. Mäkelä** Department of Bioproducts and Biosystems, Aalto University, Espoo, Finland

**Francis Martin** Université de Lorraine, INRAE, UMR Interactions Arbres/Microorganismes, Centre INRAE Grand Est - Nancy, Nancy, France

**Megan C. McDonald** School of Biosciences, Institute of Microbiology and Infection, University of Birmingham, Birmingham, West Midlands, UK

**Shingo Miyauchi** Evolutionary and Synthetic Biology unit, Okinawa Institute of Science and Technology, Okinawa, Japan

**Mao Peng** Westerdijk Fungal Biodiversity Institute, CT, Utrecht, The Netherlands

**Johanna Rhodes** MRC Centre for Global Infectious Disease Analysis, Imperial College London, London, UK

Department of Medical Microbiology, RadboudUMC, Nijmegen, Netherlands

**Martin Ryberg** Department of Organismal Biology, Uppsala University, Uppsala, Sweden

**Marisol Sánchez-García** Uppsala Biocentre, Department of Forest Mycology and Plant Pathology, Swedish University of Agricultural Sciences, Uppsala, Sweden

**Kaustuv Sanyal** Molecular Mycology Laboratory, Molecular Biology and Genetics Unit, Jawaharlal Nehru Centre for Advanced Scientific Research, Bengaluru, India

**Monika Schmoll** Department of Microbiology and Environmental Systems Science, Division of Terrestrial Ecosystem Research, University of Vienna, Vienna, Austria

**Jason E. Stajich** Department of Microbiology and Plant Pathology, University of California-Riverside, Riverside, CA, USA

**Peicheng Sun** Laboratory of Food Chemistry, Wageningen University, Wageningen, The Netherlands

**Andrew S. Urquhart** Institute of Organismal Biology, Uppsala University, Uppsala, Sweden

**Aaron A. Vogan** Institute of Organismal Biology, Uppsala Universit, Uppsala, Sweden

**Diwen Wang** Department of Botany and Plant Pathology, Purdue University, West Lafayette, IN, USA

**Duncan Wilson** Medical Research Council Centre for Medical Mycology at the University of Exeter, Department of Biosciences, Faculty of Health and Life Sciences, Exeter, UK

**Jin-Rong Xu** Department of Botany and Plant Pathology, Purdue University, West Lafayette, IN, USA

# Part 1

# Genome Sequences and Beyond

# *Starship* Discovery: A Collaborative Approach to Uncover Massive Transposable Elements

1

S. Lorena Ament-Velásquez, Emile Gluck-Thaler, Andrew S. Urquhart, Megan C. McDonald, and Aaron A. Vogan

**Abstract**

The ability to generate high-quality fungal genomes brought with it new frontiers of scientific discovery and paved the way for answering outstanding biological mysteries. Recently, several independent observations of large mobile elements within fungal genomes were recognized as manifestations of the same phenomenon: a newly discovered superfamily of mobile elements called *Starships*. In the hopes of inspiring further insight and encouraging other breakthroughs, we detail how individual stories of genomic exploration coalesced into the discovery of a previously hidden mechanism of fungal evolution. We end by describing important aspects of *Starship* biology and detail exciting perspectives on where we see the *Starships* taking the field of fungal biology in the future.

**Keywords**

Transposable elements · *Starships* · Comparative genomics

## 1.1 Introduction

How do scientific discoveries come to be? Much of science education is taught as a series of landmark discoveries that have led us to our current understanding of the world around us. It provides the impression that science is yanked forward with staccato strokes of brilliance by the agile minds of a few individuals. This belies the true nature of scientific endeavor and buries the contributions of countless researchers, the truth being that most scientific progress happens through the interchange of ideas between many different groups. The authors of this chapter have been fortunate enough to all take part in the unearthing of a novel phenomenon within the field of fungal genomics. Here, in hopes to inspire new insight and encourage other breakthroughs, we detail how individual stories of exploration coalesced into the discovery of a new superfamily of transposable elements, the *Starships*, that stands to reshape our understanding of the mechanisms of fungal evolution.

S. L. Ament-Velásquez
Department of Zoology, Stockholm University, Stockholm, Sweden

E. Gluck-Thaler
Department of Plant Pathology, University of Wisconsin-Madison, Madison, WI, USA

Wisconsin Institute for Discovery, Madison, WI, USA

A. S. Urquhart · A. A. Vogan (✉)
Institute of Organismal Biology, Uppsala University, Uppsala, Sweden
e-mail: aaron.vogan@ebc.uu.se

M. C. McDonald
School of Biosciences, Institute of Microbiology and Infection, University of Birmingham, Birmingham, West Midlands, UK


M. Nowrousian, J. E. Stajich (eds.), *Fungal Genomics*, The Mycota 13,
https://doi.org/10.1007/978-3-031-75666-5_1

Before we had the technological capabilities to sequence long stretches of nucleic acids, our understanding of organismal genomes was limited to functional regions associated with obvious phenotypes that could be disrupted or modified through mutation. These approaches necessitated that many groups focus on the same organism to slowly map the genetic landscape. Even early sequencing technologies, based on the Sanger method, still limited the field to producing single representative genomes for a species, as piecing together whole chromosomes from many separate assays required considerable time and energy. For example, the original *Saccharomyces cerevisiae* project took 7 years to complete, involving 35 different laboratories (Engel et al. 2014). Eventually, as short-read shotgun sequencing technologies developed, we could begin assembling multiple genomes from different members of the same species. However, genome assemblies produced with these methods are often fragmented because short sequencing reads do not span regions of repetitive DNA. This fact necessitated that additional individuals be compared to a single high-quality "reference" genome in order to scaffold their fragmented contigs into full chromosomal assemblies. Such comparisons provide an incomplete picture of genetic variation between individuals/strains, since they poorly resolve large structural variants, which typically consist of the presence/absence polymorphism of genes and larger genomic regions. Thus, we still lacked key pieces to the puzzle in terms of linking observed phenotypic differences between individuals to underlying genetic variation. However, more recently, with the ability to generate high-quality whole fungal genomes through long-read sequencing technologies, our investigations no longer need to be framed through the lens of a single reference genome because we can generate and compare reference quality assemblies from multiple individuals. As a result, we have begun to appreciate that the secrets to many fungal phenotypes lie in these previously hidden presence/absence structural variants. In other words, the pangenome now lies open before us, revealing a dynamic landscape of change, far greater than we had previously imagined.

The agents responsible for generating the most structural variation within populations are typically transposable elements (TEs). These relatively short genetic entities code for a small number of genes (often a single one) that act to mobilize and/or replicate themselves. By this action, TEs can become numerous within a genome, and in many cases occupy more nucleic acid real estate than the host machinery (Wells and Feschotte 2020). One peculiarity of many fungi is that their TEs are not evenly distributed among the chromosomes, but rather form clustered TE islands in an otherwise gene dense ocean (Castanera et al. 2016). However, most fungi are quite good at keeping TE copy number in check, and as a result usually have relatively low TE abundance (~25%) compared to plants or animals (Amselem et al. 2015). Nevertheless, TEs play important roles in fungi, comprising structural regions, such as centromeres (Guin et al. 2020), or mediating adaptation to niches like plant pathogens (Porquier et al. 2021). Understanding TE dynamics is key to elucidating the interplay between genomic architecture and organismal phenotype. Fortunately, the molecular mechanisms that TEs employ to travel the genetic milieu require specific structural features and often leave behind telling clues (Box 1.1). Thus, the endeavor to map structural variants or to quantify metrics of genomic evolution frequently requires the cataloging of TE movement. This process, while potentially tedious and exhausting, can lead to great revelations, such as those at the focus of this chapter.

**Box 1.1 Transposable Elements in Fungal Genomes**

**Fungal TE diversity and classification:** Transposable elements exist in a large variety of forms and likely have numerous independent origins. Researchers have strived to organize them into various classification systems based around shared mechanisms of replication, sequence similarity, and/or phylogenetic relatedness

(Wicker et al. 2007). The ways that TEs undergo replication are directly connected to specific structural features as well as genes with important enzymatic functions. For example, the LTR retrotransposons have long terminal repeats (hence the name) and genes such as reverse transcriptase that allow them to create RNA copies of themselves, which can then reintegrate elsewhere in the genome. The integration of LTRs requires an integrase enzyme that cuts DNA with a characteristic 4–6 bp overhang. Once inserted into the new location, the repair machinery of the cell fixes these staggered ends (one on each side of the LTR), creating a target site duplication (TSD). Integrase enzymes from most other types of TEs operate similarly and, in DNA transposons, the inserted elements are instead often flanked by terminal inverted repeats (TIRs). However, some superfamilies, like *Cryptons*, lack structural features or TSDs. Instead, there are small stretches (just a few base-pairs long) of homologous sequence at either end of an inserted element known as direct repeats (DRs). These look identical to TSDs, but represent microhomology to a target site, and are not generated as part of the insertion mechanism. All of these aforementioned features vary across the different types of TEs, providing clear guides for their identification and classification.

**Genomic defenses against TEs in fungal genomes:** One unique feature of many fungal genomes is the presence of the genome defense mechanism, repeat-induced point-mutation (RIP) (Selker 2002). This system identifies any duplicate sequence within a genome and induces C to T mutations. This occurs on both DNA strands, resulting in a rapid decline in GC content. Thus, many fungal genomes have large stretches of AT-rich sequence that are the result of the interactions between TEs and host defense mechanisms. Due to the nature of this conflict, these regions are often highly variable within populations making analyses of synteny or reconstructing evolutionary history impossible for many such regions. While RIP is widespread across Pezizomycotina, it is not universal. Additionally, RIP is known to operate during the sexual cycle, so asexual lineages lack this mechanism. Although RIP has not been described outside of Ascomycota, other fungal genomes have been observed to possess a bimodal distribution of GC content (Amselem et al. 2015), which is a characteristic of how RIP operates. Thus, other fungi may possess analogous systems that function similarly to protect against repetitive element expansion.

What follows next are four tales of genomic intrigue, whose details may, at a glance, seem unconnected. However, collectively they weave a tapestry of scientific endeavor that led to the breakthrough discovery that giant transposable elements, the *Starships*, are mobilizing vast amounts of diverse nucleotide sequences within the Pezizomycotina. Each story is written by their own protagonist, delving into what were initially "straightforward" forays into comparative genomics, until it became evident that they were far from that. Beneath the seemingly placid surfaces of fungal genomes large amounts of DNA are on the move; in the depths, Leviathans swim.

## 1.2 ToxhAT

Megan McDonald

One of the best characterized and recent horizontal gene transfer (HGT) events between pathogenic fungi is the virulence gene, *ToxA*. This small, secreted protein was first identified and characterized from the culture filtrate of the wheat pathogen *Pyrenophora tritici-repentis*, where it was shown to induce a strong necrosis in a gene-for-gene manner with wheat cultivars that carry the susceptibility gene *Tsn1* (Tuori et al. 1995). Much later, *ToxA* and the surrounding ~11 kb region were discovered in the whole genome sequence of the fungal pathogen *Parastagonospora nodorum* SN15 with over 95% nucleotide identity to the region in *P. tritici-repentis* (Friesen et al. 2006). Intriguingly, the gene neighbor to *ToxA* encoded a putative hAT transposase. Consistent with TE activity in *P. nodorum,* the edges of the horizontally transferred DNA were AT-rich. These AT-rich regions are a characteristic of repeat-rich regions of fungal genomes that have been affected by RIP and contain many fragments of inactive transposons (Box 1.1) (McDonald et al. 2018; Moolhuijzen et al. 2018). However, other structural features (i.e., TSDs, TIRs) that would have enabled annotation of a larger transposon that flanked or encompassed this horizontally transferred region were missing (Friesen et al. 2006).

This all changed when Prof. Peter Solomon and myself decided to add a new fungal pathogen to the lab's research program, the wheat and barley pathogen *Bipolaris sorokiniana.* This pathogen causes a severe leaf-spot disease in South East Asia, often in complex with *P. tritici-repentis* (Gupta et al. 2018). In order to get started with *B. sorokiniana* our natural first step was to sequence a few genomes. To do this we ordered three strains from the Queensland Plant Pathology herbarium in Australia and sequenced these isolates to get our first de novo assemblies for this pathogen (Shivas et al. 2019). Replicating what he did in the early 2000s, Peter used *ToxA* as a query for a local BLAST search in these newly assembled genomes. Much to our surprise, these hits came back with a near 100% identical match to the horizontally transferred 11 kb region reported in the original HGT paper by Friesen et al. (2006). Shortly after we reported our findings, *ToxA* was also confirmed to be present in *B. sorokiniana* in the United States and India (Friesen et al. 2018; Navathe et al. 2020).

One challenge that we faced with the new Illumina-based de novo assembly of *B. sorokiniana* was that the genome was extremely fragmented (+10,000 total contigs) and the horizontally transferred DNA (~11 kb) was found on one 15.6 kb scaffold (McDonald et al. 2018). The small size of this scaffold prevented further investigation of the HGT event in the context of the full chromosome, so we set out to generate long-read assemblies for both *B. sorokiniana* and *P. nodorum*. These assemblies enabled a complete alignment of the horizontally transferred region between all three species. This alignment was key to identifying the first clear breakpoints that precisely defined the boundaries of the HGT event. In *B. sorokiniana* there was one key difference compared to the other species, which was that these boundaries were not AT-rich and were unaffected by RIP. This enabled the identification of TIRs in *B. sorokiniana* that flanked and defined the horizontally transferred region. These TIRs were the first clear evidence that a Class II DNA transposon was responsible for the horizontal transfer. We named this transposon ToxhAT, a name derived from the putative hAT transposase gene neighbor to the *ToxA* effector (McDonald et al. 2019).

Given that both TIRs were intact in *B. sorokiniana* we began exploring the idea that this transposon might remain mobile in this species. To do this we sequenced and assembled two additional isolates with Oxford Nanopore long-reads. These assemblies confirmed that *ToxhAT* was conserved in these other isolates and also showed that *ToxhAT* appeared to be in a different chromosome when compared to our PacBio assembly with isolate CS10. This was exciting as these data supported our hypothesis that *ToxhAT* was still an active transposon in *B. sorokiniana.* To explore this further, I re-sequenced a total of nine isolates in addition to my reference PacBio strain and assembled these genomes de novo. These assemblies then sat on a hard-drive for over a year, waiting to be looked at in more detail. The global pandemic and subsequent lab shutdown unexpectedly brought a new Master's student, Hannah Wilson, onto the project, who was given my hard-drive of long-read assemblies.

We started exploring that putative movement of *ToxhAT* by aligning each chromosome where the transposon was found. Here, we expected to see sharp boundaries that corresponded to the TIRs that we used to define *ToxhAT.* These whole genome alignments did reveal sharp boundaries where near 100% nucleotide identity dropped to background levels, however these boundaries were not located at the *ToxhAT* TIRs, but were in fact much further apart spanning nearly 200 kb. We termed this larger region, which had some features of a TE, but not any that we knew of yet, the "Region of Uncertainty" (ROU). Close examination of these boundaries did not reveal TIRs but instead showed a six base-pair AT-rich sequence that resembled an imperfect direct repeat (DR). The ROU was found in a different chromosome in every single isolate of *B. sorokiniana* we sequenced, which strongly indicated that this region might be a giant transposon carrying the smaller *ToxhAT*. However, at the time no TE of that size had ever been described.

Our discovery of the ROU coincided with the posting of a new preprint describing a novel giant transposon known as *Enterprise*, where we started to draw many parallels.

## 1.3 Enterprise

S. Lorena Ament-Velásquez

For over a century, the coprophilous fungus *P. anserina* has served as a model organism for genetics and molecular biology research (Silar 2020). Although many aspects of its biology are peculiar and fascinating, the presence of multiple genetic elements with non-Mendelian inheritance has been particularly notorious. One of these elements, a prion, is now well-understood (Saupe 2007), but the remaining elements confused "podosporologists" for decades. Different research groups reported segregation distortion in crosses of strains from France, the Netherlands, and Germany (Padieu and Bernet 1967; van der Gaag et al. 2000; Hamann and Osiewacz 2004). However, it was unclear if there was a common underlying cause or if there were multiple, independent genetic elements disrupting the production of spores in their favor. In time, these elements were collectively termed the "spore-killers," due to their overt phenotypes.

A major breakthrough in the spore-killing scene came from the discovery of the *Spok* gene family, whose members were each capable of inducing meiotic drive on their own (Grognet et al. 2014). At first only two paralogs were known, one in *P. anserina* and another in its sibling species *Podospora comata*. That is when Prof. Hanna Johannesson, a professor in Uppsala University, became interested in the topic. She turned her attention to a well-studied *P. anserina* population from Wageningen, the Netherlands. Multiple spore-killers had been previously described in Wageningen and a rich strain collection was actively maintained since the early 90s (van der Gaag et al. 2000; Debets et al. 2012). She established collaboration with the group in Wageningen University & Research, and I later joined her group as a PhD student in 2015. Aaron A. Vogan was hired as a postdoc a year later. At that time, long-read sequencing technologies were on the rise but still relatively expensive. Nonetheless, *Podospora* fungi were an easy target for population sequencing due to their relatively small genomes (~36 Mb) and low repeat content (~5%). Eventually, we managed to sequence multiple representative spore-killer strains (Vogan et al. 2019, 2021), producing mostly telomere-to-telomere genome assemblies and opening a treasure trove of *Podospora* genomics.

The main finding coming from this data was the existence of two novel *Spok* paralogs that, unlike the original *Spok* copies, were part of a much larger region present in different genomic locations depending on the strain: the *Spok* block. These blocks of sequence represented rather massive structural variants, from 110 kb to 247 kb, in an otherwise extremely collinear genomic background (Vogan et al. 2019, 2021). The block content itself could vary, having one or both novel *Spok* paralogs. Ultimately, the different combinations of *Spok* genes and positions of the *Spok* block were producing different meiotic drive phenotypes, and hence were acting as independent spore-killers (Vogan et al. 2019). The mechanistic puzzle of meiotic drive was finally solved, but a big question remained: how is the *Spok* block moving?

The answer came late on a Friday night of February 2018. I was observing the breakpoints of the structural variants across multiple strains, trying to understand the history of the *Spok* block and what I thought were regular translocations. Unexpectedly, the ends of the *Spok* blocks were very conserved and flanked by a directed repeat (DR) of six base pairs (RGGTAG). This motif was suspiciously similar to the TSD produced by DNA transposons, giving us the idea that the whole *Spok* block might be a gigantic transposable element. However, there were no other obvious TE attributes, such as TIRs or LTRs. Hence, we reasoned that conservation of any features in this putative gigantic transposable element should give an indication of what components are important for movement. By comparing all known *Spok* block copies within *P. anserina*, as well as smaller versions of the element in *P. comata*, we narrowed down our focus to a putative protein coding gene. Located at the very beginning of the *Spok* block, this open reading frame harbored a few domains: a zinc finger domain, a CHROMO domain, and most importantly, a domain of unknown function called DUF3435. The first two domains are often found in TEs, while the latter domain has a remote similarity to tyrosine recombinases (YRs) (Vogan et al. 2021). YRs, in turn, are enzymes involved in the DNA integration of various selfish genetic elements, including *Cryptons* (Kojima and Jurka 2011). Like the *Spok* block, *Cryptons* have no TIRs or LTRs but produce short DRs flanking the element. As a result, this open reading frame emerged as our leading candidate for a mobilizing protein and was designated with the name *Kirc*, which stands for "spore killing related crypton."

At this point, we were pretty convinced that the *Spok* block was a new kind of transposable element. In the fictional original series Star Trek, the character Spock is a crew member of the USS Enterprise starship, piloted by its Captain J. T. Kirk. Hence, the *Spok* block and closely related elements got the name *Enterprise*, and the passengers received slightly misspelled names to follow this theme. Ultimately, the *Enterprise* element in *Podospora* had managed to obtain a formidable cargo, the *Spok* genes, upgrading its selfish capabilities from just movement within a single genome to also transmission advantage within a population. We experimentally found evidence that different *Spok* block copies have deleterious effects on its host (Vogan et al. 2021). However, the fact that the spore-killing phenomenon first detected in the late 60s still prevailed by the time of the last sampling event in 2017, suggests that the *P. anserina* genome has not been able to purge this two-times-selfish genetic element.

By early 2020, we were finally polishing the last details of the *Enterprise* manuscript, when our colleague Alex Suh spotted an intriguing new preprint. He sent it to Aaron, insisting it was relevant to our work. The preprint was describing a new element called *Hephaestus*, a huge TE with cargo genes that conferred a particular phenotype to the host. Conceptually, it was so similar to the *Enterprise*! Aaron paid special attention to the gene at the very beginning of this new TE. Superficially, this gene looked nothing like *Kirc*, but was reported to contain a DUF3435 domain. In other words, *Hephaestus* also had a captain! We made sure to at least mention *Hephaestus* (and *ToxhAT*) in our discussion, and submitted the *Enterprise* manuscript to bioRxiv a few weeks later.

## 1.4 Hephaestus

Andrew Urquhart

During my PhD studies with Assoc. Prof. Alexander Idnurm at the University of Melbourne, I was focused on the plant pathogen *Leptosphaeria maculans*. In the first year of my PhD, as is typical, things were not going well in the lab, my difficulty in getting *L. maculans* to grow with consistent morphology being one issue. By chance or by poor aseptic technique, one day in 2016 I noticed that I had a colony of *Paecilomyces variotii* growing on the edge of one of my cultures. I was instantly enamored with its fast, consistent, and vibrant yellow growth. Taking this as a sign from the universe as to what species I really should be working on, and with Alex's guidance, a side project was born.

A logical first step was to sequence the genome of the two *P. variotii* strains (my initial contaminant and a strain of opposite mating type) and to develop the techniques to work with this organism. Fortunately, *P. variotii* proved to be very amenable to genetic manipulation and this work was published as our first *P. variotii* paper (Urquhart et al. 2018). The question then was what to do next with this new "model species?" During peer review of this first *Paecilomyces* paper it was suggested to us that we should compare the gene content of our two genomes, and as a result of doing this we noticed that one of the two strains had more genes for metal resistance. What was initially intriguing is that these metal resistance genes were physically clustered together in the genome in a large region that was absent in the genome of the other strain. Typically, metal resistance loci are dispersed across the genome (Bazzicalupo et al. 2020). We named this region *Hephaestus* after the Greek god of metalworking (Urquhart et al. 2022). The region seemed like an easy direction to pursue given that we could initially test the metal resistances of the two wildtype strains without needing to do any genetic manipulations. We found that of our two strains, the strain containing *Hephaestus* was more resistant to cadmium, lead, arsenic, and copper ions. To expand this correlation we obtained another four strains of *P. variotii* and found that three out of four strains had *Hephaestus.* Examining the metal resistances of these additional strains supported the correlation between *Hephaestus* and the four metal resistances, which we subsequently confirmed using gene knockouts. However, what was unexpected was that in each case *Hephaestus* was located in a different place in the genome. We also noticed that *Hephaestus* was flanked by TIRs, a characteristic feature of some TEs. We then realized we were dealing with a massive transposable element carrying host-beneficial genes, the first ever discovered in eukaryotes.

Given that it had been previously hypothesized that giant transposons would mobilize gene cargo between species of eukaryotes (Arkhipova and Yushenova 2019), we searched for evidence of HGT in *Hephaestus*. BLAST searches uncovered a very similar, albeit fragmented, element in the genome of *Penicillium fuscoglaucum*—a clear HGT event. Subsequently, we have found two more HGT events of *Hephaestus* within the genus *Paecilomyces,* providing exciting evidence that *Hephaestus* might be an agent of horizontal transfer.

The *Hephaestus* preprint was quickly followed by the preprint version of the *Enterprise* paper (Vogan et al. 2021), as mentioned above, which described a similarly large transposable element. This awakened us to the fact that *Hephaestus* was likely to be the tip of the iceberg of a greater diversity and spurred us to look for similar elements. Excitingly, we found a great diversity of elements closely related to *Hephaestus* containing similar Captain genes and flanking sequences. These elements included a range of different putative resistance genes including those for mercury and sodium resistance—implying that the original *Hephaestus* transposon was not an isolated phenomenon.

## 1.5 Voyager

Emile Gluck-Thaler

Several serendipitous events set the stage for revealing these giant TEs to be a widespread phenomenon across the Pezizomycotina. As conferences transitioned to virtual formats in the first summer of the COVID-19 pandemic in 2020, Aaron A. Vogan and I both were able to present our postdoctoral research at the first CanFunNet Fungal Biology Conference, even though we were physically located thousands of kilometers away from each other. Although we had never met before, we began messaging immediately after seeing each other talk because we realized that we had a unique opportunity to combine our seemingly unrelated fungal genomic datasets into something much larger. Through an ongoing collaboration with Horacio Lopez-Nicora at the Ohio State University, I assembled a dataset of 12 long-read, near-chromosome quality assemblies from the fungal pathogen and endophyte *Macrophomina phaseolina* and was finding large and dynamic genomic hotspots for gene gain and loss. Aaron took one look at that data and wondered if a giant TE like the *Enterprise* could be responsible for producing those patterns. In fact, one of the advantages of being at an online conference was that Aaron was able to BLAST a publicly available *M. phaesolina* genome for homologs of the *Enterprise*'s DUF3435-containing *Kirc* as I was talking, yielding several promising hits.

The *M. phaseolina* dataset proved ideal for resolving the boundaries of giant TEs at a population scale because it enabled high-quality TE annotations through pairwise genome alignments. Fairly soon after our initial meeting, on new year's day in 2021, Aaron found evidence of a giant TE, which we named *Voyager,* that shared a common feature with *Enterprise*, a predicted DUF3435-domain containing tyrosine recombinase "captain" at the 5′ end of the element. At this point, we began to wonder: are *Enterprise, Hephaestus,* and *Voyager* outliers, or are they archetypes of a common phenomenon?

In an attempt to recover more giant TEs to test these alternative hypotheses, I used the cargo carried by *Voyager* to search a fungal kingdom-wide database for evidence that at least some of the cargo genes had remained clustered together with the tyrosine recombinase captain over long periods of evolutionary time. The maintenance of physical linkage is a powerful signature of natural selection acting to maintain particular combinations of gene functions and has been successfully used to identify gene clusters encoding specialized metabolic pathways (Gluck-Thaler and Slot 2018; Gluck-Thaler et al. 2020; Konkel et al. 2023). By sheer luck, I found that *Voyager* carries four genes that frequently co-occurred with each other in close proximity to tyrosine recombinases across hundreds of different fungal species: DUF3723-domain containing genes, ferric reductases, patatin-like phosphatases, and NOD-like receptors (Gluck-Thaler et al. 2022). Upon manual annotation, many of the higher quality "giant TE-like" regions containing these conserved, co-localized genes were indeed found to be bonafide giant TEs with flanking direct repeats and occasionally terminal inverted repeats (Fig. 1.1).

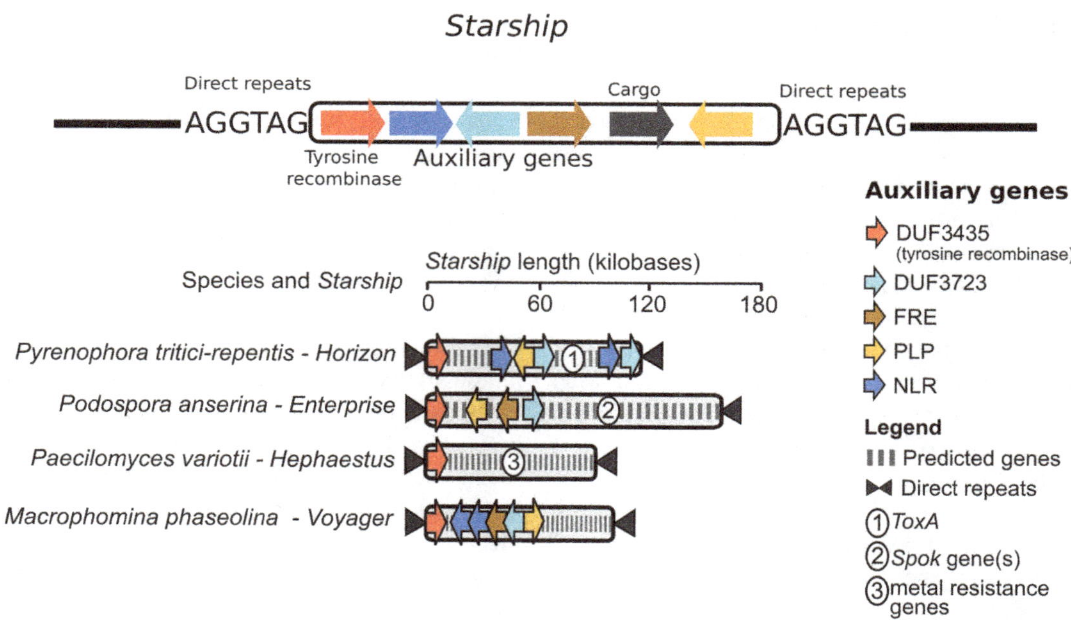


**Fig. 1.1** Schematic diagrams of the *Starship* transposable elements. (Top) A generalized depiction including important genes and an example of what the DRs may look like (AGGTAG). (Bottom) Specific *Starships* described in the text. Relevant cargo is indicated with numbers, see the legend. Note that the predicted genes are illustrative and do not represent accurate models. *DUF* domain of unknown function, *FRE* ferric reductase, *PLP* patatin-like phosphatases, *NLR* NOD-like receptors

*Voyager* thus provided a powerful template for revealing shared patterns of gene organization, even in poorer quality genomes, and a means to identify giant TE-like regions through a guilt-by-association approach independent of sequence alignments. Now, clusters of tyrosine recombinases and its four "auxiliary" genes could be used as a signature for finding giant TE-like regions. This advance dramatically increased the number of sequences available for a comparative analysis and provided an opportunity to comprehensively assess whether they all shared a common origin. I reconstructed the evolutionary history of the tyrosine recombinases from giant TE-like regions, as well as those in the *ToxhAT* region, *Hephaestus, Enterprise*, and *Voyager*, and found that all examined tyrosine recombinases indeed share a single origin and are phylogenetically distinct from all other families of eukaryotic and prokaryotic TEs (Gluck-Thaler et al. 2022). This was the first indication that all of the giant TEs identified to date are indeed manifestations of the same biological process, and thus the *Starship* superfamily was born. Aaron and I soon began contacting and meeting virtually with the other groups studying these elements to share our results ahead of publication so we could collectively coordinate and orient our research agendas.

## 1.6 *Starships* and Perspectives

The discovery of *Starships* as a widespread, previously unaccounted-for phenomenon fundamentally changes how we should interpret large structural variants in fungal genomes and challenges our current expectations about the mode and tempo of fungal evolution. Indeed, part of the motivation for searching for a deeper mechanism connecting these phenomena were numerous observations in the literature of large structural variants and horizontal gene transfers in fungal genomes whose underlying mechanisms were obscure. These included horizontally transferred regions such as those in *Penicillium* species domesticated for cheese making (Ropars et al. 2015), multiple biosynthetic gene clusters (Wisecaver and Rokas 2015), regions important for plant pathogenicity (Qiu et al. 2016), and large "genomic islands" whose presence/absence varied across individuals (Fedorova et al. 2008). It should be noted that important shifts in sequencing technologies were instrumental in all of our work finding the *Starships*. Cheap and/or easy access to long-read technologies (i.e., Oxford Nanopore Technologies, PacBio) allowed us to assemble several individual strain genomes from the same species. Having multiple assemblies provides the opportunity to observe "empty" sites, where the *Starships* are variably present or absent among strains, while their high quality reveals the TE signatures that were obscured in previous, fragmented assemblies.

Although a relatively short amount of time has elapsed between the description of the *Starships* and the writing of this chapter, considerable progress has been made toward understanding the functions and diversity of these enigmatic elements. It has since been confirmed that the movement of *Starships* can be observed experimentally and that the captain tyrosine recombinases are the sole gene within the *Starships* necessary and sufficient for their transposition (Urquhart et al. 2023). This was achieved by "sandwiching" the *Hephaestus* element within an antibiotic resistance cassette. When *Hephaestus* jumped out of the antibiotic marker the strain became resistant to antibiotics and in most cases *Hephaestus* was reintegrated elsewhere in the genome. Interestingly, the captain protein could be supplied in trans, raising the possibility of non-autonomous *Starship* elements. A further development in the field is that instead of relying on expert inspections of manual sequence alignments, researchers with no prior experience can now systematically annotate *Starships* using "starfish," a newly developed *Starship* annotation workflow (Gluck-Thaler and Vogan 2024). The first application of starfish yielded a torrent of newly identified *Starships* whose immense sequence diversity necessitated the development of a classification system. We have now organized extant *Starship* diversity into 11 families, which are further divided into groups of more closely related elements (called "naves," latin for "ships"), and below that into individual haplotypes. Investigations of these data have led to the realization that specific elements and families target unique genomic niches, including AT-rich regions and the 5S rDNA gene (Gluck-Thaler and Vogan 2024). As more elements continue to be described, we expect manifold opportunities for comparative analyses to further reveal the biology of these elements.

In conjunction with understanding the diversity of these elements, there is also a growing appreciation for the contributions of *Starships* to organismal phenotypes. Previous studies have identified large presence/absence polymorphisms linked to stress tolerance and aflatoxin production in *Aspergillus flavus* (Fountain et al. 2020), and Sorbic acid resistance in *Penicillium roqueforti* (Punt et al. 2022). After examining these regions, we can now see that both are in fact *Starships*. The *ToxA* region has been firmly identified as a *Starship*, now named *Horizon* (Gourlie et al. 2022). Other research groups have begun to identify suites of elements with compelling phenotypes. The presence of multiple *Starships* from fungi involved in the fermentation of sausage, *P. nalgiovense* and *P. salamii*, correlates with proteolytic and lipolytic activity and have clear signature of HGT (Lo et al. 2023). A putatively degraded *Starship* is evidence that the T-toxin of *Cochliobolus heterostrophus* was transferred into the species via *Starship* mediated HGT (Haridas et al. 2023). Lastly, a gene associated with climate adaptation in the wheat pathogen *Zymoseptoria tritici* is carried on the *Starship Swordfish*, which although no longer active, still segregates within natural populations and contributes to the genetic diversity of this species (Moser Tralamazza et al. 2023).

Since the delineation of the *Starship* superfamily, there has also been a steady increase in the characterization of *Starships* that have signatures of recent transfer between species, suggesting that these elements are a long-sought after active mechanism of fungal horizontal gene transfer. In many cases, the successful transfer of these elements and their cargo has resulted in the rapid evolution of adaptive phenotypes (Urquhart et al. 2022, 2023; Gourlie et al. 2022). In a particularly illustrative example, a gene cluster responsible for the detoxification of formaldehyde has been horizontally transferred by four different families of *Starships* in total at least nine times. The recurrent nature of these horizontal transfers implies that *Starships* are true drivers of HGT in fungi (Urquhart et al. 2022, 2023; Gourlie et al. 2022). In other cases, the elements have proven to mediate large-scale chromosomal rearrangements, the insertions of large genomic regions, and the transposition of "mobile" biosynthetic gene clusters (Fountain et al. 2020; Gluck-Thaler et al. 2022, 2024; Bucknell et al. 2024). Not only is there now a common mechanism to explain how previously disparate genomic variation likely arises in fungal genomes, but the beginnings of a new framework for evaluating and predicting fungal adaptation.

The *Starships* provide an emblematic example of how scientific progress is achieved. The story began with the independent identification of individual *Starships* by multiple groups, followed by the recognition that these elements represented a single new phenomenon, and eventually advancing with the experimental demonstration of *Starship* movement. Progression of the *Starship* story, at each stage, was enabled through collaboration and open science policies, like sharing unpublished data and methods at conferences, reaching out to strangers, meeting with each other, and publishing preprints. Thus, through a combination of independent discoveries, collaborative sharing, and technological advances, the *Starships* were revealed. Now, the recognition of *Starships* as distinct entities with evolvable features has revealed myriad research opportunities to delve deeper into the genetic bases of fungal adaptation. Neither the short- nor long-term evolutionary consequences of *Starship*–fungal interactions are currently understood, but based on analogous nested symbioses between bacteria and their mobile elements, such interactions have vast potential to mediate ecological outcomes through both additive and non-additive effects (Batstone 2022; Weisberg et al. 2022). Building on the strong foundation of how *Starships* were first identified, we expect the next decade of research into *Starship* biology to continue fueling collaborative opportunities for discovery that will together reveal new insights into the principals of fungal ecology and evolution.

**Acknowledgments** This work was supported by the Office of the Vice Chancellor for Research and Graduate Education at the University of Wisconsin-Madison with funding from the Wisconsin Alumni Research Foundation to E.G.-T; by funding from start-up funds from the Department of Plant Pathology at the University of Wisconsin-Madison; by the Swedish Research Council Formas (grant number 2019-01227) and the Swedish Research Council VR to A.A.V (grant number 2021-04290) and to S.L.A.-V. (grant number 2022-00341); by the Sun Foundation Peer Prize for Women in Science to M.C.M.

## References

Amselem J, Lebrun M-H, Quesneville H (2015) Whole genome comparative analysis of transposable elements provides new insight into mechanisms of their inactivation in fungal genomes. BMC Genomics 16:141

Arkhipova IR, Yushenova IA (2019) Giant transposons in eukaryotes: is bigger better? Genome Biol Evol 11:906–918

Batstone RT (2022) Genomes within genomes: nested symbiosis and its implications for plant evolution. New Phytol 234:28–34

Bazzicalupo AL, Ruytinx J, Ke Y-H et al (2020) Fungal heavy metal adaptation through single nucleotide polymorphisms and copy-number variation. Mol Ecol 29:4157–4169

Bucknell A, Wilson HM, Gonçalves do Santos KC, et al (2024) Sanctuary: a starship transposon facilitating the movement of the virulence factor ToxA in fungal wheat pathogens. bioRxiv

Castanera R, López-Varas L, Borgognone A et al (2016) Transposable elements versus the fungal genome: impact on whole-genome architecture and transcriptional profiles. PLoS Genet 12:e1006108

Debets AJM, Dalstra HJP, Slakhorst M et al (2012) High natural prevalence of a fungal prion. Proc Natl Acad Sci USA 109:10432–10437

Engel SR, Dietrich FS, Fisk DG et al (2014) The reference genome sequence of *Saccharomyces cerevisiae*: then and now. G3 4:389

Fedorova ND, Khaldi N, Joardar VS et al (2008) Genomic islands in the pathogenic filamentous fungus *Aspergillus fumigatus*. PLoS Genet 4:e1000046

Fountain JC, Clevenger JP, Nadon B et al (2020) Two new reference genomes reveal a large insertion potentially contributing to isolate stress tolerance and aflatoxin production. G3 10:3515–3531

Friesen TL, Holmes DJ, Bowden RL, Faris JD (2018) *ToxA* is present in the U.S. *Bipolaris sorokiniana* population and is a significant virulence factor on wheat harboring *Tsn1*. Plant Dis 102:2446–2452

Friesen TL, Stukenbrock EH, Liu Z et al (2006) Emergence of a new disease as a result of interspecific virulence gene transfer. Nat Genet 38:953–956

Gluck-Thaler E, Haridas S, Binder M et al (2020) The architecture of metabolism maximizes biosynthetic diversity in the largest class of fungi. Mol Biol Evol 37:2838–2856

Gluck-Thaler E, Ralston T, Konkel Z et al (2022) Giant *Starship* elements mobilize accessory genes in fungal genomes. Mol Biol Evol 39. https://doi.org/10.1093/molbev/msac109

Gluck-Thaler E, Slot JC (2018) Specialized plant biochemistry drives gene clustering in fungi. ISME J 12:1694–1705

Gluck-Thaler E, Vogan AA (2024) Systematic identification of cargo-mobilizing genetic elements reveals new dimensions of eukaryotic diversity. Nucleic Acids Res 52:5496. https://doi.org/10.1093/nar/gkae327

Gluck-Thaler E, Forsythe A, Puerner C, Stajich JE, Croll D, Cramer RA, Vogan AA (2024) Giant transposons promote strain heterogeneity in a major fungal pathogen. bioRxiv [Preprint]. https://doi.org/10.1101/2024.06.28.601215

Gourlie R, McDonald M, Hafez M et al (2022) The pangenome of the wheat pathogen *Pyrenophora tritici-repentis* reveals novel transposons associated with necrotrophic effectors *ToxA* and *ToxB*. BMC Biol 20:239

Grognet P, Lalucque H, Malagnac F, Silar P (2014) Genes that bias Mendelian segregation. PLoS Genet 10:e1004387

Guin K, Sreekumar L, Sanyal K (2020) Implications of the evolutionary trajectory of centromeres in the fungal kingdom. Ann Rev Microbiol 74:835–853

Gupta PK, Chand R, Vasistha NK et al (2018) Spot blotch disease of wheat: the current status of research on genetics and breeding. Plant Pathol 67:508–531

Hamann A, Osiewacz HD (2004) Genetic analysis of spore killing in the filamentous ascomycete *Podospora anserina*. Fungal Genet Biol 41:1088–1098

Haridas S, González JB, Riley R et al (2023) T-toxin virulence genes: unconnected dots in a sea of repeats. MBio 14:e0026123

Kojima KK, Jurka J (2011) *Crypton* transposons: identification of new diverse families and ancient domestication events. Mob DNA 2:12

Konkel Z, Kubatko L, Slot JC (2023) *CLOCI:*unveiling cryptic gene clusters with generalized detection. bioRxiv

Lo Y-C, Bruxaux J, Rodríguez de la Vega RC et al (2023) Domestication in dry-cured meat fungi: convergent specific phenotypes and horizontal gene transfers without strong genetic subdivision. Evol Appl 16:1637–1660

McDonald MC, Ahren D, Simpfendorfer S et al (2018) The discovery of the virulence gene *ToxA* in the wheat and barley pathogen *Bipolaris sorokiniana*. Mol Plant Pathol 19:432–439

McDonald MC, Taranto AP, Hill E et al (2019) Transposon-mediated horizontal transfer of the host-specific virulence protein *ToxA* between three fungal wheat pathogens. MBio 10. https://doi.org/10.1128/mBio.01515-19

Moolhuijzen P, See PT, Hane JK et al (2018) Comparative genomics of the wheat fungal pathogen *Pyrenophora tritici-repentis* reveals chromosomal variations and genome plasticity. BMC Genomics 19:279

Moser Tralamazza S, Gluck-Thaler E, Feurtey A, Croll D (2023) Copy number variation introduced by a massive mobile element underpins global thermal adaptation in a fungal wheat pathogen. bioRxiv

Navathe S, Yadav PS, Chand R et al (2020) ToxA-Tsn1 interaction for spot blotch susceptibility in Indian wheat: an example of inverse gene-for-gene relationship. Plant Dis 104:71–81

Padieu E, Bernet J (1967) Mode of action of the genes responsible for abortion of certain products of meiosis in the ascomycete, *Podospora anserina*. C R Acad Hebd Seances Acad Sci D 264:2300–2303

Porquier A, Tisserant C, Salinas F et al (2021) Retrotransposons as pathogenicity factors of the plant pathogenic fungus *Botrytis cinerea*. Genome Biol 22:225

Punt M, Seekles SJ, van Dam JL et al (2022) High sorbic acid resistance of *Penicillium roqueforti* is mediated by the SORBUS gene cluster. PLoS Genet 18:e1010086

Qiu H, Cai G, Luo J et al (2016) Extensive horizontal gene transfers between plant pathogenic fungi. BMC Biol 14:41

Ropars J, Rodríguez de la Vega RC, López-Villavicencio M et al (2015) Adaptive horizontal gene transfers between multiple cheese-associated fungi. Curr Biol 25:2562–2569

Saupe SJ (2007) A short history of small s: a prion of the fungus *Podospora anserina*. Prion 1:110–115

Selker EU (2002) Repeat-induced gene silencing in fungi. Adv Genet 46:439–450

Shivas RG, Beasley DR, Bransgrove K et al (2019) Biodiscovery and the Queensland plant pathology herbarium. Microbiol Aust 40:134

Silar P (2020) Podospora anserina. Hal Open Sci

Tuori RP, Wolpert TJ, Ciuffetti LM (1995) Purification and immunological characterization of toxic components from cultures of *Pyrenophora tritici-repentis*. Mol Plant-Microbe Interact 8:41–48

Urquhart AS, Chong NF, Yang Y, Idnurm A (2022) A large transposable element mediates metal resistance in the fungus *Paecilomyces variotii*. Curr Biol 32:937–950.e5

Urquhart AS, Mondo SJ, Mäkelä MR et al (2018) Genomic and genetic insights into a cosmopolitan fungus, *Paecilomyces variotii* (Eurotiales). Front Microbiol 9:3058

Urquhart AS, Vogan AA, Gardiner DM, Idnurm A (2023) *Starships* are active eukaryotic transposable elements mobilized by a new family of tyrosine recombinases. Proc Natl Acad Sci USA 120:e2214521120

van der Gaag M, Debets AJ, Oosterhof J et al (2000) Spore-killing meiotic drive factors in a natural population of the fungus *Podospora anserina*. Genetics 156:593–605

Vogan AA, Ament-Velásquez SL, Bastiaans E et al (2021) The *Enterprise*, a massive transposon carrying *Spok* meiotic drive genes. Genome Res 31:789–798

Vogan AA, Ament-Velásquez SL, Granger-Farbos A et al (2019) Combinations of *Spok* genes create multiple meiotic drivers in *Podospora*. elife 8:e46454

Weisberg AJ, Rahman A, Backus D et al (2022) Pangenome evolution reconciles robustness and instability of Rhizobial Symbiosis. MBio 13:e0007422

Wells JN, Feschotte C (2020) A field guide to eukaryotic transposable elements. Annu Rev Genet 54:539–561

Wicker T, Sabot F, Hua-Van A et al (2007) A unified classification system for eukaryotic transposable elements. Nat Rev Genet 8:973–982

Wisecaver JH, Rokas A (2015) Fungal metabolic gene clusters-caravans traveling across genomes and environments. Front Microbiol 6:161

# 2 The Three-Dimensional Chromatin Architecture in Fungi

Satya Dev Polisetty and Kaustuv Sanyal

**Abstract**

Three-dimensional genome organization regulates gene expression of key cellular processes, including DNA replication, transcription, repair, and recombination. In fungal species like *Saccharomyces cerevisiae*, *Candida albicans*, *Schizosaccharomyces pombe*, and *Neurospora crassa,* chromosomes are organized in a polarized fashion to adopt the Rabl configuration. In this mode of chromosomal arrangement, all centromeres form a single cluster at the nuclear envelope near the spindle pole bodies. Telomeres, on the other hand, are present at the opposite end toward the nucleolus. In metazoans, the genome is organized into distinct chromosome territories and adopts non Rabl-like configuration. During the cell cycle, the chromatin undergoes various morphological changes as the cells progress through the cell division. Understanding the key processes involved in modulating the chromatin architecture to suit the needs of the cell is essential. This chapter discusses the principles of fungal genome organization and how the chromatin is folded to facilitate chromosome segregation and its role in DNA replication timing.

**Keywords**

Three-dimensional genome organization · Rabl · Non-Rabl · Centromere clustering · Hi-C · Chromosome territories · Replication timing · Cohesins · Condensins

## 2.1 Introduction

Typical hallmarks of a eukaryotic chromosome include the presence of several replication origins, one centromere, and two telomeres. DNA replication origins are the sites where DNA replication initiates, thereby ensuring the completion of the genome duplication within a specified time frame (Fragkos et al. 2015). The centromere is a specific locus on the eukaryotic chromosome that acts as a platform for the assembly of a multi-protein structure, the kinetochore, that binds to microtubules (MTs) and facilitates the equal segregation of duplicated chromosomes (McAinsh and Marston 2022; Guin et al. 2020b). Telomeres are the repetitive DNA regions that constitute the ends of the chromosomes. Telomeric repeats, in association with telomere-associated proteins, protect the DNA ends (Kupiec 2014). During the mitotic cell cycle,

S. D. Polisetty · K. Sanyal (✉)
Molecular Mycology Laboratory, Molecular Biology and Genetics Unit, Jawaharlal Nehru Centre for Advanced Scientific Research, Bengaluru, India
e-mail: pvssdev@jncasr.ac.in; sanyal@jncasr.ac.in


M. Nowrousian, J. E. Stajich (eds.), *Fungal Genomics*, The Mycota 13,
https://doi.org/10.1007/978-3-031-75666-5_2

duplicated chromosomes are segregated equally into daughter cells.

The mitotic cell cycle is a coordinated set of events that results in the generation of two cells. In eukaryotes, the typical cell cycle consists of four stages G1 (Gap1), S (Synthesis), G2 (Gap2), and M (Mitosis) phase. Interphase constitutes G1, G2, and S phases. G1 and G2 comprise the gap phases, where cells undergo changes to suit the needs of a cell division event. During S phase, a cell duplicates its genome, and in M phase, the duplicated genome is separated into two daughter cells. In multicellular organisms like metazoans, cell cycle phases have defined boundaries. However, in certain unicellular organisms, such as the budding yeast *Saccharomyces cerevisiae*, there is no defined G2 phase between S and M phases, and it is referred to as the G2/M phase (for a comprehensive review about cell cycle, refer (Morgan 2006)). M phase consists of four stages: prophase, metaphase, anaphase, and telophase. As the cells enter prophase, chromatin is structurally reorganized to form compact, dense structures in form of chromosomes (Batty and Gerlich 2019). Cells showing defects in chromatin condensation result in the mis-segregation of chromosomes during mitosis (Hirano 2012). Hence, proper folding and compaction of interphase chromatin is essential for the faithful segregation of chromosomes. During metaphase, compacted chromosomes are aligned along the metaphase plate in metazoans. The kinetochores assembled onto centromere DNA facilitate the attachment of chromosomes to MTs emanating from the MT organizing centers (MTOCs) known as centrosomes or spindle pole bodies (SPBs) (Bloom and Joglekar 2010; Winey and Bloom 2012). Once both the sister kinetochores of the duplicated chromatids are attached to the MTs and are oriented toward the poles, the sister chromatids are segregated to the daughter cells during anaphase. Chromatin is decondensed as the cells enter telophase, and the interphase chromatin organization is restored (Batty and Gerlich 2019). Therefore, to ensure genome stability, a cell must fold its chromatin correctly, and the folded, condensed chromosomes must be properly attached to the MTs and segregated equally into daughter cells.

## 2.2 General Features of the 3D Genome Organization in the Nucleus

The eukaryotic DNA inside the nucleus is wrapped around histone octamers or nucleosomes to form the 11-nm beads-on-a-string structure. This structure is further folded to form a 30-nm chromatin fiber (Ou et al. 2017). A combination of molecular biology and DNA sequencing techniques has led to the development of the chromosome conformation capture (3C) technology that improved our understanding of the organization features of eukaryotic genomes significantly (Dekker et al. 2002) (For more information, read Box 2.1). 3C-based studies in metazoans revealed several features of genome folding (Fig. 2.1). Within the nucleus, at the kilobase scale, chromatin is organized in discrete domains where chromatin loops have a higher tendency to interact among themselves. These structures are referred to as topologically associating domains (TADs). Apart from mammalian genomes, TAD-like structures are observed in many organisms and range from ten to hundreds of kilobases (Dixon et al. 2012; Nora et al. 2012; Sexton et al. 2012; Hou et al. 2012). TADs of similar epigenetic states are further arranged to form megabase scale self-interacting regions known as A and B compartments. The A compartment comprises transcriptionally active genes (euchromatic), and the B compartment is generally enriched with relatively poorly transcribed genes (heterochromatic) (Lieberman-Aiden et al. 2009; Rao et al. 2014; Wang et al. 2016). The transcriptionally silent heterochromatin regions are present toward the periphery of the nucleus, and the transcriptionally active euchromatin regions are more often located toward the center of the nucleus. On a whole genome scale, the individual chromosomes occupy their own space known as chromosome territories (CTs) (Cremer and Cremer 2010). The chromosome with the ribosomal DNA (rDNA) locus is organized in the form of a membrane-less structure known as the nucleolus and is located toward the center of the nucleus in mammals and at the nuclear periphery in most yeast species. The heterochromatin loci, such as telomeres and centromeres, are present at the nuclear periphery in yeasts, plants, and animals (Mekhail and Moazed 2010).

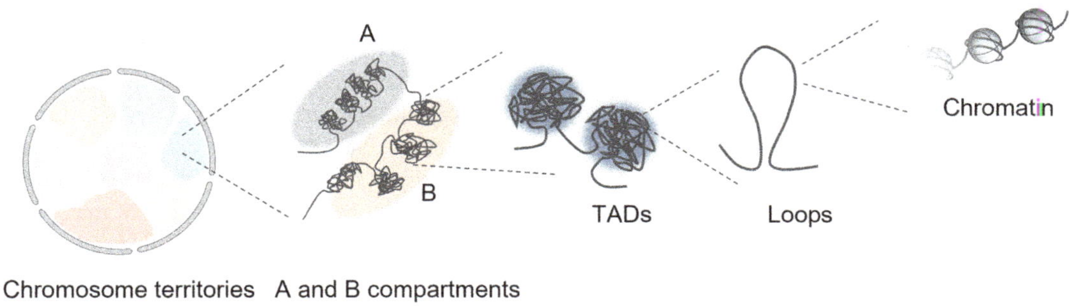


**Fig. 2.1** Hierarchical organization of chromatin within the nucleus. In metazoans like humans, chromosomes are present in defined spaces within the nucleus known as chromosome territories (CTs). Within CTs, compartments of different epigenetic states known as A and B compartments are present. The A and B compartments are composed of several topologically associating domains (TADs). TADs are structures that are enriched with self-interacting chromatin loops. On the finest scale, chromatin loops are formed of DNA wrapped around octameric nucleosomes

**Box 2.1 Hi-C**

Hi-C (High-throughput chromosome conformation capture) is a DNA sequencing-based technique that is a modification of the 3C technique and allows genome-wide identification of 3D chromatin interactions in the nucleus (Lieberman-Aiden et al. 2009). In Hi-C, the protein–DNA interactions are fixed using cross-linkers such as formaldehyde. The cross-linked nuclei are subjected to digestion with restriction enzymes such as *DpnII*. The digested DNA ends are filled in with biotinylated nucleotide (dATP or dCTP), and the ends are ligated. This leads to the formation of chimeric ligated fragments. Both digestion and ligation are performed inside the nucleus to minimize non-specific ligations arising from random interactions among chromatin fragments. Ligated DNA fragments are decrosslinked and sonicated to break the circularized ligated products. The sonicated fragments are enriched for ligation products by pulling down biotinylated DNA fragments. These fragments are sonicated to sizes suitable for high-throughput sequencing and are sequenced using the paired-end sequencing method. The sequencing reads are mapped onto the genome, and whole genome interaction maps are obtained.

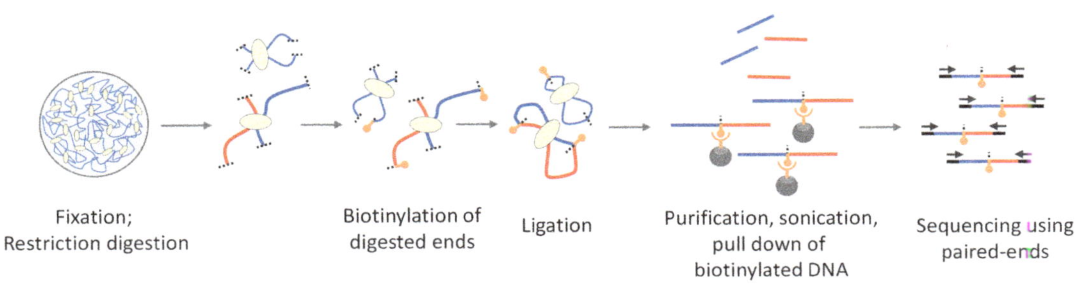


**Box Fig. 2.1** Key steps in Hi-C protocol

## 2.3 Centromere Clustering and its Role in Nuclear Architecture

In most organisms studied, microscopic observations identified that the centromeres have a strong propensity to cluster, and they are preferentially found to be localized at the nuclear periphery (Muller et al. 2019). Many species of fungi, animals, plants, and protozoans studied show clustering of centromeres inside the nucleus. Centromere clustering can be constitutive or stage-specific (Fig. 2.2). Centromere clustering is nearly constitutive in most fungal species studied (Guin et al. 2020b; Torres et al. 2023). Apart from constitutive clustering, cell cycle or developmental stage-specific centromere clustering is also present in certain species of fungi, animals, and plants (Fig. 2.2a) (Muller et al. 2019; Torres et al. 2023; Guin et al. 2020b). Centromere clustering also imparts a profound effect on the positioning and organization of the chromosomes inside the nucleus. The arrangement of chromosomes inside the nucleus determines the nuclear architecture, and it varies across organisms, depending on the cell cycle and developmental stages (Muller et al. 2019). Based on the spatial positioning of the centromeres and telomeres within the nucleus, nuclear architecture is classified into Rabl and non-Rabl configurations (Fig. 2.2b, c, and d) (Hoencamp et al. 2021). In the Rabl configuration, centromeres and telomeres are arranged in a polarized manner within the nucleus. In this configuration, centromeres are clustered at one end of the nucleus, and the telomeres are at the opposite end of the nucleus, away from the centromere cluster. In the non-Rabl configuration, the centromeres and telomeres are scattered inside the nucleus without any polarized arrangement.

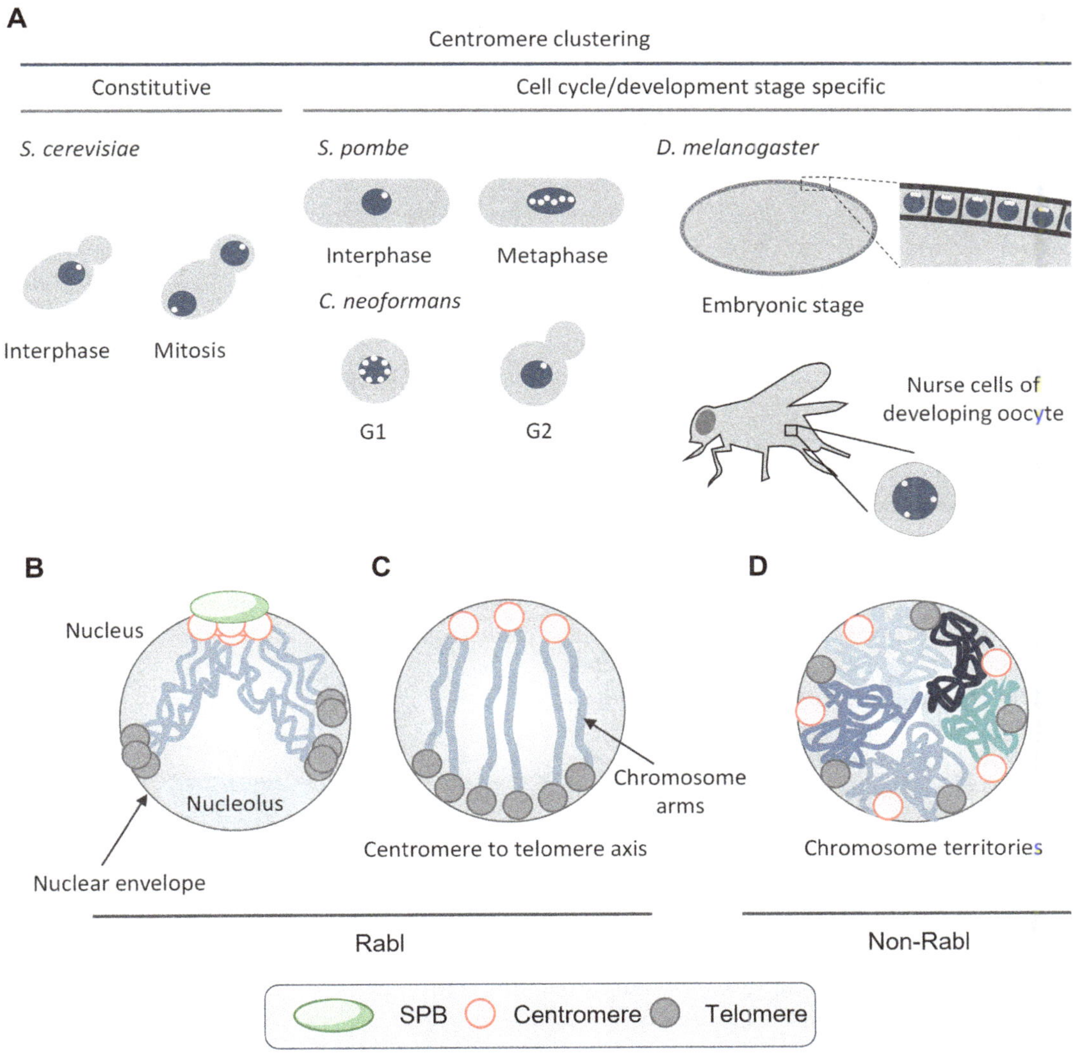


**Fig. 2.2** Centromere clustering and types of nuclear architecture. (**a**) Schematic showing constitutive or cell cycle/developmental stage-specific clustering of centromeres at the nuclear periphery. The nucleus is represented as a dark gray circle, and the centromeres as a light pink circle. Left, constitutive clustering of centromeres throughout the cell cycle as seen in various fungal species such as *S. cerevisiae*, *C. albicans*, *C. tropicalis*, and *N. crassa*. Center, cell cycle stage-specific clustering of centromeres as seen in certain fungal species such as *S. pombe* and *C. neoformans*. Right, metazoan species such as *Drosophila* exhibit developmental stage-specific clustering of centromeres. Embryonic stages have clustered centromeres, and nurse cells of adult flies display unclustered centromeres. (**b**) Schematic of the Rabl configuration representing centromeres clustered beneath the SPBs or MTOCs and telomeres clustered away from the centromere cluster near the nucleolus toward the nuclear periphery, as seen in *S. cerevisiae* and *S. pombe*. (**c**) Schematic of the Rabl arrangement of centromeres and telomeres as observed in some of the plant species such as maize and barley (Mascher et al. 2017; Dong et al. 2017). In this configuration, the chromosomes are arranged along the centromere-to-telomere axis such that the arms of a chromosome tend to interact more frequently. (**d**) Scattered arrangement of centromeres and telomeres representing the non-Rabl arrangement of chromosomes

## 2.4 Fungi

In various fungal species studied, such as *S. cerevisiae* (Guacci et al. 1997; Jin et al. 2000; Duan et al. 2010), *S. pombe* (Funabiki et al. 1993; Mizuguchi et al. 2014), *Neurospora crassa* (Galazka et al. 2016), *C. albicans* (Thakur and Sanyal 2012; Burrack et al. 2016), *Candida tropicalis* (Guin et al. 2020a), *Malassezia sympodialis* (Sankaranarayanan et al. 2020), and *M. oryzae* (Yadav et al. 2019) chromosomes are organized in the Rabl configuration (Fig. 2.2b). Centromeres show robust clustering and appear as a single punctum inside the nucleus. Microscopic studies in *S. cerevisiae* and *S. pombe* revealed that centromeres cluster near the SPBs during interphase. Telomeres are located at the periphery of the nucleus and lie closer to the nucleolus in these organisms (Funabiki et al. 1993; Guacci et al. 1997; Jin et al. 2000). Hi-C studies in several fungal species belonging to major phyla such as ascomycetes and basidiomycetes indicated that the chromosomes are organized in the Rabl configuration (Torres et al. 2023). Though most studies show that fungal genomes are predominantly organized in the Rabl configuration, certain fungi deviate from this trend. For example, the centromeres are unclustered during interphase in *Cryptococcus neoformans* and are located near the periphery of the nucleus (Kozubowski et al. 2013) in interphase. A recent Hi-C-based study in the glomeromycete fungus *Rhizophagus irregularis* indicated the absence of centromere clustering but prominent telomere clustering was evident (Yildirir et al. 2022).

## 2.5 Features of the Genome Organization in Different Organisms

### 2.5.1 *S. cerevisiae* and *S. pombe*

The three-dimensional organization of the genome was studied using high-throughput chromatin conformation capture (Hi-C) techniques, which has revealed the organization principles of genomes in many organisms. *S. cerevisiae* and *S. pombe* have similar genome sizes (12 and 14 Mb, respectively). Extensive studies in these two model yeast species have shed light on how the genomes are organized in these fungal species. High-resolution Hi-C contact maps of these fungal species revealed several striking features of the genome organization (For more information on interpreting Hi-C contact maps, read Box 2.2). As mentioned previously, the most prominent feature visible in the Hi-C contact maps in both organisms is the presence of strong centromere clustering, which appears as dark spots highlighting inter-centromeric interactions (Fig. 2.3). Similarly, inter-telomere interactions are also observed as dark spots on the maps, indicating telomere clustering. In *S. cerevisiae*, it is shown that the chromosome arm length influences the intra-telomeric interactions within the chromosome arms. Chromosomes with equal arm lengths show enhanced intra-telomeric inter-

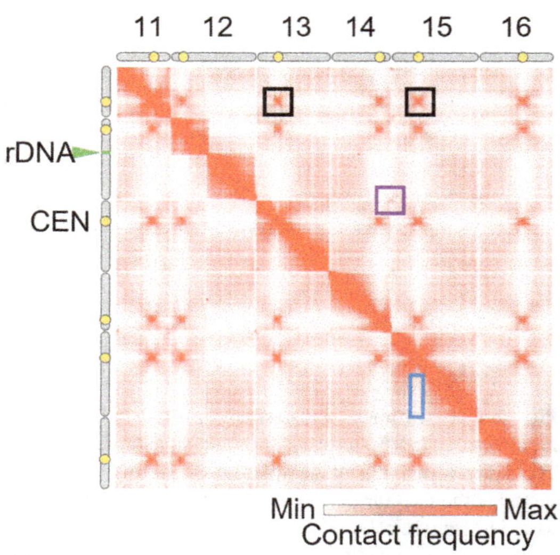


**Fig. 2.3** Hi-C contact map of asynchronously grown *S. cerevisiae*. Contact map of chromosomes 11, 12, 13, 14, 15, and 16 of *S. cerevisiae*. The rDNA locus is indicated with a green arrowhead and centromeres are represented as yellow circles. Black squares represent inter-centromeric interactions, violet squares represent inter-telomeric interactions, and the blue rectangle represents the region devoid of centromeric interactions with the arms of the chromosome (HiC data source: (Rutledge et al. 2015), processed Hi-C data available in the Juicebox software (Durand et al. 2016) was used for generating the contact maps)

action compared to unequal arm lengths (Schober et al. 2008; Duan et al. 2010; Therizols et al. 2010). The centromere-proximal inter-arm interactions are enriched as the chromosomes are arranged in the Rabl configuration. As a result, a cross-like pattern of inter-centromeric interactions appears (Fig. 2.3, black squares). Other than inter-centromeric interactions, the spatial confinement of centromeres and telomeres within the nucleus prevents their interactions with chromosomal arms. As a result, centromeres act as a strong barrier that compartmentalizes the chromosomal arms into two distinct regions. This leads to higher intra-arm interaction frequencies than the inter-arm interactions observed within the chromosome. In *S. cerevisiae*, similar to centromeres, the nucleolus-associated ribosomal DNA (rDNA) locus on chromosome 12 also acts as a barrier similar to the centromeres. As a result, the rDNA locus and the centromere compartmentalize chromosome 12 into three distinct regions (Duan et al. 2010) (Fig. 2.3, green arrowhead). A similar feature is not evident in *S. pombe* as the rDNA locus is located close to the telomeres of chromosome 3. One striking feature in *S. pombe* Hi-C maps is the long-range inter-chromosomal interaction between the mating type (*mat*) locus on chromosome 2 and the right telomere of chromosome 1. Microscopy analysis has revealed that only a certain subset of cells shows this interaction (Mizuguchi et al. 2014).

**Box 2.2 Interpreting Hi-C Contact Maps**

The genome-wide DNA-DNA contacts captured from a Hi-C experiment are represented as a two-dimensional heatmap/contact map. A typical contact map represents the strength of interactions between different genomic loci. Shown below on the left is a simplified schematic of the Hi-C contact map for three chromosomes, and on the right is the schematic showing two chromosomes highlighting the interactions in the Hi-C contact map. The dark-to-light red color gradient represents higher to lower interaction frequency/strength. The x- and y-axes represent the genomic coordinates of the three chromosomes. The inter-centromere interactions between two chromosomes (black arrow) are highlighted as a black box in the map. The inter-telomeric interaction (green arrows) is highlighted as a green box in the map. The intra-chromosomal interactions within the chromosomal arms are represented as a yellow box. The strong diagonal along the matrix represents enriched interaction among the neighboring regions within the chromosome. The intra-chromosomal arm interactions (violet arrows) and the inter-chromosomal arm interactions (blue arrows) are highlighted as violet and blue boxes, respectively, in the map.

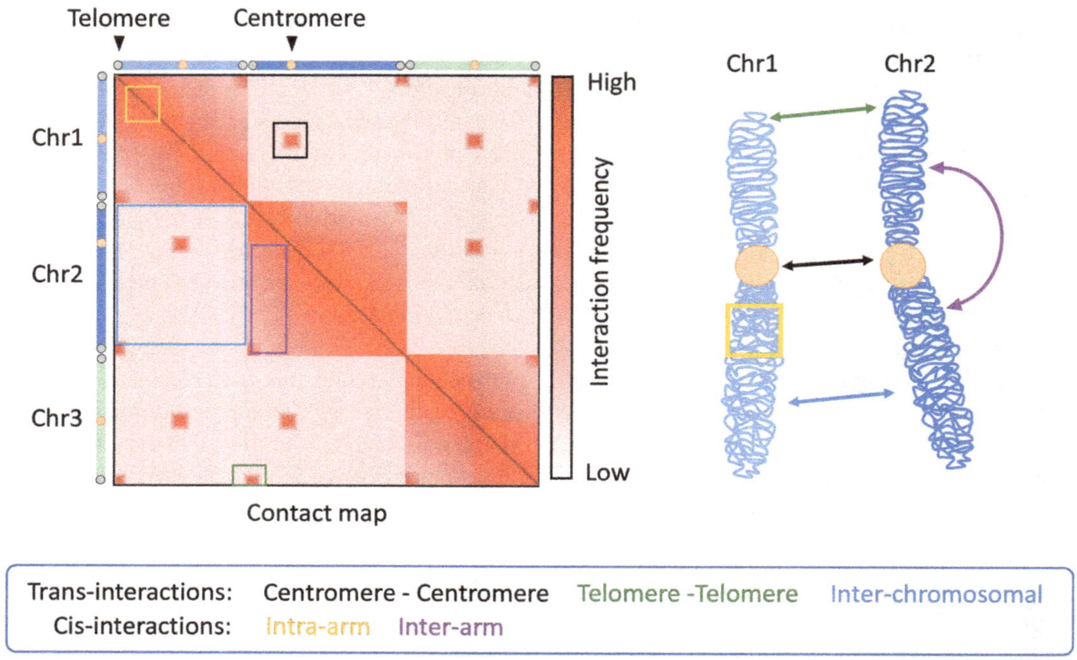


**Box Fig. 2.2** Schematic of Hi-C contact map highlighting genomic interactions

## 2.5.2 Other Fungal Model Organisms

In *N. crassa*, on a global scale, the contact maps exhibit prominent interaction among centromeres and telomeres similar to that of *S. cerevisiae* and *S. pombe*. The genome of *N. crassa* is characterized by the presence of interspersed heterochromatic regions that are AT-rich (Lewis et al. 2009). These regions range from <1 to 400 kb in size (Klocko et al. 2016; Galazka et al. 2016). The AT-rich regions or isochores exhibit strong interactions within and as well as across the chromosomes. As a result of these interactions, chromatin forms "globules" of self-interacting regions of sizes approximately 60–80 kb. These globules flank the gene-rich regions along the chromosomes (Rodriguez et al. 2022).

Similar to *N. crassa*, in fungal species such as *Epichloe festucae*, the genome is interspersed with long blocks of AT-rich gene-poor regions. The AT-rich isochores harbor a tightly packed chromatin structure and exhibit higher levels of interactions with other AT-rich isochores of the genome (Winter et al. 2018). Although TADs are not readily distinguishable in *S. cerevisiae* and *S. pombe*, the plant pathogen *Verticillium dahliae* exhibits distinct TADs across the genome where the TAD boundaries are well-defined by gene-poor and less transcriptionally active genes (Torres et al. 2024).

As mentioned previously, in several organisms, the A and B compartments are majorly composed of euchromatin and heterochromatin, respectively. The A compartment is more gene-dense than the B compartment and corresponds to transcriptionally active and repressed genomic regions, respectively (Szabo et al. 2019). A/B compartments are absent in the genomes of most fungal species studied. One study reported the compartmentalization of the genome in *R. irregularis*, The HiC contact maps of *R. irregularis* exhibit a checkerboard pattern indicating the presence of distinct A and B compartments (Yildirir et al. 2022). Strikingly, the transcription state of A/B compartments in *R. irregularis* depends on the growth conditions, where in planta growth shows higher gene expression in B compartment as compared to spore germination conditions (Yildirir et al. 2022).

## 2.6 The Dynamic Architecture of Chromatin: How Chromatin Is Modulated to Suit Cell Division

During the transition from interphase to metaphase, chromosomes gradually become condensed to form units that can be segregated easily. In this process, the interphase chromatin architecture is modified to suit the requirements of different cell cycle stages and to ensure proper chromosome segregation. Hi-C studies revealed the presence of conserved principles of chromatin organization across several organisms studied. Though the overall chromatin architecture principles are conserved, the mechanisms by which chromatin folding is achieved during a cell cycle differ across organisms and genome sizes (Kakui et al. 2017; Lazar-Stefanita et al. 2017; Nagano et al. 2017; Schalbetter et al. 2017; Naumova et al. 2013; Gibcus et al. 2018; Abramo et al. 2019).

## 2.7 Cohesins and Condensins: Molecular Rings that Organize Chromosomes

Cohesins and condensins constitute the structural maintenance of chromosomes (SMC) protein complexes. Multiple subunits of SMC associate together to form a ring-like structure and function as chromatin linkers. These SMC proteins play an important role in organizing the genome.

### 2.7.1 Cohesins

Cohesins are heterotrimeric complexes that hold replicated DNA molecules (Ivanov and Nasmyth 2005; Haering et al. 2008). The core complex consists of Smc1, Smc3, and the α-kleisin subunit Rad21/Scc1/Mcd1 (Scc1 is replaced by the Rec8 subunit in meiosis). The Smc1 and Smc3 heterodimerize to form a V-shaped structure. The α-kleisin subunit closes the ring by connecting Smc1 and Smc3 (Peters et al. 2008; Nasmyth and Haering 2009). The heterotrimeric ring (core complex) is associated with accessory proteins, such as Pds5, Scc1, and Scc3, that regulate the cohesin function (Fig. 2.4a) (Mirkovic and Oliveira 2017). The components of the core and accessory complex are mostly conserved across eukaryotes. *S. cerevisiae* and *S. pombe* genomes encode all the core complex components and most of the accessory proteins (Makrantoni and Marston 2018). Cohesins play a significant role in sister chromatid cohesion during cell division. Apart from sister chromatid cohesion, cohesins also play a role in interphase chromatin organization. Cohesins contribute to transcription regulation by means of DNA looping, which brings distal regulatory elements such as promoters and enhancers together to control transcription (Rollins et al. 1999). Cohesins also play a role in TAD formation, and disruption of cohesin subunits such as Rad21 results in the disappearance of TADs (Rao et al. 2017; Wutz et al. 2017).

### 2.7.2 Condensins

Similar to cohesins, the core components of condensins are made up of SMC subunits. Condensins play an essential role in chromatin compaction. In eukaryotes, at least two different types of condensin complexes exist: condensin I and II. Both complexes have identical SMC subunits, SMC2 and SMC4 (Hirano 2006). The two complexes differ in the type of non-SMC subunits that associate with the SMC components. Condensin I associates with CAP-D2, CAP-G, and CAP-H subunits, whereas condensin II associates with CAP-D2, CAP-G2, and CAP-H2 subunits (Fig. 2.4b) (Hirano 2012). The genome of several yeast species such as *S. cerevisiae* and *S. pombe* encodes only condensin I subunits (Hirano 2012). Unlike yeasts, vertebrates encode two different condensin complexes, condensin I and condensin II (Ono et al. 2003). Two condensin complexes act in different ways to shape the mitotic chromosomes. The action of condensin II leads to axial compaction of chromosomes, whereas condensin I promotes lateral compaction of chromosomes (Shintomi and Hirano 2011; Green et al. 2012). A recent comparative study across several species

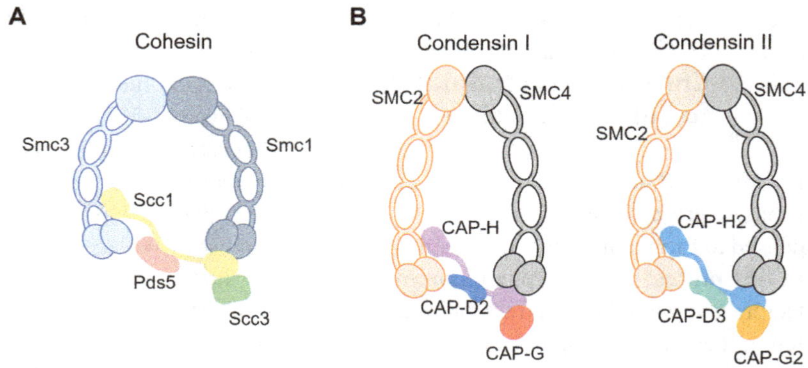


**Fig. 2.4** Cohesins and condensins. Schematic of the various subunits of (**a**) cohesin and (**b**) condensin I and II

of eukaryotes revealed that condensin II plays an important role in shaping the genome architecture across different species (Hoencamp et al. 2021). In metazoans cells lacking condensin II, the genome is organized in a Rabl-like configuration with a loose territorial organization of chromosomes. Most fungi lack the condensin II subunits, and thus they most likely favor the Rabl configuration (Hoencamp et al. 2021).

## 2.8 Cell Cycle-Associated Changes in Chromatin Architecture

### 2.8.1 *S. cerevisiae*

In budding yeast *S. cerevisiae,* chromatin in G1 shows the presence of small self-interacting domains of size 5-kb resembling TAD-like structures observed in metazoans. These 5-kb domains are enriched with one to five genes and the domain boundaries are flanked with promoters of highly expressed genes (Hsieh et al. 2015). Unlike in metazoans, no distinct A and B compartments exist in budding yeast. The G1 phase chromatin exhibits increased inter-chromosomal interactions as compared to chromatin in mitosis. Moreover, intermingling among chromosomes indicates that distinct chromosome territories are absent in budding yeast. During S phase, the long-range intra-chromosomal interactions (>20 kb) are enriched, and gradual individualization of chromosomes is observed. During mitosis, chromatin shows an increase in short-range (<20 kb) and reduced long-range interactions within a chromosome (Lazar-Stefanita et al. 2017). Chromatin of the mitotic chromosomes in budding yeast is organized into loops along the length of the chromatid and promotes compaction of the chromosomes (Dauban et al. 2020). In budding yeast, cohesins play a genome-wide role in modulating the abovementioned changes during the cell cycle. In contrast, condensins are shown to play a focused role in condensing the rDNA locus and in regulating the extent of centromere clustering during mitosis (Lazar-Stefanita et al. 2017; Schalbetter et al. 2017). A recent study in *S. cerevisiae* has also shown that chromatin remodelers play an important role in affecting the 3D genome architecture depending on the cell cycle stage and chromosomal context (Jo et al. 2021).

### 2.8.2 *S. pombe*

In the fission yeast *S. pombe*, G1 chromatin is characterized by numerous small cohesin-dependent self-interacting domains called globules, which range from 50 to 100 kb in size. During mitosis, the reorganization of interphase chromatin takes place. During this process, the appearance of small globules diminishes, and large condensin-dependent globules of size 300 kb to 1 Mb start to appear (Tanizawa et al. 2017). This leads to compaction of the chromatin during mitosis. Unlike *S. cerevisiae*, in *S. pombe*, cohesins and condensins play important roles in genome organization at different cell cycle stages.

## 2.9 Functional Role of the 3D Genome Organization and Nuclear Architecture

In the interphase nucleus, the positioning of one chromosomal region with respect to neighboring regions and nuclear space has implications on the genome function. The spatial organization of the genome affects cellular processes such as transcription, DNA repair, and DNA replication. In the following sections, the role of nuclear architecture in DNA replication timing will be discussed.

### 2.9.1 Genome Organization and DNA Replication Timing

DNA within a cell must be duplicated to be distributed faithfully to daughter cells via mitosis. DNA replication initiation occurs at multiple sites across the genome, known as the origins of replication. Replication factors assembled at the origins are required to replicate the genome. Activation of these origins involves two events: replication licensing and initiation of DNA replication. Though multiple origins can be licensed for DNA replication during G1, only a few become activated during S phase. The blueprint defining the activation of origins and their activation timeframe is known as the replication timing (RT) program. The 3D genome organization and the nuclear architecture regulate the replication timing during S phase, or it is possible that replication timing itself can shape the three-dimensional genome organization.

In mammals, the replication timing is linked to chromatin folding. The replication timing of genomic regions closely correlates with the A/B compartments and TADs. The early replicating DNA is positioned in the A compartment, and the late replicating DNA is positioned in the B compartment. During the S phase, certain regions of the genome activate their origins simultaneously and replicate synchronously. Such regions are referred to as replication domains (RD). Studies show that replication domains overlap with the TAD structures (Fig. 2.5). As mentioned previously, TADs are formed during early G1, and the formation of TADs coincides with establishing the replication timing program during a similar time frame (Ryba et al. 2010; Pope et al. 2014; Xiang et al. 2018). Although there is a striking similarity between the TADs and replication domains, perturbing the TAD organization by depletion of CCCTC-binding factor (CTCF, a conserved zinc finger transcription factor) or cohesins or deletion of TAD boundary sequences did not affect the replication program within a RD (Oldach and Nieduszynski 2019). This indicates that even though RDs colocalize with TADs, the organization of the TAD itself does not influence the replication program of an RD.

Similar to mammals, in yeasts such as *S. cerevisiae,* and to a certain extent in *S. pombe,* replication timing follows a spatial pattern that corresponds to the chromatin topology. A study in *S. cerevisiae* revealed that the early and late replicating regions correlate with the TADs, which span around 200 kb (Fig. 2.5). Similar to mammals, the boundaries of these budding yeast TADs are enriched with transcription activity and cohesins (Eser et al. 2017). Unlike *S. cerevisiae*, the replication origins are fired randomly in *S. pombe* (Heichinger et al. 2006). Even though the origin firing is stochastic, clustering of origins similar to replication foci is observed, and a subset of origin clusters follow spatial patterns in replication timing (Kaykov and Nurse 2015), indicating that chromatin conformation possibly influence origin firing in *S. pombe*.

### 2.9.2 Nuclear Architecture and DNA Replication Timing

Apart from genome folding influencing DNA replication timing, the nuclear positioning of the genome also affects replication timing. As discussed previously, in mammals, heterochromatin is present toward the periphery and around the nucleolus, whereas euchromatin is more toward the inside core of the nucleus. Chromatin positioned toward the periphery is shown to replicate late compared to chromatin positioned inside the core of the nucleus (Fig. 2.5, left panel). During

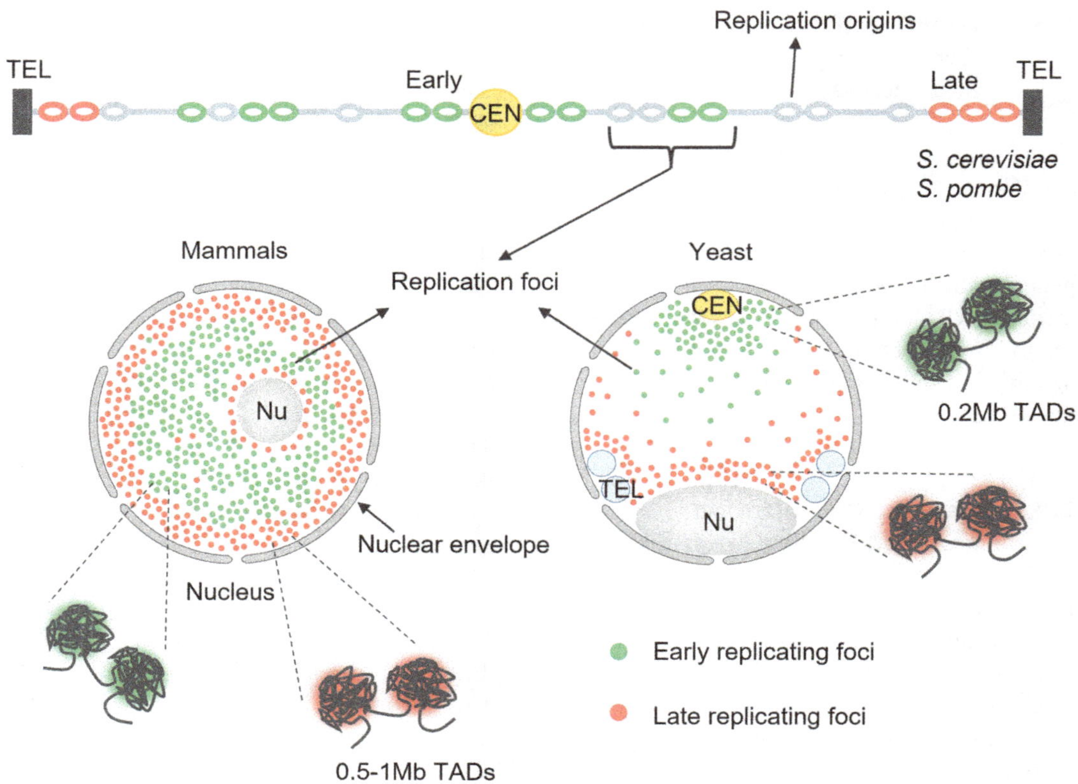


**Fig. 2.5** Role of 3D genome folding and nuclear positioning of chromatin in regulating replication timing. Top, linear representation of a chromosome with centromeres, telomeres, and replication origins and their replication timing as observed in *S. cerevisiae* and *S. pombe*. Below is a schematic of mammalian and yeast nuclei depicting the spatiotemporal dynamics of DNA replication. Replication foci consist of origins that fire synchronously during DNA replication and represent the replication domains. Replication foci can contain inactive origins that can be passively replicated by neighboring active origins (green ovals—active or early replicating origins; light gray oval—passively replicated or inactive origins; and red ovals—late replicating origins). The replication foci or domains correspond to the TADs within the nucleus. In mammals, the early replicating domains are primarily toward the interior, and late replicating foci occupy the periphery of the nucleus and around the nucleolus (Nu). The replication domains span 0.5–1 Mb in mammals. In budding yeast, the early replicating domains are located toward the centromeres, and the late replicating foci are present more toward the telomeres and nucleolus (Nu). The TADs in budding correspond to replication domains that span 0.2 Mb in length

cellular differentiation, changes in the rate of transcription and replication timing correlate with nuclear relocalization. Shifting from late to early replication timing correlates with the activation of transcription and repositioning of chromatin from the periphery to the nuclear interior (Hiratani et al. 2010). Based on two contrasting studies published, it appears that position alone does not dictate the replication timing for at least some genomic loci; instead, the transcription activation plays a vital role in influencing replication timing (Therizols et al. 2014; Heinz et al. 2018).

In yeasts, chromosomes are arranged in the Rabl configuration. This arrangement also results in spatial segregation of the early and late replicating regions for the chromosomes. In budding and fission yeast, centromeres replicate early, and the early replicating regions are proximal to the centromeres. Similarly, telomeres are late replicating, and the late replicating regions are closer to the telomeres and nucleolus (Fig. 2.5, right panel). Even though chromatin (other than the centromeric regions of fungi) localized to the periphery of the nucleus is late replicating, dissociating chromatin from the nuclear envelope does not alter the replication timing. Studies in the fission yeast indicate that the origins of other chromatin regions that are in proximity to telomeres replicate late compared to the origins close to the nuclear envelope (Ogawa et al. 2018). The telomere-associated shelterin components influence the replication timing of origins close to telomeres (Ogawa et al. 2018). Recent studies indicate that the nuclear envelope creates a microenvironment that facilitates the ease of replication of heterochromatic regions, such as telomeres, that are otherwise difficult to replicate (Ebrahimi et al. 2018). It has been proposed that the Swi6/HP1-mediated recruitment of Dbf4-dependent kinase (DDK, responsible for replication initiation) (Hayashi et al. 2009) creates a microenvironment of enriched DDK activity at centromeres which can result in origin firing at centromere-proximal loci or the nuclear space. Whereas at the telomeres, the presence of PP1 (phosphatase activity) creates an environment that is refractory for the initiation of replication (Dave et al. 2014) which will result in late replication of telomeres and telomere-proximal regions. Similarly, a recent report in *C. albicans* has shown that Orc4, a component of origin recognition complex (ORC) which binds to DNA replication origins is reported to be enriched near the centromere cluster and the Orc4 level gradually reduces toward the telomeres. This contributes to the early and late replication of centromeres and telomeres (Sreekumar et al. 2021).

## 2.10 Conclusions

The Rabl and non-Rabl arrangement of chromosomes shows significant differences in the 3D genome organization. Both configurations differ in terms of how the centromeres are arranged, the rate at which the genomic interactions decay with increasing distance, the ratio of inter-/intra-chromosomal interactions, the length of the chromatin loops, and structural proteins (such as condensin II) that associate with chromatin (Hoencamp et al. 2021; Alvarez-Gonzalez et al. 2022). A thorough understanding of how chromosomes are organized in the Rabl and non-Rabl configurations and the changes that occur during the switching between the configurations will enable us to understand the evolutionary dynamics of the 3D chromatin organization and its effect on genome organization. The same organism may even exhibit—either Rabl or non-Rabl configuration in a developmental stage-specific manner. While most fungal species studied are known to arrange chromosomes in the Rabl configuration and the hierarchical organization of chromatin is conserved from fungi to humans there are few fungal species (*C. neoformans* and *R. irregularis*) which deviate from the Rabl arrangement of chromosomes. These deviations warrant studying of other poorly explored fungal species to get a comprehensive understanding of the genome organization in fungi.

## References


Abramo K, Valton AL, Venev SV, Ozadam H, Fox AN, Dekker J (2019) A chromosome folding intermediate at the condensin-to-cohesin transition during telophase. Nat Cell Biol 21(11):1393–1402. https://doi.org/10.1038/s41556-019-0406-2

Alvarez-Gonzalez L, Arias-Sarda C, Montes-Espuna L, Marin-Gual L, Vara C, Lister NC, Cuartero Y, Garcia F, Deakin J, Renfree MB, Robinson TJ, Marti-Renom MA, Waters PD, Farre M, Ruiz-Herrera A (2022) Principles of 3D chromosome folding and evolutionary genome reshuffling in mammals. Cell Rep 41(12):111839. https://doi.org/10.1016/j.celrep.2022.111839

Batty P, Gerlich DW (2019) Mitotic chromosome mechanics: how cells segregate their genome. Trends Cell Biol 29(9):717–726. https://doi.org/10.1016/j.tcb.2019.05.007

Bloom K, Joglekar A (2010) Towards building a chromosome segregation machine. Nature 463(7280):446–456. https://doi.org/10.1038/nature08912

Burrack LS, Hutton HF, Matter KJ, Clancey SA, Liachko I, Plemmons AE, Saha A, Power EA, Turman B, Thevandavakkam MA, Ay F, Dunham MJ, Berman J (2016) Neocentromeres provide chromosome segregation accuracy and centromere clustering to multiple loci along a Candida albicans chromosome. PLoS Genet 12(9):e1006317. https://doi.org/10.1371/journal.pgen.1006317

Cremer T, Cremer M (2010) Chromosome territories. Cold Spring Harb Perspect Biol 2(3):a003889. https://doi.org/10.1101/cshperspect.a003889

Dauban L, Montagne R, Thierry A, Lazar-Stefanita L, Bastie N, Gadal O, Cournac A, Koszul R, Beckouet F (2020) Regulation of Cohesin-mediated chromosome folding by Eco1 and other partners. Mol Cell 77(6):1279–1293. e1274. https://doi.org/10.1016/j.molcel.2020.01.019

Dave A, Cooley C, Garg M, Bianchi A (2014) Protein phosphatase 1 recruitment by Rif1 regulates DNA replication origin firing by counteracting DDK activity. Cell Rep 7(1):53–61. https://doi.org/10.1016/j.celrep.2014.02.019

Dekker J, Rippe K, Dekker M, Kleckner N (2002) Capturing chromosome conformation. Science 295(5558):1306–1311. https://doi.org/10.1126/science.1067799

Dixon JR, Selvaraj S, Yue F, Kim A, Li Y, Shen Y, Hu M, Liu JS, Ren B (2012) Topological domains in mammalian genomes identified by analysis of chromatin interactions. Nature 485(7398):376–380. https://doi.org/10.1038/nature11082

Dong P, Tu X, Chu PY, Lu P, Zhu N, Grierson D, Du B, Li P, Zhong S (2017) 3D chromatin architecture of large plant genomes determined by local a/B compartments. Mol Plant 10(12):1497–1509. https://doi.org/10.1016/j.molp.2017.11.005

Duan Z, Andronescu M, Schutz K, McIlwain S, Kim YJ, Lee C, Shendure J, Fields S, Blau CA, Noble WS (2010) A three-dimensional model of the yeast genome. Nature 465(7296):363–367. https://doi.org/10.1038/nature08973

Durand NC, Robinson JT, Shamim MS, Machol I, Mesirov JP, Lander ES, Aiden EL (2016) Juicebox provides a visualization system for hi-C contact maps with unlimited zoom. Cell Syst 3(1):99–101. https://doi.org/10.1016/j.cels.2015.07.012

Ebrahimi H, Masuda H, Jain D, Cooper JP (2018) Distinct 'safe zones' at the nuclear envelope ensure robust replication of heterochromatic chromosome regions. elife 7. https://doi.org/10.7554/eLife.32911

Eser U, Chandler-Brown D, Ay F, Straight AF, Duan Z, Noble WS, Skotheim JM (2017) Form and function of topologically associating genomic domains in budding yeast. Proc Natl Acad Sci USA 114(15):E3061–E3070. https://doi.org/10.1073/pnas.1612256114

Fragkos M, Ganier O, Coulombe P, Mechali M (2015) DNA replication origin activation in space and time. Nat Rev Mol Cell Biol 16(6):360–374. https://doi.org/10.1038/nrm4002

Funabiki H, Hagan I, Uzawa S, Yanagida M (1993) Cell cycle-dependent specific positioning and clustering of centromeres and telomeres in fission yeast. J Cell Biol 121(5):961–976. https://doi.org/10.1083/jcb.121.5.961

Galazka JM, Klocko AD, Uesaka M, Honda S, Selker EU, Freitag M (2016) Neurospora chromosomes are organized by blocks of importin alpha-dependent heterochromatin that are largely independent of H3K9me3. Genome Res 26(8):1069–1080. https://doi.org/10.1101/gr.203182.115

Gibcus JH, Samejima K, Goloborodko A, Samejima I, Naumova N, Nuebler J, Kanemaki MT, Xie L, Paulson JR, Earnshaw WC, Mirny LA, Dekker J (2018) A pathway for mitotic chromosome formation. Science 359(6376). https://doi.org/10.1126/science.aao6135

Green LC, Kalitsis P, Chang TM, Cipetic M, Kim JH, Marshall O, Turnbull L, Whitchurch CB, Vagnarelli P, Samejima K, Earnshaw WC, Choo KH, Hudson DF (2012) Contrasting roles of condensin I and condensin II in mitotic chromosome formation. J Cell Sci 125(Pt 6):1591–1604. https://doi.org/10.1242/jcs.097790

Guacci V, Hogan E, Koshland D (1997) Centromere position in budding yeast: evidence for anaphase a. Mol Biol Cell 8(6):957–972. https://doi.org/10.1091/mbc.8.6.957

Guin K, Chen Y, Mishra R, Muzaki SRB, Thimmappa BC, O'Brien CE, Butler G, Sanyal A, Sanyal K (2020a) Spatial inter-centromeric interactions facilitated the emergence of evolutionary new centromeres. elife 9. https://doi.org/10.7554/eLife.58556

Guin K, Sreekumar L, Sanyal K (2020b) Implications of the evolutionary trajectory of centromeres in the fungal kingdom. Ann Rev Microbiol 74:835–853. https://doi.org/10.1146/annurev-micro-011720-122512

Haering CH, Farcas AM, Arumugam P, Metson J, Nasmyth K (2008) The cohesin ring concatenates sister DNA molecules. Nature 454(7202):297–301. https://doi.org/10.1038/nature07098

Hayashi MT, Takahashi TS, Nakagawa T, Nakayama J, Masukata H (2009) The heterochromatin protein Swi6/HP1 activates replication origins at the pericentromeric region and silent mating-type locus. Nat Cell Biol 11(3):357–362. https://doi.org/10.1038/ncb1845

Heichinger C, Penkett CJ, Bahler J, Nurse P (2006) Genome-wide characterization of fission yeast DNA replication origins. EMBO J 25(21):5171–5179. https://doi.org/10.1038/sj.emboj.7601390

Heinz KS, Casas-Delucchi CS, Torok T, Cmarko D, Rapp A, Raska I, Cardoso MC (2018) Peripheral re-localization of constitutive heterochromatin advances its replication timing and impairs maintenance of silencing marks. Nucleic Acids Res 46(12):6112–6128. https://doi.org/10.1093/nar/gky368

Hirano T (2006) At the heart of the chromosome: SMC proteins in action. Nat Rev Mol Cell Biol 7(5):311–322. https://doi.org/10.1038/nrm1909

Hirano T (2012) Condensins: universal organizers of chromosomes with diverse functions. Genes Dev 26(15):1659–1678. https://doi.org/10.1101/gad.194746.112

Hiratani I, Ryba T, Itoh M, Rathjen J, Kulik M, Papp B, Fussner E, Bazett-Jones DP, Plath K, Dalton S, Rathjen PD, Gilbert DM (2010) Genome-wide dynamics of replication timing revealed by in vitro models of mouse embryogenesis. Genome Res 20(2):155–169. https://doi.org/10.1101/gr.099796.109

Hoencamp C, Dudchenko O, Elbatsh AMO, Brahmachari S, Raaijmakers JA, van Schaik T, Sedeno Cacciatore A, Contessoto VG, van Heesbeen R, van den Broek

B, Mhaskar AN, Teunissen H, St Hilaire BG, Weisz D, Omer AD, Pham M, Colaric Z, Yang Z, Rao SSP, Mitra N, Lui C, Yao W, Khan R, Moroz LL, Kohn A, St Leger J, Mena A, Holcroft K, Gambetta MC, Lim F, Farley E, Stein N, Haddad A, Chauss D, Mutlu AS, Wang MC, Young ND, Hildebrandt E, Cheng HH, Knight CJ, Burnham TLU, Hovel KA, Beel AJ, Mattei PJ, Kornberg RD, Warren WC, Cary G, Gomez-Skarmeta JL, Hinman V, Lindblad-Toh K, Di Palma F, Maeshima K, Multani AS, Pathak S, Nel-Themaat L, Behringer RR, Kaur P, Medema RH, van Steensel B, de Wit E, Onuchic JN, Di Pierro M, Lieberman Aiden E, Rowland BD (2021) 3D genomics across the tree of life reveals condensin II as a determinant of architecture type. Science 372(6545):984–989. https://doi.org/10.1126/science.abe2218

Hou C, Li L, Qin ZS, Corces VG (2012) Gene density, transcription, and insulators contribute to the partition of the drosophila genome into physical domains. Mol Cell 48(3):471–484. https://doi.org/10.1016/j.molcel.2012.08.031

Hsieh TH, Weiner A, Lajoie B, Dekker J, Friedman N, Rando OJ (2015) Mapping nucleosome resolution chromosome folding in yeast by micro-C. Cell 162(1):108–119. https://doi.org/10.1016/j.cell.2015.05.048

Ivanov D, Nasmyth K (2005) A topological interaction between cohesin rings and a circular minichromosome. Cell 122(6):849–860. https://doi.org/10.1016/j.cell.2005.07.018

Jin QW, Fuchs J, Loidl J (2000) Centromere clustering is a major determinant of yeast interphase nuclear organization. J Cell Sci 113(Pt 11):1903–1912. https://doi.org/10.1242/jcs.113.11.1903

Jo H, Kim T, Chun Y, Jung I, Lee D (2021) A compendium of chromatin contact maps reflecting regulation by chromatin remodelers in budding yeast. Nat Commun 12(1):6380. https://doi.org/10.1038/s41467-021-26629-6

Kakui Y, Rabinowitz A, Barry DJ, Uhlmann F (2017) Condensin-mediated remodeling of the mitotic chromatin landscape in fission yeast. Nat Genet 49(10):1553–1557. https://doi.org/10.1038/ng.3938

Kaykov A, Nurse P (2015) The spatial and temporal organization of origin firing during the S-phase of fission yeast. Genome Res 25(3):391–401. https://doi.org/10.1101/gr.180372.114

Klocko AD, Ormsby T, Galazka JM, Leggett NA, Uesaka M, Honda S, Freitag M, Selker EU (2016) Normal chromosome conformation depends on subtelomeric facultative heterochromatin in Neurospora crassa. Proc Natl Acad Sci USA 113(52):15048–15053. https://doi.org/10.1073/pnas.1615546113

Kozubowski L, Yadav V, Chatterjee G, Sridhar S, Yamaguchi M, Kawamoto S, Bose I, Heitman J, Sanyal K (2013) Ordered kinetochore assembly in the human-pathogenic basidiomycetous yeast Cryptococcus neoformans. mBio 4(5):e00614–e00613. https://doi.org/10.1128/mBio.00614-13

Kupiec M (2014) Biology of telomeres: lessons from budding yeast. FEMS Microbiol Rev 38(2):144–171. https://doi.org/10.1111/1574-6976.12054

Lazar-Stefanita L, Scolari VF, Mercy G, Muller H, Guerin TM, Thierry A, Mozziconacci J, Koszul R (2017) Cohesins and condensins orchestrate the 4D dynamics of yeast chromosomes during the cell cycle. EMBO J 36(18):2684–2697. https://doi.org/10.15252/embj.201797342

Lewis ZA, Honda S, Khlafallah TK, Jeffress JK, Freitag M, Mohn F, Schubeler D, Selker EU (2009) Relics of repeat-induced point mutation direct heterochromatin formation in Neurospora crassa. Genome Res 19(3):427–437. https://doi.org/10.1101/gr.086231.108

Lieberman-Aiden E, van Berkum NL, Williams L, Imakaev M, Ragoczy T, Telling A, Amit I, Lajoie BR, Sabo PJ, Dorschner MO, Sandstrom R, Bernstein B, Bender MA, Groudine M, Gnirke A, Stamatoyannopoulos J, Mirny LA, Lander ES, Dekker J (2009) Comprehensive mapping of long-range interactions reveals folding principles of the human genome. Science 326(5950):289–293. https://doi.org/10.1126/science.1181369

Makrantoni V, Marston AL (2018) Cohesin and chromosome segregation. Curr Biol 28(12):R688–R693. https://doi.org/10.1016/j.cub.2018.05.019

Mascher M, Gundlach H, Himmelbach A, Beier S, Twardziok SO, Wicker T, Radchuk V, Dockter C, Hedley PE, Russell J, Bayer M, Ramsay L, Liu H, Haberer G, Zhang XQ, Zhang Q, Barrero RA, Li L, Taudien S, Groth M, Felder M, Hastie A, Simkova H, Stankova H, Vrana J, Chan S, Munoz-Amatriain M, Ounit R, Wanamaker S, Bolser D, Colmsee C, Schmutzer T, Aliyeva-Schnorr L, Grasso S, Tanskanen J, Chailyan A, Sampath D, Heavens D, Clissold L, Cao S, Chapman B, Dai F, Han Y, Li H, Li X, Lin C, McCooke JK, Tan C, Wang P, Wang S, Yin S, Zhou G, Poland JA, Bellgard MI, Borisjuk L, Houben A, Dolezel J, Ayling S, Lonardi S, Kersey P, Langridge P, Muehlbauer GJ, Clark MD, Caccamo M, Schulman AH, Mayer KFX, Platzer M, Close TJ, Scholz U, Hansson M, Zhang G, Braumann I, Spannagl M, Li C, Waugh R, Stein N (2017) A chromosome conformation capture ordered sequence of the barley genome. Nature 544(7651):427–433. https://doi.org/10.1038/nature22043

McAinsh AD, Marston AL (2022) The four causes: the functional architecture of centromeres and kinetochores. Annu Rev Genet 56:279–314. https://doi.org/10.1146/annurev-genet-072820-034559

Mekhail K, Moazed D (2010) The nuclear envelope in genome organization, expression and stability. Nat Rev Mol Cell Biol 11(5):317–328. https://doi.org/10.1038/nrm2894

Mirkovic M, Oliveira RA (2017) Centromeric Cohesin: molecular glue and much more. Prog Mol Subcell Biol 56:485–513. https://doi.org/10.1007/978-3-319-58592-5_20

Mizuguchi T, Fudenberg G, Mehta S, Belton JM, Taneja N, Folco HD, FitzGerald P, Dekker J, Mirny L,

Barrowman J, Grewal SIS (2014) Cohesin-dependent globules and heterochromatin shape 3D genome architecture in S. Pombe. Nature 516(7531):432–435. https://doi.org/10.1038/nature13833

Morgan DO (2006) The cell cycle: principles of control, 1st edn. New Science Press

Muller H, Gil J Jr, Drinnenberg IA (2019) The impact of centromeres on spatial genome architecture. Trends Genet 35(8):565–578. https://doi.org/10.1016/j.tig.2019.05.003

Nagano T, Lubling Y, Varnai C, Dudley C, Leung W, Baran Y, Mendelson Cohen N, Wingett S, Fraser P, Tanay A (2017) Cell-cycle dynamics of chromosomal organization at single-cell resolution. Nature 547(7661):61–67. https://doi.org/10.1038/nature23001

Nasmyth K, Haering CH (2009) Cohesin: its roles and mechanisms. Annu Rev Genet 43:525–558. https://doi.org/10.1146/annurev-genet-102108-134233

Naumova N, Imakaev M, Fudenberg G, Zhan Y, Lajoie BR, Mirny LA, Dekker J (2013) Organization of the mitotic chromosome. Science 342(6161):948–953. https://doi.org/10.1126/science.1236083

Nora EP, Lajoie BR, Schulz EG, Giorgetti L, Okamoto I, Servant N, Piolot T, van Berkum NL, Meisig J, Sedat J, Gribnau J, Barillot E, Bluthgen N, Dekker J, Heard E (2012) Spatial partitioning of the regulatory landscape of the X-inactivation Centre. Nature 485(7398):381–385. https://doi.org/10.1038/nature11049

Ogawa S, Kido S, Handa T, Ogawa H, Asakawa H, Takahashi TS, Nakagawa T, Hiraoka Y, Masukata H (2018) Shelterin promotes tethering of late replication origins to telomeres for replication-timing control. EMBO J 37(15). https://doi.org/10.15252/embj.201898997

Oldach P, Nieduszynski CA (2019) Cohesin-mediated genome architecture does not define DNA replication timing domains. Genes (Basel) 10(3). https://doi.org/10.3390/genes10030196

Ono T, Losada A, Hirano M, Myers MP, Neuwald AF, Hirano T (2003) Differential contributions of condensin I and condensin II to mitotic chromosome architecture in vertebrate cells. Cell 115(1):109–121. https://doi.org/10.1016/s0092-8674(03)00724-4

Ou HD, Phan S, Deerinck TJ, Thor A, Ellisman MH, O'Shea CC (2017) ChromEMT: visualizing 3D chromatin structure and compaction in interphase and mitotic cells. Science 357(6349). https://doi.org/10.1126/science.aag0025

Peters JM, Tedeschi A, Schmitz J (2008) The cohesin complex and its roles in chromosome biology. Genes Dev 22(22):3089–3114. https://doi.org/10.1101/gad.1724308

Pope BD, Ryba T, Dileep V, Yue F, Wu W, Denas O, Vera DL, Wang Y, Hansen RS, Canfield TK, Thurman RE, Cheng Y, Gulsoy G, Dennis JH, Snyder MP, Stamatoyannopoulos JA, Taylor J, Hardison RC, Kahveci T, Ren B, Gilbert DM (2014) Topologically associating domains are stable units of replication-timing regulation. Nature 515(7527):402–405. https://doi.org/10.1038/nature13986

Rao SS, Huntley MH, Durand NC, Stamenova EK, Bochkov ID, Robinson JT, Sanborn AL, Machol I, Omer AD, Lander ES, Aiden EL (2014) A 3D map of the human genome at kilobase resolution reveals principles of chromatin looping. Cell 159(7):1665–1680. https://doi.org/10.1016/j.cell.2014.11.021

Rao SSP, Huang SC, Glenn St Hilaire B, Engreitz JM, Perez EM, Kieffer-Kwon KR, Sanborn AL, Johnstone SE, Bascom GD, Bochkov ID, Huang X, Shamim MS, Shin J, Turner D, Ye Z, Omer AD, Robinson JT, Schlick T, Bernstein BE, Casellas R, Lander ES, Aiden EL (2017) Cohesin loss eliminates all loop domains. Cell 171(2):305–320 e324. https://doi.org/10.1016/j.cell.2017.09.026

Rodriguez S, Ward A, Reckard AT, Shtanko Y, Hull-Crew C, Klocko AD (2022) The genome organization of Neurospora crassa at high resolution uncovers principles of fungal chromosome topology. G3 (Bethesda) 12(5). https://doi.org/10.1093/g3journal/jkac053

Rollins RA, Morcillo P, Dorsett D (1999) Nipped-B, a drosophila homologue of chromosomal adherins, participates in activation by remote enhancers in the cut and Ultrabithorax genes. Genetics 152(2):577–593. https://doi.org/10.1093/genetics/152.2.577

Rutledge MT, Russo M, Belton JM, Dekker J, Broach JR (2015) The yeast genome undergoes significant topological reorganization in quiescence. Nucleic Acids Res 43(17):8299–8313. https://doi.org/10.1093/nar/gkv723

Ryba T, Hiratani I, Lu J, Itoh M, Kulik M, Zhang J, Schulz TC, Robins AJ, Dalton S, Gilbert DM (2010) Evolutionarily conserved replication timing profiles predict long-range chromatin interactions and distinguish closely related cell types. Genome Res 20(6):761–770. https://doi.org/10.1101/gr.099655.109

Sankaranarayanan SR, Ianiri G, Coelho MA, Reza MH, Thimmappa BC, Ganguly P, Vadnala RN, Sun S, Siddharthan R, Tellgren-Roth C, Dawson TLJ, Heitman J, Sanyal K (2020) Loss of centromere function drives karyotype evolution in closely related Malassezia species. elife 9. https://doi.org/10.7554/eLife.53944

Schalbetter SA, Goloborodko A, Fudenberg G, Belton JM, Miles C, Yu M, Dekker J, Mirny L, Baxter J (2017) SMC complexes differentially compact m itotic chromosomes according to genomic context. Nat Cell Biol 19(9):1071–1080. https://doi.org/10.1038/ncb3594

Schober H, Kalck V, Vega-Palas MA, Van Houwe G, Sage D, Unser M, Gartenberg MR, Gasser SM (2008) Controlled exchange of chromosomal arms reveals principles driving telomere interactions in yeast. Genome Res 18(2):261–271. https://doi.org/10.1101/gr.6687808

Sexton T, Yaffe E, Kenigsberg E, Bantignies F, Leblanc B, Hoichman M, Parrinello H, Tanay A, Cavalli G (2012) Three-dimensional folding and functional organization principles of the drosophila genome. Cell 148(3):458–472. https://doi.org/10.1016/j.cell.2012.01.010

Shintomi K, Hirano T (2011) The relative ratio of condensin I to II determines chromosome shapes. Genes Dev 25(14):1464–1469. https://doi.org/10.1101/gad.2060311

Sreekumar L, Kumari K, Guin K, Bakshi A, Varshney N, Thimmappa BC, Narlikar L, Padinhateeri R, Siddharthan R, Sanyal K (2021) Orc4 spatiotemporally stabilizes centromeric chromatin. Genome Res 31(4):607–621. https://doi.org/10.1101/gr.265900.120

Szabo Q, Bantignies F, Cavalli G (2019) Principles of genome folding into topologically associating domains. Sci Adv 5(4):eaaw1668. https://doi.org/10.1126/sciadv.aaw1668

Tanizawa H, Kim KD, Iwasaki O, Noma KI (2017) Architectural alterations of the fission yeast genome during the cell cycle. Nat Struct Mol Biol 24(11):965–976. https://doi.org/10.1038/nsmb.3482

Thakur J, Sanyal K (2012) A coordinated interdependent protein circuitry stabilizes the kinetochore ensemble to protect CENP-A in the human pathogenic yeast Candida albicans. PLoS Genet 8(4):e1002661. https://doi.org/10.1371/journal.pgen.1002661

Therizols P, Duong T, Dujon B, Zimmer C, Fabre E (2010) Chromosome arm length and nuclear constraints determine the dynamic relationship of yeast subtelomeres. Proc Natl Acad Sci USA 107(5):2025–2030. https://doi.org/10.1073/pnas.0914187107

Therizols P, Illingworth RS, Courilleau C, Boyle S, Wood AJ, Bickmore WA (2014) Chromatin decondensation is sufficient to alter nuclear organization in embryonic stem cells. Science 346(6214):1238–1242. https://doi.org/10.1126/science.1259587

Torres DE, Kramer HM, Tracanna V, Fiorin GL, Cook DE, Seidl MF, Thomma B (2024) Implications of the three-dimensional chromatin organization for genome evolution in a fungal plant pathogen. Nat Commun 15(1):1701. https://doi.org/10.1038/s41467-024-45884-x

Torres DE, Reckard AT, Klocko AD, Seidl MF (2023) Nuclear genome organization in fungi: from gene folding to Rabl chromosomes. FEMS Microbiol Rev 47:3. https://doi.org/10.1093/femsre/fuad021

Wang S, Su JH, Beliveau BJ, Bintu B, Moffitt JR, Wu CT, Zhuang X (2016) Spatial organization of chromatin domains and compartments in single chromosomes. Science 353(6299):598–602. https://doi.org/10.1126/science.aaf8084

Winey M, Bloom K (2012) Mitotic spindle form and function. Genetics 190(4):1197–1224. https://doi.org/10.1534/genetics.111.128710

Winter DJ, Ganley ARD, Young CA, Liachko I, Schardl CL, Dupont PY, Berry D, Ram A, Scott B, Cox MP (2018) Repeat elements organise 3D genome structure and mediate transcription in the filamentous fungus Epichloe festucae. PLoS Genet 14(10):e1007467. https://doi.org/10.1371/journal.pgen.1007467

Wutz G, Varnai C, Nagasaka K, Cisneros DA, Stocsits RR, Tang W, Schoenfelder S, Jessberger G, Muhar M, Hossain MJ, Walther N, Koch B, Kueblbeck M, Ellenberg J, Zuber J, Fraser P, Peters JM (2017) Topologically associating domains and chromatin loops depend on cohesin and are regulated by CTCF, WAPL, and PDS5 proteins. EMBO J 36(24):3573–3599. https://doi.org/10.15252/embj.201798004

Xiang W, Roberti MJ, Heriche JK, Huet S, Alexander S, Ellenberg J (2018) Correlative live and super-resolution imaging reveals the dynamic structure of replication domains. J Cell Biol 217(6):1973–1984. https://doi.org/10.1083/jcb.201709074

Yadav V, Yang F, Reza MH, Liu S, Valent B, Sanyal K, Naqvi NI (2019) Cellular dynamics and genomic identity of centromeres in cereal blast fungus. mBio 10(4). https://doi.org/10.1128/mBio.01581-19

Yildirir G, Sperschneider J, Malar CM, Chen ECH, Iwasaki W, Cornell C, Corradi N (2022) Long reads and hi-C sequencing illuminate the two-compartment genome of the model arbuscular mycorrhizal symbiont Rhizophagus irregularis. New Phytol 233(3):1097–1107. https://doi.org/10.1111/nph.17842

# Fungi from Extreme Environments: Genome Sequences and Beyond

3

Cene Gostinčar, Claudia Coleine, Nina Gunde-Cimerman, and Jason E. Stajich

**Abstract**

Fungi have been documented to inhabit nearly all habitats on earth including some of the most extreme environmental and physico-chemical conditions. The extremophilic and extremotolerant fungi which thrive in harsh conditions include specialists which inhabit a narrow though extreme range of temperature, pH, or aridity, and generalists which are documented in often broader environmental conditions but are successfully in the extremes. Studies of these fungi have been driven by culture-based work documenting the phenotypes and capabilities in different growth conditions, genotyping and genomic sequencing to catalog gene content and genetic variation. Ongoing research is working to uncover the basic biology, unravel novel molecular adaptations to understand their evolution, and develop biotechnological applications from these resources and knowledge.

C. Gostinčar · N. Gunde-Cimerman
Biotechnical Faculty, Department of Biology, University of Ljubljana, Ljubljana, Slovenia

C. Coleine
Department of Ecological and Biological Sciences, University of Tuscia, Viterbo, Italy

J. E. Stajich (✉)
Department of Microbiology and Plant Pathology, University of California-Riverside, Riverside, CA, USA
e-mail: jason.stajich@ucr.edu

**Keywords**

Extremophile · Halophile · Black yeast · Meristematic · Genomics

## 3.1 Extremophilic and Extremotolerant Fungi

Extreme environments on Earth are characterized, by an anthropocentric point of view, by challenging physico-chemical conditions such as high concentrations of solutes, extreme temperature fluctuations, high radiation levels, nutrient scarcity, and extreme pH values (Gostinčar et al. 2010). There are commonalities in the challenges faced by organisms inhabiting seemingly very different environments. These include low water activity ($a_w$), which limits the amount of water accessible by the cells. In hypersaline waters, for example, the biologically available water is limited due to the high salt concentration, while in polar environments, e.g. in glacial ice, it is scarce due to binding of water molecules into ice and the concentration of solutes in the remaining liquid water (Gostinčar et al. 2010; Gostinčar and Gunde-Cimerman 2023). Yet, in arid regions, water availability is limited due to low precipitation and high evaporation rates (Coleine et al. 2020b; Coleine et al. 2022; Gostinčar et al. 2023). Fungi that colonize all these low water $a_w$ have to protect themselves from desiccation and often


M. Nowrousian, J. E. Stajich (eds.), *Fungal Genomics*, The Mycota 13,
https://doi.org/10.1007/978-3-031-75666-5_3

also from the detrimental effects of extreme temperatures. In extreme environments, fungi play important roles, some of which are poorly understood. Fungi colonizing hot desert crusts are crucial for soil stabilization and nutrient cycling (Coleine et al. 2020b; Coleine et al. 2022; Gostinčar et al. 2023).

Fungi that grow on and within Antarctic rocks, for instance, are adapted to extreme cold and drought, form biofilms, and contribute to nutrient cycling in these environments. Fungi in hypersaline environments are involved in the formation of microbial mats and in the precipitation of halite (Cantrell et al. 2013; Gostinčar and Gunde-Cimerman 2023). Fungi inhabiting glaciers also interact with bacteria supporting biogeochemical cycles that occur under retreating glaciers (Butinar et al. 2007; Anesio and Laybourn-Parry 2012; Anesio et al. 2017).

The growing interest in fungi living in extreme environments has spurred many surveys and enabled a relatively complete understanding of the mycobiota of some of these environments. At the same time, there are other environments, which have been much less investigated. These include alkaline environments with low water activity, such as alkaline soda lakes with high concentration of sodium carbonate, or acidic environments, such as acidic hypersaline lakes (Ghori et al. 2021), soils, acid mine drainage streams, and volcanic lakes (Gross and Robbins 2000).

Extreme environments are also found in the built environment, and these have been successfully colonized by fungi as well. Fungi growing indoors occupy unique niches, shaped by artificial, man-made materials and conditions, and therefore distinct from niches occupied by the same fungi in nature (Gümral et al. 2016; Novak Babič et al. 2020). The built environment often selects for specific groups of fungi, with adaptations such as growth under nutrient-poor conditions and oxidative stress. On artificial surfaces, these fungi can form biofilms together with bacteria (Zupančič et al. 2018). Some indoor environments favor opportunistic pathogenic fungi (Raghupathi et al. 2018) of which some species and strains are capable of survival even in strictly regulated environments such as spaceships, threatening astronauts' health on long space travels (Novikova et al. 2006; Vesper et al. 2008; Schuerger et al. 2021; Parker et al. 2022; Averesch et al. 2022; Horne et al. 2022).

Extremophilic fungi with their diverse adaptations offer insights not only into the adaptability of life on Earth, both in natural and indoor settings, but also into the potential for life beyond our planet. Many extreme habitats provide analogs for extraterrestrial habitats (ten Kate and Preston 2015; Heinicke and Arnhof 2021; Biddanda et al. 2021). Fungal metabolic products and cellular structures in such sites can be studied as potential biosignatures to detect past or present extraterrestrial life (Scheller et al. 2022; Gibson and Benison 2023), possibly even after hundreds of millions of years (Horne et al. 2022).

Fungal species found in extreme environments can be divided into generalists, often categorized as polyextremotolerant, and specialists (Gostinčar et al. 2010; Gostinčar and Gunde-Cimerman 2023). The main characteristic of specialist species is their high level of adaptation to specific extreme environments and occupation of a limited range of habitats. They have evolved specific adaptations tailored to their specific extreme environment. They display limited tolerance to other conditions and may struggle to survive outside their specialized niche. Despite their limited distribution, specialists can play essential roles within their ecosystems, contributing to specific ecological functions (Gostinčar et al. 2010; Gostinčar and Gunde-Cimerman 2023). Examples of specialist species, presented in more details later, include the ascomycetous melanized psychrotolerant black yeast *Aureobasidium subglaciale* (Zajc et al. 2022; Gostinčar and Gunde-Cimerman 2023), the extremely halotolerant ascomycetous black yeast *Hortaea werneckii* (Gunde-Cimerman et al. 2018; Gostinčar and Gunde-Cimerman 2023), the psychrophilic basidiomycetous yeast *Mrakia psychrophila* (Xin and Zhou 2007; Su et al. 2016), and the halophilic basidiomycetous filamentous fungus *Wallemia ichthyophaga* (Zajc et al. 2013; Gunde-Cimerman et al. 2018).

In contrast, generalist and polyextremotolerant species display a much broader ecological range and can adapt to a wide range of extreme conditions. They are versatile and when polyextremotolerant, they can tolerate multiple stress factors simultaneously, such as extreme salinity, freezing temperatures, UV radiation, nutrient scarcity, and more. They are more widely distributed and can be found in various extreme and moderate environments worldwide. They have flexible ecological roles, contributing to nutrient cycling, organic matter decomposition, and different ecosystem processes. They can exploit different resources and habitats and display various growth forms (Gostinčar et al. 2010; Gostinčar and Gunde-Cimerman 2023). Due to their multiple adaptations many have found different biotechnological uses. Examples of generalist and polyextremotolerant species include the filamentous ascomycetous *Aspergillus sydowii* (Pérez-Llano et al. 2020)*),* the black yeast *Aureobasidium pullulans* (Gostinčar et al. 2014; Gostinčar and Gunde-Cimerman 2023), the yeast *Debaryomyces hansenii* (Seiler and Busse 1990; Saldanha-da-Gama et al. 1997; Prista et al. 2005), and the basidiomycetous yeast *Rhodotorula mucilaginosa* (Gunde-Cimerman et al. 2005; Collins and Margesin 2019), among others.

Among the listed examples of specialist and generalist fungi, three species belong to black, yeast-like fungi that are referred to as black yeasts, meristematic fungi, or microcolonial fungi. These names derive from phenotypic classifications relating to growth. In most cases, sexual structures have not been observed in these fungi, limiting the morphological characteristics to variation in asexual growth forms, which can appear as single cell yeasts with budding structures and chains of incompletely separated cells, but also as highly resilient clusters of meristematically growing cells (Gostinčar et al. 2012; Mitchison-Field et al. 2019; El Baidouri et al. 2021; Gostinčar and Gunde-Cimerman 2024). Black yeasts are a polyphyletic group of extremophilic and extremotolerant fungi, with unique characteristics. They are characterized by thick dark melanized cell walls, providing protection against UV radiation, oxidative stress, and other environmental stressors. The cell wall is often covered by extracellular polysaccharides. They typically grow slowly, either as yeast cells, hyphae, or meristematic clumps, and can form biofilms, all according to exterior conditions. Black yeasts are not a monophyletic evolutionary clade but occupy two lineages in the Ascomycota phylum. The first lineage encompasses members of Dothideales (Dothideomycetes) which are related to plant pathogens such as the genus *Zymoseptoria*. Notable members also include the whiskey fungus *Baudoinia compniacensis*, the extremely halotolerant *Hortaea werneckii*, an emerging model system for rock-inhabiting fungi *Knufia petricola*, and a ubiquitous yeast *Aureobasidium pullulans* with multiple growth forms (Scott et al. 2007; Voigt et al. 2020; Erdmann et al. 2022; Masi et al. 2024; Wirshing et al. 2024). The other main lineage of black yeasts are members of the Chaetothyriales (Eurotiomycetes) that is a sister group to Eurotiales, which include *Aspergillus* and *Penicillium*. Notable members include species of *Exophiala*, which are found in soil, rock environments, and are associated with human disease and high radiation resistance (Zakharova et al. 2013; Coleine et al. 2019; Schultzhaus et al. 2020; Sun et al. 2020; Yuzon et al. 2023; Kurbessoian et al. 2024; Ravenel et al. 2024). They are found in various extreme environments, from hypersaline waters, deep-sea, glacial ice, as rock-inhabiting fungi ranging from marble statues to Antarctic rocks and radiation exposed sites (Réblová et al. 2016; Liu et al. 2022; Sastoque et al. 2024; Gostinčar and Gunde-Cimerman 2024). Some are associated with animals, plants, foods, and even humans, causing different conditions, from neurotropic infections to tinea nigra (Reid 1998; Bonifaz et al. 2008; Kirchhoff et al. 2020; Barrantes Murillo et al. 2023; Kurbessoian et al. 2023). They produce enzymes and metabolites active at extreme physico-chemical conditions (Selbmann et al. 2020).

A textbook example of a generalist black yeast species is *A. pullulans* (Dothideomycetes, Dothideales). It can be found in a wide variety of environments, spanning temperate, polar, and tropical regions. It can thrive in association with other organisms and directly on inorganic substances in environments both indoors and out doors. It is commonly associated with plants, animals, and isolated from rocks, soil, aqueous environments, house dust, household surfaces, and many more. *A. pullulans* is considered poly-extremotolerant, since it has a high tolerance for a wide range of stress factors and extreme conditions. It can tolerate high salinity, high UV radiation, heavy metal contamination, and even gamma irradiation. *A. pullulans* and related sub-subspecies are metabolically versatile and can utilize various carbon sources. It exhibits great phenotypic plasticity, resulting in various morphological and physiological traits (Slepecky and Starmer 2009). The fungus is also known for its many biotechnological uses, from the production of pullulan, an extracellular polysaccharide with applications in cosmetics, medicine, and the food industry, to the production of a variety of enzymes, including amylases, lipases, and hemicellulose- and xylan-degrading enzymes, with different industrial applications. *A. pullulans* is also used as a registered biocontrol agent in agriculture to combat several plant pathogens and as a producer of the antifungal peptide aureobasidin A (Chi et al. 2009; Di Francesco et al. 2020; Gostinčar and Gunde-Cimerman 2023).

Generalists are not just found among black yeasts, however. *Debaryomyces hansenii* (Saccharomycetes, Saccharomycetales), often referred to as *Hansenula anomala* or *Candida famata* is a non-pigmented, oval, or spherical cell occurring singly or in pairs. *D. hansenii* can be found in many environments, including soil, marine water, sea sediments, salted dairy products, on the surface of fruits and vegetables, and notably in polar glaciers (Seiler and Busse 1990; Saldanha-da-Gama et al. 1997; Prista et al. 2005; Jacques et al. 2015). It is known for its halotolerance, osmotolerance, and thermotolerance, for efficient nitrogen metabolism and to produce antifungal and antibacterial compounds. Combination of these characteristics enables its frequent use in food industry, to produce salted foods, such as cheese and fish based. Due to its versatility and adaptability, *D. hansenii* is used also in various biotechnological processes, including the production of enzymes, biosurfactants, and bioactive compounds, such as riboflavin (Navarrete et al. 2022).

Another example of a non-melanized generalist yeast is basidiomycetous *Rhodotorula mucilaginosa* (Microbotryomycetes, Sporidiobolales). The color of this red pigmented polyextremotolerant yeast is due to carotenoid pigments. Cells are typically unicellular and oval-shaped and form creamy or mucilaginous colonies and mixed biofilms. *R. mucilaginosa* is globally isolated from soil, water, air, and different extreme environments, from glaciers to industrially polluted hypersaline waters and artificial environments (Martínez-Ávila et al. 2021). It can tolerate a wide range of environmental stress factors, including high concentrations of salts and sugars, UV radiation, extreme pH levels, a wide range of temperatures. Additionally, it is nutritionally versatile and capable of metabolizing different substrates. In rare cases, it has been associated with opportunistic human infections (Wirth and Goldani 2012). Biotechnologically it is mainly used to produce carotenoids, which have applications in food, cosmetics, and pharmaceutical industries (Li et al. 2022).

*Aspergillus sydowii* (Eurotiomycetes, Eurotiales) is an example of a generalist filamentous fungus. *A. sydowii* has a broad ecological distribution and can be found in a broad range of natural and artificial environments in marine and terrestrial habitats. Lately, it has gained importance as a pathogen of corals, and hypersaline marine habitats have been hypothesized as potential reservoirs for its spread (Butinar et al. 2011; Soler-Hurtado et al. 2016). In marine environments it can be isolated from sea sediments, algae, seawater, and from coastal sites, such as mangrove ecosystems. Terrestrial sources of *A. sydowii* include soil and decaying plant and other organic matter, from near-Arctic to tropical areas. In both terrestrial and marine ecosystems, it contributes to the breakdown of complex organic compounds (Cong et al. 2017). Terrestrial isolates of *A. sydowii* have been shown to be non-pathogenic to corals, suggesting the existence of distinct ecotypes or strains with different ecological roles and host interacting factors (Alker et al. 2001). Some strains of *A. sydowii* have been studied for their biotechnological potential, in particular as a potential source of lignin and cellulose—degrading enzymes and as a producer of bioactive compounds (Madhu et al. 2009; Wang et al. 2014; Cong et al. 2017).

Specialist species range from fungi with a very limited distribution to those with a strong preference for one main selective environmental constraint or abiotic factor. The first example is the black yeast *Hortaea werneckii* (Dothideomycetes, Capnodiales), known for its adaptability to different salinities and temperatures. *H. werneckii* is exceptionally halotolerant and has been documented to grow from 0% to almost NaCl saturation at 32%, with optimal growth occurring between 6% and 14% NaCl. Some strains can grow at human body temperature, in the presence of NaCl, while others have been isolated from cold deep-sea water and glacial ice (Gunde-Cimerman et al. 2018; Gostinčar and Gunde-Cimerman 2023). Although *H. werneckii* is found primarily in thalassohaline, eutrophic hypersaline waters of man-made salterns around the world, it can be isolated with a lesser frequency in a wide array of other marine-related environments, Mg-rich bitterns that remain after halite precipitation, from salted foods and many other salty surfaces, including human skin, where they can cause a condition named tinea nigra (Reid 1998).

Another specialist black yeast species is *Aureobasidium subglaciale* (Dothideomycetes, Dothideales). This black yeast is found in only a very narrow range of habitats. It is specialized to low temperatures and oligotrophic conditions encountered in polar subglacial ice and refrigerators. It is also adapted to occasional high salinity and high UV radiation (Gostinčar et al. 2014; Zajc et al. 2022).

A third example of a specialist species is the filamentous basidiomycetous *Wallemia ichthyophaga* (Wallemiales, Wallemiaceae), an obligately halophilic fungus, with an interesting phylogenetic history. Genus *Wallemia* includes some of the most xerophilic fungi that belong to the class Wallemiomycetes within the 500-million-year-old subphylum Wallemiomycotina (Zajc et al. 2013; Zhao et al. 2017). *Wallemia* spp. have a global distribution and are found in extremely dry and saline environments. All species exhibit slow growth rates and some have a requirement for low $a_w$ (Gunde-Cimerman et al. 2018). The most specialized within the genus is the rarely found *W. ichthyophaga*, which has been observed in a very limited type of environment. It can be found primarily in thalassohaline hypersaline waters and bitterns of salterns and on salted meat products (Gunde-Cimerman et al. 2018; Gostinčar and Gunde-Cimerman 2023). It is the most halophilic fungus known to date, since it can grow in solutions saturated with NaCl, requiring a minimum of around 10% NaCl in its growth medium. It can even grow in solutions with high concentrations of other salts like KCl and $MgSO_4$, as well as at 2.1 M $MgCl_2$ (Zajc et al. 2014). The halophilic specialization of *W. ichthyophaga* can be recognized also by the increase of its production of secondary metabolites, including toxins, at high NaCl concentrations. This contrasts with the general trend of reduced secondary metabolite synthesis at low water activity in other fungi (Jančič et al. 2016).

## 3.2 From Genotype to Phenotype

Studying molecular adaptations of fungi to extreme environments has been of interest since the discovery of their consistent appearance as part of extreme microbiota. Understanding the remarkable survival strategies of these species can help us better understand life as it has evolved on our planet, predict the impact of climate change, identify the limit of life on Earth, and guess the possibilities of extraterrestrial life (Coleine et al. 2022; Gostinčar and Gunde-Cimerman 2023). Some of the more immediate impacts of mycology of extremes are the use of extremotolerant and extremophilic fungi in biotechnology (e.g., enzymes with unusual optima, industrial microorganism that can be cultivated under extreme conditions), agriculture (e.g., crops with improved resilience to stress), and astrobiology (e.g., in exploration for the search for life on acidic and hot planets such as our sister planet Venus).

However, research of fungi from extreme environments is not trivial. The lack of efficient methodology for genetic manipulation, slow and often polymorphic growth, and the complexity of adaptations have long stymied the progress of investigation. The initial isolation and deciphering individual mechanisms that laid the groundwork for later studies have lagged the work performed with well-studied model organisms such as *Saccharomyces cerevisiae*. While the work on mesophilic species exposed to environmental shocks has provided valuable data, it is a less than ideal substitute for extremotolerant and extremophilic species grown under constant extreme conditions (Gostinčar and Gunde-Cimerman 2023). Therefore, the first genome sequences of species from extreme environments were hailed as a fresh opportunity for mycology of extremes.

The genome of the halotolerant yeast *Debaryomyces hansenii* was published in 2004 (Dujon et al. 2004), followed by the genome of the extremely halotolerant black yeast *Hortaea werneckii* as well as the extremophilic *Wallemia ichthyophaga* in 2013 (Lenassi et al. 2013; Zajc et al. 2013), and the polyextremotolerant *Aureobasidium pullulans* and the psychrotolerant *Aureobasidium subglaciale* in 2014 (Gostinčar et al. 2014). However, with only limited information on what to look for and few genomes to work with, these studies remained mostly limited to confirming the presence of genes already known to be involved in response to stress in other organisms (Prista et al. 2005; Lenassi et al. 2013; Gostinčar et al. 2014). This included the identification of genes for the components of the high-osmolarity glycerol (HOG) pathway, for enzymes in the biosynthesis of compatible solutes, and for synthesis of melanin (Lenassi et al. 2013; Gostinčar et al. 2014). Genes for aquaporins and for membrane transporters involved in maintaining the homeostasis of metal cations, particularly $Na^+$ and $K^+$, have also been found (Gostinčar et al. 2014). Besides confirming their presence, genome analysis also discovered a great redundance of these genes (e.g., three different types of $K^+$ importers in *A. pullulans*) (Gostinčar et al. 2014). While logical at first sight, the role of this redundance is still not entirely explained, since it appears to be neither sufficient nor necessary for halotolerance (Gostinčar and Gunde-Cimerman 2023).

Recently, the first genomes of Antarctic rock-inhabiting black yeasts were published (Coleine et al. 2019; Coleine et al. 2020a). For instance, the stress-tolerant genome of *Friedmanniomyces endolithicus* (Tesei et al. 2012) was explored to identify genomic patterns that could be attributed to its specialization. The genome was compared to phylogenetically close melanized fungi from extreme environments, including *F. simplex, Baudoinia panamericana, Acidomyces acidophilus, Hortaea thailandica*, and *H. werneckii.* It was reported that genome encoded copies of genes (i.e., Gene Ontology [GO] terms) for stress tolerance such as and UV irradiation, DNA damage repair after damage, cold adaptation, and salt stress responses were particularly enriched in *F. endolithicus* (Coleine et al. 2020a). More recently, 118 black yeast genomes of Dothideomycetes and Eurotiomycetes, encompassing different lifestyles and ecologies, were used to investigate the traits explaining the differences and adaptability of these two classes. This study focused on key genomic traits associated with the main metabolic competences for their extremophilic behavior such as ability to withstand low or high temperatures, UV radiation, efficiency in DNA repair, and degradation of monoaromatic toxins and pollutants (Coleine et al. 2024).

Furthermore, a significant advancement of our understanding of the genomics of extreme fungi is sought through the DOE Joint Genome Project-funded initiative "Shed lighT in the daRk lineagES of the Fungal tree of life" (STRES). The main aim of this project is sequencing the reference genome of around 100 black yeasts (spanning most unsampled lineages in Dothideomycetes and Eurotiomycetes) to clarify the relationship between stress response, ecology and phylogeny and track transcripts and metabolites expressed genes under different stress conditions (i.e., high salinity, desiccation, UV radiation, and oligotrophy) (Selbmann et al. 2020).

Some other interesting genes were also found, such as a large number of genes for extracellular enzymes and major facilitator superfamily (MFS) transporters as well as several copies that encode for type 1 microbial rhodopsins in *Aureobasidium* spp. (Gostinčar et al. 2014). The latter were subsequently confirmed to be efficient light-driven proton transporters (Panzer et al. 2021). Fungal rhodopsins have been known for some time for their ability to build the transmembrane proton gradient (Waschuk et al. 2005) and were recently suggested to provide the cell with additional energy even when heterologously expressed in *S. cerevisiae* (Peterson et al. 2023). Their potential role in halotolerant fungi, which often grow under conditions of high solar irradiation and increased energy requirements (Oren 1999), is a good example of an early speculation based on genomic data, which is given further relevance with later research.

In the last decade, also transcriptomics was applied to investigate the adaptive strategies of extremotolerant fungi. For instance, the analyses of the transcriptome of *Cladophialophora immunda* (Chaetothyriales, Eurotiomycetes), a black fungus typically associated with hydrocarbon polluted environments, revealed that exposure to toluene activated degradation genes which likely protects the fungus (Blasi et al. 2017). Transcriptomic analysis revealed that a series of genes in the *Acidiella bohemica* genome were related to its metabolic pathways of C, N, S, and Fe as well as its adaptation mechanisms, including the response to acid stress and the resistance to heavy metals (Ou et al. 2021). Transcriptomics also revealed an altered response to ionizing radiation exposure in *Exophiala dermatitidis,* which showed an overlap in response between the control conditions and reactive oxygen species conditions, whereas exposure to an alpha particle source resulted in a robust downregulation of metabolic processes and upregulation of DNA replication and repair genes, and RNA metabolic processes (Malo et al. 2021). The transcriptome of members of *Aspergillus*, *Penicillium*, *Fodinomyces*, *Microascus*, *Trichoderma*, and *Verticillium* exposed to different stresses (e.g., extreme pH, salinity and temperature) was also investigated; findings showed that *Aspergillus terreus* NTOU4989 was the only fungus that was metabolically active at 45 °C, pH 3, and 30% salinity (Pang et al. 2020). Transcriptional profiling of the halophilic *Aspergillus sydowii* under similar stresses found the fungus upregulates genes related to starvation, cell wall remodeling, degradation and metabolism of xenobiotics, DNA/RNA metabolism, energy generation, signaling and general stress responses (Peidro-Guzmán et al. 2021).

## 3.3 From One Genome to Many

Several comparative genomic studies have extended the analyses of genomes to include either mesophilic relatives or other fungi from extreme environments. These studies often focus on large functional gene groups, for example reporting enrichment in amino acid uptake genes in *Aspergillus sydowii*, *Aspergillus versicolor*, and *Hyphopichia* spp. (Pérez-Llano et al. 2020; Lee et al. 2021), genes for MFS transporters in *Aspergillus salisburgensis*, *Aspergillus penicillioides*, *Aspergillus sclerotialis* and *Mrakia psychrophila* (Nazareth and Gonsalves 2014; Su et al. 2016; Tafer et al. 2019), and genes for hydrophobins in *W. ichthyophaga* (Zajc et al. 2013), pointing to the role of small metabolite accumulation and cell wall in adaptation to extreme conditions. Genome sequencing of *H. werneckii* grown for 7 years at constant high salinity indicated adaptations on the level of cell wall and morphology, signaling pathways, and the pentose phosphate cycle (Gostinčar et al. 2021). These studies provide a good starting point for further research, but mostly avoid mechanistic explanations of relatively general observations. More precise comparative and genetic studies might in the future pinpoint loci that are linked to extremotolerance, but such studies are hampered by the limited repertoires of strains and genetic methods, as discussed below. While homologous genes are easily compared, annotations of different quality, functionally analogous solutions, differences reflecting phylogenetic distances rather than adaptations to extreme conditions, and the lack of clear boundaries between some mesophilic and extremophilic species can obfuscate the results.

Sequencing tens or hundreds of genomes and employing population genomics offer another view of the lifestyle of extremotolerant and extremophilic fungi. Moreover, Single Nucleotide Polymorphism (SNP) calling could be also employed to allow the simultaneous identification of a genome-wide average and outliers for any given statistic to help to identify genomic regions contributing to local adaptation or even speciation. While not yet employed for the identification of genetic determinants of extremotolerance, these studies have so far uncovered an unexpected diversity of the reproductive strategies in fungi from extreme environments. Three species that were previously considered asexual have been shown to recombine: *A. pullulans* (Gostinčar et al. 2019b), *W. mellicola*, and *W. ichthyophaga* (Gostinčar et al. 2019a). Furthermore, *A. pullulans* was confirmed to be a true generalist. Despite the unusually wide variety of environments it can inhabit, it showed no population structure that would indicate cryptic specialization (Gostinčar et al. 2019b). Uncovering signs of recombination in fungi previously believed to be asexual is a common outcome of population genetic analyses and most fungi are now believed to be capable of sexual reproduction (Taylor et al. 2015; Nieuwenhuis and James 2016). For example, population genomics studies now possible with strain sequencing has found rates of recombination and LD decay to be extremely high in the previously regarded asexual fungus *Aspergillus fumigatus* (Lofgren et al. 2022; Auxier et al. 2023). Nevertheless, several clonal species have been confirmed among fungi from extreme environments with population genomics, among them *A. subglaciale* (Zajc et al. 2022), *A. melanogenum* (Černoša et al. 2021), and *H. werneckii* (Gostinčar et al. 2022a). Clonal reproduction has long been proposed as an attractive option for extremotolerant species, enabling them to redirect energy otherwise required for sexual reproduction into energetically costly adaptations and protecting efficient genome configurations from recombination load (Gostinčar et al. 2022b). The long-term success of this strategy is unclear, considering Muller's ratchet and other disadvantages of strictly asexual lifestyle. Interestingly, despite their strict clonality, wild populations of *A. melanogenum* and *H. werneckii* are dominated by diploid strains. In *H. werneckii* these diploids were initially thought to be a consequence of whole-genome duplication (Lenassi et al. 2013; Sinha et al. 2017). However, later analyses suggested they are the products of intraspecific hybridization between highly heterozygous haploids and are stable enough to spread over large geographical distances without any obvious loss of ploidy (Gostinčar et al. 2018). The process has been termed "stable parasexuality" (Gostinčar et al. 2022a) to distinguish it from the otherwise well-known parasexual cycle in fungi, which typically encompasses a gradual return to the initial ploidy through random chromosome loss (Pontecorvo 1956; Clutterbuck 1996). Hybridization in fungi is by no means limited to extremotolerant species (Gabaldón 2020), but there is increasing evidence for the large variety in ploidy in extreme conditions. For example, recent investigation of *Laburnicola rhizohalophila*, an endophyte of a halophyte, uncovered intraspecific diploid-like hybrids with increased fitness at high salinity in at least one of the hybrids (Li et al. 2024). Importantly, stable parasexuality and other variations in ploidy are easily missed during data analysis; therefore, the possible role of these phenomena in extremotolerance remains to be investigated.

The role of Whole-Genome Duplication (WGD) and other large-scale gene duplication events requires more study in extreme fungi. Duplications events may have enabled rapid evolution of proteins and consequently enhanced the metabolic plasticity or increasing the fitness during the colonization of hostile ecological niches. Studies of genomes and gene redundancy, copy number expansion, or the role of WGD in the history of these fungi can highlight whether any of these mechanisms have been important in the adaptation and evolution as extremophiles.

If comparing genomes of different species is complicated by the noise of differences unrelated to extremotolerance and comparing different strains alleviates that concern but is hampered by the lack of phenotypic differences between the strains—as discussed below—sequencing strains acquired by experimental evolution provides an attractive alternative to both strategies. Even if differences in extremotolerance between strains are smaller than between wild strains, the signal-to-noise ratio is smaller due to short evolutionary times and can be further decreased by biological replicates and the search for cases of convergent evolution. The number of experimental evolution studies on fungi from extreme environments is limited, but they provide some encouraging results. During 7 years of growth at constant high salinity, a diploid strain of *H. werneckii* accumulated aneuploidies, all duplications rather than deletions, enriched in genes related with cell wall and morphology, signaling pathways, and the pentose phosphate cycle (Gostinčar et al. 2021). A similar but shorter experiment with *A. pullulans* produced strains with improved growth at high salinity and mutations associated with genes for voltage-gated potassium channels, while the mutations introduced during the evolution of *W. ichthyophaga* at low salinity were linked to hydrophobins (Hariri Akbari et al. 2024).

Despite the wealth of genomic data, predicting complex phenotypes such as extremotolerance from genomes remains challenging. However, the quickly accelerating sequencing of fungal genomes presents an unprecedented resource that can be used by increasingly sophisticated analytical tools and genetic methods. For instance, an unprecedented advance would be the inclusion of additional species in the already existing Fungal Stress Response Database (Karányi et al. 2013) and the *Saccharomyces cerevisiae*- and *Aspergillus nidulans*-based stress response databases (Orosz et al. 2018), which currently incorporate filamentous fungi and yeasts but do not specifically address stress-adapted species. This would represent a necessary expansion of tool for scientists interested in comparative analyses of fungi across different extreme ecosystems.

The development of these new genomic resources has expanded the genomes and gene information available for cultured isolates of extremophiles and their relatives. As sequencing technology has improved however, it is now possible to generate nearly complete genome assemblies of fungi as Metagenome-Assembled Genomes (MAGs) from metagenome sequencing of environmental samples (Peng et al. 2021). These metagenome surveys are also supporting additional perspective on the abundance and taxonomic diversity as well as predicted functional diversity of the organisms based on analysis of predicted gene sequences of fungal metagenomes.

## 3.4 Perspectives

Genome sequencing of fungi from extreme environments has provided a welcome addition to the limited toolbox of extreme mycology. However, the increasingly powerful and accessible high-throughput sequencing methods may offer many further opportunities. Future research on extremotolerant and extremophilic fungi and their genomes should focus on several key directions—ecology, astrobiology, multi-omics, phylogenomics, comparative genomics and population genomics, and transcriptomics.

The main research gap in the ecology of extremophilic and extremotolerant fungi is in assessing the size of uncultivable fungal microbiota within various extreme environments. Tailored approaches are essential for the extraction of the total DNA, since methods used for mesophilic fungi (or even bacteria) often do not produce DNA of sufficient quality and quantity and can introduce major biases. The main reason is that extremophilic fungi, particularly those that are constitutively melanized, typically have very thick cell walls, often covered by substantial quantities of protective extracellular polymeric substances (EPS) (Gostinčar and Gunde-Cimerman 2023). If homogenization of the sample is only partially successful and the resulting DNA is used for amplicon sequencing of taxonomic markers or metagenomic sequencing, the abundance of species with the most recalcitrant cells can be substantially underestimated (Lindahl et al. 2013; Frau et al. 2019; Tiew et al. 2020).

Advanced metagenomics is needed to estimate the diversity and abundance of these uncultivable fungal species, to distinguish between native species and sporadic inputs and to enhance our understanding of the roles of extremophilic fungi in nutrient cycling, interactions of extremophilic fungi with other microorganisms, and overall contributions to ecosystem dynamics in extreme environments. This research area lags considerably behind our understanding of prokaryotes. For instance, while a large number of prokaryotic Metagenome-Assembled Genomes (MAGs) have been published, only a few examples of eukaryotic and fungi in particular MAGs have been reported; this is in part due to the absence of dedicated and easy-to-use processing pipelines or curated databases.

Research should also focus on how extremophilic fungi disperse, colonize, and connect within small and fragmented extreme habitats. So far fungal distribution has been mainly studied in relation to Aeolian deposits, while ocean currents, clouds, aerial black carbon, and animal vectors have been mainly overlooked. These investigations can reveal the mechanisms governing their distribution and diversity by carefully designed metagenomic studies or population genomics. The latter can also uncover the prevalence of stable diploid or polyploid hybrids in relation to haploid strains in the global population dynamics of specialist and generalist species.

Phylogenomics can be used to better understand the evolutionary relationships of extremotolerant and extremophilic fungi and to delineate species with complex taxonomy—genomics in fungal taxonomy remains a significantly underutilized tool (Hibbett et al. 2013; Naranjo-Ortiz and Gabaldón 2019; Li et al. 2021).

The analysis of multiple genomes within a population allows the identification of genetic variations, patterns of adaptation, and evolutionary trajectories. Genetic background of extremotolerance is perhaps the most obvious target of such studies. Other topics include questions about metabolism and energy allocation to understand how extremotolerant and extremophilic fungi balance their energy expenditure between growth and stress survival mechanisms and how phenotypic plasticity and the ability to switch between growth forms contribute to stress tolerance and survival strategies. Examination of genetic differences and similarities among different species can provide insights into conserved adaptive mechanisms and species-specific strategies. Identification of Quantitative Trait Loci (QTL) and Genome-Wide Association Studies (GWAS) provide opportunities to determine genetic variants related to extremotolerance. However, application of these methods to extremotolerant and extremophilic fungi is not trivial (Gostinčar and Gunde-Cimerman 2023). Approaches such as segregant analysis would be difficult in species with largely unknown sexual cycle and impossible in clonal species. Recent discoveries provide hope that these obstacles can be surmounted at least in some species, for example in *A. pullulans*, which appears to recombine (even if the process itself has not yet been directly observed) (Gostinčar et al. 2019b), can be grown as a yeast (Zalar et al. 2008), and is genetically tractable (Zhang et al. 2019; Kreuter et al. 2022; Wirshing et al. 2024). The second concern is whether the phenotypes of strains included in the analysis differ sufficiently to provide enough statistical power to any analysis. For example, while the genomic variability between strains of *H. werneckii* and *A. pullulans* is considerable (Gostinčar et al. 2019b; Gostinčar et al. 2022a), the difference in halotolerance between strains is relatively minor (our unpublished data). In some species close relatives can be used, but these usually differ from the investigated species in more than one trait. For example, *Aureobasidium subglaciale* and *Aureobasidium melanogenum* are certainly less halotolerant than *A. pullulans* (Zalar et al. 2008), but are also clonal and have different temperature preferences (Zalar et al. 2008; Černoša et al. 2021; Zajc et al. 2022).

Genomics of extremotolerant and extremophilic fungi hold future promise in various biotechnological applications. An important area is bioremediation, where they can play a vital role in cleaning up polluted environments, such as saline or extremely cold ones (Martínez-Ávila et al. 2021; Peidro-Guzmán et al. 2021). Another significant but relatively unexplored area involves the identification of specific stress-tolerance genes within extremotolerant fungi. These genes can potentially be introduced into other organisms, including microorganisms and plants, to enhance their capacity to withstand challenging environmental conditions (Gašparič et al. 2013). This is particularly important in the face of ongoing global climate change that disproportionately affects agriculturally important plants. CRISPR/Cas9 techniques, which have been underutilized in the context of extremophilic fungi so far, can serve as a valuable tool to investigate their adaptations and transfer it to other organisms.

The genomics of extremophilic fungi could also contribute to additional valuable applications, such as using extremotolerant fungi as effective biocontrol agents (Zajc et al. 2022), as sources of enzymes in specialized industrial processes (Porter et al. 2023), and for studying radiation resistance (Pulschen et al. 2015; Romsdahl et al. 2021; Horne et al. 2022; Yuzon et al. 2023).

Genomic studies have propelled our understanding of how fungi adapt to extreme environments. While sequencing provides a foundation, translating genomic information into phenotypic predictions necessitates addressing the complexities inherent in gene–environment interactions. Nonetheless, the advances made in studying extremotolerant and extremophilic fungi genomes offer promising avenues for harnessing their unique adaptations in various practical applications.

Ultimately, studying extremotolerant and extremophilic fungi is also integral to providing the important information to help to predict what sort of Earth-like organisms we might find elsewhere in the Solar System and beyond, and to understand the limit of life on Earth. Unearthing these aspects is the subject of astromycology, a recent field within astrobiology (Simões et al. 2023) and the origins of life on Earth. We further suggest that multi omics-based techniques, including genomics, epigenomics, transcriptomics, proteomics, metabolomics, and phenomics, should be used to untangle the mechanisms underlying adaptations, which allow fungi to persist in the extremes, while synthetic biology might represent a tool to increase the fitness of organisms and to confer new functions and allow them to potentially survive in extraterrestrial environments (Coleine and Delgado-Baquerizo 2022).

## References


Alker AP, Smith GW, Kim K (2001) Characterization of *Aspergillus sydowii* (Thom et Church), a fungal pathogen of Caribbean Sea fan corals. Hydrobiologia 460:105–111

Anesio AM, Laybourn-Parry J (2012) Glaciers and ice sheets as a biome. Trends Ecol Evol 27:219–225

Anesio AM, Lutz S, Chrismas NAM, Benning LG (2017) The microbiome of glaciers and ice sheets. NPJ Biofilms Microbiomes 3:10

Auxier B, Debets AJM, Stanford FA, Rhodes J, Becker FM, Reyes Marquez F, Nijland R, Dyer PS, Fisher MC, van den Heuvel J, Snelders E (2023) The human fungal pathogen *Aspergillus fumigatus* can produce the highest known number of meiotic crossovers. PLoS Biol 21:e3002278

Averesch NJH, Shunk GK, Kern C (2022) Cultivation of the dematiaceous fungus *Cladosporium sphaerospermum* aboard the International Space Station and effects of ionizing radiation. Front Microbiol 13:877625

Barrantes Murillo DF, Anderson S, Capobianco C, Lewbart GA, Wiederhold NP, Cañete-Gibas CF, Negrão Watanabe TT (2023) Systemic *Exophiala equina* infection in an eastern box turtle (*Terrapene carolina carolina*): a case report and literature review. Front Vet Sci 10:1158393

Biddanda B, Weinke A, Stone I, Kendall S, Hartmeyer P, Lusardi W, Gandulla S, Bright J, Ruberg S (2021) Extant earthly microbial Mats and Microbialites as models for exploration of life in extraterrestrial mat worlds. Life (Basel) 11:883

Blasi B, Tafer H, Kustor C, Poyntner C, Lopandic K, Sterflinger K (2017) Genomic and transcriptomic analysis of the toluene degrading black yeast *Cladophialophora immunda*. Sci Rep 7:11436

Bonifaz A, Badali H, de Hoog GS, Cruz M, Araiza J, Cruz MA, Fierro L, Ponce RM (2008) Tinea nigra by *Hortaea werneckii*, a report of 22 cases from Mexico. Stud Mycol 61:77–82

Butinar L, Frisvad JC, Gunde-Cimerman N (2011) Hypersaline waters–a potential source of foodborne toxigenic aspergilli and penicillia. FEMS Microbiol Ecol 77:186–199

Butinar L, Spencer-Martins I, Gunde-Cimerman N (2007) Yeasts in high Arctic glaciers: the discovery of a new habitat for eukaryotic microorganisms. Antonie Van Leeuwenhoek 91:277–289

Cantrell SA, Tkavc R, Gunde-Cimerman N, Zalar P, Acevedo M, Báez-Félix C (2013) Fungal communities of young and mature hypersaline microbial mats. Mycologia 105:827–836

Černoša A, Sun X, Gostinčar C, Fang C, Gunde-Cimerman N, Song Z (2021) Virulence traits and population genomics of the black yeast *Aureobasidium melanogenum*. J Fungi (Basel) 7:665

Chi Z, Wang F, Chi Z, Yue L, Liu G, Zhang T (2009) Bioproducts from *Aureobasidium pullulans*, a biotechnologically important yeast. Appl Microbiol Biotechnol 82:793–804

Clutterbuck AJ (1996) Parasexual recombination in fungi. J Genet 75:281–286

Coleine C, Delgado-Baquerizo M (2022) Unearthing terrestrial extreme microbiomes for searching terrestrial-like life in the solar system. Trends Microbiol 30:1101–1115

Coleine C, Kurbessoian T, Calia G, Delgado-Baquerizo M, Cestaro A, Pindo M, Armanini F, Asnicar F, Isola D, Segata N, Donati C, Stajich J, de Hoog S, Selbmann L (2024) Class-wide genomic tendency throughout specific extremes in black fungi. Fungal Divers 125:121–138

Coleine C, Masonjones S, Sterflinger K, Onofri S (2020a) Peculiar genomic traits in the stress-adapted cryptoendolithic Antarctic fungus *Friedmanniomyces endolithicus*, vol 124. Fungal Biol, p 458

Coleine C, Selbmann L, Masonjones S, Onofri S, Zucconi L, Stajich JE (2019) Draft genome sequence of an Antarctic isolate of the black yeast fungus *Exophiala mesophila*. Microbiol Resour Announc 8. https://doi.org/10.1128/mra.00142-19

Coleine C, Stajich JE, de Los RA, Selbmann L (2020b) Beyond the extremes: rocks as ultimate refuge for fungi in drylands. Mycologia 113:108–133

Coleine C, Stajich JE, Selbmann L (2022) Fungi are key players in extreme ecosystems. Trends Ecol Evol 37:P517–P528

Collins T, Margesin R (2019) Psychrophilic lifestyles: mechanisms of adaptation and biotechnological tools. Appl Microbiol Biotechnol 103:2857–2871

Cong B, Wang N, Liu S, Liu F, Yin X, Shen J (2017) Isolation, characterization and transcriptome analysis of a novel Antarctic *Aspergillus sydowii* strain MS-19 as a potential lignocellulosic enzyme source. BMC Microbiol 17:129

Di Francesco A, Di Foggia M, Zajc J, Gunde-Cimerman N, Baraldi E (2020) Study of the efficacy of Aureobasidium strains belonging to three different species: *A. pullulans*, *A. subglaciale* and *A. melanogenum* against *Botrytis cinerea* of tomato. Ann Appl Biol 177:266–275

Dujon B, Sherman D, Fischer G, Durrens P, Casaregola S, Lafontaine I, De Montigny J, Marck C, Neuvéglise C, Talla E, Goffard N, Frangeul L, Aigle M, Anthouard V, Babour A, Barbe V, Barnay S, Blanchin S, Beckerich J-M, Beyne E, Bleykasten C, Boisramé A, Boyer J, Cattolico L, Confanioleri F, De Daruvar A, Despons L, Fabre E, Fairhead C, Ferry-Dumazet H, Groppi A, Hantraye F, Hennequin C, Jauniaux N, Joyet P, Kachouri R, Kerrest A, Koszul R, Lemaire M, Lesur I, Ma L, Muller H, Nicaud J-M, Nikolski M, Oztas S, Ozier-Kalogeropoulos O, Pellenz S, Potier S, Richard G-F, Straub M-L, Suleau A, Swennen D, Tekaia F, Wésolowski-Louvel M, Westhof E, Wirth B, Zeniou-Meyer M, Zivanovic I, Bolotin-Fukuhara M, Thierry A, Bouchier C, Caudron B, Scarpelli C, Gaillardin C, Weissenbach J, Wincker P, Souciet J-L (2004) Genome evolution in yeasts. Nature 430:35–44

El Baidouri F, Zalar P, James TY, Gladfelter AS, Amend AS (2021) Evolution and physiology of amphibious yeasts. Ann Rev Microbiol 75:337–357

Erdmann EA, Nitsche S, Gorbushina AA, Schumacher J (2022) Genetic engineering of the rock inhabitant *Knufia petricola* provides insight into the biology of Extremotolerant black fungi. Front Fungal Biol 3:862429

Frau A, Kenny JG, Lenzi L, Campbell BJ, Ijaz UZ, Duckworth CA, Burkitt MD, Hall N, Anson J, Darby AC, Probert CSJ (2019) DNA extraction and amplicon production strategies deeply influence the outcome of gut mycobiome studies. Sci Rep 9:9328

Gabaldón T (2020) Hybridization and the origin of new yeast lineages. FEMS Yeast Res 20:foaa040

Gašparič MB, Lenassi M, Gostinčar C, Rotter A, Plemenitaš A, Gunde-Cimerman N, Gruden K, Zel J (2013) Insertion of a specific fungal 3′-phosphoadenosine-5′-phosphatase motif into a plant homologue improves halotolerance and drought tolerance of plants. PLoS One 8:e81872

Ghori N-U-H, Wise MJ, Whiteley AS (2021) Temporal microbial community dynamics within a unique acid saline Lake. Front Microbiol 12:649594

Gibson ME, Benison KC (2023) It's a trap!: modern and ancient halite as Lagerstätten. J Sediment Res 93:642–655

Gostinčar C, Grube M, de Hoog S, Zalar P, Gunde-Cimerman N (2010) Extremotolerance in fungi: evolution on the edge. FEMS Microbiol Ecol 71:2–11

Gostinčar C, Gunde-Cimerman N (2023) Understanding fungi in glacial and hypersaline environments. Ann Rev Microbiol 77:89–109

Gostinčar C, Gunde-Cimerman N (2024) Black yeasts in hypersaline conditions. Appl Microbiol Biotechnol 108:252

Gostinčar C, Muggia L, Grube M (2012) Polyextremotolerant black fungi: oligotrophism, adaptive potential, and a link to lichen symbioses. Front Microbiol 3:390

Gostinčar C, Ohm RA, Kogej T, Sonjak S, Turk M, Zajc J, Zalar P, Grube M, Sun H, Han J, Sharma A, Chiniquy J, Ngan CY, Lipzen A, Barry K, Grigoriev IV, Gunde-Cimerman N (2014) Genome sequencing of four *Aureobasidium pullulans* varieties: biotechnological potential, stress tolerance, and description of new species. BMC Genomics 15:549

Gostinčar C, Stajich JE, Gunde-Cimerman N (2023) Extremophilic and extremotolerant fungi. Curr Biol 33:R752–R756

Gostinčar C, Stajich JE, Kejžar A, Sinha S, Nislow C, Lenassi M, Gunde-Cimerman N (2021) Seven years at high salinity-experimental evolution of the extremely halotolerant black yeast *Hortaea werneckii*. J Fungi (Basel) 7:723

Gostinčar C, Stajich JE, Zupančič J, Zalar P, Gunde-Cimerman N (2018) Genomic evidence for intraspecific hybridization in a clonal and extremely halotolerant yeast. BMC Genomics 19:364

Gostinčar C, Sun X, Černoša A, Fang C, Gunde-Cimerman N, Song Z (2022a) Clonality, inbreeding, and hybridization in two extremotolerant black yeasts. Gigascience 11:giac095

Gostinčar C, Sun X, Zajc J, Fang C, Hou Y, Luo Y, Gunde-Cimerman N, Song Z (2019a) Population genomics of an obligately halophilic basidiomycete *Wallemia ichthyophaga*. Front Microbiol 10:2019

Gostinčar C, Turk M, Zajc J, Gunde-Cimerman N (2019b) Fifty *Aureobasidium pullulans* genomes reveal a recombining polyextremotolerant generalist. Environ Microbiol 21:3638–3652

Gostinčar C, Zalar P, Gunde-Cimerman N (2022b) No need for speed: slow development of fungi in extreme environments. Fungal Biol Rev 39:1–14

Gross S, Robbins EI (2000) Acidophilic and acid-tolerant fungi and yeasts. Hydrobiologia 433:91–109

Gümral R, Özhak-Baysan B, Tümgör A, Saraçlı MA, Yıldıran ŞT, Ilkit M, Zupančič J, Novak-Babič M, Gunde-Cimerman N, Zalar P, de Hoog GS (2016) Dishwashers provide a selective extreme environment for human-opportunistic yeast-like fungi. Fungal Divers 76:1–9

Gunde-Cimerman N, Oren A, Plemenitaš A (eds) (2005) Adaptation to life at high salt concentrations in Archaea, Bacteria, and Eukarya. Springer, Netherlands

Gunde-Cimerman N, Plemenitaš A, Oren A (2018) Strategies of adaptation of microorganisms of the three domains of life to high salt concentrations. FEMS Microbiol Rev 42:353–375

Hariri Akbari F, Song Z, Turk M, Gunde-Cimerman N, Gostinčar C (2024) Experimental evolution of extremotolerant and extremophilic fungi under osmotic stress. IUBMB Life 76:617. https://doi.org/10.1002/iub.2825

Heinicke C, Arnhof M (2021) A review of existing analog habitats and lessons for future lunar and Martian habitats. Reach Out 21–22:100038

Hibbett DS, Stajich JE, Spatafora JW (2013) Toward genome-enabled mycology. Mycologia 105:1339–1349

Horne WH, Volpe RP, Korza G, DePratti S, Conze IH, Shuryak I, Grebenc T, Matrosova VY, Gaidamakova EK, Tkavc R, Sharma A, Gostinčar C, Gunde-Cimerman N, Hoffman BM, Setlow P, Daly MJ (2022) Effects of desiccation and freezing on microbial ionizing radiation survivability: considerations for Mars sample return. Astrobiology 22:1337–1350

Jacques N, Zenouche A, Gunde-Cimerman N, Casaregola S (2015) Increased diversity in the genus Debaryomyces from Arctic glacier samples. Antonie Van Leeuwenhoek 107:487–501

Jančič S, Frisvad JC, Kocev D, Gostinčar C, Džeroski S, Gunde-Cimerman N (2016) Production of secondary metabolites in extreme environments: food- and airborne *Wallemia spp.* produce toxic metabolites at hypersaline conditions. PLoS One 11:e0169116

Karányi Z, Holb I, Hornok L, Pócsi I, Miskei M (2013) FSRD: fungal stress response database. Database 2013:bat037

Kirchhoff L, Weisner A-K, Schrepffer M, Hain A, Scharmann U, Buer J, Rath P-M, Steinmann J (2020) Phenotypical characteristics of the black yeast *Exophiala dermatitidis* are affected by *Pseudomonas aeruginosa* in an artificial sputum medium mimicking cystic fibrosis-like conditions. Front Microbiol 11:471

Kreuter J, Stark G, Mach RL, Mach-Aigner AR, Zimmermann C (2022) Fast and efficient CRISPR-mediated genome editing in *Aureobasidium* using Cas9 ribonucleoproteins. J Biotechnol 350:11–16

Kurbessoian T, Murante D, Crocker A, Hogan DA, Stajich JE (2023) In host evolution of *Exophiala dermatitidis* in cystic fibrosis lung micro-environment. G3 (Bethesda) 13:jkad126

Kurbessoian T, Stevenson D, Haro R, Ying S, Stajich JE (2024) Draft genome sequence of a black yeast fungus *Exophiala xenobiotica* isolated from La Brea tar pits. Microbiol Resour Announc 13:e0091323

Lee DW, Hong CP, Thak EJ, Park S-G, Lee CH, Lim JY, Seo J-A, Kang HA (2021) Integrated genomic and transcriptomic analysis reveals unique mechanisms for high osmotolerance and halotolerance in *Hyphopichia* yeast. Environ Microbiol 23:3499–3522

Lenassi M, Gostinčar C, Jackman S, Turk M, Sadowski I, Nislow C, Jones S, Birol I, Cimerman NG, Plemenitaš A (2013) Whole genome duplication and enrichment of metal cation transporters revealed by de novo genome sequencing of extremely halotolerant black yeast *Hortaea werneckii*. PLoS One 8:e71328

Li Y, Steenwyk JL, Chang Y, Wang Y, James TY, Stajich JE, Spatafora JW, Groenewald M, Dunn CW, Hittinger CT, Shen X-X, Rokas A (2021) A genome-scale phylogeny of the kingdom Fungi. Curr Biol 31:1653–1665

Li Z, Li C, Cheng P, Yu G (2022) *Rhodotorula mucilaginosa*-alternative sources of natural carotenoids, lipids, and enzymes for industrial use. Heliyon 8:e11505

Li Z, Zhu Z, Qian K, Tang B, Han B, Zhong Z, Fu T, Zhou P, Stukenbrock EH, Martin FM, Yuan Z (2024) Intraspecific diploidization of a halophyte root fungus drives heterosis. Nat Commun 15:5872

Lindahl BD, Nilsson RH, Tedersoo L, Abarenkov K, Carlsen T, Kjøller R, Kõljalg U, Pennanen T, Rosendahl S, Stenlid J, Kauserud H (2013) Fungal community analysis by high-throughput sequencing of amplified markers—a user's guide. New Phytol 199:288–299

Liu B, Fu R, Wu B, Liu X, Xiang M (2022) Rock-inhabiting fungi: terminology, diversity, evolution and adaptation mechanisms. Mycology 13:1–31

Lofgren LA, Ross BS, Cramer RA, Stajich JE (2022) The pan-genome of *Aspergillus fumigatus* provides a high-resolution view of its population structure revealing high levels of lineage-specific diversity driven by recombination. PLoS Biol 20:e3001890

Madhu KM, Beena PS, Chandrasekaran M (2009) Extracellular β-glucosidase production by a marine *Aspergillus sydowii* BTMFS 55 under solid state fermentation using statistical experimental design. Biotechnol Bioprocess Eng 14:457–466

Malo ME, Schultzhaus Z, Frank C, Romsdahl J, Wang Z, Dadachova E (2021) Transcriptomic and genomic changes associated with radioadaptation in *Exophiala dermatitidis*. Comput Struct Biotechnol J 19:196–205

Martínez-Ávila L, Peidro-Guzmán H, Pérez-Llano Y, Moreno-Perlín T, Sánchez-Reyes A, Aranda E, Ángeles de Paz G, Fernández-Silva A, Folch-Mallol JL, Cabana H, Gunde-Cimerman N, Batista-García RA (2021) Tracking gene expression, metabolic profiles, and biochemical analysis in the halotolerant basidiomycetous yeast *Rhodotorula mucilaginosa* EXF-1630 during benzo[a]pyrene and phenanthrene biodegradation under hypersaline conditions. Environ Pollut 271:116358

Masi A, Wögerbauer K, Mach RL, Mach-Aigner AR (2024) Genomic deletions in *Aureobasidium pullulans* by an AMA1 plasmid for gRNA and CRISPR/Cas9 expression. Fungal Biol Biotechnol 11:6

Mitchison-Field LMY, Vargas-Muñiz JM, Stormo BM, Vogt EJD, Van Dierdonck S, Pelletier JF, Ehrlich C, Lew DJ, Field CM, Gladfelter AS (2019) Unconventional cell division cycles from marine-derived yeasts. Curr Biol 29:3439–3456.e5

Naranjo-Ortiz MA, Gabaldón T (2019) Fungal evolution: diversity, taxonomy and phylogeny of the fungi. Biol Rev Camb Philos Soc 94:2101–2137

Navarrete C, Estrada M, Martínez JL (2022) *Debaryomyces hansenii*: an old acquaintance for a fresh start in the era of the green biotechnology. World J Microbiol Biotechnol 38:99

Nazareth S, Gonsalves V (2014) *Aspergillus penicillioides*—a true halophile existing in hypersaline and polyhaline econiches. Ann Microbiol 64:397–402

Nieuwenhuis BPS, James TY (2016) The frequency of sex in fungi. Philos Trans R Soc Lond Ser B Biol Sci 371:20150540

Novak Babič M, Gostinčar C, Gunde-Cimerman N (2020) Microorganisms populating the water-related indoor biome. Appl Microbiol Biotechnol 104:6443–6462

Novikova N, De Boever P, Poddubko S, Deshevaya E, Polikarpov N, Rakova N, Coninx I, Mergeay M (2006) Survey of environmental biocontamination on board the International Space Station. Res Microbiol 157:5–12

Oren A (1999) Bioenergetic aspects of halophilism. Microbiol Mol Biol Rev 63:334–348

Orosz E, van de Wiele N, Emri T, Zhou M, Robert V, de Vries RP, Pócsi I (2018) Fungal Stress Database (FSD)–a repository of fungal stress physiological data. Database 2018:bay009

Ou S-N, Liang J-L, Jiang X-M, Liao B, Jia P, Shu W-S, Li J-T (2021) Physiological, genomic and transcriptomic analyses reveal the adaptation mechanisms of *Acidiella bohemica* to extreme acid mine drainage environments. Front Microbiol 12:705839

Pang K-L, Chiang MW-L, Guo S-Y, Shih C-Y, Dahms HU, Hwang J-S, Cha H-J (2020) Growth study under combined effects of temperature, pH and salinity and transcriptome analysis revealed adaptations of *Aspergillus terreus* NTOU4989 to the extreme conditions at Kueishan Island hydrothermal vent Field, Taiwan. PLoS One 15:e0233621

Panzer S, Zhang C, Konte T, Bräuer C, Diemar A, Yogendran P, Yu-Strzelczyk J, Nagel G, Gao S, Terpitz U (2021) Modified Rhodopsins from *Aureobasidium pullulans* excel with very high proton-transport rates. Front Mol Biosci 8:750528

Parker CW, de Teixeira MM, Singh NK, Raja HA, Cank KB, Spigolon G, Oberlies NH, Barker BM, Stajich JE, Mason CE, Venkateswaran K (2022) Genomic Characterization of *Parengyodontium torokii* sp. nov., a Biofilm-Forming Fungus Isolated from Mars 2020 Assembly facility. J Fungi (Basel) 8:66

Peidro-Guzmán H, Pérez-Llano Y, González-Abradelo D, Fernández-López MG, Dávila-Ramos S, Aranda E, Hernández DRO, García AO, Lira-Ruan V, Pliego OR, Santana MA, Schnabel D, Jiménez-Gómez I, Mouriño-Pérez RR, Aréchiga-Carvajal ET, Del Rayo S-CM, Folch-Mallol JL, Sánchez-Reyes A, Vaidyanathan VK, Cabana H, Gunde-Cimerman N, Batista-García RA (2021) Transcriptomic analysis of polyaromatic hydrocarbon degradation by the halophilic fungus *Aspergillus sydowii* at hypersaline conditions. Environ Microbiol 23:3435–3459

Peng X, Wilken SE, Lankiewicz TS, Gilmore SP, Brown JL, Henske JK, Swift CL, Salamov A, Barry K, Grigoriev IV, Theodorou MK, Valentine DL, O'Malley MA (2021) Genomic and functional analyses of fungal and bacterial consortia that enable lignocellulose breakdown in goat gut microbiomes. Nat Microbiol 6:499–511

Pérez-Llano Y, Rodríguez-Pupo EC, Druzhinina IS, Chenthamara K, Cai F, Gunde-Cimerman N, Zalar P, Gostinčar C, Kostanjšek R, Folch-Mallol JL, Batista-García RA, Sánchez-Carbente MDR (2020) Stress reshapes the physiological response of halophile fungi to salinity. Cells 9:525

Peterson A, Baskett C, Ratcliff WC, Burnetti A (2023) Using light for energy: examining the evolution of phototrophic metabolism through synthetic construction bioRxiv 2022.12.06.519405

Pontecorvo G (1956) The parasexual cycle in fungi. Ann Rev Microbiol 10:393–400

Porter R, Černoša A, Fernández-Sanmartín P, Cortizas AM, Aranda E, Luo Y, Zalar P, Podlogar M, Gunde-Cimerman N, Gostinčar C (2023) Degradation of polypropylene by fungi *Coniochaeta hoffmannii* and *Pleurostoma richardsiae*. Microbiol Res 277:127507

Prista C, Loureiro-Dias MC, Montiel V, García R, Ramos J (2005) Mechanisms underlying the halotolerant way of *Debaryomyces hansenii*. FEMS Yeast Res 5:693–701

Pulschen AA, Rodrigues F, Duarte RTD, Araujo GG, Santiago IF, Paulino-Lima IG, Rosa CA, Kato MJ, Pellizari VH, Galante D (2015) UV-resistant yeasts isolated from a high-altitude volcanic area on the Atacama Desert as eukaryotic models for astrobiology. Microbiology 4:574–588

Raghupathi PK, Zupančič J, Brejnrod AD, Jacquiod S, Houf K, Burmølle M, Gunde-Cimerman N, Sørensen SJ (2018) Microbial diversity and putative opportunistic pathogens in dishwasher biofilm communities. Appl Environ Microbiol 84:e02755–e02717

Ravenel K, Guegan H, Gastebois A, Bouchara J-P, Gangneux J-P, Giraud S (2024) Fungal colonization of the Airways of Patients with cystic fibrosis: the role of the environmental reservoirs. Mycopathologia 189:19

Réblová M, Hubka V, Thureborn O, Lundberg J, Sallstedt T, Wedin M, Ivarsson M (2016) From the tunnels into the treetops: new lineages of black yeasts from biofilm in the Stockholm metro system and their relatives among ant-associated fungi in the Chaetothyriales. PLoS One 11:e0163396

Reid BJ (1998) *Exophiala werneckii* causing tinea nigra in Scotland. Br J Dermatol 139:157–158

Romsdahl J, Schultzhaus Z, Chen A, Liu J, Ewing A, Hervey J, Wang Z (2021) Adaptive evolution of a melanized fungus reveals robust augmentation of radiation resistance by abrogating non-homologous end-joining. Environ Microbiol 23:3627–3645

Saldanha-da-Gama A, Malfeito-Ferreira M, Loureiro V (1997) Characterization of yeasts associated with Portuguese pork-based products. Int J Food Microbiol 37:201–207

Sastoque AP, Cano-Lira JF, Stchigel AM (2024) Diversity of rock-inhabiting fungi in Tarragona Province, Spain. J Fungi (Basel) 10:170

Scheller EL, Razzell Hollis J, Cardarelli EL, Steele A, Beegle LW, Bhartia R, Conrad P, Uckert K, Sharma S, Ehlmann BL, Abbey WJ, Asher SA, Benison KC, Berger EL, Beyssac O, Bleefeld BL, Bosak T, Brown AJ, Burton AS, Bykov SV, Cloutis E, Fairén AG, DeFlores L, Farley KA, Fey DM, Fornaro T, Fox AC, Fries M, Hickman-Lewis K, Hug WF, Huggett JE, Imbeah S, Jakubek RS, Kah LC, Kelemen P, Kennedy MR, Kizovski T, Lee C, Liu Y, Mandon L, McCubbin FM, Moore KR, Nixon BE, Núñez JI, Rodriguez Sanchez-Vahamonde C, Roppel RD, Schulte M, Sephton MA, Sharma SK, Siljeström S, Shkolyar S, Shuster DL, Simon JI, Smith RJ, Stack KM, Steadman K, Weiss BP, Werynski A, Williams AJ, Wiens RC, Williford KH, Winchell K, Wogsland B, Yanchilina A, Yingling R, Zorzano M-P (2022) Aqueous alteration processes in Jezero crater, Mars-implications for organic geochemistry. Science 378:1105–1110

Schuerger AC, Amaradasa BS, Dufault NS, Hummerick ME, Richards JT, Khodadad CL, Smith TM, Massa GD (2021) *Fusarium oxysporum* as an opportunistic fungal pathogen on *Zinnia hybrida* plants grown on board the International Space Station. Astrobiology 21:1029–1048

Schultzhaus Z, Chen A, Shuryak I, Wang Z (2020) The transcriptomic and phenotypic response of the Melanized yeast *Exophiala dermatitidis* to ionizing particle exposure. Front Microbiol 11:609996

Scott JA, Untereiner WA, Ewaze JO, Wong B, Doyle D (2007) Baudoinia, a new genus to accommodate *Torula compniacensis*. Mycologia 99:592–601

Seiler H, Busse M (1990) The yeasts of cheese brines. Int J Food Microbiol 11:289–303

Selbmann L, Benkő Z, Coleine C, de Hoog S, Donati C, Druzhinina I, Emri T, Ettinger CL, Gladfelter AS, Gorbushina AA, Grigoriev IV, Grube M, Gunde-Cimerman N, Karányi ZÁ, Kocsis B, Kubressoian T, Miklós I, Miskei M, Muggia L, Northen T, Novak-Babič M, Pennacchio C, Pfliegler WP, Pòcsi I, Prigione V, Riquelme M, Segata N, Schumacher J, Shelest E, Sterflinger K, Tesei D, U'Ren JM, Varese GC, Vázquez-Campos X, Vicente VA, Souza EM, Zalar P, Walker AK, Stajich JE (2020) Shed light in the DaRk Lineag ES of the fungal tree of life-STRES. Life 10:362

Simões MF, Cortesão M, Azua-Bustos A, Bai FY, Canini F, Casadevall A, Cassaro A, Cordero RJB, Fairén AG, González-Silva C, Gunde-Cimerman N, Koch SM, Liu XY, Onofri S, Pacelli C, Selbmann L, Tesei D, Waghmode A, Wang T, Zucconi L, Antunes A (2023) The relevance of fungi in astrobiology research—Astromycology. Mycosphere 14:1190–1253

Sinha S, Flibotte S, Neira M, Formby S, Plemenitaš A, Cimerman NG, Lenassi M, Gostinčar C, Stajich JE, Nislow C (2017) Insight into the recent genome duplication of the halophilic yeast *Hortaea werneckii*: combining an improved genome with gene expression and chromatin structure. G3 7:2015–2022

Slepecky RA, Starmer WT (2009) Phenotypic plasticity in fungi: a review with observations on *Aureobasidium pullulans*. Mycologia 101:823–832

Soler-Hurtado MM, Sandoval-Sierra JV, Machordom A, Diéguez-Uribeondo J (2016) *Aspergillus sydowii* and other potential fungal pathogens in gorgonian Octocorals of the Ecuadorian Pacific. PLoS One 11:e0165992

Su Y, Jiang X, Wu W, Wang M, Hamid MI, Xiang M, Liu X (2016) Genomic, transcriptomic, and proteomic analysis provide insights into the cold adaptation mechanism of the obligate psychrophilic fungus *Mrakia psychrophila*. G3 6:3603–3613

Sun W, Su L, Yang S, Sun J, Liu B, Fu R, Wu B, Liu X, Cai L, Guo L, Xiang M (2020) Unveiling the hidden diversity of rock-inhabiting fungi: Chaetothyriales from China. J Fungi (Basel) 6. https://doi.org/10.3390/jof6040187

Tafer H, Poyntner C, Lopandic K, Sterflinger K, Piñar G (2019) Back to the salt mines: genome and transcriptome comparisons of the halophilic fungus *Aspergillus salisburgensis* and its halotolerant relative *Aspergillus sclerotialis*. Genes 10:381

Taylor JW, Hann-Soden C, Branco S, Sylvain I, Ellison CE (2015) Clonal reproduction in fungi. Proc Natl Acad Sci USA 112:8901–8908

ten Kate IL, Preston LJ (2015) Earth analogues. In: Beysens DA, van Loon JJWA (eds) Generation and applications of extra-terrestrial environments on earth. River Publishers, New York, pp 165–171

Tesei D, Marzban G, Zakharova K, Isola D, Selbmann L, Sterflinger K (2012) Alteration of protein patterns in black rock inhabiting fungi as a response to different temperatures. Fungal Biol 116:932–940

Tiew PY, Mac Aogain M, Ali NABM, Thng KX, Goh K, Lau KJX, Chotirmall SH (2020) The mycobiome in health and disease: emerging concepts, methodologies and challenges. Mycopathologia 185:207–231

Vesper SJ, Wong W, Kuo CM, Pierson DL (2008) Mold species in dust from the International Space Station identified and quantified by mold-specific quantitative PCR. Res Microbiol 159:432–435

Voigt O, Knabe N, Nitsche S, Erdmann EA, Schumacher J, Gorbushina AA (2020) An advanced genetic toolkit for exploring the biology of the rock-inhabiting black fungus *Knufia petricola*. Sci Rep 10:22021

Wang J-F, Lin X-P, Qin C, Liao S-R, Wan J-T, Zhang T-Y, Liu J, Fredimoses M, Chen H, Yang B, Zhou X-F, Yang X-W, Tu ZC, Liu Y-H (2014) Antimicrobial and antiviral sesquiterpenoids from sponge-associated fungus, *Aspergillus sydowii* ZSDS1-F6. J Antibiot 67:581–583

Waschuk SA, Bezerra AG Jr, Shi L, Brown LS (2005) Leptosphaeria rhodopsin: bacteriorhodopsin-like proton pump from a eukaryote. Proc Natl Acad Sci USA 102:6879–6883

Wirshing ACE, Petrucco CA, Lew DJ (2024) Chemical transformation of the multibudding yeast, *Aureobasidium pullulans*. J Cell Biol 223:e202402114

Wirth F, Goldani LZ (2012) Epidemiology of Rhodotorula: an emerging pathogen. Interdiscip Perspect Infect Dis 2012:465717

Xin M-X, Zhou P-J (2007) *Mrakia psychrophila* sp. nov., a new species isolated from Antarctic soil. J Zhejiang Univ Sci B 8:260–265

Yuzon JD, Schultzhaus Z, Wang Z (2023) Transcriptomic and genomic effects of gamma-radiation exposure on strains of the black yeast *Exophiala dermatitidis* evolved to display increased ionizing radiation resistance. Microbiol Spectr 11:e0221923

Zajc J, Černoša A, Sun X, Fang C, Gunde-Cimerman N, Song Z, Gostinčar C (2022) From glaciers to refrigerators: the population genomics and biocontrol potential of the black yeast *Aureobasidium subglaciale*. Microbiol Spectr 10:e0145522

Zajc J, Džeroski S, Kocev D, Oren A, Sonjak S, Tkavc R, Gunde-Cimerman N (2014) Chaophilic or chaotolerant fungi: a new category of extremophiles? Front Microbiol 5:708

Zajc J, Liu Y, Dai W, Yang Z, Hu J, Gostinčar C, Gunde-Cimerman N (2013) Genome and transcriptome sequencing of the halophilic fungus *Wallemia ichthyophaga*: haloadaptations present and absent. BMC Genomics 14:617

Zakharova K, Tesei D, Marzban G, Dijksterhuis J, Wyatt T, Sterflinger K (2013) Microcolonial fungi on rocks: a life in constant drought? Mycopathologia 175:537–547

Zalar P, Gostinčar C, de Hoog GS, Ursic V, Sudhadham M, Gunde-Cimerman N (2008) Redefinition of *Aureobasidium pullulans* and its varieties. Stud Mycol 61:21–38

Zhang Y, Feng J, Wang P, Xia J, Li X, Zou X (2019) CRISPR/Cas9-mediated efficient genome editing via protoplast-based transformation in yeast-like fungus *Aureobasidium pullulans*. Gene 709:8–16

Zhao R-L, Li G-J, Sánchez-Ramírez S, Stata M, Yang Z-L, Wu G, Dai Y-C, He S-H, Cui B-K, Zhou J-L, Wu F, He M-Q, Moncalvo J-M, Hyde KD (2017) A six-gene phylogenetic overview of *Basidiomycota* and allied phyla with estimated divergence times of higher taxa and a phyloproteomics perspective. Fungal Divers 84:43–74

Zupančič J, Raghupathi PK, Houf K, Burmølle M, Sørensen SJ, Gunde-Cimerman N (2018) Synergistic interactions in microbial biofilms facilitate the establishment of opportunistic pathogenic fungi in household dishwashers. Front Microbiol 9:21

# Part 2

# Cell and Developmental Biology

4

# Recent Advances in High-Throughput Genetics in Fungi

Joshua W. Lyon and R. Blake Billmyre

**Abstract**

Large-scale genetic screens deal with two competing challenges: generating mutants and mapping them. Traditional forward and reverse genetics each solve one problem but not the other. Recent innovations in sequencing-coupled genetic approaches have enabled solutions that help to solve both issues, either by rapidly mapping random mutants (TN-seq) or by generating targeted mutants at extremely high frequencies (CRISPR). In this review, we will discuss advances in high-throughput genetic screens in fungi, starting with arrayed mutant libraries and insertional mutagenesis before moving on to discuss the recent developments in sequencing-coupled TN-seq and CRISPR-based approaches. Finally, we will discuss two other useful approaches that can be coupled with these new approaches: knockout Sudoku and RH-seq. Taken together, these approaches have the potential to drive more rapid exploration of fungal biology.

**Keywords**

TN-seq · Transposons · CRISPR · RH-seq · Insertional mutagenesis · Fungal genetics

## 4.1 Introduction

At the core, genetic approaches perturb gene function and study the resulting phenotypes. While there are many ways to perturb gene function, genetic approaches often utilize loss-of-function mutations. Traditional or "forward" genetic approaches generate a large set of mutants and then screen those for a phenotype of interest. Once interesting mutants are identified, the mutation linked to that phenotype must be identified. Even with modern sequencing approaches, identifying or mapping the causative mutation is often a slow and laborious process.

As a result, modern approaches often instead start from predetermined libraries of mutants (reverse genetics). These libraries can then be individually screened for a phenotype of interest. While forward genetic screens move from phenotype to gene, reverse genetic screens move from gene to phenotype. Unlike in forward genetics,

J. W. Lyon · R. B. Billmyre (✉)
Department of Pharmaceutical and Biological Sciences, College of Pharmacy, University of Georgia, Athens, GA, USA

Department of Infectious Diseases, College of Veterinary Medicine, University of Georgia, Athens, GA, USA
e-mail: blake.billmyre@uga.edu


M. Nowrousian, J. E. Stajich (eds.), *Fungal Genomics*, The Mycota 13,
https://doi.org/10.1007/978-3-031-75666-5_4

where construction of mutants is often quite easy but mapping the specific mutation of interest responsible for a given phenotype is difficult, in reverse genetics identification of the mutation of interest is simple but the construction of the mutants and mutant libraries in the first place is often quite difficult. Many reverse mutant collections are built collaboratively by communities over long periods of time. These collections often become key resources for individual fields, but their construction is time and labor-intensive, such that building new collections to answer specific questions is frequently impractical. In essence, forward and reverse genetic approaches deal with a balance between the difficulty of making mutants and the difficulty of mapping them.

Recent advances in sequencing-coupled approaches have generated solutions to these competing problems. While many originated outside of fungi, they are now being broadly applied to fungal systems as well. Like bacteria, fungi are typically far more experimentally tractable than multicellular animals. However, because they are evolutionarily more closely related to animals than bacteria, fungi are powerful models for understanding basic and conserved human-relevant biology using simple microbes. This evolutionary similarity also makes fungal pathogens of humans difficult to treat, creating a need for better tools to examine specific and often non-model species of fungi. In this review, we will discuss high-throughput genetic approaches in fungi, with an emphasis on the recent development of transposon-based and CRISPR-based mutagenesis systems that enable screening of enormous numbers of pooled mutants. Many of these techniques blur the line between forward genetics, reverse genetics, and functional genomics.

## 4.2 Classic Mutant Collections

Traditional reverse genetic approaches involve large libraries of mutants, which are often cryopreserved in 96 well plates (Giaever et al. 2002; Baba et al. 2006; Colot et al. 2006; Liu et al. 2008; Kim et al. 2010; Son et al. 2011). These mutants are typically constructed manually, using targeted genetic approaches. In organisms with efficient homologous recombination, like *Saccharomyces cerevisiae*, repair templates can be constructed using short homology regions (~50 bp) synthesized as part of the oligonucleotide primers used to amplify the drug resistance marker. In organisms with less efficient homologous recombination, this process may require hundreds of bases of homology on either side of deleted sequences (Nelson et al. 2003). In these organisms, construction of deletion cassettes requires time and labor-intensive steps like overlap or stitch PCR prior to transformation, increasing the cost and time to produce deletion libraries.

These libraries can be screened for phenotypes of interest, often by either manual pinning or using automated robotic approaches. Barcodes can also be incorporated during library construction in order to enable pooled screen approaches (reviewed in (Mazurkiewicz et al. 2006)). Because both the contents of the wells and the barcode sequence are already known, these arrayed collections require minimal effort to link interesting screen hits back to the causative gene. Outside of screening, deletion collections also enable individual labs to quickly access deletions of specific strains for specific targeted experiments.

However, there are a few distinct limitations to arrayed mutants collections. First, phenotypes are assumed to be linked to the deletion allele. However, the long time frame in which these libraries are constructed means that many mutants within a library are, by necessity, constructed at different times or even by different labs, which can introduce unlinked mutations. For example, an entire plate of the *S. cerevisiae* mutant collection was constructed from a starting culture that carried a de novo mutation resulting in a defective allele of *msh3*. All 96 isolates on this plate were found as hits in a screen for elevated mutation rate, but the phenotype was unlinked to the genes deleted in those strains (Lehner et al. 2007). Likewise, deletion collections are prone to handling errors, such as cross contamination between wells or simply freezing the wrong isolate in a well. Multiple screens of deletion collections have found mistakes upon validation that led to false positives. While tools like barcoded mutations can help identify these events, this type of mistake is hard to avoid with projects of this scale and the best quality collections are typically those that are iteratively maintained by an invested community. Finally, deletion collections do not contain deletions of essential genes, leaving a subset of genes that cannot be screened using this approach.

Essential genes are an organism's set of genes that are absolutely required for it to grow. Loss of an essential gene causes cells to die, meaning that it is impossible to make a null mutant of a truly essential gene. Traditionally, identification of essential genes has taken advantage of diploid genetics and meiosis. A heterozygous deletion can be made in a diploid isolate and then the diploid can be put through meiosis. If two of the four meiotic products are nonviable and the viable products always contain the wildtype allele of the gene of interest, then that gene can be inferred to be essential for growth. This process is slow and also prone to false positives, such as identifying genes that are required for spore germination. In pathogens, identification of essential genes is important in and of itself in order to prioritize drug development. However, the inability to construct viable null alleles of essential genes means that traditional genetic approaches struggle to determine the function of these genes. Death is an uninformative phenotype.

There are two key solutions to this issue of essentiality: Libraries of inducible/repressible promoters and libraries of disruption alleles. The first type of collection uses systematic replacement of the promoter region of genes with a regulatable promoter. When expression is turned off, the mutant phenotype can be observed. As a result, regulatable promoters can be constructed for even essential alleles and linkage between mutation and phenotype can be directly tested by assaying with or without the addition of the compound the promoter responds to. This approach was used in *Candida albicans* in the gene replacement and conditional expression (GRACE) collection (Roemer et al. 2003). One allele is replaced with a drug marker and the other allele is tightly regulated by a transactivator. Ideally, the compound used to induce or repress the promoter should not participate in metabolism or it may cause effects that are distinct from the activity of the target gene. Even relatively innocuous compounds do have some known issues. For example, the GRACE library uses a tetracycline repressible promoter, but tetracycline is a strong iron chelator and may cause proteotoxic stress in the mitochondria. Screens of the GRACE library missed three single gene knockdown mutants where the null mutants demonstrated azole sensitivity without tetracycline (Chen et al. 2018).

A second approach is construction of perturbation alleles that affect the amount of protein produced, but do not eliminate it completely. The DAmP collection in *S. cerevisiae* uses disruption of the 3' UTR by a drug resistance marker to create hypomorphic alleles (Breslow et al. 2008). This mRNA perturbation results in a decrease of protein synthesis by ~2 to ~10 times. DAmP mutants in low stress environments will be able to survive, but will become sick when exposed to selective pressure. The DAmP approach can be complemented by the addition of a targeted degron system to reduce protein levels by an additional ~2.5 to >10 times (Breslow et al. 2008).

## 4.3 Insertional Mutagenesis

For any mutational analysis system there are two main challenges: making mutants and identifying them. For example, chemical mutagens like Ethyl methanesulfonate (EMS) are extremely efficient at generating mutations. However, identifying the causative mutation within a strain is far more difficult. Even with whole genome sequencing, a particular mutation of interest is often masked by significant additional background mutations. Homologous recombination with specific repair templates solves the identification issue but produces mutants slowly. Improving the speed and throughput of whole genome screens requires solving these competing challenges.

One solution is the use of insertional mutagens. Rather than deleting gene sequences, insertional mutagenesis interrupts those sequences with a large piece of exogenous DNA. This mutagenesis can be random or targeted and thus can be used for forward or reverse genetics, depending on the system. Insertional mutagenesis is particularly advantageous for random mutagenesis because it introduces a known exogenous DNA sequence. That known sequence can then be used as a seed to aid in mapping the specific disrupted gene via either sequencing or PCR-based approaches.

In fungi, *Agrobacterium tumefaciens* (now also called *A. radiobacter*) has been used to effectively generate mutations. *A. tumefaciens* inserts a transfer DNA (T-DNA) fragment into plant cells to alter host biology and enable bacteria growth. However, *A. tumefaciens* is capable of transforming a variety of other organisms, including many fungi. *Agrobacterium*-mediated transformations avoid the background mutations associated with EMS mutagenesis and even the preliminary strain construction associated with some of the methods discussed below. While in some species *A. tumefaciens* can drive homologous gene replacement (Ianiri et al. 2016), in others like *Cryptococcus neoformans* and *Candida auris*, it is highly effective in generating random insertions (Idnurm et al. 2004; Santana and O'Meara 2021).

One of the advantages of *Agrobacterium*-mediated transformation is that because the insertion sequence is known, the insertions can be mapped. Classical approaches used a variety of PCR-based techniques such as inverse PCR to map the genomic DNA associated with T-DNA insertions. Briefly, inverse PCR involves digesting genomic DNA with a restriction enzyme that does not cut within the T-DNA sequence. The resulting fragment library can then be re-ligated, with the goal of creating a circular fragment containing the T-DNA piece. Inverse PCR relies on molecules self-ligating. If there are too many molecules in close proximity (i.e., high DNA concentration), then molecules are more likely to cross-ligate instead of self-ligating. Performing multiple ligations with a gradient of DNA concentrations can help to balance the tradeoff between cross-ligation at high DNA concentration, and low reaction yield at low DNA concentrations. PCR primers within the T-DNA can then be used to amplify outwards from the fragment and into the genomic DNA and the resulting PCR product can be Sanger sequenced to identify the location. *Agrobacterium*-mediated insertions have been used to map a variety of phenotypes in various fungi, including the identification of essential genes in *Cryptococcus neoformans* (Ianiri and Idnurm 2015), oxidative stress response in *Aspergillus fumigatus* (Fan et al. 2016), filamentous morphology in *Candida auris* (Santana and O'Meara 2021), fatty acid catabolism and lipid accumulation in *Rhodosporidium toruloides* (Coradetti et al. 2018), and intracellular survival of *Histoplasma capsulatum* (Hilty et al. 2008).

Unfortunately, inverse PCR is often technically challenging and as a result, some mutants of interest can be difficult to map. For example, if an insert is distant from the restriction sites used for digestion, the PCR amplicon may be too large to efficiently amplify. To solve this issue, many labs use sequencing-based approaches. For example, *Agrobacterium*-mediated insertional mutagenesis sequencing (AIM-SEQ) uses untargeted whole genome sequencing to map pools of insertion mutants and then follow up PCR to identify which strain within each pool contains which insertion (Esher et al. 2015). Similar approaches have been used to identify insertions in *C. auris* (Santana and O'Meara 2021). These approaches are highly effective, but not generally scalable for very large numbers of mutants, where both the cost of whole genome sequencing and the number of disentangling PCRs becomes too high.

## 4.4 TN-SEQ

Transposon sequencing (TN-seq) is a variant of insertional mutagenesis that combines random transposon insertions with high-throughput next-generation sequencing (NGS). Transposons are natural selfish elements that self-mobilize. The transposons that are used in TN-seq are DNA transposons which mobilize via a cut and paste mechanism, whereas retrotransposons use a mechanism that functions more like copy and paste. TN-seq works by randomly inserting a transposon throughout the genome within a pool, such that every cell within the pool has a single random insertion somewhere in the genome.

Similar to *Agrobacterium*-mediated mutagenesis, insertions will interrupt any gene sequences they land in, resulting in loss-of-function alleles. Through a variety of molecular biology methods, the boundaries between the transposon insertions and the genome can be amplified and deep sequenced. This enables high-throughput identification of the disrupted genes in a single experiment. The targeted sequencing approach also allows TN-seq to scale to very large numbers of inserts, unlike methods that rely on whole genome sequencing such as AIM-seq (Fig. 4.1).

TN-seq allows researchers to determine gene essentiality (Fig. 4.2a) and to systematically

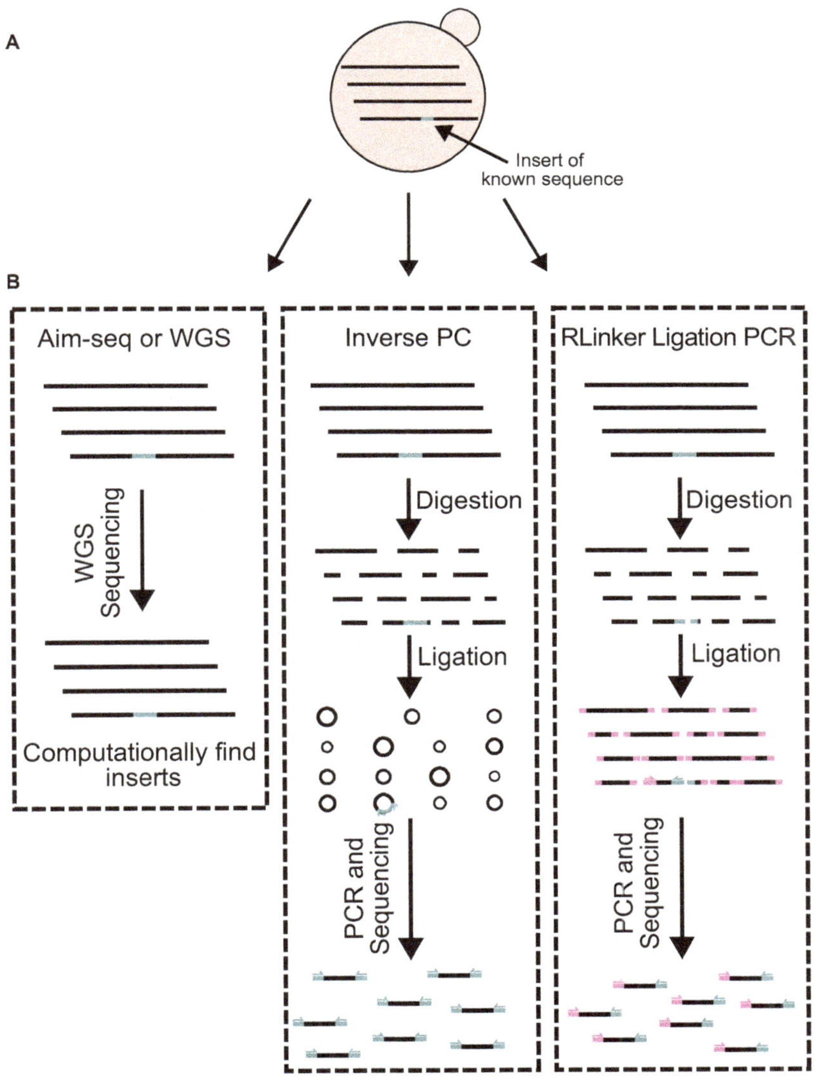


**Fig. 4.1** Strategies for identifying location of insert. (**a**) Insertional mutagenesis introduces a known sequence into an unknown location within the genome. (**b**) Three methods of identifying the location of the insertion within the genome are shown. Left: WGS (whole genome sequencing) or Aim-seq sequence the entire genome and then find the location of the insert computationally. Middle: In inverse PCR the genome is digested and the digested fragments are self-ligated. PCR is performed using primers facing out from the known sequence of the insertion to generate amplicons which include the unknown genomic sequence surrounding the insert. Right: In linker ligation the genome is digested and the fragments are ligated to linkers with a known sequence. Primers from both the linker and the known insertion will amplify the sequence between them. For both inverse PCR and linker ligation PCR, the amplicon is sequenced to identify the insert location

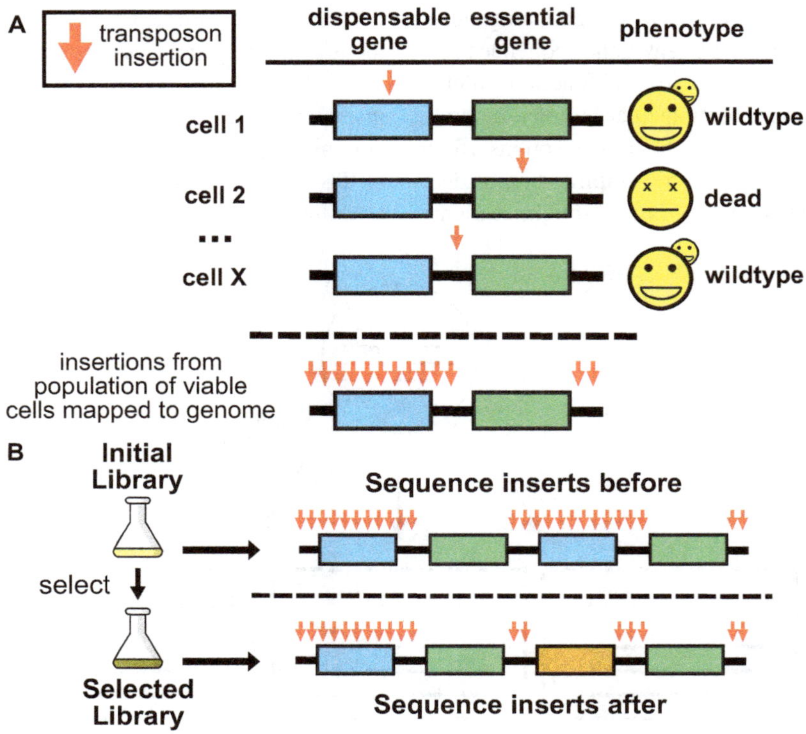


**Fig. 4.2** TN-seq insertional mutagenesis enables essential gene identification and whole genome screens. (**a**) TN-seq randomly inserts a single transposon per cell in a random location. If that transposon interrupts an essential gene, it will kill the cell and not be recoverable. Otherwise, the pooled cells, each with unique insertion locations, can be collected and transposon insertion sites can be mapped using next-generation sequencing to produce maps of essential regions (green) and nonessential regions (blue). (**b**) Once transposon insertion pools are built, they can be selected and resequenced to identify genes important in that given selection (orange)

quantify differential fitness of gene deletions (Fig. 4.2b). Genes that do not tolerate transposon insertions are assumed to be essential genes. TN-seq is commonly applied in bacteria to determine the essential gene set. In addition, TN-seq mutant libraries can be selected as a pool and the transposon insertion frequencies can be assayed before and after selection to give quantitative readout of the importance of individual genes in the genome. Pooled assays have the advantage of controlling for variation in growth conditions. Every mutant within a pool is in direct competition for resources with every other mutant in the pool with growth rate being the ultimate read out. TN-seq systems in fungi have been used to identify essential genes (Segal et al. 2018; Gale et al. 2020; Fu et al. 2021), genes involved in heterochromatin formation (Lee et al. 2020), general fitness (Grech et al. 2019), sexual reproduction (Billmyre et al. 2022), low-density growth (Billmyre et al. 2022), and drug resistance (Michel et al. 2019; Gale et al. 2020), across multiple species including *S. pombe*, *S. cerevisiae*, *C. albicans*, and *C. glabrata*.

TN-seq was originally introduced in bacteria in 2009 (as TN-seq, HITS, TraDIS, and INSeq) and revolutionized our ability to study bacterial genomes (Goodman et al. 2009; Gawronski et al. 2009; van Opijnen et al. 2009; Langridge et al. 2009). In bacteria, TN-seq is often made simpler by the fact that many bacteria are intrinsically highly competent for transformation. As a result, many bacterial TN-seq approaches utilize in vitro transposition reactions to generate a diverse library of transposon insert locations. Because high transformation efficiency and high rates of homologous recombination are possible for many bacteria, this library can then be transformed into bacteria, and a drug resistance marker within the transposon can be used to select for successful homologous integration of transposon inserts (Guo and Mekalanos 2001; Rosconi et al. 2022). This approach has three main benefits. First, transposon inserts in vitro are not constrained by fitness. Thus, bacterial geneticists can sequence their libraries both prior to and after transformation to more directly detect which inserts are selected against in vivo. Second, there is less direct bias in transposon insertion outside the context of active transcription or genome organization. Third, a pre-existing set of transposon mutants reduces the likelihood of jackpot mutations that can dominate a transposon insert library.

However, most fungi are not nearly as genetically labile as bacteria. Consequently, the transposition step for fungal TN-seq must be performed in vivo. Fungal TN-seq systems require two parts. First, a non-autonomous DNA transposon is used to generate insertional mutants. This transposon is incapable of movement on its own and relies on the second component, an inducible transposase, to mobilize it and insert it randomly across the genome. These two elements can be integrated into the genome or harbored on one or two plasmids. In any case, transposition events must be detectable and selectable because otherwise transposons in the original location will dominate sequencing libraries.

Plasmid-based systems can use counterselection to eliminate the original transposon carrying plasmid. For example, the first TN-seq system established in fungi is in the fission yeast *Schizosaccharomyces pombe* and uses two plasmids, one carrying a G418-marked transposon on a *ura4* marked vector and the other carrying the transposase (Guo et al. 2013). After mobilization, the transposon is selected with G418 while the transposon plasmid is simultaneously counterselected using 5FOA, which selects against *ura4* function. As a result, the only surviving cells are those where the transposon has moved from the plasmid to the genome. This strategy has been used successfully in *C. glabrata* as well (Gale et al. 2020). The saturated transposon analysis in yeast (SATAY) system in *S. cerevisiae* instead selects for transposon excision events from the plasmid (Michel et al. 2017). The transposon interrupts an *ade2* gene on the plasmid. Excision and appropriate repair events regenerate an *ADE2* allele and so cells can be grown on media lacking adenine to select for transposon movement events.

However, many fungal systems lack stably maintained plasmids and so these approaches can be impractical. To solve this issue, the TN-seq system in *Candida albicans* instead integrates the original transposon into the genome (Segal et al. 2018). Like the SATAY system described above, transposition events are detected by selecting for the excision event. The *C. albicans* system also uses a transposon integrated into the *ADE2* locus. Unlike SATAY, in *C. albicans* the transposon is integrated into the endogenous *ADE2* locus rather than an ectopic copy on a plasmid, but both are selected the same way using media lacking adenine.

Genome integration strategies require careful planning to ensure that the components required to make the system work do not themselves cause unwanted changes in phenotype, such as through local changes in gene expression. In TN-seq, this means that introduction of the transposase should be fitness neutral. One way to ensure this is to use known neutral or "safe haven" regions. These regions are known to tolerate gene insertion without alterations in cell fitness or virulence. In multiple species of *Cryptococcus*, intergenic regions of more than 2 kb between convergent genes that have very low levels of transcription have been used successfully as safe haven regions (Arras et al. 2015; Fan and Lin 2020; Li et al. 2022). This same system of assessing annotations and transcription levels was used to effectively discover two safe haven regions in *A. fumigatus.* Importantly, these safe haven regions did not change virulence in murine and insect models (Furukawa et al. 2022).

TN-seq systems also vary in their choice of transposon. In fungi, three transposons have been used thus far: PiggyBac, Ac/Ds, and Hermes. All of these systems have insertional bias, which does generate some uneven distribution of insertion events. Ac/Ds prefers to insert near the point of excision and into intergenic regions but does not have a strong sequence-based target preference. The Ac/Ds system has more unique insertions and more insertions per gene than libraries made using Hermes and PiggyBac (Levitan et al. 2020). While Hermes does have some insertion sequence bias, it is relatively minimal ("NTNNNNAN") and the high density and randomness of insertions provides strong results. PiggyBac is the least effective for identifying essential genes due to strong specificity of transposon target sequence ("TTAA"). There are only approximately 120,000 and 111,000 PiggyBac target sequences in *C. albicans* and *S. pombe*, respectively; which generates insufficient insert density to accurately gauge essentiality of genes (Levitan et al. 2020).

TN-seq methods have several potential complications. First, TN-seq assumes that transposons are capable of insertion into all sequences. If this assumption is not valid, some dispensable but not insertable genes may be inappropriately scored as essential. For the most part, essentiality predictions are consistent between classical genetic approaches and TN-seq but there are exceptions that may be explained by insertability. One alternate explanation is that TN-seq struggles to differentiate essential genes from those that are not absolutely required but whose mutants are extremely sick. TN-seq experiments are conducted in pools by necessity and slow growing mutants are often diluted out over time such that they are indistinguishable from mutants that were never present in the pool. Many transposons also have some bias in insertion sequence based either on nucleotide sequence or DNA accessibility (discussed above). These biases can be compensated for by generating larger mutant pools so that the low insertion parts of the genome are also testable. However, larger mutant pools come with increased library complexity and sequencing costs. Finally, induction of transposition typically is driven by a regulatable promoter. Many non-model fungal systems have few options for inducible promoters and those options are frequently linked to metabolism. For example, the *S. cerevisiae* SATAY system used galactose induction, the *C. albicans* system used maltose, and the *S. pombe* system used thiamine poor conditions. Genes that are required for growth in any of these conditions will not tolerate transposon insertions and will both be scored as essential and invisible to future screens using that library.

## 4.5 CRISPR

TN-seq solves the mutation mapping problem by using more efficient mapping of random insertions. An alternative approach is to more efficiently generate targeted insertions. CRISPR-Cas9 approaches have recently become common in fungi and enable more efficient targeted transformations in many systems. CRISPR uses a ribonucleoprotein to make cuts in the DNA at a specific location that is designated by the sequence of the guide RNA component of the complex. Cells must repair double strand breaks to continue growing. Ideally, those cuts are then repaired to generate null alleles, either through a homologous repair template, through nonhomologous sequence deletion, or through nonhomologous sequence insertion. Because the guide RNA is inserted as an expression construct into the cells for mutagenesis, researchers can efficiently genotype the mutants by sequencing the specific guide RNA expression cassette, which is flanked by known sequences (Fig. 4.3a).

CRISPR approaches can be used with pools of guide RNA constructs to achieve similar results to TN-seq (Fig. 4.3b). Like transposon insertion frequencies in TN-seq, the frequency of each guide RNA insertion within the pool can be measured via sequencing as a proxy for the frequency of mutations in the targeted gene. Further, the ability of CRISPR to quickly mutate both alleles in a diploid makes this approach particularly useful in organisms like *C. albicans* where diploidy is the default state. CRISPR has even been used creatively to generate gene drive models where two individual mutants with mutations in different genes can be mated together to generate a

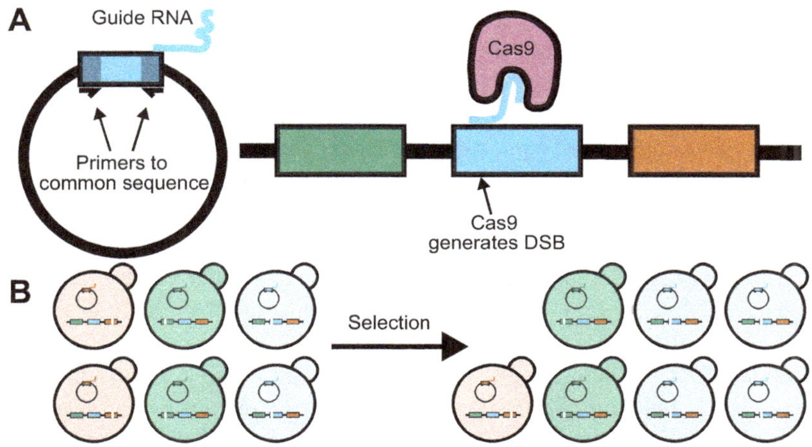


**Fig. 4.3** CRISPR screens allow pooled mutational analysis. (**a**) Guide RNAs for specific genes can be synthesized and added to cells in a pool. By PCR amplifying the guide RNA sequence (left) using common sequences, the frequency of guide RNA containing cells for each specific gene can be determined using next-generation sequencing. (**b**) Pools of cells can be selected and changes in the frequency of guide RNAs to specific target genes can be assayed before and after selection via sequencing

double mutant diploid through continued expression of the guide RNAs and the Cas9 protein (Shapiro et al. 2018; Rosiana et al. 2021).

This type of approach has been used genome-wide outside of fungi, including in mouse cells (Chen et al. 2015), toxoplasma (Sidik et al. 2016), and bacteria, but is not yet common in fungi. However, many individual fungal systems have CRISPR mutagenesis protocols, including *C. albicans* (Min et al. 2016), *C. neoformans* (Fan and Lin 2018; Huang et al. 2022), *A. fumigatus* (van Rhijn et al. 2020), *C. glabrata* (Enkler et al. 2016), *S. cerevisiae* (DiCarlo et al. 2013), and *S. pombe* (Jacobs et al. 2014; Torres-Garcia et al. 2020). Large-scale use of these protocols to generate high-throughput pooled mutagenesis tools is likely to be highly fruitful in the future and could provide an alternative to TN-seq approaches.

One complication of genome-wide CRISPR approaches is the requirement for guide RNA libraries for the entire genome. Synthesizing guide RNA libraries at whole genome scale is expensive even for short oligos. Further, differences in strain background will lead to reduced success of a subset of the guide RNAs. This can become prohibitive as the number of potential strains to be tested increases and fewer guides are shared between the strains or species of interest. Rather than synthesize new libraries for each species of interest, one solution developed to this problem was to generate the guide RNA libraries through a series of enzymatic digestions of the genomic DNA to yield appropriate small fragments. When coupled with *Agrobacterium* mutagenesis of a strain endogenously expressing Cas9, this approach enables rapid construction of a genome-wide mutation library in a process the authors named RELATe (Li and Kim 2020). The specific mutations can then be mapped by sequencing the guide RNA constructs that were inserted into the genome.

In addition to Cas9 based deletion assays, there are a number of CRISPR activation (CRISPRa) (Roux et al. 2020; Gervais et al. 2023; Maroc et al. 2023) and CRISPR inactivation (CRISPRi) (Zhang et al. 2023; Billerbeck et al. 2023) protocols in fungi. These assays enable activation or inhibition of specific genes using the same guide RNAs used for typical deletion assays. Both approaches allow screens to assay more subtle changes in gene function, either by increasing gene expression or by decreasing, but not eliminating, it. Like CRISPR-based deletion assays, whole genome assays using these approaches will likely become more common in the next few years.

Like TN-seq, there are some issues with CRISPR-based mutagenesis. Many of the CRISPR-Cas9 approaches make three assumptions. First, they assume that Cas9 mutagenesis will result in null mutations. Not all double strand breaks that eliminate the recognition site will necessarily result in null alleles. Further, some repair events that do generate null alleles may also result in regional deletions and loss of neighboring gene functions or even cause translocations (Adikusuma et al. 2018; Cullot et al. 2019). Second, CRISPR systems assume that phenotypes are linked to mutation of the gene targeted by the guide RNA. Off target effects are not infrequent (Höijer et al. 2022), and while these events can be reduced by computationally predicting off target potential for guides, this problem likely still exists at some level. Third, many systems utilize constitutive expression of Cas9 and make the assumption that this is unlikely to affect their phenotype of interest or cause additional mutations. Many of these issues can be mitigated by increased density of guide RNAs, although with a concomitant increase in cost.

## 4.6 Other Applications of High-Throughput Genetics

### 4.6.1 Knockout Sudoku

One significant downside of the pooled TN-seq or pooled CRISPR-Cas9 approaches is that neither approach generates individual mutants that can be used for downstream validation experiments. Traditional deletion collections have significant value for many labs because of the ability to quickly retrieve a specific mutant in a type strain background. While pooled TN-seq or CRISPR sequencing data may identify a gene of interest, that specific mutant cannot be practically retrieved from the millions of other mutants within the pool. In theory the entire pooled library could be plated to single colonies and then arrayed in plates and tested one at a time. In practice, this approach is too slow and inefficient to be productive and sequencing-based approaches to speed it up are too expensive for this scale.

Fortunately, a solution has already been developed for bacteria called "Knockout Sudoku" (Anzai et al. 2017). This approach starts with isolation of individual mutant clones from a pool. Those mutants are then arrayed in plates and a series of pools are made (i.e., all the mutants in one row of the plate or all the mutants in one column of the plate) so that individual mutants are contained within a number of distinct pools. By sequencing all of those pools, the content of any individual well can then be back calculated by a process similar to the well-known Sudoku game. The plates can then be rearrayed to produce a traditional deletion collection. This process dramatically reduces the cost and time required to produce this type of classical arrayed collection.

### 4.6.2 RH-seq

Both TN-seq and CRISPR approaches primarily aim to assay the effects of null mutations on gene function, although there are some applications of each that can test over or underexpression. However, neither TN-seq or CRISPR-based approaches directly test the consequences of nucleotide sequence variation. Variation in nucleotide sequence accumulates rapidly between species and often within species and drives differences in numerous complex traits. Traditionally, this variation has been assayed by a process called quantitative trait locus analysis or QTL analysis. This process involves crossing two isolates from a single species with distinct phenotypes to generate an F1 offspring generation with variation in both the phenotype and the inherited nucleotide sequences. Traits can then be tested for association with particular nucleotide sequences or markers to identify a set of candidate genes or variants associated with that trait (quantitative trait loci).

Validation of these candidate quantitative trait loci in yeast is often carried out via reciprocal hemizygosity testing. This process involves creating a hybrid of the two parents and disrupting one of the parental alleles within the locus at a time. If parent A grows better under a condition and disrupting the allele from parent A impairs growth while disrupting the allele from parent B does not, this confirms that allele is a causative allele for that trait. This process can be labor-intensive, especially for large candidate regions, as it requires construction of numerous mutants within a hybrid strain.

TN-seq can be used in an alternative approach called RH-seq to solve this issue. Rather than mapping quantitative trait loci, RH-seq skips directly to the reciprocal hemizygosity test and uses TN-seq to turn it into a high-throughput assay. Essentially, RH-seq conducts a TN-seq mutagenesis in a heterozygous diploid background, such that one allele from each parent is disrupted at a time on a genome-wide scale (Fig. 4.4a, b). Because a heterozygous hybrid strain is all that is required for this test, RH-seq can be performed across species boundaries where crosses produce viable hybrids but sterile offspring. This greatly expands the breadth of organisms that can be directly compared. RH-seq also has the advantage that the size of quantitative trait loci is not limited by recombination (Fig. 4.4c).

RH-seq was pioneered in *S. cerevisiae* (Weiss et al. 2018; Abrams et al. 2022) but has also been successfully used in nematodes (Wang et al. 2022). *Saccharomyces* spp. are particularly amenable to RH-seq because they readily form stable diploid hybrids. The ability to mate and form stable diploid hybrids may limit how broadly RH-seq can be used in fungi. For example, *C. glabrata* lacks a known mating cycle and *Cryptococcus* diploids need to be maintained at 37 °C to prevent sporulation (Sia et al. 2000), which limits the testable conditions. However, broader use of RH-seq across fungi will likely generate significant advances in the future.

RH-seq could also potentially be adapted to use CRISPR approaches. The challenge is that guide RNAs must be allele specific to avoid inactivating both alleles in the diploid hybrid. This requires sufficient sequence variation to design allele-specific guide RNAs. As a result, guide libraries would likely need to be synthesized for specific hybrid pairs and tools like RELATe would not be able to replace synthesis.

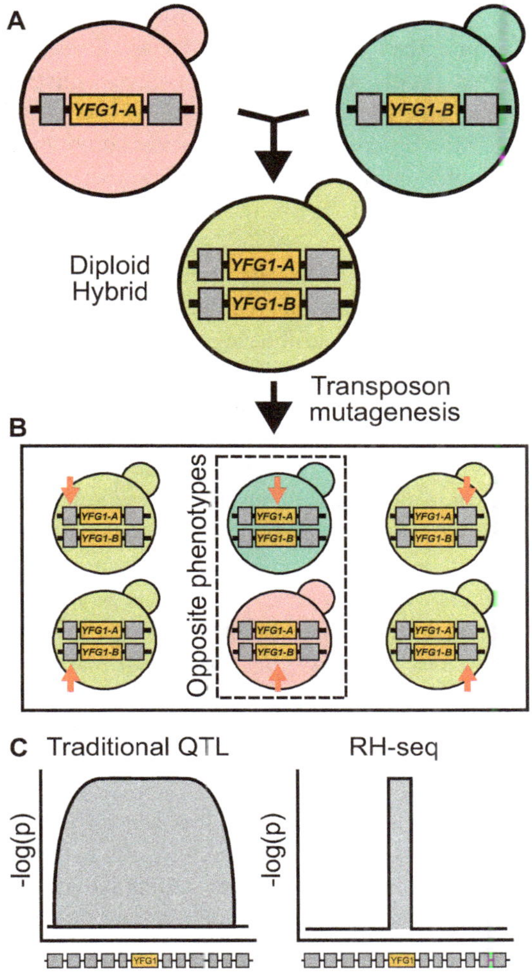


**Fig. 4.4** Schematic of RH-seq. (**a**) Two haploid strains with divergent drug resistance phenotypes are fused to make a heterozygous diploid hybrid. (**b**) Mutagenesis is carried out in a diploid to make a pooled library of hemizygous diploids. Those diploids are then selected on drug and transposon insertions are sequenced. (**c**) RH-seq generates peaks centered on single genes instead of broad peaks limited by recombination frequency

## 4.7 Conclusions

High-throughput genetic approaches have been incredibly impactful in model yeasts like *S. cerevisiae*. However, while model organism communities benefit from large knowledge bases and resources, these are often unavailable for many non-model microbes, including many pathogens. New high-throughput approaches are particularly valuable in these organisms because they reach many of the same experimental goals but require very few existing organism specific resources. TN-seq and CRISPR-based screen approaches both have the potential to drive more rapid genetic analysis in a broad range of fungal species and isolates.

Each approach has its strengths and weaknesses. TN-seq does not require upfront synthesis of large guide RNA libraries and thus is initially less expensive than CRISPR. However, CRISPR has the advantage of customizability. The inability to directly control transposon insertion location in TN-seq, in combination with some unavoidable biases in insert distribution mean that TN-seq libraries must be sequenced more deeply than an equivalent CRISPR library to guarantee coverage of all genes. Over large numbers of experiments within a single strain, the higher cost of sequencing for TN-seq will eventually surpass the initial cost of guide RNA synthesis. For some experiments however, the high transposon insertion density is beneficial because it reveals differential effects across insertions that may be invisible to cheaper, less dense CRISPR experiments.

TN-seq is particularly advantageous in experiments that explore strain and species variation. Even within individual species, there is substantial sequence variation with phenotypic consequences. Further, recent work to establish pangenomes in fungi is revealing differences in gene content between isolates of fungal species (Barber et al. 2021; Lofgren et al. 2022). Historically, many communities work on a single type strain, meaning that genetic variation found within species is unexplored. TN-seq can be applied to many strains or species without new synthesis or substantial upfront work. This approach was successfully applied in *S. cerevisiae* to explore differences in essential genes across the species (Chen et al. 2022). For strains with sufficient genetic variation, CRISPR guide libraries may need to be redesigned and resynthesized to be used effectively. This becomes even more important when different species are compared.

CRISPR also has the advantage of direct access to gene up or down regulation through CRISPRa and CRISPRi, respectively. These tools can utilize the same guide libraries and thus simply require swapping out the protein component in the cells. TN-seq primarily generates null alleles, although some inserts into regulatory regions have been observed to alter expression of neighboring genes (Michel and Kornmann 2022) (Michel and Kornmann 2022). Likewise, alterations in the cargo of these transposons may enable future TN-seq experiments to assay gene expression changes in similar ways to CRISPR.

Genetic analysis has been incredibly impactful in studying fungi, but most of the fungi used to establish and develop these high-throughput techniques can grow as yeasts. Developing strategies that allow these approaches to be used in fungi that are more challenging to grow in lab settings is an important next step. New high-throughput genetic approaches have the potential to rapidly expand our ability to study non-model fungi. This is important particularly in the context of emerging pathogens that may have no pre-existing research community. Further, examination of larger sets of isolates and species will enable us to answer questions about the impact of genome variation on numerous phenotypes.

**Acknowledgments** The authors thank members of the Billmyre Lab for helpful comments ahead of publication.

## References

Abrams MB, Chuong JN, AlZaben F et al (2022) Barcoded reciprocal hemizygosity analysis via sequencing illuminates the complex genetic basis of yeast thermotolerance. G3 12:12. https://doi.org/10.1093/g3journal/jkab412

Adikusuma F, Piltz S, Corbett MA et al (2018) Large deletions induced by Cas9 cleavage. Nature 560:E8–E9. https://doi.org/10.1038/s41586-018-0380-z

Anzai IA, Shaket L, Adesina O et al (2017) Rapid curation of gene disruption collections using knockout Sudoku. Nat Protoc 12:2110–2137. https://doi.org/10.1038/nprot.2017.073

Arras SDM, Chitty JL, Blake KL et al (2015) A genomic safe haven for mutant complementation in Cryptococcus neoformans. PLoS One 10:e0122916. https://doi.org/10.1371/journal.pone.0122916

Baba T, Ara T, Hasegawa M et al (2006) Construction of Escherichia coli K-12 in-frame, single-gene knockout mutants: the Keio collection. Mol Syst Biol 2(2006):0008. https://doi.org/10.1038/msb4100050

Barber AE, Sae-Ong T, Kang K et al (2021) Aspergillus fumigatus pan-genome analysis identifies genetic variants associated with human infection. Nat Microbiol 6:1526–1536. https://doi.org/10.1038/s41564-021-00993-x

Billerbeck S, Prins RC, Marquardt M (2023) A modular cloning toolkit including CRISPRi for the engineering of the human fungal pathogen and biotechnology host Candida glabrata. ACS Synth Biol 12:1358–1363. https://doi.org/10.1021/acssynbio.2c00560

Billmyre RB, Eickbush MT, Craig CJ et al (2022) Genome-wide quantification of contributions to sexual fitness identifies genes required for spore viability and health in fission yeast. PLoS Genet 18:e1010462. https://doi.org/10.1371/journal.pgen.1010462

Breslow DK, Cameron DM, Collins SR et al (2008) A comprehensive strategy enabling high-resolution functional analysis of the yeast genome. Nat Methods 5:711–718. https://doi.org/10.1038/nmeth.1234

Chen P, Michel AH, Zhang J (2022) Transposon insertional mutagenesis of diverse yeast strains suggests coordinated gene essentiality polymorphisms. Nat Commun 13:1490. https://doi.org/10.1038/s41467-022-29228-1

Chen S, Sanjana NE, Zheng K et al (2015) Genome-wide CRISPR screen in a mouse model of tumor growth and metastasis. Cell 160:1246–1260. https://doi.org/10.1016/j.cell.2015.02.038

Chen Y, Mallick J, Maqnas A et al (2018) Chemogenomic profiling of the fungal pathogen Candida albicans. Antimicrob Agents Chemother 62. https://doi.org/10.1128/aac.02365-17

Colot HV, Park G, Turner GE et al (2006) A high-throughput gene knockout procedure for Neurospora reveals functions for multiple transcription factors. Proc Natl Acad Sci USA 103:10352–10357. https://doi.org/10.1073/pnas.0601456103

Coradetti ST, Pinel D, Geiselman GM et al (2018) Functional genomics of lipid metabolism in the oleaginous yeast Rhodosporidium toruloides. eLife 7:e32110. https://doi.org/10.7554/eLife.32110

Cullot G, Boutin J, Toutain J et al (2019) CRISPR-Cas9 genome editing induces megabase-scale chromosomal truncations. Nat Commun 10:1136. https://doi.org/10.1038/s41467-019-09006-2

DiCarlo JE, Norville JE, Mali P et al (2013) Genome engineering in Saccharomyces cerevisiae using CRISPR-Cas systems. Nucleic Acids Res 41:4336–4343. https://doi.org/10.1093/nar/gkt135

Enkler L, Richer D, Marchand AL et al (2016) Genome engineering in the yeast pathogen Candida glabrata using the CRISPR-Cas9 system. Sci Rep 6:35766. https://doi.org/10.1038/srep35766

Esher SK, Granek JA, Alspaugh JA (2015) Rapid mapping of insertional mutations to probe cell wall regulation in Cryptococcus neoformans. Fungal Genet Biol FG B 82:9–21. https://doi.org/10.1016/j.fgb.2015.06.003

Fan Y, Lin X (2020) An intergenic "safe haven" region in Cryptococcus neoformans serotype D genomes. Fungal Genet Biol 144:103464. https://doi.org/10.1016/j.fgb.2020.103464

Fan Y, Lin X (2018) Multiple applications of a transient CRISPR-Cas9 coupled with electroporation (TRACE) system in the Cryptococcus neoformans species complex. Genetics 208:1357–1372. https://doi.org/10.1534/genetics.117.300656

Fan Z, Yu H, Guo Q et al (2016) Identification and characterization of an anti-oxidative stress-associated mutant of *Aspergillus fumigatus* transformed by *Agrobacterium tumefaciens*. Mol Med Rep 13:2367–2376. https://doi.org/10.3892/mmr.2016.4839

Fu C, Zhang X, Veri AO et al (2021) Leveraging machine learning essentiality predictions and chemogenomic interactions to identify antifungal targets. Nat Commun 12:6497. https://doi.org/10.1038/s41467-021-26850-3

Furukawa T, Van Rhijn N, Chown H et al (2022) Exploring a novel genomic safe-haven site in the human pathogenic mould *Aspergillus fumigatus*. Fungal Genet Biol 161:103702. https://doi.org/10.1016/j.fgb.2022.103702

Gale AN, Sakhawala RM, Levitan A et al (2020) Identification of essential genes and fluconazole susceptibility genes in *Candida glabrata* by profiling *Hermes* transposon insertions. G3 10:3859–3870. https://doi.org/10.1534/g3.120.401595

Gawronski JD, Wong SMS, Giannoukos G et al (2009) Tracking insertion mutants within libraries by deep sequencing and a genome-wide screen for Haemophilus genes required in the lung. Proc Natl Acad Sci USA 106:16422–16427. https://doi.org/10.1073/pnas.0906627106

Gervais NC, La Bella AA, Wensing LF et al (2023) Development and applications of a CRISPR activation system for facile genetic overexpression in Candida albicans. G3 13:jkac301. https://doi.org/10.1093/g3journal/jkac301

Giaever G, Chu AM, Ni L et al (2002) Functional profiling of the *Saccharomyces cerevisiae* genome. Nature 418:387–391. https://doi.org/10.1038/nature00935

Goodman AL, McNulty NP, Zhao Y et al (2009) Identifying genetic determinants needed to establish a human gut symbiont in its habitat. Cell Host Microbe 6:279–289. https://doi.org/10.1016/j.chom.2009.08.003

Grech L, Jeffares DC, Sadée CY et al (2019) Fitness landscape of the fission yeast genome. Mol Biol Evol 36:1612–1623. https://doi.org/10.1093/molbev/msz113

Guo BP, Mekalanos JJ (2001) Helicobacter pylori mutagenesis by mariner in vitro transposition. FEMS Immunol Med Microbiol 30:87–93. https://doi.org/10.1111/j.1574-695X.2001.tb01554.x

Guo Y, Park JM, Cui B et al (2013) Integration profiling of gene function with dense maps of transposon integration. Genetics 195:599–609. https://doi.org/10.1534/genetics.113.152744

Hilty J, Smulian AG, Newman SL (2008) The Histoplasma capsulatum vacuolar ATPase is required for iron homeostasis, intracellular replication in macrophages and virulence in a murine model of histoplasmosis. Mol Microbiol 70:127–139. https://doi.org/10.1111/j.1365-2958.2008.06395.x

Höijer I, Emmanouilidou A, Östlund R et al (2022) CRISPR-Cas9 induces large structural variants at on-target and off-target sites in vivo that segregate across generations. Nat Commun 13:627. https://doi.org/10.1038/s41467-022-28244-5

Huang MY, Joshi MB, Boucher MJ et al (2022) Short homology-directed repair using optimized Cas9 in the pathogen Cryptococcus neoformans enables rapid gene deletion and tagging. Genetics 220:iyab180. https://doi.org/10.1093/genetics/iyab180

Ianiri G, Averette AF, Kingsbury JM et al (2016) Gene function analysis in the ubiquitous human commensal and pathogen Malassezia genus. mBio 7:e01853–e01816. https://doi.org/10.1128/mBio.01853-16

Ianiri G, Idnurm A (2015) Essential gene discovery in the basidiomycete *Cryptococcus neoformans* for antifungal drug target prioritization. MBio 6. https://doi.org/10.1128/mbio.02334-14

Idnurm A, Reedy JL, Nussbaum JC, Heitman J (2004) *Cryptococcus neoformans* virulence gene discovery through insertional mutagenesis. Eukaryot Cell 3:420–429. https://doi.org/10.1128/EC.3.2.420-429.2004

Jacobs JZ, Ciccaglione KM, Tournier V, Zaratiegui M (2014) Implementation of the CRISPR-Cas9 system in fission yeast. Nat Commun 5:5344. https://doi.org/10.1038/ncomms6344

Kim D-U, Hayles J, Kim D et al (2010) Analysis of a genome-wide set of gene deletions in the fission yeast Schizosaccharomyces pombe. Nat Biotechnol 28:617–623. https://doi.org/10.1038/nbt.1628

Langridge GC, Phan M-D, Turner DJ et al (2009) Simultaneous assay of every Salmonella Typhi gene using one million transposon mutants. Genome Res 19:2308–2316. https://doi.org/10.1101/gr.097097.109

Lee SY, Hung S, Esnault C et al (2020) Dense transposon integration reveals essential cleavage and polyadenylation factors promote heterochromatin formation. Cell Rep 30:2686–2698.e8. https://doi.org/10.1016/j.celrep.2020.01.094

Lehner KR, Stone MM, Farber RA, Petes TD (2007) Ninety-Six haploid yeast strains with individual disruptions of open Reading frames between YOR097C and YOR192C, constructed for the saccharomyces genome deletion project, have an additional mutation in the mismatch repair gene MSH3. Genetics 177:1951–1953. https://doi.org/10.1534/genetics.107.079368

Levitan A, Gale AN, Dallon EK et al (2020) Comparing the utility of in vivo transposon mutagenesis approaches in yeast species to infer gene essentiality. Curr Genet 66:1117–1134. https://doi.org/10.1007/s00294-020-01096-6

Li Y, Pham T, Xie X, Lin X (2022) Identification and characterization of an intergenic "safe haven" region in human fungal pathogen *Cryptococcus gattii*. J Fungi Basel Switz 8:178. https://doi.org/10.3390/jof8020178

Li Z, Kim KS (2020) RELATe enables genome-scale engineering in fungal genomics. Sci Adv 6:eabb8783. https://doi.org/10.1126/sciadv.abb8783

Liu OW, Chun CD, Chow ED et al (2008) Systematic genetic analysis of virulence in the human fungal pathogen Cryptococcus neoformans. Cell 135:174–188. https://doi.org/10.1016/j.cell.2008.07.046

Lofgren LA, Ross BS, Cramer RA, Stajich JE (2022) The pan-genome of *Aspergillus fumigatus* provides a high-resolution view of its population structure revealing high levels of lineage-specific diversity driven by recombination. PLoS Biol 20:e3001890. https://doi.org/10.1371/journal.pbio.3001890

Maroc L, Shaker H, Shapiro RS (2023) Functional genetic characterization of stress tolerance and biofilm formation in Nakaseomyces glabrata via a novel CRISPR activation system. 2023.12.07.569977

Mazurkiewicz P, Tang CM, Boone C, Holden DW (2006) Signature-tagged mutagenesis: barcoding mutants for genome-wide screens. Nat Rev Genet 7:929–939. https://doi.org/10.1038/nrg1984

Michel AH, Hatakeyama R, Kimmig P et al (2017) Functional mapping of yeast genomes by saturated transposition. eLife 6:e23570. https://doi.org/10.7554/eLife.23570

Michel AH, Kornmann B (2022) SAturated Transposon Analysis in Yeast (SATAY) for deep functional mapping of yeast genomes. In: Devaux F (ed) Yeast functional genomics: methods and protocols. Springer, US, New York, NY, pp 349–379

Michel AH, Schie S van, Mosbach A, et al (2019) Exploiting homologous recombination increases SATAY efficiency for loss- and gain-of-function screening 866483

Min K, Ichikawa Y, Woolford CA, Mitchell AP (2016) *Candida albicans* gene deletion with a transient CRISPR-Cas9 system. mSphere 1:e00130–e00116. https://doi.org/10.1128/mSphere.00130-16

Nelson RT, Pryor BA, Lodge JK (2003) Sequence length required for homologous recombination in Cryptococcus neoformans. Fungal Genet Biol 38:1–9. https://doi.org/10.1016/S1087-1845(02)00510-8

Roemer T, Jiang B, Davison J et al (2003) Large-scale essential gene identification in Candida albicans and applications to antifungal drug discovery. Mol Microbiol 50:167–181. https://doi.org/10.1046/j.1365-2958.2003.03697.x

Rosconi F, Rudmann E, Li J et al (2022) A bacterial pan-genome makes gene essentiality strain-dependent and evolvable. Nat Microbiol 7:1580–1592. https://doi.org/10.1038/s41564-022-01208-7

Rosiana S, Zhang L, Kim GH et al (2021) Comprehensive genetic analysis of adhesin proteins and their role in virulence of Candida albicans. Genetics 217:iyab003. https://doi.org/10.1093/genetics/iyab003

Roux I, Woodcraft C, Hu J et al (2020) CRISPR-mediated activation of biosynthetic gene clusters for bioactive molecule discovery in filamentous fungi. ACS

Synth Biol 9:1843–1854. https://doi.org/10.1021/acssynbio.0c00197
Santana DJ, O'Meara TR (2021) Forward and reverse genetic dissection of morphogenesis identifies filament-competent Candida auris strains. Nat Commun 12:7197. https://doi.org/10.1038/s41467-021-27545-5
Segal ES, Gritsenko V, Levitan A et al (2018) Gene essentiality analyzed by in vivo transposon mutagenesis and machine learning in a stable haploid isolate of Candida albicans. MBio 9. https://doi.org/10.1128/mbio.02048-18
Shapiro RS, Chavez A, Porter CBM et al (2018) A CRISPR–Cas9-based gene drive platform for genetic interaction analysis in Candida albicans. Nat Microbiol 3:73–82. https://doi.org/10.1038/s41564-017-0043-0
Sia RA, Lengeler KB, Heitman J (2000) Diploid strains of the pathogenic basidiomycete Cryptococcus neoformans are thermally dimorphic. Fungal Genet Biol 29:153–163. https://doi.org/10.1006/fgbi.2000.1192
Sidik SM, Huet D, Ganesan SM et al (2016) A genome-wide CRISPR screen in toxoplasma identifies essential apicomplexan genes. Cell 166:1423–1435.e12. https://doi.org/10.1016/j.cell.2016.08.019
Son H, Seo Y-S, Min K et al (2011) A phenome-based functional analysis of transcription factors in the cereal head blight fungus, *Fusarium graminearum*. PLOS Pathog 7:e1002310. https://doi.org/10.1371/journal.ppat.1002310
Torres-Garcia S, Di Pompeo L, Eivers L et al (2020) SpEDIT: a fast and efficient CRISPR/Cas9 method for fission yeast. Wellcome Open Res 5:274. https://doi.org/10.12688/wellcomeopenres.16405.1
van Opijnen T, Bodi KL, Camilli A (2009) Tn-seq: high-throughput parallel sequencing for fitness and genetic interaction studies in microorganisms. Nat Methods 6:767–772. https://doi.org/10.1038/nmeth.1377
van Rhijn N, Furukawa T, Zhao C et al (2020) Development of a marker-free mutagenesis system using CRISPR-Cas9 in the pathogenic mould *Aspergillus fumigatus*. Fungal Genet Biol 145:103479. https://doi.org/10.1016/j.fgb.2020.103479
Wang W, Flury AG, Rodriguez AT et al (2022) A role for worm cutl-24 in background- and parent-of-origin-dependent ER stress resistance. BMC Genomics 23:842. https://doi.org/10.1186/s12864-022-09063-w
Weiss CV, Roop JI, Hackley RK et al (2018) Genetic dissection of interspecific differences in yeast thermotolerance. Nat Genet 50:1501–1504. https://doi.org/10.1038/s41588-018-0243-4
Zhang Y-M, Zheng L, Xie K (2023) CRISPR/dCas9-mediated gene silencing in two plant fungal pathogens. mSphere 8:e0059422. https://doi.org/10.1128/msphere.00594-22

# Genome-Wide A-to-I RNA Editing During Sexual Reproduction in Filamentous Ascomycetes

5

Zhuyun Bian, Diwen Wang, and Jin-Rong Xu

**Abstract**

Like plants, fungi lack the Adenosine Deaminase that Acts on RNA (ADAR) enzymes that are unique to animals for A-to-I mRNA editing. However, genome-wide RNA editing has been reported to occur specifically during sexual reproduction in a number of filamentous ascomycetes. Unlike ADAR-mediated editing, fungal editing events have a strong preference for U at the −1 position and hairpin loops. Furthermore, majority of editing events in fungi are in the open reading frame and non-synonymous. For *PUK1* and other 70 genes with premature stop codons, mRNA editing is essential for their functions in ascosporogenesis. For conserved nonsynonymous editing sites, editing provides heterozygous advantages with both edited and unedited transcripts. Furthermore, recent studies have showed that genome-wide mRNA editing is mediated by adenosine deaminases acting on tRNA (ADATs) and stage-specific cofactors during sexual reproduction in *Fusarium graminearum*.

**Keywords**

Ascosporogenesis · mRNA editing · ADAT · ADAR · *Gibberella zeae*

## 5.1 Introduction

Although both A-to-I and C-to-U editing occur to nuclear-encoded mRNA in metazoans, the predominant form is the editing of A to I, which involves the deamination of adenosine (A), transforming it into inosine (I). The recognition of inosine as guanosine during translation or cDNA synthesis is facilitated by their similar base-pairing properties. The enzyme catalyzing A-to-I editing is Adenosine Deaminase that Acts on RNA (ADAR), which is unique to metazoans. Typically, ADAR proteins contain 1–3 double-stranded-RNA binding domains (dsRBDs) and a deaminase domain. In humans, millions of A-to-I editing sites have been identified and deficiencies in A-to-I RNA editing can lead to health problems and diseases (Slotkin and Nishikura 2013; Goncharov et al. 2019).

Like plants, fungi lack ADAR homologs. To date, none of the sequenced fungal genomes contains genes that are predicted to encode proteins with both dsRBD and deaminase domains. However, stage-specific A-to-I RNA editing during sexual reproduction has been reported in a number of filamentous fungi (Bian et al. 2019).

Z. Bian · D. Wang · J.-R. Xu (✉)
Department of Botany and Plant Pathology, Purdue University, West Lafayette, IN, USA
e-mail: jinrong@purdue.edu



M. Nowrousian, J. E. Stajich (eds.), *Fungal Genomics*, The Mycota 13,
https://doi.org/10.1007/978-3-031-75666-5_5

In this chapter, we will focus on unique features and functions of fungal A-to-I RNA editing.

## 5.2 Stage-Specific A-to-I RNA Editing During Sexual Reproduction in Fungi

### 5.2.1 Discovery of RNA Editing Events in PUK1

A-to-I editing was first observed in *Fusarium graminearum*, a major causal agent of Fusarium head blight (FHB) of wheat and barley. It is a highly fertile homothallic ascomycete that forms abundant darkly pigmented perithecia on plant debris or artificial media (Karlsson et al. 2021). Mature perithecia can release four-celled ascospores by physical ejection or oozing out from the ostiole as ascospore cirrhi (Keller et al. 2014). Because airborne ascospores are the primary inoculum, sexual reproduction plays a critical role in the FHB disease cycle. In a study to systematically characterize protein kinase genes in *F. graminearum*, *PUK1* (named after its specific expression in perithecia as perithecium unique kinase 1) was identified as one of the kinase genes that are unique to filamentous ascomycetes and have stage-specific functions during sexual reproduction (Liu et al. 2016). The *puk1* deletion mutant is normal in vegetative growth, asexual reproduction, and plant infection but is defective in ascosporogenesis and ascospore ejection. Ascospores formed by the *puk1* mutant often have morphological defects (Liu et al. 2016). In a follow-up study to generate a *PUK1* bait construct for yeast two-hybrid assays, the third intron in its predicted ORF (open reading frame) was found to be artificially introduced by automated gene prediction to avoid two tandem stop codons ($UA^{1831}GUA^{1834}G$) located inside the kinase domain. In repeated attempts, all the cDNA clones of *PUK1* contained this incorrectly predicted intron. However, sequence analysis showed that the TAGTAG sequence in genomic DNA was changed to TGGTGG in amplified cDNA fragments, indicating the occurrence of A-to-I RNA editing (Liu et al. 2016).

### 5.2.2 Genome-Wide RNA Editing During Sexual Reproduction

Strand-specific RNA-seq analysis with RNA isolated from vegetative hyphae, conidia, and perithecia collected at 8 days post-fertilization (dpf) showed that A-to-I editing specifically occurs during sexual reproduction in *F. graminearum*. A total of 26,056 editing events were identified in RNA isolated from perithecia (Liu et al. 2016). On average, the editing level is 14.8% for the 4594 genes that have at least one editing site, which is similar to the average editing level in humans (Brummer et al. 2017). Over 300 genes have 10 or more editing events in their transcripts. For most of the genes with two or more editing events, the editing levels vary significantly between different editing sites. However, for $A^{1831}$ and $A^{1834}$ of *PUK1*, the editing level is 91% and 99%, respectively, in RNA-seq data, which is significantly higher than the average.

Genome-wide RNA editing during sexual reproduction has also been reported in several other Sordariomycetes, including *Neurospora crassa*, *Sordaria macrospora*, *F. verticillioides*, and *N. tetrasperma* (Liu et al. 2017; Teichert et al. 2017). In *N. crassa*, over 41,000 A-to-I RNA editing sites were identified by RNA-seq analysis with perithecia collected at 3 to 6 dpf. Interestingly, whereas some of these editing sites are common among 3–6 dpf perithecia, others are unique to perithecia at 3, 4, 5, or 6 dpf, suggesting that there are differences in RNA editing events among different sexual differentiation stages. Furthermore, 454 editing sites identified in *N. crassa* were found to be edited at the same sites in their orthologs in *F. graminearum* and *N. tetrasperma*, suggesting that some A-to-I editing events may have conserved functions among these fungi. One of these genes with conserved editing sites is *PUK1*. Whereas editing of $A^{1831}$ is conserved in *N. crassa*, editing of $A^{1834}$ is conserved in *F. verticillioides* (Liu et al. 2017). In *S. macrospora*, A-to-I editing events were detected in protoperithecia but more abundantly in young perithecia formed by the wild type (Teichert et al. 2017). Vegetative hyphae and the *pro1* and *nox1* mutants blocked in the development of fruiting bodies had no RNA editing, confirming that A-to-I editing occurs specifically during sexual reproduction in Sordariomycetes (Teichert et al. 2017).

A-to-I RNA editing is not observed in the budding and fission yeasts that form naked asci and belong to Saccharomycotina and Taphrinomycotina, respectively. However, *Pyronema confluens*, a Pezizomycete that forms apothecia, also has sexual stage-specific RNA editing. Although absent in vegetative hyphae, 2772 A-to-I editing sites were detected in fruiting bodies. Furthermore, 15 of the 16 editing events selected for verification by PCR amplification and sequencing were confirmed (Teichert et al. 2017). Nevertheless, the gray mold fungus *Botrytis cinerea*, a member of Leotiomycetes that also forms apothecia, appears to lack A-to-I editing in fruiting bodies (Bian et al. 2019).

For Basidiomycetes, putative RNA-editing events were reported in *Ganoderma lucidum* during fruiting body formation and vegetative hyphae of *Fomitopsis pinicola* and other four *Polyporales* species although A-to-G single nucleotide polymorphisms (SNPs) were not enriched in comparison with other types of SNPs in their RNA-seq data (Wu et al. 2018, 2019). A careful reanalysis showed that these fungi do not have A-to-I RNA editing and un-enriched A-to-G SNPs observed in RNA-seq data are likely related to sequencing errors and/or problems with genome assemblies (Min et al. 2021). Taken together, stage-specific genome-wide A-to-I editing during sexual reproduction has only been observed in filamentous ascomycetes although it remains to be determined which fungal classes in Pezizomycotina have RNA editing. To date, all the fungi with confirmed A-to-I RNA editing, except *P. confluens* belonging to Pezizomycetes, are Sordariomycetes species.

## 5.3 Distinct Features of Fungal A-to-I Editing

RNA editing catalyzed by animal ADARs has only a weak preference for the flanking sequences of the edited adenosines, with the depletion of G at the −1 position and an enrichment of G at the +1 position (Eggington et al. 2011, Porath et al. 2017). In contrast, A-to-I editing in fungi strongly favors U at the −1 position (>70% in *F. graminearum* and *N. crassa*) and prefers A and G at +1 and + 3 positions (Fig. 5.1a) (Liu et al. 2016, 2017). Furthermore, adenosines in the preferred flanking sequences tend to have higher editing levels than those in non-preferred flanking sequences, suggesting that nucleotides surrounding the editing site influence both the specificity and efficiency of A-to-I editing in fungi.

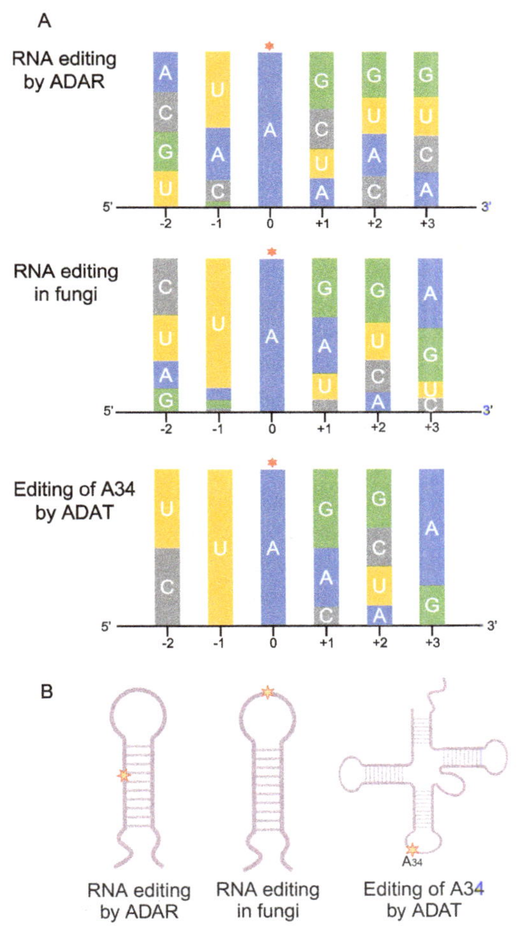


**Fig. 5.1** Unique features of A-to-I RNA editing in fungi. (**a**) Preference of nucleotides flanking the edited adenosines (A*) in ADAR-mediated RNA editing, RNA editing in fungi, and editing of A34 in the anticodon loop of tRNAs. (**b**) Preferential distribution of the edited adenosines (A*) in predicted secondary structures of mRNA or tRNA

Besides their differences in the flanking nucleotide sequences of edited adenosines, RNA editing in fungi also differs from ADAR-mediated editing in metazoans in the preference of secondary structures. Whereas ADAR-mediated editing preferentially occurs in the stem region in metazoans (Zhang et al. 2017), A-to-I RNA editing in fungi displays a preference for adenosines in hairpin loops (Fig. 5.1b) of predicted secondary mRNA structures (Liu et al. 2016, 2017). Furthermore, edited adenosines in hairpin loops tend to have higher editing levels than those in predicted stem regions of mRNA in *F. graminearum* and *N. crassa*, indicating that secondary structures of mRNA also influence the occurrence of A-to-I editing in fungi. These observations are consistent with the fact that fungi lack ADAR-like enzymes and likely involve a novel mechanism for mRNA editing.

Another significant difference between fungi and animals is the effects of A-to-I editing on protein complexity. In humans, over 98% of the editing sites are in the non-coding regions, mainly in the 3'-UTRs and introns. Only less than 0.02% of the editing events result in changes in amino acid sequences in translated proteins (Brummer et al. 2017). Nonsynonymous A-to-I editing is also less than 0.02% in *C. elegans* and only 10% in Drosophila (Zhao et al. 2015; Yablonovitch et al. 2017). However, 70.5% of the editing sites are in the coding regions and 64% of them cause changes in the amino acid sequences of encoding proteins in *F. graminearum* (Liu et al. 2016). In *N. crassa*, 64% of the editing sites are in the coding regions and 52% of them affect protein sequences (Liu et al. 2017). With majority of editing events being nonsynonymous and the average editing level of approximately 15%, A-to-I RNA editing significantly increases the proteome complexity during sexual reproduction in filamentous ascomycetes.

## 5.4 Importance of Fungal mRNA Editing

Because inosines can pair with cytosines or be recognized as guanosines, like in metazoans, A-to-I RNA editing can affect various processes affected by changes in nucleotide sequences, including intron splicing, RNA folding, and RNA–RNA or protein interaction. Below we will summarize only editing of stop codons and abundant nonsynonymous editing events that are somewhat unique to RNA editing in fungi.

### 5.4.1 Editing of Stop Codons in Genes Important for Ascosporogenesis

Because of the strong preference for U at −1 position, it is not surprising that editing of stop codons occurs frequently in fungi. Whereas editing of the premature stop codons (PSCs) in the coding region is essential to synthesize full-length proteins, such as the editing of $UA^{1831}GUA^{1834}G$ in *PUK1*, editing of the canonical stop codon of ORFs (stop-loss editing) will convert the UAG stop codon to a tryptophan (TRP) codon and lead to an extension at the C-terminus.

*F. graminearum* has a total of 71 genes with PSC editing events, including *AMD1, FgRID, FgAMA1*, and *FgBUD14* (Liu et al. 2016; Cao et al. 2017; Hao et al. 2019; Liang et al. 2021). Similar to *PUK1*, *AMD1* and *FgAMA1* are specifically expressed during sexual reproduction. Expressing the non-editable alleles of *AMD1* or *FgAMA1* with the PSC (TAG) edited to TAA failed to complement the defect of the *amd1* or *Fgama1* mutant in ascospore morphology and/or discharge, indicating that editing of the PSCs in these two genes is essential for their functions during sexual reproduction. *FgBUD14* is constitutively expressed in vegetative hyphae and perithecia. Both editing of $UA^{1334}G$ to UGG and alternative splicing of intron 2 containing the PSC editing site occur in the *FgBUD14* transcripts during sexual reproduction (Liang et al. 2021). Interestingly, *FgRID* is orthologous to the *RID* gene that is essential for repeat-induced point mutation (RIP) pre-meiotically in *N. crassa* (Freitag et al. 2002; Selker et al. 2002). RIP has been experimentally confirmed in *F. graminearum* and predicted to occur in other Fusarium species (Van Wyk et al. 2020). However, *FgRID* and its orthologs from other

Hypocreales species all have a PSC (TAG) editing site at the $W^{145}$ position that is conserved among RID orthologs (Wang et al. 2016). Therefore, RNA editing is essential for RID orthologs to encode full-length functional DNA methyltransferase proteins in *F. graminearum* and other Hypocreales species. Recently, functional characterization of the other genes with PSC editing events showed that restorative editing in 16 of them is important for fruiting body development (Qi et al. 2024). Interestingly, for some of these genes, restorative editing of the PSCs appears to have adaptive advantages for balancing the survival–reproduction trade-offs.

Both *F. graminearum* and *N. crassa* have over hundreds of genes with stop-loss editing events (Liu et al. 2016, 2017) that will result in an addition of extra amino acid sequences at the C-terminus (until the next in-frame stop codon) although ribosome stalling may occur beyond the canonical stop codon UAG that is edited to UGG. Another type of editing events at canonical stop codons is stop-retaining editing that changes UAA to UGA. Whereas it is unlike for stop-retaining editing to be important because the similarity between UAA and UGA, some of these stop-loss editing events may affect protein expression, folding, stabilities, or functions. Unfortunately, to date, there is no report on functional characterization of the importance of stop-loss editing events in filamentous ascomycetes.

### 5.4.2 Importance of Nonsynonymous Editing

As noted above, the prevalence of nonsynonymous or missense editing in fungi significantly increases the proteome complexity (Liu et al. 2016, 2017; Teichert et al. 2017). In addition to many genes associated with ascosporogenesis and highly expressed in perithecia, nonsynonymous editing may exert a global effect on gene expression during sexual development because genes involved in chromatin organization and modification, RNA transcription and processing, and protein transport and localization are enriched for nonsynonymous editing sites (Liu et al. 2016). Furthermore, in a recent study with editing events conserved between *F. graminearum* and *N. crassa*, Xin and colleagues provided experimental evidence for the adaptive advantage of nonsynonymous editing (Xin et al. 2023). Similar to the *cme5* deletion mutant, the under-edited mutant of *CME5*, one of the genes with conserved missense editing sites, has severe defects in ascus and ascospore formation. However, the pre-editing residue of *CME5* is evolutionarily conserved across diverse classes of Ascomycota, while the post-editing one is rarely hardwired into the genome, implying that editing at this site leads to higher fitness than a genomic A-to-G mutation.

Because the average editing level is less than 15% in fungi, RNA editing at a specific site is nearly binary, resulting the co-expression of both edited and unedited versions (Gommans et al. 2009). In *F. graminearum*, the co-expression of unedited and edited versions of *CME11* is required for its full function during sexual reproduction (Xin et al. 2023), which shows the importance of nonsynonymous editing and co-existence of resulting protein variants. In addition, nonsynonymous editing in genes under functional constraints or at residues at critical sites that are inaccessible to genetic mutations in the genome due to potential fitness costs (Xu and Zhang 2014) can render stage-specific functions or slightly altered functions during sexual reproduction without detrimental effects because only a fraction of transcripts is edited to generate protein variants.

## 5.5 Mechanisms of Stage-Specific RNA Editing

Although lack of ADAR-like genes, the *F. graminearum* genome has 18 *ACD* genes that encode proteins with predicted adenosine/cytosine deaminase domains. Systematical characterization of these *ACD* genes showed that *ACD17* and *ACD18* are essential genes but deletion of the other 16 *ACD* genes has no effect on A-to-I editing although some *ACD* deletion mutants are defective in purine metabolism (Sun et al. 2021). *ACD17* and *ACD18* are orthologous to *TAD2* and *TAD3*, two essential ADAT (adenosine deaminase acting on tRNA) genes in the budding yeast *Saccharomyces cerevisiae* (Torres et al. 2014). ADATs catalyze the editing of A34 in the anticodon loop of tRNA, which is structurally similar to hairpin loops favored by RNA editing in fungi. Furthermore, the −1 position of A34 is U33 in fungal tRNAs, which is similar to the strong preference of U at −1 position for fungal RNA editing. In addition, +1 to +3 positions also share similar preferences between RNA editing in fungi and editing of A34 in tRNA by ADATs. Similar preferences for the primary sequence and secondary structures between fungal RNA editing and ADAT-mediated tRNA editing further indicate that ADATs orthologous to yeast Tad2 and Tad3 may be responsible for A-to-I RNA editing during sexual reproduction in filamentous ascomycetes, possibly involving stage-specific cofactors. In *S. cerevisiae*, Tad3 has the E-to-V mutation in the catalytic core of the deaminase domain and is enzymatically inactive but it is essential for tRNA editing by forming heterodimers with Tad2. FgTad3 (Acd18) and its orthologs from other fungi have the same E-to-V mutation as in yeast Tad3 and likely play only a facilitatory role in enzymatic editing of adenosines. Therefore, the ortholog of yeast *TAD2* is likely responsible for catalyzing the deamination of adenosines in mRNA in *F. graminearum* and other filamentous fungi.

In *F. graminearum*, RNA-seq analysis with RNA isolated from mating cultures and perithecia collected at 1–8 dpf showed that editing events increase significantly after 3 dpf and kept on increasing thereafter. Interestingly, both *FgTAD2* and *FgTAD3* have two transcript isoforms. Whereas the long isoforms of both ADAT genes are constitutively expressed, the short ones are specifically expressed during sexual reproduction and their expression levels increased rapidly after 3 dpf. In fact, the shorter isoforms of *FgTAD2* and *FgTAD3* became more abundant than the longer transcripts in perithecia after 4 dpf (Bian et al. 2024).

To further characterize the roles of FgTad2 and FgTad3 in A-to-I mRNA editing, nonfunctional *FgTAD2* and *FgTAD3* fragments were integrated ectopically into the wild-type strain PH-1. Ascospore progeny with RIP mutations in these two ADAT genes that specifically affect mRNA editing and ascosporogenesis was identified (Bian et al. 2024) (Feng et al. 2024). Among these RIP mutations, the H352Y mutation in *FgTAD2* was verified for its effect on editing of mRNA in an in vitro assays with affinity purified proteins from perithecia (Bian et al. 2024). Results from these genetic and biochemical studies confirm that A-to-I mRNA editing is mediated by ADATs in *F. graminearum.*

To identify stage-specific cofactors that facilitate the editing of mRNA by ADATs during sexual reproduction, putative FgTad2-ineracting proteins (FIPs) were identified by affinity purification (Bian et al. 2024). Deletion of *FIP5* that encodes a protein with the RNA recognition motif had defects in mRNA editing and specifically affected perithecium development and ascospore formation (Bian et al. 2024). The *AME1* (activator of mRNA editing 1) gene was identified by functional characterization of genes that are specifically expressed in perithecia and conserved between *F. graminearum* and *N. crassa*. Ame1 interacts with the N-terminal region of FgTad2 and it is required for A-to-I mRNA editing during sexual reproduction (Feng et al. 2024). Constitutive expression of *AME1* results in A-to-I editing at over hundreds of sites in vegetative hyphae. Co-expression of *AME1* together with the short isoforms of *FgTAD2* and *FgTAD3* also causes A-to-I RNA editing events in the budding yeast, *Escherichia coli*, or HEK 293 T cells (Feng et al. 2024). Therefore, the FgTad2 and TgTad2 ADATs catalyze the editing of A34 of tRNA in vegetative hyphae but they are responsible for A-to-I mRNA editing during sexual reproduction with the short isoforms interacting with their stage-specific cofactors for mRNA editing in *F. graminearum.*

## 5.6 Conclusions and Perspectives

To date, genome-wide A-to-I mRNA editing has only been reported in a limited number of filamentous ascomycetes. It is likely that more and more Sordariomycetes species will be found to have genome-wide RNA editing during sexual reproduction. However, it is difficult to predict which fungal groups outside Sordariomycetes have sexual specific genome-wide RNA editing. It is also not clear whether A-to-I RNA editing occurs in other developmental or infection stages (other than sexual reproduction) in filamentous fungi, such as during plant infection and sclerotium formation. Considering the importance of fungal effectors during infectious growth, editing of effector mRNAs, if occurs, will allow plant pathogenic fungi to increase the complexity of effectors without increasing the number of effector genes.

Regarding the roles of ADATs and stage-specific cofactors required for genome-wide A-to-I RNA editing, it is not clear what the mechanism is for mRNA recognition and binding. The roles of Fip5 or other proteins with RNA binding domains need to be further characterized. Furthermore, other FgTad2- and FgTad3-interacting proteins remain to be functionally characterized for their roles in ascosporogenesis and mRNA editing. Whereas some of them may be responsible for developmental stage-specific editing events during different stages of sexual reproduction, others may play a role in determining editing efficiencies or levels at different editing sites, which is not even clear in metazoans (Eisenberg and Levanon 2018).

## References

Bian Z, Ni Y, Xu JR, Liu H (2019) A-to-I mRNA editing in fungi: occurrence, function, and evolution. Cell Mol Life Sci 76:329–340

Bian Z, Wang Z, Wang DW, Xu JR (2024) Sexual stage–specific A-to-I mRNA editing is mediated by tRNA-editing enzymes in fungi. Proc Natl Acad Sci 121(12):e2319235121

Brummer A, Yang Y, Chan TW, Xiao X (2017) Structure-mediated modulation of mRNA abundance by A-to-I editing. Nat Commun 8(1):1255

Cao S, He Y, Hao C, Xu Y, Zhang H, Wang C, Liu H, Xu JR (2017) RNA editing of the AMD1 gene is important for ascus maturation and ascospore discharge in *Fusarium graminearum*. Sci Rep 7(1):4617

Eggington JM, Greene T, Bass BL (2011) Predicting sites of ADAR editing in double-stranded RNA. Nat Commun 2(1):1–9

Eisenberg E, Levanon EY (2018) A-to-I RNA editing—immune protector and transcriptome diversifier. Nat Rev 19(8):473–490

Feng C, Xin K, Du Y, Zou J, Xing X, Xiu Q, Zhang B, Zhang Y, Huang W, Wang QH, Jiang C, Wang X, Kang Z, Xu JR, Liu H (2024) Unveiling the A-to-I mRNA editing machinery and its regulation and evolution in fungi. Nat Commun 15(1):3934

Freitag M, Williams RL, Kothe GO, Selker EU (2002) A cytosine methyltransferase homologue is essential for repeat-induced point mutation in. Neurospora crassa 99(13):8802–8807

Gommans WM, Mullen SP, Maas S (2009) RNA editing: a driving force for adaptive evolution? BioEssays 31(10):1137–1145

Goncharov A, Kliuchnikova A, Nasaev S, Moshkovskii S (2019) RNA editing by ADAR adenosine deaminases: from molecular plasticity of neural proteins to the mechanisms of human cancer. Biochemistry (Mosc) 84:896–904

Hao C, Yin J, Sun M, Wang Q, Liang J, Bian Z, Liu H, Xu JR (2019) The meiosis-specific APC activator FgAMA1 is dispensable for meiosis but important for ascosporogenesis in *Fusarium graminearum*. Mol Microbiol 111(5):1245–1262

Karlsson I, Persson P, Friberg H (2021) Fusarium head blight from a microbiome perspective. Front Microbiol 12:628373

Keller MD, Bergstrom GC, Shields EJ (2014) The aerobiology of *Fusarium graminearum*. Aerobiologia 30:123–136

Liang J, Fu X, Hao C, Bian Z, Liu H, Xu JR, Wang G (2021) FgBUD14 is important for ascosporogenesis and involves both stage-specific alternative splicing and RNA editing during sexual reproduction. Environ Microbiol 23(9):5052–5068

Liu H, Li Y, Chen D, Qi Z, Wang Q, Wang J, Jiang C, Xu J-RJPNAS (2017) A-to-I RNA editing is developmentally regulated and generally adaptive for sexual reproduction in *Neurospora crassa*. Proc Natl Acad Sci U S A 114(37):E7756–E7765

Liu H, Wang Q, He Y, Chen L, Hao C, Jiang C, Li Y, Dai Y, Kang Z, Xu JR (2016) Genome-wide A-to-I RNA editing in fungi independent of ADAR enzymes. Genome Res 26(4):499–509

Min B, Wu B, Gaskell J, Zhang J, Toapanta C, Ahrendt S, Blanchette RA, Master E, Cullen D, Hibbett DS, Grigoriev IV (2021) RNA-editing in Basidiomycota, revisited. ISME Commun 1(1):70

Porath HT, Knisbacher BA, Eisenberg E, Levanon EY (2017) Massive A-to-I RNA editing is common across the Metazoa and correlates with dsRNA abundance. Genome Biol 18:1–12

Qi Z, Lu P, Long X, Cao X, Wu M, Xin K, Xue T, Gao X, Huang Y, Wang Q (2024) Adaptive advantages of restorative RNA editing in fungi for resolving survival-reproduction trade-offs. Sci Adv 10(1):eadk6130

Selker EU, Freitag M, Kothe GO, Margolin BS, Rountree MR, Allis CD, Tamaru H (2002) Induction and maintenance of nonsymmetrical DNA methylation in *Neurospora*. PNAS 99(suppl_4):16485–16490

Slotkin W, Nishikura K (2013) Adenosine-to-inosine RNA editing and human disease. Genome Med 5(11):1–13

Sun M, Bian Z, Luan Q, Chen Y, Wang W, Dong Y, Chen L, Hao C, Xu JR, Liu H (2021) Stage-specific regulation of purine metabolism during infectious growth and sexual reproduction in *Fusarium graminearum*. New Phytol 230(2):757–773

Teichert I, Dahlmann TA, Kuck U, Nowrousian M (2017) RNA editing during sexual development occurs in distantly related filamentous Ascomycetes. Genome Biol Evol 9(4):855–868

Torres AG, Pineyro D, Filonava L, Stracker TH, Batlle E, Ribas de Pouplana L (2014) A-to-I editing on tRNAs: biochemical, biological and evolutionary implications. FEBS Lett 588(23):4279–4286

Van Wyk S, Wingfield BD, De Vos L, van der Merwe NA, Steenkamp ET (2020) Genome-wide analyses of repeat-induced point mutations in the Ascomycota. Front Microbiol 11:622368

Wang C, Xu JR, Liu H (2016) A-to-I RNA editing independent of ADARs in filamentous fungi. RNA Biol 13(10):940–945

Wu B, Gaskell J, Held BW, Toapanta C, Vuong T, Ahrendt S, Lipzen A, Zhang J, Schilling JS, Master E (2018) Substrate-specific differential gene expression and RNA editing in the brown rot fungus *Fomitopsis pinicola*. AEM 84(16):e00991–e00918

Wu B, Gaskell J, Zhang J, Toapanta C, Ahrendt S, Grigoriev IV, Blanchette RA, Schilling JS, Master E, Cullen D (2019) Retracted article: evolution of substrate-specific gene expression and RNA editing in brown rot wood-decaying fungi. ISME J 13(6):1391–1403

Xin K, Zhang Y, Fan L, Qi Z, Feng C, Wang Q, Jiang C, Xu JR, Liu H (2023) Experimental evidence for the functional importance and adaptive advantage of A-to-I RNA editing in fungi. PNAS 120(12):e2219029120

Xu G, Zhang J (2014) Human coding RNA editing is generally nonadaptive. PNAS 111(10):3769–3774

Yablonovitch AL, Deng P, Jacobson D, Li JB (2017) The evolution and adaptation of A-to-I RNA editing. PLoS Genet 13(11):e1007064

Zhang R, Deng P, Jacobson D, Li JB (2017) Evolutionary analysis reveals regulatory and functional landscape of coding and non-coding RNA editing. PLoS Genet 13(2):e1006563

Zhao HQ, Zhang P, Gao H, He X, Dou Y, Huang AY, Liu XM, Ye AY, Dong MQ, Wei L (2015) Profiling the RNA editomes of wild-type C. elegans and ADAR mutants. Genome Res 25(1):66–75

# A Symphony of Roles for Codon Usage in Fungal Genomics

6

Abigail Leavitt LaBella

**Abstract**

Codon usage bias, or the uneven use of synonymous codons, is a universal feature of genes and genomes. Although previously considered silent, synonymous mutations can alter cellular translational and even transcriptional dynamics. Fungi have served as model organisms for many groundbreaking discoveries in codon usage. Moreover, surveys of fungi repeatedly find that selection has shaped codon usage—suggesting a functional role for synonymous sites. Here, we review how codon usage plays an active role throughout the flow of information from DNA to RNA to protein while highlighting findings from fungi. We will also discuss leveraging information encoded in fungal synonymous codon usage to address biological and phylogenetic hypotheses.

**Keywords**

Fungus · Codon usage · tRNA · Ribosome · Reverse ecology

A. L. LaBella (✉)
Department of Bioinformatics and Genomics, The North Carolina Research Campus,
Kannapolis, NC, USA
e-mail: alabell3@charlotte.edu

## 6.1 Introduction

Before the genetic code was cracked, amino acids were hypothesized to be encoded by more than one synonymous codon (Crick 1955). Mutations between synonymous codons were known as "silent" mutations because researchers could not observe them using protein electrophoresis (Bowman 1959). The genetic code is, in fact, degenerate, with amino acids encoded by up to 6 different synonymous codons (Nirenberg and Leder 1964). Pioneering DNA sequencing revealed that the usage of synonymous codons varied widely between organisms (Goel et al. 1972) and that the relative frequency of synonymous codons was proportional to the number of tRNAs with a corresponding anticodon (Fig. 6.1; Grantham 1978). In addition to variation between species, codon usage was found to be biased within genomes (Sharp et al. 1986). In this case, transcripts containing codons decoded by abundant tRNAs were expressed higher (Sharp et al. 1986). Despite early evidence suggesting a functional role for codon usage bias, the moniker "silent" was expanded to define these mutations as evolving neutrally or nearly neutrally (Kimura 1977). Research has revealed numerous functional roles for codon usage—inspiring the frequent use of The Sounds of Silence lyrics. There are many great reviews on the various functional roles of codon usage (Liu et al. 2021; Komar 2016; Iriarte et al. 2021; Novoa and Ribas de


M. Nowrousian, J. E. Stajich (eds.), *Fungal Genomics*, The Mycota 13,
https://doi.org/10.1007/978-3-031-75666-5_6

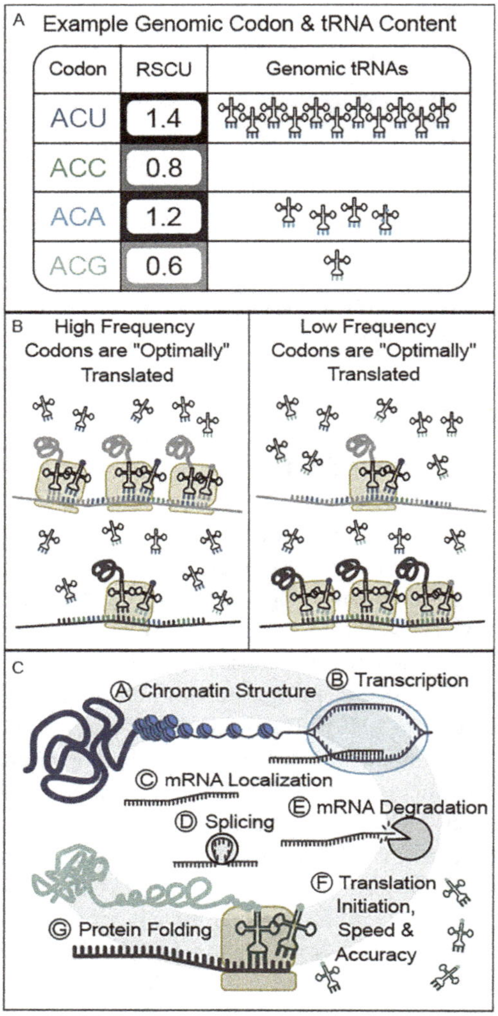


**Fig. 6.1** Overview of the principles of codon usage bias. (**a**) Sample codon usage metrics from *S. cerevisiae*. The synonymous codons encoding threonine and their respective Relative Synonymous Codon Usage (RSCU) in the genome. ACU and ACA are preferred, while ACC and ACG are avoided. There are three isoacceptors that decode the four synonymous codons. The preferred codons are decoded by isoacceptors with high genomic copy numbers. All eleven of the genomic tRNAs that decode ACU (anticodon AGU) have an identical sequence. There are two unique isodecoder sequences (one with three copies and one with a single copy) that decode ACA (anticodon UGU). The sequence differences are shown in blue dots. The codon ACC is decoded by wobble base pairing. (**b**) Codons decoded by abundant tRNAs, especially those that do not require wobble base pairing, are optimally translated leading to higher protein output. The codon that is considered optimal is dependent on the tRNA context and may change in different regulatory states. On the left side, the high-frequency blue codon is optimal. On the right side, the low-frequency green codon is optimal. (**c**) The many points along the central dogma impacted by codon usage

Pouplana 2012). Fungi have played a critical role in the research conducted on codon usage bias. Here, we will use fungal research, when available, as a lens into the growing field of codon usage bias.

### 6.1.1 Measures of Codon Usage

Codon usage bias refers to the non-random usage of codons within a protein-coding sequence. These measurements fall into two distinct categories: (1) those that use only frequency information and (2) those that aim to capture codon usage that optimizes gene expression. We will introduce the frequency-based metrics here and then define the functional metrics after exploring codon optimization.

The most basic measures of codon usage are absolute number and frequency. The relative abundance of amino acids highly influences these frequency-based metrics. Amino acid usage varies across genes with different functions (Karlin and Bucher 1992) and species with different lifestyles (Singer and Hickey 2003). Codon frequencies are, therefore, often reported within amino acids (i.e., among synonymous codons). The most frequently used codons for an amino acid are often referred to as "preferred" codons. To measure how preferred a codon is, researchers often turn to the Relative Synonymous Codon Usage (RSCU) (Fig. 6.1a; Sharp et al. 1986). RSCU is the observed number of occurrences of a synonymous codon divided by the expected number if the synonymous codons were random. Codon usage indistinguishable from random (or for non-degenerate amino acids) has a value of 1. Codon usage greater than random has a value greater than 1. RSCU can be applied to gene regions, genes, or genomes. Frequently, the RSCU for each individual gene will be averaged for a genome-wide RSCU average.

Another measure of the deviation of codon usage from random is the effective number of codons (ENC or $N_c$). ENC is a single value representing how many codons are used within each amino acid and, therefore, ranges from 20 (only one codon per amino acid) to 61 (equal use of all codons) (Fig. 6.2; Wright 1990). A new formula-

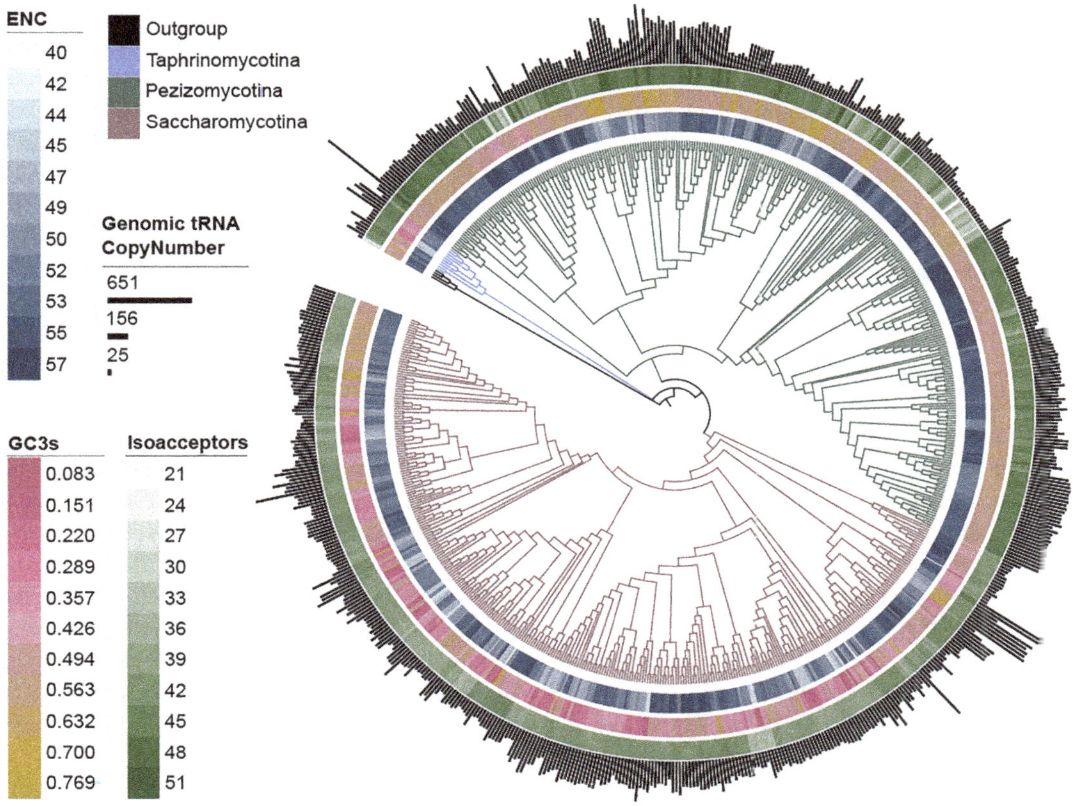


**Fig. 6.2** Variation in codon usage associated metrics across a subset of Ascomycota. The average number of effective codons (blue bars) vary widely and nearly reach the theoretical maximum of 61. The synonymous GC component of the third codon position (GC3s) exhibits major shifts between the Pezizomycotina and the Saccharomycotina. The underlying phylogenetic tree was obtained from Shen et al. (2020) and the genomes and annotations for the specific strains used were downloaded from NCBI. The ENC was calculated using DAMBE 7.3.32 (Xia 2018), tRNAs were obtained using tRNA-scan-SE, and GC3s was calculated using codonW (https://codonw.sourceforge.net/). Information about the genomes used for the analysis can be found under https://doi.org/10.6084/m9.figshare.c.7063571.v1

tion of ENC has been introduced that better accounts for issues caused by six-fold degenerate amino acids (Sun et al. 2013). ENC values measure bias for a gene or genome accounting for all synonymous codons.

We are also often interested in the role of GC bias in codon usage. Several metrics of GC content are frequently used. The first is the global GC content, often used as a background metric reflecting genome-wide forces. The second is GC content at the third codon position (GC3). Finally, we can examine the bias only at synonymous third codon positions (GC3s; Fig. 6.2). This removes the bias introduced by amino acids encoded by 1 or 3 codons. All these metrics help us describe a bias toward a preferred set of high-frequency synonymous codons and a bias against low-frequency rare synonymous codons.

### 6.1.2 Codon Usage Bias in Fungi

Fungi are an exceptional system for studying codon usage due to the incredible diversity within this kingdom. Codon usage bias in fungi was initially noted in six genes within *Saccharomyces cerevisiae* (Bennetzen and Hall 1982). Across the fungal kingdom, differences in RSCU are sufficient to distinguish subkingdoms due to differences in the usage of $AUA^{Ile}$ and

GGG$^{Gly}$ (Wint et al. 2022). Within the subphylum Saccharomycotina, RSCU alone is sufficiently informative to classify species into the proper order using a random forest algorithm (unpublished data). In addition to differences in codon preference, fungi exhibit both high (ENC of 32.8) and low (ENC of 56.9) bias in their genomic codon usage (Fig. 6.2; Wint et al. 2022). Ancestral reconstruction suggests that early diverging lineages have more biased (lower ENC) genomes—especially within the *Neocallimastigomycota*.

Studies of codon usage in fungi are also of particular interest as there are three independent nuclear reassignments of the CUG codon within the Saccharomycotina (Muhlhausen and Kollmar 2014; Shen et al. 2018; Krassowski et al. 2018). In the order Serinales, the CUG codon is completely reassigned from leucine to serine (Ohama et al. 1993; Groenewald et al. 2023). Similarly, the CUG codon is stochastically translated as both serine and leucine in the Ascoideales (Muhlhausen et al. 2018). In the Alaninales, the CUG codon is decoded as alanine (Riley et al. 2016; Muhlhausen et al. 2016). These codon reassignments are hypothesized to be associated with extreme avoidance and subsequent recapture of the CUG codon (Kollmar and Muhlhausen 2017). They present a unique research opportunity but also require modification of standard codon metrics to account for changes in the number of synonymous codons (Choudhuri and Sau 2023).

### 6.1.3 Forces Shaping Codon Bias Variation in Fungi

The variation of codon bias within fungi naturally leads to the question, why does variation exist? Much of this variation is likely closely related to variation in tRNAs. tRNA sequences can vary at the anticodon and along other regions of the sequence. tRNA isoacceptors decode the same amino acid but contain different anticodons, while tRNAs that contain the same anticodon but vary in other regions are isodecoders. Across fungi, the total number of tRNAs with unique isoacceptors ranges from 58 in *Ascobolus immersus* to 18 in *Sporobolomyces linderae* (Wint et al. 2022)—which is below the theoretical minimum of 25–30 tRNAs required to decode all codons in the standard genetic code (Marck and Grosjean 2002; Alkatib et al. 2012). Species that contain fewer than the theoretical minimums likely rely on tRNA modifications to extend wobble base pairing. There is also the possibility that some codons are not used within fungal genomes—in the >1200 sequences available on GenBank for *Colletotrichum godetiae* (taxID 1,209,918) there are no TTA (leucine), ATA (isoleucine), or TCA (serine) codons (data from CoCoPUTS https://dnahive.fda.gov/ (Alexaki et al. 2019)). The total number of tRNA genes within the genome (known as tRNAome) also varies from 24 to 4547 (Wint et al. 2022). The tRNAs found at the highest number in the genome are typically found at the highest concentration within the cell and decode the most frequently used codons (Bulmer 1987). This suggests that codon usage and tRNAs have co-evolved.

The co-evolution of tRNAs and codon usage suggests that selection has acted to shape codon usage. Below, we will outline many of the functional roles for codon usage that have been experimentally tested. If selection on these functions is sufficient, it may outweigh neutral and mutational forces traditionally thought to govern synonymous site evolution (Kimura 1977). Despite the abundance of clonal population structures, fungi have evolved mechanisms like accessory chromosomes and high rates of molecular evolution that increase the efficacy of selection sufficiently to act on synonymous codon usage (Stukenbrock and Croll 2014; Stajich 2017). Numerous studies have found that most fungi exhibit signatures of selection on codon usage within their genomes (LaBella et al. 2019; Wint et al. 2022; Cope and Shah 2022).

There are, however, some fungi that lack this strong selection signal on codon usage. In those cases, codon usage patterns may be driven by drift, mutational biases, and GC-biased gene conversion (GCB). Drift is likely to be dominant in fungi with small effective population sizes associated with specific lifestyles. A lack of selection on codon usage has been shown in specialist parasites such as *Zymoseptoria tritici* and *Blumeria graminis* (Badet et al. 2017). Mutational biases vary in both their direction and degree—genomic GC content across fungi ranges from 62% (*Ceratocystiopsis minuta*) to 25% (*Ascoidea asiatica*) (Shen et al. 2020). These biases may be associated with mutation rates, rates of transitions versus transversions (Zhu et al. 2014), or processes like repeat-induced point mutation (RIP) found in fungi like *Neurospora crassa* (Selker et al. 1987). Biases in GC composition can also be impacted by GC-biased gene conversion, or the non-reciprocal exchange of alleles resulting from double-strand breaks favors GC alleles (Bengtsson 1986). Recent work, however, has called into question the GC bias of gene conversion in *S. cerevisiae* and *N. crassa* (Liu et al. 2018). Given the incredibly diversity of fungi, we cannot make any universal statements about the amount of selection acting on codon usage within this kingdom. Nevertheless, fungi have been instrumental in identifying and characterizing the role of codons in the central dogma.

## 6.2 Functional Role of Codon Usage Bias Across the Central Dogma

At nearly every step between DNA and protein, a specific functional role for codon usage has been identified. We will discuss the various roles starting with translation because this was the first and most well recognized function of codon usage. During the process of protein translation, codon usage bias also plays a major role in protein folding and stability. More recently, a critical role for codon usage in transcription and even epigenetics has emerged. Given its numerous roles it is clear that codon usage can be used to fine-tune gene expression (Quax et al. 2015). As a caveat, many of the functions of codon usage we describe can be extrapolated to other organisms, but their relative importance in species such as humans is still under debate.

### 6.2.1 Translation

From the very beginning, codon usage bias was found to be correlated in *S. cerevisiae* with cognate tRNA abundance and transcript expression level (Ikemura 1982). This led to the hypothesis that mRNA transcripts containing codons decoded by abundant tRNAs are translated at higher rate than those with codons decoded by rare tRNAs (Sharp et al. 1986). Highly expressed genes, therefore, would be subject to translational selection that would optimize codon usage for high expression. The specific codons decoded by abundant cognate tRNAs are referred to as "optimal codons." The evidence supporting a role for translational selection in *S. cerevisiae* has been followed up with numerous studies aimed at characterizing the functional dynamics occurring during translation and the overall role of translational selection in fungal evolution.

#### 6.2.1.1 Translation Initiation

Translational initiation is an important rate-limiting step for protein synthesis. Ribosomes must be loaded with enough distance to prevent collisions or traffic jams that could result in aborted protein synthesis and wasted energy (Tuller et al. 2010). In highly expressed genes, which require many ribosomes, there are ~30–50 non-optimal codons at the beginning of the ORF that are associated with rarer tRNAs (Tuller et al. 2010). This set of non-optimal codons is known as a ramp and was identified in the transcripts of highly expressed genes in fungi including *Yarrowia lipolytica*, *Schizosaccharomyces pombe*, *Candida albicans*, and *Aspergillus nidulans* (Tuller et al. 2010). However, studies in *S. cerevisiae* (Chu et al. 2014) and *Drosophila* (Barrington et al. 2023) cells suggest that the rate of translation initiation is impacted by codons found throughout the whole mRNA and not specifically at the beginning. This is in line with an alternative hypothesis that the low codon optimization of the 5′ region of mRNA is due to their rapid evolutionary turnover (Sejour et al. 2023). Translation initiation is also impacted by mRNA secondary structure which is affected by synonymous codon usage (Osterman et al. 2020).

Additional work is needed to disentangle the mechanisms and specific contexts in which codon usage impacts translation initiation.

#### 6.2.1.2 Elongation Rate

Due to low availability of corresponding tRNAs, non-optimal codons may wait longer for a tRNA to become available and slow the rate of translation elongation (Gardin et al. 2014). This hypothesis was originally tested by asking if rare codons had longer ribosomal occupancy as measured by ribosomal profiling. Results from *S. cerevisiae* (Qian et al. 2012) and other organisms (Ingolia et al. 2011) suggested that elongation speed was not impacted by rare codons. Reanalysis of this data, however, demonstrated that this was due to both analysis methods (Dana and Tuller 2014) and limitations with earlier ribosomal profiling methods (Yu et al. 2015). Ribosomes do have longer dwell times on codons decoded by rarer tRNAs in *S. cerevisiae* (Weinberg et al. 2016). *Neurospora crassa* has been a critical model for elucidating the roles of codon usage in elongation. In *Neurospora* rare codons are translated more slowly than their synonymous counterparts and can trigger ribosomal stalling (Yu et al. 2015). This can lead to premature termination of translation mediated by translation termination factor eRF1 (Yang et al. 2019). Work in *Neurospora* has also identified components, such as CPC-3, involved in the feedback loop that connects transcript length, codon usage, and elongation speed (Lyu et al. 2021). CPC-3, the *Neurospora* homolog to GCN2, is a kinase that phosphorylates eIF2$\alpha$ (Karki et al. 2020). During elongation, ribosomal pausing at rare codons activates CPC-3 which prevents translation initiation by phosphorylating eIF2$\alpha$. The rate at which CPC-3 is activated is also dependent on the length of the transcript (Lyu et al. 2021). In addition to rare codons, wobble base pairing may decrease translation speed. Wobble-decoding of the CGA codon in *S. cerevisiae* is so inefficient that it can halt translation (Letzring et al. 2010). Ribosomal profiling in *S. cerevisiae* found a slower rate of translation codons requiring wobble base pairing (Wang et al. 2017).

These codon-dependent elongation rates are typically attributed to tRNA abundances (Rak et al. 2018). Interestingly, codon-specific elongation rates were identified the bacterium *Mycoplasma pneumoniae* but the rates were not associated with tRNA abundance (Weber et al. 2023). This could be due to tRNA charging rates (Parker et al. 2020), differential tRNA modification (Lyu et al. 2020), competition between tRNAs (Vieira et al. 2016), and local codon usage biases. In *S. cerevisiae* the ordering of specific codons impacts translational speeds—sequences that optimize tRNA reuse are translated more quickly (Cannarozzi et al. 2010). It is difficult to disentangle all the ways in which codon usage impacts translational elongation, but new methodologies may allow us to directly measure elongation rates (Livingston et al. 2023) to tackle outstanding questions.

#### 6.2.1.3 Co-translational Protein Folding

The rate of elongation has been implicated in co-translational protein folding—structures that are difficult to fold or are intrinsically disordered may require more time to fold (Liu et al. 2021). Rare non-optimal codons, therefore, are proposed to slow translation to allow for proper folding. Analysis of 10 Ascomycota yeast genomes (*S. cerevisiae, Candida glabrata, Debaryomyces hansenii, Kluyveromyces lactis, Saccharomyces bayanus, Saccharomyces kluyveri, Saccharomyces mikatae, Saccharomyces paradoxus, S. pombe*, and *Yarrowia lipolytica*) found conserved patterns of non-optimal codons associated with secondary protein structures (Pechmann and Frydman 2013). In vitro assays in *Neurospora* showed that optimizing codon usage in the luciferase gene increased translation speed but altered the protein structure which reduced its activity (Yu et al. 2015). This has been expanded to in vivo studies showing the Frequency (FRQ) (Zhou et al. 2015) and Cross Pathway Control Protein 1 (CPC-1) (Lyu and Liu 2020) genes contain non-optimal codons that are critical to proper protein folding and function. In the yeast *Pichia pastoris* some genes exhibit a negative correlation between codon bias and protein structural disorder tendency (Xu et al. 2021). Experiments to codon-optimize 4 of these transcripts did not result in increased expression but had an effect on the secondary structure (Liu et al. 2022).

Across the tree of life, proteins with codon-sensitive protein structures have been identified (Moss et al. 2023). These specific findings, however, may not be generalizable across organisms, codons, or protein structures. Analysis of *S. cerevisiae* and *E. coli* did not find strong selection on codon usage acting across classes of secondary structures or organisms when accounting for biases in amino acids and gene expression (Cope and Gilchrist 2022). Other work found that conserved rare codon clusters are not primarily involved in co-translational folding but act to separate small protein structural motifs within domains (Chaney et al. 2017). This is complemented by similar work that found no enrichment in conserved rare codons at protein domain boundaries in *Escherichia coli* (Jacobs and Shakhnovich 2017). Efforts are underway to synthesize the large-scale evolutionary studies and experimental work to produce a "folding-code" for mRNA (Komar et al. 2023).

#### 6.2.1.4 Protein Modification and Transport

Decreasing the rate of protein elongation may also allow for extended detection of signals encoded in the protein sequence. Translation rate at the 5′ end of the mammalian γ-actin gene is decreased by rare codons leading to longer exposure of a lysine residue that increases arginylation and protein degradation (Jacobs and Shakhnovich 2017). Slowing translation can also promote co-translational protein translocation by increasing the time available for the signal recognition particle (SRP) to recognize the signal sequence at the N-terminus (Zhang and Shan 2012). In vivo experiments in *S. cerevisiae* found that a cluster of rare non-optimal codons downstream from the SRP-binding site slows elongation just as the site emerges from the ribosome (Pechmann et al. 2014). In *Aspergillus nidulans* a pair of rare non-optimal codons at the boundary between the N-terminus and the first transmembrane segment were identified as critical to biogenesis of the protein UreA at 37 °C (Sanguinetti et al. 2019). While their exact function is unclear, they may contribute to proper signal recognition functioning. By modulating translational dynamics, codon usage can also modulate other co-translational processes such as glycosylation (Sun et al. 2023) and chaperone recruitment (Rojano-Nisimura et al. 2020).

#### 6.2.1.5 Translation Accuracy

Wobbling at the third codon position allows species to exist without a complete set of cognate tRNAs. Codon usage is typically biased toward cognate tRNAs with codons requiring wobble-decoding being used at a lower rate. This has long been hypothesized to be a result of decreased translational accuracy when wobble base pairing is used (Akashi 1994). Evolutionary analysis in multiple species including *S. cerevisiae* suggested that codon usage is tuned to enhance translational accuracy to reduce protein misfolding (Drummond and Wilke 2008). Experiments in *E. coli* showed that some, but not all, preferred codons have a lower rate of mistranslation (Kramer and Farabaugh 2007). Analysis of proteome-wide probe of mistranslation in *E. coli* (Mordret et al. 2019) found that preferred synonymous codons are translated more accurately (Sun and Zhang 2022). Additional high throughput proteomics work is needed to confirm these findings in eukaryotes like yeasts.

Translational accuracy is also influenced by the speed of elongation. If translation occurs too quickly, accuracy levels can decrease leading to mistranslation (Zhang et al. 2015). In *S. cerevisiae,* mRNA secondary structure modulated by codon usage was found to tune translational speeds to balance efficiency and accuracy (Yang et al. 2014). Codon usage bias is optimized in a manner that accounts for the trade-off between accuracy and speed.

#### 6.2.1.6 Global Translational Dynamics

Codon usage can be tuned within a transcript to optimize translational initiation, elongation, accuracy, protein folding, and more. A single transcript, however, is only one of many expressed at any given time. The cell has a limited number of ribosomes, tRNAs, amino acids, and other translational components. Changing the codon usage of a single gene can alter the availability of these components. In *S. cerevisiae*, a whole cell simulation found that increasing the abundance of codons associated with abundant tRNAs in a highly expressed transgene impacted global protein production by altering the free ribosome pool (Shah et al. 2013). Due to this *trans* effect, codon usage is likely tuned to optimize translation across all transcripts. One genomic analysis study found that global *S. cerevisiae* codon usages are tuned to cognate tRNA concentrations resulting in uniform translation speed across transcripts (Qian et al. 2012). The global impacts of altering codon usage were experimentally tested in *E. coli* (Qian et al. 2012). In this study, they introduced 60 instances of a rare codon into highly expressed genes and found that this led to decreased translation efficiency in the other transcripts containing this rare codon. This effect could be mediated by altering the tRNA levels within the cell—suggesting these expression changes are directly related to tRNA abundances in the cell. Therefore, optimal codon usage is tuned to gene-level and global translational requirements.

## 6.2.2 Codon Optimization Metrics

At any position within a transcript, we can ask what codon will give us the optimal output? Given the wide variety of forces listed above, the optimal codon choice is context dependent. This makes optimal codon choice difficult to define. One of the earliest definitions of optimal codons refers to specific codons that induce "an 'optimal' energy of stabilization for the codon anticodon complex" (Grosjean et al. 1978). In this case, optimal codons are simply those that do not require wobble pairing for decoding. Since then, over 50 different metrics have been developed to identify the optimal codon choice considering the influence of codon usage on translation (for a review see Bahiri-Elitzur and Tuller (2021)). We will briefly review some of the most popular codon optimization metrics and the importance of contextualizing codon optimization.

### 6.2.2.1 Frequency of Optimal Codons (Fop)

In the original formulation of this metric in *E. coli*, each codon was assigned to the categories optimal or non-optimal based on observed tRNA abundances (Ikemura 1981). For each transcript, therefore, the frequency of optimal codon is simply the proportion of codons that are defined as optimal. Categorizing codons into these two categories is not always clear-cut and the original formulation in *S. cerevisiae* identified only 19 optimal codons across 13 amino acids (Ikemura 1985).

### 6.2.2.2 Codon Adaptation Index (CAI)

The CAI was created to account for each codon's relative degree of optimization (Sharp and Li 1987). The first step was to calculate the RSCU of codons found in highly expressed transcripts under the assumption that these genes experienced the highest level of selection for translational efficiency. These per-codon RSCU values were then used to calculate a mean RSCU value for a transcript ($CAI_{obs}$) and the maximum possible mean RSCU value assuming the same amino acid composition ($CAI_{max}$). The degree of codon optimality in a transcript is the ratio of the $CAI_{obs}$ to the $CAI_{max}$. The CAI was found to be highly correlated with gene expression levels in bacteria and yeast (Sharp and Li 1987).

Since its original formulation there have been many iterations of the CAI. Some methods calculate CAI without expression data by using ribosomal genes (under the assumption they are highly expressed) (Roymondal et al. 2009) or iterative processes (Carbone et al. 2003). Others have used global transcriptional levels in *S. cerevisiae* to parameterize the CAI (Jansen et al. 2003). Changes to the formulation of CAI address issues such as rare or nonexistent codons in the reference set (Xia 2007) and account for background (genome-wide) codon usage (Lee et al. 2010). All these formulations of CAI aim to predict expression level using codon frequency to identify which codons are highly adapted to the tRNA pool.

#### 6.2.2.3 tRNA Adaptation Index (tAI)

The tAI is another metric that aims to describe gene expression based on optimal codon usage based on cellular tRNA abundance and tRNA–codon interaction efficiency (dos Reis et al. 2003). The tRNA abundances can be either directly measured or inferred from genomic copy number. Optimal codons are those with efficient interactions with abundant tRNAs. The original formulation of tAI measured the efficiency of interactions in *S. cerevisiae.* The species-specific tAI (stAI) was introduced to allow for the calculation of tRNA efficiencies for individual species (Sabi and Tuller 2014). This led to the formulation of the genetic tAI (gtAI) which further improved the algorithm for computing the codon–tRNA interaction efficiencies by using a genetic algorithm (Anwar et al. 2023). Genetic algorithms are metaheuristic search algorithms that were inspired by the adage survival of the fittest (Katoch et al. 2021). Each iteration of the tAI has improved its correlation with empirical protein abundances (Anwar et al. 2023).

#### 6.2.2.4 Function Specific Codon Metrics

The metrics discussed above focus primarily on identifying and measuring codons that are optimal for high gene expression. Other metrics aim to identify optimal codons for specific functions. To specifically study the effect of codon usage on translation elongation, the index of translation elongation ($I_{TE}$) was developed (Xia 2015). In addition to elongation, initiation rates and ramps can be examined using codon frequencies using programs like ExtRamp (Miller et al. 2019). Codon usage metrics are also being incorporated into protein folding prediction models (Hallee et al. 2023).

#### 6.2.2.5 Complex Optimization Metrics

In the earliest heterologous protein expression experiments, codons were chosen to try and match the host organism. The codons in the human somatostatin gene were replaced with those most frequently used by the bacteriophage that infects *E. coli* which was the target (Itakura et al. 1977). We now know, however, that altering one or more codons to increase the optimization of translational elongation speed may have unintended consequences for the transcript's mRNA structure, protein function, or even global transcription rates. These unintended consequences are a primary driver in the common result that efforts to increase expression of heterologous proteins through codon optimization have varied success (Gustafsson et al. 2004). In experiments designed to compare codon optimized and wild-type sequences the results range from no change in expression (Alexeyev and Winkler 1999) to over one thousand fold increase (Zhou et al. 1999). These results vary across donor species, protein features, and host species. A study in *E. coli* tested multiple codon optimization methods to increase the expression of KRas4B and found that the tested optimization software produced highly varied sequences and each algorithm almost as likely to decrease recombinant yields as increase it (Ranaghan et al. 2021). This has prompted the development of methods to optimize codon usage considering the varied roles of codons.

If codon usage has been tuned over evolutionary time to optimize codon usage, it may be best to learn the rules directly from the organisms. Several methods have been developed to learn and replicate codon usage. First, are complex algorithms that leverage evolutionary or statistical methods to predict gene expression. The ribosome overhead costs Stochastic Evolutionary Model of Protein Production Rate (ROC SEMPPR) incorporates phylogenetic models of protein evolution to better account for mutational biases in the estimation of gene expression from codon usage data (Gilchrist et al. 2015). Maximum likelihood approaches, inspired by statistical physics, have also been deployed in SCUMBLE to better capture codon usage variation associated with expression (Kloster and Tang 2008). Other methods examine extensive transcriptomics data to learn patterns of codon usage that predict expression levels (Boel et al. 2016).

Second is "codon harmonization" (Angov et al. 2008). The goal of codon harmonization is to replace codons based on the relative optimization or frequency of those codons in the original gene or organism. For example, a codon with a low CAI in the original organism is then replaced with a synonymous codon of similar CAI. Experimental work in *E. coli* found that codon harmonization improved heterologous protein expression likely by preserving codon patterns that influence protein folding (Angov et al. 2011). The development of open-source codon harmonization methods, such as CHARMING (Wright et al. 2022), will allow researchers to deploy this method in fungi and other organisms.

Finally, machine learning has also been leveraged to produce better predictions of gene expression based on codon usage and gene sequences (Nikolados and Oyarzun 2023). The use of machine learning to optimize codons has shown promise in *E. coli* (Fu et al. 2020; Jain et al. 2023) and *S. cerevisiae* (Wang et al. 2022a). Some of these methods are unsupervised, such as ChimeraARS (Zur and Tuller 2015) and ChimeraUGEM (Diament et al. 2019), and they predict gene expression based on genomic data. There are also deep learning models that have been trained on gene expression data, including *S. cerevisiae,* which predicted up to 82% of the variation in mRNA abundance from DNA sequences (Zrimec et al. 2020). Interestingly, they found that models trained on codon frequencies alone could explain over half of mRNA abundance variation and that this information was highly overlapping with that produced from regulatory region models. These methods are highly adaptable to many fungal species. The Codon2Vec neural network was developed in fungi and can predict high and low gene expression when trained on the codon usage of highly expressed genes (Wint et al. 2022). This work examined 450 fungi from six of the eight fungal phyla with genomic and transcriptomic data. This included early diverging fungi from phyla such as Chytridiomycota, Blastocladiomycota, Zoopagomycota, and Mucoromycota. Other methods, like FUN-PROSE which was deployed in *S. cerevisiae, N. crassa*, and *Issatchenkia orientalis*, are trained directly on DNA sequences but are likely learning codon usage data (Nambiar et al. 2023). Machine learning methods will continue to help us interpret the complex patterns and variations of codon usage across the fungal kingdom.

#### 6.2.2.6 Measuring the Degree of Codon Optimization

The last set of codon metrics aim to understand how much the observed codon usage patterns are driven by selection or other forces. In species where codon usage is not heavily shaped by selection, it may be difficult to identify patterns of codon usage associated with gene expression. Methods have been developed that attempt to separate and quantify the effects of translational selection, mutational bias, and drift on genomes. One metric devised to identify the strength of translational selection across the genome is the S-value (dos Reis et al. 2004). The S-value is constructed to identify the proportion of codon bias variance that cannot be explained by mutational bias or random factors alone. Specifically, it measures the correlation between the tRNA adaptation index (tAI) and the combined effects of the selection effect of the codon usage of a gene and uncontrollable random factors. The S-value has been deployed in non-model fungi to estimate translational selection. High levels of translational selection were found across the majority of 332 budding yeast species (LaBella et al. 2019) and 22 *Aspergillus* species belonging to section *Flavi* (Hugaboom et al. 2023). An analysis of 40 fungal pathogens found higher levels of translational selection in fungi with higher host ranges, like *Botrytis cinerea* (host range of 556), than in species with small host ranges, like *Ophiocordyceps unilateralis* (host range of 1) (Badet et al. 2017).

Methods like ROC SEMPPR can also be used to identify the effects of mutation and selection to better predict gene expression and can therefore be used to generate codon-specific estimates of these forces (Gilchrist et al. 2015). The models of

ROC SEMPPR have recently been expanded to explicitly test for natural selection on codon usage in a package called AnaCoDa (Landerer et al. 2018). Analysis of 49 budding yeast genomes (a subset of those noted above) using AnaCoDa revealed that intergenomic variation caused both over- and under-estimation of the strength of selection in these fungi (Cope and Shah 2022).

Phylogenetic methods are also shedding light on the influence of selection on codon usage bias. Evolutionary analysis of 450 species across fungi found that adaptive radiation to tRNA pools, as opposed to drift, is the best explanation for the variation of interspecific codon bias (Wint et al. 2022). The level of translational selection on codon usage is also important to consider when conducting evolutionary analyses using nonsynonymous to synonymous codon substitution rates (*dN/dS*) (Mugal et al. 2014). Recent work in bacteria showed that selection at synonymous sites can underestimate levels of selection and implement a new model that includes both neutral and selected synonymous substitutions (Rahman et al. 2021). Implementing *dN/dS* methods that account for translational selection on codon usage will improve our ability to detect selection associated with important fungal phenotypes.

### 6.2.3 Transcription

Codon usage bias has been found to be correlated with both protein levels and mRNA transcript levels (Zhou et al. 2016; Gouy and Gautier 1982). We know that protein levels are impacted by both the rate of translation and the total number of transcripts available in the cell. Over evolutionary time, genes encoding highly expressed proteins have likely been subject to both translational selection on codon usage and selection for increased transcription. This begs the question, is the correlation between codon usage and mRNA due only to a shared selection for increased protein expression? Foundational work in *S. cerevisiae* (Coghlan and Wolfe 2000) and *E. coli* (dos Reis et al. 2003) found that codon usage was correlated with transcript levels in a manner that could not be fully accounted for by invoking only translational selection. This spurred investigation into the role of codon usage in mRNA stability, transcription, and more.

#### 6.2.3.1 mRNA Stability

Levels of mRNA within a cell are modulated both by producing new mRNAs through transcription and destroying existing mRNAs through degradation mechanisms (Dori-Bachash et al. 2011). RNA surveillance mechanisms, like No-Go, Nonsense-Mediated, and Non-Stop Decay, can target specific transcripts for degradation (Powers et al. 2020). In *S. cerevisiae* introduction of a run of rare codons in a transcript resulted in degradation of that transcript (Hu et al. 2009) in a manner not associated with the RNA surveillance mechanisms mentioned above (Shoemaker and Green 2012). Indeed, mRNA half-life correlates with optimal codon content through transcript-specific translation elongation rates determined by codon usage in *S. cerevisiae* (Presnyak et al. 2015) and *Schizosaccharomyces pombe* (Harigaya and Parker 2016). This genome-wide phenomenon contributes to the global differences in mRNA half-lives—*S. cerevisiae* mRNA half-lives range from less than a minute to over an hour (Coller and Parker 2004). The mechanism by which transcripts containing non-optimal codons are targeted for decay in *S. cerevisiae* involves Dhh1p (DDX6) which is a conserved and abundant DEAD-box protein (Radhakrishnan et al. 2016). The strength of the effect of codon usage on mRNA stability, however, may not be universal. In *Neurospora crassa* there is not a consistent impact of codon usage on mRNA stability (Zhou et al. 2016).

#### 6.2.3.2 Translation Independent Effects on Transcription

Genomes encode a variety of sequences that direct transcription levels. Some of this information overlaps with coding regions. In plants and animals, transcription factor recognition sites often reside in exons (Stergachis et al. 2013; Sullivan et al. 2014) and therefore constrain local codon usage. More broadly, codon usage correlated with transcription levels independently of mRNA decay in transcripts with GC-biased codon usage in *Neurospora* (Zhou et al. 2016) and mammals (Kudla et al. 2006). Follow-up studies in *Neurospora* showed that rare codons induced premature transcription termination and prevented the transcription of full-length mRNAs (Zhou et al. 2018). Moreover, RNA levels in the nucleus of *Neurospora* are highly correlated with codon usage biases (Zhou et al. 2018). A knockout screen identified 18 genes that disrupted the correlation between transcripts and codon usage (Zhao et al. 2021). All but one of these 18 genes have known functional roles in transcription including transcription factors and chromatin modifiers. These results support a role for codon usage in chromatin accessibility. For example, SET-2 and FKH1 both altered the correlation between codon usage and transcript levels. These genes, respectively, have the function of methylating H3K36 co-transcriptionally (Carrozza et al. 2005) and regulating Poll II carboxy-terminal domain phosphorylation state (Morillon et al. 2003). Additional work is needed to fully understand the role of codon usage bias in regulating chromatin accessibility.

#### 6.2.3.3 Other Effects of Codon Usage on mRNAs

Between the initiation of transcription and translation, mRNAs are processed and moved throughout the cell. Mature mRNA transcripts contain a polyA tail whose construction is triggered by a polyadenylation signal which is typically an A-rich sequence element (Graber et al. 1999). There may also be additional cryptic signals impacting polyadenylation encoded within exons. This is supported by the premature polyadenylation associated with rare codons in *Neurospora* (Zhou et al. 2018) and with altered GC content (and therefore codon bias) of heterologous genes in *Aspergillus oryzae* (Tanaka et al. 2014). Codon context, therefore, is likely important in determining proper polyadenylation.

Premature mRNAs also undergo splicing to remove introns. This process must be extremely precise to prevent frame-shift mutations. There are specific exonic sequences that enhance splicing (McCullough and Berget 1997). Comparative studies have identified selective pressure on codon usage to maintain splicing enhancers and avoid unintended splice sites (Eskesen et al. 2004; Willie and Majewski 2004). The biased codon usage at intron–exon junctions, however, may also be associated with other forces (Chamary and Hurst 2005). Splicing and mRNA transport may also be intertwined. Analysis of synonymous reporter gene splicing in human cells found that codon usage which increased GC content increased protein yield and increased movement of mRNA from the nucleus to the cytoplasm (Mordstein et al. 2020). This work suggests that codon usage is also tuned to ensure proper splicing and transport.

## 6.3 Codon Usage in Regulation

The transcriptome and proteome of a cell are not static and dynamically change throughout the cell cycle, with the circadian rhythm and in response to external stimuli. During these responses the codon usage landscape can change—codon frequencies in expressed genes change while tRNAs are differentially expressed, modified, and charged. This raises the question, does codon usage play an active and functional role in regulation?

### 6.3.1 Codon Usage during Stress Response

Proteins are the functional workhorses of the cell and protein synthesis requires amino acids. Under amino acid starvation cells undergo reprogramming to recycle amino acids already incorporated into proteins to maintain essential functions (He and Klionsky 2009). This requires the selective upregulation of degradation and autophagy pathways in the face of amino acid shortages. Computational and experimental work in *E. coli* found that amino acid starvation induces selective charging of rare tRNA isoacceptors (Elf et al. 2003; Dittmar et al. 2005). These rare tRNAs maintain the translation of rare codons during amino acid starvation leading to differential mRNA translation (Saikia et al. 2016). In *S. cerevisiae* and *N. crassa* loss of tRNA modifications triggered starvation responses suggesting these modifications play a key role in triggering stress responses (Bruch et al. 2020; Klassen et al. 2020; Lyu et al. 2020). The change in the tRNA pool and corresponding codon dynamics during amino acid starvation may also trigger other responses. A decrease in translation efficiency is hypothesized to lead to protein aggregation and a starvation response. Similarly, deletion of non-essential single copy tRNA genes in *S. cerevisiae* invoked proteotoxic stress responses (Bloom-Ackermann et al. 2014). Other stressors can also induce global decreases in protein synthesis (Spriggs et al. 2010) and are likely utilizing this codon-mediated mechanism.

Evidence for a stress-induced codon response is also found in *S. cerevisiae*. Under multiple stressors, the transcriptome of *S. cerevisiae* exhibits over-representation of low-efficiency codons (Gingold et al. 2012). This effect was robust to species-specific differences in tRNA and codon usage including in *S. pombe*. Altered transcriptional codon usage has also been found in response to heat stress in *Arabidopsis thaliana* (Chu and Wei 2019) and rice (*Oryza sativa*) (Tyagi et al. 2023). While the codon usage of the transcriptome changes, does the optimization of these codons change? If a rare codon is now over-expressed, this could place additional strain on the tRNA pool, further decreasing their optimization (Fig. 6.1c) Conversely, if over-expression of a rare codon is coordinated with an increase in availability of the corresponding tRNA, the codon could transition from non-optimal to optimal expression.

There is mixed evidence for changes in tRNA expression, charging, and modification in response to stressors beyond amino acid starvation. Experimental work in *S. cerevisiae* found that oxidative, osmotic, and temperature stress-induced changes in some tRNA molecules (Torrent et al. 2018). Conversely, other experiments found that the relative amount of each isotype is constant within cells (Kutter et al. 2011) and that differential modification of isodecoders is the primary mechanism by which tRNAs and codons coordinate a variety of stress responses (Gao et al. 2024; Advani and Ivanov 2019; Dedon and Begley 2014; Chan et al. 2012). While not discussed here, the degradation of tRNAs and the production of tRNA fragments are another potential regulatory mechanism (Saikia et al. 2012). Recent advances in tRNA sequencing, including sequencing of tRNA modifications (Lucas et al. 2023; Zhang et al. 2022), will continue to improve our ability to infer condition-specific codon optimization. For example, commercial tRNA modification detection was recently used to characterize a tRNA modification which, along with codon usage, tunes the secretion of unconventional protein effectors in *Magnaporthe oryzae* (Li et al. 2023a). Overall, current evidence supports an active role for tRNA and codons in stress response (Chan et al. 2018).

### 6.3.2 Codon Usage Across the Cell Cycle

The cell cycle is an essential component of cellular life. Mutations that over-stimulate cell proliferation (proto-oncogenes) or inhibit the cell cycle (tumor suppressors) can result in uncontrolled growth such as cancers and other invasive phenotypes (Aktipis et al. 2015). In fungi the cell cycle is critical to proper distribution of DNA to daughter cells and the vegetative growth of multicellular filamentous fungi (Dorter and Momany 2016). Each stage of the cell cycle requires the expression of stage-specific factors and progression from one stage to another is tightly controlled (Vermeulen et al. 2003). In *S. cerevisiae* approximately 14% of the genome is periodically transcribed across the cell cycle (Spellman et al. 1998).

Analysis of cell-cycle regulated genes in *S. pombe* and *S. cerevisiae* found these genes contained an excess of rare codons (Frenkel-Morgenstern et al. 2012). They hypothesized that the oscillation of these cell-cycle genes is enhanced by a corresponding oscillation in tRNA pool (differential charging, modification, or expression) which would increase translation only during the proper time. Indeed, they found that several tRNA synthetases increased during G2/M phase and this corresponded with increased expression of mRNAs with rare codons. There is also evidence for tRNA modifications playing a role in cell cycle regulation: 16 tRNA modifications in *S. cerevisiae* were found to oscillate during the cell cycle (Patil et al. 2012). This group further demonstrated that these modifications interact with specific codons to regulate ribonucleotide reductase (RNR). The RNR complex catalyzes the biosynthesis of deoxyribonucleotides (dNTPs) (Yao et al. 2003) and is therefore required during DNA repair (Hakansson et al. 2006) and cell division (Turner et al. 1968). Evidence for differences in tRNA expression during the cell cycle is mixed. Measurement of the tRNA pool in hundreds of cell lines found tRNAs induced in proliferating cells decode rare codons associated with proliferation-related genes (Gingold et al. 2014). Other studies find increased translation rates of proliferation-related genes with rare codons but do not attribute this to differential tRNA expression (Guimaraes et al. 2020; Frenkel-Morgenstern et al. 2012). As with stress response, new tRNA sequencing methodologies will enable us to build a clearer view of the dynamics of tRNAs during the cell cycle. This will allow us to pinpoint the mechanism by which rare codons are optimally translated during the cell cycle.

### 6.3.3 A Role for Non-optimal Codon Usage

Non-optimal codon usage is associated with lower expression levels, which for many proteins is the "optimal" expression level. Maintaining a precise protein level is critical for many processes including the circadian rhythm which is made up of many negative-feedback loops. In *N. crassa* (Zhou et al. 2013), the cyanobacterium *Synechococcus elongatus* (Xu et al. 2013) and *Drosophila melanogaster* (Fu et al. 2016) non-optimal codon usage in key circadian regulators is critical to proper functionality. When the optimal codon composition of the circadian gene FRQ is increased in *N. crassa,* the protein is degraded likely due to erroneous folding caused by increased ribosomal speed. In *S. elongatus* the non-optimal codon usage is associated with suppression of the circadian rhythm at cool temperatures where growth is not robust. Collectively this work supports both a specific and general role for non-optimal codon usage in circadian regulation.

### 6.3.4 Other Regulatory Processes

Working in concert, codon usage and tRNA abundances can drastically alter the expression of specific sets of genes. Organisms face many environmental changes, some stressful and some not, that elicit regulatory changes. Is codon usage leveraged to respond to a wide diversity of environmental changes beyond stress responses? One indication that this is possible is that co-expressed and interacting proteins have similar codon usage patterns in *S. cerevisiae* (Najafabadi and Salavati 2008). This has also been found in *E. coli* where regions of the protein–protein interaction network containing interacting proteins had similar codon usage biases (Dilucca et al. 2015). This principle has been expanded to a wide range of bacterial species (Dilucca et al. 2021). Codon usage is also shared across genes that have shared functions when examined across organisms including *S. cerevisiae* (Najafabadi et al. 2009). These findings suggest that codon usage has been differentially tuned across groups of co-expressed genes and that changes in the tRNA pool could selectively regulate these genes.

## 6.4 Codon Usage in Organelles and Viruses

Codon usage bias is not restricted to the nuclear genome. Codon usage bias has been observed in organelles (Bonitz et al. 1980) and viruses (Karlin et al. 1990). For a more comprehensive discussion, we refer the reader to additional resources on mitochondrial (Jia and Higgs 2008), chloroplast (Liu and Xue 2005) and viral codon usage (Belalov and Lukashev 2013; Simon et al. 2021). We will briefly review codon usage bias in mitochondria and viruses in the context of fungi.

### 6.4.1 Mitochondrial Codon Usage

Codon usage bias has been found in mitochondrial genomes of many organisms including fungi: *S. cerevisiae*, *Candida* species (Wang et al. 2023), *Aspergillus* section *Flavi* (Hugaboom et al. 2023), Amanita (Li et al. 2023b), Ganoderma (Wu et al. 2023), *Hyaloraphidium curvatum* (Forget et al. 2002), and more. There is some evidence that selection can act on codon usage in the mitochondrial genome despite the high mutational pressure (Jia and Higgs 2008; Hugaboom et al. 2023). This suggests that in some cases the codon usage of mitochondrial genes can be subjected to translational pressures induced by the tRNA pool. The tRNA pool of the mitochondria, however, is difficult to estimate. While human mitochondria contain the minimal set of required tRNAs to complete translation, other species lack a complete set. For example, the mitochondrial genome of the protozoa *Trypanosoma brucei* contains no tRNAs (Salinas et al. 2008) and the mitochondrial genomes of *Aspergillus* lack tRNAs decoding for multiple codons without substantial modification or extended wobble hypotheses (Hugaboom et al. 2023).

In species without a complete set of tRNAs, nuclear import of tRNAs is required (Schneider and Marechal-Drouard 2000). Even in species with sufficient mitochondrial tRNAs, import of nuclear produced tRNAs still occurs (Kamenski et al. 2007). In *S. cerevisiae* import of a cytosolic tRNA with a CUU anticodon is required for translation at high temperatures when the mitochondrial UUU anticodon loses a modification and the ability to decode AAG codons (Kamenski et al. 2007). Beyond model yeasts, there is little known about the diversity and rate of tRNA import into the mitochondria which in turn obscures calculations of codon optimization. Given the importance of mitochondria to the development of antifungal resistance (Shingu-Vazquez and Traven 2011), further research on tRNA abundance and codon usage in the mitochondria could prove critical.

### 6.4.2 Viral Codon Usage

Like in the case of mitochondria, viral codon usage is subject to high levels of mutational pressure as opposed to natural selection (Gu et al. 2004). Viruses, as obligate parasites, are also subject to selective pressures from the host that may influence codon usage (Simon et al. 2017). It may be advantageous for a virus to adapt to the codon usage of its host. Conversely, infection with viruses can alter tRNA abundances by impacting pol III (Fradkin et al. 1987) and potentially modulate the environment to the viral codon usage. Analysis of thousands of viral genomes found that phage codon usage is dependent on the host when the host is a prokaryote, fungi, or protist but not when the host is an animal or plant (Simon et al. 2021). Unlike other eukaryotic viruses, mycoviruses exhibited a positive correlation between host and viral codon usage (Simon et al. 2021). This finding was replicated in analysis of 98 RNA mycoviruses infecting phytopathogenic fungi (Wang et al. 2022b). The role of codon usage in viral infection has also been demonstrated by codon de-optimization experiments that result in virus attenuation (Groenke et al. 2020).

## 6.5 Applications of Codon Usage in Fungal Genomics

Patterns of codon usage are shaped by the numerous functional roles that codons play within the cell. Examining patterns of codon usage, therefore, can reveal insights into organismal evolution, ecology, and more. We will use fungal biology to showcase the many ways in which codon usage can be decoded into information about biology, ecology, and evolution.

### 6.5.1 Codon Usage in Phylogenetic Inference

Given the functional role of codon usage it is unsurprising that closely related species have similar codon usage bias patterns that reflect phylogenetic relationships (Miller et al. 2017). Analysis of 12,337 species across domains of life demonstrated that codon aversion and codon use are phylogenetically informative (Miller et al. 2020) and supports the same resolution of controversial nodes as recent high powered studies (Shen et al. 2017). Phylogenetic information in codon usage has also been mined to improve the extraction of metagenomic contigs for metagenome assembly (Yu et al. 2018). Similarly, codon usage information can be informative for identifying horizontally transferred genes (Waack et al. 2006). Although this method of detection may be biased as successful horizontal transfer events may be more frequent between organisms with similar codon usage and horizontal gene transfer events can include tRNA genes (Tuller 2013). Finally, codon substitution models are especially powerful for the study of protein-coding genes as they can measure selection at the nucleotide and amino acid level simultaneously (Gupta and Vadde 2023).

### 6.5.2 Reverse Ecology

One of the most exciting arenas in which codon usage is being leveraged is in reverse ecology. Reverse ecology is a framework in which adaptive phenotypes and the underlying genetic variation associated with those phenotypes are inferred directly from genomic information (Levy and Borenstein 2012). In the context of codon usage, adaptation that results in increased gene expression will also result in codon optimization of these genes. This hypothesis has been used to identify adaptive phenotypes in multiple fungal species. In generalist fungal pathogens, virulence genes are enriched in optimal codons suggesting that selection has optimized expression of these genes (Badet et al. 2017). In the budding yeasts of the subphylum Saccharomycotina, codon usage has been used to study galactose and xylose metabolism. In the galactose metabolism pathway (also known as the Leloir pathway), high growth rate on galactose media is correlated with the degree of codon optimization in the galactose metabolism (GAL) genes (LaBella et al. 2021). Optimal codon usage in the GAL pathway was also enriched in yeasts from dairy and human ecological niches. This suggests that the galactose metabolism pathway is adaptive in these two environments. A similar analysis of the xylose metabolism genes in yeasts found that codon optimization could differentiate between species with the xylose metabolism genes that do or do not grow on xylose (Nalabothu et al. 2023).

Future studies of codon usage may help to illuminate features of the hundreds of thousands of fungi that have yet to be cultured within the lab (Abarenkov et al. 2022). We are now able to sequence complete or near-complete metagenome-assembled genomes from environmental samples (Chivian et al. 2023). For example, the genomes of fungal symbionts from lichen have recently been generated from metagenomic sequencing (Tagirdzhanova et al. 2021). From these genomes, we can identify genes which have undergone translational selection for increased gene expression. In prokaryotes, codon usage data has also been used to estimate growth rate in common and uncultivated species (Weissman et al. 2021). Codon usage can also be used to infer community level metabolism as was demonstrated in a study of prokaryotic metagenomics where they identified enriched metabolic pathways associated with the microbiome (Fabijanić and Vlahoviček 2016).

## 6.6 Conclusions

Codon usage is an important genetic feature that shapes gene expression across the central dogma. Model fungi have been instrumental in identifying the specific mechanisms by which codons interact with tRNAs, ribosomes, and other partners (Xu et al. 2013; Radhakrishnan et al. 2016). There are, however, many open questions in the field of codon usage. There are currently contradictory reports of the relative fitness effects of codon usage in *S. cerevisiae*—there are reports that synonymous mutations are strongly non-neutral (Shen et al. 2022) and work that calls these findings into question (Kruglyak et al. 2023). While we have captured specific instances in which codon usage is critical to fungal gene function, such as in circadian rhythm proteins (Xu et al. 2013), the generalized and global impact of codon usage on fitness is still being explored.

There are also outstanding hypotheses to be tested in the temporal dynamics of codon usage bias. For example, while we see differences in codon usage between species, little work has been done to measure codon usage bias within or between populations. Numerous fungal species such as *A. fumigatus* (Rhodes et al. 2022), *S. cerevisiae* (Peter et al. 2018), and *C. albicans* (Hirakawa et al. 2015) now have population level sampling that will enable us to test for codon usage changes. Expanded genome sampling will enable studies of codon usage bias across multiple evolutionary timescales. On a cellular timescale, we lack a clear understanding of how dynamic the mRNA codon usage and tRNA landscapes are. There is some evidence that optimal codon identity changes in accordance with differences in tRNA abundance (Patil et al. 2012). New advancements in tRNA sequencing (Lucas et al. 2023) will enable more rigorous studies of codon dynamics within a cell.

Patterns of codon usage contain more information than simply the amino acid sequence of proteins. These patterns contain instructions for chromatin modification, transcription rates, mRNA localization, splicing, mRNA stability, translation, and protein folding. Studying fungal codon usage bias can help us understand the fundamental principles of codon usage and is an opportunity to generate ecological and biological hypotheses directly from genomes.

## References


Abarenkov K, Kristiansson E, Ryberg M, Nogal-Prata S, Gomez-Martinez D, Stuer-Patowsky K, Jansson T, Polme S, Ghobad-Nejhad M, Corcoll N, Scharn R, Sanchez-Garcia M, Khomich M, Wurzbacher C, Nilsson RH (2022) The curse of the uncultured fungus. MycoKeys 86:177–194. https://doi.org/10.3897/mycokeys.86.76053

Advani VM, Ivanov P (2019) Translational control under stress: reshaping the Translatome. BioEssays 41(5):e1900009. https://doi.org/10.1002/bies.201900009

Akashi H (1994) Synonymous codon usage in Drosophila melanogaster: natural selection and translational accuracy. Genetics 136(3):927–935. https://doi.org/10.1093/genetics/136.3.927

Aktipis CA, Boddy AM, Jansen G, Hibner U, Hochberg ME, Maley CC, Wilkinson GS (2015) Cancer across the tree of life: cooperation and cheating in multicellularity. Philos Trans R Soc Lond Ser B Biol Sci 370(1673):20140219. https://doi.org/10.1098/rstb.2014.0219

Alexaki A, Kames J, Holcomb DD, Athey J, Santana-Quintero LV, Lam PVN, Hamasaki-Katagiri N, Osipova E, Simonyan V, Bar H, Komar AA, Kimchi-Sarfaty C (2019) Codon and codon-pair usage tables (CoCoPUTs): facilitating genetic variation analyses and recombinant Gene Design. J Mol Biol 431(13):2434–2441. https://doi.org/10.1016/j.jmb.2019.04.021

Alexeyev MF, Winkler HH (1999) Gene synthesis, bacterial expression and purification of the *Rickettsia prowazekii* ATP/ADP translocase. Biochim Biophys Acta 1419(2):299–306. https://doi.org/10.1016/s0005-2736(99)00078-4

Alkatib S, Scharff LB, Rogalski M, Fleischmann TT, Matthes A, Seeger S, Schottler MA, Ruf S, Bock R (2012) The contributions of wobbling and superwobbling to the reading of the genetic code. PLoS Genet 8(11):e1003076. https://doi.org/10.1371/journal.pgen.1003076

Angov E, Hillier CJ, Kincaid RL, Lyon JA (2008) Heterologous protein expression is enhanced by harmonizing the codon usage frequencies of the target gene with those of the expression host. PLoS One 3(5):e2189. https://doi.org/10.1371/journal.pone.0002189

Angov E, Legler PM, Mease RM (2011) Adjustment of codon usage frequencies by codon harmonization improves protein expression and folding. Methods Mol Biol 705:1–13. https://doi.org/10.1007/978-1-61737-967-3_1

Anwar AM, Khodary SM, Ahmed EA, Osama A, Ezzeldin S, Tanios A, Mahgoub S, Magdeldin S (2023) gtAI: an improved species-specific tRNA adaptation index using the genetic algorithm. Front Mol Biosci 10:1218518. https://doi.org/10.3389/fmolb.2023.1218518

Badet T, Peyraud R, Mbengue M, Navaud O, Derbyshire M, Oliver RP, Barbacci A, Raffaele S (2017) Codon optimization underpins generalist parasitism in fungi. elife 6. https://doi.org/10.7554/eLife.22472

Bahiri-Elitzur S, Tuller T (2021) Codon-based indices for modeling gene expression and transcript evolution. Comput Struct Biotechnol J 19:2646–2663. https://doi.org/10.1016/j.csbj.2021.04.042

Barrington CL, Galindo G, Koch AL, Horton ER, Morrison EJ, Tisa S, Stasevich TJ, Rissland OS (2023) Synonymous codon usage regulates translation initiation. Cell Rep 42(12):113413. https://doi.org/10.1016/j.celrep.2023.113413

Belalov IS, Lukashev AN (2013) Causes and implications of codon usage bias in RNA viruses. PLoS One 8(2):e56642. https://doi.org/10.1371/journal.pone.0056642

Bengtsson BO (1986) Biased conversion as the primary function of recombination. Genet Res 47(1):77–80. https://doi.org/10.1017/s001667230002454x

Bennetzen JL, Hall BD (1982) Codon selection in yeast. J Biol Chem 257(6):3026–3031

Bloom-Ackermann Z, Navon S, Gingold H, Towers R, Pilpel Y, Dahan O (2014) A comprehensive tRNA deletion library unravels the genetic architecture of the tRNA pool. PLoS Genet 10(1):e1004084. https://doi.org/10.1371/journal.pgen.1004084

Boel G, Letso R, Neely H, Price WN, Wong KH, Su M, Luff J, Valecha M, Everett JK, Acton TB, Xiao R, Montelione GT, Aalberts DP, Hunt JF (2016) Codon influence on protein expression in *E. coli* correlates with mRNA levels. Nature 529(7586):358–363. https://doi.org/10.1038/nature16509

Bonitz SG, Berlani R, Coruzzi G, Li M, Macino G, Nobrega FG, Nobrega MP, Thalenfeld BE, Tzagoloff A (1980) Codon recognition rules in yeast mitochondria. Proc Natl Acad Sci USA 77(6):3167–3170. https://doi.org/10.1073/pnas.77.6.3167

Bowman BH (1959) Inheritance of protein variation in human serum. In: Bier M (ed) Electrophoresis: theory, methods, and applications, 2nd edn. Academic Press Inc., New York, N.Y, p 157

Bruch A, Laguna T, Butter F, Schaffrath R, Klassen R (2020) Misactivation of multiple starvation responses in yeast by loss of tRNA modifications. Nucleic Acids Res 48(13):7307–7320. https://doi.org/10.1093/nar/gkaa455

Bulmer M (1987) Coevolution of codon usage and transfer RNA abundance. Nature 325(6106):728–730. https://doi.org/10.1038/325728a0

Cannarozzi G, Schraudolph NN, Faty M, von Rohr P, Friberg MT, Roth AC, Gonnet P, Gonnet G, Barral Y (2010) A role for codon order in translation dynamics. Cell 141(2):355–367. https://doi.org/10.1016/j.cell.2010.02.036

Carbone A, Zinovyev A, Kepes F (2003) Codon adaptation index as a measure of dominating codon bias. Bioinformatics 19(16):2005–2015. https://doi.org/10.1093/bioinformatics/btg272

Carrozza MJ, Li B, Florens L, Suganuma T, Swanson SK, Lee KK, Shia WJ, Anderson S, Yates J, Washburn MP, Workman JL (2005) Histone H3 methylation by Set2 directs deacetylation of coding regions by Rpd3S to suppress spurious intragenic transcription. Cell 123(4):581–592. https://doi.org/10.1016/j.cell.2005.10.023

Chamary JV, Hurst LD (2005) Biased codon usage near intron-exon junctions: selection on splicing enhancers, splice-site recognition or something else? Trends Genet 21(5):256–259. https://doi.org/10.1016/j.tig.2005.03.001

Chan C, Pham P, Dedon PC, Begley TJ (2018) Lifestyle modifications: coordinating the tRNA epitranscriptome with codon bias to adapt translation during stress responses. Genome Biol 19(1):228. https://doi.org/10.1186/s13059-018-1611-1

Chan CT, Pang YL, Deng W, Babu IR, Dyavaiah M, Begley TJ, Dedon PC (2012) Reprogramming of tRNA modifications controls the oxidative stress response by codon-biased translation of proteins. Nat Commun 3:937. https://doi.org/10.1038/ncomms1938

Chaney JL, Steele A, Carmichael R, Rodriguez A, Specht AT, Ngo K, Li J, Emrich S, Clark PL (2017) Widespread position-specific conservation of synonymous rare codons within coding sequences. PLoS Comput Biol 13(5):e1005531. https://doi.org/10.1371/journal.pcbi.1005531

Chivian D, Jungbluth SP, Dehal PS, Wood-Charlson EM, Canon RS, Allen BH, Clark MM, Gu T, Land ML, Price GA, Riehl WJ, Sneddon MW, Sutormin R, Zhang Q, Cottingham RW, Henry CS, Arkin AP (2023) Metagenome-assembled genome extraction and analysis from microbiomes using KBase. Nat Protoc 18(1):208–238. https://doi.org/10.1038/s41596-022-00747-x

Choudhuri S, Sau K (2023) CodonU: a python package for codon usage analysis. IEEE/ACM Trans Comput Biol Bioinform 21:36. https://doi.org/10.1109/TCBB.2023.3335823

Chu D, Kazana E, Bellanger N, Singh T, Tuite MF, von der Haar T (2014) Translation elongation can control translation initiation on eukaryotic mRNAs. EMBO J 33(1):21–34. https://doi.org/10.1002/embj.201385651

Chu D, Wei L (2019) Characterizing the heat response of Arabidopsis thaliana from the perspective of codon usage bias and translational regulation. J Plant Physiol 240:153012. https://doi.org/10.1016/j.jplph.2019.153012

Coghlan A, Wolfe KH (2000) Relationship of codon bias to mRNA concentration and protein length in Saccharomyces cerevisiae. Yeast 16(12):1131–1145. https://doi.org/10.1002/1097-0061(20000915)16:12<1131::AID-YEA609>3.0.CO;2-F

Coller J, Parker R (2004) Eukaryotic mRNA decapping. Annu Rev Biochem 73:861–890. https://doi.org/10.1146/annurev.biochem.73.011303.074032

Cope AL, Gilchrist MA (2022) Quantifying shifts in natural selection on codon usage between protein regions: a population genetics approach. BMC Genomics 23(1):408. https://doi.org/10.1186/s12864-022-08635-0

Cope AL, Shah P (2022) Intragenomic variation in non-adaptive nucleotide biases causes underestimation of selection on synonymous codon usage. PLoS Genet 18(6):e1010256. https://doi.org/10.1371/journal.pgen.1010256

Crick F (1955) On degenerate templates and the adaptor hypothesis: a note for the RNA tie club. Francis Harry Compton Crick Papers

Dana A, Tuller T (2014) The effect of tRNA levels on decoding times of mRNA codons. Nucleic Acids Res 42(14):9171–9181. https://doi.org/10.1093/nar/gku646

Dedon PC, Begley TJ (2014) A system of RNA modifications and biased codon use controls cellular stress response at the level of translation. Chem Res Toxicol 27(3):330–337. https://doi.org/10.1021/tx400438d

Diament A, Weiner I, Shahar N, Landman S, Feldman Y, Atar S, Avitan M, Schweitzer S, Yacoby I, Tuller T (2019) ChimeraUGEM: unsupervised gene expression modeling in any given organism. Bioinformatics 35(18):3365–3371. https://doi.org/10.1093/bioinformatics/btz080

Dilucca M, Cimini G, Forcelloni S, Giansanti A (2021) Co-evolution between codon usage and protein-protein interaction in bacteria. Gene 778:145475. https://doi.org/10.1016/j.gene.2021.145475

Dilucca M, Cimini G, Semmoloni A, Deiana A, Giansanti A (2015) Codon bias patterns of E. coli's interacting proteins. PLoS One 10(11):e0142127. https://doi.org/10.1371/journal.pone.0142127

Dittmar KA, Sorensen MA, Elf J, Ehrenberg M, Pan T (2005) Selective charging of tRNA isoacceptors induced by amino-acid starvation. EMBO Rep 6(2):151–157. https://doi.org/10.1038/sj.embor.7400341

Dori-Bachash M, Shema E, Tirosh I (2011) Coupled evolution of transcription and mRNA degradation. PLoS Biol 9(7):e1001106. https://doi.org/10.1371/journal.pbio.1001106

Dorter I, Momany M (2016) Fungal Cell Cycle: A Unicellular versus Multicellular Comparison. Microbiol Spectr 4(6). https://doi.org/10.1128/microbiolspec.FUNK-0025-2016

dos Reis M, Savva R, Wernisch L (2004) Solving the riddle of codon usage preferences: a test for translational selection. Nucleic Acids Res 32(17):5036–5044. https://doi.org/10.1093/nar/gkh834

dos Reis M, Wernisch L, Savva R (2003) Unexpected correlations between gene expression and codon usage bias from microarray data for the whole Escherichia coli K-12 genome. Nucleic Acids Res 31(23):6976–6985. https://doi.org/10.1093/nar/gkg897

Drummond DA, Wilke CO (2008) Mistranslation-induced protein misfolding as a dominant constraint on coding-sequence evolution. Cell 134(2):341–352. https://doi.org/10.1016/j.cell.2008.05.042

Elf J, Nilsson D, Tenson T, Ehrenberg M (2003) Selective charging of tRNA isoacceptors explains patterns of codon usage. Science 300(5626):1718–1722. https://doi.org/10.1126/science.1083811

Eskesen ST, Eskesen FN, Ruvinsky A (2004) Natural selection affects frequencies of AG and GT dinucleotides at the 5′ and 3′ ends of exons. Genetics 167(1):543–550. https://doi.org/10.1534/genetics.167.1.543

Fabijanić M, Vlahoviček K (2016) Big data, evolution, and metagenomes: predicting disease from gut microbiota codon usage profiles. Data mining techniques for the life sciences:509–531

Forget L, Ustinova J, Wang Z, Huss VA, Lang BF (2002) Hyaloraphidium curvatum: a linear mitochondrial genome, tRNA editing, and an evolutionary link to lower fungi. Mol Biol Evol 19(3):310–319. https://doi.org/10.1093/oxfordjournals.molbev.a004084

Fradkin LG, Yoshinaga SK, Berk AJ, Dasgupta A (1987) Inhibition of host cell RNA polymerase III-mediated transcription by poliovirus: inactivation of specific transcription factors. Mol Cell Biol 7(11):3880–3887. https://doi.org/10.1128/mcb.7.11.3880-3887.1987

Frenkel-Morgenstern M, Danon T, Christian T, Igarashi T, Cohen L, Hou YM, Jensen LJ (2012) Genes adopt non-optimal codon usage to generate cell cycle-dependent oscillations in protein levels. Mol Syst Biol 8:572. https://doi.org/10.1038/msb.2012.3

Fu H, Liang Y, Zhong X, Pan Z, Huang L, Zhang H, Xu Y, Zhou W, Liu Z (2020) Codon optimization with deep learning to enhance protein expression. Sci Rep 10(1):17617. https://doi.org/10 1038/s41598-020-74091-z

Fu J, Murphy KA, Zhou M, Li YH, Lam VH, Tabuloc CA, Chiu JC, Liu Y (2016) Codon usage affects the structure and function of the Drosophila circadian clock protein PERIOD. Genes Dev 30(15):1761–1775. https://doi.org/10.1101/gad.281030.116

Gao L, Behrens A, Rodschinka G, Forcelloni S Wani S, Strasser K, Nedialkova DD (2024) Selective gene expression maintains human tRNA anticodon pools during differentiation. Nat Cell Biol 26(1):100–112. https://doi.org/10.1038/s41556-023-01317-3

Gardin J, Yeasmin R, Yurovsky A, Cai Y, Skiena S, Futcher B (2014) Measurement of average decoding rates of the 61 sense codons in vivo. elife 3. https://doi.org/10.7554/eLife.03735

Gilchrist MA, Chen WC, Shah P, Landerer CL, Zaretzki R (2015) Estimating gene expression and codon-specific translational efficiencies, mutation biases, and selection coefficients from genomic data alone. Genome Biol Evol 7(6):1559–1579. https://doi.org/10.1093/gbe/evv087

Gingold H, Dahan O, Pilpel Y (2012) Dynamic changes in translational efficiency are deduced from codon usage of the transcriptome. Nucleic Acids Res 40(20):10053–10063. https://doi.org/10.1093/nar/gks772

Gingold H, Tehler D, Christoffersen NR, Nielsen MM, Asmar F, Kooistra SM, Christophersen NS, Christensen LL, Borre M, Sorensen KD, Andersen LD, Andersen CL, Hulleman E, Wurdinger T, Ralfkiaer E, Helin K, Gronbaek K, Orntoft T, Waszak SM, Dahan O, Pedersen JS, Lund AH, Pilpel Y (2014) A dual program for translation regulation in cellular proliferation and differentiation. Cell 158(6):1281–1292. https://doi.org/10.1016/j.cell.2014.08.011

Goel NS, Rao GS, Ycas M, Bremermann HJ King L (1972) A method for calculating codon frequencies in DNA. J Theor Biol 35(3):399–457. https://doi.org/10.1016/0022-5193(72)90143-9

Gouy M, Gautier C (1982) Codon usage in bacteria: correlation with gene expressivity. Nucleic Acids Res 10(22):7055–7074. https://doi.org/10.1093/nar/10.22.7055

Graber JH, Cantor CR, Mohr SC, Smith TF (1999) Genomic detection of new yeast pre-mRNA 3′-end-processing signals. Nucleic Acids Res 27(3):888–894. https://doi.org/10.1093/nar/27.3.888

Grantham R (1978) Viral, prokaryote and eukaryote genes contrasted by mRNA sequence indexes. FEBS Lett 95(1):1–11. https://doi.org/10.1016/0014-5793(78)80041-6

Groenewald M, Hittinger C, Bensch K, Opulente D, Shen X-X, Li Y, Liu C, LaBella A, Zhou X, Limtong S (2023) A genome-informed higher rank classification of the biotechnologically important fungal subphylum Saccharomycotina. Studies. in Mycology

Groenke N, Trimpert J, Merz S, Conradie AM, Wyler E, Zhang H, Hazapis OG, Rausch S, Landthaler M, Osterrieder N, Kunec D (2020) Mechanism of virus attenuation by codon pair Deoptimization. Cell Rep 31(4):107586. https://doi.org/10.1016/j.celrep.2020.107586

Grosjean H, Sankoff D, Jou WM, Fiers W, Cedergren RJ (1978) Bacteriophage MS2 RNA: a correlation between the stability of the codon: anticodon interaction and the choice of code words. J Mol Evol 12(2):113–119. https://doi.org/10.1007/BF01733262

Gu W, Zhou T, Ma J, Sun X, Lu Z (2004) Analysis of synonymous codon usage in SARS coronavirus and other viruses in the Nidovirales. Virus Res 101(2):155–161. https://doi.org/10.1016/j.virusres.2004.01.006

Guimaraes JC, Mittal N, Gnann A, Jedlinski D, Riba A, Buczak K, Schmidt A, Zavolan M (2020) A rare codon-based translational program of cell proliferation. Genome Biol 21(1):44. https://doi.org/10.1186/s13059-020-1943-5

Gupta MK, Vadde R (2023) Next-generation development and application of codon model in evolution. Front Genet 14:1091575. https://doi.org/10.3389/fgene.2023.1091575

Gustafsson C, Govindarajan S, Minshull J (2004) Codon bias and heterologous protein expression. Trends Biotechnol 22(7):346–353. https://doi.org/10.1016/j.tibtech.2004.04.006

Hakansson P, Hofer A, Thelander L (2006) Regulation of mammalian ribonucleotide reduction and dNTP pools after DNA damage and in resting cells. J Biol Chem 281(12):7834–7841. https://doi.org/10.1074/jbc.M512894200

Hallee L, Rafailidis N, Gleghorn JP (2023) cdsBERT–extending protein language models with codon awareness. bioRxiv. https://doi.org/10.1101/2023.09.15.558027

Harigaya Y, Parker R (2016) Analysis of the association between codon optimality and mRNA stability in Schizosaccharomyces pombe. BMC Genomics 17(1):895. https://doi.org/10.1186/s12864-016-3237-6

He C, Klionsky DJ (2009) Regulation mechanisms and signaling pathways of autophagy. Annu Rev Genet 43:67–93. https://doi.org/10.1146/annurev-genet-102808-114910

Hirakawa MP, Martinez DA, Sakthikumar S, Anderson MZ, Berlin A, Gujja S, Zeng Q, Zisson E, Wang JM, Greenberg JM, Berman J, Bennett RJ, Cuomo CA (2015) Genetic and phenotypic intra-species variation in Candida albicans. Genome Res 25(3):413–425. https://doi.org/10.1101/gr.174623.114

Hu W, Sweet TJ, Chamnongpol S, Baker KE, Coller J (2009) Co-translational mRNA decay in Saccharomyces cerevisiae. Nature 461(7261):225–229. https://doi.org/10.1038/nature08265

Hugaboom M, Hatmaker EA, LaBella AL, Rokas A (2023) Evolution and codon usage bias of mitochondrial and nuclear genomes in Aspergillus section Flavi. G3 (Bethesda) 13(1). https://doi.org/10.1093/g3journal/jkac285

Ikemura T (1981) Correlation between the abundance of *Escherichia coli* transfer RNAs and the occurrence of the respective codons in its protein genes: a proposal for a synonymous codon choice that is optimal for the *E. coli* translational system. J Mol Biol 151(3):389–409. https://doi.org/10.1016/0022-2836(81)90003-6

Ikemura T (1982) Correlation between the abundance of yeast transfer RNAs and the occurrence of the respective codons in protein genes. Differences in synonymous codon choice patterns of yeast and Escherichia coli with reference to the abundance of isoaccepting transfer RNAs. J Mol Biol 158(4):573–597. https://doi.org/10.1016/0022-2836(82)90250-9

Ikemura T (1985) Codon usage and tRNA content in unicellular and multicellular organisms. Mol Biol Evol 2(1):13–34. https://doi.org/10.1093/oxfordjournals.molbev.a040335

Ingolia NT, Lareau LF, Weissman JS (2011) Ribosome profiling of mouse embryonic stem cells reveals the complexity and dynamics of mammalian proteomes. Cell 147(4):789–802. https://doi.org/10.1016/j.cell.2011.10.002

Iriarte A, Lamolle G, Musto H (2021) Codon usage bias: An endless tale. J Mol Evol 89(9–10):589–593. https://doi.org/10.1007/s00239-021-10027-z

Itakura K, Hirose T, Crea R, Riggs AD, Heyneker HL, Bolivar F, Boyer HW (1977) Expression in Escherichia coli of a chemically synthesized gene for the hormone somatostatin. Science 198(4321):1056–1063. https://doi.org/10.1126/science.412251

Jacobs WM, Shakhnovich EI (2017) Evidence of evolutionary selection for cotranslational folding. Proc Natl Acad Sci USA 114(43):11434–11439. https://doi.org/10.1073/pnas.1705772114

Jain R, Jain A, Mauro E, LeShane K, Densmore D (2023) ICOR: improving codon optimization with recurrent neural networks. Bmc Bioinformatics 24(1):132. https://doi.org/10.1186/s12859-023-05246-8

Jansen R, Bussemaker HJ, Gerstein M (2003) Revisiting the codon adaptation index from a whole-genome perspective: analyzing the relationship between gene expression and codon occurrence in yeast using a variety of models. Nucleic Acids Res 31(8):2242–2251. https://doi.org/10.1093/nar/gkg306

Jia W, Higgs PG (2008) Codon usage in mitochondrial genomes: distinguishing context-dependent mutation from translational selection. Mol Biol Evol 25(2):339–351. https://doi.org/10.1093/molbev/msm259

Kamenski P, Kolesnikova O, Jubenot V, Entelis N, Krasheninnikov IA, Martin RP, Tarassov I (2007) Evidence for an adaptation mechanism of mitochondrial translation via tRNA import from the cytosol. Mol Cell 26(5):625–637. https://doi.org/10.1016/j.molcel.2007.04.019

Karki S, Castillo K, Ding Z, Kerr O, Lamb TM, Wu C, Sachs MS, Bell-Pedersen D (2020) Circadian clock control of eIF2alpha phosphorylation is necessary for rhythmic translation initiation. Proc Natl Acad Sci USA 117(20):10935–10945. https://doi.org/10.1073/pnas.1918459117

Karlin S, Blaisdell BE, Schachtel GA (1990) Contrasts in codon usage of latent versus productive genes of Epstein-Barr virus: data and hypotheses. J Virol 64(9):4264–4273. https://doi.org/10.1128/JVI.64.9.4264-4273.1990

Karlin S, Bucher P (1992) Correlation analysis of amino acid usage in protein classes. Proc Natl Acad Sci USA 89(24):12165–12169. https://doi.org/10.1073/pnas.89.24.12165

Katoch S, Chauhan SS, Kumar V (2021) A review on genetic algorithm: past, present, and future. Multimed Tools Appl 80(5):8091–8126. https://doi.org/10.1007/s11042-020-10139-6

Kimura M (1977) Preponderance of synonymous changes as evidence for the neutral theory of molecular evolution. Nature 267(5608):275–276. https://doi.org/10.1038/267275a0

Klassen R, Bruch A, Schaffrath R (2020) Induction of protein aggregation and starvation response by tRNA modification defects. Curr Genet 66(6):1053–1057. https://doi.org/10.1007/s00294-020-01103-w

Kloster M, Tang C (2008) SCUMBLE: a method for systematic and accurate detection of codon usage bias by maximum likelihood estimation. Nucleic Acids Res 36(11):3819–3827. https://doi.org/10.1093/nar/gkn288

Kollmar M, Muhlhausen S (2017) Nuclear codon reassignments in the genomics era and mechanisms behind their evolution. BioEssays 39(5). https://doi.org/10.1002/bies.201600221

Komar AA (2016) The Yin and Yang of codon usage. Hum Mol Genet 25(R2):R77–R85. https://doi.org/10.1093/hmg/ddw207

Komar AA, Samatova E, Rodnina MV (2023) Translation rates and protein folding. J Mol Biol 436:168384. https://doi.org/10.1016/j.jmb.2023.168384

Kramer EB, Farabaugh PJ (2007) The frequency of translational misreading errors in *E. coli* is largely determined by tRNA competition. RNA 13(1):87–96. https://doi.org/10.1261/rna.294907

Krassowski T, Coughlan AY, Shen XX, Zhou X, Kominek J, Opulente DA, Riley R, Grigoriev IV, Maheshwari N, Shields DC, Kurtzman CP, Hittinger CT, Rokas A, Wolfe KH (2018) Evolutionary instability of CUG-Leu in the genetic code of budding yeasts. Nat Commun 9(1):1887. https://doi.org/10.1038/s41467-018-04374-7

Kruglyak L, Beyer A, Bloom JS, Grossbach J, Lieberman TD, Mancuso CP, Rich MS, Sherlock G, Kaplan CD (2023) Insufficient evidence for non-neutrality of synonymous mutations. Nature 616(7957):E8–E9. https://doi.org/10.1038/s41586-023-05865-4

Kudla G, Lipinski L, Caffin F, Helwak A, Zylicz M (2006) High guanine and cytosine content increases mRNA levels in mammalian cells. PLoS Biol 4(6):e180. https://doi.org/10.1371/journal.pbio.0040180

Kutter C, Brown GD, Goncalves A, Wilson MD, Watt S, Brazma A, White RJ, Odom DT (2011) Pol III binding in six mammals shows conservation among amino acid isotypes despite divergence among tRNA genes. Nat Genet 43(10):948–955. https://doi.org/10.1038/ng.906

LaBella AL, Opulente DA, Steenwyk JL, Hittinger CT, Rokas A (2019) Variation and selection on codon usage bias across an entire subphylum. PLoS Genet 15(7):e1008304. https://doi.org/10.1371/journal.pgen.1008304

LaBella AL, Opulente DA, Steenwyk JL, Hittinger CT, Rokas A (2021) Signatures of optimal codon usage in metabolic genes inform budding yeast ecology. PLoS Biol 19(4):e3001185. https://doi.org/10.1371/journal.pbio.3001185

Landerer C, Cope A, Zaretzki R, Gilchrist MA (2018) AnaCoDa: analyzing codon data with Bayesian mixture models. Bioinformatics 34(14):2496–2498. https://doi.org/10.1093/bioinformatics/bty138

Lee S, Weon S, Lee S, Kang C (2010) Relative codon adaptation index, a sensitive measure of codon usage bias. Evol Bioinformatics Online 6:47–55. https://doi.org/10.4137/ebo.s4608

Letzring DP, Dean KM, Grayhack EJ (2010) Control of translation efficiency in yeast by codon–anticodon interactions. RNA 16(12):2516–2528. https://doi.org/10.1261/rna.2411710

Levy R, Borenstein E (2012) Reverse ecology: from systems to environments and back. Adv Exp Med Biol 751:329–345. https://doi.org/10.1007/978-1-4614-3567-9_15

Li G, Dulal N, Gong Z, Wilson RA (2023a) Unconventional secretion of Magnaporthe oryzae effectors in rice cells is regulated by tRNA modification and codon usage control. Nat Microbiol 8(9):1706–1716. https://doi.org/10.1038/s41564-023-01443-6

Li Q, Luo Y, Sha A, Xiao W, Xiong Z, Chen X, He J, Peng L, Zou L (2023b) Analysis of synonymous codon usage patterns in mitochondrial genomes of nine Amanita species. Front Microbiol 14:1134228. https://doi.org/10.3389/fmicb.2023.1134228

Liu H, Huang J, Sun X, Li J, Hu Y, Yu L, Liti G, Tian D, Hurst LD, Yang S (2018) Tetrad analysis in plants and fungi finds large differences in gene conversion rates but no GC bias. Nat Ecol Evol 2(1):164–173. https://doi.org/10.1038/s41559-017-0372-7

Liu K, Ouyang Y, Lin R, Ge C, Zhou M (2022) Strong negative correlation between codon usage bias and protein structural disorder impedes protein expression after codon optimization. J Biotechnol 343:15–24. https://doi.org/10.1016/j.jbiotec.2021.11.001

Liu Q, Xue Q (2005) Comparative studies on codon usage pattern of chloroplasts and their host nuclear genes in four plant species. J Genet 84(1):55–62. https://doi.org/10.1007/BF02715890

Liu Y, Yang Q, Zhao F (2021) Synonymous but not silent: the codon usage code for gene expression and protein folding. Annu Rev Biochem 90:375–401. https://doi.org/10.1146/annurev-biochem-071320-112701

Livingston NM, Kwon J, Valera O, Saba JA, Sinha NK, Reddy P, Nelson B, Wolfe C, Ha T, Green R, Liu J, Wu B (2023) Bursting translation on single mRNAs in live cells. Mol Cell 83(13):2276–2289 e2211. https://doi.org/10.1016/j.molcel.2023.05.019

Lucas MC, Pryszcz LP, Medina R, Milenkovic I, Camacho N, Marchand V, Motorin Y, Ribas de Pouplana L, Novoa EM (2023) Quantitative analysis of tRNA abundance and modifications by nanopore RNA sequencing. Nat Biotechnol 42:72. https://doi.org/10.1038/s41587-023-01743-6

Lyu X, Liu Y (2020) Nonoptimal codon usage is critical for protein structure and function of the master general amino acid control regulator CPC-1. MBio 11(5). https://doi.org/10.1128/mBio.02605-20

Lyu X, Yang Q, Li L, Dang Y, Zhou Z, Chen S, Liu Y (2020) Adaptation of codon usage to tRNA I34 modification controls translation kinetics and proteome landscape. PLoS Genet 16(6):e1008836. https://doi.org/10.1371/journal.pgen.1008836

Lyu X, Yang Q, Zhao F, Liu Y (2021) Codon usage and protein length-dependent feedback from translation elongation regulates translation initiation and elongation speed. Nucleic Acids Res 49(16):9404–9423. https://doi.org/10.1093/nar/gkab729

Marck C, Grosjean H (2002) tRNomics: analysis of tRNA genes from 50 genomes of Eukarya, Archaea, and Bacteria reveals anticodon-sparing strategies and domain-specific features. RNA 8(10):1189–1232. https://doi.org/10.1017/s1355838202022021

McCullough AJ, Berget SM (1997) G triplets located throughout a class of small vertebrate introns enforce intron borders and regulate splice site selection. Mol Cell Biol 17(8):4562–4571. https://doi.org/10.1128/MCB.17.8.4562

Miller JB, Brase LR, Ridge PG (2019) ExtRamp: a novel algorithm for extracting the ramp sequence based on the tRNA adaptation index or relative codon adaptiveness. Nucleic Acids Res 47(3):1123–1131. https://doi.org/10.1093/nar/gky1193

Miller JB, Hippen AA, Belyeu JR, Whiting MF, Ridge PG (2017) Missing something? Codon aversion as a new character system in phylogenetics. Cladistics 33(5):545–556. https://doi.org/10.1111/cla.12183

Miller JB, McKinnon LM, Whiting MF, Ridge PG (2020) Codon use and aversion is largely phylogenetically conserved across the tree of life. Mol Phylogenet Evol 144:106697. https://doi.org/10.1016/j.ympev.2019.106697

Mordret E, Dahan O, Asraf O, Rak R, Yehonadav A, Barnabas GD, Cox J, Geiger T, Lindner AB, Pilpel Y (2019) Systematic detection of amino acid substitutions in proteomes reveals mechanistic basis of ribosome errors and selection for translation Fidelity. Mol Cell 75(3):427–441 e425. https://doi.org/10.1016/j.molcel.2019.06.041

Mordstein C, Savisaar R, Young RS, Bazile J, Talmane L, Luft J, Liss M, Taylor MS, Hurst LD, Kudla G (2020) Codon usage and splicing jointly influence mRNA localization. Cell Syst 10(4):351–362 e358. https://doi.org/10.1016/j.cels.2020.03.001

Morillon A, O'Sullivan J, Azad A, Proudfoot N, Mellor J (2003) Regulation of elongating RNA polymerase II by forkhead transcription factors in yeast. Science 300(5618):492–495. https://doi.org/10.1126/science.1081379

Moss MJ, Chamness LM, Clark PL (2023) The effects of codon usage on protein structure and folding. Annu Rev Biophys 53:87. https://doi.org/10.1146/annurev-biophys-030722-020555

Mugal CF, Wolf JB, Kaj I (2014) Why time matters: codon evolution and the temporal dynamics of dN/dS. Mol Biol Evol 31(1):212–231. https://doi.org/10.1093/molbev/mst192

Muhlhausen S, Findeisen P, Plessmann U, Urlaub H, Kollmar M (2016) A novel nuclear genetic code alteration in yeasts and the evolution of codon reassignment in eukaryotes. Genome Res 26(7):945–955. https://doi.org/10.1101/gr.200931.115

Muhlhausen S, Kollmar M (2014) Molecular phylogeny of sequenced Saccharomycetes reveals polyphyly of the alternative yeast codon usage. Genome Biol Evol 6(12):3222–3237. https://doi.org/10.1093/gbe/evu152

Muhlhausen S, Schmitt HD, Pan KT, Plessmann U, Urlaub H, Hurst LD, Kollmar M (2018) Endogenous stochastic decoding of the CUG codon by competing Ser- and Leu-tRNAs in Ascoidea asiatica. Curr Biol 28(13):2046–2057. e2045. https://doi.org/10.1016/j.cub.2018.04.085

Najafabadi HS, Goodarzi H, Salavati R (2009) Universal function-specificity of codon usage. Nucleic Acids Res 37(21):7014–7023. https://doi.org/10.1093/nar/gkp792

Najafabadi HS, Salavati R (2008) Sequence-based prediction of protein-protein interactions by means of codon usage. Genome Biol 9(5):R87. https://doi.org/10.1186/gb-2008-9-5-r87

Nalabothu RL, Fisher KJ, LaBella AL, Meyer TA, Opulente DA, Wolters JF, Rokas A, Hittinger CT (2023) Codon optimization improves the prediction of xylose metabolism from gene content in budding yeasts. Mol Biol Evol 40. https://doi.org/10.1093/molbev/msad111

Nambiar A, Dubinkina V, Liu S, Maslov S (2023) FUNPROSE: a deep learning approach to predict condition-specific gene expression in fungi. PLoS Comput Biol 19(11):e1011563. https://doi.org/10.1371/journal.pcbi.1011563

Nikolados EM, Oyarzun DA (2023) Deep learning for optimization of protein expression. Curr Opin Biotechnol 81:102941. https://doi.org/10.1016/j.copbio.2023.102941

Nirenberg M, Leder P (1964) Rna Codewords and protein synthesis. The effect of trinucleotides upon the binding of Srna to ribosomes. Science 145(3639):1399–1407. https://doi.org/10.1126/science.145.3639.1399

Novoa EM, Ribas de Pouplana L (2012) Speeding with control: codon usage, tRNAs, and ribosomes. Trends Genet 28(11):574–581. https://doi.org/10.1016/j.tig.2012.07.006

Ohama T, Suzuki T, Mori M, Osawa S, Ueda T, Watanabe K, Nakase T (1993) Non-universal decoding of the leucine codon CUG in several Candida species. Nucleic Acids Res 21(17):4039–4045. https://doi.org/10.1093/nar/21.17.4039

Osterman IA, Chervontseva ZS, Evfratov SA, Sorokina AV, Rodin VA, Rubtsova MP, Komarova ES, Zatsepin TS, Kabilov MR, Bogdanov AA, Gelfand MS, Dontsova OA, Sergiev PV (2020) Translation at first sight: the influence of leading codons. Nucleic Acids Res 48(12):6931–6942. https://doi.org/10.1093/nar/gkaa430

Parker DJ, Lalanne JB, Kimura S, Johnson GE, Waldor MK, Li GW (2020) Growth-optimized aminoacyl-tRNA Synthetase levels prevent maximal tRNA charging. Cell Syst 11(2):121–130 e126. https://doi.org/10.1016/j.cels.2020.07.005

Patil A, Dyavaiah M, Joseph F, Rooney JP, Chan CT, Dedon PC, Begley TJ (2012) Increased tRNA modification and gene-specific codon usage regulate cell cycle progression during the DNA damage response. Cell Cycle 11(19):3656–3665. https://doi.org/10.4161/cc.21919

Pechmann S, Chartron JW, Frydman J (2014) Local slowdown of translation by nonoptimal codons promotes nascent-chain recognition by SRP in vivo. Nat Struct Mol Biol 21(12):1100–1105. https://doi.org/10.1038/nsmb.2919

Pechmann S, Frydman J (2013) Evolutionary conservation of codon optimality reveals hidden signatures of cotranslational folding. Nat Struct Mol Biol 20(2):237–243. https://doi.org/10.1038/nsmb.2466

Peter J, De Chiara M, Friedrich A, Yue JX, Pflieger D, Bergstrom A, Sigwalt A, Barre B, Freel K, Llored A, Cruaud C, Labadie K, Aury JM, Istace B, Lebrigand K, Barbry P, Engelen S, Lemainque A, Wincker P, Liti G, Schacherer J (2018) Genome evolution across 1,011 Saccharomyces cerevisiae isolates. Nature 556(7701):339–344. https://doi.org/10.1038/s41586-018-0030-5

Powers KT, Szeto JA, Schaffitzel C (2020) New insights into no-go, non-stop and nonsense-mediated mRNA decay complexes. Curr Opin Struct Biol 65:110–118. https://doi.org/10.1016/j.sbi.2020.06.011

Presnyak V, Alhusaini N, Chen YH, Martin S, Morris N, Kline N, Olson S, Weinberg D, Baker KE, Graveley BR, Coller J (2015) Codon optimality is a major determinant of mRNA stability. Cell 160(6):1111–1124. https://doi.org/10.1016/j.cell.2015.02.029

Qian W, Yang JR, Pearson NM, Maclean C, Zhang J (2012) Balanced codon usage optimizes eukaryotic translational efficiency. PLoS Genet 8(3):e1002603. https://doi.org/10.1371/journal.pgen.1002603

Quax TE, Claassens NJ, Soll D, van der Oost J (2015) Codon bias as a means to fine-tune gene expression. Mol Cell 59(2):149–161. https://doi.org/10.1016/j.molcel.2015.05.035

Radhakrishnan A, Chen YH, Martin S, Alhusaini N, Green R, Coller J (2016) The DEAD-box protein Dhh1p couples mRNA decay and translation by monitoring codon optimality. Cell 167(1):122–132 e129. https://doi.org/10.1016/j.cell.2016.08.053

Rahman S, Pond SLK, Webb A, Hey J (2021) Weak selection on synonymous codons substantially inflates dN/dS estimates in bacteria. P Natl Acad Sci USA 118(20):e2023575118. https://doi.org/10.1073/pnas.2023575118

Rak R, Dahan O, Pilpel Y (2018) Repertoires of tRNAs: the couplers of genomics and proteomics. Annu Rev Cell Dev Biol 34:239–264. https://doi.org/10.1146/annurev-cellbio-100617-062754

Ranaghan MJ, Li JJ, Laprise DM, Garvie CW (2021) Assessing optimal: inequalities in codon optimization algorithms. BMC Biol 19(1):36. https://doi.org/10.1186/s12915-021-00968-8

Rhodes J, Abdolrasouli A, Dunne K, Sewell TR, Zhang Y, Ballard E, Brackin AP, van Rhijn N, Chown H, Tsitsopoulou A, Posso RB, Chotirmall SH, McElvaney NG, Murphy PG, Talento AF, Renwick J, Dyer PS, Szekely A, Bowyer P, Bromley MJ, Johnson EM, Lewis White P, Warris A, Barton RC, Schelenz S, Rogers TR, Armstrong-James D, Fisher MC (2022) Population genomics confirms acquisition of drug-resistant *Aspergillus fumigatus* infection by humans from the environment. Nat Microbiol 7(5):663–674. https://doi.org/10.1038/s41564-022-01091-2

Riley R, Haridas S, Wolfe KH, Lopes MR, Hittinger CT, Goker M, Salamov AA, Wisecaver JH, Long TM, Calvey CH, Aerts AL, Barry KW, Choi C, Clum A, Coughlan AY, Deshpande S, Douglass AP, Hanson SJ, Klenk HP, LaButti KM, Lapidus A, Lindquist EA, Lipzen AM, Meier-Kolthoff JP, Ohm RA, Otillar RP, Pangilinan JL, Peng Y, Rokas A, Rosa CA, Scheuner C, Sibirny AA, Slot JC, Stielow JB, Sun H, Kurtzman CP, Blackwell M, Grigoriev IV, Jeffries TW (2016) Comparative genomics of biotechnologically important yeasts. Proc Natl Acad Sci USA 113(35):9882–9887. https://doi.org/10.1073/pnas.1603941113

Rojano-Nisimura AM, Haning K, Janovsky J, Vasquez KA, Thompson JP, Contreras LM (2020) Codon selection affects recruitment of ribosome-associating factors during translation. ACS Synth Biol 9(2):329–342. https://doi.org/10.1021/acssynbio.9b00344

Roymondal U, Das S, Sahoo S (2009) Predicting gene expression level from relative codon usage bias: an application to Escherichia coli genome. DNA Res 16(1):13–30. https://doi.org/10.1093/dnares/dsn029

Sabi R, Tuller T (2014) Modelling the efficiency of codon-tRNA interactions based on codon usage bias. DNA Res 21(5):511–526. https://doi.org/10.1093/dnares/dsu017

Saikia M, Krokowski D, Guan BJ, Ivanov P, Parisien M, Hu GF, Anderson P, Pan T, Hatzoglou M (2012) Genome-wide identification and quantitative analysis of cleaved tRNA fragments induced by cellular stress. J Biol Chem 287(51):42708–42725. https://doi.org/10.1074/jbc.M112.371799

Saikia M, Wang XY, Mao YH, Wan J, Pan T, Qian SB (2016) Codon optimality controls differential mRNA translation during amino acid starvation. RNA 22(11):1719–1727. https://doi.org/10.1261/rna.058180.116

Salinas T, Duchene AM, Marechal-Drouard L (2008) Recent advances in tRNA mitochondrial import. Trends Biochem Sci 33(7):320–329. https://doi.org/10.1016/j.tibs.2008.04.010

Sanguinetti M, Iriarte A, Amillis S, Marin M, Musto H, Ramon A (2019) A pair of non-optimal codons are necessary for the correct biosynthesis of the *Aspergillus nidulans* urea transporter, UreA. R Soc Open Sci 6(11):190773. https://doi.org/10.1098/rsos.190773

Schneider A, Marechal-Drouard L (2000) Mitochondrial tRNA import: are there distinct mechanisms? Trends Cell Biol 10(12):509–513. https://doi.org/10.1016/s0962-8924(00)01854-7

Sejour R, Leatherwood J, Yurovsky A, Futcher B (2023) No ramp needed: spandrels, statistics, and a slippery slope. elife 12

Selker EU, Cambareri EB, Jensen BC, Haack KR (1987) Rearrangement of duplicated DNA in specialized cells of Neurospora. Cell 51(5):741–752. https://doi.org/10.1016/0092-8674(87)90097-3

Shah P, Ding Y, Niemczyk M, Kudla G, Plotkin JB (2013) Rate-limiting steps in yeast protein translation. Cell 153(7):1589–1601. https://doi.org/10.1016/j.cell.2013.05.049

Sharp PM, Li WH (1987) The codon Adaptation Index--a measure of directional synonymous codon usage bias, and its potential applications. Nucleic Acids Res 15(3):1281–1295. https://doi.org/10.1093/nar/15.3.1281

Sharp PM, Tuohy TM, Mosurski KR (1986) Codon usage in yeast: cluster analysis clearly differentiates highly and lowly expressed genes. Nucleic Acids Res 14(13):5125–5143. https://doi.org/10.1093/nar/14.13.5125

Shen X, Song S, Li C, Zhang J (2022) Synonymous mutations in representative yeast genes are mostly strongly non-neutral. Nature 606(7915):725–731. https://doi.org/10.1038/s41586-022-04823-w

Shen XX, Hittinger CT, Rokas A (2017) Contentious relationships in phylogenomic studies can be driven by a handful of genes. Nat Ecol Evol 1(5):126. https://doi.org/10.1038/s41559-017-0126

Shen XX, Opulente DA, Kominek J, Zhou X, Steenwyk JL, Buh KV, Haase MAB, Wisecaver JH, Wang M, Doering DT, Boudouris JT, Schneider RM, Langdon QK, Ohkuma M, Endoh R, Takashima M, Manabe R, Cadez N, Libkind D, Rosa CA, DeVirgilio J, Hulfachor AB, Groenewald M, Kurtzman CP, Hittinger CT, Rokas A (2018) Tempo and mode of genome evolution in the budding yeast subphylum. Cell 175:1533. https://doi.org/10.1016/j.cell.2018.10.023

Shen XX, Steenwyk JL, LaBella AL, Opulente DA, Zhou X, Kominek J, Li Y, Groenewald M, Hittinger CT, Rokas A (2020) Genome-scale phylogeny and contrasting modes of genome evolution in the fungal phylum Ascomycota. Sci Adv 6(45). https://doi.org/10.1126/sciadv.abd0079

Shingu-Vazquez M, Traven A (2011) Mitochondria and fungal pathogenesis: drug tolerance, virulence, and potential for antifungal therapy. Eukaryot Cell 10(11):1376–1383. https://doi.org/10.1128/EC.05184-11

Shoemaker CJ, Green R (2012) Translation drives mRNA quality control. Nat Struct Mol Biol 19(6):594–601. https://doi.org/10.1038/nsmb.2301

Simon D, Cristina J, Musto H (2021) Nucleotide composition and codon usage across viruses and their respective hosts. Front Microbiol 12:646300. https://doi.org/10.3389/fmicb.2021.646300

Simon D, Fajardo A, Sonora M, Delfraro A, Musto H (2017) Host influence in the genomic composition of flaviviruses: a multivariate approach. Biochem Biophys Res Commun 492(4):572–578. https://doi.org/10.1016/j.bbrc.2017.06.088

Singer GA, Hickey DA (2003) Thermophilic prokaryotes have characteristic patterns of codon usage, amino acid composition and nucleotide content. Gene 317(1–2):39–47. https://doi.org/10.1016/s0378-1119(03)00660-7

Spellman PT, Sherlock G, Zhang MQ, Iyer VR, Anders K, Eisen MB, Brown PO, Botstein D, Futcher B (1998) Comprehensive identification of cell cycle-regulated genes of the yeast Saccharomyces cerevisiae by microarray hybridization. Mol Biol Cell 9(12):3273–3297. https://doi.org/10.1091/mbc.9.12.3273

Spriggs KA, Bushell M, Willis AE (2010) Translational regulation of gene expression during conditions of cell stress. Mol Cell 40(2):228–237. https://doi.org/10.1016/j.molcel.2010.09.028

Stajich JE (2017) Fungal genomes and insights into the evolution of the kingdom. Microbiol Spectr 5(4). https://doi.org/10.1128/microbiolspec.FUNK-0055-2016

Stergachis AB, Haugen E, Shafer A, Fu W, Vernot B, Reynolds A, Raubitschek A, Ziegler S, LeProust EM, Akey JM, Stamatoyannopoulos JA (2013) Exonic transcription factor binding directs codon choice and affects protein evolution. Science 342(6164):1367–1372. https://doi.org/10.1126/science.1243490

Stukenbrock EH, Croll D (2014) The evolving fungal genome. Fungal Biol Rev 28(1):1–12

Sullivan AM, Arsovski AA, Lempe J, Bubb KL, Weirauch MT, Sabo PJ, Sandstrom R, Thurman RE, Neph S, Reynolds AP, Stergachis AB, Vernot B, Johnson AK, Haugen E, Sullivan ST, Thompson A, Neri FV 3rd, Weaver M, Diegel M, Mnaimneh S, Yang A, Hughes TR, Nemhauser JL, Queitsch C, Stamatoyannopoulos JA (2014) Mapping and dynamics of regulatory DNA and transcription factor networks in *A. thaliana*. Cell Rep 8(6):2015–2030. https://doi.org/10.1016/j.celrep.2014.08.019

Sun L, Kallolimath S, Palt R, Eminger F, Strasser R, Steinkellner H (2023) Codon optimization regulates IgG3 and IgM expression and glycosylation in N. Benthamiana. Front Bioeng. Biotechnol 11:1320586. https://doi.org/10.3389/fbioe.2023.1320586

Sun M, Zhang J (2022) Preferred synonymous codons are translated more accurately: proteomic evidence, among-species variation, and mechanistic basis. Sci Adv 8(27):eabl9812. https://doi.org/10.1126/sciadv.abl9812

Sun X, Yang Q, Xia X (2013) An improved implementation of effective number of codons (nc). Mol Biol Evol 30(1):191–196. https://doi.org/10.1093/molbev/mss201

Tagirdzhanova G, Saary P, Tingley JP, Diaz-Escandon D, Abbott DW, Finn RD, Spribille T (2021) Predicted input of uncultured fungal symbionts to a lichen Symbiosis from metagenome-assembled genomes. Genome Biol Evol 13(4). https://doi.org/10.1093/gbe/evab047

Tanaka M, Tokuoka M, Gomi K (2014) Effects of codon optimization on the mRNA levels of heterologous genes in filamentous fungi. Appl Microbiol Biotechnol 98(9):3859–3867. https://doi.org/10.1007/s00253-014-5609-7

Torrent M, Chalancon G, de Groot NS, Wuster A, Madan Babu M (2018) Cells alter their tRNA abundance to selectively regulate protein synthesis during stress conditions. Sci Signal 11(546). https://doi.org/10.1126/scisignal.aat6409

Tuller T (2013) The effect of codon usage on the success of horizontal gene transfer. In: Lateral gene transfer in evolution. Springer, pp. 147–158

Tuller T, Carmi A, Vestsigian K, Navon S, Dorfan Y, Zaborske J, Pan T, Dahan O, Furman I, Pilpel Y (2010) An evolutionarily conserved mechanism for controlling the efficiency of protein translation. Cell 141(2):344–354. https://doi.org/10.1016/j.cell.2010.03.031

Turner MK, Abrams R, Lieberman I (1968) Levels of ribonucleotide reductase activity during the division cycle of the L cell. J Biol Chem 243(13):3725–3728

Tyagi S, Kabade PG, Gnanapragasam N, Singh UM, Gurjar AKS, Rai A, Sinha P, Kumar A, Singh VK (2023) Codon usage provide insights into the adaptation of Rice genes under stress condition. Int J Mol Sci 24(2). https://doi.org/10.3390/ijms24021098

Vermeulen K, Van Bockstaele DR, Berneman ZN (2003) The cell cycle: a review of regulation, deregulation and therapeutic targets in cancer. Cell Prolif 36(3):131–149. https://doi.org/10.1046/j.1365-2184.2003.00266.x

Vieira JP, Racle J, Hatzimanikatis V (2016) Analysis of translation elongation dynamics in the context of an Escherichia coli cell. Biophys J 110(9):2120–2131. https://doi.org/10.1016/j.bpj.2016.04.004

Waack S, Keller O, Asper R, Brodag T, Damm C, Fricke WF, Surovcik K, Meinicke P, Merkl R (2006) Score-based prediction of genomic islands in prokaryotic genomes using hidden Markov models. Bmc Bioinformatics 7:142. https://doi.org/10.1186/1471-2105-7-142

Wang H, McManus J, Kingsford C (2017) Accurate Recovery of Ribosome Positions Reveals Slow Translation of Wobble-Pairing Codons in Yeast Journal of Computational Biology 24(6):486–500. https://doi.org/10.1089/cmb.2016.0147

Wang C, Zhang W, Tian R, Zhang J, Zhang L, Deng Z, Lv X, Li J, Liu L, Du G, Liu Y (2022a) Model-driven design of synthetic N-terminal coding sequences for regulating gene expression in yeast and bacteria. Biotechnol J 17(5):e2100655. https://doi.org/10.1002/biot.202100655

Wang F, Zhang N, Zhao C, Song Z, Caiyan X (2023) Codon usage bias analysis of mitochondrial protein-coding genes in 12 species of Candida. J Genet 102

Wang Q, Lyu X, Cheng J, Fu Y, Lin Y, Abdoulaye AH, Jiang D, Xie J (2022b) Codon usage provides insights into the adaptive evolution of Mycoviruses in their associated fungi host. Int J Mol Sci 23(13). https://doi.org/10.3390/ijms23137441

Weber M, Sogues A, Yus E, Burgos R, Gallo C, Martinez S, Lluch-Senar M, Serrano L (2023) Comprehensive quantitative modeling of translation efficiency in a genome-reduced bacterium. Mol Syst Biol 19(10):e11301. https://doi.org/10.15252/msb.202211301

Weinberg DE, Shah P, Eichhorn SW, Hussmann JA, Plotkin JB, Bartel DP (2016) Improved ribosome-footprint and mRNA measurements provide insights into dynamics and regulation of yeast translation. Cell Rep 14(7):1787–1799. https://doi.org/10.1016/j.celrep.2016.01.043

Weissman JL, Hou S, Fuhrman JA (2021) Estimating maximal microbial growth rates from cultures, metagenomes, and single cells via codon usage patterns. Proc Natl Acad Sci USA 118(12). https://doi.org/10.1073/pnas.2016810118

Willie E, Majewski J (2004) Evidence for codon bias selection at the pre-mRNA level in eukaryotes. Trends Genet 20(11):534–538. https://doi.org/10.1016/j.tig.2004.08.014

Wint R, Salamov A, Grigoriev IV (2022) Kingdom-wide analysis of fungal protein-coding and tRNA genes reveals conserved patterns of adaptive evolution. Mol Biol Evol 39(2). https://doi.org/10.1093/molbev/msab372

Wright F (1990) The 'effective number of codons' used in a gene. Gene 87(1):23–29. https://doi.org/10.1016/0378-1119(90)90491-9

Wright G, Rodriguez A, Li J, Milenkovic T, Emrich SJ, Clark PL (2022) CHARMING: harmonizing synonymous codon usage to replicate a desired codon usage pattern. Protein Sci 31(1):221–231. https://doi.org/10.1002/pro.4223

Wu P, Xiao W, Luo Y, Xiong Z, Chen X, He J, Sha A, Gui M, Li Q (2023) Comprehensive analysis of codon bias in 13 Ganoderma mitochondrial genomes. Front Microbiol 14:1170790. https://doi.org/10.3389/fmicb.2023.1170790

Xia X (2007) An improved implementation of codon adaptation index. Evol Bioinformatics Online 3:53–58

Xia X (2015) A major controversy in codon-anticodon adaptation resolved by a new codon usage index. Genetics 199(2):573–579. https://doi.org/10.1534/genetics.114.172106

Xia X (2018) DAMBE7: New and Improved Tools for Data Analysis in Molecular Biology and Evolution Molecular Biology and Evolution 35(6):1550–1552. https://doi.org/10.1093/molbev/msy073

Xu Y, Liu K, Han Y, Xing Y, Zhang Y, Yang Q, Zhou M (2021) Codon usage bias regulates gene expression and protein conformation in yeast expression system P. pastoris. Microb Cell Factories 20(1):91. https://doi.org/10.1186/s12934-021-01580-9

Xu Y, Ma P, Shah P, Rokas A, Liu Y, Johnson CH (2013) Non-optimal codon usage is a mechanism to achieve circadian clock conditionality. Nature 495(7439):116–120. https://doi.org/10.1038/nature11942

Yang JR, Chen X, Zhang J (2014) Codon-by-codon modulation of translational speed and accuracy via mRNA folding. PLoS Biol 12(7):e1001910. https://doi.org/10.1371/journal.pbio.1001910

Yang Q, Yu CH, Zhao F, Dang Y, Wu C, Xie P, Sachs MS, Liu Y (2019) eRF1 mediates codon usage effects on mRNA translation efficiency through premature termination at rare codons. Nucleic Acids Res 47(17):9243–9258. https://doi.org/10.1093/nar/gkz710

Yao R, Zhang Z, An X, Bucci B, Perlstein DL, Stubbe J, Huang M (2003) Subcellular localization of yeast ribonucleotide reductase regulated by the DNA replication and damage checkpoint pathways. Proc Natl Acad Sci USA 100(11):6628–6633. https://doi.org/10.1073/pnas.1131932100

Yu CH, Dang Y, Zhou Z, Wu C, Zhao F, Sachs MS, Liu Y (2015) Codon usage influences the local rate of translation elongation to regulate co-translational protein folding. Mol Cell 59(5):744–754. https://doi.org/10.1016/j.molcel.2015.07.018

Yu G, Jiang Y, Wang J, Zhang H, Luo H (2018) BMC3C: binning metagenomic contigs using codon usage, sequence composition and read coverage. Bioinformatics 34(24):4172–4179. https://doi.org/10.1093/bioinformatics/bty519

Zhang D, Shan SO (2012) Translation elongation regulates substrate selection by the signal recognition particle. J Biol Chem 287(10):7652–7660. https://doi.org/10.1074/jbc.M111.325001

Zhang J, Ieong KW, Johansson M, Ehrenberg M (2015) Accuracy of initial codon selection by aminoacyl-tRNAs on the mRNA-programmed bacterial ribosome. Proc Natl Acad Sci USA 112(31):9602–9607. https://doi.org/10.1073/pnas.1506823112

Zhang W, Foo M, Eren AM, Pan T (2022) tRNA modification dynamics from individual organisms to metaepitranscriptomics of microbiomes. Mol Cell 82(5):891–906. https://doi.org/10.1016/j.molcel.2021.12.007

Zhao F, Zhou Z, Dang Y, Na H, Adam C, Lipzen A, Ng V, Grigoriev IV, Liu Y (2021) Genome-wide role of codon usage on transcription and identification of potential regulators. Proc Natl Acad Sci USA 118(6). https://doi.org/10.1073/pnas.2022590118

Zhou J, Liu WJ, Peng SW, Sun XY, Frazer I (1999) Papillomavirus capsid protein expression level depends on the match between codon usage and tRNA availability. J Virol 73(6):4972–4982. https://doi.org/10.1128/JVI.73.6.4972-4982.1999

Zhou M, Guo J, Cha J, Chae M, Chen S, Barral JM, Sachs MS, Liu Y (2013) Non-optimal codon usage affects expression, structure and function of clock protein FRQ. Nature 495(7439):111–115. https://doi.org/10.1038/nature11833

Zhou M, Wang T, Fu J, Xiao G, Liu Y (2015) Nonoptimal codon usage influences protein structure in intrinsically disordered regions. Mol Microbiol 97(5):974–987. https://doi.org/10.1111/mmi.13079

Zhou Z, Dang Y, Zhou M, Yuan H, Liu Y (2018) Codon usage biases co-evolve with transcription termination machinery to suppress premature cleavage and polyadenylation. elife 7. https://doi.org/10.7554/eLife.33569

Zhou ZP, Dang YK, Zhou M, Li L, Yu CH, Fu JJ, Chen S, Liu Y (2016) Codon usage is an important determinant of gene expression levels largely through its effects on transcription. P Natl Acad Sci USA 113(41):E6117–E6125. https://doi.org/10.1073/pnas.1606724113

Zhu YO, Siegal ML, Hall DW, Petrov DA (2014) Precise estimates of mutation rate and spectrum in yeast. Proc Natl Acad Sci USA 111(22):E2310–E2318. https://doi.org/10.1073/pnas.1323011111

Zrimec J, Borlin CS, Buric F, Muhammad AS, Chen R, Siewers V, Verendel V, Nielsen J, Topel M, Zelezniak A (2020) Deep learning suggests that gene expression is encoded in all parts of a co-evolving interacting gene regulatory structure. Nat Commun 11(1):6141. https://doi.org/10.1038/s41467-020-19921-4

Zur H, Tuller T (2015) Exploiting hidden information interleaved in the redundancy of the genetic code without prior knowledge. Bioinformatics 31(8):1161–1168. https://doi.org/10.1093/bioinformatics/btu797

# Epigenetic Regulation in Early-Diverging Fungi

# 7

Carlos Lax and Victoriano Garre

**Abstract**

Epigenetics, defined as the study of heritable change in gene function independent of changes in DNA sequence, offers a fascinating approach to the dynamic regulation of biological processes. Beyond the static information encoded within the genome, epigenetic modifications create a dynamic and intricate layer that regulates gene expression, shaping the phenotypes in response to environmental and endogenous cues. The exploration of epigenetic regulation and its underlying mechanisms has been extensive in animals and plants, incorporating significant contributions from relevant discoveries in yeasts. In addition, studies in asco- and basidiomycetes have also advanced, unveiling several mechanisms primarily dedicated to safeguarding the genome from transposable elements (TEs). However, early-diverging fungi (EDF), which encompass the highest diversity among the fungal kingdom, have received limited attention. We anticipate a forthcoming shift in this scenario, as initial studies have surfaced, and many more are in progress. This transformation is propelled by the accessibility of affordable high-throughput sequencing technologies and the development of new procedures for genome manipulation. This chapter aims to summarize recent discoveries in EDF epigenetics, emphasizing unique features such as the prevalence of 6-methyladenine (6mA) in DNA, a characteristic found at very low levels in most eukaryotes. In addition, we discuss known epigenetics phenomena in EDF, including the control of TEs through RNA interference (RNAi) and constitutive heterochromatin, and the resistance to antifungal drugs by RNAi-dependent generation of epimutants.

**Keywords**

RNAi · Transposable elements · Genome stability · Antifungal resistance · Heterochromatin · 6-methyladenine · 5-methylcytosine · Epimutants

Carlos Lax and Victoriano Garre contributed equally.

C. Lax · V. Garre (✉)
Departamento de Genética y Microbiología, Facultad de Biología, Universidad de Murcia, Murcia, Spain
e-mail: vgarre@um.es


M. Nowrousian, J. E. Stajich (eds.), *Fungal Genomics*, The Mycota 13,
https://doi.org/10.1007/978-3-031-75666-5_7

## 7.1 Introduction

The main components of epigenetic regulation are DNA methylation, histone modification, and non-coding RNAs (Duncan et al. 2014). The first two can alter transcriptional activity by the addition or removal of chemical tags to nucleotides or histones, respectively. Non-coding RNA can interact directly with DNA and recruit enzymes to modify bases and histones. Regarding DNA methylation, bacteria rely on the use of 6-methyladenine (6mA) to regulate processes such as the discrimination between self and foreign DNA, DNA replication, and repair (Casadesús and Low 2006; Sánchez-Romero et al. 2015; Adhikari and Curtis 2016; Zhao et al. 2023). On the contrary, eukaryotes have shifted to the use of 5-methylcytosine (5mC) to regulate critical biological processes, including gene expression regulation, genome stability maintenance, and X chromosome inactivation, among others (Zemach et al. 2010; Smith and Meissner 2013). In the case of histone modifications, their functions vary depending on the type of modification (methylation, acetylation, phosphorylation, and ubiquitination) and the residues carrying those modifications (Bannister and Kouzarides 2011; Morgan and Shilatifard 2020). Moreover, different histone variants such as H2A.Z and H2A.X also play an important role in the shaping of the epigenetic landscape as they are involved in the regulation of chromatin accessibility and DNA repair (Kurumizaka et al. 2021; Talbert and Henikoff 2021). Classified according to their length and secondary structure (LncRNAs, miRNAs, CircRNAs, etc.) (Hombach and Kretz 2016), there is also a wide repertoire of non-coding RNA molecules that participate in epigenetic regulation at different levels (Peschansky and Wahlestedt 2014; Hombach and Kretz 2016; Mattick et al. 2023).

Fungi have not only consistently kept up with epigenetics research but have also been the protagonists of some of the most relevant discoveries in this field. The *Saccharomyces cerevisiae* Rpd3 transcriptional regulator was the first histone deacetylase ever cloned (Vidal and Gaber 1991; Rundlett et al. 1996; Taunton et al. 1996), whereas studies with *Schizosaccharomyces pombe* allowed the discovery of a conserved eukaryotic mechanism for the recognition of H3K9 methylation and heterochromatin regulation as well as the hereditary transmission of this epigenetic modification (Nakayama et al. 2001; Ragunathan et al. 2015). For its part, the filamentous fungus *Neurospora crassa* served to identify Dim-5, a histone methyltransferase that also controls DNA methylation (Tamaru and Selker 2001). In addition, *N. crassa* was the first organism where a gene, QDE-1, involved in what was later termed RNAi, was cloned (Cogoni and Macino 1999a). At that time, the process was referred to as quelling, and the complete set of genes coding RNAi core components was identified through a well-designed mutant screening and targeted gene deletion (Cogoni and Macino 1999b; Catalanotto et al. 2000, 2004).

As with most eukaryotic organisms, the predominant regulatory modification in the DNA of Dikarya fungi is 5mC (Bewick et al. 2019). This epigenetic modification participates mainly in the maintenance of genome stability by regulating the activity of transposable elements (TEs) (Antequera et al. 1984; Bender 1998; Bewick et al. 2019; He et al. 2020). 5mC is frequent in Basidiomycetes and exceptionally in some Ascomycetes and the early-diverging fungus *Phycomyces blakesleeanus* (Antequera et al. 1984; Bewick et al. 2019). For its part, histone modifications play a crucial role in the development and production of secondary metabolites in filamentous fungi such as *Trichoderma reesei* and *Aspergillus nidulans* (Millar and Grunstein 2006; Brosch et al. 2008; Dubey and Jeon 2017; Osorio-Concepción et al. 2017; Wang et al. 2022) as well as in other relevant processes, including regulation of the sexual cycle in the fission yeast *S. pombe* (Huang 2002). Furthermore, histone modification dynamics have also been reported to be important for fungal virulence; histone acetylation and deacetylation have been shown to affect antifungal resistance, biofilm formation, and dimorphic transition in *Candida* (Robbins et al. 2012; Hnisz et al. 2012; Nobile et al. 2014). Beyond particular examples, it is important to emphasize the high variability that exists between epigenomes and epigenetic regulatory mechanisms within the fungal kingdom (Madhani 2021). This marked variability highlights the need to approach the study of epigenetics in fungi in the broadest possible way, including representatives of the different fungal groups.

Unfortunately, the earlier branches of the fungal kingdom have been traditionally neglected in fungal research, including epigenetics characterization. Nevertheless, EDF comprises most of the variability within the fungal kingdom, with several representatives included in this group of increasing biotechnological and clinical interest (Voigt et al. 2021; Strassert and Monaghan 2022). Interestingly, EDF have retained some common characteristics with metazoans, such as the presence of flagella and actin structures for movement, the utilization and requirement for vitamins, and the presence of cholesterol in cell membranes (Weete et al. 2010; Orłowska et al. 2021; Prostak et al. 2021). These distinct aspects also extend to epigenetic regulation. With considerable variability within different phyla and even among species within the same phylum, EDF display an extensive variability at an epigenetic level, which will be presented in this Chapter. Investigating epigenetic mechanisms in EDF holds significance beyond solely elucidating their biology. It can also provide a crucial avenue to comprehending the broader principles governing epigenetic function and evolution across the entire eukaryotic domain.

## 7.2 Epigenetic Modifications in EDF

### 7.2.1 DNA Methylation

Instead of having a predominant DNA epigenetic modification, the EDF encompass organisms with high levels of 5mC and low or almost absent 6mA, and vice versa. In addition, there are organisms in which high levels of both epigenetic modifications coexist, while others exhibit undetectable levels of neither. This extensive variability, found among the different EDF representatives, highlights the suitability of this group of fungi for the study of epigenetic variability and its evolution in eukaryotic organisms (Fig. 7.1).

In fungi, genomic 5mC levels correlate with TE content and the catalog of 5mC methyltransferases present in each species (Bewick et al. 2019). While this modification is rare over gene bodies and has a questioned role in gene expression regulation, it shows a clear enrichment over repetitive regions where it plays a role in preventing transposon activity (Selker et al. 1988; Selker 1990; Lewis et al. 2009, 2010; Mondo et al. 2017; Bewick et al. 2019; Dallaire et al. 2021) (Fig. 7.1 and Table 7.1), similar to what has also been described in other eukaryotes (Zemach et al. 2010). Although 5mC levels in EDF are generally reduced (Bewick et al. 2019), two notable exceptions are *Rhizophagus irregularis* and *P. blakesleeanus*. The DNA epigenetic landscape of *R. irregularis*, which belongs to the Glomeromycota, a sister phylum of Mucoromycota, is dominated by 5mC with a very reduced presence of symmetric (both strands methylated) 6mA sites (Dallaire et al. 2021; Chaturvedi et al. 2021). Both in *R. irregularis* and *P. blakesleeanus*, 5mC is enriched in TEs and is probably involved in controlling their activity.

Interestingly, 5mC is not the only epigenetic modification present in the *P. blakesleeanus* DNA as this fungus also possesses high 6mA content. Unlike most EDF displaying high 6mA levels, *P. blakesleeanus* stands as a unique exception, being rich in both 5mC and 6mA (Mondo et al. 2017; Lax et al. 2024). Despite being traditionally considered a prokaryotic DNA modification, 6mA is abundant in the genome of three groups of unicellular eukaryotic organisms: protists, green algae, and several EDF (Mondo et al. 2017; Bochtler and Fernandes 2021; Lax et al. 2024). Despite some discrepancies regarding the role of 6mA in eukaryotes, this modification appears in high and reliable levels in these three eukaryotic groups (Kong et al. 2022). When first discovered, 6mA in EDF was found to be clustered downstream of the transcriptional start sites (TSS) of actively expressed genes, suggesting a role for this modification in gene expression regulation

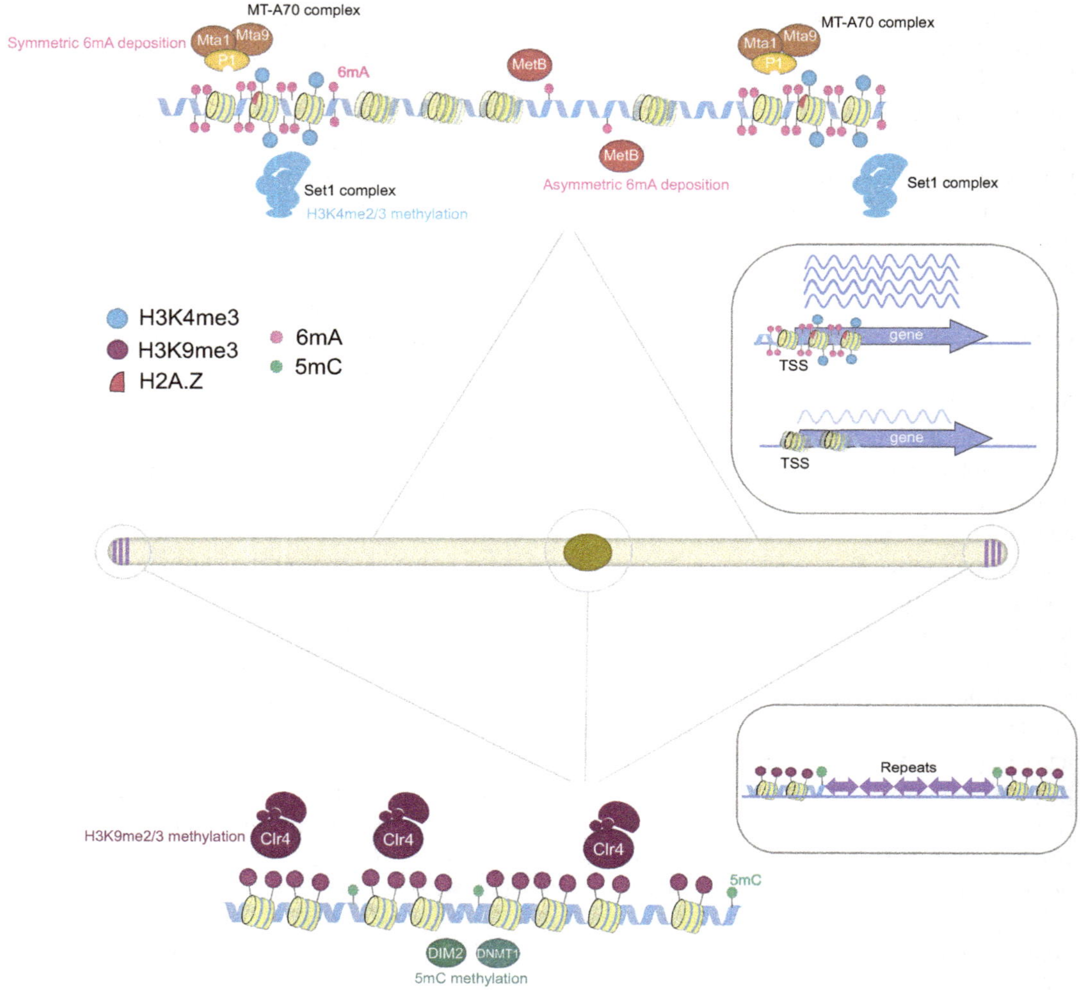


**Fig. 7.1** Epigenetic landscape in early-diverging fungi. The figure shows the main known DNA and histone modifications and the machinery involved in their regulation. Subtelomeric, pericentromeric, and repeat-rich regions are populated with H3K9me3 and 5mC, which mark constitutive heterochromatin regions. Euchromatin regions can be marked with dense clusters of symmetric 6mA and dense H3K4me3 deposition over nucleosomes that can contain the histone variant H2A.Z. These regions are associated with actively expressed genes, whereas the absence of this modification determines open chromatin regions with poorly expressed genes

(Mondo et al. 2017). Methylated sites in 6mA-rich EDF are mostly found in ApT context, following a symmetric distribution (Mondo et al. 2017; Chaturvedi et al. 2021; Lax et al. 2024, our unpublished data) (Fig. 7.1 and Table 7.1). The distribution of 6mA is inversely correlated with 5mC distribution in these fungi, indicating that they serve different roles. Moreover, 6mA is frequent in EDF with low 5mC levels, whereas those with higher 5mC content usually display reduced or very scarce 6mA presence (Mondo et al. 2017). As mentioned above, *P. blakesleeanus* stands as an extraordinary exception to this observation as this fungus displays high levels of both epigenetic modifications (Lax et al. 2024). In this model organism, 1–1.10% of adenines are methylated, and these sites display a similar distribution to what has been described in other EDF (Lax et al. 2024). Here, 6mA is also associated with highly expressed genes and is absent in

**Table 7.1** Epigenetic modifications described in EDF

| Epigenetic modification | Enzymes | Distribution | Roles | References |
|---|---|---|---|---|
| 5mC | DIM2, DNMT1 | Transposons/repetitive regions | TEs control. Genome stability maintenance | Antequera et al. (1985); Bewick et al. (2019); Dallaire et al. (2021); Chaturvedi et al. (2021); Lax et al. (2024) |
| 6mA | Mta1c (Mta1, Mta9, P1), MetB | Symmetric sites are clustered downstream the TSS of actively expressed genes. | Gene expression regulation. Marking active euchromatin regions | Mondo et al. (2017); Chaturvedi et al. (2021); Lax et al., (2024, our unpublished data) |
| H3K4me | Set1 complex (Set1, Swd1, Swd2, Swd3, Sdc1, Spp1, Bre2) | Associated with actively expressed genes and preferentially located downstream of the TSS showing a similar distribution to symmetric 6mA sites | Gene expression regulation. DNA repair | Osorio-Concepción et al. (2024); Lax et al. (our unpublished data) |
| H3K9me | Clr4 complex (Clr4, Cul4, Rik1, Raf1, Swi6) | Transposons/repetitive regions. Found in pericentromeric and subtelomeric regions | TEs control. Genome stability maintenance. Essential | Navarro-Mendoza et al. (2023); Lax et al. (our unpublished data) |
| H2A.Z | | Frequently found in the +1 nucleosome of actively expressed genes marked with 6mA and H3K4me | Gene expression regulation | Lax et al. (our unpublished data) |
| sRNAs | Dcl1<br>Dcl2<br>Ago1<br>RdRP1<br>RdRP2<br>RdRP3<br>R3B2 | Genomic repeats and regulated genes | TEs control. Genome stability maintenance. Gene expression regulation. Antifungal resistance | Ruiz-Vázquez et al. (2015); Torres-Martínez and Ruiz-Vázquez (2017); Pérez-Arques et al. (2020); Navarro-Mendoza et al. (2023) |

silent genes. Moreover, differentially expressed genes during different developmental stages and environmental stimuli also showed different methylation states, reinforcing the direct association with the presence of 6mA clusters downstream of the TSS and gene expression activation (Lax et al. 2024).

The rich diversity encompassed within the EDF group is also evident in the utilization of 6mA as the predominant DNA epigenetic mark. Extensive characterizations of the distribution of this modification in EDF have revealed marked differences between different phyla, as well as between species belonging to the same phylum (Mondo et al. 2017; Lax et al. 2024). One example of these exceptions is *Mucor lusitanicus*, another basal fungus that belongs to the Mucoromycota phylum and the Mucorales order, like *P. blakesleeanus*, but presents very low levels of both 5mC and 6mA in its genome. The few 6mA sites in the *M. lusitanicus* genome are mostly asymmetric (only DNA strand is methylated) and do not appear concentrated downstream of the TSS, nor do they show a clear association with active genes (Lax et al. 2024). Nevertheless, the functions associated with the most densely methylated genes include many that overlap with those also associated with highly methylated genes in *R. irregularis*, another EDF representative with low 6mA levels (Chaturvedi et al. 2021; Lax et al. 2024). This suggests that although 6mA is scarce in some EDF, it appears to have highly conserved functions in these organisms.

However, despite being found at low levels, the molecular tools available in *M. lusitanicus* have allowed the identification of the methyltransferases involved in the deposition of 6mA in EDF (Fig. 7.1 and Table 7.1). On the one hand, asymmetric methylation, which supposes 99% of the total 6mA in the *Mucor* genome, depends on an atypical methyltransferase called MetB (PF02384) (Lax et al. 2024). This methyltransferase only appears in some terrestrial EDF belonging to the Mucoromycota, Mortierellomycota and Glomeromycota phyla and shows a high similarity to the methyltransferases of type I and IC bacterial restriction systems. Conservation analyses of this methyltransferase allowed the detection of a high similarity between the methyltransferases of some Glomeromycetes and the sequences of their respective endosymbiotic bacteria, suggesting that EDF could have acquired this methyltransferase from symbiotic relationships with bacteria as a result of horizontal gene transfer (Lax et al. 2024). On the other hand, the deposition of symmetric 6mA is regulated by an MT-A70 type methylation complex in which the methyltransferase Mta1 has the catalytic activity and is assisted by another methyltransferase, Mta9, and a DNA binding protein, called P1 (Lax et al. 2024). *M. lusitanicus* mutants for subunits of this MT-A70 complex show a reduction in growth, indicating that despite the low levels of symmetric 6mA in this fungus, it plays an important role (Lax et al. 2024). Interestingly, this complex is conserved in 6mA-rich unicellular eukaryotes, such as green algae, protists, and basal fungi (Beh et al. 2019; Wang et al. 2019; our unpublished data). The functionality of this symmetric methylation complex has also been characterized in *Rhizopus microsporus* (our unpublished data), which also belong to the order Mucorales. This fungus has high levels of 6mA, which are associated with transcriptionally active genes transcribed by RNA polymerase II. Moreover, the characterization of the chromatin structure in this fungus revealed that the presence of 6mA is associated with regions of strongly positioned nucleosomes, while in those regions devoid of methylation, nucleosomes show a fuzzier positioning (our unpublished data). The development of genetic manipulation tools in this fungus has allowed the functional study of genes of interest (Lax et al. 2021, 2022). Thus, it was possible to establish that the components of the symmetric methylation complex MT-A70 are essential for the survival of this fungus (our unpublished data), which highlights the central role of symmetric 6mA in those EDF with high levels of this modification.

Beyond methyltransferases, 6mA demethylases have also been characterized in EDF. A study with *M. lusitanicus* revealed the existence of three demethylases named Dmt1, Dmt2, and Dmt3 (Osorio-concepción et al. 2021). Among them, the Dmt1 demethylase was shown to be involved in the dynamic regulation of 6mA levels during the yeast-to-mycelium dimorphic transition. A *M. lusitanicus* mutant in the *dmt1* gene displayed higher levels of 6mA during this transition, resulting in a reduced ability to germinate after the yeast growth stage and to escape phagocytosis by macrophages (Osorio-concepción et al. 2021).

### 7.2.2 Histone Modifications

Along with DNA epigenetic modifications, histone modifications in EDF have also gained interest in recent years. *M. lusitanicus* possesses four copies of histone H3 and three copies of histone H4 genes. As in most eukaryotes, these histones are highly conserved and contain the residues targeted by the major histone modifying enzymes (Navarro-Mendoza et al. 2019). This fungus has served for the study of the distribution of histone 3 lysine 9 methylation (H3K9me), a typical mark of constitutive heterochromatin. This modification (H3K9me2/me3) covers transcriptional deserts enriched in repeat sequences, including TEs, in the *M. lusitanicus* genome and constitutes the main repressive epigenetic modification in this fungus (Navarro-Mendoza et al. 2023) (Fig. 7.1 and Table 7.1). Noteworthy, the methyltransferase Clr4, involved in the deposition of this modification, is essential for cell viability in *M.*

*lusitanicus*, highlighting the importance of constitutive heterochromatin regions in this fungus (Navarro-Mendoza et al. 2023).

*M. lusitanicus* has also been instrumental in characterizing the function of Set1, the methyltransferase responsible for histone 3 lysine 4 methylation (H3K4me) (Osorio-Concepción et al. 2024). The loss-of-function mutation of Set1 led to the disappearance of the H3K4me1/2/3 marks, although the mutant strain obtained was viable, indicating that, unlike H3K9me, H3K4me is not essential in this fungus (Osorio-Concepción et al. 2024). Nevertheless, the *set1* mutant strain showed clear growth and sporulation defects. As in many other eukaryotic representatives, including fungi, this modification is associated with actively expressed genes (Bernstein et al. 2005; Liu et al. 2005; Zhang et al. 2009). In addition, loss of *set1* leads to defects in DNA repair, implicating H3K4me1/2/3 marks in the maintenance of genome integrity, a function previously described in budding yeast (Faucher and Wellinger 2010; Mushtaq et al. 2021).

Histone epigenetic modifications have also been studied in the model fungus *R. microsporus*. The combined analysis of DNA and histone epigenetic modifications in this fungus revealed the existence of a compartmentalized chromatin (Fig. 7.1). H3K9me3 covers constitutive heterochromatin regions devoid of 6mA, which are transcriptional deserts (our unpublished data). Conversely, as in *Mucor* and other EDF, the facultative heterochromatin mark H3K27me is absent (our unpublished data). Active euchromatin regions, on the other hand, are enriched in H3K4me3 and 6mA. Both modifications follow a similar distribution with respect to the TSS. Furthermore, the loss of 6mA clusters leads to the loss of H3K4me3, highlighting the interplay between these two epigenetic modifications (our unpublished data). The presence of both modifications is associated with strongly positioned nucleosomes, which could facilitate the entry of RNA polymerase II and promote expression (our unpublished data).

In addition to histone epigenetic modifications, the distribution of the histone H2A.Z variant has also been studied in this fungus. The H2A.Z variant is typically found in the nucleosome located at position +1 (from the TSS) of genes that are actively expressed and that are marked by H3K4me3 and 6mA (our unpublished data).

### 7.2.3 Non-coding RNAs

RNA-based epigenetic mechanisms in EDF have been minimally explored, with Mucorales, specifically *M. lusitanicus*, being the sole exception. The non-coding RNAs (ncRNAs) involved in these mechanisms are categorized into small RNAs (sRNAs) and long non-coding RNAs (lncRNAs) based on their length and secondary structure. While the diversity and functions of sRNAs have been extensively investigated in *M. lusitanicus* (Torres-Martínez and Ruiz-Vázquez 2017), other EDF remain understudied, despite the conservation of core RNAi machinery components across all EDF phyla (Navarro-Mendoza et al. 2023). Conversely, lncRNAs have not been studied in any EDF, although they have recently been identified in the Mucorales *M. lusitanicus* and *R. microsporus* (Ghizlane Tahiri et al., personal communication). Therefore, the rest of this section will focus only on sRNAs.

The sRNAs are generated by the silencing or RNA interference (RNAi) machinery, which in its basic form consists of an RNase III protein (Dicer) that processes double-stranded RNA (dsRNA) precursors into sRNAs, and an Argonaute protein (Ago) that uses these sRNAs as a guide to selectively repress gene expression. Post-transcriptional gene silencing (PTGS) and transcriptional gene silencing (TGS) mechanisms are distinguished, with the former involving translation by degradation or blocking of specific target messenger RNAs (mRNAs) and the latter mediating heterochromatin formation and deposition of 5mC (Ghildiyal and Zamore 2009). In plants, nematodes, and fungi, including most EDF (Navarro-Mendoza et al. 2023), RNA-dependent RNA polymerases (RdRPs) complement the basic machinery, participating in RNAi induction or amplification by generating dsRNA from single-stranded RNA (Ghildiyal and

Zamore 2009; Torres-Martínez and Ruiz-Vázquez 2017). *M. lusitanicus* and all EDF, except Umbelopsidales, have more than one gene encoding Dicer, Ago, or RdRP proteins (Navarro-Mendoza et al. 2023). In *M. lusitanicus*, only PTGS has been observed, but no TGS, with at least three RNAi pathways resulting from the combination of different RNAi components (Calo et al. 2017).

The first RNAi pathway identified in *M. lusitanicus* was a canonical RNAi mechanism because it uses Dicer and Ago and it is involved in the defense system against transgenes, plasmids, and TEs (Torres-Martínez and Ruiz-Vázquez 2017; Navarro-Mendoza et al. 2019; Pérez-Arques et al. 2020; Navarro-Mendoza et al. 2023). However, this RNAi pathway also generates four classes of endogenous small RNAs (esRNAs) derived from exons (ex-siRNAs) that act as riboregulators of endogenous functions (Nicolas et al. 2010). The second canonical RNAi pathway, named the epimutational pathway, generates transient drug resistance by silencing the gene encoding the target proteins or genes related to drug activation (Calo et al. 2014; Chang et al. 2019a). The third is a non-canonical RNAi pathway (NCRIP) that requires RdRPs, but not Dicer, which is replaced by R3B2, an atypical RNase III conserved in Mucorales that shows an unusual catalytic activity against single-stranded RNA (Cánovas-Márquez et al. 2021). NCRIP regulates gene expression by degrading specific mRNAs (Trieu et al. 2015; Pérez-Arques et al. 2020) and negatively regulates the canonical pathways involved in the generation of epimutants (Calo et al. 2017). The existence of multiple RNAi pathways in other EDF is possible due to the presence of multiple copies of genes encoding RNAi components (Navarro-Mendoza et al. 2023).

## 7.3 Biological Processes Under Epigenetic Control in EDF

The studies in *M. lusitanicus* have unveiled that several cellular processes are regulated by sRNAs and chromatin epigenetic marks. Mutants in key components of the RNAi machinery show defects in both asexual and sexual reproduction, response to stressors such as nutrient deprivation and oxidative stress, and virulence (Ruiz-Vázquez et al. 2015; Torres-Martínez and Ruiz-Vázquez 2017; Pérez-Arques et al. 2020). Mutants lacking H3K4me1/2/3 exhibit a significant reduction in vegetative growth, sporulation, DNA repair activity, and virulence, along with cell wall alterations (Osorio-Concepción et al. 2024). Similar to methylation on H3K4, the absence of enzymes involved in 6mA deposition results in defects in vegetative growth, lipid accumulation, dimorphism, and virulence (Osorio-concepción et al. 2021). Furthermore, the indispensability of the H3K9 methyltransferase Clr4 in this fungus indicates a crucial role for constitutive heterochromatin (Navarro-Mendoza et al. 2023). The best-known epigenetic phenomena in EDF through the study of *M. lusitanicus* are the defense against TEs and the development of acquired resistance to antifungal drugs.

### 7.3.1 Genome Defense

The study of epigenetic phenomena in filamentous fungi has unearthed several pathways that are primarily dedicated to limiting the expansion and expression of non-unique genomic sequences, mostly TEs. These include chromatin- and RNA-based epigenetic mechanisms, although these are not two completely independent entities, as sRNAs may mediate the establishment of repressive heterochromatin at loci occupied by TEs in some fungi (Smith et al. 2012). Chromatin-based mechanisms include the repeat-induced point mutation (RIP), which operates in ascomycete species, and the related methylation induced premeiotically (MIP). Although they differ in initiation, both lead to 5mC deposition of repeat sequences they target (Gladyshev 2017). In addition, in the case of RIP, 5mC is associated with heterochromatin formation (Aramayo and Selker 2013).

RNA-based mechanisms protecting the genome from TEs in filamentous fungi include the meiotic silencing by unpaired DNA (MSUD)

(Wang et al. 2015) found in filamentous ascomycetes (Shiu et al. 2001; Son et al. 2011; Tsuji et al. 2022), sex-induced silencing (SIS) of *Cryptococcus neoformans* (Wang et al. 2010), and other mechanisms that operate in vegetative cells such as quelling in *N. crassa* (Gladyshev 2017). The common feature of these different mechanisms is that the silencing takes place at the post-transcriptional level by RNAi (Gladyshev 2017).

Although the mechanisms safeguarding EDF genomes from TEs and other invading nucleic acids have been scarcely studied, recent studies in fungi belonging to the Mucoromycota and Glomeromycota have provided incisive hints. The study of TEs in *M. lusitanicus* has clearly linked its canonical PTGS pathway with repressing the movement of TEs. The complete set of TEs in the genome of this species have recently been characterized, identifying both DNA or class II elements (Tc1/Mariner, PIF/Harbinger, and RC/Helitron) and RNA or class I elements, which include non-LTR GremLINE1 (Genomic Retrotransposable Element of Mucoromycotina LINE1-like) and other LINE, as well as LTR Ty3/Gypsy (Navarro-Mendoza et al. 2023). The most extensively studied TEs in *M. lusitanicus* are the GremLINE1 elements. These elements occupy transcriptionally silent regions devoid of genes flanking the centromeres (Navarro-Mendoza et al. 2019), whereas the remaining TE families show interspersed distribution throughout the genome (Navarro-Mendoza et al. 2023). Although families of retrotransposons are controlled by PTGS pathways, differences exist among families. Thus, the control of GremLINE1 elements is entirely dependent on the canonical pathway, while an interaction between canonical RNAi pathway and NCRIP repressed expression of other LINE and Gypsy elements (Navarro-Mendoza et al. 2023).

Surprisingly, both the canonical RNAi pathway and NCRIP are involved in the production of sRNAs derived from the majority of DNA TEs, even though they use a DNA intermediate for jumping. Furthermore, the canonical pathway represses Mariner and one Helitron subfamily, similar to its repression of Gypsy and LINE RNA TEs. The enrichment of sRNA toward the transposase-encoding transcript suggests the existence of an uncharacterized mechanism that guides the canonical PTGS machinery to this transcript, preventing DNA transposition events as part of a genome defense mechanism. The remaining copies of DNA TEs are likely transcriptionally silenced by heterochromatin formation (Navarro-Mendoza et al. 2023).

In the glomeromycete *R. irregularis*, TEs promote genetic variability and gene expansion of specific families (Dallaire et al. 2021). To maintain a stable balance between adaptation mediated by genetic variability and the defense against the deleterious effects of uncontrolled TE insertions, 5mC coupled with RNAi is proposed to regulate transposon activity in this fungus (Dallaire et al. 2021). Transposon families with lower transcriptional activities are more frequently methylated than those with higher transcriptional activities and an important proportion of sRNAs bound to Ago is derived from TEs (Dallaire et al. 2021). *P. blakesleeanus* also displays high 5mC levels, which change dynamically through its life cycle, with higher levels detected in the sporangiospore stage (Antequera et al. 1984, 1985). In this fungus, 5mC (primarily found in the CG context) also spans repetitive elements, especially over DNA and LTR TEs (Lax et al. 2024). Interestingly, TEs are more active in the sporangiophore stage of *P. blakesleeanus* than in the mycelial stage (in terms of transcriptional activity and novel insertions), which coincides with a reduced expression of the DNMT1 5mC methyltransferase (Lax et al. 2024). Moreover, when analyzing sets of the most differentially expressed TEs during different growth stages and environmental stimuli, a correlation was found between an increase in the transcriptional activity and a reduction in the 5mC content (Lax et al. 2024), reinforcing the role of this modification in the genomic stability maintenance during the life cycle of this fungus. Observations from *R. irregularis* and *P. blakesleeanus* suggest that 5mC is involved in controlling TEs in EDF with high levels of this epigenetic mark.

Heterochromatin plays a crucial architectural role in eukaryotic chromosomes maintaining genomic stability by suppressing the activity of TEs, sequestering DNA repair in repetitive regions to the periphery of heterochromatin compartments, and ensuring precise chromosome segregation (Allshire and Madhani 2018). As mentioned earlier, EDF lack facultative heterochromatin and only show constitutive heterochromatin (Navarro-Mendoza et al. 2023). Significant insights into the function and formation of constitutive heterochromatin in filamentous fungi have emerged through insightful research in *N. crassa* (Freitag 2017), where it plays an important role in the control of TEs alongside RIP and 5mC (Smith et al. 2012). In *M. lusitanicus* and *R. microsporus*, constitutive heterochromatin predominantly resides in genomic regions enriched with repeats, including TEs, suggesting its potential contribution to TE control (Navarro-Mendoza et al. 2023; our unpublished data). In *M. lusitanicus*, most regions of constitutive heterochromatin are also enriched with sRNAs, and the canonical PTGS pathway is pivotal for controlling TE movement. Mutants in key genes of this pathway increase TE expression, leading to new insertions of centromeric GremLINE1 elements (Navarro-Mendoza et al. 2023). Interestingly, these RNAi mutants maintain constitutive heterochromatin indicating that the RNAi is not required for the maintenance of constitutive heterochromatin in *M. lusitanicus*. This contrasts with observations in several eukaryotes, where sRNAs generated by RNAi pathways play a crucial role in the formation and maintenance of constitutive heterochromatin (Volpe and Martienssen 2011). On the other hand, the H3K9 methyltransferases Clr4, found in the majority of EDF, has a chromodomain (Navarro-Mendoza et al. 2023) that may bind methylated H3K9 to maintain and spread constitutive heterochromatin, as reported in fission yeast and animals (Zhang et al. 2008; Shirai et al. 2017).

The observation that certain repeats in the *M. lusitanicus* genome are located in regions enriched with H3K9me2/3, but not sRNAs suggests that RNAi and heterochromatin formation function as independent genome defense in this fungus (Navarro-Mendoza et al. 2023). The conservation of H3K9 methyltransferases (Clr4/KMT1) and the RNAi machinery in EDF (Navarro-Mendoza et al. 2023) further suggests that these two defense mechanisms could also operate in other EDF.

### 7.3.2 Acquired Resistance to Antifungal Drugs

Although the number of examples is limited, both RNA- and chromatin-based epigenetic mechanisms of resistance to antifungal drugs have been described in fungi. In *S. pombe*, the *nc-tgp1* lncRNA represses the transcription of the adjacent *tgp1* gene, which encodes a transporter for glycerophosphodiester, resulting in tolerance to a broad spectrum of drugs. This repression is mediated by increasing nucleosome density at the *tgp1* promoter, preventing transcription factor access, without the need for RNAi or heterochromatin formation (Ard et al. 2014). A second mechanism of drug resistance in this yeast was discovered by the isolation of unstable epimutants resistant to caffeine. In this case, the mechanism is based on the establishment of heterochromatin islands with reduced expression of embedded genes, particularly genes that confer sensitivity to caffeine (Torres-Garcia et al. 2020). Additionally, histone acetylation has been shown to be involved in antifungal resistance in several fungi (Chang et al. 2019b).

Mucorales fungi show an intrinsic resistance to most commercially available antifungal drugs, but they also have an adaptative mechanism for generating epimutants that exhibit transient and reversible resistance to drugs (Calo et al. 2014; Chang et al. 2019a). This was the first mechanism of epimutant generation described in fungi and the only one known in EDF. The mechanism relies on the epimutation RNAi pathway, which post-transcriptionally silences genes involved in drug sensitivity (Calo et al. 2014; Chang et al. 2019a). This mechanism was discovered in screening for spontaneous mutants of *M. lusitanicus* resistant to the antifungal agent FK506, which binds to the peptidylprolyl isomerase FKBP12 and then inhibits calcineurin, resulting in the enforced yeast growth of the fungus. In addition to stable resistant isolates having mutations in the *fkbA* gene encoding FKBP12, epimutants showing unstable resistance were isolated. Resistance in these epimutants results from the production of endogenous sRNAs against *fkbA*, leading to degradation of *fkbA* mRNA by the epimutation pathway. Fulfilling the requirement for being considered an epigenetic phenomenon, this phenotype of resistance is transient and reversible when the antifungal drug is removed (Calo et al. 2014). Subsequently, it was demonstrated that this mechanism is not exclusive to resistance to FK506, but also for other antifungal agents. Epimutants were also isolated for 5-fluoroorotic acid (5-FOA), which silenced the expression of either *pyrF* or *pyrG* encoding enzymes in the pyrimidine biosynthetic pathway that convert 5-FOA from an inactive prodrug into a toxic nucleotide analog. The silencing is mediated by the epimutation RNAi pathway and involved the production of sRNAs against these genes (Chang et al. 2019a). This mechanism seems to be purely post-transcriptional, and an association with changes in the chromatin structure in the target *loci* needs further investigations. The unanswered question about this phenomenon is why some specific genes are targeted by the epimutation RNAi pathway.

The RNAi-based epimutation mechanism of *Mucor* represents a bet-hedging strategy that allows transient adaptation to insults while maintaining an intact genome. Interestingly, this mechanism could be modulated by certain environmental conditions, such as mammalian infection, as strains recovered from organs, with the brain being the most important, showed an increased rate of epimutation when they were exposed to FK506 in vitro (Chang and Heitman 2019).

## 7.4 Conclusions

The epigenetics of EDF is largely unexplored, although it deserves a special attention due to its peculiarities compared to other fungi. In addition to the extensive absence of facultative heterochromatin, in contrast to asco- and basidiomycetes, the majority of these fungi show 6mA as the major epigenetic modification of the DNA, with an almost complete absence of 5mC. Interestingly, some EDF fungi have high levels of both 6mA and 5mC, representing key organisms for understanding the evolution of DNA epigenetic modifications and associated epigenetic regulatory mechanisms in eukaryotes. We feel fortunate because we are probably witnessing the beginning of an outburst of studies characterizing the epigenomes of EDF, provoked by the decreasing cost of high-throughput sequencing technologies and the increasing number of sequenced genomes available. In a second step, we will probably see the functional characterization and the discovery of new epigenetic mechanisms that will contribute to a better understanding of the epigenetic regulation in eukaryotes.

**Acknowledgments** The researchers are funded by Fundación Séneca-Agencia de Ciencia y Tecnología de la Región de Murcia, Spain (21969/PI/22) and grants PGC2018-097452-B-I00 and PID2021-124674NB-I00 funded by MCIN/AEI/ 10.13039/501100011033 by "ERDF A way of making Europe," by the "European Union."

## References

Adhikari S, Curtis PD (2016) DNA methyltransferases and epigenetic regulation in bacteria. FEMS Microbiol Rev 40:575–591. https://doi.org/10.1093/femsre/fuw023

Allshire RC, Madhani HD (2018) Ten principles of heterochromatin formation and function. Nat Rev Mol Cell Biol 19:229–244. https://doi.org/10.1038/nrm.2017.119

Antequera F, Tamame M, Villanueva JR, Santos T (1984) DNA methylation in the fungi. J Biol Chem 259:8033–8036. https://doi.org/10.1016/s0021-9258(17)39681-3

Antequera F, Tamame M, Villanueva JR, Santos T (1985) Developmental modulation of DNA methylation in the fungus *Phycomyces blakesleeanus*. Nucleic Acids Res 13:6545–6558. https://doi.org/10.1093/nar/13.18.6545

Aramayo R, Selker EU (2013) *Neurospora crassa*, a model system for epigenetics research. Cold Spring Harb Perspect Biol 5:a017921. https://doi.org/10.1101/cshperspect.a017921

Ard R, Tong P, Allshire RC (2014) Long non-coding RNA-mediated transcriptional interference of a permease gene confers drug tolerance in fission yeast. Nat Commun 5:5576. https://doi.org/10.1038/ncomms6576

Bannister AJ, Kouzarides T (2011) Regulation of chromatin by histone modifications. Cell Res 21:381–395. https://doi.org/10.1038/cr.2011.22

Beh LY, Debelouchina GT, Clay DM, Thompson RE, Lindblad KA, Hutton ER, Bracht JR, Sebra RP, Muir TW, Landweber LF (2019) Identification of a DNA N6-adenine methyltransferase complex and its impact on chromatin organization. Cell 177:1781–1796.e25. https://doi.org/10.1016/j.cell.2019.04.028

Bender J (1998) Cytosine methylation of repeated sequences in eukaryotes: the role of DNA pairing. Trends Biochem Sci 23:252–256. https://doi.org/10.1016/s0968-0004(98)01225-0

Bernstein BE, Kamal M, Lindblad-Toh K, Bekiranov S, Bailey DK, Huebert DJ, McMahon S, Karlsson EK, Kulbokas EJ, Gingeras TR, Schreiber SL, Lander ES (2005) Genomic maps and comparative analysis of histone modifications in human and mouse. Cell 120:169–181. https://doi.org/10.1016/j.cell.2005.01.001

Bewick AJ, Hofmeister BT, Powers RA, Mondo SJ, Grigoriev IV, James TY, Stajich JE, Schmitz RJ (2019) Diversity of cytosine methylation across the fungal tree of life. Nat Ecol Evol 3:479–490. https://doi.org/10.1038/s41559-019-0810-9

Bochtler M, Fernandes H (2021) DNA adenine methylation in eukaryotes: enzymatic mark or a form of DNA damage? BioEssays 43:1–15. https://doi.org/10.1002/bies.202000243

Brosch G, Loidl P, Graessle S (2008) Histone modifications and chromatin dynamics: a focus on filamentous fungi. FEMS Microbiol Rev 32:409–439. https://doi.org/10.1111/j.1574-6976.2007.00100.x

Calo S, Nicolás FE, Lee SC, Vila A, Cervantes M, Torres-Martinez S, Ruiz-Vazquez RM, Cardenas ME, Heitman J (2017) A non-canonical RNA degradation pathway suppresses RNAi-dependent epimutations in the human fungal pathogen *Mucor circinelloides*. PLoS Genet 13:e1006686. https://doi.org/10.1371/journal.pgen.1006686

Calo S, Shertz-Wall C, Lee SC, Bastidas RJ, Nicolás FE, Granek JA, Mieczkowski P, Torres-Martínez S, Ruiz-Vázquez RM, Cardenas ME, Heitman J (2014) Antifungal drug resistance evoked via RNAi-dependent epimutations. Nature 513:555–558. https://doi.org/10.1038/nature13575

Cánovas-Márquez JT, Falk S, Nicolás FE, Padmanabhan S, Zapata-Pérez R, Sánchez-Ferrer Á, Navarro E, Garre V (2021) A ribonuclease III involved in virulence of Mucorales fungi has evolved to cut exclusively single-stranded RNA. Nucleic Acids Res 49:5294–5307. https://doi.org/10.1093/nar/gkab238

Casadesús J, Low D (2006) Epigenetic gene regulation in the bacterial world. Microbiol Mol Biol Rev 70:830–856. https://doi.org/10.1128/mmbr.00016-06

Catalanotto C, Azzalin G, Macino G, Cogoni C (2000) Gene silencing in worms and fungi. Nature 404:245–245. https://doi.org/10.1038/35005169

Catalanotto C, Pallotta M, ReFalo P, Sachs MS, Vayssie L, Macino G, Cogoni C (2004) Redundancy of the two dicer genes in transgene-induced posttranscriptional gene silencing in *Neurospora crassa*. Mol Cell Biol 24:2536–2545. https://doi.org/10.1128/MCB.24.6.2536-2545.2004

Chang Z, Billmyre RB, Lee SC, Heitman J (2019a) Broad antifungal resistance mediated by RNAi-dependent epimutation in the basal human fungal pathogen *Mucor circinelloides*. PLoS Genet 15:e1007957. https://doi.org/10.1371/journal.pgen.1007957

Chang Z, Heitman J (2019) Drug-resistant epimutants exhibit organ-specific stability and induction during murine infections caused by the human fungal pathogen *Mucor circinelloides*. mBio 10:e02579–e02519. https://doi.org/10.1128/mBio.02579-19

Chang Z, Yadav V, Chan Lee S, Heitman J (2019b) Epigenetic mechanisms of drug resistance in fungi. Fungal Genet Biol 132:103253. https://doi.org/10.1016/j.fgb.2019.103253

Chaturvedi A, Cruz Corella J, Robbins C, Loha A, Menin L, Gasilova N, Masclaux FG, Lee S-J, Sanders IR (2021) The methylome of the model arbuscular mycorrhizal fungus, *Rhizophagus irregularis*, shares characteristics with early diverging fungi and Dikarya. Commun Biol 4:901. https://doi.org/10.1038/s42003-021-02414-5

Cogoni C, Macino G (1999a) Gene silencing in *Neurospora crassa* requires a protein homologous to RNA-dependent RNA polymerase. Nature 399:166–169. https://doi.org/10.1038/20215

Cogoni C, Macino G (1999b) Posttranscriptional gene silencing in *Neurospora* by a RecQ DNA helicase. Science 286:2342–2344. https://doi.org/10.1126/science.286.5448.2342

Dallaire A, Manley BF, Wilkens M, Bista I, Quan C, Evangelisti E, Bradshaw CR, Ramakrishna NB, Schornack S, Butter F, Paszkowski U, Miska EA (2021) Transcriptional activity and epigenetic regulation of transposable elements in the symbiotic fungus *Rhizophagus irregularis*. Genome Res 31:2290–2302. https://doi.org/10.1101/gr.275752.121

Dubey A, Jeon J (2017) Epigenetic regulation of development and pathogenesis in fungal plant pathogens. Mol Plant Pathol 18:887–898. https://doi.org/10.1111/mpp.12499

Duncan EJ, Gluckman PD, Dearden PK (2014) Epigenetics, plasticity, and evolution: how do we link epigenetic change to phenotype? J Exp Zoolog B Mol Dev Evol 322:208–220. https://doi.org/10.1002/jez.b.22571

Faucher D, Wellinger RJ (2010) Methylated H3K4, a transcription-associated histone modification, is involved in the DNA damage response pathway. PLoS Genet 6:e1001082. https://doi.org/10.1371/journal.pgen.1001082

Freitag M (2017) Histone methylation by SET domain proteins in fungi. Ann Rev Microbiol 71:413–439. https://doi.org/10.1146/annurev-micro-102215-095757

Ghildiyal M, Zamore PD (2009) Small silencing RNAs: an expanding universe. Nat Rev Genet 10:94–108. https://doi.org/10.1038/nrg2504

Gladyshev E (2017) Repeat-induced point mutation and other genome defense mechanisms in fungi. Microbiol Spectr 5. https://doi.org/10.1128/microbiolspec.funk-0042-2017

He C, Zhang Z, Li B, Tian S (2020) The pattern and function of DNA methylation in fungal plant pathogens. Microorganisms 8:227. https://doi.org/10.3390/microorganisms8020227

Hnisz D, Bardet AF, Nobile CJ, Petryshyn A, Glaser W, Schöck U, Stark A, Kuchler K (2012) A histone deacetylase adjusts transcription kinetics at coding sequences during *Candida albicans* morphogenesis. PLoS Genet 8:e1003118. https://doi.org/10.1371/journal.pgen.1003118

Hombach S, Kretz M (2016) Non-coding RNAs: classification, biology and functioning. In: Slaby O, Calin GA (eds) Non-coding RNAs in colorectal cancer. Springer International Publishing, Cham, pp 3–17

Huang Y (2002) Transcriptional silencing in *Saccharomyces cerevisiae* and *Schizosaccharomyces pombe*. Nucleic Acids Res 30:1465–1482. https://doi.org/10.1093/nar/30.7.1465

Kong Y, Cao L, Deikus G, Fan Y, Mead EA, Lai W, Zhang Y, Yong R, Sebra R, Wang H, Zhang X-S, Fang G (2022) Critical assessment of DNA adenine methylation in eukaryotes using quantitative deconvolution. Science 375:515–522. https://doi.org/10.1126/science.abe7489

Kurumizaka H, Kujirai T, Takizawa Y (2021) Contributions of histone variants in nucleosome structure and function. J Mol Biol 433:166678. https://doi.org/10.1016/j.jmb.2020.10.012

Lax C, Mondo SJ, Osorio-Concepción M, Muszewska A, Corrochano-Luque M, Gutiérrez G, Riley R, Lipzen A, Guo J, Hundley H, Amirebrahimi M, Ng V, Lorenzo-Gutiérrez D, Binder U, Yang J, Song Y, Cánovas D, Navarro E, Freitag M, Gabaldón T, Grigoriev IV, Corrochano LM, Nicolás FE, Garre V (2024) Symmetric and asymmetric DNA N6-adenine methylation regulates different biological responses in Mucorales. Nat Commun 15:6066. https://doi.org/10.1038/s41467-024-50365-2

Lax C, Navarro-Mendoza MI, Pérez-Arques C, Navarro E, Nicolás FE, Garre V (2021) Stable and reproducible homologous recombination enables CRISPR-based engineering in the fungus *Rhizopus microsporus*. Cell Rep Methods 1:100124. https://doi.org/10.1016/J.CRMETH.2021.100124

Lax C, Navarro-Mendoza MI, Pérez-Arques C, Navarro E, Nicolás FE, Garre V (2022) Transformation and CRISPR-Cas9-mediated homologous recombination in the fungus *Rhizopus microsporus*. STAR Protoc 3:101237. https://doi.org/10.1016/j.xpro.2022.101237

Lewis ZA, Adhvaryu KK, Honda S, Shiver AL, Knip M, Sack R, Selker EU (2010) DNA methylation and normal chromosome behavior in *Neurospora* depend on five components of a histone methyltransferase complex, DCDC. PLoS Genet 6:e1001196. https://doi.org/10.1371/journal.pgen.1001196

Lewis ZA, Honda S, Khlafallah TK, Jeffress JK, Freitag M, Mohn F, Schübeler D, Selker EU (2009) Relics of repeat-induced point mutation direct heterochromatin formation in *Neurospora crassa*. Genome Res 19:427–437. https://doi.org/10.1101/gr.086231.108

Liu CL, Kaplan T, Kim M, Buratowski S, Schreiber SL, Friedman N, Rando OJ (2005) Single-nucleosome mapping of histone modifications in *S. cerevisiae*. PLoS Biol 3:e328. https://doi.org/10.1371/journal.pbio.0030328

Madhani HD (2021) Unbelievable but true: epigenetics and chromatin in fungi. Trends Genet 37:12–20. https://doi.org/10.1016/j.tig.2020.09.016

Mattick JS, Amaral PP, Carninci P, Carpenter S, Chang HY, Chen L-L, Chen R, Dean C, Dinger ME, Fitzgerald KA, Gingeras TR, Guttman M, Hirose T, Huarte M, Johnson R, Kanduri C, Kapranov P, Lawrence JB, Lee JT, Mendell JT, Mercer TR, Moore KJ, Nakagawa S, Rinn JL, Spector DL, Ulitsky I, Wan Y, Wilusz JE, Wu M (2023) Long non-coding RNAs: definitions, functions, challenges and recommendations. Nat Rev Mol Cell Biol 24:430–447. https://doi.org/10.1038/s41580-022-00566-8

Millar CB, Grunstein M (2006) Genome-wide patterns of histone modifications in yeast. Nat Rev Mol Cell Biol 7:657–666. https://doi.org/10.1038/nrm1986

Mondo SJ, Dannebaum RO, Kuo RC, Louie KB, Bewick AJ, LaButti K, Haridas S, Kuo A, Salamov A, Ahrendt SR, Lau R, Bowen BP, Lipzen A, Sullivan W, Andreopoulos BB, Clum A, Lindquist E, Daum C, Northen TR, Kunde-Ramamoorthy G, Schmitz RJ, Gryganskyi A, Culley D, Magnuson J, James TY, O'Malley MA, Stajich JE, Spatafora JW, Visel A, Grigoriev IV (2017) Widespread adenine N6-methylation of active genes in fungi. Nat Genet 49:964–968. https://doi.org/10.1038/ng.3859

Morgan MAJ, Shilatifard A (2020) Reevaluating the roles of histone-modifying enzymes and their associated chromatin modifications in transcriptional regulation. Nat Genet 52:1271–1281. https://doi.org/10.1038/s41588-020-00736-4

Mushtaq A, Mir US, Hunt CR, Pandita S, Tantray WW, Bhat A, Pandita RK, Altaf M, Pandita TK (2021) Role of histone methylation in maintenance of genome integrity. Genes 12:1000. https://doi.org/10.3390/genes12071000

Nakayama J, Rice JC, Strahl BD, Allis CD, Grewal SIS (2001) Role of histone H3 lysine 9 methylation in epigenetic control of heterochromatin assembly. Science 292:110–113. https://doi.org/10.1126/science.1060118

Navarro-Mendoza MI, Pérez-Arques C, Heitman J (2023) Heterochromatin and RNAi act independently to ensure genome stability in Mucorales human fungal pathogens. Proc Natl Acad Sci 120:e2220475120. https://doi.org/10.1073/pnas.2220475120

Navarro-Mendoza MI, Pérez-Arques C, Panchal S, Nicolás FE, Mondo SJ, Ganguly P, Pangilinan J, Grigoriev IV, Heitman J, Sanyal K, Garre V (2019) Early diverging fungus *Mucor circinelloides* lacks centromeric histone CENP-A and displays a mosaic of point and regional centromeres. Curr Biol 29:3791–3802.e6. https://doi.org/10.1016/j.cub.2019.09.024

Nicolas FE, Moxon S, de Haro JP, Calo S, Grigoriev IV, Torres-Martínez S, Moulton V, Ruiz-Vázquez RM, Dalmay T (2010) Endogenous short RNAs generated by Dicer 2 and RNA-dependent RNA polymerase 1 regulate mRNAs in the basal fungus *Mucor circinelloides*. Nucleic Acids Res 38:5535–5541. https://doi.org/10.1093/nar/gkq301

Nobile CJ, Fox EP, Hartooni N, Mitchell KF, Hnisz D, Andes DR, Kuchler K, Johnson AD (2014) A histone deacetylase complex mediates biofilm dispersal and drug resistance in *Candida albicans*. mBio 5:e01201–e01214. https://doi.org/10.1128/mBio.01201-14

Orłowska M, Steczkiewicz K, Muszewska A (2021) Utilization of cobalamin is ubiquitous in early-branching fungal phyla. Genome Biol Evol 13:evab043. https://doi.org/10.1093/gbe/evab043

Osorio-Concepción M, Cristóbal-Mondragón GR, Gutiérrez-Medina B, Casas-Flores S (2017) Histone deacetylase HDA-2 regulates *Trichoderma atroviride* growth, conidiation, blue light perception, and oxidative stress responses. Appl Environ Microbiol 83:1–19. https://doi.org/10.1128/AEM.02922-16

Osorio-Concepción M, Lax C, Lorenzo-Gutiérrez D, Cánovas-Márquez JT, Tahiri G, Navarro E, Binder U, Nicolás FE, Garre V (2024) H3K4 methylation regulates development, DNA repair, and virulence in Mucorales. IMA Fungus 15:6. https://doi.org/10.1186/s43008-023-00136-3

Osorio-concepción M, Lax C, Navarro E, Nicolás FE, Garre V (2021) DNA methylation on N6-adenine regulates the hyphal development during dimorphism in the early-diverging fungus *Mucor lusitanicus*. J Fungi 7:738. https://doi.org/10.3390/jof7090738

Pérez-Arques C, Navarro-Mendoza MI, Murcia L, Navarro E, Garre V, Nicolás FE (2020) A non-canonical RNAi pathway controls virulence and genome stability in Mucorales. PLoS Genet 16:e1008611. https://doi.org/10.1371/journal.pgen.1008611

Peschansky VJ, Wahlestedt C (2014) Non-coding RNAs as direct and indirect modulators of epigenetic regulation. Epigenetics 9:3–12. https://doi.org/10.4161/epi.27473

Prostak SM, Robinson KA, Titus MA, Fritz-Laylin LK (2021) The actin networks of chytrid fungi reveal evolutionary loss of cytoskeletal complexity in the fungal kingdom. Curr Biol 31:1192–1205.e6. https://doi.org/10.1016/j.cub.2021.01.001

Ragunathan K, Jih G, Moazed D (2015) Epigenetic inheritance uncoupled from sequence-specific recruitment. Science 348:1258699. https://doi.org/10.1126/science.1258699

Robbins N, Leach MD, Cowen LE (2012) Lysine deacetylases Hda1 and Rpd3 regulate Hsp90 function thereby governing fungal drug resistance. Cell Rep 2:878–888. https://doi.org/10.1016/j.celrep.2012.08.035

Ruiz-Vázquez RM, Nicolás FE, Torres-Martínez S, Garre V (2015) Distinct RNAi pathways in the regulation of physiology and development in the fungus *Mucor circinelloides*. In: Advances in genetics, pp 55–102

Rundlett SE, Carmen AA, Kobayashi R, Bavykin S, Turner BM, Grunstein M (1996) HDA1 and RPD3 are members of distinct yeast histone deacetylase complexes that regulate silencing and transcription. Proc Natl Acad Sci 93:14503–14508. https://doi.org/10.1073/pnas.93.25.14503

Sánchez-Romero MA, Cota I, Casadesús J (2015) DNA methylation in bacteria: from the methyl group to the methylome. Curr Opin Microbiol 25:9–16. https://doi.org/10.1016/j.mib.2015.03.004

Selker EU (1990) DNA methylation and chromatin structure: a view from below. Trends Biochem Sci 15:103–107. https://doi.org/10.1016/0968-0004(90)90193-f

Selker EU, Cambareri EB, Garrett PW, Haack KR, Jensen BC, Schabtach E (1988) DNA methylation and control of genome organization in *Neurospora crassa*. Gene 74:109–111. https://doi.org/10.1016/0378-1119(88)90264-8

Shirai A, Kawaguchi T, Shimojo H, Muramatsu D, Ishida-Yonetani M, Nishimura Y, Kimura H, Nakayama J, Shinkai Y (2017) Impact of nucleic acid and methylated H3K9 binding activities of Suv39h1 on its heterochromatin assembly. eLife 6:e25317. https://doi.org/10.7554/eLife.25317

Shiu PKT, Raju NB, Zickler D, Metzenberg RL (2001) Meiotic silencing by unpaired DNA. Cell 107:905–916

Smith KM, Phatale PA, Bredeweg EL, Connolly LR, Pomraning KR, Freitag M (2012) Epigenetics of filamentous fungi. In: Meyers RA (ed) Encyclopedia of molecular cell biology and molecular medicine. Wiley-VCH Verlag GmbH & Co. KGaA, Weinheim, Germany, p mcb.201100035

Smith ZD, Meissner A (2013) DNA methylation: roles in mammalian development. Nat Rev Genet 14:204–220. https://doi.org/10.1038/nrg3354

Son H, Min K, Lee J, Raju NB, Lee Y-W (2011) Meiotic silencing in the homothallic fungus *Gibberella zeae*. Fungal Biol 115:1290–1302. https://doi.org/10.1016/j.funbio.2011.09.006

Strassert JFH, Monaghan MT (2022) Phylogenomic insights into the early diversification of fungi. Curr Biol 32:3628–3635.e3. https://doi.org/10.1016/j.cub.2022.06.057

Talbert PB, Henikoff S (2021) Histone variants at a glance. J Cell Sci 134:jcs244749. https://doi.org/10.1242/jcs.244749

Tamaru H, Selker EU (2001) A histone H3 methyltransferase controls DNA methylation in *Neurospora crassa*. Nature 414:277–283. https://doi.org/10.1038/35104508

Taunton J, Hassig CA, Schreiber SL (1996) A mammalian histone deacetylase related to the yeast transcriptional regulator Rpd3p. Science 272:408–411. https://doi.org/10.1126/science.272.5260.408

Torres-Garcia S, Yaseen I, Shukla M, Audergon PNCB, White SA, Pidoux AL, Allshire RC (2020) Epigenetic gene silencing by heterochromatin primes fungal resistance. Nature 585:453–458. https://doi.org/10.1038/s41586-020-2706-x

Torres-Martínez S, Ruiz-Vázquez RM (2017) The RNAi universe in fungi: a varied landscape of small RNAs and biological functions. Ann Rev Microbiol 71:371–391. https://doi.org/10.1146/annurev-micro-090816-093352

Trieu TA, Calo S, Nicolás FE, Vila A, Moxon S, Dalmay T, Torres-Martínez S, Garre V, Ruiz-Vázquez RM (2015) A non-canonical RNA silencing pathway promotes mRNA degradation in basal fungi. PLoS Genet 11:e1005168. https://doi.org/10.1371/journal.pgen.1005168

Tsuji K, Kitade Y, Yoshimi A, Tanaka C (2022) Meiotic silencing in Dothideomycetous *Bipolaris maydis*. Front Fungal Biol 3:931888. https://doi.org/10.3389/ffunb.2022.931888

Vidal M, Gaber RF (1991) RPD3 encodes a second factor required to achieve maximum positive and negative transcriptional states in *Saccharomyces cerevisiae*. Mol Cell Biol 11:6317–6327. https://doi.org/10.1128/mcb.11.12.6317-6327.1991

Voigt K, James TY, Kirk PM, de Santiago ALCM, Waldman B, Griffith GW, Fu M, Radek R, Strassert JFH, Wurzbacher C, Jerônimo GH, Simmons DR, Seto K, Gentekaki E, Hurdeal VG, Hyde KD, TTT N, Lee HB (2021) Early-diverging fungal phyla: taxonomy, species concept, ecology, distribution, anthropogenic impact, and novel phylogenetic proposals. Fungal Divers 109:59–98. https://doi.org/10.1007/s13225-021-00480-y

Volpe T, Martienssen RA (2011) RNA interference and heterochromatin assembly. Cold Spring Harb Perspect Biol 3:a003731. https://doi.org/10.1101/cshperspect.a003731

Wang L, Liu J, Li X, Lyu X, Liu Z, Zhao H, Jiao X, Zhang W, Xie J, Liu W (2022) A histone H3K9 methyltransferase Dim5 mediates repression of sorbicillinoid biosynthesis in *Trichoderma reesei*. Microb Biotechnol 15:2533–2546. https://doi.org/10.1111/1751-7915.14103

Wang X, Hsueh Y-P, Li W, Floyd A, Skalsky R, Heitman J (2010) Sex-induced silencing defends the genome of *Cryptococcus neoformans* via RNAi. Genes Dev 24:2566–2582. https://doi.org/10.1101/gad.1970910

Wang Y, Sheng Y, Liu Y, Zhang W, Cheng T, Duan L, Pan B, Qiao Y, Liu Y, Gao S (2019) A distinct class of eukaryotic MT-A70 methyltransferases maintain symmetric DNA N6-adenine methylation at the ApT dinucleotides as an epigenetic mark associated with transcription. Nucleic Acids Res 47:11771–11789. https://doi.org/10.1093/nar/gkz1053

Wang Y, Smith KM, Taylor JW, Freitag M, Stajich JE (2015) Endogenous small RNA mediates meiotic silencing of a novel DNA transposon. G3, 5:1949–1960. https://doi.org/10.1534/g3.115.017921

Weete JD, Abril M, Blackwell M (2010) Phylogenetic distribution of fungal sterols. PLoS One 5:e10899. https://doi.org/10.1371/journal.pone.0010899

Zemach A, McDaniel IE, Silva P, Zilberman D (2010) Genome-wide evolutionary analysis of eukaryotic DNA methylation. Science 328:916–919. https://doi.org/10.1126/science.1186366

Zhang K, Mosch K, Fischle W, Grewal SIS (2008) Roles of the Clr4 methyltransferase complex in nucleation, spreading and maintenance of heterochromatin. Nat Struct Mol Biol 15:381–388. https://doi.org/10.1038/nsmb.1406

Zhang X, Bernatavichute YV, Cokus S, Pellegrini M, Jacobsen SE (2009) Genome-wide analysis of mono-, di- and trimethylation of histone H3 lysine 4 in *Arabidopsis thaliana*. Genome Biol 10:R62. https://doi.org/10.1186/gb-2009-10-6-r62

Zhao J, Zhang M, Hui W, Zhang Y, Wang J, Wang S, Kwok L-Y, Kong J, Zhang H, Zhang W (2023) Roles of adenine methylation in the physiology of *Lacticaseibacillus paracasei*. Nat Commun 14:2635. https://doi.org/10.1038/s41467-023-38291-1

# Part 3

# Biotechnology

# Fungal Diversity Related to Plant Biomass Degradation

**8**

Miia R. Mäkelä, Peicheng Sun, Mao Peng, and Ronald P. de Vries

**Abstract**

Plant biomass degradation is of major importance for many fungi, as this is one of the most abundant and renewable carbon sources on earth. With a global push toward a bio-based economy based on plant biomass conversion, research in this area has obtained enormous momentum. In this chapter, recent insights obtained from fungal genomes and post-genomic studies related to plant biomass degradation will be discussed, focusing in particular on the diversity of the approaches used by different fungi. In addition, an updated overview will be presented of the enzymes currently considered to be involved in the degradation of the plant biomass polymers.

M. R. Mäkelä
Department of Bioproducts and Biosystems, Aalto University, Espoo, Finland
e-mail: miia.makela@aalto.fi

P. Sun
Laboratory of Food Chemistry, Wageningen University, Wageningen, The Netherlands
e-mail: p.sun@wi.knaw.nl

M. Peng · R. P. de Vries (✉)
Westerdijk Fungal Biodiversity Institute, Utrecht, The Netherlands
e-mail: m.peng@wi.knaw.nl; r.devries@wi.knaw.nl

**Keywords**

Diversity · Plant polymer degradation · Enzymes · (post-)genomics · Bio-based economy

## 8.1 Introduction

Degradation of plant biomass has become one of the focal points of the bio-based economy, mainly because plant biomass is the most abundant renewable biomass on earth that already is extensively used in agriculture and industry. The challenges for the bio-based economy are to use that part of plant biomass that is not used for the production of food or other valuable products, and convert these into biochemicals and bio-based materials. A major step in these processes is the depolymerization of the polymeric components of the plant cell wall, and for that detailed understanding of the composition of plant biomass is required.

### 8.1.1 Plant Biomass Composition

Plant biomass consists mainly of polymeric components, such as polysaccharides, the aromatic polymer lignin and proteins. In the polysaccharide fraction, a distinction can be made between cell wall polysaccharides (cellulose, hemicellulose, pectin) and storage polysaccharides (starch, inulin,


M. Nowrousian, J. E. Stajich (eds.), *Fungal Genomics*, The Mycota 13,
https://doi.org/10.1007/978-3-031-75666-5_8

different gums). Cell wall polysaccharides form a tight network together with lignin that provides strength for the plant, allowing it to grow upright, and is also a major defense against microbial attack. The foundation of this network is formed by the cellulose microfibrils, which are long cables of parallel cellulose polymers. Hemicelluloses and pectin form non-covalent linkages with the microfibrils, creating the structural integrity required by the plant. Hemicelluloses and pectin can be covalently linked to each other by the formation of diferulic-ester bonds, providing additional structure to the network. This network is further strengthened by ester-linkages between hemicellulose and lignin, either through ferulic acid or through (4-O-methyl-)glucuronic acid.

## 8.1.2 Cellulose

Cellulose is a linear polymer of β-1,4-linked D-glucopyranose residues, which are non-covalently organized into a cable-like structure called microfibril (Kolpak and Blackwell 1976). These microfibrils contain crystalline (highly ordered) and more amorphous regions, the latter of which are more susceptible to enzymatic attack. Degradation of cellulose requires the combined action of four types of enzymes, endoglucanases, cellobiohydrolases/exo-glucanases, β-glucosidases, and lytic polysaccharide dehydrogenases (Cragg et al. 2015). Different families of these enzymes have been identified in fungi and other organisms, which are classified in the Carbohydrate Active enZyme database (CAZy, www.cazy.org) (Drula et al. 2022). Cellulose is the most abundant polymer in plant biomass and also the most abundant organic compound on earth. It is already of major importance for several industries, such as paper and pulp, but is currently of particular interest as a substrate for biomaterials and biochemicals (Meyer et al. 2020).

## 8.1.3 Hemicelluloses

Three main types of hemicellulose are found in plant cell walls: xylan, xyloglucan, and galactomannan. These are briefly described below.

### 8.1.3.1 Xylan

Xylan is the most abundant hemicellulose and is found in a wide range of plant species. It consists of a backbone of β-1,4-linked D-xylopyranose units that can be decorated with a range of different residues (de Vries and Visser 2001). The most abundant residues attached to the xylan backbone are L-arabinose and (4-O-methyl-)D-glucuronic acid, but the relative amount of these decorations differs strongly depending on the source. In cereals and grasses, L-arabinose is the most abundant decoration and up to 50% of the xylose residues can contain an L-arabinose side group. These xylans are therefore often referred to as arabinoxylans. In contrast, in hardwood (4-O-methyl)-D-glucuronic acid is the most abundant side group and these xylans are therefore often named glucuronoxylans. Other side groups in xylan are α- and β-linked D-galactose and acetyl groups that are linked to the main chain xylose residues, and feruloyl and *p*-coumaroyl residues that are linked to the L-arabinose side groups (de Vries and Visser 2001). The high variation of the xylan structure makes it a very challenging substrate to degrade for fungi and also requires a much larger number of enzymes (Table 8.1) (de Vries and Visser 2001). The main chain of xylan is degraded by endoxylanases and β-xylosidases, while the side groups are removed by arabinoxylan arabinofuranohydrolases, α-arabinofuranosidases, α-glucuronidases, α- and β-galactosidases, feruloyl esterases, and acetyl xylan esterases.

### 8.1.3.2 Xyloglucan

The backbone of xyloglucan consists of β-1,4-linked D-glucopyranose units, just like the backbone of cellulose. However, these glucose residues can be decorated with α-1,4-linked D-xylose residues in two distinct repeating patterns, XXGG or XXXG, in which two or three decorated glucose residues are separated by two or one undecorated glucose residues, respectively (Vincken et al. 1997). The xylose residues can be further decorated with L-arabinose, D-galactose, L-fucose, or acetyl residues, creating a further level of variation in the xyloglucan structure (Hantus et al. 1997; Hisamatsu et al. 1991, 1992; Huisman et al. 2000; Maruyama et al. 1996). Degradation of the backbone of xyloglucan

**Table 8.1** Fungal enzymatic activities involved in plant polysaccharide degradation

| Enzyme code | Activity | EC number | Substrate | CAZy families |
|---|---|---|---|---|
| ABF | α-L-Arabinofuranosidase | 3.2.1.55 | Xyloglucan, xylan, pectin | GH51, GH54 |
| ABN | Endoarabinanase | 3.2.1.99 | Pectin | GH43 |
| ABX | Exoarabinanase | 3.2.1.-* | Pectin | GH93 |
| AFC | α-L-fucosidase | 3.2.1.51, 3.2.1.63 | Xyloglucan | GH29, GH95, **GH139, GH141** |
| AGD | α-1,4-D-glucosidase | 3.2.1.20 | Starch | GH31 |
| AGL | α-1,4-D-galactosidase | 3.2.1.22 | Xyloglucan, xylan, galactomannan | GH27, GH36 |
| AGU | α-Glucuronidase | 3.2.1.131, 3.2.1.139 | Xylan | GH67, GH115 |
| AMY | α-Amylase | 3.2.1.1 | Starch | GH13 |
| AXE | Acetyl xylan esterase | 3.1.1.72 | Xylan | CE1 |
| AXH | Arabinoxylan arabinofuranohydrolase | 3.2.1.55 | Xylan | GH62 |
| AXL | α-D-xylosidase | 3.2.1.177 | Xyloglucan | GH31 |
| BGL | β-1,4-D-glucosidase | 3.2.1.21 | Cellulose, xyloglucan | GH1, GH3, GH9 |
| BXL | β-1,4-D-xylosidase | 3.2.1.37 | Xylan, pectin | GH3, GH39, GH43 |
| CBH | Cellobiohydrolase | 3.2.1.91 | Cellulose, xyloglucan | GH6, GH7, GH9 |
| EGL | β-1,4-D-endoglucanase | 3.2.1.4 | Cellulose, xyloglucan | GH5, GH7, GH9, GH12, GH44, GH45 |
| FAE | Feruloyl esterase | 3.1.1.73 | Xylan, pectin | CE1 |
| GAL | β-1,4-endogalactanase | 3.2.1.89 | Pectin | GH53 |
| GE | Glucuronoyl esterase | 3.1.1.117 | Xylan | CE15 |
| GLA | Glucoamylase | 3.2.1.3 | Starch | GH15 |
| GLN | β-1,6-endogalactanase | 3.2.1.164 | Pectin | GH5, GH30 |
| **HAE** | **Hemicellulose acetyl esterase** | **3.1.1.6** | **Xylan, galactomannan, xyloglucan** | **CE16** |
| INU | Endo-inulinase | 3.2.1.7 | Inulin | GH32 |
| INX | Exo-inulinase | 3.2.1.80 | Inulin | GH32 |
| LAC | β-1,4-D-galactosidase | 3.2.1.23 | Xyloglucan, xylan, pectin, galactomannan | GH2, GH35, GH42 |
| **LPMO** | **Lytic polysaccharide monooxygenase** | **1.14.99.54/56** | **Cellulose, hemicellulose** | **AA9** |
| **LPMO** | **Lytic polysaccharide monooxygenase** | **1.14.99.55** | **Starch** | **AA13** |
| **LPMO** | **Lytic polysaccharide monooxygenase** | **1.14.99.-** | **Xylan (only when associated with cellulose)** | **AA14** |
| **LPMO** | **Lytic polysaccharide monooxygenase** | **1.14.99.53/54** | **Chitin, cellulose** | **AA15** |
| **LPMO** | **Lytic polysaccharide monooxygenase** | **1.14.99.54** | **Cellulose** | **AA16** |
| MAN | β-1,4-D-endomannanase | 3.2.1.78 | Galactomannan | GH5, GH26, **GH134** |
| MND | β-1,4-D-mannosidase | 3.2.1.25 | Galactomannan | GH2 |
| PAE | Pectin acetyl esterase | 3.2.1.- | Pectin | CE12? |
| PEL | Pectin lyase | 4.2.2.10 | Pectin | PL1 |
| RGL | Rhamnogalacturonan lyase | 4.2.2.23/24 | Pectin | PL4, PL11, **PL26** |
| PGA | Endopolygalacturonase | 3.2.1.15 | Pectin | GH28 |
| PGX | Exopolygalacturonase | 3.2.1.82 | Pectin | GH28 |

(continued)

**Table 8.1** (continued)

| Enzyme code | Activity | EC number | Substrate | CAZy families |
|---|---|---|---|---|
| PLY | Pectate lyase | 4.2.2.2 | Pectin | PL1, PL3, PL9 |
| PME | Pectin methyl esterase | 3.1.1.11 | Pectin | CE8 |
| RGAE | Rhamnogalacturonan acetyl esterase | 3.1.1.86 | Pectin | CE12 |
| RGX | Rhamnogalacturonan galaturonohydrolase / exorhamnogalacturonase | 3.2.1.173 | Pectin | GH28 |
| RHA | α-Rhamnosidase / rhamnogalacturonan rhamnohydrolase | 3.2.1.40 | Pectin | GH78 |
| RHG | Rhamnogalacturonan hydrolase / endorhamnogalacturonase | 3.2.1.171 | Pectin | GH28 |
| SUC | Invertase / fructofuranosidase | 3.2.1.26 | Inulin | GH32 |
| UGH | D-4,5 unsaturated-glucuronyl hydrolase | 3.2.1.- | Pectin | GH88 |
| URH | Unsaturated rhamnogalacturonan hydrolase | 3.2.1.172 | Pectin | GH105 |
| **XBH** | **Xylobiohydrolase** | **3.2.1.-** | **Xylan** | **GH30** |
| XEG | Xyloglucan-active β-1,4-D-endoglucanase | 3.2.1.151 | Xyloglucan | GH12, **GH44**, GH74 |
| XGH | Xylogalacturonan hydrolase | 3.2.1.- | Pectin | GH28 |
| XLN | β-1,4-D-endoxylanase | 3.2.1.8 | Xylan | GH10, GH11, **GH30** |
| XTG | β-1,3-exogalactanase | 3.2.1.145 | Pectin | GH43 |

The enzymes are sorted alphabetically based on the enzyme code. Source: ww.cazy.org (Drula et al. 2022), updated from (Mäkelä et al. 2014). New enzyme activities or families are indicated in bold. These families typically do not have extensive functional analysis but are strongly suggested to be involved in the degradation of the associated polysaccharides. They are discussed in more detail in this chapter

[a]If the last digit of the EC number is missing, this means that this enzyme does not have an EC number yet, but that based on its activity we can already predict in which group it will be classified

requires similar activities as cellulose and some cellulose active enzymes are also active on xyloglucan (Table 8.1). However, xyloglucan-specific endoglucanases and exoglucanases have also been described (Hasper et al. 2002; Master et al. 2008; Pauly et al. 1999). The side groups are removed by α-xylosidases (Matsushita et al. 1985; Matsushita et al. 1987; Yoshikawa et al. 1993a, b), α-arabinofuranosidases (Saha 2000), β-galactosidases, α-fucosidases, and xyloglucan acetyl esterases (Table 8.1).

#### 8.1.3.3 Galactomannan

Galactomannan is mainly found in the cell walls of woody plants and is the most abundant hemicellulose in softwood (Jacobs et al. 2002). Galactomannan consists of a backbone of β-1,4-linked D-mannose residues which can be decorated with α-1,4-linked D-galactose residues and acetyl groups (Sjöström 1993). In some plants, the backbone is interrupted by D-glucose residues in which case the polysaccharide is referred to as galactoglucomannan. Low levels of β-1,4-linked D-galactose residues have also been found attached to the mannose residues. The complete degradation of galacto(gluco)mannan involves the action of endomannases, β-mannosidases, α- and β-galactosidases, β-glucosidases, and galactomannan acetyl esterases (Gilbert et al. 2008) (Table 8.1). Galactomannans are also found as storage polysaccharides, such as locust bean gum and guar gum, but then do not contain glucose residues in the backbone (Srivastava and Kapoor 2005).

### 8.1.4 Pectin

Pectin is a highly variable polysaccharide that consists of four distinct substructures: homogalacturonan, xylogalacturonan, and rhamnogalacturonan I and II (Carpita and Gibeaut 1993; Mohnen 1999; Ridley et al. 2001; Willats et al. 2001). Their characteristics and the enzymes that act on them are briefly discussed below.

#### 8.1.4.1 Homogalacturonan and Xylogalacturonan

Homogalacturonan (also called the pectin smooth region) is a linear polymer of α-1,4-linked D-galacturonic acid residues, which can be methylated or acetylated. Xylogalacturonan is a modified form of homogalacturonan in which some of the galacturonic acid residues are decorated with β-1,3-linked D-xylose residues. The degradation of these polymers involves the action of endo- and exo-polygalacturonases, pectin and pectate lyases, pectin acetyl and pectin methyl esterases, and β-xylosidases (Table 8.1).

#### 8.1.4.2 Rhamnogalacturonan I and II

Rhamnogalacturonan I (RG-I) consists of a backbone of alternating D-galacturonic acid and L-rhamnose residues that are linked through α-1,2- and α-1,4-linkages, respectively. It is also referred to as the hairy region of pectin, since the rhamnose residues can be decorated with arabinan or arabinogalactan I side chains. The arabinan side chains consist of α-1,5-linked L-arabinose residues, which can be decorated with α-1,2- and α-1,3-linked L-arabinose residues (Schols and Voragen 1996). Arabinogalactan I consists of a β-1,4-linked D-galactose backbone that is decorated with α-1,3-linked L-arabinose residues (Mohnen 1999; Ridley et al. 2001). Both arabinan and arabinogalactan I can contain terminal feruloyl residues, and the galacturonic acid residues of the RG-I backbone can be acetylated. The degradation of RG-I requires the combined action of a large set of enzymes, such as rhamnogalacturonan hydrolases and lyases, α-rhamnosidases, endo- and exo-arabinanases, α-arabinofuranosidases, endogalactanases, β-galactosidases, feruloyl esterases, and rhamnogalacturonan acetyl esterases (Table 8.1).

Rhamnogalacturonan II (RG-II) is a minor component of pectin and consists of a modified homogalacturonan that contains side chains consisting of 11 uncommon sugars (O'Neill et al. 1996; Vidal et al. 2000). Due to its unusual structure, it is highly resistant to degradation by common pectinolytic enzymes.

### 8.1.5 Lignin

Lignin is an aromatic heteropolymer that is present in the cell walls of wood and grasses. It is covalently linked to hemicelluloses and cross-linked to other plant cell wall components, thus rigidifying, and mechanically strengthening the plant cell walls. Due to its hydrophobic nature, lignin also enhances water transport in plants. Importantly, the highly irregular structure of lignin polymers, synthesized from phenylpropanoid monomers, i.e., *p*-coumaryl, coniferyl, and sinapyl alcohol (Boerjan et al. 2003; Higuchi 1997), endows the plant cell walls with resistance to microbial attack. The plant enzymes, such as laccases and peroxidases, catalyze the oxidative radical coupling of the lignin monomers to *p*-hydroxyphenyl (H-units), guaiacyl (G-units), and syringyl (S-units) type of subunits, and further to lignin polymers (Higuchi 2006). Other phenolic compounds that do not originate from the canonical phenylpropanoid biosynthetic pathway of lignin, including flavonoids, hydroxystilbenes, and hydroxycinnamic amides, are also considered as lignin monomers and in certain plant species they have been shown to be integrated into lignin molecules (del Rio et al. 2020).

Several linkages exist between the lignin subunits, of which *β*-O-4 aryl ether is the most abundant (Rinaldi et al. 2016). A variety of carbon-carbon bonds, including β-5′ (phenylcoumaran), β-β' (resinol), and 5–5′ (biphenyl) bonds, are also present. The composition and structure of lignin are highly variable between plant species, which makes the resolution of the chemical structure of lignin molecules challenging. In addition, the amount of lignin in plant biomass differs between plant species, being highest in softwoods (25–33% of dry weight). Hardwoods contain 20–25% lignin of the wood dry weight, while a lower amount of lignin, approx. 5–24% of the dry weight, is present in grasses (Buranov and Mazza 2008).

The irregular structure of lignin, which lacks hydrolyzable interunit linkages, needs to be degraded by oxidative, unspecific enzymes. The so-called lignin-modifying enzymes required for efficient degradation and mineralization of native lignin molecules include lignin peroxidases (LiPs), manganese peroxidases (MnPs), and versatile peroxidases (VP) as well as laccases. Of these, LiPs, MnPs, and VPs are unique for the basidiomycete white-rot fungi that are renowned for their capacity to degrade lignin. As these peroxidase enzymes use $H_2O_2$ as an electron acceptor in their catalysis, also several $H_2O_2$-generating oxidases play an important role in lignin degradation. All these enzymes are classified into different families of auxiliary activities (AA) in the CAZy database (Drula et al. 2022) and were discussed in detail in the previous version of this chapter (Mäkelä et al. 2014). Brown-rot fungi do not possess lignin-modifying peroxidases, and therefore they mainly demethoxylate lignin (Hatakka and Hammel 2010).

## 8.2 Plant Biomass Degrading Enzymes of Fungi

In the previous edition of this book, a chapter (Mäkelä et al. 2014) already discussed many of the enzymes involved in plant biomass degradation in detail, and this will therefore not be repeated here. A complete overview of the enzymes is provided in Table 8.1. However, since that publication, additional enzymes, specificities, and enzyme families have been described, which will be briefly described here. One of these novel enzyme classes is the lytic polysaccharide monooxygenases (LPMOs) that have received significant attention due to their high industrial potential. These enzymes will therefore be described in more detail below.

### 8.2.1 Novel Enzymes Activities and Families Involved in Plant Biomass Degradation

Several studied reported endoxylanase (XLN) activity GH30 enzymes of several fungi (Biely et al. 2014b; Espinoza and Eyzaguirre 2018; Luo et al. 2010; Nakamichi et al. 2019a, b, 2020), but also enzymes with exo-acting activity (Katsimpouras et al. 2019; Tenkanen et al. 2013). A comparative study of a wider set of GH30 enzymes revealed distinct subfamilies acting on different polysaccharides (Li et al. 2022c). The subset acting on xylan showed high diversity ranging from enzymes releasing only xylobiose (xylobiohydrolase, XBH) to endoxylanases with different cleavage patterns. The GH30 XLNs showed distinct product profiles compared to GH10 and GH11 XLNs. However, also within those well-studied families more diverse specificities have been demonstrated. This was especially obvious when the GH10 and GH11 enzymes of *Penicillium subrubescens* were compared (Li et al. 2022b). This *Penicillium* species was previously reported to have a specific expansion of inulinases, hemicellulases, and pectinases compared to other Penicillia (de Vries and Mäkelä 2020; Peng et al. 2017b). When the product profiles of the GH10 and GH11 members were compared distinct subsets with different oligosaccharide product profiles were identified (Li et al. 2022b), indicating that the expansion of members of these families resulted in functional diversification of the enzymes.

This appears to be a common phenomenon when the number of genes in an enzyme family expands, as this was also demonstrated for two other expanded enzyme activities of this fungal species: α-L-arabinofuranosidases (ABF/AXH; GH51, GH53, GH62) and α-D-galactosidases (AGL; GH27, GH36). Comparison of the AGLs from *P. subrubescens* revealed clear differences in their substrate specificity and dependence on endomannanase activity (Coconi Linares et al. 2020), while comparison of the ABFs and AXHs showed strong differences in specific activity and substrate specificity (Coconi Linares et al. 2022).

Some other novel hemicellulose activities were also described, such as the hemicellulose acetyl esterase (HAE; CE16). Initially, this family was described as a novel class of acetyl xylan esterases (Biely et al. 2014a; Koutaniemi et al. 2013; Puchart et al. 2015, 2016). However, a comparison of the four CE16 enzymes of *Aspergillus niger* revealed that at least one of them has broader substrate specificity, with activity on acetylated xylan, galactomannan, and xyloglucan (Venegas et al. 2022).

Detailed analysis of fungal genomes revealed the presence of GH44 enzymes in part of the basidiomycete genomes, but not in ascomycete genomes (Sun et al. 2022b). This family had already been described in bacteria as xyloglucanase (XEG) (Warner et al. 2011) but had likely been missed in fungal genomes due to its absence in the best-studied fungi related to plant biomass degradation (*A. niger*, *Trichoderma reesei*). Characterization of two basidiomycete GH44 enzymes demonstrated their activity on xyloglucan, as well as a distinct cleavage pattern when compared to GH12 and GH74 XEGs (Sun et al. 2022b).

Additional families were also described for two other endo-acting enzymes. A novel family of endomannanases (MAN) was identified in both bacteria (You et al. 2018) and fungi (Sakai et al. 2017; Shimizu et al. 2015). The enzyme from *Aspergillus nidulans* released mannobiose, mannotriose, and mannotetraose, but not higher oligosaccharides, suggesting a different mechanism than that described for GH5 and GH26 endomannanases (Shimizu et al. 2015). In *Penicillium chrysogenum*, a novel rhamnogalacturonan lyase (RGL) family (PL26) was described that is structurally distinct from other RGL families (PL4, PL11) (Iwai et al. 2015).

In light of the exponentially increasing number of fungal genomes, it can be expected that additional enzyme families and specificities will be discovered. However, the current speed of genome sequencing is not matched by an increase in efforts in enzyme characterization, which likely means that incorrect specificities will be assigned to enzymes of established CAZy families, and that several novel enzyme families may remain undescribed.

### 8.2.2 Lytic Polysaccharide Monooxygenases

In 1950, Reese et al. suggested that some unknown non-hydrolytic proteins may play a role in cellulose degradation in addition to hydrolytic enzymes (Reese et al. 1950). One type of these unknown non-hydrolytic proteins from fungi, which was classified into the GH61 family, has been shown to increase cellulose degradation when supplemented to cellulosic enzyme cocktails, although the activity of sole GH61 proteins toward cellulosic substrates was not detected (Harris et al. 2010; Karkehabadi et al. 2008). Until 2010, the catalytic role and function of GH61 members were still unknown, even after the first structure of GH61 protein (Cel61A from *T. reesei*) was published (Karkehabadi et al. 2008). Finally, in 2010, Vaaje-Kolstad et al demonstrated for the first time that a bacterium-derived GH61-like protein (CBM33) cleaves the glycosidic bonds via oxidative mechanisms using molecular oxygen ($O_2$) and external electron donors (Vaaje-Kolstad et al. 2010). Later, GH61 proteins were demonstrated to oxidatively cleave cellulose (Beeson et al. 2012; Forsberg et al. 2011; Quinlan et al. 2011), and renamed as polysaccharide monooxygenases (PMOs) or lytic polysaccharide monooxygenases (LPMOs). In 2013, the GH61 was reclassified as AA9 and AA10, respectively (Lombard et al. 2014). Next, AA11 LPMOs were found in fungi, and characterized to be only active toward chitin (Hemsworth et al. 2014). AA13 members are only present in fungi and characterized to be starch-active LPMOs (Lo Leggio et al. 2015). In the beginning of 2018, AA14 LPMOs were identified in fungi, showing activity toward xylan, but only when associated with cellulose (Couturier et al. 2018). One month later, AA15 LPMOs were discovered from the genomes of marine and terrestrial invertebrates, fungi, algae, and oomycetes (Sabbadin et al. 2018). AA15 LPMOs were shown to be active toward both cellulose and chitin (Sabbadin et al. 2018). AA16 has also been described as LPMO family, of which one protein was reported to cleave cellulose, although LPMO activity of other AA16 members was not confirmed

(Filiatrault-Chastel et al. 2019; Sun et al. 2023). The newest LPMO family is AA17, which only contains proteins from oomycetes and is shown to oxidatively cleave pectin substrate (demethylated homogalacturonan) (Sabbadin et al. 2021).

The first structure of a fungal AA9 LPMO was published in 2008 (Karkehabadi et al. 2008). Subsequently, more structures of AA9–AA11 and AA13–AA17 LPMOs have been elucidated (Couturier et al. 2018; Hemsworth et al. 2014; Sabbadin et al. 2018, 2021; Sun et al. 2023). Although generally, LPMOs exhibit a low amino acid sequence similarity, both in between and within AA families, overall they display a common fibronectin–/immunoglobulin-like β-sandwich core structure (Frandsen et al. 2016; Vaaje-Kolstad et al. 2017). Different from GH enzymes, which have a surface cleft or groove, the active site of LPMOs is located at a relatively flat, solvent (and substrate) exposed surface. LPMOs have a single copper atom in the active site, which is coordinated by an N-terminal histidine, a second histidine, and a third aromatic amino acid. This structural coordination is recognized as a histidine-brace motif, which is conserved in all LPMOs (Meier et al. 2017; Quinlan et al. 2011; Tandrup et al. 2018).

Although LPMOs were discovered more than a decade ago, catalytic mechanism of LPMOs is still not fully clear. Quinlan and coworkers reported that an AA9 LPMO *Ta*GH61A is a mono copper-dependent monooxygenase, which can release oxidized cello-oligosaccharides from cellulose in the presence of $O_2$ and reductant, such as gallate that was used in the study (Quinlan et al. 2011). Other studies further confirmed these results, and also showed that extracellular redox enzyme-based electron-donating systems (e.g., cellobiose dehydrogenases (CDHs)) can also activate the AA9 LPMOs in addition to small molecular reductants (Kracher et al. 2016; Loose et al. 2014, 2016; Tan et al. 2015). Subsequent work and studies focused on insights into the mechanism of the LPMO-catalyzed reaction, and several routes were proposed (Horn et al. 2012; Johansen 2016; Span and Marletta 2015; Walton and Davies 2016; Wang et al. 2020). It is well accepted that the LPMO catalytic cycle starts with a priming one-electron reduction, meaning that a Cu(II) is reduced to Cu(I) by external electron donors (Fig. 8.1). After this priming reduction, the catalytic reaction can follow two routes depending on the co-substrate, which can either be $H_2O_2$ or $O_2$ (Fig. 8.1) (Beeson et al. 2015; Bissaro et al. 2017; Hangasky et al. 2018). The mechanistic details of these routes are still under debate as extensively reviewed elsewhere (Manavalan et al. 2021; Wang et al. 2020). In the absence of a carbohydrate substrate, superoxide (or hydroperoxide) is released, resulting in the production of $H_2O_2$ and regeneration of the Cu(II) state (Fig. 8.1) (Kittl et al. 2012). Although $H_2O_2$ has been shown to enhance the LPMO catalytic reaction, overdosed $H_2O_2$ results in oxidative damage of the LPMO active site and further leads to the self-inactivation of LPMOs (Bissaro et al. 2017).

One intriguing feature is that LPMOs display selective oxidation of carbon position in the glycosidic linkages, which is also referred to as regioselectivity (Fig. 8.1). Regarding fungal AA9 LPMOs, Quinlan et al. were the first to determine that *Ta*GH61A oxidatively cleaved cellulose at both C1- and C4-carbon position (Quinlan et al. 2011). Afterwards, more and more AA9 LPMOs have been characterized for their regioselectivity, which was extensively reviewed elsewhere (Frommhagen et al. 2018). Oxidation at the C6-carbon position was suggested to occur for *Pa*GH61B from the ascomycete fungus *Podospora anserina* (Bey et al. 2013), however, the C6-oxidation was not confirmed by other studies. It is also proposed that double C4- and C6-oxidation occurred (Sun et al. 2022a). Interestingly, many fungal species, such as *P. anserina, Neurospora crassa* (*Nc*) or *Myceliophthora thermophila* (*Mt*) produce a variety of LPMOs differing in C1- or C4- or both C1- and C4-oxidizing abilities (Frommhagen et al. 2018). The reason why LPMOs exhibit regioselectivities has not been elucidated yet. It is proposed that the regioselectivity reflects how the LPMOs interact with the substrates (e.g., exposed cellulose-fiber surface) (Frommhagen et al. 2018). Borisova and coworkers suggested that

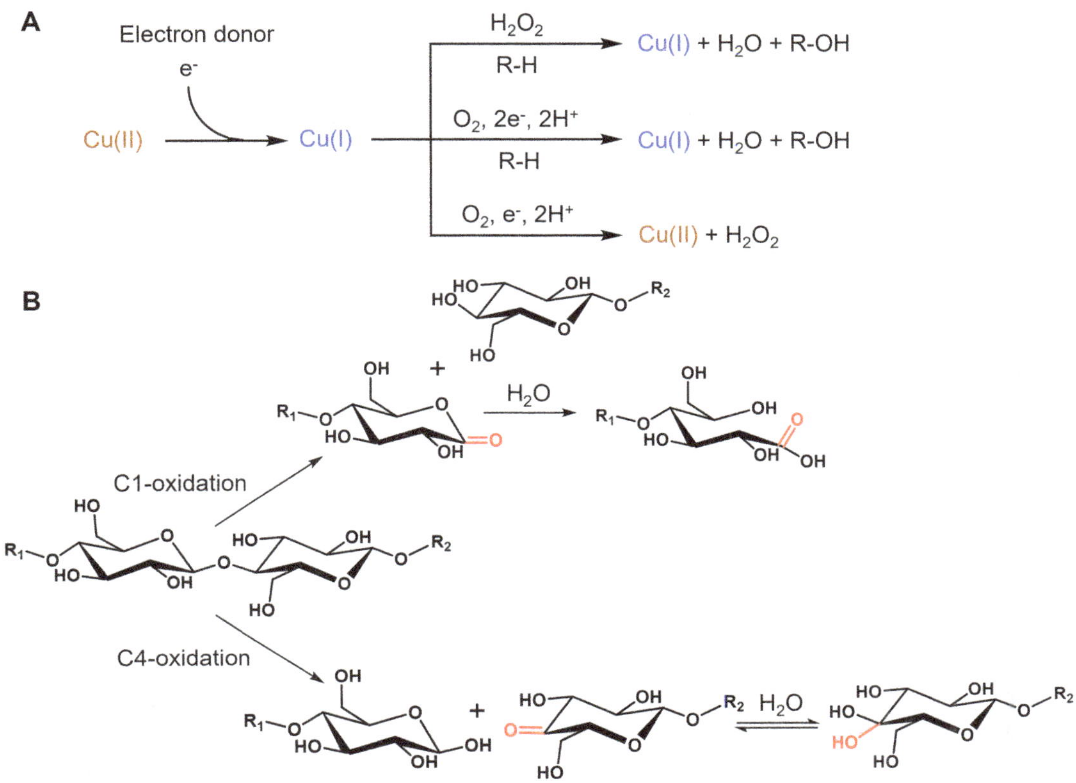


**Fig. 8.1** (**a**) Proposed catalytic routes for LPMO reactions using $H_2O_2$ or $O_2$. In both $H_2O_2$ and $O_2$-dependent routes, the C1- and/or C4-carbon position of the carbohydrate (R-H) substrate is hydroxylated (R-OH) and a Cu(I) is ready for the next catalytic cycle. (**b**) The oxidation at the C1-carbon leads to the formation of δ-lactone, which is converted to an aldonic acid in water. The oxidation at C4-carbon results in the generation of 4-ketoaldose, which is in pH-dependent equilibrium with the geminal diol. Adapted from (Sun et al. 2023) and Wang et al. (2019)

differences in the accessibility of the solvent-facing axial position around the active site may explain the regioselectivity of LPMOs (Borisova et al. 2015). Vu and coworkers performed amino acid sequence analysis of 497 (putative) AA9 LPMOs with their published regioselectivity, and four groups (PMO1, PMO2, PMO3, and PMO3*) were classified (Vu et al. 2014). This study showed that regioselectivity might be correlated to the configuration of loops that were renamed as active site segments in later studies. This correlation is further strengthened by Laurent et al. who identified a more extended structural determination of regioselectivity based on a structure-based multiple sequence alignment (Laurent et al. 2019; Vu et al. 2014). However, with more and more published AA9 LPMO data regarding regioselectivity and amino acid sequences, an increased number of exceptions was shown in each LPMO group, indicating that understanding of LPMO structure-regioselectivity correlation is still inadequate. The structure-regioselectivity correlation for other fungal LPMO families has not been studied yet.

Until now, fungal AA9 LPMOs have obtained the most diverse substrate specificity of all characterized LPMOs. Initially, AA9 LPMOs were shown to only oxidize cellulose. In 2014, Isaksen and coworkers showed that *Nc*LPMO9C is able to oxidatively cleave soluble cello-oligosaccharides (Isaksen et al. 2014). Later, Borisova and coworkers found that *Nc*LPMO9C is also active toward xyloglucan, glucomannan, and mixed β-(1 → 3, 1 → 4)-linked glucan (Borisova et al. 2015). From this time onward, oxidative cleavage of hemicellulosic substrates has been observed for various AA9 LPMOs (Agger et al. 2014; Bennati-Granier et al. 2015; Cannella et al. 2016; Fanuel et al. 2017; Frommhagen et al. 2016; Kojima et al. 2016; Ladeveze et al. 2017; Nekiunaite et al. 2016; Simmons et al. 2017; Sun et al. 2020). Frommhagen et al. reported that *Mt*LPMO9A is able to oxidatively cleave xylan, but only when associated with cellulose (Frommhagen et al. 2015). Hüttner et al. found that multiple AA9 LPMOs from *Malbranchea cinnamomea* (*Mc*) display diverse hemicellulose specificity including xylan, glucomannan, and xyloglucan (Huttner et al. 2019). Notably, *Mc*LPMO9H (*Mc*AA9H) preferentially cleaves xylan when exposed to cellulose-xylan blends to release (oxidized) xylo-oligosaccharides (Huttner et al. 2019). A similar preference for xylan cleavage when associated with cellulose was also found for *Nc*LPMO9F and *Nc*LPMO9L (Hegnar et al. 2021). So far only one LPMO (*Ls*LPMO9A) from the basidiomycete fungus *Lentinus similis* has been shown to solely act on glucuronoxylan (Simmons et al. 2017). In a recent study, Martinez-D'Alto and coworkers reported that *Mo*LPMO9A from the plant pathogen *Magnaporthe oryzae* (*Pyricularia oryzae*) is only active toward mixed β-(1 → 3, 1 → 4)-linked glucan, which reflects the still poor understanding of substrate specificity and biological functions of LPMOs (Martinez-D'Alto et al. 2023). The substrate specificity of other fungal LPMO families is not currently understood.

The above-described LPMO diversity of structures, catalytic routes, regioselectivity, and substrate specificity reflect the functional diversity of the biological processes in nature that LPMOs are involved in. It can be expected that the understanding of LPMO diversity helps to develop LPMO-supplemented application-tailored enzyme products or cocktails. For instance, cellulose is associated with hemicelluloses such as xylan, xyloglucan, and glucomannan even after pretreatment, and the additional hemicellulosic side-activities from cellulases might contribute to the overall plant cell wall polysaccharide degradation. Therefore, LPMOs with broad substrate specificity, especially active on xylan, such as *Mt*LPMO9A, *Ls*LPMO9A, and *Mc*LPMO9H, or active on glucomannan such as *Nc*LPMO9C, are interesting to study in relation to cellulose degradation from real plant biomass.

### 8.2.3 Current Status of Fungal Genomes and Post-Genomics in Relation to Plant Biomass Degradation

The number of available fungal genomes is rapidly increasing, as evidenced by the number in the Mycocosm database (https://mycocosm.jgi.doe.gov/mycocosm/home) (Grigoriev et al. 2014). While the relevance of a species for industry or society has been a major driver in the selection of species for genome sequencing, the ongoing 1000 fungal genome project (http://genome.jgi.doe.gov/pages/fungi-1000-projects-user-guidelines.jsf) has been working toward reference genome sequences for every clade in the fungal tree of life, and therefore has resulted in the reduction of the heavy bias observed in the initial set of available fungal genomes.

CAZy annotation is now a standard aspect of these genomes, facilitating quick comparison of the plant biomass degradation potential of fungi. It also revealed the high diversity between fungi with respect to the number of candidate genes per CAZy family in their genomes. To illustrate this, a comparison of the variation in gene numbers of the ten plant biomass degradation-related CAZy families with the highest average gene number was performed across the fungal kingdom (Fig. 8.2). The error bars show the high variations for them, even when this was analyzed within a fungal genus, *Aspergillus* (source: https://mycocosm.jgi.doe.gov/mycocosm/home).

The advancement in genome sequencing technologies has resulted in a huge increase in species with a public genome sequence and has expanded the genome-driven studies in fungi, including those addressing plant biomass degradation. Already in 2012, a study in which genomes of 18 Dothidiomycetes were compared was published that revealed significant differences in CAZy numbers in subgroups of these

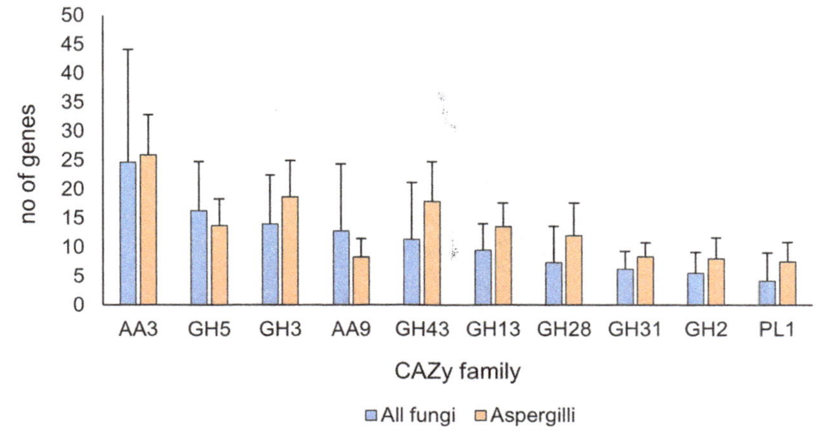


**Fig. 8.2** Diversity of gene numbers in CAZy families related to plant biomass degradation in fungi and in the genus *Aspergillus*. The error bars reflect the standard deviation of the gene number per family. Only the ten families with the highest average number of genes are displayed

fungi, particularly for CAZy families related to cellulose degradation (Ohm et al. 2012). Since then, comparisons of larger groups of fungi have become more common, although not all of these studies addressed the plant biomass degradation potential of fungi. Some examples of studies that address this topic are mentioned here.

The earlier mentioned Dothidiomycete study of 2012 was expanded to 101 species in a new study in 2020 (Haridas et al. 2020), which revealed that saprobic species had in general larger sets of CAZy genes involved in plant biomass degradation than phytopathogenic species. It also concluded that the ancestral Dothidiomycete had a saprobic lifestyle, suggesting gene deletions in the descendants with a pathogenic lifestyle.

Basidiomycete mushroom-forming fungi contain many ectomycorrhizal species and two studies explored the evolutionary transition to this lifestyle using genome sequences of species with ectomycorrhizal and other lifestyles (Kohler et al. 2015; Looney et al. 2022). This revealed that among other features ectomycorrhizal lifestyles are accompanied by a reduction in genes involved in plant biomass degradation. Evolutionary patterns in CAZy gene content related to plant biomass degradation were also observed when 50 basidiomycete Polyporales genomes were compared (Hage et al. 2021), supporting the important role that carbon source acquisition plays in evolution.

A higher sampling of individual and sets of related genera has resulted in better insights into the diversity within fungal genera. Analysis of 24 genomes of the genus *Lentinula* revealed high conservation of their set of CAZy genes involved in plant biomass degradation, reflecting their conserved ecological lifestyle as saprobes on hardwood (Sierra-Patev et al. 2023). Currently, three fungal whole-genus genome sequencing projects are at various stages of completion through projects with the Joint Genome Institute, addressing the ascomycete genera *Aspergillus*, *Penicillium*, and *Trichoderma*. While the latter two are still in progress, the *Aspergillus* project is close to completion, with already several papers published (Kjaerbolling et al. 2020; Vesth et al. 2018). This revealed high variation in the plant biomass-degrading potential of *Aspergillus* species, and therefore confirmed earlier studies on smaller sets from this genus (Benoit et al. 2015; de Vries et al. 2017; Mäkelä et al. 2018).

In the years since the last version of this Mycota volume, functional genomics of fungi has become a mature field with many transcriptomic and proteomic studies having been published. Many of these studies address the utilization of plant biomass and its components due to the relevance of carbon utilization for fungal survival in its natural biotopes and for fungal biotechnology. However, the division of such studies is clearly biased to some fungal species with a long history of use in research and bio-

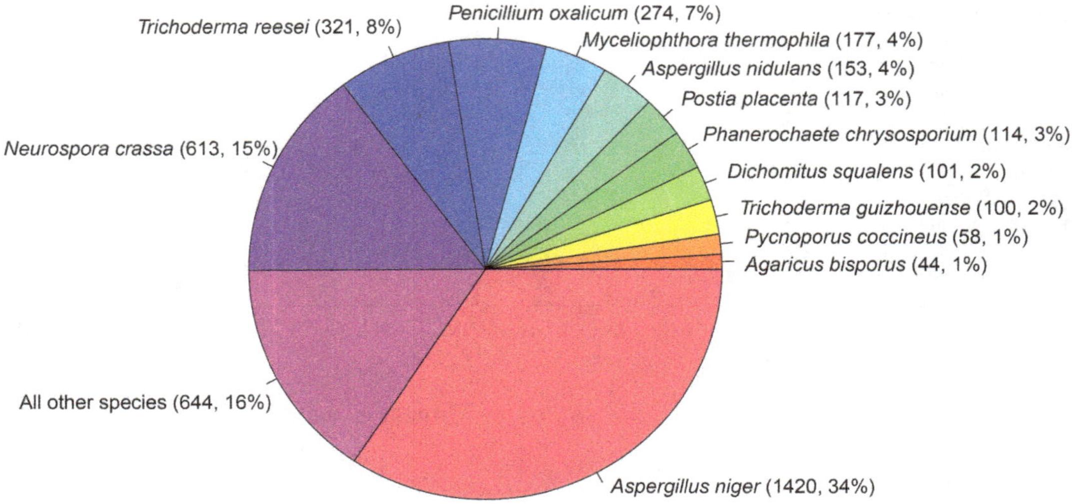


**Fig. 8.3** Distribution of transcriptome samples related to plant biomass conversion in different fungi. The survey is based on search of corresponding data in NCBI GEO (https://www.ncbi.nlm.nih.gov/geo/) and SRA databases (https://www.ncbi.nlm.nih.gov/sra)

technology, in particular *A. niger* (Fig. 8.3). This bias in part reflects the size of the community studying these species for plant biomass degradation but is also strongly affected by the approach used by some groups in this community that employs large sets of transcriptomic samples in their studies. While many of these studies remain species centric, more comparative studies are appearing, enabling a more detailed view on the post-genomic diversity of plant biomass degradation in fungi. A recent study compared the transcriptomic response of five diverse fungi to the same plant biomass substrates and revealed strong diversity, suggesting different preferences with respect to the plant biomass component they use as a carbon source (Li et al. 2023). High diversity was also observed in the transcriptomic response of white-rot and brown-rot species during growth on wood wafers (Zhang et al. 2019), which was further supported by a proteomic analysis of secreted proteins of several species (Presley et al. 2018). However, a more global comparison of basidiomycete transcriptomes revealed that despite high diversity, there is a core gene set that appears to be consistently expressed during growth on plant biomass (Peng et al. 2017a). Similarly, a conserved set of CAZy genes was expressed during growth *in planta* of six phytopathogenic fungi (Mesny et al. 2021).

The combined insights of all the species-specific and comparative post-genomic studies revealed that the diversity in the plant biomass-degrading approach of fungi goes far beyond the diversity observed in fungal genome sequences. It appears that regulation of gene expression may have a dominant influence and may even cause strong differences between isolates of the same species.

## 8.3 Conclusions

The continued increase in fungal genomes, and transcriptomics and proteomics datasets is demonstrating the high diversity of the approaches used by fungi to obtain their carbon from plant biomass. However, many challenges remain to be solved to apply the full width of the fungal kingdom in biotechnology. Some of these challenges are technical, such as the high percentage of incorrect gene models in fungal genomes and the poor biochemical support of most of the CAZy families. However, while variation within fungal genera is being addressed, very few of these studies address the intraspecies variation. In addition, most of the fungal tree of life has poor support from post-genomic studies to be able to better evaluate the potential encoded in their genomes.

The (post-)genomic data and resources are enabling a more system-based approach to the study of plant biomass degradation by fungi. Rather than a main focus on the extracellular enzymes specifically involved in the degradation of the plant biomass polymers, studies are now also addressing the other aspects required to use plant biomass as a carbon source such as sugar transport (Gu et al. 2023; Wang et al. 2022; Xu et al. 2024) and metabolism (Li et al. 2022a; Steindorff et al. 2021) and the regulatory systems controlling them.

It can be expected that the coming years will see a further increase in data related to plant biomass degradation, especially in fungi already applied in biotechnology. These species will therefore remain forerunners and references for scientists studying other species.

## References


Agger JW, Isaksen T, Várnai A, Vidal-Melgosa S, Willats WG, Ludwig R, Horn SJ, Eijsink VG, Westereng B (2014) Discovery of LPMO activity on hemicelluloses shows the importance of oxidative processes in plant cell wall degradation. Proc Natl Acad Sci USA 111:6287–6292. https://doi.org/10.1073/pnas.1323629111

Beeson WT, Phillips CM, Cate JH, Marletta MA (2012) Oxidative cleavage of cellulose by fungal copper-dependent polysaccharide monooxygenases. J Am Chem Soc 134:890–892. https://doi.org/10.1021/ja210657t

Beeson WT, Vu VV, Span EA, Phillips CM, Marletta MA (2015) Cellulose degradation by polysaccharide monooxygenases. Annu Rev Biochem 84:923–946. https://doi.org/10.1146/annurev-biochem-060614-034439

Bennati-Granier C, Garajova S, Champion C, Grisel S, Haon M, Zhou S, Fanuel M, Ropartz D, Rogniaux H, Gimbert I, Record E, Berrin J-G (2015) Substrate specificity and regioselectivity of fungal AA9 lytic polysaccharide monooxygenases secreted by *Podospora anserina*. Biotechnol Biofuels 8:90. https://doi.org/10.1186/s13068-015-0274-3

Benoit I, Culleton H, Zhou M, DiFalco M, Aguilar-Osorio G, Battaglia E, Bouzid O, Brouwer CPJM, El-Bushari HBO, Coutinho PM, Gruben BS, Hildén KS, Houbraken J, Barboza LAJ, Levasseur A, Majoor E, Mäkelä MR, Narang H-M, Trejo-Aguilar B, van den Brink J, vanKuyk PA, Wiebenga A, McKie V, McCleary B, Tsang A, Henrissat B, de Vries RP (2015) Closely related fungi employ diverse enzymatic strategies to degrade plant biomass. Biotechnol Biofuels 8(1):107. https://doi.org/10.1186/s13068-015-0285-0

Bey M, Zhou S, Poidevin L, Henrissat B, Coutinho PM, Berrin J-G, Sigoillot JC (2013) Cello-oligosaccharide oxidation reveals differences between two lytic polysaccharide monooxygenases (family GH61) from *Podospora anserina*. Appl Environ Microbiol 79(2):488–496. https://doi.org/10.1128/AEM.02942-12

Biely P, Cziszarova M, Agger JW, Li XL, Puchart V, Vrsanska M, Eijsink VG, Westereng B (2014a) *Trichoderma reesei* CE16 acetyl esterase and its role in enzymatic degradation of acetylated hemicellulose. Biochim Biophys Acta 1840(1):516–525. https://doi.org/10.1016/j.bbagen.2013.10.008

Biely P, Puchart V, Stringer MA, Morkeberg Krogh KB (2014b) *Trichoderma reesei* XYN VI–a novel appendage-dependent eukaryotic glucuronoxylan hydrolase. FEBS J 281(17):3894–3903. https://doi.org/10.1111/febs.12925

Bissaro B, Rohr AK, Muller G, Chylenski P, Skaugen M, Forsberg Z, Horn SJ, Vaaje-Kolstad G, Eijsink VGH (2017) Oxidative cleavage of polysaccharides by monocopper enzymes depends on $H_2O_2$. Nat Chem Biol 13:1123–1128. https://doi.org/10.1038/nchembio.2470

Boerjan W, Ralph J, Baucher M (2003) Lignin biosynthesis. Annu Rev Plant Biol 54:519–546

Borisova AS, Isaksen T, Dimarogona M, Kognole AA, Mathiesen G, Várnai A, Rohr AK, Payne CM, Sorlie M, Sandgren M, Eijsink VGH (2015) Structural and functional characterization of a lytic polysaccharide monooxygenase with broad substrate specificity. J Biol Chem 290:22955–22969. https://doi.org/10.1074/jbc.M115.660183

Buranov AU, Mazza G (2008) Lignin in straw of herbaceous crops. Ind Crop Prod 28:237–259

Cannella D, Mollers KB, Frigaard NU, Jensen PE, Bjerrum MJ, Johansen KS, Felby C (2016) Light-driven oxidation of polysaccharides by photosynthetic pigments and a metalloenzyme. Nat Commun 7:11134. https://doi.org/10.1038/ncomms11134

Carpita NC, Gibeaut DM (1993) Structural models of primary cell walls in flowering plants: consistency of molecular structure with the physical properties of the walls during growth. Plant J 3:1–30

Coconi Linares N, Dilokpimol A, Stålbrand H, Mäkelä MR, de Vries RP (2020) Recombinant production and characterization of six novel GH27 and GH36 α-galactosidases from *Penicillium subrubescens* and their synergism with a commercial mannanase during the hydrolysis of lignocellulosic biomass. Bioresour Technol 295:122258. https://doi.org/10.1016/j.biortech.2019.122258

Coconi Linares N, Li X, Dilokpimol A, de Vries RP (2022) Comparative characterization of nine novel GH51, GH54 and GH62 α-L-arabinofuranosidases from *Penicillium subrubescens*. FEBS Lett 596(3):360–368. https://doi.org/10.1002/1873-3468.14278

Couturier M, Ladeveze S, Sulzenbacher G, Ciano L, Fanuel M, Moreau C, Villares A, Cathala B, Chaspoul F, Frandsen KE, Labourel A, Herpoël-Gimbert I, Grisel S, Haon M, Lenfant N, Rogniaux H, Ropartz D, Davies GJ, Rosso MN, Walton PH, Henrissat B, Berrin J-G (2018) Lytic xylan oxidases from wood-decay fungi unlock biomass degradation. Nat Chem Biol 14:306–310. https://doi.org/10.1038/nchembio.2558

Cragg SM, Beckham GT, Bruce NC, Bugg TD, Distel DL, Dupree P, Etxabe AG, Goodell BS, Jellison J, McGeehan JE, McQueen-Mason SJ, Schnorr K, Walton PH, Watts JE, Zimmer M (2015) Lignocellulose degradation mechanisms across the tree of life. Curr Opin Chem Biol 29:108–119. https://doi.org/10.1016/j.cbpa.2015.10.018

de Vries RP, Mäkelä MR (2020) Genomic and postgenomic diversity of fungal plant biomass degradation approaches. Trends Microbiol 28(6):487–499. https://doi.org/10.1016/j.tim.2020.01.004

de Vries RP, Riley R, Wiebenga A, Aguilar-Osorio G, Amillis S, Uchima CA, Anderluh G, Asadollahi M, Askin M, Barry K, Battaglia E, Bayram O, Benocci T, Braus-Stromeyer SA, Caldana C, Canovas D, Cerqueira GC, Chen F, Chen W, Choi C, Clum A, Dos Santos RA, Damasio AR, Diallinas G, Emri T, Fekete E, Flipphi M, Freyberg S, Gallo A, Gournas C, Habgood R, Hainaut M, Harispe ML, Henrissat B, Hildén KS, Hope R, Hossain A, Karabika E, Karaffa L, Karanyi Z, Krasevec N, Kuo A, Kusch H, LaButti K, Lagendijk EL, Lapidus A, Levasseur A, Lindquist E, Lipzen A, Logrieco AF, MacCabe A, Mäkelä MR, Malavazi I, Melin P, Meyer V, Mielnichuk N, Miskei M, Molnar AP, Mule G, Ngan CY, Orejas M, Orosz E, Ouedraogo JP, Overkamp KM, Park HS, Perrone G, Piumi F, Punt PJ, Ram AF, Ramon A, Rauscher S, Record E, Riano-Pachon DM, Robert V, Rohrig J, Ruller R, Salamov A, Salih NS, Samson RA, Sandor E, Sanguinetti M, Schutze T, Sepcic K, Shelest E, Sherlock G, Sophianopoulou V, Squina FM, Sun H, Susca A, Todd RB, Tsang A, Unkles SE, van de Wiele N, van Rossen-Uffink D, Oliveira JV, Vesth TC, Visser J, Yu JH, Zhou M, Andersen MR, Archer DB, Baker SE, Benoit I, Brakhage AA, Braus GH, Fischer R, Frisvad JC, Goldman GH, Houbraken J, Oakley B, Pocsi I, Scazzocchio C, Seiboth B, vanKuyk PA, Wortman J, Dyer PS, Grigoriev IV (2017) Comparative genomics reveals high biological diversity and specific adaptations in the industrially and medically important fungal genus *Aspergillus*. Genome Biol 18(1):28. https://doi.org/10.1186/s13059-017-1151-0

de Vries RP, Visser J (2001) *Aspergillus* enzymes involved in degradation of plant cell wall polysaccharides. Microb Mol Biol Rev 65:497–522

del Rio JC, Rencoret J, Gutiérrez A, Elder T, Kim H, Ralph J (2020) Lignin monomers from beyond the canonical monolignol biosynthetic pathway: another brick in the wall. ACS Sust Chem Eng 8(13):4997–5012

Drula E, Garron ML, Dogan S, Lombard V, Henrissat B, Terrapon N (2022) The carbohydrate-active enzyme database: functions and literature. Nucleic Acids Res 50(D1):D571–D577. https://doi.org/10.1093/nar/gkab1045

Espinoza K, Eyzaguirre J (2018) Identification, heterologous expression and characterization of a novel glycoside hydrolase family 30 xylanase from the fungus *Penicillium purpurogenum*. Carbohydr Res 468:45–50. https://doi.org/10.1016/j.carres.2018.08.006

Fanuel M, Garajova S, Ropartz D, McGregor N, Brumer H, Rogniaux H, Berrin J-G (2017) The *Podospora anserina* lytic polysaccharide monooxygenase PaLPMO9H catalyzes oxidative cleavage of diverse plant cell wall matrix glycans. Biotechnol Biofuels 10:63. https://doi.org/10.1186/s13068-017-0749-5

Filiatrault-Chastel C, Navarro D, Haon M, Grisel S, Herpoël-Gimbert I, Chevret D, Fanuel M, Henrissat B, Heiss-Blanquet S, Margeot A, Berrin J-G (2019) AA16, a new lytic polysaccharide monooxygenase family identified in fungal secretomes. Biotechnol Biofuels 12:55. https://doi.org/10.1186/s13068-019-1394-y

Forsberg Z, Vaaje-Kolstad G, Westereng B, Bunaes AC, Stenstrom Y, MacKenzie A, Sorlie M, Horn SJ, Eijsink VGH (2011) Cleavage of cellulose by a CBM33 protein. Prot Sci 20:1479–1483. https://doi.org/10.1002/pro.689

Frandsen KEH, Simmons TJ, Dupree P, Poulsen JCN, Hemsworth GR, Ciano L, Johnston EM, Tovborg M, Johansen KS, von Freiesleben P, Marmuse L, Fort S, Cottaz S, Driguez H, Henrissat B, Lenfant N, Tuna F, Baldansuren A, Davies GJ, Lo Leggio L, Walton PH (2016) The molecular basis of polysaccharide cleavage by lytic polysaccharide monooxygenases. Nat Chem Biol 12(4):298. https://doi.org/10.1038/Nchembio.2029

Frommhagen M, Koetsier MJ, Westphal AH, Visser J, Hinz SW, Vincken JP, van Berkel WJ, Kabel MA, Gruppen H (2016) Lytic polysaccharide monooxygenases from *Myceliophthora thermophila* C1 differ in substrate preference and reducing agent specificity. Biotechnol Biofuels 9(1):186. https://doi.org/10.1186/s13068-016-0594-y

Frommhagen M, Sforza S, Westphal AH, Visser J, Hinz SW, Koetsier MJ, van Berkel WJ, Gruppen H, Kabel MA (2015) Discovery of the combined oxidative cleavage of plant xylan and cellulose by a new fungal polysaccharide monooxygenase. Biotechnol Biofuels 8:101. https://doi.org/10.1186/s13068-015-0284-1

Frommhagen M, Westphal AH, Van Berkel WJ, Kabel MA (2018) Distinct substrate specificities and electron-donating systems of fungal lytic polysaccharide monooxygenases. Front Microbiol 9:1080

Gilbert HJ, Stalbrand H, Brumer H (2008) How the walls come crumbling down: recent structural biochemistry of plant polysaccharide degradation. Curr Opin Plant Biol 11:338–348

Grigoriev IV, Nikitin R, Haridas S, Kuo A, Ohm R, Otillar R, Riley R, Salamov A, Zhao X, Korzeniewski F, Smirnova T, Nordberg H, Dubchak I, Shabalov I (2014) MycoCosm portal: gearing up for 1000 fungal genomes. Nucleic Acids Res 42(Database issue):D699. https://doi.org/10.1093/nar/gkt1183

Gu S, Zhao Z, Xue F, Liu D, Liu Q, Li J, Tian C (2023) The arabinose transporter MtLat-1 is involved in hemicellulase repression as a pentose transceptor in *Myceliophthora thermophila*. Biotechnol Biofuels Bioprod 16(1):51. https://doi.org/10.1186/s13068-023-02305-3

Hage H, Miyauchi S, Viragh M, Drula E, Min B, Chaduli D, Navarro D, Favel A, Norest M, Lesage-Meessen L, Balint B, Merenyi Z, de Eugenio L, Morin E, Martinez AT, Baldrian P, Stursova M, Martinez MJ, Novotny C, Magnuson JK, Spatafora JW, Maurice S, Pangilinan J, Andreopoulos W, LaButti K, Hundley H, Na H, Kuo A, Barry K, Lipzen A, Henrissat B, Riley R, Ahrendt S, Nagy LG, Grigoriev IV, Martin F, Rosso MN (2021) Gene family expansions and transcriptome signatures uncover fungal adaptations to wood decay. Environ Microbiol 23(10):5716–5732. https://doi.org/10.1111/1462-2920.15423

Hangasky JA, Iavarone AT, Marletta MA (2018) Reactivity of $O_2$ versus $H_2O_2$ with polysaccharide monooxygenases. Proc Natl Acad Sci USA 115:4915–4920. https://doi.org/10.1073/pnas.1801153115

Hantus S, Pauly M, Darvill AG, Albersheim P, York WS (1997) Structural characterization of novel L-galactose-containing oligosaccharide subunits of jojoba seed xyloglucans. Carbohydr Res 304:11–20

Haridas S, Albert R, Binder M, Bloem J, LaButti K, Salamov A, Andreopoulos B, Baker SE, Barry K, Bills G, Bluhm BH, Cannon C, Castanera R, Culley DE, Daum C, Ezra D, Gonzalez JB, Henrissat B, Kuo A, Liang C, Lipzen A, Lutzoni F, Magnuson J, Mondo SJ, Nolan M, Ohm RA, Pangilinan J, Park HJ, Ramirez L, Alfaro M, Sun H, Tritt A, Yoshinaga Y, Zwiers LH, Turgeon BG, Goodwin SB, Spatafora JW, Crous PW, Grigoriev IV (2020) 101 Dothideomycetes genomes: a test case for predicting lifestyles and emergence of pathogens. Stud Mycol 96:141–153. https://doi.org/10.1016/j.simyco.2020.01.003

Harris PV, Welner D, McFarland K, Re E, Navarro Poulsen J-C, Brown K, Salbo R, Ding H, Vlasenko E, Merino S (2010) Stimulation of lignocellulosic biomass hydrolysis by proteins of glycoside hydrolase family 61: structure and function of a large, enigmatic family. Biochemist 49(15):3305–3316

Hasper AA, Dekkers E, van Mil M, van de Vondervoort PJ, de Graaff LH (2002) EglC, a new endoglucanase from *Aspergillus niger* with major activity towards xyloglucan. Appl Environ Microbiol 68(4):1556–1560

Hatakka A, Hammel KE (2010) Fungal biodegradation of lignocelluloses. In: Esser K, Hofrichter M (eds) The Mycota: a comprehensive treatise on fungi as experimental systems for basic and applied research industrial applications, vol X, 2nd edn. Springer, Berlin, Heidelberg, pp 319–340

Hegnar OA, Østby H, Petrović DM, Olsson L, Várnai A, Eijsink VG (2021) Quantifying oxidation of cellulose-associated glucuronoxylan by two lytic polysaccharide monooxygenases from *Neurospora crassa*. Appl Environ Microbiol 87:e01652–e01621

Hemsworth GR, Henrissat B, Davies GJ, Walton PH (2014) Discovery and characterization of a new family of lytic polysaccharide monooxygenases. Nat Chem Biol 10(2):122–126. https://doi.org/10.1038/nchembio.1417

Higuchi T (1997) Biochemistry and molecular biology of wood. Springer Series in Wood Science. Springer Verlag, Berlin

Higuchi T (2006) Look back over the studies of lignin biochemistry. J Wood Sci 52:2–8

Hisamatsu M, Impallomeni G, York WS, Albersheim P, Darvill AG (1991) The structure of plant cell walls. 31. A new undecasaccharide subunit of xyloglucans with 2 α-L-fucosyl residues. Carbohydr Res 211:117–129

Hisamatsu M, York WS, Darvill AG, Albersheim P (1992) Characterization of seven xyloglucan oligosaccharides containing from seventeen to twenty glycosyl residues. Carbohydr Res 227:45–71

Horn SJ, Vaaje-Kolstad G, Westereng B, Eijsink VG (2012) Novel enzymes for the degradation of cellulose. Biotechnol Biofuels 5:45. https://doi.org/10.1186/1754-6834-5-45

Huisman MMH, Weel KGC, Schols HA, Voragen AGJ (2000) Xyloglucan from soybean (*Glycine max*) meal is composed of XXXG-type building units. Carbohydr Pol 42:185–191

Huttner S, Várnai A, Petrovic DM, Bach CX, Anh DTK, Thanh VN, Eijsink VGH, Larsbrink J, Olsson L (2019) Specific xylan activity revealed for AA9 lytic polysaccharide monooxygenases of the thermophilic fungus *Malbranchea cinnamomea* by functional characterization. Appl Environ Microbiol 85:e01408–e01419. https://doi.org/10.1128/AEM.01408-19

Isaksen T, Westereng B, Aachmann FL, Agger JW, Kracher D, Kittl R, Ludwig R, Haltrich D, Eijsink VG, Horn SJ (2014) A C4-oxidizing lytic polysaccharide monooxygenase cleaving both cellulose and cello-oligosaccharides. J Biol Chem 289:2632–2642

Iwai M, Kawakami T, Ikemoto T, Fujiwara D, Takenaka S, Nakazawa M, Ueda M, Sakamoto T (2015) Molecular characterization of a *Penicillium chrysogenum* exo-rhamnogalacturonan lyase that is structurally distinct from other polysaccharide lyase family proteins. Appl Microbiol Biotechnol 99(20):8515–8525. https://doi.org/10.1007/s00253-015-6600-7

Jacobs A, Lundqvist J, Stalbrand H, Tjerneld F, Dahlman O (2002) Characterization of water-soluble hemicelluloses from spruce and aspen employing SEC/MALDI mass spectroscopy. Carbohydr Res 337(8):711–717

Johansen KS (2016) Discovery and industrial applications of lytic polysaccharide mono-oxygenases. Biochem Soc Trans 44:143–149. https://doi.org/10.1042/BST20150204

Karkehabadi S, Hansson H, Kim S, Piens K, Mitchinson C, Sandgren M (2008) The first structure of a glycoside hydrolase family 61 member, Cel61B from *Hypocrea jecorina*, at 1.6 Å resolution. J Mol Biol 383(1):144–154

Katsimpouras C, Dedes G, Thomaidis NS, Topakas E (2019) A novel fungal GH30 xylanase with xylobiohydrolase auxiliary activity. Biotechnol Biofuels 12:120. https://doi.org/10.1186/s13068-019-1455-2

Kittl R, Kracher D, Burgstaller D, Haltrich D, Ludwig R (2012) Production of four *Neurospora crassa* lytic polysaccharide monooxygenases in *Pichia pastoris* monitored by a fluorimetric assay. Biotechnol Biofuels 5:79. https://doi.org/10.1186/1754-6834-5-79

Kjaerbolling I, Vesth T, Frisvad JC, Nybo JL, Theobald S, Kildgaard S, Petersen TI, Kuo A, Sato A, Lyhne EK, Kogle ME, Wiebenga A, Kun RS, Lubbers RJM, Mäkelä MR, Barry K, Chovatia M, Clum A, Daum C, Haridas S, He G, LaButti K, Lipzen A, Mondo S, Pangilinan J, Riley R, Salamov A, Simmons BA, Magnuson JK, Henrissat B, Mortensen UH, Larsen TO, de Vries RP, Grigoriev IV, Machida M, Baker SE, Andersen MR (2020) A comparative genomics study of 23 *Aspergillus* species from section Flavi. Nat Commun 11(1):1106. https://doi.org/10.1038/s41467-019-14051-y

Kohler A, Kuo A, Nagy LG, Morin E, Barry KW, Buscot F, Canbäck B, Choi C, Cichocki N, Clum A, Colpaert J, Copeland A, Costa MD, Doré J, Floudas D, Gay G, Girlanda M, Henrissat B, Herrmann S, Hess J, Högberg N, Johansson T, Khouja H-R, LaButti K, Lahrmann U, Levasseur A, Lindquist EA, Lipzen A, Marmeisse R, Martino E, Murat C, Ngan CY, Nehls U, Plett JM, Pringle A, Ohm RA, Perotto S, Peter M, Riley R, Rineau F, Ruytinx J, Salamov A, Shah F, Sun H, Tarkka M, Tritt A, Veneault-Fourrey C, Zuccaro A, Consortium MGI, Tunlid A, Grigoriev IV, Hibbett DS, Martin F (2015) Convergent losses of decay mechanisms and rapid turnover of symbiosis genes in mycorrhizal mutualists. Nat Genet 47:410–415. https://doi.org/10.1038/ng.3223

Kojima Y, Várnai A, Ishida T, Sunagawa N, Petrovic DM, Igarashi K, Jellison J, Goodell B, Alfredsen G, Westereng B, Eijsink VG, Yoshida M (2016) A lytic polysaccharide monooxygenase with broad xyloglucan specificity from the brown-rot fungus *Gloeophyllum trabeum* and its action on cellulose-xyloglucan complexes. Appl Environ Microbiol 82:6557–6572. https://doi.org/10.1128/AEM.01768-16

Kolpak FJ, Blackwell J (1976) Determination of the structure of cellulose II. Macromolecules 9:273–278

Koutaniemi S, van Gool MP, Juvonen M, Jokela J, Hinz SW, Schols HA, Tenkanen M (2013) Distinct roles of carbohydrate esterase family CE16 acetyl esterases and polymer-acting acetyl xylan esterases in xylan deacetylation. J Biotechnol 168(4):684–692. https://doi.org/10.1016/j.jbiotec.2013.10.009

Kracher D, Scheiblbrandner S, Felice AK, Breslmayr E, Preims M, Ludwicka K, Haltrich D, Eijsink VG, Ludwig R (2016) Extracellular electron transfer systems fuel cellulose oxidative degradation. Science 352:1098–1101. https://doi.org/10.1126/science.aaf3165

Ladeveze S, Haon M, Villares A, Cathala B, Grisel S, Herpoël-Gimbert I, Henrissat B, Berrin J-G (2017) The yeast *Geotrichum candidum* encodes functional lytic polysaccharide monooxygenases. Biotechnol Biofuels 10:215. https://doi.org/10.1186/s13068-017-0903-0

Laurent CVFP, Sun P, Scheiblbrandner S, Csarman F, Cannazza P, Frommhagen M, van Berkel WJH, Oostenbrink C, Kabel MA, Ludwig R (2019) Influence of lytic polysaccharide monooxygenase active site segments on activity and affinity. Int J Mol Sci 20:6219. https://doi.org/10.3390/ijms20246219

Li J, Chroumpi T, Garrigues S, Kun RS, Meng J, Salazar-Cerezo S, Aguilar-Pontes MV, Zhang Y, Tejomurthula S, Lipzen A, Ng V, Clendinen CS, Tolic N, Grigoriev IV, Tsang A, Mäkelä MR, Snel B, Peng M, de Vries RP (2022a) The sugar metabolic model of *Aspergillus niger* can only be reliably transferred to fungi of its phylum. J Fungi 8(12). https://doi.org/10.3390/jof8121315

Li J, Wiebenga A, Lipzen A, Ng V, Tejomurthula S, Zhang Y, Grigoriev IV, Peng M, de Vries RP (2023) Comparative genomics and transcriptomics analyses reveal divergent plant biomass-degrading strategies in fungi. J Fungi 9(8). https://doi.org/10.3390/jof9080860

Li X, Kouzounis D, Kabel MA, de Vries RP (2022b) GH10 and GH11 endoxylanases in *Penicillium subrubescens*: comparative characterization and synergy with GH51, GH54, GH62 α-L-arabinofuranosidases from the same fungus. New Biotechnol 70:84–92. https://doi.org/10.1016/j.nbt.2022.05.004

Li X, Kouzounis D, Kabel MA, de Vries RP, Dilokpimol A (2022c) Glycoside hydrolase family 30 harbors fungal subfamilies with distinct polysaccharide specificities. New Biotechnol 67:32–41. https://doi.org/10.1016/j.nbt.2021.12.004

Lo Leggio L, Simmons TJ, Poulsen JC, Frandsen KE, Hemsworth GR, Stringer MA, von Freiesleben P, Tovborg M, Johansen KS, De Maria L, Harris PV, Soong CL, Dupree P, Tryfona T, Lenfant N, Henrissat B, Davies GJ, Walton PH (2015) Structure and boosting activity of a starch-degrading lytic polysaccharide monooxygenase. Nat Commun 6:5961. https://doi.org/10.1038/ncomms6961

Lombard V, Ramulu HG, Drula E, Coutinho PM, Henrissat B (2014) The carbohydrate-active enzymes database (CAZy) in 2013. Nucleic Acids Res 42:D490–D495. https://doi.org/10.1093/nar/gkt1178

Looney B, Miyauchi S, Morin E, Drula E, Courty PE, Kohler A, Kuo A, LaButti K, Pangilinan J, Lipzen A, Riley R, Andreopoulos W, He G, Johnson J, Nolan M, Tritt A, Barry KW, Grigoriev IV, Nagy LG, Hibbett D, Henrissat B, Matheny PB, Labbe J, Martin FM (2022) Evolutionary transition to the ectomycorrhizal habit in the genomes of a hyperdiverse lineage of mushroom-forming fungi. New Phytol 233(5):2294–2309. https://doi.org/10.1111/nph.17892

Loose JS, Forsberg Z, Fraaije MW, Eijsink VG, Vaaje-Kolstad G (2014) A rapid quantitative activity assay shows that the *Vibrio cholerae* colonization factor GbpA is an active lytic polysaccharide monooxygenase. FEBS Lett 588(18):3435–3440. https://doi.org/10.1016/j.febslet.2014.07.036

Loose JS, Forsberg Z, Kracher D, Scheiblbrandner S, Ludwig R, Eijsink VG, Vaaje-Kolstad G (2016) Activation of bacterial lytic polysaccharide monooxygenases with cellobiose dehydrogenase. Prot Sci 25(12):2175–2186. https://doi.org/10.1002/pro.3043

Luo H, Yang J, Li J, Shi P, Huang H, Bai Y, Fan Y, Yao B (2010) Molecular cloning and characterization of the novel acidic xylanase XYLD from *Bispora sp*. MEY-1 that is homologous to family 30 glycosyl hydrolases. Appl Microbiol Biotechnol 86(6):1829–1839. https://doi.org/10.1007/s00253-009-2410-0

Mäkelä MR, DiFalco M, McDonnell E, Nguyen TTM, Wiebenga A, Hildén K, Peng M, Grigoriev IV, Tsang A, de Vries RP (2018) Genomic and exoproteomic diversity in plant biomass degradation approaches among Aspergilli. Stud Mycol 91:79–99. https://doi.org/10.1016/j.simyco.2018.09.001

Mäkelä MR, Hildén KS, de Vries RP (2014) Degradation and modification of plant biomass by fungi. In: Nowrousian M (ed) Mycota vol XIII Fungal Genomics, 2nd edn. Springer-Verlag, Berlin, pp 175–208

Manavalan T, Stepnov AA, Hegnar OA, Eijsink VGH (2021) Sugar oxidoreductases and LPMOs–two sides of the same polysaccharide degradation story? Carbohydr Res 505:108350. https://doi.org/10.1016/j.carres.2021.108350

Martinez-D'Alto A, Yan X, Detomasi TC, Sayler RI, Thomas WC, Talbot NJ, Marletta MA (2023) Characterization of a unique polysaccharide monooxygenase from the plant pathogen *Magnaporthe oryzae*. Proc Natl Acad Sci USA 120(8):e2215426120. https://doi.org/10.1073/pnas.2215426120

Maruyama K, Goto C, Numata M, Suzuki T, Nakagawa Y, Hoshino T, Uchiyama T (1996) *O*-acetylated xyloglucan in extracellular polysaccharides from cell-suspension cultures of *Mentha*. Phytochemistry 41:1309–1314

Master ER, Zheng Y, Storms R, Tsang A, Powlowski J (2008) A xyloglucan-specific family 12 glycosyl hydrolase from *Aspergillus niger*: recombinant expression, purification and characterization. Biochem J 411(1):161–170. https://doi.org/10.1042/BJ20070819

Matsushita J, Kato Y, Matsuda K (1985) Purification and properties of an α-D-xylosidase from *Aspergillus niger*. J Biochem 98(3):825–832

Matsushita J, Kato Y, Matsuda K (1987) Characterization of α-D-xylosidase II from *Aspergillus niger*. Agric Biol Chem 51(7):2

Meier KK, Jones SM, Kaper T, Hansson H, Koetsier MJ, Karkehabadi S, Solomon EI, Sandgren M, Kelemen B (2017) Oxygen activation by Cu LPMOs in recalcitrant carbohydrate polysaccharide conversion to monomer sugars. Chem Rev 118:2593. https://doi.org/10.1021/acs.chemrev.7b00421

Mesny F, Miyauchi S, Thiergart T, Pickel B, Atanasova L, Karlsson M, Huttel B, Barry KW, Haridas S, Chen C, Bauer D, Andreopoulos W, Pangilinan J, LaButti K, Riley R, Lipzen A, Clum A, Drula E, Henrissat B, Kohler A, Grigoriev IV, Martin FM, Hacquard S (2021) Genetic determinants of endophytism in the *Arabidopsis* root mycobiome. Nat Commun 12(1):7227. https://doi.org/10.1038/s41467-021-27479-y

Meyer V, Basenko EY, Benz JP, Braus GH, Caddick MX, Csukai M, de Vries RP, Endy D, Frisvad JC, Gunde-Cimerman N, Haarmann T, Hadar Y, Hansen K, Johnson RI, Keller NP, Krasevec N, Mortensen UH, Perez R, Ram AFJ, Record E, Ross P, Shapaval V, Steiniger C, van den Brink H, van Munster J, Yarden O, Wosten HAB (2020) Growing a circular economy with fungal biotechnology: a white paper. Fungal Biol Biotechnol 7:5. https://doi.org/10.1186/s40694-020-00095-z

Mohnen D (1999) Biosynthesis of pectins and galactomannans. In: Barton D, Nakanishi K, Meth-Cohn O (eds) Comprehensive natural products chemistry, Elsevier, vol 3. Dordrecht, The Netherlands, pp 497–527

Nakamichi Y, Fouquet T, Ito S, Matsushika A, Inoue H (2019a) Mode of action of GH30-7 reducing-end xylose-releasing exoxylanase A (Xyn30A) from the filamentous fungus *Talaromyces cellulolyticus*. Appl Environ Microbiol 85(13). https://doi.org/10.1128/AEM.00552-19

Nakamichi Y, Fouquet T, Ito S, Watanabe M, Matsushika A, Inoue H (2019b) Structural and functional characterization of a bifunctional GH30-7 xylanase B from the filamentous fungus *Talaromyces cellulolyticus*. J Biol Chem 294(11):4065–4078. https://doi.org/10.1074/jbc.RA118.007207

Nakamichi Y, Fujii T, Watanabe M, Matsushika A, Inoue H (2020) Crystal structure of GH30-7 endoxylanase C from the filamentous fungus *Talaromyces cellulolyticus*. Acta crystallographica Section F. Struct Biol Comm 76(Pt 8):341–349. https://doi.org/10.1107/S2053230X20009024

Nekiunaite L, Petrovic DM, Westereng B, Vaaje Kolstad G, Hachem MA, Várnai A, Eijsink VG (2016) FgLPMO9A from *Fusarium graminearum* cleaves xyloglucan independently of the backbone substitution pattern. FEBS Lett 590:3346–3356. https://doi.org/10.1002/1873-3468.12385

O'Neill MA, Warrenfeltz D, Kates K, Pellerin P, Doci T, Darvill AG, Albersheim P (1996) Rhamnogalacturonan-II, a pectic polysaccharide in the walls of growing plant cell, forms a dimer that is covalently-linked by a borate ester. J Biol Chem 271:22923–22930

Ohm RA, Feau N, Henrissat B, Schoch CL, Horwitz BA, Barry KW, Condon BJ, Copeland AC Dhillon B, Glaser F, Hesse CN, Kosti I, Labutti K, Lindquist EA, Lucas S, Salamov AA, Bradshaw RE, Ciuffetti L, Hamelin RC, Kema GH, Lawrence C, Scott JA, Spatafora JW, Turgeon BG, de Wit PJ, Zhong S, Goodwin SB, Grigoriev IV (2012) Diverse lifestyles and strategies of plant pathogenesis encoded in the genomes of eighteen dothideomycetes fungi. PLoS Pathog 8(12):e1003037. https://doi.org/10.1371/journal.ppat.1003037

Pauly M, Andersen LN, Kauppinen S, Kofod LV, York WS, Albersheim P, Darvill A (1999) A xyloglucan-specific *endo*-β-1,4-glucanase from *Aspergillus aculeatus*: expression cloning in yeast, purification and characterization of the recombinant enzyme. Glycobiol 9(1):93–100

Peng M, Aguilar-Pontes MV, Hainaut M, Henrissat B, Hildén K, Mäkelä MR, de Vries RP (2017a) Comparative analysis of basidiomycete transcriptomes reveals a core set of expressed genes encoding plant biomass degrading enzymes. Fung Genet Biol 112:40–46

Peng M, Dilokpirnol A, Mäkelä MR, Hilden K, Bervoets S, Riley R, Grigoriev IV, Hainaut M, Henrissat B, de Vries RP, Granchi Z (2017b) The draft genome sequence of the ascomycete fungus *Penicillium subrubescens* reveals a highly enriched content of plant biomass related CAZymes compared to related fungi. J Biotechnol 246:1–3

Presley GN, Panisko E, Purvine SO, Schilling JS (2018) Coupling secretomics with enzyme activities to compare the temporal processes of wood metabolism among white and brown rot fungi. Appl Environ Microbiol 84(16). https://doi.org/10.1128/AEM.00159-18

Puchart V, Agger JW, Berrin J-G, Várnai A, Westereng B, Biely P (2016) Comparison of fungal carbohydrate esterases of family CE16 on artificial and natural substrates. J Biotechnol 233:228–236. https://doi.org/10.1016/j.jbiotec.2016.07.003

Puchart V, Berrin J-G, Haon M, Biely P (2015) A unique CE16 acetyl esterase from *Podospora anserina* active on polymeric xylan. Appl Microbiol Biotechnol 99(24):10515–10526. https://doi.org/10.1007/s00253-015-6934-1

Quinlan RJ, Sweeney MD, Leggio LL, Otten H, Poulsen J-CN, Johansen KS, Krogh KB, Jørgensen CI, Tovborg M, Anthonsen A (2011) Insights into the oxidative degradation of cellulose by a copper metalloenzyme that exploits biomass components. Proc Nat Acad Sci U S A 108:15079–15084

Reese ET, Siu RG, Levinson HS (1950) The biological degradation of soluble cellulose derivatives and its relationship to the mechanism of cellulose hydrolysis. J Bact 59(4):485

Ridley BL, O'Neill MA, Mohnen D (2001) Pectins: structure, biosynthesis, and oligogalacturonide-related signaling. Phytochemistry 57:929–967

Rinaldi R, Jastrzebski R, Clough MT, Ralph J, Kennema M, Bruijnincx PC, Weckhuysen BM (2016) Paving the way for lignin valorisation: recent advances in bioengineering, biorefining and catalysis. Angew Chem Int Ed Engl 55(29):8164–8215. https://doi.org/10.1002/anie.201510351

Sabbadin F, Hemsworth GR, Ciano L, Henrissat B, Dupree P, Tryfona T, Marques RDS, Sweeney ST, Besser K, Elias L, Pesante G, Li Y, Dowle AA, Bates R, Gomez LD, Simister R, Davies GJ, Walton PH, Bruce NC, McQueen-Mason SJ (2018) An ancient family of lytic polysaccharide monooxygenases with roles in arthropod development and biomass digestion. Nat Commun 9(1):756. https://doi.org/10.1038/s41467-018-03142-x

Sabbadin F, Urresti S, Henrissat B, Avrova AO, Welsh LR, Lindley PJ, Csukai M, Squires JN, Walton PH, Davies GJ (2021) Secreted pectin monooxygenases drive plant infection by pathogenic oomycetes. Science 373:774–779

Saha BC (2000) α-L-Arabinofuranosidases: biochemistry, molecular biology and application in biotechnology. Biotechnol Adv 18(5):403–423

Sakai K, Mochizuki M, Yamada M, Shinzawa Y, Minezawa M, Kimoto S, Murata S, Kaneko Y, Ishihara S, Jindou S, Kobayashi T, Kato M, Shimizu M (2017) Biochemical characterization of thermostable β-1,4-mannanase belonging to the glycoside hydrolase family 134 from *Aspergillus oryzae*. Appl Microbiol Biotechnol 101(8):3237–3245. https://doi.org/10.1007/s00253-017-8107-x

Schols HA, Voragen AGJ (1996) Complex pectins: structure elucidation using enzymes. In: Visser J, Voragen AGJ (eds) Pectin and pectinases. Progress in Biotechnology, vol 14. Elsevier Science, Amsterdam, pp 793–798

Shimizu M, Kaneko Y, Ishihara S, Mochizuki M, Sakai K, Yamada M, Murata S, Itoh E, Yamamoto T, Sugimura Y, Hirano T, Takaya N, Kobayashi T, Kato M (2015) Novel β-1,4-mannanase belonging to a new glycoside hydrolase family in *Aspergillus nidulans*. J Biol Chem 290(46):27914–27927. https://doi.org/10.1074/jbc.M115.661645

Sierra-Patev S, Min B, Naranjo-Ortiz M, Looney B, Konkel Z, Slot JC, Sakamoto Y, Steenwyk JL, Rokas A, Carro J, Camarero S, Ferreira P, Molpeceres G, Ruiz-Duenas FJ, Serrano A, Henrissat B, Drula E, Hughes KW, Mata JL, Ishikawa NK, Vargas-Isla R, Ushijima S, Smith CA, Donoghue J, Ahrendt S, Andreopoulos W, He G, LaButti K, Lipzen A, Ng V, Riley R, Sandor L, Barry K, Martinez AT, Xiao Y, Gibbons JG, Terashima K, Grigoriev IV, Hibbett D (2023) A global phylogenomic analysis of the shiitake genus *Lentinula*. Proc Natl Acad Sci USA 120(10):e2214076120. https://doi.org/10.1073/pnas.2214076120

Simmons TJ, Frandsen KEH, Ciano L, Tryfona T, Lenfant N, Poulsen JC, Wilson LFL, Tandrup T, Tovborg M, Schnorr K, Johansen KS, Henrissat B, Walton PH, Lo Leggio L, Dupree P (2017) Structural and electronic determinants of lytic polysaccharide monooxygenase reactivity on polysaccharide substrates. Nat Commun 8:1064. https://doi.org/10.1038/s41467-017-01247-3

Sjöström E (1993) Wood chemistry, fundamentals and applications, 2nd edn. Academic press, San Diego

Span EA, Marletta MA (2015) The framework of polysaccharide monooxygenase structure and chemistry. Curr Opin Struct Biol 35:93–99. https://doi.org/10.1016/j.sbi.2015.10.002

Srivastava M, Kapoor VP (2005) Seed galactomannans: an overview. Chem Biodivers 2:295–317

Steindorff AS, Serra LA, Formighieri EF, de Faria FP, Pocas-Fonseca MJ, de Almeida JRM (2021) Insights into the lignocellulose-degrading enzyme system of *Humicola grisea* var. *thermoidea* based on genome and transcriptome analysis. Microbiol Spectr 9(2):e0108821. https://doi.org/10.1128/Spectrum.01088-21

Sun P, Huang Z, Banerjee S, Kadowaki MA, Veersma RJ, Magri S, Hilgers R, Muderspach SJ, Laurent CV, Ludwig R (2023) AA16 oxidoreductases boost cellulose-active AA9 lytic polysaccharide monooxygenases from *Myceliophthora thermophila*. ACS Catal 13:4454–4467

Sun P, Laurent C, Scheiblbrandner S, Frommhagen M, Kouzounis D, Sanders MG, van Berkel WJH, Ludwig R, Kabel MA (2020) Configuration of active site segments in lytic polysaccharide monooxygenases steers oxidative xyloglucan degradation. Biotechnol Biofuels 13:95. https://doi.org/10.1186/s13068-020-01731-x

Sun P, Laurent CVFP, Boerkamp VJP, van Erven G, Ludwig R, van Berkel WJH, Kabel MA (2022a) Regioselective C4 and C6 double oxidation of cellulose by lytic polysaccharide monooxygenases. ChemSusChem 15:e202102203

Sun P, Li X, Dilokpimol A, Henrissat B, de Vries RP, Kabel MA, Mäkelä MR (2022b) Fungal glycoside hydrolase family 44 xyloglucanases are restricted to the phylum Basidiomycota and show a distinct xyloglucan cleavage pattern. iScience 25(1):103666. https://doi.org/10.1016/j.isci.2021.103666

Tan TC, Kracher D, Gandini R, Sygmund C, Kittl R, Haltrich D, Hallberg BM, Ludwig R, Divne C (2015) Structural basis for cellobiose dehydrogenase action during oxidative cellulose degradation. Nat Commun 6:7542. https://doi.org/10.1038/ncomms8542

Tandrup T, Frandsen KEH, Johansen KS, Berrin J-G, Lo Leggio L (2018) Recent insights into lytic polysaccharide monooxygenases (LPMOs). Biochem Soc Trans 46:1431. https://doi.org/10.1042/BST20170549

Tenkanen M, Vrsanska M, Siika-aho M, Wong DW, Puchart V, Penttilä M, Saloheimo M, Biely P (2013) Xylanase XYN IV from *Trichoderma reesei* showing exo- and endo-xylanase activity. FEBS J 280(1):285–301. https://doi.org/10.1111/febs.12069

Vaaje-Kolstad G, Forsberg Z, Loose JS, Bissaro B, Eijsink VG (2017) Structural diversity of lytic polysaccharide monooxygenases. Curr Opin Struct Biol 44:67–76

Vaaje-Kolstad G, Westereng B, Horn SJ, Liu Z, Zhai H, Sørlie M, Eijsink VG (2010) An oxidative enzyme boosting the enzymatic conversion of recalcitrant polysaccharides. Science 330:219–222

Venegas FA, Koutaniemi S, Langeveld SMJ, Bellemare A, Chong SL, Dilokpimol A, Lowden MJ, Hilden KS, Leyva-Illades JF, Mäkelä MR, My Pham TT, Peng M, Hancock MA, Zheng Y, Tsang A, Tenkanen M, Powlowski J, de Vries RP (2022) Carbohydrate esterase family 16 contains fungal hemicellulose acetyl esterases (HAEs) with varying specificity. New Biotechnol 70:28–38. https://doi.org/10.1016/j.nbt.2022.04.003

Vesth TC, Nybo JL, Theobald S, Frisvad JC, Larsen TO, Nielsen KF, Hoof JB, Brandl J, Salamov A, Riley R, Gladden JM, Phatale P, Nielsen MT, Lyhne EK, Kogle ME, Strasser K, McDonnell E, Barry K, Clum A, Chen C, LaButti K, Haridas S, Nolan M, Sandor L, Kuo A, Lipzen A, Hainaut M, Drula E, Tsang A, Magnuson JK, Henrissat B, Wiebenga A, Simmons BA, Mäkelä MR, de Vries RP, Grigoriev IV, Mortensen UH, Baker SE, Andersen MR (2018) Investigation of inter- and intraspecies variation through genome sequencing of *Aspergillus* section *Nigri*. Nat Genet 50:1688–1695. https://doi.org/10.1038/s41588-018-0246-1

Vidal S, Doco T, Williams P, Pellerin P, York WS, O'Neill MA, Glushka J, Darvill AG, Albersheim P (2000) Structural characterization of the pectic polysaccharide rhamnogalacturonan II: evidence for the backbone location of the aceric acid-containing oligoglycosyl side chain. Carbohydr Res 326:227–294

Vincken J-P, York WS, Beldman G, Voragen AGJ (1997) Two general branching patterns of xyloglucan, XXXG and XXGG. Plant Physiol 114:9–13

Vu VV, Beeson WT, Phillips CM, Cate JH, Marletta MA (2014) Determinants of regioselective hydroxylation in the fungal polysaccharide monooxygenases. J Am Chem Soc 136:562–565. https://doi.org/10.1021/ja409384b

Walton PH, Davies GJ (2016) On the catalytic mechanisms of lytic polysaccharide monooxygenases. Curr Opin Chem Biol 31:195–207. https://doi.org/10.1016/j.cbpa.2016.04.001

Wang BJ, Walton PH, Rovira C (2019) Molecular mechanisms of oxygen activation and hydrogen peroxide formation in lytic polysaccharide monooxygenases. ACS Catal 9:4958–4969. https://doi.org/10.1021/acscatal.9b00778

Wang BJ, Wang ZF, Davies GJ, Walton PH, Rovira C (2020) Activation of $O_2$ and $H_2O_2$ by lytic polysaccharide monooxygenases. ACS Catal 10:12760–12769. https://doi.org/10.1021/acscatal.0c02914

Wang H, Pang AP, Wang W, Li B, Li C, Wu FG, Lin F (2022) Discovery of ER-localized sugar transporters for cellulase production with *lac1* being essential. Biotechnol Biofuels Bioprod 15(1):132. https://doi.org/10.1186/s13068-022-02230-x

Warner CD, Go RM, Garcia-Salinas C, Ford C, Reilly PJ (2011) Kinetic characterization of a glycoside hydrolase family 44 xyloglucanase/endoglucanase from *Ruminococcus flavefaciens* FD-1. Enzyme Microbial Technol 48(1):27–32. https://doi.org/10.1016/j.enzmictec.2010.08.009

Willats WGT, McCartney L, MacKie W, Knox JP (2001) Pectin: cell biology and prospects for functional analysis. Plant Mol Biol 47:9–27

Xu L, Li J, Gonzalez Ramos VM, Lyra C, Wiebenga A, Grigoriev IV, de Vries RP, Mäkelä MR, Peng M (2024) Genome-wide prediction and transcriptome analysis of sugar transporters in four ascomycete fungi. Bioresour Technol 391(Pt B):130006. https://doi.org/10.1016/j.biortech.2023.130006

Yoshikawa K, Yamamoto K, Okada S (1993a) Isolation of *Aspergillus flavus* MO-5 producing two types of intracellular α-D-xylosidases: purification and characterization of α-D-xylosidase I. Biosci Biotech Biochem 57:1275–1280

Yoshikawa K, Yamamoto K, Okada S (1993b) Purification and characterization of an intracellular α-D-xylosidase II from *Aspergillus flavus* MO-5. Biosci Biotechnol Biochem 57(8):1281–1285

You X, Qin Z, Li YX, Yan QJ, Li B, Jiang ZQ (2018) Structural and biochemical insights into the substrate-binding mechanism of a novel glycoside hydrolase family 134 β-mannanase. Biochim Biophys Acta Gen Subj 1862(6):1376–1388. https://doi.org/10.1016/j.bbagen.2018.03.016

Zhang J, Silverstein KAT, Castano JD, Figueroa M, Schilling JS (2019) Gene regulation shifts shed light on fungal adaption in plant biomass decomposers. mBio 10(6). https://doi.org/10.1128/mBio.02176-19

# Genomes and Genomics of the Genus *Trichoderma*

9

Monika Schmoll

**Abstract**

The genus *Trichoderma* comprises more than 400 species and fungi from diverse ecological niches as well as model fungi for different phenomena, such as plant cell wall degradation, biocontrol, or light response. The widespread uses that arise from these characteristics initiated applications in industry and agriculture, but also development of a plethora of molecular tools for genome manipulation and strain improvement with *Trichoderma*. Analysis of evolution with high-quality genome sequences revealed the acquisition of an important share of enzymes by horizontal gene transfer (HGT). The "core genome" of *Trichoderma* was narrowed down to 7000 genes and repeat-induced point mutation (RIP) was discovered as one mechanism for evolutionary genome modification. Genome sequencing studies further revealed numerous mitochondrial genomes and the presence of a surprising diversity of mycoviruses in *Trichoderma* genomes. Moreover, discoveries on the relevance of long non-coding RNA (lncRNA) and small RNAs as well as selected transcriptome and proteome studies are discussed. Besides new insights from comparative analysis of genomes, genome mining yielded new enzymes, pathways, and progress in the investigation of secondary metabolism of diverse species of the genus.

**Keywords**

*Trichoderma* · Biocontrol · Enzyme production · Mycoparasitism · Plant protection · Biofuel production · Cellulase gene expression · Transcriptome · Biosynthetic gene cluster · Secondary metabolism

## 9.1 Introduction

The genus *Trichoderma* includes some of the most beneficial strains for industry and agriculture currently used commercially and has hence been subject to extensive research efforts aimed at both basic and application-oriented research (Guzmán-Guzmán et al. 2023; Bischof et al. 2016; Schmoll et al. 2016; Woo et al. 2022). According to the NCBI taxonomy database (accessed March 2024) this genus currently comprises 418 species, additionally, seven species related to *Trichoderma* (species affinis/aff.), 128 species with uncertain classification (conferatur/cf.), and 1813 unclassified *Trichoderma* species/strains. The JGI mycocosm database as the largest for fungi in general, lists 88 sequenced

M. Schmoll (✉)
Centre for Microbiology and Environmental Systems Science, Division of Terrestrial Ecosystem Research, University of Vienna, Vienna, Austria
e-mail: monika.schmoll@univie.ac.at


M. Nowrousian, J. E. Stajich (eds.), *Fungal Genomics*, The Mycota 13,
https://doi.org/10.1007/978-3-031-75666-5_9

genomes of the genus *Trichoderma*, although in several cases multiple strains of the same species were sequenced, for example, wildtypes and mutants of industrial relevance or isolates from different habitats. *Trichoderma* spp. can have very diverse lifestyles and preferences for various habitats, ranging from mycoparasitism (Karlsson et al. 2017) to plant saprotrophy to life as endophytes in plants (Chaverri and Samuels 2013; Woo et al. 2022), where they show potential for fighting phytopathogens (Zachow et al. 2016). In the majority of cases, *Trichoderma* spp. exert a positive effect on their plant hosts, protecting them from pathogens by mycoparasitism, but also by initiating the innate immune response of the plants and by priming plants (also called "green vaccination") against pathogens (Guzmán-Guzmán et al. 2023). This priming was recently even shown to be heritable in plants (Tiwari et al. 2022; Moran-Diez et al. 2021).

Strains of the genus *Trichoderma* play major roles in biotechnology (Arnau et al. 2020; Paloheimo et al. 2016), plant protection (Tyskiewicz et al. 2022), but also as human pathogens (Hatvani et al. 2013) and a plethora of molecular and genetic methods facilitates physiological analysis and genetic tractability of these fungi (Mach-Aigner and Martzy 2021). With respect to functional genomics, genome modification by CRISPR/Cas9 was developed and optimized for several *Trichoderma* spp. and is now broadly applied (Wang et al. 2022a), both for improvement of enzyme production (Zhang et al. 2024) and for studying biocontrol agents (Luo et al. 2023; Cortes et al. 2023). Besides application-oriented research, which substantially drives research with *Trichoderma*, several species of the genus also have become model systems for basic research, for example, in the field of plant cell wall degradation (Glass et al. 2013; Yan et al. 2021), biocontrol (Harman et al. 2004, 2021; Guzman-Guzman et al. 2019), effector-like molecules (Ramirez-Valdespino et al. 2019), and light response (Carreras-Villaseñor et al. 2012; Schmoll et al. 2010, 2018a, b). In recent years, natural products of *Trichoderma* spp. and particularly secondary metabolites with biological activity came into focus of research (Guo et al. 2023; Bai et al. 2023; Zeilinger et al. 2016), also enabled by enhanced bioinformatic screening tools (Rush et al. 2021).

Generally, species of the genus *Trichoderma* are considered safe for applications in industry (Nevalainen et al. 1994; Frisvad et al. 2018) and agriculture and have a long history of safe use, also due to cost-efficient production in low-income countries, where biocontrol agents are more affordable than classic chemical pesticides. However, safe application requires knowledge-based development and consideration of potential risks to avoid deleterious outcomes.

Already quite some time ago, *Trichoderma* spp. were mentioned as emerging fungal pathogens of immunocompromised patients (Walsh et al. 2004; Hatvani et al. 2013), for which treatment is often problematic due to a high resistance to antimicrobials (Sal et al. 2022). With better identification methods and awareness of the threat posed by emerging pathogenic fungi, also the number of identified *Trichoderma* strains isolated from patients has increased (Dos Santos and Dos Santos 2023). Although a causal connection to disease has not been confirmed in all cases, it is relevant that strains of *T. viride*, *T. hamatum*, *T. gamsii*, *T. asperellum*, and *T. reesei* (Dos Santos and Dos Santos 2023) were reported from such samples besides the known opportunistic pathogens *T. longibrachiatum* and *T. citrinoviride* (Hatvani et al. 2013; Kredics et al. 2003). In part, the appearance of *Trichoderma* as a human pathogen is attributed to its presence in agriculture and increased biocontrol applications (Hatvani et al. 2019). Hence, research toward genomic characteristics facilitating opportunistic pathogenicity is highly timely to enable safe and beneficial application of *Trichoderma* in the environment.

As for agricultural use, *T. afroharzianum* was recently found to cause maize ear rot disease (Pfordt et al. 2020; Sanna et al. 2022), although this species is close to other, very potent biocontrol species, which have long been used without known problems. Since awareness of the toxicity of chemical pesticides in agriculture is increasing and some of these chemicals are already banned in the European Union, screening efforts for biological or integrated pest management have become a focus of industry. Accordingly, the market value of *Trichoderma* biocontrol agents globally is reported to have the highest performance in terms of value, even exceeding that of bacterial biocontrol agents together (https://www.marketsandmarkets.com/Market-Reports/biofungicide-market-8734417.html). As these biocontrol agents are distributed in nature, particular care has to be taken for species identification as well as investigation of genome content to avoid the propagation of harmful strains. Additionally, the policy to apply fungal strains even from different continents as biocontrol agents without consideration of the local microbial ecosystem should be reconsidered (Averill et al. 2022).

In order to fully leverage the benefits provided by strains of the genus *Trichoderma*, while avoiding adversities due to potential human or plant pathogenicity, omics research is of utmost importance to elucidate risk factors, support species identification and characterization, and discover novel pathways, biosynthetic clusters, and enzymes. Particularly in the last 1–2 decades, significant progress has been made, both regarding tools and techniques (Mach-Aigner and Martzy 2021; Schmoll and Dattenböck 2016) and knowledge of the genomic diversity of *Trichoderma*, which is fundamental for functional genomics. Besides the development of CRISPR/Cas9 protocols for *Trichoderma* (Wang et al. 2022a), newer developments in this direction include the TrichoGate cloning strategy (Nogueira-Lopez et al. 2019) for efficient cloning in *Trichoderma* or the SES (synthetic expression system), which applies a synthetic transcription factor and enables strong and tunable expression (Rantasalo et al. 2016, 2018, 2019).

In the following, I provide an overview of the progress made with the investigation of genomes of *Trichoderma* species along with some insight into more general omics research inspired by genome mining or new research approaches enabled by new developments in sequencing technologies. Due to the breadth of the topics in which *Trichoderma* strains are relevant, only a selection of topics and research work can be covered, which is not meant to imply a judgement on importance or novelty.

## 9.2 Genome Research with *Trichoderma*

### 9.2.1 Aspects of Evolution

Availability of multiple genomes of *Trichoderma* species, for some of them in high quality, allowed for a more accurate assessment of their evolution. Around 66 million years ago, the genus *Trichoderma* evolved and subsequently speciation led to formation of the sections Longibrachiatum and Trichoderma followed by the Harzianum/Virens clade (Kubicek et al. 2019). The most closely related genera include *Escovopsis*, *Metarhizium*, *Cordyceps*, and *Beauveria*. In the course of evolution, species of the genus acquired a considerable number of genes (for example, transcription factor genes, HET domain genes, and ankyrin domain-containing genes), but also gene losses. Ultimately, this resulted in the detected core genome of 7000 genes, among which there are mainly genes involved in metabolism and such assigned to the KOG families "posttranslational modification, protein turnover and chaperones," "transcription," and "carbohydrate transport and metabolism." Additionally, numerous genes encoding glycoside hydrolases and fungal-specific transcription factors were detected (Kubicek et al. 2019). The finding of 105 orphan genes unique to *Trichoderma* among the 7000 core genes reflects the flexibility of the genus in adaptation to new habitats and environmental conditions. As most of these orphan genes are under strong purifying selection, this evolutionary optimization of the core genome is currently still ongoing (Kubicek et al. 2019).

Considering the evolution of the most prominent cellulases of *T. reesei*, CEL6a and CEL7a, investigation of their structural features shed new light on processivity (Uchiyama et al. 2020). Family 7 glycoside hydrolases are a group of enzymes with similar characteristics. In both organism groups, the combination of highly processive and less processive cellulases is required to avoid "traffic jams" during degradation of cellulose fibers and hence decreased efficiency. Consequently, convergent evolution of cellulases for optimization of processivity and interplay on the cellulose fibers occurred in both bacteria and fungi (Uchiyama et al. 2020).

## 9.2.2 Horizontal Gene Transfer (HGT) in the Genus *Trichoderma*

Horizontal or lateral gene transfer, meaning the acquisition of novel genes from a different species is well known to play an important role in evolution, predominantly in bacteria, but also in fungi (Soucy et al. 2015). However, the relevance and even the occurrence of this process in eukaryotes is subject to debates (Leger et al. 2018; Martin 2017) and careful analyses have to be done to exclude artifacts due to sequencing or assembly errors. In *Trichoderma*, the most prominent example of HGT is the SOR (sorbicillin) cluster, which was in part acquired from Eurotiomycetes and is subject to selection (Druzhinina et al. 2016). For an overview of the characteristics of this cluster see Schalamun and Schmoll (2022) and references therein. More recent findings extended the knowledge on regulators of sorbicillin production and specific modification of sorbicillinoid components (Schalamun et al. 2023a, 2024) and revealed the production of bisorbibutenolide in *T. reesei* (Schalamun et al. 2023a).

With respect to secondary metabolite production, also a terpene synthase gene (*tri5*), predicted to be involved in trichothecene biosynthesis was obviously acquired by HGT i (Malmierca et al. 2013; Vicente et al. 2020; Gutierrez et al. 2021). However, the genomic localization in *Trichoderma* is not syntenic to other fungi, as could be expected as *tri5* is not part of a cluster (Lindo et al. 2018; Proctor et al. 2018).

Further cases of gene acquisition by HGT include ceratoplatanins (Gaderer et al. 2014; Gao et al. 2020), likely from Basidiomycetes; more than 40% of plant cell wall degrading enzyme genes encoding CAZymes (Carbohydrate Active enZymes) (Druzhinina et al. 2018). Interestingly, these enzyme genes were acquired from even closely related Ascomycetes, such as *Pestaliopsis fici* or *Talaromyces stipitatus*, but not from Basidiomycetes, which can also be host to mycoparasitic *Trichoderma* (Druzhinina et al. 2018). In contrast to the relatively close relationship of *Trichoderma* to donor species of plant cell wall degrading enzyme genes, acquisition of the auxiliary protein swollenin encoding gene was traced back to green plants as donors (Druzhinina et al. 2018; Saloheimo et al. 2002). In contrast to this high number of enzyme-encoding genes, the transcription factors mainly responsible for their regulation, like XYR1, ACE2, and ACE3 (Benocci et al. 2017) were not acquired by HGT (Druzhinina et al. 2018).

This extensive acquisition of CAZyme encoding genes set *Trichoderma* spp. apart from other hypocrealean fungi. Very likely, the mycoparasitic competences of *Trichoderma* spp. were beneficial for HGT and enabled them to optimize their metabolic competences by "sampling" their prey and selecting useful components during evolution.

Besides genes contributing to the nutrient versatility of *Trichoderma*, 123 genes were predicted to result from previous HGT and which are in part still with unknown function (Druzhinina et al. 2018), including GPCRs with potential relevance for cellulase regulation (Stappler et al. 2017), the small unique protein OOC1 with a function in nitrogen metabolism and potentially plant cell wall degradation (Schmoll and Kubicek 2005; Pang et al. 2021), the dehydrogenase GRD1, which impacts cellulose degradation (Schuster et al. 2011), the high-affinity nitrate transporter NIT10 and CIP1, a protein important for efficient cellulose degradation (Lehmann et al. 2016).

## 9.2.3 Telomere-to-Telemere Sequences of *Trichoderma* Strains (Gold Standard Genomes)

Omics analyses require high-quality genomes for optimal interpretation and consequently, novel assembly methods were applied to *T. reesei* QM6a (Jourdier et al. 2017; Marie-Nelly et al. 2014). The advent of third-generation sequencing facilitated the assembly of telomere to telomore genome sequences for several *Trichoderma* species, initially for *T. reesei* QM6a (Li et al. 2017), which revealed numerous sequencing errors in the earlier assemblies. Moreover, a comparison of the complete genome of *T. reesei* QM6a (Li et al. 2017) with that of the high cellulase producer RutC30 genome indicates that the translocations in this mutant are less extensive than previously postulated (Seidl et al. 2008; Peterson and Nevalainen 2012).

In the course of the analyses of the complete genomes of seven *Trichoderma* species, telomere sequences were found to be highly conserved in *Trichoderma* (Li et al. 2021c). Analysis of the now available AT-rich sequences which were not complete in previous genome sequences based on short read sequencing showed that the percentages of AT-rich blocks of the *T. reesei* strains QM6a and CBS999.97 (7–8%) are lower than those of *T. virens* Gv29.8, *T. virens* FT-333, *T. atroviride* P1, and *T. asperellum* FT-101 (11–13%) (Li et al. 2021c). Analysis of these high-quality genomes further confirmed earlier studies (Kubicek et al. 2019) showing that these genomes are highly divergent with 2000–3000 species-specific genes each. Also, remnants of numerous transposons were detected along with the operation of RIP as well as considerable genome shuffling (Li et al. 2021c). Major differences in gene content were observed for transcription factor genes, especially in the subfamilies of fungal zinc-binuclear cluster domains and fungal-specific transcription factor domains. Also, the hypothesis of physically linked CAZyme gene clusters in *Trichoderma* genomes is further supported (Li et al. 2021c). Fourteen of them are shared between *T. reesei* and *T. virens*, eighteen are shared between *T. atroviride* and *T. asperellum* (Li et al. 2021c). Considering secondary metabolite clusters, 27 and 37 clusters, respectively, are common in these species (Li et al. 2021c). In this respect, in-depth annotation of Cytochrome P450 encoding genes revealed 12 families of the encoded proteins and hence added additional biosynthetic genes in some of the clusters (Chadha et al. 2018).

High-quality genomes of the fully fertile *T. reesei* strains CBS999.97 (MAT1–1) and CBS999.97 (MAT1–2) (Seidl et al. 2009) were of high interest for investigation of recombination and meiosis. A very high number of SNPs and indels (ca. 1 Mio) between the CBS999.97 strains and QM6a is in agreement with the unexpectedly high phenotypic diversity of progeny from their crosses (Li et al. 2021b). Between the two different mating types of CBS999.97, only around 2700 SNPs and indels were found (Li et al. 2021b). In the course of this investigation, the specifically developed software tool TSETA (Liu et al. 2021a) further increased accuracy of detection of such alterations.

### 9.2.4 Random Mutants Demystified: New Insights for Protein Production

After the discovery of the high potential of *T. reesei* for enzyme production at an industrial scale, numerous mutant strains were created and selected for improved characteristics (Seidl and Seiboth 2010). For decades, the molecular basis of the physiology of these strains remained unknown until genetic tools combined with an increasingly higher performance of sequencing technology enabled genome-wide identification of mutations and subsequent functional genomics (Bischof et al. 2016; Peterson and Nevalainen 2012).

*T. reesei* QM9978 is a strain, which cannot be induced to express cellulases, but only produces them at a very low, basic level (Torigoi et al. 1996). Hence, this strain was frequently used to narrow down genes responsible for induction and high-level cellulase production (for example, Schmoll et al. (2004), Zeilinger et al. (2003)). A comparison of the genome of QM9978 with that of QM6a revealed 43 mutations, of which a translocation upstream of the transcription factor gene *vib1* was identified as responsible for the defect in cellulase regulation (Ivanova et al. 2017). VIB1 serves as a link between glucose signaling and carbon catabolite repression in *N. crassa* (Xiong et al. 2014) and its transcript level is strongly regulated by light (Hitzenhammer et al. 2019).

Another, less frequently investigated cellulase-negative *T. reesei* strain, QM9136 (Mandels et al. 1971; Nevalainen and Palva 1978) was found to have suffered a frameshift mutation causing a truncation in the gene-encoding XYR1, the major positive cellulase transcription factor (Lichius et al. 2015). The *T. reesei* strain PC-3-7 (Kawamori et al. 1985) is a descendant of the cellulase overproducer QM9414 and contains 260 SNPs, mostly in non-coding regions (Porciuncula Jde et al. 2013). However, an SNP in the *cre1*-gene led to a decreased binding affinity of the carbon catabolite repressor CRE1 to the *cbh1* promotor and consequently to increased cellulase production (Porciuncula Jde et al. 2013).

## 9.3 Sexual Development and Recombination

Sexual development plays an important role in the propagation of a species (Bennett and Turgeon 2016; Wallen and Perlin 2018), especially if the habitat of a species changes or the environment requires adaptation of physiological competences of a species (Debuchy et al. 2010; McDonald et al. 2016). For the genus *Trichoderma* with its multiple applications in industry and agriculture, sexual development could also be seen as a tool for strain improvement or research toward physiology in case of QTL mapping of progeny of crosses. However, despite the isolation of many strains from sexual fruiting bodies in nature (Jaklitsch and Voglmayr 2015), sexual development under laboratory conditions was so far only achieved for *T. reesei* (Seidl et al. 2009; Hinterdobler et al. 2020b).

Investigation of sexual development in *T. reesei* revealed that the parental strain for most industrial applications, QM6a (Martinez et al. 2008), is female sterile, meaning it cannot assume the female role in a cross and hence even introduction of the other mating type into this strain is not sufficient to enable mating (Seidl et al. 2009). Later on, the fertility defect of QM6a could be traced back to the scaffolding protein HAM5, which shows several mutations resulting in a frameshift (Linke et al. 2015; Tisch et al. 2017). While unfortunate for strain improvement with QM6a-derived strains, this defect renders QM6a a tester strain to assess the role of genes and/or mutations in female fertility (Hinterdobler et al. 2020b): If a mutant strain is able to mate with a fully fertile MAT1-2 strain, for example, CBS999.97 (MAT1–2), but not with QM6a (MAT1-2), which cannot assume the female role, then the tested mutation or gene is very likely responsible for female sterility of this strain.

### 9.3.1 Repeat Induced Point Mutation (RIP)

Analysis of the first sequenced genome of the genus *Trichoderma*, that of *T. reesei* (Martinez et al. 2008), revealed a strikingly low number of gene duplications and transposon residues. Hence, repeat-induced point mutation (RIP), which was first discovered in *Neurospora* (Selker et al. 1987) was likely to be operative as similar characteristics were found for *N. crassa* (Borkovich et al. 2004; Krumlauf and Marzluf 1980). RIP was found for most Ascomycota (van Wyk et al. 2020) and serves as a mechanism protecting the genome from transposable elements and mobile DNA (Gladyshev 2017). During RIP, repetitive DNA sequences larger than 400 bp become subject to C-5 cytosine methylation and deamination by the methyltransferases RID1 and DIM2 prior to meiosis (Kouzminova and Selker 2001; Li et al. 2018).

As also observed for *T. reesei*, if RIP is operative in a genome, only a low number of repetitive sequences and transposon remnants are found (Gladyshev 2017). Finally, 10 years after the hypothesis was made, RIP was indeed confirmed to be operative in *T. reesei* (Li et al. 2018). However, RIP in *T. reesei* has a different dinucleotide preference compared to *N. crassa* (CpA > CpT > CpG > CpC in *N. crassa* (Margolin et al. 1998) versus CpG ≥ CpA ≫ CpT > CpC in *T. reesei* (Li et al. 2017). Moreover, RID1 and

DIM2 are required for normal sexual development in *T. reesei* but not in *N. crassa* (Li et al. 2018). Analysis of a full-length genome of *T. reesei* QM6a along with a new assembly then revealed several transposon sequences, mostly located in AT-rich regions, which were not detected previously (Li et al. 2017).

It is reasonable to assume that RIP was of high importance for the evolution of *Trichoderma* considering the extensive occurrence of HGT causing gene gain (in case of similar copies integrated), which needs to be balanced with gene loss. Evolution of *Trichoderma* genomes was hence likely impacted by RIP by counteracting gene duplication in sexually propagating populations of *Trichoderma*.

### 9.3.2 Segmental Aneuploidy in *Trichoderma*

Already early after the achievement of sexual crosses of QM6a with the wild, fertile isolate CBS999.97 (Lieckfeldt et al. 2000; Seidl et al. 2009) under laboratory conditions, it became obvious that the resulting progeny had unexpectedly diverse phenotypes. In-depth analysis of sexual development of *T. reesei* subsequently revealed that not all of the 16 ascospores resulting from meiosis and two rounds of postmeiotic mitosis were viable. In fact, 4–8 of them did not germinate and segmentally aneuploid ascospores were found as well (Chuang et al. 2015; Schmoll and Wang 2016). Interestingly, segmental duplication and deletion on a chromosome led to the loss of the polyketide synthase gene *pks4* (Chuang et al. 2015), which is responsible for spore coloration (Atanasova et al. 2013). Moreover, a xylanase gene resides in the concerned genomic region and is lost as well in case of segmental aneuploidy, which leads to altered capacity of lignocellulosic biomass degradation in the respective progeny (Chuang et al. 2015). Unfortunately, the process leading to segmental aneuploidy is not readily predictable, which would be desirable for applications in strain improvement.

## 9.4 Regulatory Non-Coding RNAs in *Trichoderma*

Besides investigation of the protein-coding transcriptome, also non-coding RNAs (Cech and Steitz 2014) are highly relevant for gene regulation and hence physiology. These RNAs exert a considerable influence on gene regulation (Li et al. 2021a; Song et al. 2021), targeting diverse cellular functions. Moreover, the different versions of non-coding RNAs such as long non-coding RNAs and microRNAs interact and lncRNAs can, for example, act as precursors for microRNAs (Ulitsky 2018).

### 9.4.1 microRNA-like RNAs

In recent years, omics analysis beyond protein-coding genes and transcripts came into focus also for research with *Trichoderma*. Micro RNAs or micro RNA-like RNAs are non-coding RNAs with a usually negative effect on gene regulation of target genes (Cech and Steitz 2014). Already 10 years ago microRNA-like RNAs were detected in *T. reesei* (Kang et al. 2013). Comparing cellulase-inducing with non-inducing conditions, 726 and 631 small RNAs were detected, respectively, of which 13 were differentially expressed. Moreover, functions for milRNAs were proposed for enzyme expression and cellulase regulation (Kang et al. 2013). Thereafter especially the relevance of small RNA-mediated functions in fungus–plant interaction (Cai et al. 2018; Weiberg et al. 2015; Weiberg and Jin 2015) initiated research with *Trichoderma* biocontrol species. On the plant side, regulation of milRNAs was shown to be specific to pathogens versus beneficial *Trichoderma* species in several cases (Wang et al. 2021, 2023; Olmo et al. 2024).

For *T. asperellum*, three milRNA candidates were reported to be differentially expressed in the presence of tomato roots with potential roles in *Trichoderma*–plant interaction. Additionally, transboundary milRNAs targeting plant proteins relevant for growth, development, and disease resistance were identified (Wang et al. 2021).

Hundreds of target genes were identified for *T. breve* milRNAs during mycoparasitism and intriguingly, also 41 transboundary milRNAs originating from *T. breve* and acting on genes of *B. cinerea* were found (Liu et al. 2023). These milRNAs targeted genes with functions in signaling, secondary metabolism, and carbon metabolism, indicating an important influence of *T. breve* on its prey via manipulation of gene regulation using milRNAs.

Differential expression of milRNAs was also shown for the interaction between interspecies *Trichoderma* fusants (originating from protoplast fusion) confronted with *Sclerotium rolfsii* (Hirpara et al. 2023). This interaction involved alterations in amino acid metabolism and biosynthesis of secondary metabolism as controlled by several milRNAs (Hirpara and Gajera 2023).

Besides the relevance of milRNAs in plant fungus interactions, also a function for hyphal regeneration after injury and for conidiation was recently shown for *T. atroviride* (Villalobos-Escobedo et al. 2022). The three main milRNAs subject to regulation by Dcr2 are also induced during injury response, hence revealing a post-transcriptional signaling process in hyphal regeneration. A relevance of the small RNA-mediated gene silencing machinery was discovered for *T. atroviride* interacting with Arabidopsis and is required for growth, systemic resistance, and proper epigenetic modification of the gene NSP4, which is crucial for fungus–plant interaction (Rebolledo-Prudencio et al. 2022).

### 9.4.2 Long Non-Coding RNAs (lncRNAs)

LncRNAs represent transcripts which do not give rise to an associated protein, but have extensive regulatory functions (Till et al. 2018a). Research with lncRNAs is still very limited in *Trichoderma* spp. However, the discovery of one lncRNA, HAX1, which impacts cellulase gene expression (Till et al. 2018b) opened up a new topic of omics research toward strain improvement. Interestingly, the hypercellulolytic strains of *T. reesei* (QM9414 and RutC30) comprise longer versions of HAX1, which are crucial for the function in cellulase regulation (Till et al. 2018b). The respective positive effect of HAX1 on cellulase expression was found to be due to the interference of HAX1 with the negative feedback loop of the cellulase activator XYR1 (Till et al. 2020).

Importantly, there is now also a bioinformatic tool available, which enables genome-wide identification of non-coding RNAs, which also works for fungi (Schmal et al. 2022).

## 9.5 Mitochondrial Genomes

Mitochondria are known as the powerhouse of eukaryotic cells. They are involved in production of ATP and the aerobic citric acid cycle as well as biosynthesis of metabolites, but also in senescence and maintenance of iron homeostasis (Medina et al. 2020; Basse 2010; Chatre and Ricchetti 2013). In fungi, mitochondria are further relevant as targets for fungicides due to their importance for fungal virulence (Medina et al. 2020). Mitochondria have their own genome and their own codon usage. Moreover, the mitochondrial genome can be present in multiple copies, depending on environmental conditions, and the mitochondrial genome can replicate and be inherited independently (Medina et al. 2020; Burger et al. 2003). In Hypocreales, an alternative mechanism for intron evolution in mitochondrial genomes was postulated and duplication of mitochondrial genes is common (Fonseca et al. 2020) and several cases of HGT of genes were detected from bacterial genera to fungi (Megarioti and Kouvelis 2020). Also, initially endosymbiotic genes were transferred to the nuclear genome in the course of evolution (Burger et al. 2003).

The sizes of the circular mitochondrial genomes in the Hypocreales range from about 24 kb to more than 200 kb (Fonseca et al. 2020) and until recently, only a few mitochondrial genome sequences were published for *Trichoderma* spp. (Kwak 2020, 2021).

Physiological studies on *T. reesei* revealed the importance of mitochondria for cellulase production (Abrahao-Neto et al. 1995) and in particular, the 5′ region of the major cellulase gene *cbh1* was found to be crucial for sensitivity to the functional state of the mitochondria (El-Dorry et al. 1996). Later on, mitochondria were reported to play a role in oxidative stress resistance in *T. reesei* (Wang et al. 2015). With the recent increase in genome sequencing projects, often using third-generation sequencing, also more mitochondrial genomes became available. For *T. reesei*, sequencing of the nature isolates of CBS999.97 MAT1-1 and CBS999.97 MAT1-2 (Seidl et al. 2009) showed that the mitochondrial genomes of these strains are significantly different from that of *T. reesei* QM6a, but that the available genes required for mitochondrial functions are nevertheless similar (Li et al. 2017, 2021b). Sequences of the mitochondrial genomes of the biocontrol strains *T. atroviride* P1, *T. asperellum* FT-101, *T. virens* Gv29-8, and *T. virens* FT-333 are variable in size, and detection of remnants of mobile genetic elements suggests that they were relevant for shaping the mitochondrial genomes in *Trichoderma* (Li et al. 2021c). Analyses of further mitochondrial genomes of *T. harzianum, T. afroharzianum, T. atroviride*, the poisonous *T. cornu-damae*, and other strains (Fonseca et al. 2020; Medina et al. 2020; Kwak 2020, 2021; Chung et al. 2022; Özkale et al. 2023) confirmed the variability of mitochondrial genomes in size, with a highly conserved gene order, while intergenic regions, nucleotide composition bias, number of protein coding sequences and the size of mitochondrial genomes is rather divergent (Kwak 2020). With 27.5 kb, the mitogenome of *T. koningiopsis* represents the smallest of the genus so far (Castrillo et al. 2023).

## 9.6 Mycoviruses in *Trichoderma*

Viruses infecting fungi—so-called mycoviruses (Kondo et al. 2022) have been known for decades and more often than not such infection remains without influence on the phenotype or just somewhat decreased fitness revealed by slower growth (Ghabrial et al. 2015; Ghabrial and Suzuki 2009; Son et al. 2015). The detection of viruses in *Trichoderma* species happened only quite recently (Jom-in and Akarapisan 2009). Even if the infection often does not change the phenotype, there are intriguing cases, in which the virus assumes an important function in the host, which should be considered when screening for mycoviruses in *Trichoderma*. For example, a highly transmissible, but defective version of a mycovirus was found to negatively impact biocontrol capabilities in *T. harzianum* and *T. koningiopsis* (You et al. 2019). A small DNA mycovirus of *Sclerotinia sclerotiorum* can turn this fungus from a pathogen to a beneficial endophyte of *Brassica* spp. (Zhang et al. 2020). Similarly interesting is the case of a three-way interaction between the plant *Dichanthelium lanuginosum*, its endophytic fungus *Curvularia protuberata*, and a mycovirus, which enables the plant to grow at high temperatures (Marquez et al. 2007). In *Cryphonectria parasitica*, an influence of a mycovirus on the production of cellulases was shown (Wang and Nuss 1995). Mycoviruses were assumed to have a rather narrow host range, but it was also shown that they can be transmitted between species (Nerva and Chitarra 2021; Wu et al. 2012) and that mycoviruses from endophytes can even replicate in the host plant (Nerva et al. 2017). Propagation predominantly occurs vertically via conidiospores (Ghabrial et al. 2015).

In *Trichoderma* spp. screening of more than 300 isolates responsible for green mold disease indicated that the presence of mycoviruses of diverse groups is common in the genus (Yun et al. 2016). Moreover, investigation of the viromes of different *Trichoderma* isolates indicates a complex combination of a range of virus types, including negative-sense RNA viruses (Pagnoni et al. 2023).

## 9.7 Gene Regulation for Heterologous Protein Expression: New Promotors

Regulation of cellulase gene expression is among the most frequently targeted topics of functional genomics—on the one hand because of the importance of plant cell wall degradation, but on the other hand also because the strong cellulase promotors are frequently used for heterologous protein production in *T. reesei* (Gudynaite-Savitch and White 2016; Paloheimo et al. 2016; Arbige et al. 2019). Nevertheless, alternative promotors are still of interest and some with new characteristics were identified (Beier et al. 2022; Fitz et al. 2018), including tunable promotors responding to methionine (Bischof et al. 2015), a bidirectional promotor derived from the SOR cluster for dual protein expression (Wu et al. 2022; Lv et al. 2015), a copper-controlled promotor (Lv et al. 2015; Zheng et al. 2017), a synthetic hybrid promotor (Wang et al. 2022b) and a light switchable bidirectional expression system (Zhang et al. 2020). Aiming at gene knockdown by RNAi, a copper-responsive promoter was used in *T. reesei* for tunable expression (Wang et al. 2018).

## 9.8 Discoveries in Genome Wide Gene Regulation at the Transcript Level

Omics research on gene regulation patterns with strains of the genus *Trichoderma* is a very active scientific field, particularly with the aim of strain improvement for industry (Adnan et al. 2022; Novy et al. 2019; Jadhav et al. 2024; Gupta et al. 2016). Recent studies combining different omics approaches now pave the way toward more general conclusions as to plant biomass degradation strategies in fungi (Novy et al. 2019; Li et al. 2023). Also in the field of biocontrol and mycoparasitism, transcriptome studies provided important insights into the determinants for efficient plant protection mechanisms operative in *Trichoderma* (Guzman-Guzman et al. 2019). While these topics are continuously summarized in excellent specialized review articles, some selected new discoveries from rather rarely studied characteristics or particularly promising research directions are highlighted here.

The interrelationship between circadian oscillations with metabolism, which was confirmed for *N. crassa* previously (Hurley et al. 2016) was also shown in *Trichoderma* (Henriquez-Urrutia et al. 2022) and the impact of light on regulation of enzyme expression and metabolism was supported further (Pola-Sanchez et al. 2021; Schalamun et al. 2023a, c; Schmoll 2018b).

Looking at the very early events after a spore transitions from dormancy to germination and growth, tremendous changes in transcript abundance in *T. asperelloides* were detected (Gortikov et al. 2022). Chitinase and glucanase genes are especially important during this phase, which is essentially in accordance with the earlier finding that CAZyme gene expression is associated with sporulation in *T. reesei* (Metz et al. 2011) considering the transition from spore to mycelium and vice versa. Intriguingly, recognition of *Rhizoctonia solani*, which can be parasitized by *T. asperelloides*, already happens during these very early steps of development (Gortikov et al. 2022).

Recent transcriptome studies delve into the requirements and consequences of the interkingdom interaction between *Trichoderma* and plants. Colonization of roots is among the most important prerequisites for successful beneficial interaction (Macias-Rodriguez et al. 2020). For interaction of endophytic *T. virens* with maize roots, considerable metabolic changes were found, for example, from the shikimate pathway, which was in agreement with gene regulation under these conditions (Schweiger et al. 2021). For the same interaction system, the small, secreted cysteine-rich protein SIR1 was identified as crucial for *Trichoderma*-maize interaction, regulating the response of *T. virens* to the plant (Taylor et al. 2021). Recognition of maize by *T. virens* was further found to result in initial repression of transcription and subsequently, biosynthesis of phytohormone-like compounds, secretion of specific cell wall degrading enzymes and secondary metabolites, accompanied by significant shifts in metabolism (Malinich et al. 2019). Investigation of chemotropic sensing of a plant by *T. harzianum* led to the identification of a recognition specific, putative secondary metabolite cluster required for colonization of soy plant roots by the fungus (Schalamun et al. 2023b). Upon investigation of the interaction of *T. asperellum* with *Populus*, indications for alternative splicing associated with the plant–fungus interaction were detected (Ji et al. 2019), which opens up an interesting direction for research in this field.

## 9.9 Genomics for Discovery of Novel Enzymes

The diverse applications of enzymes in industry (Meyer et al. 2020; Arnau et al. 2020) still require bioprospecting for stable proteins with desired or novel properties (Cologna et al. 2018; Filiatrault-Chastel et al. 2021), which can be applied under diverse conditions—from fermentors to chemical reactors to washing machines in households (Arnau et al. 2020; Wilson 2009; Singh et al. 2015). For example, in *T. harzianum* B13–1, 50 putative lipases were identified, which are in part inducible with olive oil (Canseco-Perez et al. 2018). In *T. simonsii*, seven predicted asparaginase-encoding genes were detected, of which three show higher transcript levels under conditions enhancing asparaginase activity (Chung et al. 2021). In the mycoparasite *T. parareesei*, a novel glucuronan lyase system was discovered (Pilgaard et al. 2022) hence adding another highlight to the versatility of polysaccharide degradation mechanisms in the genus *Trichoderma*. This glucuronan lyase system, for example, enables complete degradation of *Agaricus bisporus* glucuronan, a fungus, which could serve as prey for *T. parareesei*. Discovery of a novel metal-dependent arabinofuranosidase (glycoside hydrolase family 54) in *T. harzianum* revealed its broad substrate specificity (Motta et al. 2021). Regarding the functionality of enzymes under unusual, challenging conditions, a novel, acidophilic, halophilic, and ethanol-tolerant beta-1,4-endoglucanase (glycoside hydrolase family 12) and an acidophilic beta-mannanase (glycoside hydrolase family 5), and an acidophilic beta-xylanase of *T. asperellum* were characterized and shown to be active on lignocellulosic biomass (Zheng et al. 2022, 2023, 2024). With respect to degradation of diverse carbohydrates, the function of lytic polysaccharide monooxygenases (LPMOs) (Forsberg et al. 2019) gained importance recently. For example, a novel LPMO of class AA11 was characterized from *T. guizhouense* and is targeted at chitin (Ma et al. 2021). In *T. reesei* an epoxid hydrolase was detected (de Oliveira et al. 2016, 2019), which is particularly active on epoxides from docosahexaenoic acid. Such epoxid hydrolases are particularly important in the production of chiral drug intermediates and enantiopure compounds and hence these enzymes are highly valuable for pharmaceuticals, agrochemicals, and diverse fine chemicals in industry (Choi 2009; Saini and Sareen 2017).

Considering bioremediation-applications of *Trichoderma* for contaminations with toxic organic compounds, hydrocarbons, or heavy metals (Zafra and Cortes-Espinosa 2015; Andreolli et al. 2016; Tagyan et al. 2023; Yue et al. 2022), the hidden competences of this genus are increasingly gaining appreciation and come into focus of research. For example, two laccases were identified in *T. harzianum*, which are suitable for degradation of bisphenol A (BPA) over a broad pH range (Elsayed et al. 2023). In *T. atroviride*, a paraoxonase-like enzyme was identified, which is active on organophosphate oxons and lactone and could play an important role in degradation of the pesticide Dichlorvos (DDVP) and other organophosphate pesticides (Sun et al. 2019).

## 9.10 Genomics for Discovery of Novel Secondary Metabolite Clusters and Other Genes Contributing to Production of Bioactive Metabolites

Fungi, including those of the genus *Trichoderma*, are valuable sources of bioactive molecules (Karwehl and Stadler 2016), including compounds to fight cancer or serve as alternative antibiotics in case of microbial resistance against currently common pharmaceuticals (Viglas et al. 2021). Generally, association of produced secondary metabolites with the genes responsible for their production can be challenging (Cox 2024; Kjærbølling et al. 2019). Often, these genes are assembled in so-called biosynthetic gene clusters, albeit in many cases additional genes encoding proteins or regulators reside elsewhere in the genome. Hence, elucidation of a biosynthetic pathway of a given bioactive compound is ideally done using a combination of approaches.

A recent example of such a comprehensive study is the biosynthesis of aspinolides in *T. arundinaceum*, a known biocontrol agent (Cardoza et al. 2022). Comparative genomics with *Aspergillus ochraceus* and 35 additional *Trichoderma* species, gene regulation analysis, gene deletion, and biochemical analysis were applied to narrow down the genes responsible for aspinolide production to the two polyketide synthase genes *asp1* and *asp2* (Cardoza et al. 2022).

In *T. longibrachiatum*, regulation of the production of the pharmaceutically relevant sesquiterpene alkaloid dendrobin was investigated by studying an overproducing mutant (Qian et al. 2021). Moreover, dendrobine accumulation in this fungus was investigated in response to the signaling molecule methyl jasmonate (Jia et al. 2022). Subsequently, testing of diverse cultivation conditions revealed a positive effect of CoCl2 on the production of dendrobine-type alkaloids and gene regulation associated with the TCA cycle and biosynthesis of some amino acids (Qian et al. 2024). Dendrobine was initially characterized as a pharmaceutical produced by a slow-growing orchid, the production of which is (also) accomplished by an endophytically growing fungus. Hence, studies like this one contribute to a high-level production of promising pharmaceuticals in genetically tractable fungi to ultimately enable commercial production.

Another biosynthetic gene cluster that received considerable attention in recent years is the SOR cluster in *T. reesei* (Derntl et al. 2017a; Monroy et al. 2017), which is responsible for the yellow pigments (sorbicillinoids) produced under diverse cultivation conditions. Intriguingly, this cluster has an evolutionary history of birth and death evolution and purifying selection, in other words, the genes of this cluster were in part acquired from other fungi and they are under selective pressure in *T. reesei* (Druzhinina et al. 2016). Besides the two crucial polyketide synthases, several more biosynthetic genes were so far associated with sorbicillinoid production (Monroy et al. 2017; Hinterdobler et al. 2020a) and the two transcription factors in the cluster, YPR1 and YPR2 exert light dependent influence on the cluster genes (Hitzenhammer et al. 2019). For YPR2 a function beyond SOR cluster regulation, targeting secondary metabolism and enzyme expression was revealed as well (Hitzenhammer et al. 2019). Moreover, the genes of this cluster are regulated by light, carbon source, and the regulators SUB1 (Bazafkan et al. 2015, 2017), VEL1 (Bazafkan et al. 2015), CRE1 (Monroy et al. 2017), XPP1 (Derntl et al. 2017b) as well as genes involved in signal transduction like the protein kinase USK1 (Beier et al. 2020), the MAPkinase pathway (Schalamun et al. 2023a) and the G-protein coupled receptor GPR8 (Hinterdobler et al. 2020a). Both the connection of the genes regulating the cluster to enzyme expression as well as the biological functions of the sorbicillinoids, for example, in defense (Derntl et al. 2017a; Hou et al. 2022) warrant further investigations of the SOR cluster and its importance for physiology and competition in *T. reesei*.

A more recent discovery in *T. reesei* concerns Ribosomally synthesized and Posttranscriptionally modified Peptides (RiPPs) (Vignolle et al. 2020; Ozaki et al. 2023). These peptides are encoded within a precursor protein and are processed subsequently after posttranslational modification and they include for example alpha-amanitin or phallacidin, which are formed by *Amanita* spp. The biological functions of RiPPs are still not well known, particularly not in *Trichoderma*, but are assumed to be in support of nutrient acquisition, competition, and symbiosis (Ford et al. 2022).

Although also fungal RiPPs are known, screening software was until recently focused on bacterial genomes (Kessler and Chooi 2022), which is now complemented by a tool for fungi, which was developed to screen *Trichoderma* genomes for RiPPs (Vignolle et al. 2020). The number of detected RiPP candidate sequences ranged from 6 in *T. reesei* to 222 in *T. harzianum* (Vignolle et al. 2020), indicating these compounds as a promising research field for biological function in the future.

## 9.11 Conclusions and Further Directions of Research

In recent years, we saw a dramatic increase in available genome sequences and their analysis, aimed at discovery of molecular mechanisms shaping evolution, but also to determine ecological drivers of these mechanisms. Future developments in sequencing technology and likely bioinformatics will enable us to dig even deeper into this topic and connect long-known physiological phenomena with gene content of individual species and genera.

While omics research gained increased interest also in *Trichoderma*, a comprehensive overview of the mechanisms at work at the molecular level still remains fragmentary. Especially proteomics and phosphoproteomics studies are scarce, although recently some interesting studies brought this field forward also in *Trichoderma* (Nguyen et al. 2016; Nogueira et al. 2024; Liu et al. 2021b). In the long run, integration of knowledge from the diverse aspects of *Trichoderma* genomes and omics in general, like regulatory non-coding RNAs, mycoviruses, RIP, and for example also newly discovered secondary metabolites including RiPPs, will provide intriguing insights into *Trichoderma* physiology and facilitate new strategies for strain improvement in industry. However, all these new discoveries, like the role of milRNAs, the importance of light in regulation of transcript levels and enzyme expression or posttranscriptional regulation to name just a few also pose challenges to research with this genus, as these processes can no longer be ignored and have to be considered both during experimental design and interpretation of results.

In summary, research with strains of the genus *Trichoderma* has never been more interesting and timely and the new insights of omics research open up diverse avenues for investigations for which *Trichoderma* spp. represent the ideal model fungi to accelerate physiological knowledge and even enhance the biotechnological value of this genus.

## References

Abrahao-Neto J, Rossini CH, El-Gogary S, Henrique-Silva F, Crivellaro O, El-Dorry H (1995) Mitochondrial functions mediate cellulase gene expression in *Trichoderma reesei*. Biochemistry 34:10456–10462

Adnan M, Ma X, Olsson S, Wang J, Liu G (2022) Promoter regulation and genetic engineering strategies for enhanced cellulase expression in *Trichoderma reesei*. Microbiol Res 259:127011

Andreolli M, Lampis S, Brignoli P, Vallini G (2016) *Trichoderma longibrachiatum* Evx1 is a fungal biocatalyst suitable for the remediation of soils contaminated with diesel fuel and polycyclic aromatic hydrocarbons. Environ Sci Pollut Res Int 23:9134–9143

Arbige MV, Shetty JK, Chotani GK (2019) Industrial enzymology: the next chapter. Trends Biotechnol 37:1355–1366

Arnau J, Yaver D, Hjort MC (2020) Strategies and challenges for the development of industrial enzymes using fungal cell factories. In: Nevalainen H (ed) Grand challenges in fungal biotechnology. Springer Nature Switzerland AG, Switzerland

Atanasova L, Knox BP, Kubicek CP, Druzhinina IS, Baker SE (2013) The polyketide synthase gene *pks4* of *Trichoderma reesei* provides pigmentation and stress resistance. Eukaryot Cell 12:1499–1508

Averill C, Anthony MA, Baldrian P, Finkbeiner F, van den Hoogen J, Kiers T, Kohout P, Hirt E, Smith GR, Crowther TW (2022) Defending Earth's terrestrial microbiome, vol 7. Nat Microbiol, p 1717

Bai B, Liu C, Zhang C, He X, Wang H, Peng W, Zheng C (2023) *Trichoderma* species from plant and soil: an excellent resource for biosynthesis of terpenoids with versatile bioactivities. J Adv Res 49:81–102

Basse CW (2010) Mitochondrial inheritance in fungi. Curr Opin Microbiol 13:712–719

Bazafkan H, Beier S, Stappler E, Böhmdorfer S, Oberlerchner JT, Sulyok M, Schmoll M (2017) SUB1 has photoreceptor dependent and independent functions in sexual development and secondary metabolism in *Trichoderma reesei*. Mol Microbiol 106:742–759

Bazafkan H, Dattenböck C, Böhmdorfer S, Tisch D, Stappler E, Schmoll M (2015) Mating type dependent partner sensing as mediated by VEL1 in *Trichoderma reesei*. Mol Microbiol 96:1103–1118

Beier S, Hinterdobler W, Monroy AA, Bazafkan H, Schmoll M (2020) The kinase USK1 regulates cellulase gene expression and secondary metabolite biosynthesis in *Trichoderma reesei*. Front Microbiol 11:974

Beier S, Stiegler M, Hitzenhammer E, Schmoll M (2022) Screening for genes involved in cellulase regulation by expression under the control of a novel constitutive promoter in *Trichoderma reesei*. Curr Res Biotechnol 4:238–246

Bennett RJ, Turgeon BG (2016) Fungal sex: the Ascomycota. Microbiol Spectr 4

Benocci T, Aguilar-Pontes MV, Zhou M, Seiboth B, de Vries RP (2017) Regulators of plant biomass degradation in ascomycetous fungi. Biotechnol Biofuels 10:152

Bischof RH, Horejs J, Metz B, Gamauf C, Kubicek CP, Seiboth B (2015) L-methionine repressible promoters for tuneable gene expression in *Trichoderma reesei*. Microb Cell Factories 14:120

Bischof RH, Ramoni J, Seiboth B (2016) Cellulases and beyond: the first 70 years of the enzyme producer *Trichoderma reesei*. Microb Cell Factories 15:106

Borkovich KA, Alex LA, Yarden O, Freitag M, Turner GE, Read ND, Seiler S, Bell-Pedersen D, Paietta J, Plesofsky N, Plamann M, Goodrich-Tanrikulu M, Schulte U, Mannhaupt G, Nargang FE, Radford A, Selitrennikoff C, Galagan JE, Dunlap JC, Loros JJ, Catcheside D, Inoue H, Aramayo R, Polymenis M, Selker EU, Sachs MS, Marzluf GA, Paulsen I, Davis R, Ebbole DJ, Zelter A, Kalkman ER, O'Rourke R, Bowring F, Yeadon J, Ishii C, Suzuki K, Sakai W, Pratt R (2004) Lessons from the genome sequence of *Neurospora crassa*: tracing the path from genomic blueprint to multicellular organism. Microbiol Mol Biol Rev 68:1–108

Burger G, Gray MW, Lang BF (2003) Mitochondrial genomes: anything goes. Trends Genet 19:709–716

Cai Q, Qiao L, Wang M, He B, Lin FM, Palmquist J, Huang SD, Jin H (2018) Plants send small RNAs in extracellular vesicles to fungal pathogen to silence virulence genes. Science 360:1126–1129

Canseco-Perez MA, Castillo-Avila GM, Chi-Manzanero B, Islas-Flores I, Apolinar-Hernandez MM, Rivera-Munoz G, Gamboa-Angulo M, Sanchez-Teyer F, Couoh-Uicab Y, Canto-Canche B (2018) Fungal screening on olive oil for extracellular triacylglycerol lipases: selection of a *Trichoderma harzianum* strain and genome wide search for the genes. Genes (Basel) 9

Cardoza RE, McCormick SP, Izquierdo-Bueno I, Martinez-Reyes N, Lindo L, Brown DW, Collado IG, Proctor RH, Gutierrez S (2022) Identification of polyketide synthase genes required for aspinolide biosynthesis in *Trichoderma arundinaceum*. Appl Microbiol Biotechnol 106:7153–7171

Carreras-Villaseñor N, Sanchez-Arreguin JA, Herrera-Estrella AH (2012) *Trichoderma*: sensing the environment for survival and dispersal. Microbiology 158:3–16

Castrillo ML, Bich GA, Amerio NS, Barengo MP, Zapata PD, Saparrat MCN, Villalba LL (2023) *Trichoderma koningiopsis* (Hypocreaceae) has the smallest mitogenome of the genus *Trichoderma*. Front Microbiol 14:1141087

Cech TR, Steitz JA (2014) The noncoding RNA revolution-trashing old rules to forge new ones. Cell 157:77–94

Chadha S, Mehetre ST, Bansal R, Kuo A, Aerts A, Grigoriev IV, Druzhinina IS, Mukherjee PK (2018) Genome-wide analysis of cytochrome P450s of *Trichoderma* spp.: annotation and evolutionary relationships. Fungal Biol Biotechnol 5:12

Chatre L, Ricchetti M (2013) Prevalent coordination of mitochondrial DNA transcription and initiation of replication with the cell cycle. Nucleic Acids Res 41:3068–3078

Chaverri P, Samuels GJ (2013) Evolution of habitat preference and nutrition mode in a cosmopolitan fungal genus with evidence of interkingdom host jumps and major shifts in ecology. Evolution 67:2823–2837

Choi WJ (2009) Biotechnological production of enantiopure epoxides by enzymatic kinetic resolution. Appl Microbiol Biotechnol 84:239–247

Chuang YC, Li WC, Chen CL, Hsu PW, Tung SY, Kuo HC, Schmoll M, Wang TF (2015) *Trichoderma reesei* meiosis generates segmentally aneuploid progeny with higher xylanase-producing capability. Biotechnol Biofuels 8:30

Chung D, Kwon YM, Yang Y (2021) Telomere-to-telomere genome assembly of asparaginase-producing *Trichoderma simmonsii*. BMC Genomics 22:830

Chung D, Kwon YM, Yang Y (2022) The complete mitochondrial genome of *Trichoderma simmonsii* (Hypocreales: Hypocreaceae) from the southern coast of Korea. Mitochondrial DNA B Res 7:640–641

Cologna NMD, Gómez-Mendoza DP, Zanoelo FF, Giannesi GC, Guimarães NCA, Moreira LRS, Filho EXF, Ricart CAO (2018) Exploring *Trichoderma* and *Aspergillus* secretomes: proteomics approaches for the identification of enzymes of biotechnological interest. Enzym Microb Technol 109:1–10

Cortes MVCB, Barbosa ET, de Sousa Oliveira MI, Maciel LHR, Junior ML, Contesini FJ, de Filippi MCC, da Silva-Lobo VL (2023) *Trichoderma harzianum* marker-free strain construction based on efficient CRISPR/Cas9 recyclable system: a helpful tool for the study of biological control agents. Biol Control 184:105281

Cox RJ (2024) Engineered and total biosynthesis of fungal specialized metabolites. Nat Rev Chem 8:61–78

de Oliveira GS, Adriani PP, Borges FG, Lopes AR, Campana PT, Chambergo FS (2016) Epoxide hydrolase of *Trichoderma reesei:* biochemical properties and conformational characterization. Int J Biol Macromol 89:569–574

de Oliveira GS, Adriani PP, Wu H, Morisseau C, Hammock BD, Chambergo FS (2019) Substrate and inhibitor selectivity, and biological activity of an epoxide hydrolase from *Trichoderma reesei*. Mol Biol Rep 46:371–379

Debuchy R, Berteaux-Lecellier V, Silar P (2010) Mating systems and sexual morphogenesis in ascomycetes. In: Borkovich KA, Ebbole DJ (eds) Cellular and molecular biology of filamentous fungi. ASM Press, Washington, DC

Derntl C, Guzman-Chavez F, Mello-de-Sousa TM, Busse HJ, Driessen AJM, Mach RL, Mach-Aigner AR

(2017a) In vivo study of the sorbicillinoid gene cluster in *Trichoderma reesei*. Front Microbiol 8:2037

Derntl C, Kluger B, Bueschl C, Schuhmacher R, Mach RL, Mach-Aigner AR (2017b) Transcription factor Xpp1 is a switch between primary and secondary fungal metabolism. Proc Natl Acad Sci USA 114:E560–E569

Dos Santos UR, Dos Santos JL (2023) *Trichoderma* after crossing kingdoms: infections in human populations. J Toxic Environ Health, Part B 26:97–126

Druzhinina IS, Chenthamara K, Zhang J, Atanasova L, Yang D, Miao Y, Rahimi MJ, Grujic M, Cai F, Pourmehdi S, Salim KA, Pretzer C, Kopchinskiy AG, Henrissat B, Kuo A, Hundley H, Wang M, Aerts A, Salamov A, Lipzen A, LaButti K, Barry K, Grigoriev IV, Shen Q, Kubicek CP (2018) Massive lateral transfer of genes encoding plant cell wall-degrading enzymes to the mycoparasitic fungus *Trichoderma* from its plant-associated hosts. PLoS Genet 14:e1007322

Druzhinina IS, Kubicek EM, Kubicek CP (2016) Several steps of lateral gene transfer followed by events of 'birth-and-death' evolution shaped a fungal sorbicillinoid biosynthetic gene cluster. BMC Evol Biol 16:269

El-Dorry H, Crivellaro O, Leite A, Abrahão-Neto J, Henrique-Silva F, Escobar-Vera J, Matheucci Júnior E, Carle-Urioste JC, Pereira GG, Carraro-Pereira DM, El-Gogary S (1996) Transcriptional control of the cellulase genes in *Trichoderma reesei*. Brazilian Journal of Medical and Biological Research = Revista brasileira de pesquisas medicas e biologicas 29:905–909

Elsayed AM, Mahmoud M, Abdel Karim GSA, Abdelraof M, Othman AM (2023) Purification and biochemical characterization of two laccase isoenzymes isolated from *Trichoderma harzianum* S7113 and its application for bisphenol a degradation. Microb Cell Factories 22:1

Filiatrault-Chastel C, Heiss-Blanquet S, Margeot A, Berrin JG (2021) From fungal secretomes to enzymes cocktails: the path forward to bioeconomy. Biotechnol Adv 52:107833

Fitz E, Wanka F, Seiboth B (2018) The promoter toolbox for recombinant gene expression in *Trichoderma reesei*. Front Bioeng Biotechnol 6:135

Fonseca PLC, Badotti F, De-Paula RB, Araujo DS, Bortolini DE, Del-Bem LE, Azevedo VA, Brenig B, Aguiar E, Goes-Neto A (2020) Exploring the relationship among divergence time and coding and non-coding elements in the shaping of fungal mitochondrial genomes. Front Microbiol 11:765

Ford RE, Foster GD, Bailey AM (2022) Exploring fungal RiPPs from the perspective of chemical ecology. Fungal Biol Biotechnol 9:12

Forsberg Z, Sorlie M, Petrovic D, Courtade G, Aachmann FL, Vaaje-Kolstad G, Bissaro B, Rohr AK, Eijsink VG (2019) Polysaccharide degradation by lytic polysaccharide monooxygenases. Curr Opin Struct Biol 59:54–64

Frisvad JC, Moller LLH, Larsen TO, Kumar R, Arnau J (2018) Safety of the fungal workhorses of industrial biotechnology: update on the mycotoxin and secondary metabolite potential of *Aspergillus niger*, Aspergillus oryzae, and *Trichoderma reesei* Appl Microbiol Biotechnol 102:9481–9515

Gaderer R, Bonazza K, Seidl-Seiboth V (2014) Cerato-platanins: a fungal protein family with intriguing properties and application potential. Appl Microbiol Biotechnol 98:4795–4803

Gao R, Ding M, Jiang S, Zhao Z, Chenthamara K, Shen Q, Cai F, Druzhinina IS (2020) The evolutionary and functional paradox of cerato-platanins in fungi. Appl Environ Microbiol 86

Ghabrial SA, Caston JR, Jiang D, Nibert ML, Suzuki N (2015) 50-plus years of fungal viruses. Virology 479-480:356–368

Ghabrial SA, Suzuki N (2009) Viruses of plant pathogenic fungi. Annu Rev Phytopathol 47:353–384

Gladyshev E (2017) Repeat-induced point mutation and other genome defense mechanisms in fungi. Microbiol Spectr 5

Glass NL, Schmoll M, Cate JH, Coradetti S (2013) Plant cell wall deconstruction by ascomycete fungi. Ann Rev Microbiol 67:477–498

Gortikov M, Yakubovich E, Wang Z, López-Giráldez F, Tu Y, Townsend JP, Yarden O (2022) Differential expression of cell wall remodeling genes is part of the dynamic phase-specific transcriptional program of conidial germination of *Trichoderma asperelloides*. J Fungi (Basel) 8

Gudynaite-Savitch L, White TC (2016) Fungal biotechnology for industrial enzyme production: focus on (hemi)cellulase production strategies, advances and challenges. In: Schmoll M, Dattenböck C (eds) Gene expression systems in fungi: advancements and applications. Springer International Publishing Switzerland, Switzerland

Guo Q, Shi L, Wang X, Li D, Yin Z, Zhang J, Ding G, Chen L (2023) Structures and biological activities of secondary metabolites from the *Trichoderma* genus (covering 2018-2022). J Agric Food Chem 71:13612–13632

Gupta VK, Steindorff AS, de Paula RG, Silva-Rocha R, Mach-Aigner AR, Mach RL, Silva RN (2016) The post-genomic era of *Trichoderma reesei*: What's next? Trends Biotechnol 34:970–982

Gutierrez S, McCormick SP, Cardoza RE, Kim HS, Yugueros LL, Vaughan MM, Carro-Huerga G, Busman M, Saenz de Miera LE, Jaklitsch WM, Zhuang WY, Wang C, Casquero PA, Proctor RH (2021) Distribution, function, and evolution of a gene essential for trichothecene toxin biosynthesis in *Trichoderma*. Front Microbiol 12:791641

Guzmán-Guzmán P, Kumar A, de Los S-VS, Parra-Cota FI, Orozco-Mosqueda MDC, Fadiji AE, Hyder S, Babalola OO, Santoyo G (2023) *Trichoderma* species: our best fungal allies in the biocontrol of plant diseases-a review. Plants (Basel) 12

Guzman-Guzman P, Porras-Troncoso MD, Olmedo-Monfil V, Herrera-Estrella A (2019) *Trichoderma* species: versatile plant symbionts. Phytopathology 109:6–16

Harman GE, Doni F, Khadka RB, Uphoff N (2021) Endophytic strains of *Trichoderma* increase plants' photosynthetic capability. J Appl Microbiol 130:529–546

Harman GE, Howell CR, Viterbo A, Chet I, Lorito M (2004) *Trichoderma* species–opportunistic, avirulent plant symbionts. Nat Rev Microbiol 2:43–56

Hatvani L, Homa M, Chenthamara K, Cai F, Kocsube S, Atanasova L, Mlinaric-Missoni E, Manikandan P, Revathi R, Doczi I, Bogats G, Narendran V, Buchner R, Vagvolgyi C, Druzhinina IS, Kredics L (2019) Agricultural systems as potential sources of emerging human mycoses caused by *Trichoderma*: a successful, common phylotype of *Trichoderma longibrachiatum* in the frontline. FEMS Microbiol Lett 366

Hatvani L, Manczinger L, Vagvölgyi C, Kredics L (2013) *Trichoderma* as a human pathogen. In: Mukherjee PK, Horwitz BA, Singh US, Mukherjee M, Schmoll M (eds) Trichoderma–biology and applications. CAB International, Wallingford, Oxfordshire, UK

Henriquez-Urrutia M, Spanner R, Olivares-Yanez C, Seguel-Avello A, Perez-Lara R, Guillen-Alonso H, Winkler R, Herrera-Estrella A, Canessa P, Larrondo LF (2022) Circadian oscillations in *Trichoderma atroviride* and the role of core clock components in secondary metabolism, development, and mycoparasitism against the phytopathogen *Botrytis cinerea*. elife 11

Hinterdobler W, Beier S, Monroy AA, Berger H, Dattenbock C, Schmoll M (2020a) The G-protein coupled receptor GPR8 regulates secondary metabolism in *Trichoderma reesei*. Front Bioeng Biotechnol 8:558996

Hinterdobler W, Beier S, Schmoll M (2020b) Sexual development, its determinants and regulation in *Trichoderma reesei*. In: Zeilinger S, Druzhinina I, Singh HB, Gupta VK (eds) Recent developments in Trichoderma research. Elsevier, Amsterdam, The Netherlands

Hirpara DG, Gajera HP (2023) Intracellular metabolomics and microRNAomics unveil new insight into the regulatory network for potential biocontrol mechanism of stress-tolerant *Tricho*-fusants interacting with phytopathogen *Sclerotium rolfsii* Sacc. J Cell Physiol 238:1288–1307

Hirpara DG, Gajera HP, Savaliya DD, Parakhia MV (2023) Exploring conserved and novel MicroRNA-like small RNAs from stress tolerant *Trichoderma* fusants and parental strains during interaction with fungal phytopathogen *Sclerotium rolfsii* Sacc. Pestic Biochem Physiol 191:105368

Hitzenhammer E, Büschl C, Sulyok M, Schuhmacher R, Kluger B, Wischnitzki E, Schmoll M (2019) YPR2 is a regulator of light modulated carbon and secondary metabolism in *Trichoderma reesei*. BMC Genomics 20:211

Hou X, Zhang X, Xue M, Zhao Z, Zhang H, Xu D, Lai D, Zhou L (2022) Recent advances in sorbicillinoids from fungi and their bioactivities (covering 2016-2021). J Fungi (Basel) 8

Hurley JM, Loros JJ, Dunlap JC (2016) Circadian oscillators: around the transcription-translation feedback loop and on to output. Trends Biochem Sci 41:834–846

Ivanova C, Ramoni J, Aouam T, Frischmann A, Seiboth B, Baker SE, Le Crom S, Lemoine S, Margeot A, Bidard F (2017) Genome sequencing and transcriptome analysis of *Trichoderma reesei* QM9978 strain reveals a distal chromosome translocation to be responsible for loss of *vib1* expression and loss of cellulase induction. Biotechnol Biofuels 10:209

Jadhav R, Mach RL, Mach-Aigner AR (2024) Protein secretion and associated stress in industrially employed filamentous fungi. Appl Microbiol Biotechnol 108:92

Jaklitsch WM, Voglmayr H (2015) Biodiversity of *Trichoderma* (Hypocreaceae) in southern Europe and Macaronesia. Stud Mycol 80:1–87

Ji S, Liu Z, Liu B, Wang Y (2019) Comparative analysis of biocontrol agent *Trichoderma asperellum* ACCC30536 transcriptome during its interaction with *Populus davidiana* × *P. Alba* var. *pyramidalis*. Microbiol Res 227:126294

Jia Q, Wang L, Qian X, Jin H, Shu F, Sarsaiya S, Jin L, Chen J (2022) Transcriptome analysis of dendrobine biosynthesis in *Trichoderma longibrachiatum* MD33. Front Microbiol 13:890733

Jom-in S, Akarapisan A (2009) Characterization of double stranded RNA in *Trichoderma* spp. isolates in Chiang Mai province. J Agri Technol 5:261–270

Jourdier E, Baudry L, Poggi-Parodi D, Vicq Y, Koszul R, Margeot A, Marbouty M, Bidard F (2017) Proximity ligation scaffolding and comparison of two *Trichoderma reesei* strains genomes. Biotechnol Biofuels 10:151

Kang K, Zhong J, Jiang L, Liu G, Gou CY, Wu Q, Wang Y, Luo J, Gou D (2013) Identification of microRNA-like RNAs in the filamentous fungus *Trichoderma reesei* by solexa sequencing. PLoS One 8:e76288

Karlsson M, Atanasova L, Jensen DF, Zeilinger S (2017) Necrotrophic mycoparasites and their genomes. Microbiol Spectr 5

Karwehl S, Stadler M (2016) Exploitation of fungal biodiversity for discovery of novel antibiotics. Curr Top Microbiol Immunol 398:303–338

Kawamori M, Morikawa Y, Takasawa S (1985) Inductive formation of celluases by L-sorbose in *Trichoderma reesei*. Appl Microbiol Biotechnol 22:235–236

Kessler SC, Chooi YH (2022) Out for a RiPP: challenges and advances in genome mining of ribosomal peptides from fungi. Nat Prod Rep 39:222–230

Kjærbølling I, Mortensen UH, Vesth T, Andersen MR (2019) Strategies to establish the link between biosynthetic gene clusters and secondary metabolites. Fungal Genet Biol 130:107–121

Kondo H, Botella L, Suzuki N (2022) Mycovirus diversity and evolution revealed/inferred from recent studies. Annu Rev Phytopathol 60:307–336

Kouzminova E, Selker EU (2001) *dim-2* encodes a DNA methyltransferase responsible for all known cytosine methylation in Neurospora. EMBO J 20:4309–4323

Kredics L, Antal Z, Doczi I, Manczinger L, Kevei F, Nagy E (2003) Clinical importance of the genus Trichoderma. A review. Acta Microbiol Immunol Hung 50:105–117

Krumlauf R, Marzluf GA (1980) Genome organization and characterization of the repetitive and inverted repeat DNA sequences in *Neurospora crassa*. J Biol Chem 255:1138–1145

Kubicek CP, Steindorff AS, Chenthamara K, Manganiello G, Henrissat B, Zhang J, Cai F, Kopchinskiy AG, Kubicek EM, Kuo A, Baroncelli R, Sarrocco S, Noronha EF, Vannacci G, Shen Q, Grigoriev IV, Druzhinina IS (2019) Evolution and comparative genomics of the most common *Trichoderma* species. BMC Genomics 20:485

Kwak Y (2020) Complete mitochondrial genome of the fungal biocontrol agent *Trichoderma atroviride*: genomic features, comparative analysis and insight into the mitochondrial evolution in *Trichoderma*. Front Microbiol 11:785

Kwak Y (2021) An update on *Trichoderma* mitogenomes: complete *de novo* mitochondrial genome of the fungal biocontrol agent *Trichoderma harzianum* (Hypocreales, Sordariomycetes), an ex-neotype strain CBS 226.95, and tracing the evolutionary divergences of mitogenomes in *Trichoderma*. Microorganisms 9

Leger MM, Eme L, Stairs CW, Roger AJ (2018) Demystifying eukaryote lateral gene transfer (response to Martin 2017). BioEssays 40:e1700242. https://doi.org/10.1002/bies.201700115

Lehmann L, Ronnest NP, Jorgensen CI, Olsson L, Stocks SM, Jorgensen HS, Hobley T (2016) Linking hydrolysis performance to *Trichoderma reesei* cellulolytic enzyme profile. Biotechnol Bioeng 113:1001–1010

Li J, Liu X, Yin Z, Hu Z, Zhang KQ (2021a) An overview on identification and regulatory mechanisms of long non-coding RNAs in fungi. Front Microbiol 12:638617

Li J, Wiebenga A, Lipzen A, Ng V, Tejomurthula S, Zhang Y, Grigoriev IV, Peng M, de Vries RP (2023) Comparative genomics and transcriptomics analyses reveal divergent plant biomass-degrading strategies in fungi. J Fungi (Basel) 9

Li WC, Chen CL, Wang TF (2018) Repeat-induced point (RIP) mutation in the industrial workhorse fungus *Trichoderma reesei*. Appl Microbiol Biotechnol 102:1567–1574

Li WC, Huang CH, Chen CL, Chuang YC, Tung SY, Wang TF (2017) *Trichoderma reesei* complete genome sequence, repeat-induced point mutation, and partitioning of CAZyme gene clusters. Biotechnol Biofuels 10:170

Li WC, Lee CY, Lan WH, Woo TT, Liu HC, Yeh HY, Chang HY, Chuang YC, Chen CY, Chuang CN, Chen CL, Hsueh YP, Li HW, Chi P, Wang TF (2021b) *Trichoderma reesei* Rad51 tolerates mismatches in hybrid meiosis with diverse genome sequences. Proc Natl Acad Sci USA 118

Li WC, Lin TC, Chen CL, Liu HC, Lin HN, Chao JL, Hsieh CH, Ni HF, Chen RS, Wang TF (2021c) Complete genome sequences and genome-wide characterization of *Trichoderma* biocontrol agents provide new insights into their evolution and variation in genome organization, sexual development, and fungal-plant interactions. Microbiol Spectr 9:e0066321

Lichius A, Bidard F, Buchholz F, Le Crom S, Martin J, Schackwitz W, Austerlitz T, Grigoriev IV, Baker SE, Margeot A, Seiboth B, Kubicek CP (2015) Genome sequencing of the *Trichoderma reesei* QM9136 mutant identifies a truncation of the transcriptional regulator XYR1 as the cause for its cellulase-negative phenotype. BMC Genomics 16:326

Lieckfeldt E, Kullnig CM, Samuels GJ, Kubicek CP (2000) Sexually competent, sucrose- and nitrate-assimilating strains of *Hypocrea jecorina* (*Trichoderma reesei*) from south American soils. Mycologia 92:374–380

Lindo L, McCormick SP, Cardoza RE, Brown DW, Kim HS, Alexander NJ, Proctor RH, Gutierrez S (2018) Effect of deletion of a trichothecene toxin regulatory gene on the secondary metabolism transcriptome of the saprotrophic fungus *Trichoderma arundinaceum*. Fungal Genet Biol 119:29–46

Linke R, Thallinger GG, Haarmann T, Eidner J, Schreiter M, Lorenz P, Seiboth B, Kubicek CP (2015) Restoration of female fertility in *Trichoderma reesei* QM6a provides the basis for inbreeding in this industrial cellulase producing fungus. Biotechnol Biofuels 8:155

Liu HC, Li WC, Wang TF (2021a) TSETA: a third-generation sequencing-based computational tool for mapping and visualization of SNPs, meiotic recombination products, and RIP mutations. Methods Mol Biol 2234:331–361

Liu Q, Tang S, Meng X, Zhu H, Zhu Y, Liu D, Shen Q (2021b) Proteomic analysis demonstrates a molecular dialog between *Trichoderma guizhouense* NJAU 4742 and cucumber (*Cucumis sativus* L.) roots: role in promoting plant growth. Mol Plant-Microbe Interact 34:631–644

Liu Z, Li Y, Hou J, Liu T (2023) Transboundary milRNAs: indispensable molecules in the process of *Trichoderma breve* T069 mycoparasitism of *Botrytis cinerea*. Pestic Biochem Physiol 196:105599

Luo N, Li Z, Ling J, Zhao J, Li Y, Yang Y, Mao Z, Xie B, Li H, Jiao Y (2023) Establishment of a CRISPR/Cas9-mediated efficient knockout system of *Trichoderma hamatum* T21 and pigment synthesis PKS gene knockout. J Fungi (Basel) 9

Lv X, Zheng F, Li C, Zhang W, Chen G, Liu W (2015) Characterization of a copper responsive promoter and its mediated overexpression of the xylanase regulator 1 results in an induction-independent production of cellulases in *Trichoderma reesei*. Biotechnol Biofuels 8:67

Ma L, Liu Z, Kong Z, Wang M, Li T, Zhu H, Wan Q, Liu D, Shen Q (2021) Functional characterization of a novel copper-dependent lytic polysaccharide monooxygenase TgAA11 from *Trichoderma guizhouense* NJAU 4742 in the oxidative degradation of chitin. Carbohydr Polym 258:117708

Mach-Aigner AR, Martzy R (eds) (2021) Trichoderma reesei–methods and protocols. Humana Press, New York, USA

Macias-Rodriguez L, Contreras-Cornejo HA, Adame-Garnica SG, Del-Val E, Larsen J (2020) The interactions of *Trichoderma* at multiple trophic levels: inter-kingdom communication. Microbiol Res 240:126552

Malinich EA, Wang K, Mukherjee PK, Kolomiets M, Kenerley CM (2019) Differential expression analysis of *Trichoderma virens* RNA reveals a dynamic transcriptome during colonization of *Zea mays* roots. BMC Genomics 20:280

Malmierca MG, Cardoza RE, Alexander NJ, McCormick SP, Collado IG, Hermosa R, Monte E, Gutierrez S (2013) Relevance of trichothecenes in fungal physiology: disruption of *tri5* in *Trichoderma arundinaceum*. Fungal Genet Biol 53:22–33

Mandels M, Weber J, Parizek R (1971) Enhanced cellulase production by a mutant of *Trichoderma viride*. Appl Microbiol 21:152–154

Margolin BS, Garrett-Engele PW, Stevens JN, Fritz DY, Garrett-Engele C, Metzenberg RL, Selker EU (1998) A methylated *Neurospora* 5S rRNA pseudogene contains a transposable element inactivated by repeat-induced point mutation. Genetics 149:1787–1797

Marie-Nelly H, Marbouty M, Cournac A, Flot JF, Liti G, Parodi DP, Syan S, Guillen N, Margeot A, Zimmer C, Koszul R (2014) High-quality genome (re)assembly using chromosomal contact data. Nat Commun 5:5695

Marquez LM, Redman RS, Rodriguez RJ, Roossinck MJ (2007) A virus in a fungus in a plant: three-way symbiosis required for thermal tolerance. Science 315:513–515

Martin WF (2017) Too much eukaryote LGT. BioEssays 39

Martinez D, Berka RM, Henrissat B, Saloheimo M, Arvas M, Baker SE, Chapman J, Chertkov O, Coutinho PM, Cullen D, Danchin EG, Grigoriev IV, Harris P, Jackson M, Kubicek CP, Han CS, Ho I, Larrondo LF, de Leon AL, Magnuson JK, Merino S, Misra M, Nelson B, Putnam N, Robbertse B, Salamov AA, Schmoll M, Terry A, Thayer N, Westerholm-Parvinen A, Schoch CL, Yao J, Barbote R, Nelson MA, Detter C, Bruce D, Kuske CR, Xie G, Richardson P, Rokhsar DS, Lucas SM, Rubin EM, Dunn-Coleman N, Ward M, Brettin TS (2008) Genome sequencing and analysis of the biomass-degrading fungus *Trichoderma reesei* (syn. *Hypocrea jecorina*). Nat Biotechnol 26:553–560

McDonald MJ, Rice DP, Desai MM (2016) Sex speeds adaptation by altering the dynamics of molecular evolution. Nature 531:233–236

Medina R, Franco MEE, Bartel LC, Martinez Alcantara V, Saparrat MCN, Balatti PA (2020) Fungal mitogenomes: relevant features to planning plant disease management. Front Microbiol 11:978

Megarioti AH, Kouvelis VN (2020) The coevolution of fungal mitochondrial introns and their homing endonucleases (GIY-YIG and LAGLIDADG). Genome Biol Evol 12:1337–1354

Metz B, Seidl-Seiboth V, Haarmann T, Kopchinskiy A, Lorenz P, Seiboth B, Kubicek CP (2011) Expression of biomass-degrading enzymes is a major event during conidium development in *Trichoderma reesei*. Eukaryot Cell 10:1527–1535

Meyer V, Basenko EY, Benz JP, Braus GH, Caddick MX, Csukai M, de Vries RP, Endy D, Frisvad JC, Gunde-Cimerman N, Haarmann T, Hadar Y, Hansen K, Johnson RI, Keller NP, Krasevec N, Mortensen UH, Perez R, Ram AFJ, Record E, Ross P, Shapaval V, Steiniger C, van den Brink H, van Munster J, Yarden O, Wosten HAB (2020) Growing a circular economy with fungal biotechnology: a white paper. Fungal Biol Biotechnol 7:5

Monroy AA, Stappler E, Schuster A, Sulyok M, Schmoll M (2017) A CRE1- regulated cluster is responsible for light dependent production of dihydrotrichotetronin in *Trichoderma reesei*. PLoS One e0182530

Moran-Diez ME, Martinez de Alba AE, Rubio MB, Hermosa R, Monte E (2021) *Trichoderma* and the plant heritable priming responses. J Fungi (Basel) 7

Motta MLL, Filho JAF, de Melo RR, Zanphorlin LM, Dos Santos CA, de Souza AP (2021) A novel fungal metal-dependent α-L-arabinofuranosidase of family 54 glycoside hydrolase shows expanded substrate specificity. Sci Rep 11:10961

Nerva L, Chitarra W (2021) Mycoviruses: a hidden world within fungi. In: Zaragoza Ó, Casadevall A (eds) Encyclopedia of mycology. Elsevier, Oxford

Nerva L, Varese GC, Falk BW, Turina M (2017) Mycoviruses of an endophytic fungus can replicate in plant cells: evolutionary implications. Sci Rep 7:1908

Nevalainen H, Suominen P, Taimisto K (1994) On the safety of *Trichoderma reesei*. J Biotechnol 37:193–200

Nevalainen KM, Palva ET (1978) Production of extracellular enzymes in mutants isolated from *Trichoderma viride* unable to hydrolyze cellulose. Appl Environ Microbiol 35:11–16

Nguyen EV, Imanishi SY, Haapaniemi P, Yadav A, Saloheimo M, Corthals GL, Pakula TM (2016) Quantitative site-specific phosphoproteomics of *Trichoderma reesei* signaling pathways upon induction of hydrolytic enzyme production. J Proteome Res 15:457–467

Nogueira-Lopez G, Padilla-Arizmendi F, Inwood S, Lyne S, Steyaert JM, Nieto-Jacobo MF, Stewart A, Mendoza-Mendoza A (2019) TrichoGate: an improved vector system for a large scale of functional analysis of *Trichoderma* genes. Front Microbiol 10:2794

Nogueira KMV, Mendes V, Kamath KS, Cheruku A, Oshiquiri LH, de Paula RG, Carraro C, Pedersoli WR, Pereira LMS, Vieira LC, Steindorff AS, Amirkhani A, McKay MJ, Nevalainen H, Molloy MP, Silva RN (2024) Proteome profiling of enriched membrane-associated proteins unraveled a novel sophorose and cello-oligosaccharide transporter in *Trichoderma reesei*. Microb Cell Factories 23:22

Novy V, Nielsen F, Seiboth B, Nidetzky B (2019) The influence of feedstock characteristics on enzyme production in *Trichoderma reesei:* a review on productivity, gene regulation and secretion profiles. Biotechnol Biofuels 12:238

Olmo R, Quijada NM, Morán-Diez ME, Hermosa R, Monte E (2024) Identification of tomato microRNAs in late response to *Trichoderma atroviride*. Int J Mol Sci 25

Ozaki T, Minami A, Oikawa H (2023) Recent advances in the biosynthesis of ribosomally synthesized and post-translationally modified peptides of fungal origin. J Antibiot (Tokyo) 76:3–13

Özkale E, Doğan Ö, Budak M, Mahir Korkmaz E (2023) Mitogenome evolution in *Trichoderma afroharzianum* strains: for a better understanding of distinguishing genus. Genome 67:139

Pagnoni S, Oufensou S, Balmas V, Bulgari D, Gobbi E, Forgia M, Migheli Q, Turina M (2023) A collection of *Trichoderma* isolates from natural environments in Sardinia reveals a complex virome that includes negative-sense fungal viruses with unprecedented genome organizations. Virus Evol 9:vead042

Paloheimo M, Haarmann T, Mäkinen S, Vehmaanperä J (2016) Production of industrial enzymes in *Trichoderma reesei*. In: Schmoll M, Dattenböck C (eds) Gene expression systems in fungi: advancements and applications. Springer International, Heidelberg

Pang AP, Zhang F, Hu X, Luo Y, Wang H, Durrani S, Wu FG, Li BZ, Zhou Z, Lu Z, Lin F (2021) Glutamine involvement in nitrogen regulation of cellulase production in fungi. Biotechnol Biofuels 14:199

Peterson R, Nevalainen H (2012) *Trichoderma reesei* RUT-C30–thirty years of strain improvement. Microbiology 158:58–68

Pfordt A, Schiwek S, Karlovsky P, von Tiedemann A (2020) *Trichoderma afroharzianum* ear rot–a new disease on maize in Europe. Front Agronomy 2

Pilgaard B, Vuillemin M, Munk L, Holck J, Meier S, Wilkens C, Meyer AS (2022) Discovery of a novel glucuronan lyase system in *Trichoderma parareesei*. Appl Environ Microbiol 88:e0181921

Pola-Sanchez E, Villalobos-Escobedo JM, Carreras-Villasenor N, Martinez-Hernandez P, Beltran-Hernandez EB, Esquivel-Naranjo EU, Herrera-Estrella A (2021) A global analysis of photoreceptor-mediated transcriptional changes reveals the intricate relationship between central metabolism and DNA repair in the filamentous fungus *Trichoderma atroviride*. Front Microbiol 12:724676

Porciuncula Jde O, Furukawa T, Mori K, Shida Y, Hirakawa H, Tashiro K, Kuhara S, Nakagawa S, Morikawa Y, Ogasawara W (2013) Single nucleotide polymorphism analysis of a *Trichoderma reesei* hyper-cellulolytic mutant developed in Japan. Biosci Biotechnol Biochem 77:534–543

Proctor RH, McCormick SP, Kim HS, Cardoza RE, Stanley AM, Lindo L, Kelly A, Brown DW, Lee T, Vaughan MM, Alexander NJ, Busman M, Gutierrez S (2018) Evolution of structural diversity of trichothecenes, a family of toxins produced by plant pathogenic and entomopathogenic fungi. PLoS Pathog 14:e1006946

Qian X, Jin H, Chen Z, Dai Q, Sarsaiya S, Qin Y, Jia Q, Jin L, Chen J (2021) Comparative transcriptome analysis of genes involved in sesquiterpene alkaloid biosynthesis in *Trichoderma longibrachiatum* MD33 and UN32. Front Microbiol 12:800125

Qian X, Qin Y, Sarasiya S, Chen J (2024) Transcriptomic profiling of adding cobalt chloride to improve dendrobine-type total alkaloid production. Appl Microbiol Biotechnol 108:26

Ramirez-Valdespino CA, Casas-Flores S, Olmedo-Monfil V (2019) *Trichoderma* as a model to study effector-like molecules. Front Microbiol 10:1030

Rantasalo A, Czeizler E, Virtanen R, Rousu J, Lahdesmaki H, Penttila M, Jantti J, Mojzita D (2016) Synthetic transcription amplifier system for orthogonal control of gene expression in *Saccharomyces cerevisiae*. PLoS One 11:e0148320

Rantasalo A, Landowski CP, Kuivanen J, Korppoo A, Reuter L, Koivistoinen O, Valkonen M, Penttila M, Jantti J, Mojzita D (2018) A universal gene expression system for fungi. Nucleic Acids Res 46:e111

Rantasalo A, Vitikainen M, Paasikallio T, Jantti J, Landowski CP, Mojzita D (2019) Novel genetic tools that enable highly pure protein production in *Trichoderma reesei*. Sci Rep 9:5032

Rebolledo-Prudencio OG, Estrada-Rivera M, Dautt-Castro M, Arteaga-Vazquez MA, Arenas-Huertero C, Rosendo-Vargas MM, Jin H, Casas-Flores S (2022) The small RNA-mediated gene silencing machinery is required in *Arabidopsis* for stimulation of growth, systemic disease resistance, and suppression of the nitrile-specifier gene NSP4 by *Trichoderma atroviride*. Plant J 109:873–890

Rush TA, Shrestha HK, Gopalakrishnan Meena M, Spangler MK, Ellis JC, Labbé JL, Abraham PE (2021) Bioprospecting *Trichoderma*: a systematic roadmap to screen genomes and natural products for biocontrol applications. Front Fungal Biol 2

Saini P, Sareen D (2017) An overview on the enhancement of enantioselectivity and stability of microbial epoxide hydrolases. Mol Biotechnol 59:98–116

Sal E, Stemler J, Salmanton-Garcia J, Falces-Romero I, Kredics L, Meyer E, Wurstl B, Lass-Florl C, Racil Z, Klimko N, Cesaro S, Kindo AJ, Wisplinghoff H, Koehler P, Cornely OA, Seidel D (2022) Invasive *Trichoderma* spp. infections: clinical presentation and outcome of cases from the literature and the FungiScope(R) registry. J Antimicrob Chemother 77:2850–2858

Saloheimo M, Paloheimo M, Hakola S, Pere J, Swanson B, Nyyssonen E, Bhatia A, Ward M, Penttilä M (2002) Swollenin, a *Trichoderma reesei* protein with sequence similarity to the plant expansins, exhibits disruption activity on cellulosic materials. Eur J Biochem 269:4202–4211

Sanna M, Pugliese M, Gullino ML, Mezzalama M (2022) First report of *Trichoderma afroharzianum* causing seed rot on maize in Italy. Plant Dis PDIS12212697PDN:1982

Schalamun M, Beier S, Hinterdobler W, Wanko N, Schinnerl J, Brecker L, Engl DE, Schmoll M (2023a)

MAPkinases regulate secondary metabolism, sexual development and light dependent cellulase regulation in *Trichoderma reesei*. Sci Rep 13:1912

Schalamun M, Hinterdobler W, Schinnerl J, Brecker L, Schmoll M (2024) The transcription factor STE12 influences growth on several carbon sources and production of dehydroacatic acid (DHAA) in *Trichoderma reesei*. Scientific reports in Press

Schalamun M, Li G, Hinterdobler W, Grosskinsky DK, Compant S, Dreux-Zigha A, Gerke J, Cox R, Schmoll M (2023b) Plant recognition by *Trichoderma harzianum* elicits upregulation of a novel secondary metabolite cluster required for colonization. BioRxiv 2023(04):12.536597

Schalamun M, Molin EM, Schmoll M (2023c) RGS4 impacts carbohydrate and siderophore metabolism in Trichoderma reesei. BMC Genomics 24:372

Schalamun M, Schmoll M (2022) *Trichoderma*—genomes and genomics as treasure troves for research towards biology, biotechnology and agriculture. Front Fungal Biol 3

Schmal M, Girod C, Yaver D, Mach RL, Mach-Aigner AR (2022) A bioinformatic-assisted workflow for genome-wide identification of ncRNAs. NAR Genomics and Bioinformatics 4:lqac059

Schmoll M (2018a) Light, stress, sex and carbon–the photoreceptor ENVOY as a central checkpoint in the physiology of *Trichoderma reesei*. Fungal Biol 122:479–486

Schmoll M (2018b) Regulation of plant cell wall degradation by light in *Trichoderma*. Fungal Biol Biotechnol 5:10

Schmoll M, Dattenböck C (eds) (2016) Gene expression Systems in Fungi: advancements and applications. Springer International, Heidelberg

Schmoll M, Dattenböck C, Carreras-Villasenor N, Mendoza-Mendoza A, Tisch D, Aleman MI, Baker SE, Brown C, Cervantes-Badillo MG, Cetz-Chel J, Cristobal-Mondragon GR, Delaye L, Esquivel-Naranjo EU, Frischmann A, Gallardo-Negrete Jde J, Garcia-Esquivel M, Gomez-Rodriguez EY, Greenwood DR, Hernandez-Onate M, Kruszewska JS, Lawry R, Mora-Montes HM, Munoz-Centeno T, Nieto-Jacobo MF, Nogueira Lopez G, Olmedo-Monfil V, Osorio-Concepcion M, Pilsyk S, Pomraning KR, Rodriguez-Iglesias A, Rosales-Saavedra MT, Sanchez-Arreguin JA, Seidl-Seiboth V, Stewart A, Uresti-Rivera EE, Wang CL, Wang TF, Zeilinger S, Casas-Flores S, Herrera-Estrella A (2016) The genomes of three uneven siblings: footprints of the lifestyles of three *Trichoderma* species. Microbiol Mol Biol Rev 80:205–327

Schmoll M, Esquivel-Naranjo EU, Herrera-Estrella A (2010) *Trichoderma* in the light of day–physiology and development. Fungal Genet Biol 47:909–916

Schmoll M, Kubicek CP (2005) *ooc1*, a unique gene expressed only during growth of *Hypocrea jecorina* (anamorph: *Trichoderma reesei*) on cellulose. Curr Genet 48:126–133

Schmoll M, Wang TF (2016) Sexual development in *Trichoderma*. In: Wendland J (ed) The Mycota (Vol. I): growth, differentiation and sexuality. Springer International Publishing, Switzerland

Schmoll M, Zeilinger S, Mach RL, Kubicek CP (2004) Cloning of genes expressed early during cellulase induction in *Hypocrea jecorina* by a rapid subtraction hybridization approach. Fungal Genet Biol 41:877–887

Schuster A, Kubicek CP, Schmoll M (2011) Dehydrogenase GRD1 represents a novel component of the cellulase regulon in *Trichoderma reesei* (*Hypocrea jecorina*). Appl Environ Microbiol 77:4553–4563

Schweiger R, Padilla-Arizmendi F, Nogueira-Lopez G, Rostas M, Lawry R, Brown C, Hampton J, Steyaert JM, Muller C, Mendoza-Mendoza A (2021) Insights into metabolic changes caused by the *Trichoderma virens*-maize root interaction. Mol Plant-Microbe Interact 34:524–537

Seidl V, Gamauf C, Druzhinina IS, Seiboth B, Hartl L, Kubicek CP (2008) The *Hypocrea jecorina* (*Trichoderma reesei*) hypercellulolytic mutant RUT C30 lacks a 85 kb (29 gene-encoding) region of the wild-type genome. BMC Genomics 9:327

Seidl V, Seibel C, Kubicek CP, Schmoll M (2009) Sexual development in the industrial workhorse *Trichoderma reesei*. Proc Natl Acad Sci USA 106:13909–13914

Seidl V, Seiboth B (2010) *Trichoderma reesei*: genetic approaches to improving strain efficiency. Biofuels 1:343–354

Selker EU, Cambareri EB, Jensen BC, Haack KR (1987) Rearrangement of duplicated DNA in specialized cells of *Neurospora*. Cell 51:741–752

Singh A, Taylor LE 2nd, Vander Wall TA, Linger J, Himmel ME, Podkaminer K, Adney WS, Decker SR (2015) Heterologous protein expression in *Hypocrea jecorina*: a historical perspective and new developments. Biotechnol Adv 33:142–154

Son M, Yu J, Kim KH (2015) Five questions about mycoviruses. PLoS Pathog 11:e1005172

Song Z, Lin J, Li Z, Huang C (2021) The nuclear functions of long noncoding RNAs come into focus. Noncoding RNA Res 6:70–79

Soucy SM, Huang J, Gogarten JP (2015) Horizontal gene transfer: building the web of life. Nat Rev Genet 16:472–482

Stappler E, Dattenböck C, Tisch D, Schmoll M (2017) Analysis of light- and carbon-specific transcriptomes implicates a class of G-protein-coupled receptors in cellulose sensing. mSphere 2:e00089–e00017

Sun J, Yuan X, Li Y, Wang X, Chen J (2019) The pathway of 2,2-dichlorovinyl dimethyl phosphate (DDVP) degradation by *Trichoderma atroviride* strain T23 and characterization of a paraoxonase-like enzyme. Appl Microbiol Biotechnol 103:8947–8962

Tagyan AI, Yasser MM, Mousa AM, Alkhalifah DHM, Hozzein WN, Marzouk MA (2023) A feasibility study on the recall of metallophilic fungi from Fe(III)-contaminated soil and evaluating their mycoremediation capacity: experimental and theoretical study. J Fungi (Basel) 9

Taylor JT, Wang KD, Horwitz B, Kolomiets M, Kenerley CM (2021) Early transcriptome response of *Trichoderma virens* to colonization of maize roots. Front Fungal Biol 2:718557

Till P, Derntl C, Kiesenhofer DP, Mach RL, Yaver D, Mach-Aigner AR (2020) Regulation of gene expression by the action of a fungal lncRNA on a transactivator. RNA Biol 17:47–61

Till P, Mach RL, Mach-Aigner AR (2018a) A current view on long noncoding RNAs in yeast and filamentous fungi. Appl Microbiol Biotechnol 102:7319–7331

Till P, Pucher ME, Mach RL, Mach-Aigner AR (2018b) A long noncoding RNA promotes cellulase expression in *Trichoderma reesei*. Biotechnol Biofuels 11:78

Tisch D, Pomraning KR, Collett JR, Freitag M, Baker SE, Chen CL, Hsu PW, Chuang YC, Schuster A, Dattenböck C, Stappler E, Sulyok M, Böhmdorfer S, Oberlerchner J, Wang TF, Schmoll M (2017) Omics analyses of *Trichoderma reesei* CBS999.97 and QM6a indicate the relevance of female fertility to carbohydrate-active enzyme and transporter levels. Appl Environ Microbiol 83

Tiwari M, Singh R, Jha R, Singh P (2022) Heritable priming by *Trichoderma*: a sustainable approach for wheat protection against Bipolaris *sorokiniana*. Front Plant Sci 13:1050765

Torigoi E, Henrique-Silva F, Escobar-Vera J, Carle-Urioste JC, Crivellaro O, El-Dorry H, El-Gogary S (1996) Mutants of *Trichoderma reesei* are defective in cellulose induction, but not basal expression of cellulase-encoding genes. Gene 173:199–203

Tyskiewicz R, Nowak A, Ozimek E, Jaroszuk-Scisel J (2022) *Trichoderma*: the current status of its application in agriculture for the biocontrol of fungal phytopathogens and stimulation of plant growth. Int J Mol Sci 23

Uchiyama T, Uchihashi T, Nakamura A, Watanabe H, Kaneko S, Samejima M, Igarashi K (2020) Convergent evolution of processivity in bacterial and fungal cellulases. Proc Natl Acad Sci USA 117:19896–19903

Ulitsky I (2018) Interactions between short and long noncoding RNAs. FEBS Lett 592:2874–2883

van Wyk S, Wingfield BD, De Vos L, van der Merwe NA, Steenkamp ET (2020) Genome-wide analyses of repeat-induced point mutations in the ascomycota. Front Microbiol 11:622368

Vicente I, Baroncelli R, Moran-Diez ME, Bernardi R, Puntoni G, Hermosa R, Monte E, Vannacci G, Sarrocco S (2020) Combined comparative genomics and gene expression analyses provide insights into the terpene synthases inventory in *Trichoderma*. Microorganisms 8

Viglas J, Dobiasova S, Viktorova J, Ruml T, Repiska V, Olejnikova P, Gbelcova H (2021) Peptaibol-containing extracts of *Trichoderma atroviride* and the fight against resistant microorganisms and cancer cells. Molecules 26

Vignolle GA, Mach RL, Mach-Aigner AR, Derntl C (2020) Novel approach in whole genome mining and transcriptome analysis reveal conserved RiPPs in *Trichoderma* spp. BMC Genomics 21:258

Villalobos-Escobedo JM, Martinez-Hernandez JP, Pelagio-Flores R, Gonzalez-De la Rosa PM, Carreras-Villasenor N, Abreu-Goodger C, Herrera-Estrella AH (2022) *Trichoderma atroviride* hyphal regeneration and conidiation depend on cell-signaling processes regulated by a microRNA-like RNA. Microb Genom 8

Wallen RM, Perlin MH (2018) An overview of the function and maintenance of sexual reproduction in dikaryotic fungi. Front Microbiol 9:503

Walsh TJ, Groll A, Hiemenz J, Fleming R, Roilides E, Anaissie E (2004) Infections due to emerging and uncommon medically important fungal pathogens. Clin Microbiol Infect 10(Suppl 1):48–66

Wang G, Wang H, Xiong X, Chen S, Zhang D (2015) Mitochondria thioredoxin's backup role in oxidative stress resistance in *Trichoderma reesei*. Microbiol Res 171:32–38

Wang L, Zheng F, Zhang W, Zhong Y, Chen G, Meng X, Liu W (2018) A copper-controlled RNA interference system for reversible silencing of target genes in *Trichoderma reesei*. Biotechnol Biofuels 11:33

Wang P, Nuss DL (1995) Induction of a *Cryphonectria parasitica* cellobiohydrolase I gene is suppressed by hypovirus infection and regulated by a GTP-binding-protein-linked signaling pathway involved in fungal pathogenesis. Proc Natl Acad Sci USA 92:11529–11533

Wang S, Wang X, Chen J (2023) Identification of miRNAs involved in maize–induced systemic resistance primed by *Trichoderma harzianum* T28 against *Cochliobolus heterostrophus*. J Fungi (Basel) 9

Wang W, Zhang F, Cui J, Chen D, Liu Z, Hou J, Zhang R, Liu T (2021) Identification of microRNA-like RNAs from *Trichoderma asperellum* DQ-1 during its interaction with tomato roots using bioinformatic analysis and high-throughput sequencing. PLoS One 16:e0254808

Wang Y, Chen H, Ma L, Gong M, Wu Y, Bao D, Zou G (2022a) Use of CRISPR-Cas tools to engineer Trichoderma species. Microb Biotechnol 15:2521–2532

Wang Y, Liu R, Liu H, Li X, Shen L, Zhang W, Song X, Liu W, Liu X, Zhong Y (2022b) Development of a powerful synthetic hybrid promoter to improve the cellulase system of *Trichoderma reesei* for efficient saccharification of corncob residues. Microb Cell Factories 21:5

Weiberg A, Bellinger M, Jin H (2015) Conversations between kingdoms: small RNAs. Curr Opin Biotechnol 32:207–215

Weiberg A, Jin H (2015) Small RNAs–the secret agents in the plant-pathogen interactions. Curr Opin Plant Biol 26:87–94

Wilson DB (2009) Cellulases and biofuels. Curr Opin Biotechnol 20:295–299

Woo SL, Hermosa R, Lorito M, Monte E (2022) *Trichoderma*: a multipurpose, plant-beneficial microorganism for eco-sustainable agriculture. Nat Rev Microbiol 21:312

Wu M, Jin F, Zhang J, Yang L, Jiang D, Li G (2012) Characterization of a novel bipartite double-stranded RNA mycovirus conferring hypovirulence in the phytopathogenic fungus *botrytis porri*. J Virol 86:6605–6619

Wu X, Li F, Yang R, Meng X, Zhang W, Liu W (2022) Identification of a bidirectional promoter from *Trichoderma reesei* and its application in dual gene expression. J Fungi (Basel) 8

Xiong Y, Sun J, Glass NL (2014) VIB1, a link between glucose signaling and carbon catabolite repression, is essential for plant cell wall degradation by *Neurospora crassa*. PLoS Genet 10:e1004500

Yan S, Xu Y, Yu X-W (2021) From induction to secretion: a complicated route for cellulase production in *Trichoderma reesei*. Bioresour Bioprocess 8:1–15

You J, Zhou K, Liu X, Wu M, Yang L, Zhang J, Chen W, Li G (2019) Defective RNA of a novel mycovirus with high transmissibility detrimental to biocontrol properties of *Trichoderma* spp. Microorganisms 7

Yue X, Ren X, Fu J, Wei N, Altomare C, Haidukowski M, Logrieco AF, Zhang Q, Li P (2022) Characterization and mechanism of aflatoxin degradation by a novel strain of *Trichoderma reesei* CGMCC3.5218. Front Microbiol 13:1003039

Yun SH, Lee SH, So KK, Kim JM, Kim DH (2016) Incidence of diverse dsRNA mycoviruses in *Trichoderma* spp. causing green mold disease of shiitake *Lentinula edodes*. FEMS Microbiol Lett 363

Zachow C, Berg C, Muller H, Monk J, Berg G (2016) Endemic plants harbour specific *Trichoderma* communities with an exceptional potential for biocontrol of phytopathogens. J Biotechnol 235:162–170

Zafra G, Cortes-Espinosa DV (2015) Biodegradation of polycyclic aromatic hydrocarbons by *Trichoderma* species: a mini review. Environ Sci Pollut Res Int 22:19426–19433

Zeilinger S, Gruber S, Bansal R, Mukherjee PK (2016) Secondary metabolism in *Trichoderma*–chemistry meets genomics. Fungal Biol Rev 30:74–90

Zeilinger S, Schmoll M, Pail M, Mach RL, Kubicek CP (2003) Nucleosome transactions on the *Hypocrea jecorina* ( *Trichoderma reesei*) cellulase promoter *cbh2* associated with cellulase induction. Mol Gen Genomics 270:46–55

Zhang H, Xie J, Fu Y, Cheng J, Qu Z, Zhao Z, Cheng S, Chen T, Li B, Wang Q, Liu X, Tian B, Collinge DB, Jiang D (2020) A 2-kb mycovirus converts a pathogenic fungus into a beneficial endophyte for *Brassica* protection and yield enhancement. Mol Plant 13:1420–1433

Zhang H, Xie J, Fu Y, Cheng J, Qu Z, Zhao Z, Cheng S, Chen T, Li B, Wang Q, Liu X, Tian B, Collinge DB, Jiang D (2020) A 2-kb mycovirus converts a pathogenic fungus into a beneficial endophyte for *brassica* protection and yield enhancement. Mol Plant 13:1420–1433

Zhang J, Li K, Sun Y, Yao C, Liu W, Liu H, Zhong Y (2024) An efficient CRISPR/Cas9 genome editing system based on a multiple sgRNA processing platform in *Trichoderma reesei* for strain improvement and enzyme production. Biotechnol Biofuels Bioprod 17:22

Zheng F, Basit A, Wang J, Zhuang H, Chen J, Zhang J (2023) Biochemical analyses of a novel acidophilic GH5 beta-mannanase from *Trichoderma asperellum* ND-1 and its application in mannooligosaccharides production from galactomannans. Front Microbiol 14:1191553

Zheng F, Basit A, Wang J, Zhuang H, Chen J, Zhang J (2024) Characterization of a novel acidophilic, ethanol tolerant and halophilic GH12 β-1,4-endoglucanase from *Trichoderma asperellum* ND-1 and its synergistic hydrolysis of lignocellulosic biomass. Int J Biol Macromol 254:127650

Zheng F, Basit A, Zhuang H, Chen J, Zhang J, Chen W (2022) Biochemical characterization of a novel acidophilic beta-xylanase from *Trichoderma asperellum* ND-1 and its synergistic hydrolysis of beechwood xylan. Front Microbiol 13:998160

Zheng F, Cao Y, Lv X, Wang L, Li C, Zhang W, Chen G, Liu W (2017) A copper-responsive promoter replacement system to investigate gene functions in *Trichoderma reesei:* a case study in characterizing SAGA genes. Appl Microbiol Biotechnol 101:2067–2078

# Part 4

# Interactions: Symbioses, Mutualisms, and Pathogens

# Genomic Innovation and Virulence Evolution in the Emerging Human Fungal Pathogen *Candida auris*

**10**

Hugh Gifford, Johanna Rhodes, Duncan Wilson, and Rhys Farrer

**Abstract**

*Candida auris* is an emerging human fungal pathogen and global public health threat. Composed of at least five highly clonal clades, major clinical management challenges have included nosocomial outbreaks, difficulties in identification and diagnosis, and increasing multiple drug resistance. With a crude mortality of 45% in invasive candidiasis, virulence is also accompanied by a highly adhesive, stress tolerant, and environmentally persistent phenotype. This review covers examples of the genomic innovation strategies that may drive evolution of virulence in *C. auris*. Chromosomal architecture varies between genetic clades, with evidence of major rearrangements, particularly in the less virulent clade II. As a normally haploid yeast, adaptive aneuploidy (and diploidy) has also been observed, including under drug stress, which may increase virulence. Several gene family expansions, including of hydrolase, adhesin, and transporter families, indicate diversified and polarised roles in virulence. Important genomic variation linked to differing virulence has even been identified within a single patient, including copy number variation and loss of function of the adhesin gene *ALS4* in two otherwise highly related strains. We conclude our review by suggesting research topics that need greater attention including (1) moving fungal genomic enquiry closer to patients in clinical settings, (2) targeting ecological drivers of evolution and escalating environmental discovery as a priority, and (3) interdisciplinary investigation across microbiology and immunology to leverage comparative, molecular, and population genomics to understand and control invasive candidiasis caused by *C. auris*.

**Keywords**

*Candida auris* · Fungal pathogen genomics · Population genetics · Comparative genomics · Microbial evolution · Infectious disease emergence

H. Gifford · D. Wilson · R. Farrer (✉)
Medical Research Council Centre for Medical Mycology at the University of Exeter, Department of Biosciences, Faculty of Health and Life Sciences, Exeter, UK
e-mail: hg456@exeter.ac.uk; Duncan.Wilson@exeter.ac.uk; R.Farrer@exeter.ac.uk

J. Rhodes
MRC Centre for Global Infectious Disease Analysis, Imperial College London, London, UK

Department of Medical Microbiology, RadboudUMC, Nijmegen, Netherlands
e-mail: johanna.rhodes@imperial.ac.uk


M. Nowrousian, J. E. Stajich (eds.), *Fungal Genomics*, The Mycota 13,
https://doi.org/10.1007/978-3-031-75666-5_10

## 10.1 Introduction

### 10.1.1 Emergence Detection

The emerging human fungal pathogen *Candida auris* is a global public health threat (Meis and Chowdhary 2018), an urgent antimicrobial resistance threat (CDC 2019), and a World Health Organization critical priority pathogen (WHO 2022). The earliest known isolate dates from 11th December 1996, when a challenging case of invasive candidaemia was misidentified as *Candida haemulonii* in a one-year-old girl receiving mechanical ventilation in South Korea (Lee et al. 2011). Retrospective internal transcribed spacer (ITS) and D1/D2 26S ribosomal RNA sequencing were used for species confirmation 12 years later, when a novel ascomycetous yeast was recognised and described in 2008 after isolation from the ear canal of a 70-year-old woman with otomycosis in Japan (Satoh et al. 2009). This original species description noted that *C. auris* could tolerate high salinity (up to 10% NaCl) and temperatures (up to 42 °C), with optimum growth at mammalian temperatures (37–40 °C) and phylogenetic placement close to *C. haemulonii* in the Metschnikowiaceae clade. These "first" cases illustrate the two typical diseases associated with *C. auris* in humans: (secondary) invasive candidiasis in an already critically unwell patient and (primary) epithelial infection in an otherwise healthy individual. This review will focus on the life-threatening condition of invasive candidiasis and candidaemia caused by *C. auris*.

### 10.1.2 Global Spread

Within 15 years of its initial discovery, *C. auris* rapidly transitioned from emergence to prominence—and in some settings, dominance. In the Deenanath Mangeshkar Hospital, Western India, *C. auris* was the number one cause of invasive candidiasis between December 2019 and July 2021, accounting for 43% of cases (Prayag et al. 2022). In the Chris Hani Baragwanath Hospital, Johannesburg, South Africa, *C. auris* was recently found to infect or colonise 33% of infants in the neonatal unit (Shuping et al. 2023). Across low- and middle-income countries (LMICs), *C. auris* was the third most common cause of neonatal candidiasis in the recent NeoOBS study (14% of cases) (Cook et al. 2023). A massive ongoing outbreak in the State of New York has now affected 1343 patients across 151 healthcare facilities, including 59 hospitals and 92 nursing homes (Zhu et al. 2020). Clinical cases and colonisation cases increased across the USA by 44% and over 80%, respectively, in 2019 and a further 95% and over 200%, respectively, in 2021 (Lyman et al. 2023). The rise of *C. auris* has now affected at least 42 countries across every major continent (Ragusa et al. 2023) and shows no sign of abating.

Whole-genome sequencing (WGS) was used to uncover the primary puzzle of *C. auris* emergence: its underlying population structure. After performing WGS on 47 isolates from India/Pakistan ($n$ = 31), Japan (the first identified isolate), South Africa ($n$ = 10), and Venezuela ($n$ = 5), four clades separated by at least 10,000 single nucleotide polymorphisms (SNPs) were identified and given clade assignments I–IV that appeared to correlate with geographic location (Lockhart et al. 2017). These multiple, distinct, and highly clonal clades appeared to indicate multiple simultaneous emergences of distantly related strains—or a long history of hidden transmission before detection. At this point, the apparent geographic structure led to each clade being named accordingly: South Asian (I), East Asian (II), South African (III), and South American (IV). A further keystone population genetic study then expanded WGS analyses to 304 isolates, observing a "weaker phylogeographic substructure" thought to relate to ongoing transmission of *C. auris* around the globe since its first detection in each global area (Chow et al. 2020).

Further clades have been confirmed or proposed. Isolates belonging to the more recently discovered Clade V were identified as the cause of otomycosis in four cases in Iran; Clade V has over 200,000 SNPs compared with any other clade and occupies a basal position in the phylogeny (Abastabar et al. 2019; Chow et al. 2019; Muñoz et al. 2021; Spruijtenburg et al. 2022). A sixth clade has been proposed and is currently under review, having caused three cases in Singapore and one in Bangladesh (Suphavilai et al. 2023). These isolates are over 36,000 SNPs in distance from clade IV, the "South American" clade, highlighting how phylogeographical nomenclature is misleading. Given recent findings, it is likely that additional clades will be discovered. Indeed, the authors of the clade VI paper propose a machine-learning based genomic surveillance system that could be used to detect further clade emergences (Suphavilai et al. 2023).

### 10.1.3 Clinical Management

Clinical management challenges in dealing with *C. auris* outbreaks have been overwhelming, with requirements for high technology diagnostics for identification and limited treatment options due to multiple drug resistance (MDR). Only molecular genomic techniques and mass spectrometry systems can reliably identify *C. auris*, due to the limitations of morphological and biochemical laboratory testing. Therefore, identification requires ideally WGS or partial genome sequencing such as ITS sequencing and, for routine clinical identification, Matrix-assisted laser desorption/ionisation-time-of-flight (MALDITOF), or VITEK-2 species determination with up-to-date software systems (Lone and Ahmad 2019). Concerns around high rates of resistance (fluconazole 80%, amphotericin B 27%, micafungin 7%) and MDR in 23% of cases (Chow et al. 2020) have now escalated with increasing reports of pan-resistance to all three major classes of antifungal in several clades of *C. auris* (Ostrowsky et al. 2020; O'Brien et al. 2020; Lyman 2021; Maphanga et al. 2021). Even more alarming is the ability for *C. auris* to continue to adapt to antifungal exposure. In a notable recent example, a single critically unwell patient suffered from pan-resistant invasive infection with *C. auris*, including resistance to a fourth class of antifungal, 5-flucytosine (5-FC), alongside acquired SNPs in phosphoribosyltransferase gene *FUR1* (Jacobs et al. 2022), which has been linked to 5-FC resistance in *C. auris* (Rhodes et al. 2018).

Challenges in infection control have also been extreme. Early high-profile nosocomial outbreaks in Intensive Care Units (ICUs) (Schelenz et al. 2016; Eyre et al. 2018) garnered prominent media attention with accompanying metaphors such as *C. auris* being a "creature from the black lagoon" that "bubbled up and now is everywhere" (Richtel 2019; Richtel and Jacobs 2019). In fact, drastic control measures beyond disinfection have been attempted, including hydrogen peroxide fumigation, and hospital wall and ceiling tiles being removed (Richtel 2019; Richtel and Jacobs 2019). In a "Gordian Knot" for public health services (Rapti et al. 2023), initial infection control costs have exceeded £1 million for a single facility outbreak, excluding ongoing monthly expenditure in the aftermath (Taori et al. 2019; Thoma et al. 2022). Despite clinical management challenges in diagnosis, resistance, and control, the core problem is its impact on human morbidity and mortality.

### 10.1.4 *C. auris* Pathogenicity

Current clinical observational data are limited in their ability to quantify mortality that can be attributed to *C. auris* bloodstream infection (BSI). A meta-analysis of 15 studies describing *C. auris* BSI has reported a 45% associated crude mortality (Chen et al. 2020). Invasive candidiasis tends to occur in critically unwell patients as a secondary phenomenon, e.g. developing weeks into the course of an illness such as severe community-acquired pneumonia requiring mechanical ventilation and multiple invasive procedures. Identifying the relative contribution of one species in already unwell patients suffering from multiple conditions and co-infections is complex. For *Candida* infections in general, propensity matching has been used to compare 90-day crude mortality. For example, in 269 patients with invasive candidiasis compared to 1083 matched controls, the attributable mortality was 28.4%, accounting for a 2.1-fold increase in risk of death (Mazi et al. 2022). The crude mortality rates were 42.4% (cases) and 17.1% (controls), and antifungal treatment was associated with lower mortality (35 vs. 69%, $p < 0.001$).

There are only a handful of studies comparing the mortality associated with *C. auris* BSI to other *Candida* spp., limiting generalisation across different settings. In 269 critically unwell adults across three New York hospitals with *Candida* BSI, there was no significant difference in crude mortality with *C. auris* infection ($n = 83$) compared to other *Candida* spp. at 90 days ($n = 113$, 44.6 vs. 46.9%, $p = 0.75$); the odds ratio of recurrence was higher with *C. auris* (11.9 vs. 4.0%, $p = 0.04$) (Simon et al. 2023). *C. auris* BSI cases were more likely to come from a nursing home, with a longer length of stay preceding their infection, and higher frequencies of previous infection with gram-negative and MDR bacteria (60.2 vs. 11.5%, $p < 0.001$), as well as higher rates of cerebrovascular disease, hemiplegia, chronic pulmonary disease, and use of bladder catheterisation. Rates of preceding GI surgery or parenteral nutrition use were lower in *C. auris* BSI. Propensity matching indicated an adjusted odds ratio for 90-day mortality in *C. auris* infection versus other *Candida* spp. of 0.863 (95% confidence interval 0.563–1.828), indicating no significant difference.

However, a larger propensity matched study across seven Colombian hospitals found a lower mortality associated with *C. auris* (38.1% of $n = 134$) compared to other *Candida* spp. (51.1% of $n = 378$, $p = 0.013$) (Ortiz-Roa et al. 2023). A hazard ratio (rate of mortality with *C. auris* BSI vs. with other *Candida* spp.) of 0.69 (95% CI 0.53–0.90) was calculated by propensity matching, indicating a lower risk of dying with *C. auris* for individuals with sets of matched characteristics. However, the comparator group for of non-*auris Candida* spp. was quite different between the first and second studies (26.5 vs. 49.4% *C. albicans* and 28.3 vs. 7.1% *C. glabrata*), so these results reflect different population comparisons and cannot be seen as fully representative without expanding study to many more centres.

Aside from these two studies in adult populations, this comparison was also made in newborn infants, who experience a range of different host physiologies and pathologies. The recent NeoOBS study, which collected data on neonatal candidaemia in LMICs, recorded a 22% mortality from *Candida* spp. infection overall and a hazard ratio that was not significantly different between *C. auris* and other *Candida* spp. (which was 22% overall; *C. auris* mortality univariable hazard ratio was (1.28, 95% CI 0.44–3.70) (Cook et al. 2023)). Understanding the relative contribution of *C. auris* (and other *Candida* spp.) to human morbidity and mortality will require both much broader and more granular clinical investigation, including information about both patient and causative agent(s) of invasive candidiasis, including when there is mixed infection. Overall, the mortality of *C. auris* BSI appears similar to the mortality in BSI caused by other *Candida* spp. The baseline resulting assumption is that invasive candidiasis is associated with a massive attributable mortality, including double the risk of dying in infected persons.

### 10.1.5 *C. auris* Virulence

Fundamentally, pathogenicity is an either-or characteristic describing whether an organism can cause disease (Shapiro-Ilan et al. 2005), and virulence is "the relative capacity of a microorganism to cause damage in a host" (Casadevall and Pirofski 1999; Casadevall and Pirofski 2001, 2003). In evolutionary terms, virulence can also be defined as "disease severity as assessed by reductions in host fitness following infection" (Read 1994; Cressler et al. 2016). *C. auris* has been isolated in humans from head to toe, including (i) bodily fluids and epithelial surfaces, such as skin, sputum, urinary tract, ear discharge, vagina, and stool; (ii) abiotic implants such as arterial catheters, central venous catheters, extraventricular drains, intercostal drains; and (iii) normally sterile bodily sites indicative of invasion, such as pleural fluid, cerebrospinal fluid, and blood—though each of these sterile areas can be instrumented, e.g., with intercostal drainage, extraventricular drainage, or venepuncture (Osei Sekyere 2018). The huge range of human tissues that can be colonised by *C. auris* indicates an ability to penetrate or bypass endothelial and other internal defensive barriers to enter organs and digest host macromolecules to survive, which is necessary for invasive human fungal pathogens (Köhler et al. 2017).

*C. auris* possesses a unique array of virulence factors, including adhesion, aggregation, biofilm formation, stress tolerance, secreted hydrolase production, and host immune evasion (Spivak and Hanson 2018; de Cássia Orlandi Sardi et al. 2018; Rossato and Colombo 2018; Nett 2019; de Jong and Hagen 2019; Du et al. 2020; Desoubeaux et al. 2022; Egger et al. 2022; Watkins et al. 2022; Gómez-Gaviria et al. 2023; Horton et al. 2023; Rapti et al. 2023). Genomic investigation has massive potential to explain what underlies these virulence factors (Fig. 10.1), including by uncovering the fundamental structure and regulation of the *C. auris* genome, the history of its evolution, emergence and epidemiology, the genomic drivers and signatures of virulence, drug resistance,

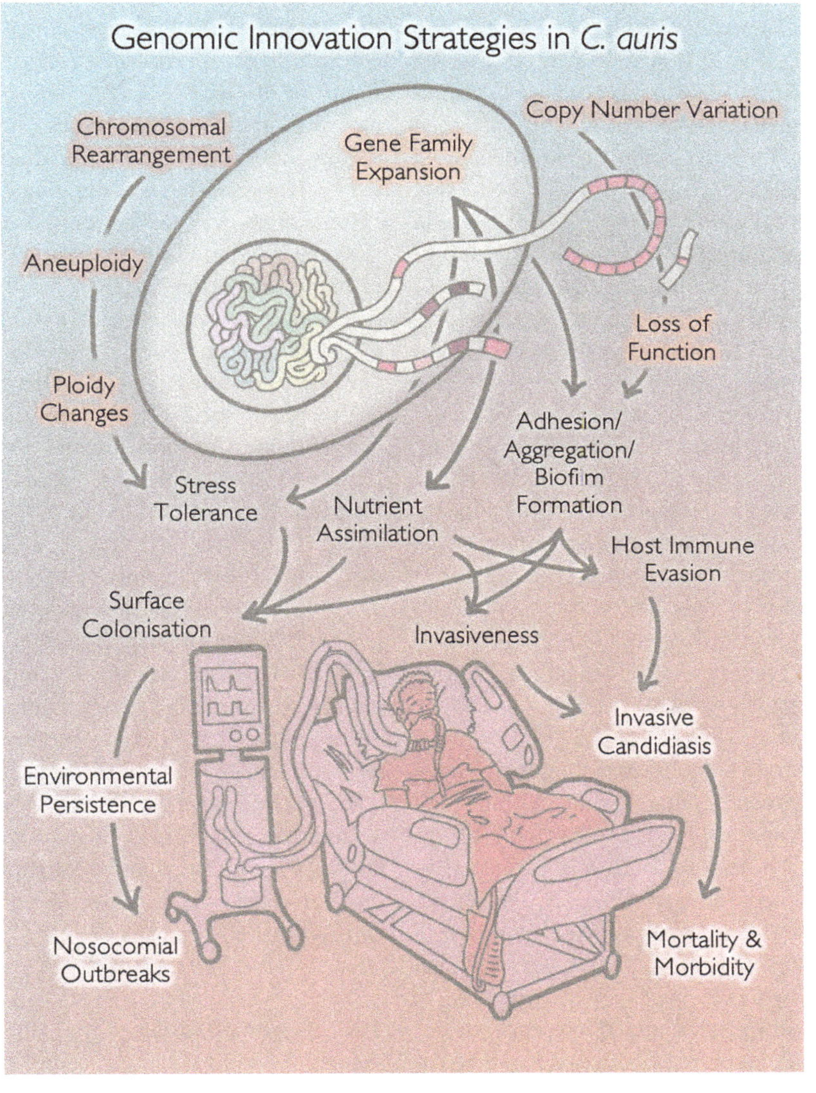


**Fig. 10.1** Genomic innovation strategies in *C. auris*. The emerging human fungal pathogen's seven-chromosome karyotype has undergone structural rearrangements between five clades linked to changes in virulence. Reversible aneuploidy under drug pressure has also been linked to resistance and stress tolerance. Diploid strains may be more virulence than haploid strains in an animal model. Gene family expansions, including of transporters, hydrolases, and adhesins, have allowed diversification and polarised expression patterns in nutrient assimilation, stress tolerance, and adhesion/aggregation/biofilm formation. These virulence factors are understood to be pivotal in host evasion and human tissue invasion, leading to the major clinical challenges of environmental persistence and nosocomial outbreaks, and invasive candidiasis leading to morbidity and mortality. Illustration by and © the authors

and targets for potential control (Chybowska et al. 2020).

## 10.2 Genomic Innovation Strategies

### 10.2.1 Structural Variation

The haploid genome of *C. auris* is 12.2–12.7 Mb consisting of seven 0.7–3.25 Mb chromosomes, together encoding 5294–5506 predicted protein-coding genes (Bravo Ruiz et al. 2019; Muñoz et al. 2021). *C. auris* belongs to the haploid Metschnikowiaceae clade of *Candida* spp. together with *C. haemulonii* and *C. lusitaniae*. In contrast, other *Candida* spp. that have received greater attention, including *C. albicans*, *C. parapsilosis*, and *C. tropicalis*, belong to the diploid Debaryomycetaceae subclade. Pulsed field gel electrophoresis has been used to identify karyotype variation in *C. auris* under thermal, osmotic, and genotoxic stress (Bravo Ruiz et al. 2019). Small segmental gains and losses occurred up to an estimated 1.7 Mb in length over five passages, and chromosomal losses were also identified. For example, some clade II isolates reduced to only five chromosomes and some clade I isolates to six. This reduction in chromosome number also manifested in a genome length as low as 10 Mb in some isolates (Bravo Ruiz et al. 2019).

Major chromosomal rearrangements have occurred between related strains within each clade, according to 12 long read assemblies from clades I–V (Muñoz et al. 2021). Given the variation in virulence between *C. auris* clades, the major structural variants in the karyotype of *C. auris* are thought to account for lower virulence for clade II, with the most genomic instability relative to the B8441 clade I reference strain. Eleven large deletion sites between 5 and 10 kb were observed, concentrated at sub-telomeric regions, and including 14 candidate adhesins (*Hyr/Iff*, *ALS* and other GPI-anchored proteins). Clade II isolates also demonstrated two to three inversions and four to eight translocations concentrated around centromeres, and even a low GC content candidate second centromere in chromosome six. One cluster of genes required for L-rhamnose assimilation is present only in clades III and V, suggesting deletion from other clades. Some chromosomal rearrangements are associated with retroelements and may be evidence for recent ectopic recombination. A further study made use of chromatin immunoprecipitation sequencing (ChIP-seq) to confirm gene sparse centromere regions with few repeats across clades I–IV and extensive chromosomal rearrangement in clade II (Narayanan et al. 2021). Such instability indicates a high likelihood of ongoing genomic innovation, supporting the ongoing use of WGS and analysis of structural variation in understanding virulence evolution.

Given the predominance of structural variants in clade II, it has been hypothesised that DNA damage response (DDR) genes may play a mechanistic role in variance generation. As many as 37 clade II-specific genes with loss-of-function mutations were identified, including the DNA Damage Checkpoint gene *DDC1* (Muñoz et al. 2021). *DDC1* was first described in *S. cerevisiae* as a critical DDR component (Longhese et al. 1997). Most DDR genes remain uncharacterised in *Candida* spp. in general (Yao et al. 2021) and are likely to be mechanistically important in generating structural variation resulting in different virulence profiles. Further genomic and molecular characterisation of DDR genes in *C. auris* could explain other forms of genomic variation, as follows, making systematic and comprehensive DDR description an important area of future research—not only for *C. auris* but for *Candida* spp. in general.

### 10.2.2 Aneuploidy

Aneuploidy is a well described mechanism of reversible stress adaptation across the fungal kingdom, and *C. auris* is no exception (Vande Zande et al. 2023). Microevolution experiments have explored genomic innovation under antifungal pressure and identified aneuploidy as a common feature (Narayanan et al. 2022–11/2022–12; Bing et al. 2020; Carolus et al. 2021; Li et al. 2022; Burrack et al. 2022). In one study, an originally susceptible clade I strain developed fluconazole resistance over multiple passages, associated with an extra copy of chromosome 5 containing the transcriptional regulator *TAC1B* (Bing et al. 2020). An additional chromosome 5 was associated with increased expression of the *ERG11* gene, which codes the target of azole antifungals, lanosterol 14, $\alpha$-demethylase (Karst and Lacroute 1977). Both these genes (*TAC1*, *ERG11*) have been implicated in *C. albicans* resistance via copy number variation (Selmecki et al. 2008). Another study used WGS and electrophoretic karyotyping to show chromosome 5 aneuploidy developing under fluconazole pressure in a drug-susceptible clade II strain and conferring resistance that was reversible without the additional chromosome (Narayanan et al. 2022–11/2022–12). However, this additional chromosome was partial, including only 638 or 675 kb of the chromosome, which is normally ~1 Mb in length (Narayanan et al. 2022–11/2022–12). Extending the range of strains studied, 17 strains from clades I and IV were exposed to fluconazole and assessed by flow cytometry and contour-clamped homogenous electric field (CHEF) electrophoresis to determine karyotype (Billmyre 2022; Burrack et al. 2022). The latter study also found that a potential chromosome 3 aneuploidy arose under azole pressure (Burrack et al. 2022). As well as an additional chromosome 5, an additional chromosome 6 has been noted during drug exposure, containing Multi-Drug Resistance gene *MDR1* (a known drug efflux pump in other *Candida* spp.), which can contribute to azole resistance independent of *ERG11* (Li et al. 2022). A study of in vitro drug exposure to all three main classes of antifungal also showed a chromosome 5 aneuploidy developed in the face of azole exposure (Carolus et al. 2021), as well as a duplication of a contig containing *ERG11* on chromosome 1. However, it is unknown how these cases of aneuploidy impact virulence.

### 10.2.3 Hypermutation

Studying genomic responses to drug pressure has demonstrated that aneuploidy and other accompanying forms of variation such as SNPs in genes encoding drug targets can rapidly increase drug resistance. Two examples of genomic change seen in the face of drug pressure with associated development of fluconazole resistance in a clade IV clinical strain include sub-telomeric deletions up to 103.8 kb, as well as a mutator phenotype (Burrack et al. 2022). The mutator phenotype in this study was confirmed by measurement of a mutation rate over ten times that of related parents (Burrack et al. 2022). It was also observed that this high mutation rate was associated with a missense mutation in the DNA mismatch repair MutL Homologue gene orthologue *MLH1* (Burrack et al. 2022). Other missense mutations were observed and included RNA-interference pathway related Dicer RNase III (*DCR1*) and Transformation/Transcription Domain Associated Protein (*TRRAP*) associated histone acetyltransferase subunit, *TRA1* (Grant et al. 1998; Bernstein et al. 2012), both of which play roles in regulation of gene expression. Such a mutator phenotype has not been described elsewhere, to our knowledge. Different mutation rates between strains and clades could be responsible for rapid acquisition of resistance in *C. auris*, but also impact branch length estimation and ancestry in phylogenetic analysis.

### 10.2.4 Polyploidy

The frequency and clinical relevance of diploid *C. auris* is currently unknown. Diploid strains have been demonstrated from clinical isolates of clade I and III using phloxine B red dye staining and flow cytometry DNA content analysis, with differing transcriptional profiles to haploid counterparts and higher virulence (measured by brain, kidney, and lung fungal burden) in a mouse model (Fan et al. 2020). It is not clear why other studies have not detected diploidy (Du et al. 2022); previous detection of exclusively haploid

*C. auris* strains used flow cytometry to compare a range of isolates from clades I–IV to haploid and diploid *S. cerevisiae* and diploid *C. albicans* (Bravo Ruiz et al. 2019). Many bioinformatic investigations do not validate or at least report ploidy and instead may simply use anticipated haploid parameters when running variant calling, missing genetic diversity. Polyploid states are thought to drive accelerated evolution within the fungal kingdom (Selmecki et al. 2015), with accelerated rates of aneuploidy and copy number variation in other species (Vande Zande et al. 2023). Characterisation of *C. auris* isolates in future should check for the possibility of diploidy to appreciate any role in *C. auris* evolution and virulence.

### 10.2.5 Hybridisation

Currently, it is unclear if *C. auris* has a cryptic sexual cycle (Ross and Lorenz 2020). Each clade appears to contain only one mating type locus (MTL): *MTL***a** in I, IV, and V, and *MTL***α** in II and III (Chow et al. 2020; Muñoz et al. 2021). Despite co-existence of clades with opposite MTL (I and III) in a single hospital in Kenya or in China (Chow et al. 2020; Bing et al. 2022) or in multiple individuals ($n = 5$) in Nevada, USA (Massic et al. 2023), there was no evidence of successful recent sexual recombination, suggesting barriers to reproduction and gene flow between clades. Based on variant calling of 1285 *C. auris* strains, several population genetic tests have implied a lack of any recombination in recent millennia (Wang and Xu 2022). Ongoing or recent mating has therefore not yet been detected in *C. auris*.

Several critical components of putative sexual or parasexual cycles appear missing or broken in *C. auris*. The *C. auris* orthologue of pheromone transporter *STE6*—originally described for its mutant conferring sterility in *Schizosaccharomyces pombe* (Girgsdies 1982)—appears dysfunctional in at least the clade I reference strain (B8441), with a two-nucleotide deletion rendering this a pseudogene (Wasi et al. 2019). Clade I isolates from the Nevada study also had a truncation in *STE6* from the 387th amino acid as well as a lack of meiosis-related α2 gene (Massic et al. 2023). At least seven meiosis-specific genes have not been detected in clades I–IV, including *DMC1*, *MEI5*, *SAE3*, *ZIP1*, *HOP1*, *RAD55*, *RAD57*, and *MSH4/5* (Muñoz et al. 2018). However, *C. lusitaniae* also lacks important meiosis-related genes but still appears to undergo meiosis during sexual reproduction (Reedy et al. 2009). Any discovery of parasexual reproduction in *C. auris* would be highly useful in advancing the study of its genomic innovation strategies. Continued asexual reproduction may however indicate a degree of genomic stability and a limitation to the adaptability of *C. auris*.

### 10.2.6 Gene Family Expansions

Several gene family expansions have been identified in *C. auris*. PFAM domain searches based on transcriptome data have identified various expanded gene families, including secreted aspartyl protease (*SAP*), lipase (*LIP*), oligopeptide transport (*OPT*), major facilitator superfamily (*MFS*), ATP-binding cassette (*ABC*) transport, siderophore iron transporters (*SIT*), and adhesin families (Muñoz et al. 2018). Gene family expansions can occur through duplication, as often as one duplication per gene per 100 million years (Lynch and Conery 2000), a rate comparable to the rate of nucleotide substitutions per site, depending on the organism (Demuth and Hahn 2009). Exact mechanisms of duplication are uncertain (Reams and Roth 2015), but the process allows a divergence of function and even development of a cast of novel proteins to facilitate host microenvironment adaptation (Hughes 1994; Innan and Kondrashov 2010). Most recently, multiple gene duplications have been linked to extremotolerance in the ecdysozoan lineage Tardigrada (Fleming et al. 2024). The following sections describe investigations into prominent gene family expansions.

#### 10.2.6.1 Hydrolases

Hydrolases and related enzymes are critical for host tissue digestion. As with other *Candida* spp.,

*C. auris* has multi-copy families of secreted hydrolases, such as at least three Secreted Aspartyl Protease (*SAP*) and eight Secreted Lipase (*LIP*) genes (Muñoz et al. 2018). A role for secreted lipases in virulence has been demonstrated using knockout experiments in *C. albicans* (Gácser et al. 2007a) and *C. parapsilosis* (Gácser et al. 2007b) gene deletion mutants. Phylogenetic analysis indicates a clear divergence of *C. auris SAP* genes from *C. albicans* and the most prominent gene identified in deletion testing for virulence and biofilm formation in *C. auris* has tentatively been named *Sapa3* (Kim et al. 2023). A multiomic approach has confirmed at least seven active phospholipases in *C. auris* (Zamith-Miranda et al. 2019). One intriguing observation is that the lipase orthologue *LIP2*, which suppresses the immune signalling molecule interleukin 17 (IL-17) via palmitic acid production and is necessary for deep tissue invasion by *C. albicans*, was expressed during *C. auris* co-incubation with macrophages (Basso et al. 2022; Miramón et al. 2023). Expanded *C. auris* protease and lipase gene families may play roles not only in nutrient assimilation but also in host immune evasion and biofilm formation.

#### 10.2.6.2 Transporters

Following hydrolytic digestion by proteases, resultant peptides must be imported to be used as a nutrient source. The oligopeptide transport (OPT) proteins are typically proton-coupled symporters that enable nitrogen assimilation from one to eight residue peptides through as many as 17 transmembrane domains (Gomolplitinant and Saier 2011). Opt1–5 have also been shown to be necessary for *C. albicans* uptake of oligopeptides when they are the sole nitrogen source (Dunkel et al. 2013). Siderophore transporters are also required for virulence in distinct types of *C. albicans* infection in differing host niches, confirmed by deletion experiments (Heymann et al. 2002). Peptide transporters can also play other roles in yeast, such as pheromone secretion and quorum sensing (Magee et al. 2002; Homer et al. 2016; Abele and Tampé 2018). Investigations into the biological roles for these transporters in other *Candida* spp. suggest that systematic functional assessment in *C. auris* will also be fruitful.

Gene family expansions allow each gene new evolutionary potential, including changes in function and regulation. A polarisation of expression patterns within the same gene family has been observed several times. For example, orthologues of both Siderophore Transporter *SIT1* and amino acid transporter dicarboxylic amino acid Permease *DIP5* had members with both a significantly higher (over four-fold change) and a lower (under four-fold change) differential expression during a macrophage phagocytosis experiment (Miramón et al. 2023).

#### 10.2.6.3 Adhesins

Expanded families of cell wall adhesins can facilitate surface colonisation, biofilm formation, host immune evasion, and tissue adhesion and invasion. The *ALS* (agglutinin-like sequence) adhesin family in *C. albicans* contains nine genes (*ALS1–9*) that play major roles in adhesion, with *ALS3* additionally mediating cellular invasion by induced endocytosis (Phan et al. 2007; Hoyer and Cota 2016); host cell invasion has not been explored or noted in *C. auris*. *C. auris* contains only three *ALS* orthologues that have been referred to inconsistently across different studies as *ALS3–5* (Pelletier et al. 2024). Another adhesin family is the *Hyr/Iff*-like (*Hil*) family, comprising eight genes in *C. auris* that each contains a range of central tandem repeat lengths that may contribute to functional variation (Smoak et al. 2023). The more recently described Surface Colonisation Factor, encoded by the *SCF1* gene, acts via cationic charge in a manner like marine bivalve adhesion, unlike other known fungal adhesins, and mediates virulence in immunocompromised mice (Santana et al. 2023). Expression of *SCF1* correlates with adhesion across various strains and, together with the hydrophobic adhesin *IFF4109*, is required to colonise in vivo murine skin and ex vivo human skin explants. *SCF1* was not seen to be necessary for adhesion in *C. haemulonii* despite the presence of an orthologue (Santana et al. 2023). The discovery of *SCF1* demonstrates the broad and nuanced range of potential adhesion mechanisms for *C. auris*.

Phenotypic diversity of *C. auris* strains has been identified even within a single patient via copy number variation or loss of function of a single gene (Bing et al. 2023). In an 81-year-old woman with a urinary infection caused by clade III *C. auris*, one strain contained a tenfold increase in copy number in a sub-telomeric location of adhesin *ALS4* (B9J08_004112) with flanking polyA/polyT motifs and was associated with aggregation, biofilm formation, and enhanced skin colonisation in a mouse skin model. A further isolate from the same patient, separated by only 38 SNPs and 90 insertions/deletions, contained a truncated *ALS4* lacking its predicted GPI anchor and fewer central repeats (8 rather than 30), and was found to be phenotypically divergent, including exhibiting reduced skin colonisation in the mouse skin model and significantly higher virulence in a *Galleria mellonella* infection model. This intriguing example shows a huge range of genomic plasticity associated with phenotypic variation that relates to both an invasive phenotype and a colonising phenotype. Of note, sub-telomeric locations of these expansions could in theory also be coupled with potential disrupted telomere maintenance, which could enable a rapid accumulation of SNPs, as has been shown in ascomycetous rice blast fungus *Pyricularia oryzae* (Rahnama et al. 2021).

Adhesins can also play a role in cellular aggregation, which has been observed in some *C. auris* isolates as a clumping together of isolates in planktonic cultures. Aggregating strains display increased biofilm formation but decreased virulence in a *G. mellonella* model (Borman et al. 2016; Sherry et al. 2017; Bing et al. 2023), but little difference in virulence in a *Drosophila melanogaster* model (Wurster et al. 2019). The mechanisms of aggregation appear to be both strain and environment dependent. Whether aggregation occurs through a stress-induced cell separation defect, *ALS*-mediated aggregation, or surface amyloid and extracellular matrix formation, aggregating forms have been hypothesised to prevent phagocytosis through physical size (Malavia-Jones et al. 2023; Pelletier et al. 2024). The adhesins thus exemplify genomic innovation through gene family expansion, copy number variation, and loss of function, and interact with a broad range of factors critical to virulence.

## 10.3 Discussion

Here we have reviewed several examples of genomic innovation strategies used by *C. auris* that are linked to its virulence (Table 10.1). However, many questions relating to the importance of genotype in pathogenicity remain unanswered including (i) a more comprehensive understanding of genomic variation (in contrast to anecdotal evidence) and its link to virulence, (ii) how evolution has generated and shapes extant and ancestral genotypes, and (iii) the molecular mechanisms governing or regulating

**Table 10.1** Genomic innovation strategies and drivers of virulence evolution in *C. auris*

| Genomic innovation strategy | Virulence attribution | Reference |
|---|---|---|
| Aneuploidy (in vitro—Chromosome three, containing *ERG11*) | Stress tolerance and azole drug pressure only; unknown virulence attribution | Burrack et al. (2022) |
| Aneuploidy (in vitro—Chromosome five, containing TAC1B, which regulates transcription of *ERG11*) | Stress tolerance and azole drug pressure only; unknown virulence attribution | Bing et al. (2020); Carolus et al. (2021); Burrack et al. (2022); Li et al. (2022); Narayanan et al. (2022) |
| Aneuploidy (in vitro—Chromosome six, containing drug efflux pump gene *MDR1*) | Stress tolerance and azole drug pressure only; unknown virulence attribution | Li et al. (2022) |
| Chromosomal rearrangement (ancestral—Sub-telomeric deletions, inversions, translocations) | Extensive rearrangement in clade II; lower virulence and tendency to cause otomycosis only | Bravo Ruiz et al. (2019), Muñoz et al. (2021), Narayanan et al. (2021) |

(continued)

**Table 10.1** (continued)

| Genomic innovation strategy | Virulence attribution | Reference |
|---|---|---|
| Copy number variation (ancestral—Gene family expansions of secreted aspartyl protease and lipase genes) | Hypothesised to enable nutrient assimilation and possibly host immune evasion via IL-17 suppression | Muñoz et al. (2018), Basso et al. (2022), Miramón et al. (2023) |
| Copy number variation (ancestral—Gene family expansions of oligopeptide and siderophore transport genes) | Polarised expression pattern between members of *SIT* and *DIP* families during survival in phagocytes | Muñoz et al. (2018), Miramón et al. (2023) |
| Copy number variation (ancestral—Gene family expansions of adhesion families *ALS* and *Hil*) | Functional differentiation within family members hypothesised to play critical roles in adhesion, aggregation, and biofilm formation | Muñoz et al. (2018); Smoak et al. (2023), Borman et al. (2016), Sherry et al. (2017) |
| Copy number variation (in vitro—Chromosome one contig including *ERG11*) | Stress tolerance and azole drug pressure only; unknown virulence attribution | Carolus et al. (2021) |
| Copy number variation (within patient—Tenfold increase of *ALS4* in sub-telomeric region in a clade III isolate) | Demonstrated in urinary infection; increased skin colonisation in a mouse model | Bing et al. (2023) |
| Diploidy (clinical strains from clade I and III) | Higher virulence in mouse model measured by fungal burden in brain, kidney, and lung | Fan et al. (2020) |
| Elevated mutation rates (in vitro—via missense mutation in DNA mismatch repair *MLH1*) | Stress tolerance and azole drug pressure only; unknown virulence attribution | Burrack et al. (2022) |
| Loss of function (within patient—*ALS4* truncation in a clade III isolate) | Demonstrated in urinary infection; increased virulence in *G. mellonella* model | Bing et al. (2023) |

each genomic innovation strategy. We therefore recommend three approaches to address *C. auris* genomic innovation. Firstly, we should consider increasing the proximity of genomic research to the clinical patient bedside to maximise the information yield and relevance to human morbidity and mortality. Secondly, we should escalate the awareness of *C. auris* and emerging fungal pathogens into integrated and One Health domains to harness the power of global environmental genomics. Thirdly, we should optimise interdisciplinary collaboration in comparative *Candida* spp. genomics across microbiology and immunology systematically and comprehensively to unveil the underlying mechanisms of genomic innovation in agents of invasive candidiasis. This is especially the case since (mainly) haploid *Candida* spp. related to *C. auris* (such as *C. haemulonii* and *C. lusitaniae*) are quite distantly related to the majority of well-researched (mainly) diploid *Candida* spp. (especially *C. albicans*, but also *C. parapsilosis* and *C. tropicalis*) (Muñoz et al. 2018).

### 10.3.1 Moving Fungal Genomic Inquiry Closer to the Patient

Identifying within host genomic diversity relevant to human virulence is critical to understanding the evolution of pathogenicity in *C. auris*. In one case previously mentioned, a single patient contained isolates differentiated by tenfold *ALS4* copy number variation, or truncation resulting in loss of function (Bing et al. 2023). Several other case reports, aided by whole-genome sequencing, have identified SNPs that may impact virulence and drug resistance. In one example of pan-resistance, 19 strains were collected and sequenced from a single patient over 72 days, during which the patient had been exposed to eight different antifungal drugs (Jacobs et al.

2022). The fact that many discoveries can be made from just one clinical case hints at a massive scope for similar work in future.

Near-patient testing is essential. It is well accepted that strains sub-cultured in vitro, away from a natural (e.g. clinical) setting, can rapidly lose their genetic and phenotypic profiles (e.g. via relaxed selection, mutations, adaptive aneuploidy, etc.). Therefore, the further cultures travel from their patient, the fewer genuine genomic innovation strategies we are likely to detect. One of the challenges of WGS is that a patient may be infected with billions of yeast cells, but only a single colony forming unit is cultured from their bloodstream, grown on culture media, and then sequenced, resulting in an under-sampling of in-host diversity. This may exaggerate how highly clonal outbreaks appear (Eyre et al. 2018). Strains isolated from non-outbreak settings have indicated a possibility that within host SNP diversity can include as many as 60 SNPs between recent clonal isolates, compared to under three SNPs from within an individual during outbreak settings (Biswas et al. 2020). Within host diversity can enable the development of clinical resistance (Diaz Caballero et al. 2023), including through bet hedging or phenotypic heterogeneity (Morawska et al. 2022). Therefore, it is important that we develop strategies to detect in patient evolution in the future, perhaps through increased partnerships between clinical settings and bioinformatics/genomics expertise.

As genomic technology continues to advance, the suggestion of moving genomic investigation to the bedside is increasingly possible. Indeed, combining host and pathogen metagenomics and metatranscriptomics at the bedside has been tested and shown to be accurate for sepsis diagnosis (Kalantar et al. 2022). One study of the metatranscriptome in clinical respiratory samples recently noted that over 51% of reads were from *Candida* spp. (Wang et al. 2023a), indicating the wealth of genomic data that could be interrogated in future to understand near-patient host–pathogen interaction. Near-patient genomic testing is at the frontier of tool development and offers the opportunity to demonstrate *C. auris* and other pathogen genomic innovation strategies in situ.

Currently, existing genomic data often lacks clinical relevance and could be a wasted resource without better clinical integration. The technology which has been made use of most extensively near the patient bedside is Illumina short read WGS, growing exponentially at the same time as the spread of *C. auris* and leading to an unprecedented wealth of data. By 1st February 2024, the National Center for Biotechnology Information Short Read Archive (NCBI SRA) contained over 10,000 paired end genomic DNA isolate datasets. Few studies have used over 1000 isolates' data from this massive source. In May 2023, one study leveraged genome-wide association statistical methods to uncover novel SNPs contributing to antifungal resistance (Wang and Xu 2024). However, only 387 of over 4000 strains with WGS data obtained from NCBI SRA in this study contained metadata denoting antifungal susceptibilities. A search for *ALS4* copy number variation indicated 2.3% of 1156 isolates with short read WGS data publicly available had more than two copies and only 0.4% had lost *ALS4* completely (Bing et al. 2023). Such studies could now be repeated or could make use of the increasing amounts of long read data that are available. To ensure public data is valuable however, relevant clinical and phenotypic information should be included. Perhaps publishing policy guidelines with reinforcement by professional societies may help.

We know little about how *C. auris* bloodstream invasion occurs in most persons—whether it is a matter of hygiene around multiple indwelling device insertion, related to translocation of microbes through broken skin and internal gastrointestinal epithelia, or an alternative route of entry (Pappas et al. 2018). A PubMed search for "*Candida auris*" and "autopsy" yields no primary literature results, and we understand that there are little to no histopathological studies in humans that systematically describe invasion. A clinical predictive model based on retrospective epidemiology of colonised individuals determined a hierarchy of risk factors, including total parenteral nutrition (TPN), a multifocal skin surface colonisation, presence of sepsis, advanced chronic kidney disease, arterial catheterisation, central venous catheterisation, and antifungal exposure (Garcia-Bustos et al. 2020). A clinical study of 27 *C. auris* colonised patients out of 117 patients admitted to an ICU during February–May 2020 (59% of whom had COVID-19) found a 25% cumulative incidence rate of *C. auris* BSI within 60 days of detection of colonisation (Briano et al. 2022). However, relating putative virulence strategies to human pathological processes is awaited, and there is a need to use genomic techniques alongside clinical pathology research to understand the underlying genomic innovation strategies that drive invasion.

Describing the host response to *C. auris* in existing clinical populations is also lacking. Pathogenesis of invasive infectious disease in general is thought to result from the interaction of host and pathogen (Casadevall and Pirofski 2003). The phenomenon of disseminated infection associated with multiple organ failure in humans has led to a recent definition of sepsis (for any pathogen) as a dysregulated host response to infection (Singer et al. 2016), and septic shock as a resulting failure of tissue perfusion in the context of circulatory failure leading to death. The relevance of host response in invasive candidiasis is illustrated by the demonstration of a high hazard ratio for mortality of 1.73 (95% CI 1.41–2.13) when the patient's presentation included septic shock in one study of *C. auris* BSI (Ortiz-Roa et al. 2023). Additionally, invasive candidiasis tends to take place in a later stage of illness defined by relative immunosuppression (Patricio et al. 2019). Therefore, dynamic host–pathogen assessment will also form a part of understanding the virulence of *C. auris*.

Studies examining the host response to *C. auris* are underway, and sorely needed. One transcriptome study found that 50% of differentially expressed genes (DEGs) in *C. auris* during macrophage phagocytosis were unique, with no orthologues in *C. albicans*, *C. tropicalis*, *C. parapsilosis*, and *C. lusitaniae* (Miramón et al. 2023). Metabolic, transport, lysis, and oxidative stress transcriptional responses to macrophage phagocytosis otherwise appeared highly conserved across these CTG clade members (Miramón et al. 2023). Given the features of *C. auris* BSI as occurring in already critically unwell patients who frequently have pre-existing medical conditions with exposure to other multi-drug resistant pathogens, considering host–pathogen interaction and the wealth of available genomic information (such as the metatranscriptome) is likely to be both a complex and profitable frontier.

### 10.3.2 Escalate Environmental Fungal Pathogen Genomics Towards Global Priority

Pathogen emergence can be defined in epidemiological terms, framed in evolutionary theory, and studied using genomic tools. A definition of emergence is "an infectious disease whose incidence is increasing following its first introduction into a new host population or…increasing in an existing host population as a result of long-term changes in its underlying epidemiology" (Woolhouse and Dye 2001). Emergence has been described as "an inevitable consequence of changes in the abundance of host populations and the contact networks that exist between them" (Cleaveland et al. 2007). An organism's genomic background enables its "jump" into new niches and its genomic innovation allows further adaptation (Woolhouse 2002; Woolhouse et al. 2005). For example, consideration of the soil-based origins of *Cryptococcus neoformans* played a role in hypothesising the similarities between phagocytosis evasion both in human hosts and when confronted with soil amoeba *Acanthamoeba castellanii* (Steenbergen et al. 2001). The discovery of an "original" source environment for *C. auris* could enable an understanding of the selection pressures that have enabled virulence to develop.

It remains unknown where *C. auris* has been hiding prior to the twenty-first century. It is also not known what the "founder" event was that caused a spill over into the global human population (Fourcade et al. 2014; Bonneaud and Longdon 2020; Snedden et al. 2021). This enigma has been reviewed in some detail (Jackson et al.

2019; Chakrabarti and Sood 2021; Zhai et al. 2021; Sharma and Kadosh 2023; Akinbobola et al. 2023; Garcia-Bustos et al. 2023; Chowdhary et al. 2023). Environmental isolates have so far included coastal wetlands of the Andaman Islands (Arora et al. 2021), the coastal waters of Columbia (Escandón 2022), and swimming pools in the Netherlands (Ekowati et al. 2018) as well as wastewater in both Eastern and Western USA (Barber et al. 2023; Rossi et al. 2023; Babler et al. 2023). *C. auris* has been isolated from picked and stored apples (Yadav et al. 2022) as well as stray dogs' ears in Delhi, India (Yadav et al. 2023) and a pet dog's mouth in Kansas, USA (White et al. 2024). None of these isolates have been proven to be so phylogenetically distant from outbreak strains as to conclusively prove an environmental "origin", but the presence of drug-susceptible isolates in coastal wetlands of the Andaman Islands is consistent with a hypothesis of a marine yeast becoming drug resistant after human exposure (Arora et al. 2021).

Ideally, searches for *C. auris* would make use of existing environmental metagenomic and metabarcoding studies. The "One Health" concept is entirely interdisciplinary, relating human health to the health of animals and ecosystems, an essential framing of the *C. auris* origins search (Garcia-Bustos et al. 2023). At least one group has begun efforts to scan public repository data for *C. auris* (Mario-Vasquez et al. 2024). *C. auris* reads were detected in an iron-reducing medium growth experiment of cultures originally taken from "contaminated soil at a former coal gasification site in Gliwice, Poland" and could be due to human contamination (Marozava et al. 2018). *Candida* spp. have been isolated from soil, e.g. from the Global Soil Mycobiome Consortium's extensive sampling of over 4000 topsoil samples with full length ITS and V9 region sequencing, but not yet explored further at a taxonomic level (Mikryukov et al. 2023). Promoting the importance of emerging fungal diseases in environmental microbial diversity may help to ensure the methods and attention required to consider and collaborate are not overlooked.

Ecological change has long been understood to explain emergence, rather than sudden genotypic or phenotypic leaps (Schrag and Wiener 1995). Anthropogenic environmental interaction, including human contact with previously unexplored natural environments leading to major disturbances, is a prime general suspect for driving emergences, such as the novel Severe Acute Respiratory Syndrome Coronavirus SARS-CoV-2 (Wilcox and Gubler 2005; Lindahl and Grace 2015; Platto et al. 2021). Several ecological drivers have been hypothesised for *C. auris* emergence, including (i) global warming as a driver of thermotolerance, (ii) widespread agricultural azole use as a driver of drug resistance, and (iii) migratory birds as a vector for transmission (Casadevall et al. 2021; Aguirre-Liguori et al. 2021). However, none of these hypotheses for *C. auris* emergence have been proven, and there is plenty of room for speculation, conjecture, and doubt. For example, one author recently argued that global warming of 1–1.5 °C may not be responsible for overcoming a thermotolerance-related limiting factor for fungal virulence, as there are tens of thousands of non-pathogenic fungal species that can already grow at mammalian temperatures (Money 2024).

Several further hypotheses exist that have not been fully explored but could be targeted via temporal analyses making use of genomic data. These include marine plastic waste as a vector of *C. auris* global transmission (Akinbobola et al. 2024), and aquaculture via attempted *C. haemulonii* seeding to promote shrimp immunity and reduce infection (Chakrabarti and Sood 2021). Bayesian analysis has been used to estimate the time to most recent common ancestor (TMRCA) of individual clades and emergent isolates in general from WGS data (Chow et al. 2020). Overall, the TMRCA for all emerging *C. auris* isolates fell around the approximate year 25,000 BCE. Excluding isolates with high drug susceptibility, the TMRCA for individual clades I and III was estimated in circa 1983 and 1984, respectively. Including all isolates, the four major clades' TMRCA dated back to circa 1869, 1833, 1658, and 1982. These global genomic epidemiological studies have provided a critical outline for the investigation of the evolution and emergence of *C. auris*. However, model parameters such as clock rate and gene choice are likely to have major influences on these results and could be analysed in future (Koch and Carmona 2024). Further tests, such as calculation of the effective population number ($N_e$) can also be performed using Bayesian phylogenomic analysis and could enrich and inform searches for *C. auris* origins. Calculation of $N_e$ has been used to link the massive increase in prevalence of Hepatitis C virus in Egypt to the period of shared needle use by misinformed public health programmes in the first half of the nineteenth century (Charlesworth 2009).

The isolation of *C. auris* from the Andaman Islands has not been accompanied by a consideration of local indigenous people groups. Andaman Islanders are made up of distinct tribes including Sentinelese, Jarawa, Great Andamanese, and Onge, who may have been isolated from other populations around the world (and even to an extent each other) for tens of thousands of years, according to genetic, historical, and linguistic features (Thangaraj et al. 2003; Thangaraj et al. 2006). A timeline for *C. auris* emergence, including a TMRCA around 25,000 BCE, falls within human prehistory and early migrations (Nielsen et al. 2017). As indigenous people groups emerge from isolation (Pringle 2014)—a phenomenon which may soon cease (Walker et al. 2016)—a prime concern has been the potential for disease introduction (such as measles and influenza) into vulnerable populations with devastating effects (Ferreira and Castro 2015). However, it is possible that *C. auris* may be an example of the opposite process, whereby *C. auris* has been present in isolated human populations until an introduction into the global population in the last 50–500 years. The sudden emergence of each clade suggested by Bayesian analysis (Chow et al. 2020) is consistent with epochs of human movement (Baker et al. 2022), including European colonisation and trans-Atlantic slave trade over the last 500 years, followed by the advent of global air travel in the last 50 years in our modern "age of emerging diseases" (Greger 2021). Sampling of indigenous populations needs extreme care (Bader et al. 2023), so the inclusion of fungal investigation in any research that combines genomics and anthropology should be considered to reduce the burden of sampling through potential repeated studies.

Despite not knowing the origins of *C. auris*, there are many human environments forming ecological niches that now influence its ongoing evolution. The reservoir of colonised and infective individuals and abiotic surfaces, particularly in hospitals, continues to rise (Lyman et al. 2023). *C. auris* appears to thrive in what has been termed "the built environment" (Ciric 2022), especially in urban areas such as, most recently, in a prison (McDougal et al. 2023). In a clinical setting, methods of decolonisation are not fully effective either in infectious patients (Elbahr et al. 2024) or infectious fomites (Nwachukwu et al. 2023), rendering robust infection control strategies critical (Ahmad and Asadzadeh 2023). Outbreaks of *C. auris*, such as during the COVID-19 pandemic, have been partially attributed to deficits in infection control practices (Thoma et al. 2022). When *C. auris* persists for long periods in patients and in healthcare environments, there is an increased potential for further genomic innovation and adaptation (Antia et al. 2003; André and Day 2005). The idea of a virulence "trade off" during evolution leading to lower pathogenicity is complex and should not be assumed (Alizon et al. 2009). The modifiable ecology of *C. auris* in ongoing transmission may be essential in preventing further evolution of virulence (Ewald 2004).

### 10.3.3 Interdisciplinary Genomics Towards Exhaustive Characterisation

*C. auris* is only one of multiple emerging infectious diseases from the fungal kingdom with major humanitarian impact (Lockhart et al. 2023). Other examples include *Batrachochytrium dendrobatidis*, a panzootic that has been responsible for massive global amphibian declines (Fisher and Garner 2020), *Magnaporthe oryzae*, the causative agent of rice blast disease threatening food security with catastrophic harvest losses (Mutiga et al. 2021), and *Trichophyton indotineae*, a zoophilic and potentially sexually transmitted drug-resistant dermatophyte (Caplan 2023). The ability to cause disease in humans has emerged in at least 16 major clades across the fungal tree of life, with no single factor obviously responsible for pathogenicity (Rokas 2022). Diversification from *Candida* clade—whether between strains and clades of *C. auris* or between other *Candida* spp.—gives an opportunity to pick apart the genomic drivers and signatures of virulence.

Comparative studies across *Candida* spp. have identified signatures of selection across multiple genes, with a focus on resistance. A study of over 2000 isolates of *C. glabrata, C. auris, C. albicans, C. tropicalis, C. parapsilosis*, and *C. orthopsilosis* identified genes associated with drug resistance and others associated with selection, but Genome Wide Association Studies (GWAS) for resistance phenotypes were probably limited again by lack of phenotypic data and annotation for *C. auris*; hence, there were no Gene Ontology (GO) biological processes significantly enriched among *C. auris* genes with signatures of selection (Schikora-Tamarit and Gabaldón 2024). The wealth of literature on *C. auris* has a clear focus on antifungal resistance (Wang et al. 2023b; Ragusa et al. 2023; Ettadili and Vural 2024), but comparative genomics will be limited until problems with WGS metadata relating to antifungal susceptibility are improved upon. Virulence-focused investigation has major potential for the discovery of disease-modifying treatment development that could lead to novel antifungal treatments (Roselletti et al. 2023). Understanding the drivers of virulence and pathogenicity in *C. auris* across different populations and compared to other *Candida* spp. will require integration of such clinical and laboratory data with recent genomic discoveries to understand what drives virulence. Such research should certainly consider prioritising research on closely related *Candida* spp. within the haploid *Candida* clade that includes *C. haemulonii* and *C. lusitaniae*, which have had far less attention than other more distantly related species, which tend to be diploid (Muñoz et al. 2018).

Careful molecular biology work is required to understand the DNA damage response (DDR) and other mechanistic bases of genomic innovation, such as via gene knockouts. Difficulty in genetic tractability has been identified alongside the inherent challenge of working with a novel strain (Bravo Ruiz and Lorenz 2021). Screening of a genome-wide mutagenesis programme in *C. auris* demonstrated a disruption of a long-coding RNA that resulted in constitutive filamentous growth (Gao et al. 2021). With deletion resulting in DNA damage, the RNA was named DNA damage-inducible non-coding RNA (DINOR) and found to be involved in co-ordination of multiple stress responses, including to macrophages, antifungals, and oxidative stress. Several direct gene knockouts have been studied during infection in a murine model, such as the Protein Kinase A regulatory subunit *BCY1* (Bypass of CYclic-AMP requirement) (Matsumoto et al. 1982), resulting in attenuated virulence for unclear reasons likely related to reduced stress tolerance (Kim et al. 2021). DDR genes have also been hypothesised to play a role in resisting immune cell oxidative stress (Yao et al. 2021). Therefore, consideration of the role of DDR genes in genomic innovation could be coupled with immunological investigation, as detailed further below.

The stress resistance profile of *C. auris* appears to be unique compared to other *Candida* spp., with a sensitivity to organic oxidative stress and an inability to grow in some anaerobic conditions or acidic conditions (Day et al. 2018). A degree of sensitivity to stress in vivo has been implied by how *Hog1* stress-activated protein kinase (SAPK) knockouts led to a reduced virulence in *C. elegans* infection alongside reduced tolerance to various stressors in vitro, including oxidative, ionic, and osmotic stress (Day et al. 2018). One recent review considered the idea of phenotypic switching (which has been clearly shown in *C. albicans*) in *C. auris* as playing a role in a switch from *C. auris* colonisation to invasion (Proctor et al. 2023). However, *C. auris* has only been reported as forming filamentous forms in rare cases, such as after passage through a mammalian body (Yue et al. 2018) or under genotoxic stress (Bravo Ruiz et al. 2020).

We suggest considering in vivo microevolution experiments as a method for detecting genomic innovation strategies relevant to virulence. In vitro microevolution experiments have demonstrated genomic diversification under drug pressure such as ploidy changes and SNPs (Carolus et al. 2021; Burrack et al. 2022). Serial passage experiments, such as with *C. neoformans* in mice, have been used for other pathogens (Hu et al. 2014) and could be replicated in *C. auris* via a range of different model organisms. An alternative would be to use host-relevant cell lines to explore microevolution in host-relevant conditions, as has been explored with *C. glabrata* during long-term co-incubation with a macrophage cell line (Brunke et al. 2014). To our knowledge, there are no microevolution studies of repeated passage in infection to examine virulence in *C. auris*, despite the potential utility (Schikora-Tamarit and Gabaldón 2022). Such experiments could unveil genomic innovation strategies that have not been previously described, such as SNPs or epigenetic regulatory mechanisms. Virulence and antifungal resistance may be co-evolving together (Pais et al. 2019), especially in human hosts, and it would be especially interesting to consider shared genomic changes in directed evolution experiments under differential and combined selection pressures towards virulence and resistance.

Integration of immunology as a discipline is also critical to understanding virulence factors in *C. auris* related to other *Candida* spp. An ex vivo whole blood model enabled transcriptional identification of host responses to *C. auris* infection (included expected neutrophil killing), with subtle differences compared to that of other *Candida* spp. including *C. albicans, C. glabrata, C. tropicalis*, and *C. parapsilosis* (Allert et al. 2022). Upregulated host genes in the whole blood model included *TLR2*, C-type lectin receptors *CLEC5a*, *MRC1*, and complement receptor *CR3*. In the whole blood model, *C. auris* differentially expressed several sets of genes related to stress tolerance, such as *HSP90*. Global transcriptional programmes need to be examined in vivo, as changes in cis- and trans-gene expression regulators can drive evolution (Hill et al. 2021). A recent hypothesis is that although genes and regulators may be conserved they can undergo rapid switching (or transcriptional rewiring) that results in accelerated evolution in *Candida* spp. (Fowler et al. 2023).

*C. auris* can survive, replicate within, and lyse human macrophages, with evidence that conserved *Candida* spp. transcriptional networks play a role. Isolates from clades I–IV can escape bone-marrow-derived macrophage (BMDM) phagocytosis within 10 h and cause lysis around 24 h without morphological change or NLRP3 inflammasome activation (Weerasinghe et al. 2023), unlike the dramatic hypha-associated lysis and candidalysin-dependent immune stimulation seen in *C. albicans* (Lewis et al. 2012; Rogiers et al. 2019). Instead, macrophages suffered glucose-starvation metabolic stress, which could be rescued by glycolytic substrate (glucose and pyruvate) provision.

Evasion of neutrophils also appears to be both highly effective and uniquely evolved in *C. auris*. Divergence of pathways relating to cell surface *β*-glucan masking may represent a highly effective immune evasion strategy compared to other *Candida* spp. Human neutrophil engagement with *C. albicans* (Johnson et al. 2018) and *C. glabrata* (Horton et al. 2021) is three times higher in vitro than with *C. auris* (clade I). Unlike with *C. albicans*, antimicrobial neutrophil extracellular traps (NETs) fail to be produced for *C. auris* both in vitro and in zebrafish hind brain injection (Johnson et al. 2018). Deletion of the orthologue of *C. albicans* cell wall mannosylation gene *PMR1* (Bates et al. 2005) increased neutrophil engagement and yeast killing in *C. auris* by over three times, with no detectable change in engagement for *C. glabrata* and *C. albicans*, in zebrafish hind brain injection (Horton et al. 2021). Despite higher lethality, fungal burden, and early pro-inflammatory cytokine release in a zebrafish swim-bladder experiment, *C. auris* infection was associated with lower expression levels of genes associated with immune recruitment, such as *MPX* and *FOXP3A/B* (Pharkjaksu et al. 2021). However, the question remains as to why *C. auris* may effectively evade neutrophil engagement, and what genomic mechanisms explain differences to other *Candida* spp.

Population genetic studies are another suggestion for genomic searches to understand virulence. Increasing the scale of phylogenetic studies could help resolve the emergence of pathogenicity and population structure of *C.*

*auris* as it continues to spread. One phylogenetic study of environmental and pathogenic *Candida* strains related to *C. auris* indicated no cladistic basis for a single emergence of pathogenicity as environmental and infecting strains were interspersed across the phylogeny (Schutz et al. 2023). However, it has not yet been possible to pick apart the roles of selection, drift, and other forces acting on the *C. auris* genome (Pouyet and Gilbert 2021). Such molecular genomic approaches for *C. auris* offer an opportunity to enrich and refine its evolutionary history. Systematic and exhaustive genomic studies of, for example, mobile genetic elements are lacking. The comparison of two differentially virulent strains in China demonstrated a large difference in numbers of the Zorro3 retrotransposon, which has unclear roles in DNA/RNA binding, with 64 copies in the more virulent (but fully drug sensitive) clade I strain compared to three copies in the less virulent (but multi-drug resistant) clade III strain (Fan et al. 2021). Zorro3 was also observed in differing numbers across strains in another study (Burrack et al. 2022), but any contribution to virulence is unknown. Large-scale genomics can help identify and quantify genomic innovations that have not been noted at a population level previously and indicate their role in virulence evolution.

## 10.4 Conclusions

We have focused our review on the genomic innovation strategies underlying virulence evolution in *C. auris*. There is evidence of structural rearrangement across seven chromosomes, particularly in the less virulent clade II. Almost all strains appear haploid, but diploid forms may increase virulence. Aneuploidy is a well described response to stress, including to drug pressures. From available strains and analyses to date, no hybridisation representative of recent mating has been identified, though further efforts are required. Gene family expansions unique to *C. auris* include transporter, hydrolase, and adhesin families. These gene families have been associated with diversification and polarised patterns of expression in virulence studies. Microevolution is an important source of rapid genomic variation. In one clinical case, an isolate underwent rapid adaptation in multiple trajectories, with tenfold copy number variation of the *ALS4* adhesin in a sub-telomeric location in one strain (associated with better skin colonisation) or a truncation and loss of function of *ALS4* in a second strain (associated with higher virulence in an experimental model).

Controlling the global public health threat of *C. auris* requires interdisciplinary integration. Clinically, understanding pathogenesis in humans requires bedside proximal research and immunological investigation. Considering *C. auris* and other fungal pathogens in global metagenomic, anthropological, and one health investigations may also help understand original niches associated with pathogenicity. Phylogenetic structures indicate a diversion of clades within human prehistory, and ancient ecological drivers of evolution and diversification are unknown. In the context of ongoing transmission in healthcare environments, the ecological factors influencing ongoing evolution in *C. auris* are modifiable by infection control practices. Comparative genomic and phenotypic studies within *C. auris* and across the *Candida* clade could explain how pathogenicity emerges in the fungal kingdom and help to reduce the global morbidity and mortality associated with invasive fungal infection.

## References

Abastabar M, Haghani I, Ahangarkani F et al (2019) *Candida auris* otomycosis in Iran and review of recent literature. Mycoses 62:101–105. https://doi.org/10.1111/myc.12886

Abele R, Tampé R (2018) Moving the cellular Peptidome by transporters. Front Cell Dev Biol 6:1–13. https://doi.org/10.3389/fcell.2018.00043

Aguirre-Liguori JA, Ramírez-Barahona S, Gaut BS (2021) The evolutionary genomics of species' responses to climate change. Nat Ecol Evol 5:1350–1360. https://doi.org/10.1038/s41559-021-01526-9

Ahmad S, Asadzadeh M (2023) Strategies to prevent transmission of *Candida auris* in healthcare settings. Curr Fungal Infect Rep 17:36–48. https://doi.org/10.1007/s12281-023-00451-7

Akinbobola AB, Kean R, Hanifi SMA, Quilliam RS (2023) Environmental reservoirs of the drug-resistant pathogenic yeast *Candida auris*. PLoS Pathog

19:e1011268. https://doi.org/10.1371/journal.ppat.1011268

Akinbobola A, Kean R, Quilliam RS (2024) Plastic pollution as a novel reservoir for the environmental survival of the drug resistant fungal pathogen *Candida auris*. Mar Pollut Bull 198:115841. https://doi.org/10.1016/j.marpolbul.2023.115841

Alizon S, Hurford A, Mideo N, Van Baalen M (2009) Virulence evolution and the trade-off hypothesis: history, current state of affairs and the future. J Evol Biol 22:245–259. https://doi.org/10.1111/j.1420-9101.2008.01658.x

Allert S, Schulz D, Kämmer P et al (2022) From environmental adaptation to host survival: attributes that mediate pathogenicity of *Candida auris*. Virulence 13:191–214. https://doi.org/10.1080/21505594.2022.2026037

André J-B, Day T (2005) The effect of disease life history on the evolutionary emergence of novel pathogens. Proc R Soc B Biol Sci 272:1949–1956. https://doi.org/10.1098/rspb.2005.3170

Antia R, Regoes RR, Koella JC, Bergstrom CT (2003) The role of evolution in the emergence of infectious diseases. Nature 426:658–661. https://doi.org/10.1038/nature02104

Arora P, Singh P, Wang Y et al (2021) Environmental isolation of *Candida auris* from the coastal wetlands of Andaman Islands, India. mBio 12:e03181–e03120. https://doi.org/10.1128/mBio.03181-20

Babler K, Sharkey M, Arenas S et al (2023) Detection of the clinically persistent, pathogenic yeast spp. *Candida auris* from hospital and municipal wastewater in Miami-Dade County, Florida. Sci Total Environ 898:165459. https://doi.org/10.1016/j.scitotenv.2023.165459

Bader AC, Van Zuylen EM, Handsley-Davis M et al (2023) A relational framework for microbiome research with indigenous communities. Nat Microbiol 8:1768–1776. https://doi.org/10.1038/s41564-023-01471-2

Baker RE, Mahmud AS, Miller IF et al (2022) Infectious disease in an era of global change. Nat Rev Microbiol 20:193–205. https://doi.org/10.1038/s41579-021-00639-z

Barber C, Crank K, Papp K et al (2023) Community-scale wastewater surveillance of *Candida auris* during an ongoing outbreak in southern Nevada. Environ Sci Technol 57:1755–1763. https://doi.org/10.1021/acs.est.2c07763

Basso P, Dang EV, Urisman A et al (2022) Deep tissue infection by an invasive human fungal pathogen requires lipid-based suppression of the IL-17 response. Cell Host Microbe 30:1589–1601.e5. https://doi.org/10.1016/j.chom.2022.10.004

Bates S, MacCallum DM, Bertram G et al (2005) *Candida albicans* Pmr1p, a secretory pathway P-type Ca2+/Mn2+-ATPase, is required for glycosylation and virulence *. J Biol Chem 280:23408–23415. https://doi.org/10.1074/jbc.M502162200

Bernstein DA, Vyas VK, Weinberg DE et al (2012) *Candida albicans* Dicer (CaDcr1) is required for efficient ribosomal and spliceosomal RNA maturation. Proc Natl Acad Sci 109:523–528. https://doi.org/10.1073/pnas.1118859109

Billmyre RB (2022) Drug resistance and Evolvability in an emerging human fungal pathogen. MBio 13:e01876–e01822. https://doi.org/10.1128/mbio.01876-22

Bing J, Guan Z, Zheng T et al (2023) Clinical isolates of *Candida auris* with enhanced adherence and biofilm formation due to genomic amplification of ALS4. PLoS Pathog 19:e1011239. https://doi.org/10.1371/journal.ppat.1011239

Bing J, Hu T, Zheng Q et al (2020) Experimental evolution identifies adaptive aneuploidy as a mechanism of fluconazole resistance in *Candida auris*. Antimicrob Agents Chemother 65:e01466–e01420. https://doi.org/10.1128/AAC.01466-20

Bing J, Wang S, Xu H et al (2022) A case of *Candida auris* candidemia in Xiamen, China, and a comparative analysis of clinical isolates in China. Mycology 13:68–75. https://doi.org/10.1080/21501203.2021.1994479

Biswas C, Wang Q, van Hal SJ et al (2020) Genetic heterogeneity of Australian *Candida auris* isolates: insights from a nonoutbreak setting using whole-genome sequencing. Open Forum Infect Dis 7:ofaa158. https://doi.org/10.1093/ofid/ofaa158

Bonneaud C, Longdon B (2020) Emerging pathogen evolution. EMBO Rep 21:e51374. https://doi.org/10.15252/embr.202051374

Borman AM, Szekely A, Johnson EM (2016) Comparative pathogenicity of United Kingdom isolates of the emerging pathogen *Candida auris* and other key pathogenic *Candida* species. mSphere 1:e00189–e00116. https://doi.org/10.1128/mSphere.00189-16

Bravo Ruiz G, Lorenz A (2021) What do we know about the biology of the emerging fungal pathogen of humans *Candida auris*? Microbiol Res 242:126621. https://doi.org/10.1016/j.micres.2020.126621

Bravo Ruiz G, Ross ZK, Gow NAR, Lorenz A (2020) Pseudohyphal growth of the emerging pathogen *Candida auris* is triggered by genotoxic stress through the S phase checkpoint. mSphere 5:e00151–e00120. https://doi.org/10.1128/mSphere.00151-20

Bravo Ruiz G, Ross ZK, Holmes E et al (2019) Rapid and extensive karyotype diversification in haploid clinical *Candida auris* isolates. Curr Genet 65:1217–1228. https://doi.org/10.1007/s00294-019-00976-w

Briano F, Magnasco L, Sepulcri C et al (2022) *Candida auris* Candidemia in critically ill, colonized patients: cumulative incidence and risk factors. Infect Dis Ther 11:1149–1160. https://doi.org/10.1007/s40121-022-00625-9

Brunke S, Seider K, Fischer D et al (2014) One small step for a yeast–microevolution within macrophages renders *Candida glabrata* hypervirulent due to a single point mutation. PLoS Pathog 10:e1004478. https://doi.org/10.1371/journal.ppat.1004478

Burrack LS, Todd RT, Soisangwan N et al (2022) Genomic diversity across *Candida auris* clinical isolates shapes rapid development of antifungal resistance in vitro and in vivo. mBio 13:e0084222. https://doi.org/10.1128/mbio.00842-22

Caplan AS (2023) Notes from the field: first reported U.S. cases of tinea caused by Trichophyton indotineae—New York City, December 2021–March 2023. Morbidity and mortality weekly report 72: https://doi.org/10.15585/mmwr.mm7219a4

Carolus H, Pierson S, Muñoz JF et al (2021) Genome-wide analysis of experimentally evolved *Candida auris* reveals multiple novel mechanisms of multidrug resistance. mBio 12:e03333–e03320. https://doi.org/10.1128/mBio.03333-20

Casadevall A, Kontoyiannis DP, Robert V (2021) Environmental *Candida auris* and the global warming emergence hypothesis. mBio 12:e00360–e00321. https://doi.org/10.1128/mBio.00360-21

Casadevall A, Pirofski L (2001) Host-pathogen interactions: the attributes of virulence. J Infect Dis 184:337–344. https://doi.org/10.1086/322044

Casadevall A, Pirofski L (2003) The damage-response framework of microbial pathogenesis. Nat Rev Microbiol 1:17–24. https://doi.org/10.1038/nrmicro732

Casadevall A, Pirofski LA (1999) Host-pathogen interactions: redefining the basic concepts of virulence and pathogenicity. Infect Immun 67:3703–3713. https://doi.org/10.1128/IAI.67.8.3703-3713.1999

CDC (2019) Centers for disease control and prevention (U.S.): antibiotic resistance threats in the United States, 2019

Chakrabarti A, Sood P (2021) On the emergence, spread and resistance of *Candida auris*: host, pathogen and environmental tipping points. J Med Microbiol 70:001318. https://doi.org/10.1099/jmm.0.001318

Charlesworth B (2009) Effective population size and patterns of molecular evolution and variation. Nat Rev Genet 10:195–205. https://doi.org/10.1038/nrg2526

Chen J, Tian S, Han X et al (2020) Is the superbug fungus really so scary? A systematic review and meta-analysis of global epidemiology and mortality of *Candida auris*. BMC Infect Dis 20:827. https://doi.org/10.1186/s12879-020-05543-0

Chow NA, de Groot T, Badali H et al (2019) Potential fifth clade of *Candida auris*, Iran, 2018. Emerg Infect Dis 25:1780–1781. https://doi.org/10.3201/eid2509.190686

Chow NA, Muñoz JF, Gade L et al (2020) Tracing the evolutionary history and global expansion of *Candida auris* using population genomic analyses. MBio 11:15

Chowdhary A, Jain K, Chauhan N (2023) *Candida auris* genetics and emergence. Ann Rev Microbiol 77:583–602. https://doi.org/10.1146/annurev-micro-032521-015858

Chybowska AD, Childers DS, Farrer RA (2020) Nine things genomics can tell us about *Candida auris*. Front Genet 11:351. https://doi.org/10.3389/fgene.2020.00351

Ciric L (2022) Microbes in the built environment. Sci Rep 12:8732. https://doi.org/10.1038/s41598-022-12254-w

Cleaveland S, Haydon DT, Taylor L (2007) Overviews of pathogen emergence: which pathogens emerge, when and why? Curr Top Microbiol Immunol 315:85–111. https://doi.org/10.1007/978-3-540-70962-6_5

Cook A, Ferreras-Antolin L, Adhisivam B et al (2023) Neonatal invasive candidiasis in low- and middle-income countries: data from the NeoOBS study. Med Mycol 61:myad010. https://doi.org/10.1093/mmy/myad010

Cressler CE, McLEOD DV, Rozins C et al (2016) The adaptive evolution of virulence: a review of theoretical predictions and empirical tests. Parasitology 143:915–930. https://doi.org/10.1017/S003118201500092X

Day AM, McNiff MM, da Silva DA et al (2018) Hog1 regulates stress tolerance and virulence in the emerging fungal pathogen *Candida auris*. mSphere 3:e00506–e00518. https://doi.org/10.1128/mSphere.00506-18

de Cássia Orlandi Sardi J, Silva DR, Soares Mendes-Giannini MJ, Rosalen PL (2018) *Candida auris*: epidemiology, risk factors, virulence, resistance, and therapeutic options. Microb Pathog 125:116–121. https://doi.org/10.1016/j.micpath.2018.09.014

de Jong AW, Hagen F (2019) Attack, defend and persist: how the fungal pathogen *Candida auris* was able to emerge globally in healthcare environments. Mycopathologia 184:353–365. https://doi.org/10.1007/s11046-019-00351-w

Demuth JP, Hahn MW (2009) The life and death of gene families. BioEssays 31:29–39. https://doi.org/10.1002/bies.080085

Desoubeaux G, Coste AT, Imbert C, Hennequin C (2022) Overview about *Candida auris*: What's up 12 years after its first description? J Med Mycol 32:101248. https://doi.org/10.1016/j.mycmed.2022.101248

Diaz Caballero J, Wheatley RM, Kapel N et al (2023) Mixed strain pathogen populations accelerate the evolution of antibiotic resistance in patients. Nat Comm 14:4083. https://doi.org/10.1038/s41467-023-39416-2

Du H, Bing J, Hu T et al (2020) *Candida auris*: epidemiology, biology, antifungal resistance, and virulence. PLoS Pathog 16:e1008921. https://doi.org/10.1371/journal.ppat.1008921

Du H, Zheng Q, Bennett RJ, Huang G (2022) Ploidy changes in human fungal pathogens: going beyond sexual reproduction. PLoS Pathog 18:e1010954. https://doi.org/10.1371/journal.ppat.1010954

Dunkel N, Hertlein T, Franz R et al (2013) Roles of different peptide transporters in nutrient acquisition in *Candida albicans*. Eukaryot Cell 12:520–528. https://doi.org/10.1128/EC.00008-13

Egger NB, Kainz K, Schulze A et al (2022) The rise of *Candida auris*: from unique traits to co-infection potential. Microbial Cell 9:141–144. https://doi.org/10.15698/mic2022.08.782

Ekowati Y, Ferrero G, Kennedy MD et al (2018) Potential transmission pathways of clinically relevant fungi in indoor swimming pool facilities. Int J Hyg Environ Health 221:1107–1115. https://doi.org/10.1016/j.ijheh.2018.07.013

Elbahr U, Khairy A, Dayyab F et al (2024) Can daily bathing with 4% chlorhexidine + daily chlorhexidine wipe for 1 week be effective in decolonizing *Candida auris* colonization? Eur J Clin Microbiol Infect Dis 43:243–247. https://doi.org/10.1007/s10096-023-04723-5

Escandón P (2022) Novel environmental niches for *Candida auris*: isolation from a coastal habitat in Colombia. J Fungi 8:748. https://doi.org/10.3390/jof8070748

Ettadili H, Vural C (2024) Current global status of *Candida auris* an emerging multidrug-resistant fungal pathogen: bibliometric analysis and network visualization. Braz J Microbiol 55:391. https://doi.org/10.1007/s42770-023-01239-0

Ewald PW (2004) Evolution of virulence. Infect Dis Clin N Am 18:1–15. https://doi.org/10.1016/S0891-5520(03)00099-0

Eyre DW, Sheppard AE, Madder H et al (2018) A *Candida auris* outbreak and its control in an intensive care setting. N Engl J Med 379:1322–1331. https://doi.org/10.1056/NEJMoa1714373

Fan S, Li C, Bing J et al (2020) Discovery of the diploid form of the emerging fungal pathogen *Candida auris*. ACS Infect Dis 6:2641–2646. https://doi.org/10.1021/acsinfecdis.0c00282

Fan S, Zhan P, Bing J et al (2021) A biological and genomic comparison of a drug-resistant and a drug-susceptible strain of *Candida auris* isolated from Beijing, China. Virulence 12:1388–1399. https://doi.org/10.1080/21505594.2021.1928410

Ferreira MU, Castro MC (2015) No longer a deadly encounter? Pathog Glob Health 109:307–308. https://doi.org/10.1080/20477724.2015.1122916

Fisher MC, Garner TWJ (2020) Chytrid fungi and global amphibian declines. Nat Rev Microbiol 18:332–343. https://doi.org/10.1038/s41579-020-0335-x

Fleming JF, Pisani D, Arakawa K (2024) The evolution of temperature and desiccation-related protein families in Tardigrada reveals a complex Acquisition of Extremotolerance. Genome Biol Evol 16:217. https://doi.org/10.1093/gbe/evad217

Fourcade Y, Keišs O, Richardson DS, Secondi J (2014) Continental-scale patterns of pathogen prevalence: a case study on the corncrake. Evol Appl 7:1043–1055. https://doi.org/10.1111/eva.12192

Fowler KR, Leon F, Johnson AD (2023) Ancient transcriptional regulators can easily evolve new pair-wise cooperativity. Proc Natl Acad Sci 120:e2302445120. https://doi.org/10.1073/pnas.2302445120

Gácser A, Stehr F, Kröger C et al (2007a) Lipase 8 affects the pathogenesis of *Candida albicans*. Infect Immun 75:4710–4718. https://doi.org/10.1128/iai.00372-07

Gácser A, Trofa D, Schäfer W, Nosanchuk JD (2007b) Targeted gene deletion in *Candida parapsilosis* demonstrates the role of secreted lipase in virulence. J Clin Invest 117:3049–3058. https://doi.org/10.1172/JCI32294

Gao J, Chow EWL, Wang H et al (2021) LncRNA DINOR is a virulence factor and global regulator of stress responses in *Candida auris*. Nat Microbiol 6:842–851. https://doi.org/10.1038/s41564-021-00915-x

Garcia-Bustos V, Cabañero-Navalon MD, Ruiz-Gaitán A et al (2023) Climate change, animals, and *Candida auris*: insights into the ecological niche of a new species from a one health approach. Clin Microbiol Infect 29:858–862. https://doi.org/10.1016/j.cmi.2023.03.016

Garcia-Bustos V, Salavert M, Ruiz-Gaitán AC et al (2020) A clinical predictive model of candidaemia by *Candida auris* in previously colonized critically ill patients. Clin Microbiol Infect 26:1507–1513. https://doi.org/10.1016/j.cmi.2020.02.001

Girgsdies O (1982) Sterile mutants of *Schizosaccharomyces pombe*: analysis by somatic hybridization. Curr Genet 6:223–227. https://doi.org/10.1007/BF00390342

Gómez-Gaviria M, Martínez-Álvarez JA, Chávez-Santiago JO, Mora-Montes HM (2023) *Candida haemulonii* complex and *Candida auris*: biology, virulence factors, immune response, and multidrug resistance. Infect Drug Resist 16:1455–1470. https://doi.org/10.2147/IDR.S402754

Gomolplitinant KM, Saier MH (2011) Evolution of the oligopeptide transporter family. J Membr Biol 240:89–110. https://doi.org/10.1007/s00232-011-9347-9

Grant PA, Schieltz D, Pray-Grant MG et al (1998) The ATM-related cofactor Tra1 is a component of the purified SAGA complex. Mol Cell 2:863–867. https://doi.org/10.1016/s1097-2765(00)80300-7

Greger M (2021) Primary pandemic prevention. Am J Lifestyle Med 15:498–505. https://doi.org/10.1177/15598276211008134

Heymann P, Gerads M, Schaller M et al (2002) The Siderophore iron transporter of *Candida albicans* (Sit1p/Arn1p) mediates uptake of Ferrichrome-type Siderophores and is required for epithelial invasion. Infect Immun 70:5246–5255. https://doi.org/10.1128/iai.70.9.5246-5255.2002

Hill MS, Vande Zande P, Wittkopp PJ (2021) Molecular and evolutionary processes generating variation in gene expression. Nat Rev Genet 22:203–215. https://doi.org/10.1038/s41576-020-00304-w

Homer CM, Summers DK, Goranov AI et al (2016) Intracellular action of a secreted peptide required for fungal virulence. Cell Host Microbe 19:849–864. https://doi.org/10.1016/j.chom.2016.05.001

Horton MV, Holt AM, Nett JE (2023) Mechanisms of pathogenicity for the emerging fungus *Candida auris*. PLoS Pathog 19:e1011843. https://doi.org/10.1371/journal.ppat.1011843

Horton MV, Johnson CJ, Zarnowski R et al (2021) *Candida auris* Cell Wall Mannosylation contributes to neutrophil evasion through pathways divergent from *Candida albicans* and *Candida glabrata*. mSphere 6:e00406–e00421. https://doi.org/10.1128/mSphere.00406-21

Hoyer LL, Cota E (2016) *Candida albicans* agglutinin-like sequence (Als) family vignettes: a review of Als protein structure and function. Front Microbiol 7:1–16. https://doi.org/10.3389/fmicb.2016.00280

Hu G, Chen SH, Qiu J et al (2014) Microevolution during serial mouse passage demonstrates FRE3 as a virulence adaptation gene in *Cryptococcus neoformans*. mBio 5:e00941–e00914. https://doi.org/10.1128/mBio.00941-14

Hughes AL (1994) The evolution of functionally novel proteins after gene duplication. Proc R Soc Lond Ser B Biol Sci 256:119–124. https://doi.org/10.1098/rspb.1994.0058

Innan H, Kondrashov F (2010) The evolution of gene duplications: classifying and distinguishing between models. Nat Rev Genet 11:97–108. https://doi.org/10.1038/nrg2689

Jackson BR, Chow N, Forsberg K et al (2019) On the origins of a species: what might explain the rise of *Candida auris*? J Fungi 5:58. https://doi.org/10.3390/jof5030058

Jacobs SE, Jacobs JL, Dennis EK et al (2022) *Candida auris* pan-drug-resistant to four classes of antifungal agents. Antimicrob Agents Chemother 66:e0005322. https://doi.org/10.1128/aac.00053-22

Johnson CJ, Davis JM, Huttenlocher A et al (2018) Emerging fungal pathogen *Candida auris* evades neutrophil attack. mBio 9:e01403–e01418. https://doi.org/10.1128/mBio.01403-18

Kalantar KL, Neyton L, Abdelghany M et al (2022) Integrated host-microbe plasma metagenomics for sepsis diagnosis in a prospective cohort of critically ill adults. Nat Microbiol 7:1805–1816. https://doi.org/10.1038/s41564-022-01237-2

Karst F, Lacroute F (1977) Ertosterol biosynthesis in *Saccharomyces cerevisiae*: mutants deficient in the early steps of the pathway. Mol Gen Genet 154:269–277. https://doi.org/10.1007/BF00571282

Kim J-S, Lee K-T, Bahn Y-S (2023) Secreted aspartyl protease 3 regulated by the Ras/cAMP/PKA pathway promotes the virulence of *Candida auris*. Front Cell Infect Microbiol 13:1257897. https://doi.org/10.3389/fcimb.2023.1257897

Kim J-S, Lee K-T, Lee MH et al (2021) Adenylyl cyclase and protein kinase a play redundant and distinct roles in growth, differentiation, antifungal drug resistance, and pathogenicity of *Candida auris*. mBio 12:e02729–e02721. https://doi.org/10.1128/mBio.02729-21

Koch NM, Carmona PM (2024) Chronospaces: an R package for the statistical exploration of divergence times reveals extreme dependence on molecular clocks and gene choice. https://doi.org/10.1101/2024.02.04.578835

Köhler JR, Hube B, Puccia R et al (2017) Fungi that infect humans. Microbiol Spectr 5. https://doi.org/10.1128/microbiolspec.FUNK-0014-2016

Lee WG, Shin JH, Uh Y et al (2011) First three reported cases of nosocomial Fungemia caused by *Candida auris*. J Clin Microbiol 49:4

Lewis LE, Bain JM, Lowes C et al (2012) Stage specific assessment of *Candida albicans* phagocytosis by macrophages identifies Cell Wall composition and morphogenesis as key determinants. PLoS Pathog 8:e1002578. https://doi.org/10.1371/journal.ppat.1002578

Li J, Coste AT, Bachmann D et al (2022) Deciphering the Mrr1/Mdr1 pathway in azole resistance of *Candida auris*. Antimicrob Agents Chemother 66:e00067–e00022. https://doi.org/10.1128/aac.00067-22

Lindahl JF, Grace D (2015) The consequences of human actions on risks for infectious diseases: a review. Infect Ecol Epidemiol 5:30048. https://doi.org/10.3402/iee.v5.30048

Lockhart SR, Chowdhary A, Gold JAW (2023) The rapid emergence of antifungal-resistant human-pathogenic fungi. Nat Rev Microbiol 1–15:818. https://doi.org/10.1038/s41579-023-00960-9

Lockhart SR, Etienne KA, Vallabhaneni S et al (2017) Simultaneous emergence of multidrug-resistant *Candida auris* on 3 continents confirmed by whole-genome sequencing and epidemiological analyses. Clin Infect Dis 64:134–140. https://doi.org/10.1093/cid/ciw691

Lone SA, Ahmad A (2019) *Candida auris*–the growing menace to global health. Mycoses 62:620–637. https://doi.org/10.1111/myc.12904

Longhese MP, Paciotti V, Fraschini R et al (1997) The novel DNA damage checkpoint protein Ddc1p is phosphorylated periodically during the cell cycle and in response to DNA damage in budding yeast. EMBO J 16:5216–5226. https://doi.org/10.1093/emboj/16.17.5216

Lyman M (2021) Notes from the field: transmission of pan-resistant and Echinocandin-resistant *Candida auris* in health care facilities–Texas and the District of Columbia, January-April 2021. Morb Mortal Wkly Rep 70:1022. https://doi.org/10.15585/mmwr.mm7029a2

Lyman M, Forsberg K, Sexton DJ et al (2023) Worsening spread of *Candida auris* in the United States, 2019 to 2021. Ann Intern Med 176:489. https://doi.org/10.7326/M22-3469

Lynch M, Conery JS (2000) The evolutionary fate and consequences of duplicate genes. Science 290:1151–1155. https://doi.org/10.1126/science.290.5494.1151

Magee BB, Legrand M, Alarco A-M et al (2002) Many of the genes required for mating in *Saccharomyces cerevisiae* are also required for mating in *Candida albicans*. Mol Microbiol 46:1345–1351. https://doi.org/10.1046/j.1365-2958.2002.03263.x

Malavia-Jones D, Farrer RA, Stappers MHT et al (2023) Strain and temperature dependent aggregation of *Candida auris* is attenuated by inhibition of surface amyloid proteins. The Cell Surface 10:100110. https://doi.org/10.1016/j.tcsw.2023.100110

Maphanga TG, Naicker SD, Kwenda S et al (2021) *In vitro* antifungal resistance of *Candida auris* isolates from bloodstream infections, South Africa. Antimicrob Agents Chemother 65:e0051721. https://doi.org/10.1128/AAC.00517-21

Mario-Vasquez JE, Bagal UR, Lowe E et al (2024) Finding *Candida auris* in public metagenomic repositories.

PLoS One 19:e0291406. https://doi.org/10.1371/journal.pone.0291406

Marozava S, Mouttaki H, Müller H et al (2018) Anaerobic degradation of 1-methylnaphthalene by a member of the *Thermoanaerobacteraceae* contained in an iron-reducing enrichment culture. Biodegradation 29:23–39. https://doi.org/10.1007/s10532-017-9811-z

Massic L, Gorzalski A, Siao DD et al (2023) Detection of five instances of dual-clade infections of *Candida auris* with opposite mating types in southern Nevada, USA. Lancet Infect Dis 23:e328–e329. https://doi.org/10.1016/S1473-3099(23)00434-6

Matsumoto K, Uno I, Oshima Y, Ishikawa T (1982) Isolation and characterization of yeast mutants deficient in adenylate cyclase and cAMP-dependent protein kinase. Proc Natl Acad Sci USA 79:2355–2359. https://doi.org/10.1073/pnas.79.7.2355

Mazi PB, Olsen MA, Stwalley D et al (2022) Attributable mortality of Candida bloodstream infections in the modern era: a propensity score analysis. Clin Infect Dis 75:1031–1036. https://doi.org/10.1093/cid/ciac004

McDougal AN, DeMaet MA, Garcia B et al (2023) A cluster investigation of *Candida auris* among hospitalized incarcerated patients. Antimicrob Steward Healthc Epidemiol 3:e244. https://doi.org/10.1017/ash.2023.520

Meis JF, Chowdhary A (2018) *Candida auris*: a global fungal public health threat. Lancet Infect Dis 18:1298–1299. https://doi.org/10.1016/S1473-3099(18)30609-1

Mikryukov V, Dulya O, Zizka A et al (2023) Connecting the multiple dimensions of global soil fungal diversity. Science. Advances 9:eadj8016. https://doi.org/10.1126/sciadv.adj8016

Miramón P, Pountain AW, Lorenz MC (2023) *Candida auris*-macrophage cellular interactions and transcriptional response. Infect Immun 91:e0027423. https://doi.org/10.1128/iai.00274-23

Money NP (2024) Fungal thermotolerance revisited and why climate change is unlikely to be supercharging pathogenic fungi (yet). Fungal Biol 128:1638–1641. https://doi.org/10.1016/j.funbio.2024.01.005

Morawska LP, Hernandez-Valdes JA, Kuipers OP (2022) Diversity of bet-hedging strategies in microbial communities—recent cases and insights. WIREs Mech Dis 14:e1544. https://doi.org/10.1002/wsbm.1544

Muñoz JF, Gade L, Chow NA et al (2018) Genomic insights into multidrug-resistance, mating and virulence in *Candida auris* and related emerging species. Nat Commun 9:5346. https://doi.org/10.1038/s41467-018-07779-6

Muñoz JF, Welsh RM, Shea T et al (2021) Clade-specific chromosomal rearrangements and loss of subtelomeric adhesins in *Candida auris*. Genetics 218:iyab029. https://doi.org/10.1093/genetics/iyab029

Mutiga SK, Rotich F, Were VM et al (2021) Integrated strategies for durable Rice blast resistance in sub-Saharan Africa. Plant Dis 105:2749–2770. https://doi.org/10.1094/PDIS-03-21-0593-FE

Narayanan A, Kumar P, Chauhan A et al (2022) Directed evolution detects supernumerary centric chromosomes conferring resistance to azoles in *Candida auris*. MBio 13:e0305222. https://doi.org/10.1128/mbio.03052-22

Narayanan A, Vadnala RN, Ganguly P et al (2021) Functional and comparative analysis of centromeres reveals clade-specific genome rearrangements in *Candida auris* and a chromosome number change in related species. MBio 12. https://doi.org/10.1128/mbio.00905-21

Nett JE (2019) *Candida auris*: an emerging pathogen "incognito"? PLoS Pathog 15:e1007638. https://doi.org/10.1371/journal.ppat.1007638

Nielsen R, Akey JM, Jakobsson M et al (2017) Tracing the peopling of the world through genomics. Nature 541:302–310. https://doi.org/10.1038/nature21347

Nwachukwu KC, Nwarunma E, David Uchenna C, Chinyere Ugbogu O (2023) Enablers of *Candida auris* persistence on medical devices and their mode of eradication. Curr Med Mycol 9:36–43. https://doi.org/10.18502/CMM.2023.150673

O'Brien B, Liang J, Chaturvedi S et al (2020) Pan-resistant *Candida auris*: New York subcluster susceptible to antifungal combinations. Lancet Microbe 1:e193–e194. https://doi.org/10.1016/S2666-5247(20)30090-2

Ortiz-Roa C, Valderrama-Rios MC, Sierra-Umaña SF et al (2023) Mortality caused by *Candida auris* bloodstream infections in comparison with other *Candida* species, a multicentre retrospective cohort. J Fungi 9:715. https://doi.org/10.3390/jof9070715

Osei Sekyere J (2018) *Candida auris*: a systematic review and meta-analysis of current updates on an emerging multidrug-resistant pathogen. Microbiol Open 7:e00578. https://doi.org/10.1002/mbo3.578

Ostrowsky B, Greenko J, Adams E et al (2020) *Candida auris* isolates resistant to three classes of antifungal medications–New York, 2019. Morb Mortal Wkly Rep 69:6–9. https://doi.org/10.15585/mmwr.mm6901a2

Pais P, Galocha M, Viana R et al (2019) Microevolution of the pathogenic yeasts *Candida albicans* and *Candida glabrata* during antifungal therapy and host infection. Microbial Cell 6:142–159. https://doi.org/10.15698/mic2019.03.670

Pappas PG, Lionakis MS, Arendrup MC et al (2018) Invasive candidiasis. Nat Rev Dis Primers 4:1–20. https://doi.org/10.1038/nrdp.2018.26

Patricio P, Paiva JA, Borrego LM (2019) Immune response in bacterial and *Candida* sepsis. Euro J Microbiol Immunol 9:105–113. https://doi.org/10.1556/1886.2019.00011

Pelletier C, Shaw S, Alsayegh S et al (2024) *Candida auris* undergoes adhesin-dependent and -independent cellular aggregation. PLoS Pathog 20:e1012076. https://doi.org/10.1371/journal.ppat.1012076

Phan QT, Myers CL, Fu Y et al (2007) Als3 is a *Candida albicans* Invasin that binds to Cadherins and induces endocytosis by host cells. PLoS Biol 5:e64. https://doi.org/10.1371/journal.pbio.0050064

Pharkjaksu S, Boonmee N, Mitrpant C, Ngamskulrungroj P (2021) Immunopathogenesis of emerging *Candida auris* and *Candida haemulonii* strains. J Fungi 7:725. https://doi.org/10.3390/jof7090725

Platto S, Zhou J, Wang Y et al (2021) Biodiversity loss and COVID-19 pandemic: the role of bats in the origin and the spreading of the disease. Biochem Biophys Res Commun 538:2–13. https://doi.org/10.1016/j.bbrc.2020.10.028

Pouyet F, Gilbert KJ (2021) Towards an improved understanding of molecular evolution: the relative roles of selection, drift, and everything in between. Peer Comm J 1. https://doi.org/10.24072/pcjournal.16

Prayag PS, Patwardhan S, Panchakshari S et al (2022) The dominance of *Candida auris*: a single-center experience of 79 episodes of Candidemia from Western India. Ind J Critical Care Med 26:560–563. https://doi.org/10.5005/jp-journals-10071-24152

Pringle H (2014) Uncontacted tribe in Brazil emerges from isolation. Science 345:125–126. https://doi.org/10.1126/science.345.6193.125

Proctor DM, Drummond RA, Lionakis MS, Segre JA (2023) One population, multiple lifestyles: commensalism and pathogenesis in the human mycobiome. Cell Host Microbe 31:539–553. https://doi.org/10.1016/j.chom.2023.02.010

Ragusa P, Prinzivalli A, Pizzini S et al (2023) *Candida auris*: a bibliometric analysis of an emerging global health threat. J Infect Public Health 16:1696–1702. https://doi.org/10.1016/j.jiph.2023.08.012

Rahnama M, Wang B, Dostart J et al (2021) Telomere roles in fungal genome evolution and adaptation. Front Genet 12. https://doi.org/10.3389/fgene.2021.676751

Rapti V, Iliopoulou K, Poulakou G (2023) The Gordian knot of *C. Auris*: if you cannot cut it, prevent it. Pathogens 12:1444. https://doi.org/10.3390/pathogens12121444

Read AF (1994) The evolution of virulence. Trends Microbiol 2:73–76. https://doi.org/10.1016/0966-842X(94)90537-1

Reams AB, Roth JR (2015) Mechanisms of gene duplication and amplification. Cold Spring Harb Perspect Biol 7:a016592. https://doi.org/10.1101/cshperspect.a016592

Reedy JL, Floyd AM, Heitman J (2009) Mechanistic plasticity of sexual reproduction and meiosis in the *Candida* pathogenic species complex. Curr Biol 19:891–899. https://doi.org/10.1016/j.cub.2009.04.058

Rhodes J, Abdolrasouli A, Farrer RA et al (2018) Genomic epidemiology of the UK outbreak of the emerging human fungal pathogen *Candida auris*. Emerg Microbes Infect 7:1–12. https://doi.org/10.1038/s41426-018-0045-x

Richtel M (2019) How a Chicago woman fell victim to *Candida auris*, a drug-resistant fungus. The New York Times

Richtel M, Jacobs A (2019) A mysterious infection, Spanning the Globe in a Climate of Secrecy. The New York Times

Rogiers O, Frising UC, Kucharíková S et al (2019) Candidalysin crucially contributes to Nlrp3 Inflammasome activation by *Candida albicans* hyphae. mBio 10:e02221–e02218. https://doi.org/10.1128/mBio.02221-18

Rokas A (2022) Evolution of the human pathogenic lifestyle in fungi. Nat Microbiol 7:607–619. https://doi.org/10.1038/s41564-022-01112-0

Roselletti E, Pericolini E, Nore A et al (2023) Zinc prevents vaginal candidiasis by inhibiting expression of an inflammatory fungal protein. Sci Transl Med 15:eadi3363. https://doi.org/10.1126/scitranslmed.adi3363

Ross ZK, Lorenz A (2020) Is *Candida auris* sexual? PLoS Pathog 16:e1009094. https://doi.org/10.1371/journal.ppat.1009094

Rossato L, Colombo AL (2018) *Candida auris*: what have we learned about its mechanisms of pathogenicity? Front Microbiol 9:3081. https://doi.org/10.3389/fmicb.2018.03081

Rossi A, Chavez J, Iverson T et al (2023) *Candida auris* discovery through community wastewater surveillance during healthcare outbreak, Nevada, USA, 2022. Emerg Infect Dis 29:422–425. https://doi.org/10.3201/eid2902.221523

Santana DJ, Anku JAE, Zhao G et al (2023) A *Candida auris*-specific adhesin, Scf1, governs surface association, colonization, and virulence. Science 381:1461–1467. https://doi.org/10.1126/science.adf8972

Satoh K, Makimura K, Hasumi Y et al (2009) *Candida auris* sp. Nov., a novel ascomycetous yeast isolated from the external ear canal of an inpatient in a Japanese hospital. Microbiol Immunol 53:41–44. https://doi.org/10.1111/j.1348-0421.2008.00083.x

Schelenz S, Hagen F, Rhodes JL et al (2016) First hospital outbreak of the globally emerging *Candida auris* in a European hospital. Antimicrob Resist Infect Control 5:35. https://doi.org/10.1186/s13756-016-0132-5

Schikora-Tamarit MÀ, Gabaldón T (2022) Using genomics to understand the mechanisms of virulence and drug resistance in fungal pathogens. Biochem Soc Trans 50:1259–1268. https://doi.org/10.1042/BST20211123

Schikora-Tamarit MÀ, Gabaldón T (2024) Recent gene selection and drug resistance underscore clinical adaptation across *Candida* species. Nat Microbiol 9:284–307. https://doi.org/10.1038/s41564-023-01547-z

Schrag SJ, Wiener P (1995) Emerging infectious disease: what are the relative roles of ecology and evolution? Trends Ecol Evol 10:319–324. https://doi.org/10.1016/S0169-5347(00)89118-1

Schutz KS, Melie T, Smith SD, Quandt CA (2023) Reassessing the origins of pathogenicity in *Candida auris* and relatives through phylogenomic analysis (bioRxiv Preprint) 2023.04.13.536682. https://doi.org/10.1101/2023.04.13.536682

Selmecki A, Gerami-Nejad M, Paulson C et al (2008) An isochromosome confers drug resistance in vivo by amplification of two genes, ERG11 and TAC1. Mol Microbiol 68:624–641. https://doi.org/10.1111/j.1365-2958.2008.06176.x

Selmecki AM, Maruvka YE, Richmond PA et al (2015) Polyploidy can drive rapid adaptation in yeast. Nature 519:349–352. https://doi.org/10.1038/nature14187

Shapiro-Ilan DI, Fuxa JR, Lacey LA et al (2005) Definitions of pathogenicity and virulence in invertebrate pathology. J Invertebr Pathol 88:1–7. https://doi.org/10.1016/j.jip.2004.10.003

Sharma C, Kadosh D (2023) Perspective on the origin, resistance, and spread of the emerging human fungal pathogen *Candida auris*. PLoS Pathog 19:e1011190. https://doi.org/10.1371/journal.ppat.1011190

Sherry L, Ramage G, Kean R et al (2017) Biofilm-forming capability of highly virulent, multidrug-resistant *Candida auris*. Emerg Infect Dis 23:328–331. https://doi.org/10.3201/eid2302.161320

Shuping L, Maphanga TG, Naicker SD et al (2023) High prevalence of *Candida auris* colonization during protracted neonatal unit outbreak, vol 29. Emerg Infect Dis, South Africa, p 1913. https://doi.org/10.3201/eid2909.230393

Simon SP, Li R, Silver M et al (2023) Comparative outcomes of *Candida auris* bloodstream infections: a multicenter retrospective case-control study. Clin Infect Dis 76:e1436–e1443. https://doi.org/10.1093/cid/ciac735

Singer M, Deutschman CS, Seymour CW et al (2016) The third international consensus definitions for sepsis and septic shock (Sepsis-3). JAMA 315:801–810. https://doi.org/10.1001/jama.2016.0287

Smoak RA, Snyder LF, Fassler JS, He BZ (2023) Parallel expansion and divergence of an Adhesin family in pathogenic yeasts. Genetics iyad024. https://doi.org/10.1093/genetics/iyad024

Snedden CE, Makanani SK, Schwartz ST et al (2021) SARS-CoV-2: cross-scale insights from ecology and evolution. Trends Microbiol 29:593–605. https://doi.org/10.1016/j.tim.2021.03.013

Spivak ES, Hanson KE (2018) *Candida auris*: an emerging fungal pathogen. J Clin Microbiol 56:e01588–e01517. https://doi.org/10.1128/JCM.01588-17

Spruijtenburg B, Badali H, Abastabar M et al (2022) Confirmation of fifth *Candida auris,* clade by whole genome sequencing. Emerg Microbes Infect 1–15:2405. https://doi.org/10.1080/22221751.2022.2125349

Steenbergen JN, Shuman HA, Casadevall A (2001) *Cryptococcus neoformans* interactions with amoebae suggest an explanation for its virulence and intracellular pathogenic strategy in macrophages. Proc Natl Acad Sci 98:15245–15250. https://doi.org/10.1073/pnas.261418798

Suphavilai C, Ko KKK, Lim KM, et al (2023) Discovery of the sixth *Candida auris* clade in Singapore (medRxiv Preprint) 2023.08.01.23293435

Taori SK, Khonyongwa K, Hayden I et al (2019) *Candida auris* outbreak: mortality, interventions and cost of sustaining control. J Infect 79:601–611. https://doi.org/10.1016/j.jinf.2019.09.007

Thangaraj K, Chaubey G, Reddy AG et al (2006) Unique origin of Andaman islanders: insight from autosomal loci. J Hum Genet 51:800–804. https://doi.org/10.1007/s10038-006-0026-0

Thangaraj K, Singh L, Reddy AG et al (2003) Genetic affinities of the Andaman islanders, a vanishing human population. Curr Biol 13:86–93. https://doi.org/10.1016/S0960-9822(02)01336-2

Thoma R, Seneghini M, Seiffert SN et al (2022) The challenge of preventing and containing outbreaks of multidrug-resistant organisms and *Candida auris* during the coronavirus disease 2019 pandemic: report of a carbapenem-resistant *Acinetobacter baumannii* outbreak and a systematic review of the literature. Antimicrob Resist Infect Control 11:12. https://doi.org/10.1186/s13756-022-01052-8

Vande Zande P, Zhou X, Selmecki A (2023) The dynamic fungal genome: polyploidy, aneuploidy and copy number variation in response to stress. Ann Rev Microbiol 77:341–361. https://doi.org/10.1146/annurev-micro-041320-112443

Walker RS, Kesler DC, Hill KR (2016) Are isolated indigenous populations headed toward extinction? PLoS One 11:e0150987. https://doi.org/10.1371/journal.pone.0150987

Wang L, Cao J-B, Xia B-B et al (2023a) Metatranscriptome of human lung microbial communities in a cohort of mechanically ventilated COVID-19 omicron patients. Signal Transduct Target Ther 8:1–12. https://doi.org/10.1038/s41392-023-01684-1

Wang Q, Cheng S, Wang Y et al (2023b) Global characteristics and trends in research on *Candida auris*. Front Microbiol 14. https://doi.org/10.3389/fmicb.2023.1287003

Wang Y, Xu J (2022) Population genomic analyses reveal evidence for limited recombination in the superbug *Candida auris* in nature. Comput Struct Biotechnol J 20:3030–3040. https://doi.org/10.1016/j.csbj.2022.06.030

Wang Y, Xu J (2024) Associations between genomic variants and antifungal susceptibilities in the archived global *Candida auris* population. J Fungi 10:86. https://doi.org/10.3390/jof10010086

Wasi M, Khandelwal NK, Moorhouse AJ et al (2019) ABC transporter genes show upregulated expression in drug-resistant clinical isolates of *Candida auris*: a genome-wide characterization of ATP-binding cassette (ABC) transporter genes. Front Microbiol 10:1445. https://doi.org/10.3389/fmicb.2019.01445

Watkins RR, Gowen R, Lionakis M, Ghannoum M (2022) Update on the pathogenesis, virulence, and treatment of *Candida auris*. Pathog Immun 7:46–65. https://doi.org/10.20411/pai.v7i2.535

Weerasinghe H, Simm C, Djajawi TM et al (2023) *Candida auris* uses metabolic strategies to escape and kill macrophages while avoiding robust activation of the NLRP3 inflammasome response. Cell Rep 42:112522. https://doi.org/10.1016/j.celrep.2023.112522

White TC, Esquivel BD, Rouse Salcido EM, et al (2024) *Candida auris* detected in the oral cavity of a dog in Kansas. MBio 0:e03080–23. doi:https://doi.org/10.1128/mbio.03080-23

WHO (2022) World Health Organisation (WHO) fungal priority pathogens list to guide research, development and public health action. https://www.who.int/publications/i/item/9789240060241

Wilcox BA, Gubler DJ (2005) Disease ecology and the global emergence of zoonotic pathogens. Environ Health Prev Med 10:263–272. https://doi.org/10.1007/BF02897701

Woolhouse MEJ (2002) Population biology of emerging and re-emerging pathogens. Trends Microbiol 10:s3–s7. https://doi.org/10.1016/S0966-842X(02)02428-9

Woolhouse MEJ, Dye C (2001) Preface. Philos Trans R Soc Lond B Biol Sci 356:981–982. https://doi.org/10.1098/rstb.2001.0899

Woolhouse MEJ, Haydon DT, Antia R (2005) Emerging pathogens: the epidemiology and evolution of species jumps. Trends Ecol Evol 20:238–244. https://doi.org/10.1016/j.tree.2005.02.009

Wurster S, Bandi A, Beyda ND et al (2019) *Drosophila melanogaster* as a model to study virulence and azole treatment of the emerging pathogen *Candida auris*. J Antimicrob Chemother 74:1904–1910. https://doi.org/10.1093/jac/dkz100

Yadav A, Jain K, Wang Y et al (2022) *Candida auris* on apples: diversity and clinical significance. MBio 13:e00518–e00522. https://doi.org/10.1128/mbio.00518-22

Yadav A, Wang Y, Jain K et al (2023) *Candida auris* in dog ears. J Fungi 9:720. https://doi.org/10.3390/jof9070720

Yao S, Feng Y, Zhang Y, Feng J (2021) DNA damage checkpoint and repair: from the budding yeast Saccharomyces cerevisiae to the pathogenic fungus *Candida albicans*. Comput Struct Biotechnol J 19:6343–6354. https://doi.org/10.1016/j.csbj.2021.11.033

Yue H, Bing J, Zheng Q et al (2018) Filamentation in *Candida auris*, an emerging fungal pathogen of humans: passage through the mammalian body induces a heritable phenotypic switch. Emerg Microbes Infect 7:1–13. https://doi.org/10.1038/s41426-018-0187-x

Zamith-Miranda D, Heyman HM, Cleare LG et al (2019) Multi-omics signature of *Candida auris*, an emerging and multidrug-resistant pathogen. mSystems 4:e00257–e00219. https://doi.org/10.1128/mSystems.00257-19

Zhai B, Rolling T, Hohl TM (2021) Exploring *Candida auris* in its habitat. Cell Host Microbe 29:150–151. https://doi.org/10.1016/j.chom.2021.01.010

Zhu Y, O'Brien B, Leach L et al (2020) Laboratory analysis of an outbreak of *Candida auris* in New York from 2016 to 2018: impact and lessons learned. J Clin Microbiol 58:e01503–e01519. https://doi.org/10.1128/JCM.01503-19

# 11 General Trends of Genomic Signatures of the Ectomycorrhizal Symbiosis in Fungi: A Comparison Across Multiple Lineages

Marisol Sánchez-García, Shingo Miyauchi, Elodie Drula, Francis Martin, and Martin Ryberg

**Abstract**

Ectomycorrhizal (ECM) symbiosis is a mutualistic association between the roots of plants, mainly tree species, and fungi, where the photosynthetic partner provides carbon to the fungus in exchange for nutrients, such as nitrogen and phosphorous. Because the majority of forest trees form this type of association, they have a tremendous impact on the maintenance and functioning of forest ecosystems. This symbiosis evolved multiple times in fungi from diverse saprotrophic ancestors. Comparative genomic studies have shown general trends in the evolution of ECM fungi, including the loss of plant cell wall degrading enzymes (PCWDEs), expansion of transposable elements and small secreted proteins, and the co-option of genes already present in saprotrophic ancestors. However, each independently evolved ECM lineage has adopted different ways of interacting with their hosts by recruiting different sets of ancestral gene families for symbiotic functions and retaining distinct sets of PCWDEs, depending on the ecology and decomposing abilities of their ancestors. Here, we review the main findings in the field of ECM fungal genomics and present the main differences among various taxonomic groups. Finally, we provide insights into new frontiers in this field.

**Keywords**

Fungal symbiosis · Comparative genomics · Transposable elements · CAZymes · Secreted proteins

M. Sánchez-García (✉)
Uppsala Biocentre, Department of Forest Mycology and Plant Pathology, Swedish University of Agricultural Sciences, Uppsala, Sweden
e-mail: marisol.sanchez.garcia@slu.se

S. Miyauchi
Evolutionary and Synthetic Biology unit, Okinawa Institute of Science and Technology, Okinawa, Japan

E. Drula
Architecture et Fonction des Macromolécules Biologiques (AFMB), CNRS, Aix-Marseille Université, Marseille, France

INRAE, UMR 1163, Biodiversité et Biotechnologie Fongiques, Marseille, France

F. Martin
Université de Lorraine, INRAE, UMR Interactions Arbres/Microorganismes, Centre INRAE Grand Est - Nancy, Nancy, France

M. Ryberg
Department of Organismal Biology, Uppsala University, Uppsala, Sweden

## 11.1 Introduction

Ectomycorrhizal (ECM) fungi form symbiotic associations with the roots of plants, primarily tree species, in forest ecosystems (Smith and


M. Nowrousian, J. E. Stajich (eds.), *Fungal Genomics*, The Mycota 13,
https://doi.org/10.1007/978-3-031-75666-5_11

Read 2008). In this symbiosis, the fungal partner provides the plant host with a range of macro- and micronutrients, including nitrogen, phosphorous, Fe, and Zn, in exchange for photosynthetic carbon (Read and Perez-Moreno 2003; Becquer et al. 2019). The exchange of nutrients occurs through the Hartig net, which is a unique structure formed by interlacing mycelia that enclose the outer cortical cells of the roots and provide a vast amount of surface area between the two symbiotic partners (Smith and Read 2008). ECM symbiosis has a significant impact on nutrient cycling and other soil processes, such as organic matter decomposition and carbon sequestration (Read and Perez-Moreno 2003; Tedersoo and Bahram 2019; Lindahl et al. 2021; Hawkins et al. 2023), plant health, and soil microbial community composition (Vik et al. 2013; Liang et al. 2020).

ECM symbiosis has evolved independently from diverse saprotrophic ancestors in up to 82 fungal lineages within Ascomycota, Basidiomycota, and Mucoromycota (Tedersoo and Smith 2017; Strullu-Derrien et al. 2018), and in 30 plant lineages within Gymnospermae and Angiospermae (Brundrett and Tedersoo 2018). Only 2% of plant species form ECM associations (Brundrett 2009), but these include the most dominant tree species in temperate and boreal ecosystems (e.g., pine, spruce, birch, and oak) and some tropical forest trees (e.g., dipterocarp) (Tedersoo et al. 2012; Corrales et al. 2018). It has been estimated that ECM fungi are associated with 60% of all tree stems (Steidinger et al. 2019). Hence, symbiosis is essential for the maintenance and functioning of forest ecosystems.

The first ECM fungal genome, *Laccaria bicolor* (Basidiomycota) was sequenced in 2008 (Martin et al. 2008), which was followed by *Tuber melanosporum* (Ascomycota; Martin et al. 2010). These pioneering studies emphasized the need for sequencing a wider diversity of ECM fungal species to gain better insights into the genomic and functional traits of ECM fungi. With the development of new sequencing tools and the launch of the JGI Fungal Genomics Program (Grigoriev et al. 2011), several ECM species were sequenced, with the aim of reflecting a broad phylogenetic diversity to gain a better understanding of the evolutionary events surrounding the origin of the ECM lifestyle. In 2015, the first comparative genomics study was published, which included 11 species of ECM fungi. The goal of this study was to identify the common features of ECM fungi with distinct evolutionary origins (Kohler et al. 2015). A second comparative genomic study in 2020 analyzed an expanded number of ECM fungal genomes, representing multiple clades across the fungal tree of life (Miyauchi et al. 2020). Together, these studies have shown that ECM fungal genomes possess the following features: (1) an expansion in genome size as a result of massive proliferation of transposable elements (TEs), (2) a decrease in the number of carbohydrate-active enzymes (CAZymes), particularly plant cell wall degrading enzymes (PCWDEs), and (3) expansion of effector-like mycorrhiza-induced secreted proteins (SSPs) (Martin et al. 2008, 2016; Labbé et al. 2012; Kohler et al. 2015; Miyauchi et al. 2020), and (4) each independent ECM lineage uses specific novel genes and members of gene families already present in saprotrophic ancestors to develop symbiotic interactions (Miyauchi et al. 2020).

Several works have also aimed to expand the knowledge within individual lineages of ECM fungi in order to distinguish common properties of ECM fungi with distinct evolutionary origins from lineage-specific evolutionary changes, and have targeted several ECM fungal taxa with a single origin of the ECM symbiosis, such as Amanitaceae, Tuberaceae, Endogonaceae, Suillaceae, Russulaceae, and Boletaceae (Hess et al. 2018; Murat et al. 2018; Chang et al. 2019; Lofgren et al. 2020; Looney et al. 2022; Lebreton et al. 2022; Wu et al. 2022; Plett et al. 2023). Here, we review the main findings on the genomics of ECM fungi, test differences among various taxonomic groups, and provide insights into new frontiers in the field.

## 11.2 Expansion of Genome Size as a Result of Proliferation of Transposable Elements

All ECM fungal genomes sequenced to date have undergone expansion of TEs, which accounts for their large genome sizes (Kohler et al. 2015; Looney et al. 2022; Fig. 11.1). TEs are mobile DNA sequences that have the ability to proliferate and expand to distant genomic regions and constitute a substantial fraction of eukaryotic genomes (Bennetzen and Wang 2014). They often range in length from 100 to 10,000 base pairs, but sometimes can be even larger (Arkhipova et al. 2019; Gluck-Thaler et al. 2022). TEs are known to have significant functional impacts on their host genome, as they can modify transcripts, interfere with gene regulation, cause mutations, and lead to chromosomal rearrangements (McCue and Slotkin 2012; Bennetzen and Wang 2014; Lorrain et al. 2021; Torres et al. 2021). TEs represent a source of genetic variation and play a major role in genomic evolution (Dubin et al. 2018).

In ECM fungi, LTR retrotransposons such as *Gypsy/DIRS1* and *Ty1/Copia* exhibit high genome coverage. A high Gypsy/DIRS1 content has been observed in *Boletus*, *Pisolithus*, *Lactarius*, and *Tuber* species (Lebreton et al. 2022; Wu et al. 2022; Plett et al. 2023; Fig. 11.2). *Tricholoma matsutake* has the largest genome in Agaricales and a higher TE content (LINEs and LTR elements) than any other ECM in Basidiomycota (Figs. 11.1 and 11.2). Notable TE expansions have been observed for the genus *Lactarius*, which also contains TE categories that are not usually found in other ECM species such as *Academ*, *Zisupton*, and *Penelope* (Lebreton et al. 2022). Ascomycota showed a distinctive pattern, with the highest TE coverage (30–70%) among the ECM fungal genomes compared (Fig. 11.1). This high TE coverage is attributed to RNA retrotransposons (LINEs and LTR elements) and DNA transposons. *PiggyBac* is only present in ECM Ascomycota (Miyauchi et al. 2020) (Fig. 11.2). A very high proportion of TEs in ECM genomes falls within the unclassified category (Fig. 11.2), for which manual curation is needed to gain a better understanding of the content and diversity of TEs in ECM fungi (Platt II et al. 2016).

The consequences of massive proliferation of TEs in ECM fungi are not fully understood. However, TE clusters are found in close proximity to MiSSPs in Russulaceae species (Looney et al. 2022), as well as to sedolisin proteases in *Lactarius* species (Lebreton et al. 2022), suggesting that the expansion of these genes is caused by TE proliferation and that TEs drive gene innovation, ECM regulation, and functional speciation (Miyauchi et al. 2020; Looney et al. 2022) as well as their possible involvement in host specificity (Tang et al. 2021; Lebreton et al. 2022). In plant pathogenic fungi, TEs facilitate adaptations to colonize host plants and tolerate environmental stresses, and they are often associated with genes that encode important pathogenicity functions and increase virulence (Hartmann and Croll 2017; Razali et al. 2019). It appears that ECM fungi utilize gene regulatory pathways and mechanisms used by other types of fungal–plant interactions, including pathogen–fungal–host interactions (Koeck et al. 2011; Plett et al. 2015; Liao et al. 2016), and because the mechanisms influencing symbiosis may be similar across different kinds of symbiotic relationships, from pathogenic to mutualistic (Veneault-Fourrey and Martin 2011), we can expect that TEs also facilitate adaptations to colonize host plants, are linked to genes involved in ECM host specificity and functional adaptation. Consequently, it drives the adaptation of ECM fungi to new hosts. Therefore, it is necessary to focus on manual curation of TEs to understand their content and diversity in ECM fungal genomes, as well as their role in the evolution of this lifestyle.

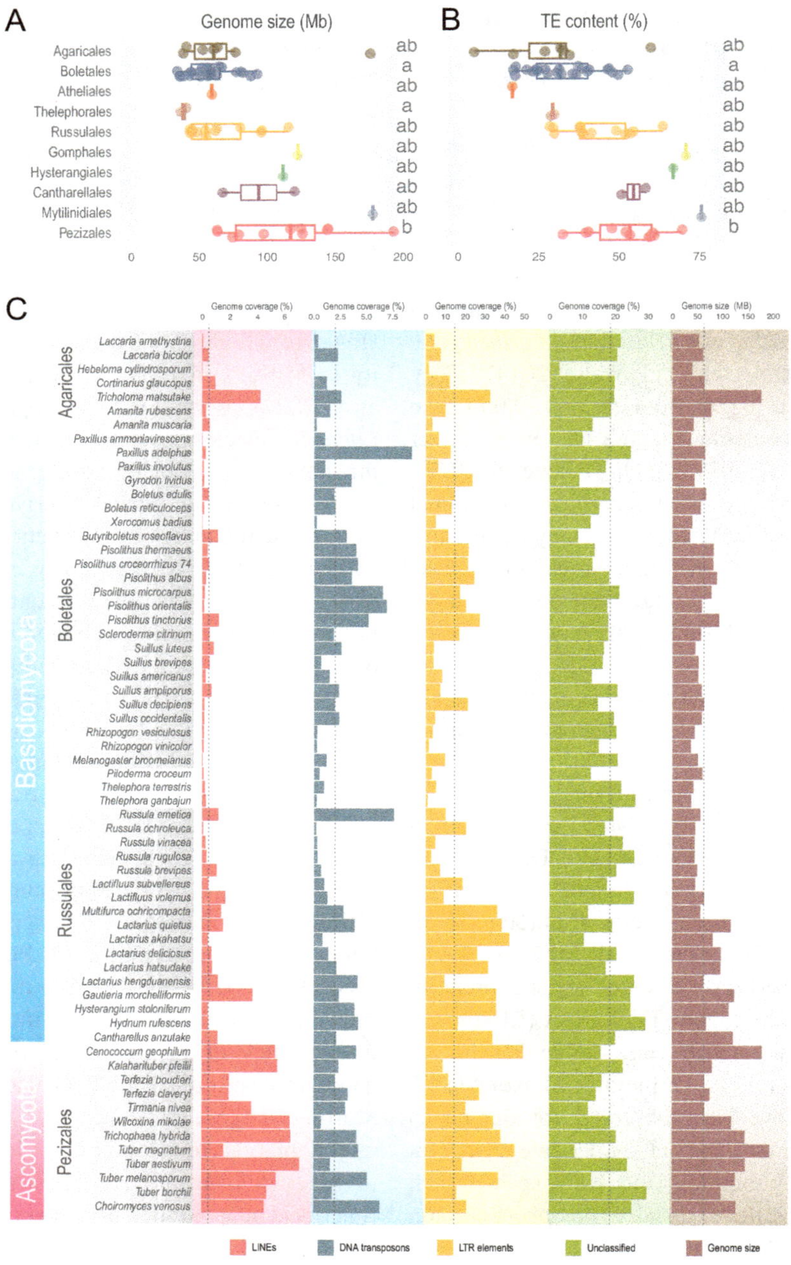


**Fig. 11.1** Overview of the genomic features of 63 ectomycorrhizal fungi. The list of ECM genomes can be found in Table S1 (https://doi.org/10.6084/m9.figshare.25992088). A. The size of genomes. B. Genome coverage of transposable elements. Colors represent different taxonomic groups. Letters indicate significant differences among taxonomic groups (FDR adjusted $p < 0.05$; Kruskal-Wallis test with post hoc Dunn test; Table S2, https://doi.org/10.6084/m9.figshare.25992088). Statistics of JGI genome assemblies were obtained from the JGI Mycocosm (https://mycocosm.jgi.doe.gov/Ectomycorrhizal_fungi/Ectomycorrhizal_fungi.info.html). Differences in variables among the groups were tested using the Kruskal-Wallis test with post hoc Dunn test using the R package DescTools (Signorell et al. 2020) C. Major TEs covering the genomes. The major TE categories are shown in color. LINEs: Long interspersed nuclear elements (non-LTR/class I retrotransposons), DNA transposons: Class II transposons, LTR elements: Long terminal repeat elements, class I retrotransposons. Unclassified: Unidentified repeats. Genome size: length of genome assemblies. The taxonomic order of the species was adopted from the JGI MycoCosm. The details and methods are shown in Fig. 11.2

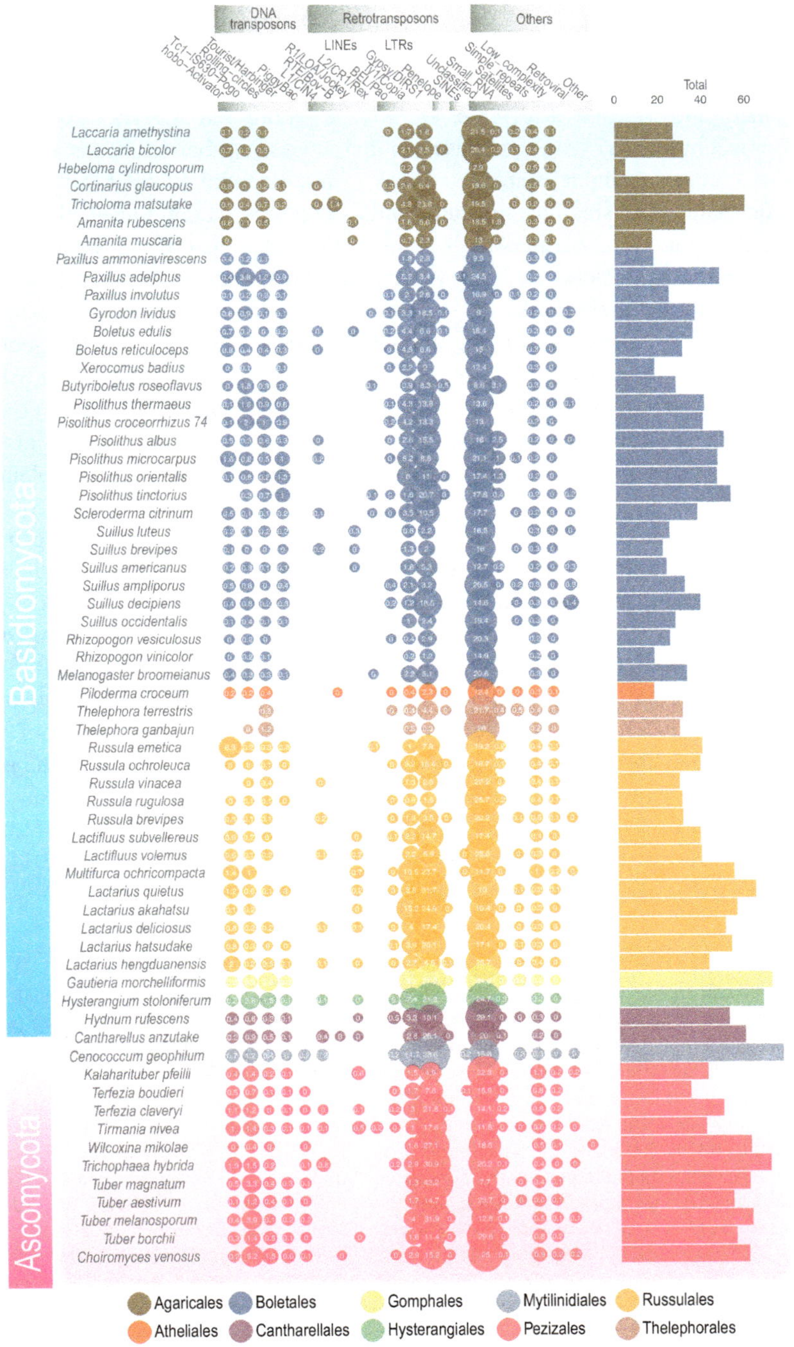


**Fig. 11.2** Genome coverage of transposable elements in 63 ectomycorrhizal fungi. LTR: Long terminal repeat retrotransposons. LINEs: Long interspersed nuclear elements. SINEs: Short interspersed nuclear elements. The bubble size is proportional to the coverage (%) of each of transposable elements (showing inside the bubbles). The right bars show the total coverage per genome. The taxonomic order of the species was adopted from the JGI MycoCosm. The genome coverage of the TEs was identified using RepeatScout and RepeatModeler (Smit and Hubley 2015; Smit et al. 2015). The information above was combined and visualized using Transposon Identification Nominative Genome Overview (TINGO; Morin et al. 2019)

## 11.3 The Secretome of Ectomycorrhizal Fungi

Fungi produce and release a diverse set of enzymes (for example, CAZymes, proteases, and lipases) into the extracellular matrix to break down organic matter into smaller molecules and interact with their microbial partners (e.g., SSPs: small secreted proteins) (Tian et al. 2009; Courty et al. 2010). Among the main classes of secreted enzymes, CAZymes have received significant attention because of their widespread reduction in all ECM fungal lineages, which will be discussed in more detail in the following section. In this section, we focus on the secreted lipases, proteases, and SSPs.

Lipases and proteases play important roles in the growth and adaptation of fungi to different environments and are particularly important for plant pathogenic fungi as they can fragment plant cell wall polymers and cuticular waxes that are found in plant epidermal cells (Subramoni et al. 2010; Podder et al. 2019). The average numbers of secreted proteases and lipases were found to be similar between ECM fungi and saprotrophs (Pellegrin et al. 2015; Miyauchi et al. 2020; Wu et al. 2022). Among the ECM fungi, Russulales contained the highest number of secreted proteases (Fig. 11.3). A large expansion of sedolisin proteases has been reported in *Lactarius* species, a feature that appears to be unique to this group, suggesting that *Lactarius* species use a protease-based toolkit to establish ECM symbiosis with their host (Lebreton et al. 2022). The ECM in Pezizales had the lowest number of secreted lipases, and the ECM in Russulales showed the highest variation in lipase number (Fig. 11.3d). Unique expansions of the secreted gene repertoire have been observed in *Russula vinacea*, in which a cluster of carboxylesterase lipases is highly expanded (Looney et al. 2022).

Fungi that interact with plants produce a relatively higher number of SSPs than free-living saprotrophic fungi do (Pellegrin et al. 2015). The expansion of mycorrhiza-induced SSPs has been observed in most ECM fungal genomes (Martin et al. 2008, 2016; Labbé et al. 2012; Kohler et al. 2015; Miyauchi et al. 2020). SSPs facilitate the establishment and functioning of ECM symbiosis (Garcia et al. 2015), some of which have been functionally characterized and appear to be indispensable for ECM symbiosis development (for example, LbMiSSP7, LbMiSSP8, and PaMiSSP10) (Plett et al. 2011, 2020; Pellegrin

**Fig. 11.3** Secretomic trends of 63 ectomycorrhizal fungi. (**a**, **b**, **c**, **d**) Taxonomic group-wise distribution of the secreted proteins. Colors represent different taxonomic groups. Letters indicate significant differences among taxonomic groups (FDR adjusted $p < 0.05$; Kruskal-Wallis test with post hoc Dunn test; Table S2, https://doi.org/10.6084/m9.figshare.25992088). E. Secretomic profiles. The first bubble plot on the left shows the number of secreted genes for CAZymes, lipases, proteases, and others (i.e., all secreted proteins that were not in the first three groups). The SSP group was a subcategory showing the number of small secreted proteins (<300 aa). The size of the bubbles corresponds to the number of genes, and the fungi are color-coded according to their taxonomic groups. The first bar plots (in the middle) represent the ratio of CAZymes, lipases, and proteases to all secreted proteins (left), and the ratio of SSPs among the entire secretome (right). The second bubble plot on the right side of the panel shows CAZymes grouped according to their functions including plant cell wall degrading enzymes (PCWDE) and fungal cell wall degrading enzymes (FCWDE), peptidoglycan (i.e., bacterial membrane) degrading enzymes (BMDE), trehalose, starch, glycogen degrading enzymes (Storage), lytic polysaccharide monooxygenase (LPMO), substrate-specific enzymes for cellulose, hemicellulose, lignin, and pectin (plant cell walls); chitin, mannan (fungal cell walls). The second bar plots (far right) show the total number of genes, including PCWDE, MCWDE, and BMDE (left), and the proportions of PCWDE, MCWDE, and BMDE (right). The taxonomic order of the species was adopted from the JGI MycoCosm. Details of the protein counts are shown in Table S3 (https://doi.org/10.6084/m9.figshare.25992088). Details of substrate-specific CAZymes are shown in Table S4 (https://doi.org/10.6084/m9.figshare.25992088). The secretomes were predicted as previously described (Pellegrin et al. 2015). CAZyme annotations were obtained from semi-manual annotation (Drula et al. 2022). We calculated, visualized, and compared the counts and ratios of secreted CAZymes, lipases, proteases, and small secreted proteins (<300 amino acids). The output files generated were combined and visualized with Proteomic Information Navigated Genomic Outlook (PRINGO; Miyauchi et al. 2020)

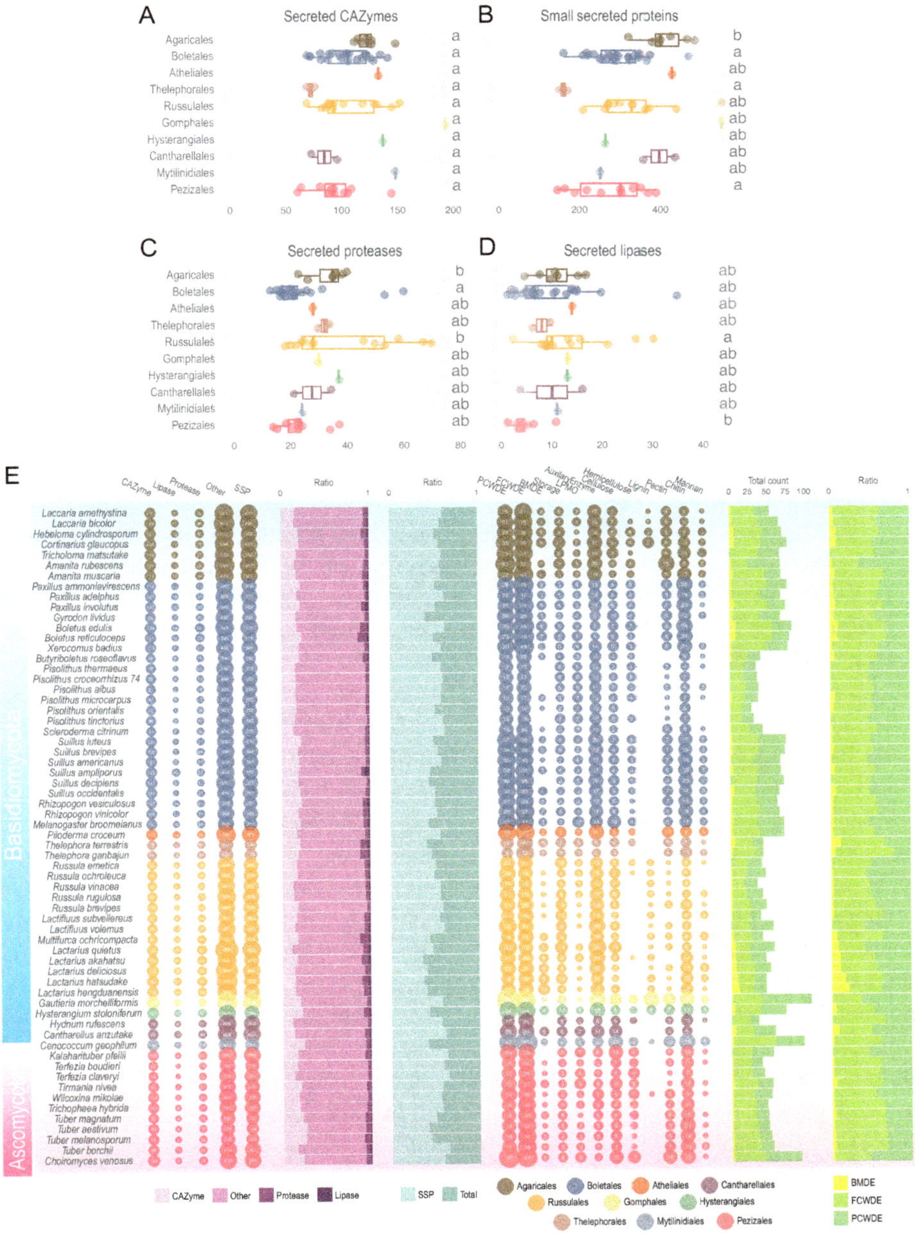


**Fig. 11.3** Continued

et al. 2019). Mycorrhiza-induced SSPs constitute 8–28% of the genes upregulated during ECM symbiosis (Martin et al. 2008; Kohler et al. 2015), and it has been hypothesized that they evolved from SSPs used in the saprotrophic ancestors of ECM fungi, and represent innovations involved in adaptations to the plant host habitat (Miyauchi et al. 2020).

A number of ECM fungal lineages do not show enrichment of SSPs when compared to their non-ECM relatives, such as members of Endogonaceae, Amanitaceae, and Russulaceae, suggesting that these groups may have evolved different ways to interact with their hosts (Hess et al. 2018; Chang et al. 2019; Looney et al. 2022).

## 11.4 Loss of Plant Cell Wall Degrading Enzymes in Ectomycorrhizal Fungi

CAZymes break down polysaccharides through various mechanisms, generating simple sugars that are used as the main source of energy for several organisms (Gavande et al. 2023). Some CAZymes specialize in the degradation of pectin, cellulose, hemicellulose, and lignin, which are the main components of plant cell walls (Cosgrove 2005), these CAZymes are known as plant cell wall degrading enzymes (PCWDEs). Saprotrophs and pathogenic fungi (necrotrophic and hemi-biotrophic) produce enzymes (e.g., celluloses and hemicelluloses) that can deconstruct and hydrolyze plant cell wall components (Kubicek et al. 2014; Ruiz-Dueñas et al. 2021). In contrast, symbiotic fungi require other mechanisms to interact with plant hosts without harming them, and they possess a reduced set of PCWDEs (Kohler et al. 2015; Miyauchi et al. 2020).

Convergent loss of PCWDEs has been observed in ECM fungal species (Kohler et al. 2015; Miyauchi et al. 2020), individual ECM fungal lineages retain unique sets of these enzymes, which reflects the ecology of their saprotrophic ancestor and their functional specialization (Kohler et al. 2015; Chang et al. 2019; Miyauchi et al. 2020; Wu et al. 2022; Looney et al. 2022). Members of Ascomycota lack the laccase AA1_1, which is involved in the oxidation of high-redox-potential lignin components (Levasseur et al. 2013), and which is present in most Basidiomycota species sequenced so far except *Thelephora* and *Hydnum* (Miyauchi et al. 2020). This lack of AA1_1 in Ascomycota accounts for the reduced set of PCWDEs compared to the ECM in Basidiomycota. On the other hand, most ECM Ascomycota possess CBM1 modules, which are involved in the degradation of cellulose, and AA1_3, a laccase-like multicopper oxidase. CBM1 is present in only a few ECM Basidiomycota, and AA1_3 is absent in ECM Basidiomycota (Fig. 11.4).

Members of Phallomycetidae (Gomphales and Hysterangiales) possess a characteristic diverse set of PCWDEs, including class II lignin-modifying peroxidases (AA2), GH6, and GH7 cellobiohydrolases (Miyauchi et al. 2020). Of particular relevance is *Gautieria morchelliformis*, which contains 25 class II Mn peroxidases that are involved in the decay of soil organic matter and phenolic aromatic substrates (Levasseur et al. 2013; Miyauchi et al. 2020), which is the only ECM species known to possess ligninolytic peroxidases (Miyauchi et al. 2020). ECM species within Russulales contain duplicated GH45 and one atypical Mn peroxidase gene, they have almost lost PCWDEs degrading hemicellulose but kept PCWDEs degrading lignin such as AA2 (Miyauchi et al. 2020; Lebreton et al. 2022; Looney et al. 2022; Fig. 11.4). GH6 and GH7 cellobiohydrolases are present in only a few ECM Basidiomycota (e.g., *Piloderma croceum* and *Hydnum rufescences*) and in several ECM Ascomycota (Fig. 11.4).

ECM fungi within Mucoromycota (Endogonaceae) lost a smaller fraction of their PCWDEs than ECM fungi within Ascomycota and Basidiomycota. This lower degree of reduction was attributed to the low number of PCWDEs present in their ancestors. The PCWDEs content of ECM Mucoromycota is characterized by the absence of AA9, which acts on cellulose- and lignin-degrading peroxidases, and CBM1 modules are absent or present in low copy numbers in Endogonaceae (Chang et al. 2019).

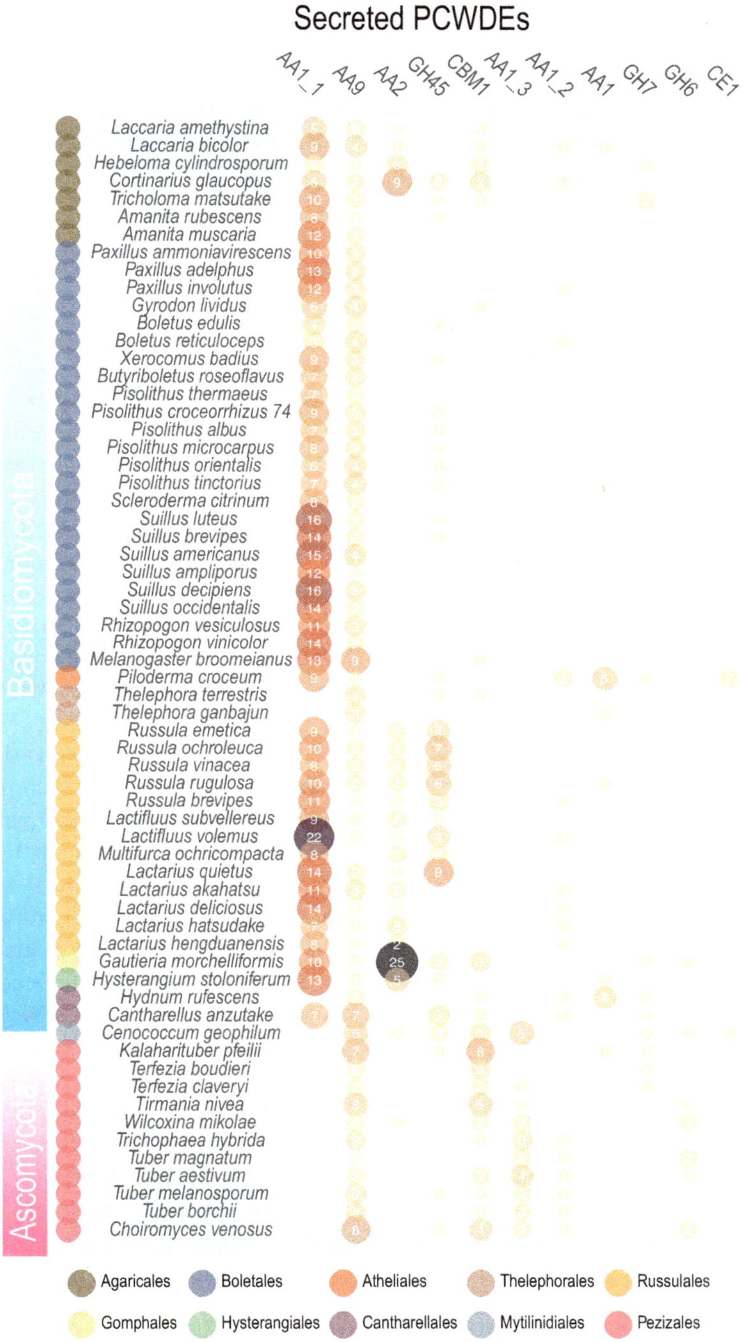


**Fig. 11.4** Distribution of key secreted PCWDEs. Bubbles with numbers contain the number of genes coding for a series of secreted PCWDEs required for cellulose and lignin degradation. Colors represent different taxonomic groups (bottom panel). The taxonomic order of the species was adopted from the JGI MycoCosm. The selection of CAZyme modules was adopted from Miyauchi et al. (2020). High loadings of variables were determined using principal components calculated with the count of secreted proteins using the R package vegan version 2.5–6 (Oksanen et al. 2019). The output files generated were combined and visualized with Proteomic Information Navigated Genomic Outlook (PRINGO; Miyauchi et al. 2020)

ECM fungi have lost most of their ability to degrade organic matter, which is evident from their reduced set of PCWDEs, but some species such as *Cortinarius* may decompose complex organic matter to mobilize N for themselves or their hosts (Bödeker et al. 2014). The genome of *Cortinarius glaucopus* contains several duplicated copies of the class II peroxidase genes (Miyauchi et al. 2020). Similarly, the genome of *Tricholoma matsutake* contains two copies of GH7 cellobiohydrolases, supporting its role in soil biomass degradation (Kusuda et al. 2008; Borisova et al. 2018; Miyauchi et al. 2020).

## 11.5 ECM Genome Features Pre-Date the Evolution of the ECM Lifestyle

Many mycorrhiza-induced genes in ECM fungal lineages have homologues in their saprotrophic ancestors (Hess et al. 2018; Miyauchi et al. 2020). In Ascomycota and Basidiomycota, approximately 70% of these genes pre-dated the evolution of the ECM lifestyle, suggesting that symbiosis-induced genes have been co-opted for ECM development during the evolution of saprotrophic ancestors (Miyauchi et al. 2020). This implies that the transition from saprotrophs to ECM does not require the origin of many new genes, but that it stems from the loss of key gene families and the co-option of already existing gene families (Hess et al. 2018). Moreover, individual ECM lineages have recruited different sets of these ancestral gene families for symbiotic functions, which together with the unique set of novel genes that evolved and diversified after the origins of ECM symbiosis, provide unique and distinguishable symbiotic gene repertoires for each independently evolved ECM lineage (Kohler et al. 2015; Miyauchi et al. 2020). For example, in *Amanita*, co-option of CAZymes involved in lignocellulose degradation facilitates interactions with the host (Hess et al. 2018).

Studies on ECM Russulaceae and Amanitaceae have shown that some of the key characteristics of ECM fungal genomes are already present in their closely related saprotrophic sister groups, such as a reduced set of PCWDEs and the expansion of TEs. These genomic changes can represent pre-adaptations for the ECM lifestyle and can explain the widespread, recurrent evolution of ECM associations (Hess et al. 2018; Looney et al. 2022).

## 11.6 Host Specificity

ECM fungi exhibit various levels of host specificity, which is a fundamental trait in the life history of this group, and is likely to play a major role in generating and maintaining diversity throughout their evolutionary history (Molina et al. 1992; Navaud et al. 2018; Kennedy et al. 2020). Some attempts have been made to understand the drivers and genetic foundations of ECM fungal host specificity. Three molecules that show differential expression during the compatible formation of mycorrhiza in *Suillus*-*Pinus* systems have received considerable attention: SSPs, secondary metabolites (SMs), and G-protein coupled receptors (GPCRs) (Liao et al. 2016). In addition, a comparative genomic study of the host specialist genus *Suillus* showed that SMs synthesized by terpene and non-ribosomal polyketide synthases (NRPS) are expanding in this genus and may play a role in determining host specificity (Lofgren et al. 2020). In *Lactarius*, another group that contains host specialist species, only a slight enrichment of terpene synthetase genes was found, in addition to the expansion of several sedolisin gene families where host switches have occurred (Lebreton et al. 2022). This suggests that these fungi use a protease-based toolkit to interact with their hosts, which is a strategy used by some plant pathogens, and thus far has not been reported for beneficial plant symbionts (Lowe et al. 2015; Lebreton et al. 2022).

## 11.7 Conclusions and Future Remarks

Despite large-scale similarities among distantly related ECM fungal genomes (e.g., expansion of TEs and reduction of PCWDEs), it is now evident that individual ECM fungal lineages have adopted different mechanisms to interact with plant hosts and establish a symbiotic relationship. Although the rate of ECM genome sequencing has dramatically increased in the last 10 years, this number represents only a small fraction of the total ECM fungal diversity, and there is still much more to be discovered. It is expected that the number of ECM lineages sequenced will continue to increase, and special attention should be paid to target species from tropical forests, which evolved under very different conditions and might reveal lineage- and biome-specific evolutionary changes.

Sequencing several species within a single ECM lineage has offered great insight into the genetic variation within ECM lineages (for example, *Russula* and *Amanita*), and has revealed high levels of functional specialization (Hess et al. 2018; Looney et al. 2022). The next step is to sequence several individuals from the same species to understand the genetic foundations of local adaptation in natural populations of ECM fungi, which is of particular importance because of the clear threat that drastic changes at the global scale pose to plant–fungal beneficial interactions (Steidinger et al. 2019).

Although there have been significant advances in our understanding of the biology and evolution of ECM fungi through the study of their genomes, as presented in this review, a number of gaps remain. Some examples are the role of TEs in host adaptation and, in general, in the evolution of symbiosis, whether functional specialization drives diversification in ECM lineages, the extent of ECM functional diversity, and within a finer level, how natural ECM fungal populations respond to environmental changes. Additionally, characterization of mycorrhiza-induced SSPs from non-model ECM fungi will expand our current knowledge of the regulation and formation of the ECM; however, this might pose a challenge because synthesizing ECM symbiosis under laboratory conditions is not trivial for most ECM fungal species.

**Supplementary Material**
Supplementary tables (S1-S4) are deposited in FigShare: https://doi.org/10.6084/m9.figshare.25992088

## References


Arkhipova IR, Yushenova IA, Angert E (2019) Giant transposons in eukaryotes: is bigger better? Genome Biol Evol 11:906–918. https://doi.org/10.1093/gbe/evz041

Becquer A, Guerrero-Galán C, Eibensteiner JL et al (2019) The ectomycorrhizal contribution to tree nutrition. Adv Bot Res 89:77–126. https://doi.org/10.1016/bs.abr.2018.11.003

Bennetzen JL, Wang H (2014) The contributions of transposable elements to the structure, function, and evolution of plant genomes. Annu Rev Plant Biol 65:505–530. https://doi.org/10.1146/annurev-arplant-050213-035811

Bödeker ITM, Clemmensen KE, de Boer W et al (2014) Ectomycorrhizal *Cortinarius* species participate in enzymatic oxidation of humus in northern forest ecosystems. New Phytol 203:245–256. https://doi.org/10.1111/nph.12791

Borisova AS, Eneyskaya EV, Jana S et al (2018) Correlation of structure, function and protein dynamics in GH7 cellobiohydrolases from *Trichoderma atroviride, T. reesei* and *T. harzianum*. Biotechnol Biofuels 11:1–22. https://doi.org/10.1186/s13068-017-1006-7

Brundrett MC (2009) Mycorrhizal associations and other means of nutrition of vascular plants: understanding the global diversity of host plants by resolving conflicting information and developing reliable means of diagnosis. Plant Soil 320:37–77. https://doi.org/10.1007/s11104-008-9877-9

Brundrett MC, Tedersoo L (2018) Evolutionary history of mycorrhizal symbioses and global host plant diversity. New Phytol 220:1108–1115. https://doi.org/10.1111/nph.14976

Chang Y, Desirò A, Na H et al (2019) Phylogenomics of Endogonaceae and evolution of mycorrhizas within Mucoromycota. New Phytol 222:511–525. https://doi.org/10.1111/nph.15613

Corrales A, Henkel TW, Smith ME (2018) Ectomycorrhizal associations in the tropics—biogeography, diversity patterns and ecosystem roles. New Phytol 220:1076–1091. https://doi.org/10.1111/nph.15151

Cosgrove DJ (2005) Growth of the plant cell wall. Nat Rev Mol Cell Biol 6:850–861. https://doi.org/10.1038/nrm1746

Courty PE, Buée M, Diedhiou AG et al (2010) The role of ectomycorrhizal communities in forest ecosystem processes: new perspectives and emerging concepts. Soil Biol Biochem 42:679–698. https://doi.org/10.1016/j.soilbio.2009.12.006

Drula E, Garron ML, Dogan S et al (2022) The carbohydrate-active enzyme database: functions and literature. Nucleic Acids Res 50:D571–D577. https://doi.org/10.1093/nar/gkab1045

Dubin MJ, Mittelsten Scheid O, Becker C (2018) Transposons: a blessing curse. Curr Opin Plant Biol 42:23–29. https://doi.org/10.1016/j.pbi.2018.01.003

Garcia K, Delaux PM, Cope KR, Ané JM (2015) Molecular signals required for the establishment and maintenance of ectomycorrhizal symbioses. New Phytol 208:79–87. https://doi.org/10.1111/nph.13423

Gavande PV, Goyal A, Fontes CMGA (2023) Chapter 1–carbohydrates and carbohydrate-active enZymes (CAZyme): an overview. In: Goyal A, Sharma K (eds) Glycoside hydrolases. Academic Press, pp 1–23

Gluck-Thaler E, Ralston T, Konkel Z et al (2022) Giant starship elements mobilize accessory genes in fungal genomes. Mol Biol Evol 39. https://doi.org/10.1093/molbev/msac109

Grigoriev IV, Cullen D, Goodwin SB et al (2011) Fueling the future with fungal genomics. Mycology 2:192–209. https://doi.org/10.1080/21501203.2011.584577

Hartmann FE, Croll D (2017) Distinct trajectories of massive recent gene gains and losses in populations of a microbial eukaryotic pathogen. Mol Biol Evol 34:2808–2822. https://doi.org/10.1093/molbev/msx208

Hawkins HJ, Cargill RIM, Van Nuland ME et al (2023) Mycorrhizal mycelium as a global carbon pool. Curr Biol 33:R560–R573. https://doi.org/10.1016/j.cub.2023.02.027

Hess J, Skrede I, De Mares MC et al (2018) Rapid divergence of genome architectures following the origin of an ectomycorrhizal symbiosis in the genus *Amanita*. Mol Biol Evol 35:2786–2804. https://doi.org/10.1093/molbev/msy179

Kennedy PG, Gagne J, Perez-Pazos E et al (2020) Does fungal competitive ability explain host specificity or rarity in ectomycorrhizal symbioses? PLoS One 15:e0234099

Koeck M, Hardham AR, Dodds PN (2011) The role of effectors of biotrophic and hemibiotrophic fungi in infection. Cell Microbiol 13:1849–1857. https://doi.org/10.1111/j.1462-5822.2011.01665.x

Kohler A, Kuo A, Nagy LG et al (2015) Convergent losses of decay mechanisms and rapid turnover of symbiosis genes in mycorrhizal mutualists. Nat Genet 47:410–415. https://doi.org/10.1038/ng.3223

Kubicek CP, Starr TL, Glass NL (2014) Plant cell wall-degrading enzymes and their secretion in plant-pathogenic fungi. Annu Rev Phytopathol 52:427–451. https://doi.org/10.1146/annurev-phyto-102313-045831

Kusuda M, Ueda M, Miyatake K, Terashita T (2008) Characterization of the carbohydrase productions of an ectomycorrhizal fungus, *Tricholoma matsutake*. Mycoscience 49:291–297. https://doi.org/10.1007/s10267-008-0423-7

Labbé J, Murat C, Morin E et al (2012) Characterization of transposable elements in the ectomycorrhizal fungus *Laccaria bicolor*. PLoS One 7:1–9. https://doi.org/10.1371/journal.pone.0040197

Lebreton A, Tang N, Kuo A et al (2022) Comparative genomics reveals a dynamic genome evolution in the ectomycorrhizal milk-cap (*Lactarius*) mushrooms. New Phytol 235:306–319. https://doi.org/10.1111/nph.18143

Levasseur A, Drula E, Lombard V et al (2013) Expansion of the enzymatic repertoire of the CAZy database to integrate auxiliary redox enzymes. Biotechnol Biofuels 6:1. https://doi.org/10.1186/1754-6834-6-41

Liang M, Johnson D, Burslem DFRP et al (2020) Soil fungal networks maintain local dominance of ectomycorrhizal trees. Nat Commun 11:1–7. https://doi.org/10.1038/s41467-020-16507-y

Liao HL, Chen Y, Vilgalys R (2016) Metatranscriptomic study of common and host-specific patterns of gene expression between pines and their symbiotic ectomycorrhizal fungi in the genus *Suillus*. PLoS Genet 12:1–24. https://doi.org/10.1371/journal.pgen.1006348

Lindahl BD, Kyaschenko J, Varenius K et al (2021) A group of ectomycorrhizal fungi restricts organic matter accumulation in boreal forest. Ecol Lett 24:1341–1351. https://doi.org/10.1111/ele.13746

Lofgren LA, Nguyen NH, Vilgalys R et al (2020) Comparative genomics reveals dynamic genome evolution in host specialist ectomycorrhizal fungi. New Phytol 230:774. https://doi.org/10.1111/nph.17160

Looney B, Miyauchi S, Morin E et al (2022) Evolutionary transition to the ectomycorrhizal habit in the genomes of a hyperdiverse lineage of mushroom-forming fungi. New Phytol 233:2294–2309. https://doi.org/10.1111/nph.17892

Lorrain C, Feurtey A, Ller MM et al (2021) Dynamics of transposable elements in recently diverged fungal pathogens: Lineage-specific transposable element content and efficiency of genome defenses. G3 11. https://doi.org/10.1093/g3journal/jkab068

Lowe RGT, McCorkelle O, Bleackley M et al (2015) Extracellular peptidases of the cereal pathogen

*Fusarium graminearum*. Front Plant Sci 6:1–13. https://doi.org/10.3389/fpls.2015.00962

Martin F, Aerts A, Ahrén D et al (2008) The genome of *Laccaria bicolor* provides insights into mycorrhizal symbiosis. Nature 452:88–92. https://doi.org/10.1038/nature06556

Martin F, Kohler A, Murat C et al (2010) Périgord black truffle genome uncovers evolutionary origins and mechanisms of symbiosis. Nature 464:1033–1038. https://doi.org/10.1038/nature08867

Martin F, Kohler A, Murat C et al (2016) Unearthing the roots of ectomycorrhizal symbioses. Nat Rev Microbiol 14:760–773. https://doi.org/10.1038/nrmicro.2016.149

McCue AD, Slotkin RK (2012) Transposable element small RNAs as regulators of gene expression. Trends Genet 28:616–623. https://doi.org/10.1016/j.tig.2012.09.001

Miyauchi S, Kiss E, Kuo A et al (2020) Large-scale genome sequencing of mycorrhizal fungi provides insights into the early evolution of symbiotic traits. Nat Commun 11:5125. https://doi.org/10.1038/s41467-020-18795-w

Molina R, Massicotte HB, Trappe JM (1992) Specificity phenomena in mycorrhizal symbiosis: community-ecological consequences and practical implications. In: Allen MF (ed) Mycorrhizal functioning. Chapman & Hall, London, UK, pp 357–423

Morin E, Miyauchi S, San Clemente H et al (2019) Comparative genomics of *Rhizophagus irregularis, R. cerebriforme, R. diaphanus* and *Gigaspora rosea* highlights specific genetic features in Glomeromycotina. New Phytol 222:1584–1598. https://doi.org/10.1111/nph.15687

Murat C, Payen T, Noel B et al (2018) Pezizomycetes genomes reveal the molecular basis of ectomycorrhizal truffle lifestyle. Nat Ecol Evol 2:1956–1965. https://doi.org/10.1038/s41559-018-0710-4

Navaud O, Barbacci A, Taylor A et al (2018) Shifts in diversification rates and host jump frequencies shaped the diversity of host range among Sclerotiniaceae fungal plant pathogens. Mol Ecol 27:1309–1323. https://doi.org/10.1111/mec.14523

Oksanen J, Guillaume B, Friendly M, et al (2019) Vegan: community ecology package

Pellegrin C, Daguerre Y, Ruytinx J et al (2019) *Laccaria bicolor* MiSSP8 is a small-secreted protein decisive for the establishment of the ectomycorrhizal symbiosis. Environ Microbiol 21:3765–3779. https://doi.org/10.1111/1462-2920.14727

Pellegrin C, Morin E, Martin FM, Veneault-Fourrey C (2015) Comparative analysis of secretomes from ectomycorrhizal fungi with an emphasis on small-secreted proteins. Front Microbiol 6:1–15. https://doi.org/10.3389/fmicb.2015.01278

Platt RN II, Blanco-Berdugo L, Ray DA (2016) Accurate transposable element annotation is vital when analyzing new genome assemblies. Genome Biol Evol 8:403–410. https://doi.org/10.1093/gbe/evw009

Plett JM, Kemppainen M, Kale SD et al (2011) A secreted effector protein of *Laccaria bicolor* is required for symbiosis development. Curr Biol 21:1197–1203. https://doi.org/10.1016/j.cub.2011.05.033

Plett JM, Miyauchi S, Morin E et al (2023) Speciation underpinned by unexpected molecular diversity in the mycorrhizal fungal genus *Pisolithus*. Mol Biol Evol 40:1–18. https://doi.org/10.1093/molbev/msad045

Plett JM, Plett KL, Wong-Bajracharya J et al (2020) Mycorrhizal effector PaMiSSP10b alters polyamine biosynthesis in *Eucalyptus* root cells and promotes root colonization. New Phytol 228:712–727. https://doi.org/10.1111/nph.16759

Plett JM, Tisserant E, Brun A et al (2015) The mutualist *Laccaria bicolor* expresses a core gene regulon during the colonization of diverse host plants and a variable regulon to counteract host-specific defenses. Mol Plant-Microbe Interact 28:261–273. https://doi.org/10.1094/MPMI-05-14-0129-FI

Podder S, Saha D, Ghosh TC (2019) Deciphering the intrinsic properties of fungal proteases in optimizing phytopathogenic interaction. Gene 711:143934. https://doi.org/10.1016/j.gene.2019.06.024

Razali NM, Cheah BH, Nadarajah K (2019) Transposable elements adaptive role in genome plasticity, pathogenicity and evolution in fungal phytopathogens. Int J Mol Sci 20. https://doi.org/10.3390/ijms20143597

Read DJ, Perez-Moreno J (2003) Mycorrhizas and nutrient cycling in ecosystems–a journey towards relevance? New Phytol 157:475–492. https://doi.org/10.1046/j.1469-8137.2003.00704.x

Ruiz-Dueñas FJ, Barrasa JM, Sánchez-García M et al (2021) Genomic analysis enlightens Agaricales lifestyle evolution and increasing peroxidase diversity. Mol Biol Evol 38:1428–1446. https://doi.org/10.1093/molbev/msaa301

Signorell A, Aho K, Alfons A, et al (2020) DescTools: tools for descriptive statistics. R Package version 09938

Smit A, Hubley R RepeatModeler open-1.0. 2008-2015. http://www.repeatmasker.org

Smit A, Hubley R, Green P RepeatMasker open-4.0. 2013-2015. http://www.repeatmasker.org

Smith SE, Read D (2008) Mycorrhizal symbiosis, 3rd. Elsevier

Steidinger BS, Crowther TW, Liang J et al (2019) Climatic controls of decomposition drive the global biogeography of forest-tree symbioses. Nature 569:404–408. https://doi.org/10.1038/s41586-019-1128-0

Strullu-Derrien C, Selosse M-A, Kenrick P, Martin FM (2018) The origin and evolution of mycorrhizal symbioses: from palaeomycology to phylogenomics. New Phytol 220:1012–1030. https://doi.org/10.1111/nph.15076

Subramoni S, Suárez-Moreno ZR, Venturi V (2010) Lipases as pathogenicity factors of plant pathogens. In: Timmis KN (ed) Handbook of hydrocarbon and lipid microbiology. Springer, Berlin Heidelberg, Berlin, Heidelberg, pp 3269–3277

Tang N, Lebreton A, Xu W et al (2021) Transcriptome profiling reveals differential gene expression of secreted proteases and highly specific gene repertoires involved in *Lactarius–Pinus* Symbioses. Front Plant Sci 12

Tedersoo L, Bahram M (2019) Mycorrhizal types differ in ecophysiology and alter plant nutrition and soil processes. Biol Rev 94:1857–1880. https://doi.org/10.1111/brv.12538

Tedersoo L, Bahram M, Toots M et al (2012) Towards global patterns in the diversity and community structure of ectomycorrhizal fungi. Mol Ecol 21:4160–4170. https://doi.org/10.1111/j.1365-294X.2012.05602.x

Tedersoo L, Smith ME (2017) Ectomycorrhizal fungal lineages: detection of four new groups and notes on consistent recognition of ectomycorrhizal taxa in high-throughput sequencing studies. In: Tedersoo L (ed) Biogeography of mycorrhizal Symbiosis, 1st edn. Springer International Publishing, pp 125–142

Tian C, Beeson WT, Iavarone AT et al (2009) Systems analysis of plant cell wall degradation by the model filamentous fungus *Neurospora crassa*. Proc Natl Acad Sci USA 106:22157–22162. https://doi.org/10.1073/pnas.0906810106

Torres DE, Thomma BPHJ, Seidl MF (2021) Transposable elements contribute to genome dynamics and gene expression variation in the fungal plant pathogen *Verticillium dahliae*. Genome Biol Evol 13:evab135. https://doi.org/10.1093/gbe/evab135

Veneault-Fourrey C, Martin F (2011) Mutualistic interactions on a knife-edge between saprotrophy and pathogenesis. Curr Opin Plant Biol 14:444–450. https://doi.org/10.1016/j.pbi.2011.03.022

Vik U, Logares R, Blaalid R et al (2013) Different bacterial communities in ectomycorrhizae and surrounding soil. Sci Rep 3. https://doi.org/10.1038/srep03471

Wu G, Miyauchi S, Morin E et al (2022) Evolutionary innovations through gain and loss of genes in the ectomycorrhizal Boletales. New Phytol 233:1383–1400. https://doi.org/10.1111/nph.17858

www.ingramcontent.com/pod-product-compliance
Lightning Source LLC
Chambersburg PA
CBHW081926120926
51338CB00005B/27
*9783031756658*

METHODS IN MOLECULAR BIOLOGY

*Series Editor*
**John M. Walker**
**School of Life and Medical Sciences,**
**University of Hertfordshire, Hatfield,**
**Hertfordshire, UK**

For further volumes:
http://www.springer.com/series/7651

For over 35 years, biological scientists have come to rely on the research protocols and methodologies in the critically acclaimed *Methods in Molecular Biology series.* The series was the first to introduce the step-by-step protocols approach that has become the standard in all biomedical protocol publishing. Each protocol is provided in readily-reproducible step-by step fashion, opening with an introductory overview, a list of the materials and reagents needed to complete the experiment, and followed by a detailed procedure that is supported with a helpful notes section offering tips and tricks of the trade as well as troubleshooting advice. These hallmark features were introduced by series editor Dr. John Walker and constitute the key ingredient in each and every volume of the *Methods in Molecular Biology series.* Tested and trusted, comprehensive and reliable, all protocols from the series are indexed in PubMed.

# Selective Autophagy

## Methods and Protocols

Edited by

**Ioannis P. Nezis**

*School of Life Sciences, University of Warwick, Coventry, UK*

Humana Press

*Editor*
Ioannis P. Nezis
School of Life Sciences
University of Warwick
Coventry, UK

ISSN 1064-3745 ISSN 1940-6029 (electronic)
Methods in Molecular Biology
ISBN 978-1-0716-4066-1 ISBN 978-1-0716-4067-8 (eBook)
https://doi.org/10.1007/978-1-0716-4067-8



This Humana imprint is published by the registered company Springer Science+Business Media, LLC, part of Springer Nature.
The registered company address is: 1 New York Plaza, New York, NY 10004, U.S.A.

# Preface

Autophagy is an evolutionarily conserved process where the cells degrade their own cellular material. It is involved in protein and organelle degradation and plays an essential role in both cellular and whole-animal homeostasis. There are various types of autophagy such as macroautophagy, microautophagy and chaperone-mediated autophagy. During macroautophagy, there is sequestration of cellular material into double-membrane vesicles called autophagosomes. The autophagosomes are subsequently fused with the lysosomes where the sequestered material is degraded by lysosomal hydrolases. The products of degradation are transported back into the cytoplasm and can be reused by the cell. Although it was initially believed that autophagy occurs randomly inside the cell, it is now established that sequestration and degradation of cytoplasmic material by autophagy can be selective. There are several types of selective autophagy described, such as mitophagy, lysophagy, aggrephagy, xenophagy, ER-phagy, ribophagy, ferritinophagy, pexophagy, Golgiphagy and nucleophagy. Comprehension of molecular and cellular mechanisms that govern the different types of selective autophagy is very critical for understanding their role in cellular and organismal physiology. To achieve this, scientists have developed a variety of methods and techniques to study selective autophagy.

This book contains a collection of 19 chapters presenting methods and techniques to study selective autophagy in a broad range of organisms, from yeast to mammals. Specifically, the chapters include methods to study selective autophagy in yeast, plants, worms, fruit flies and mammals. The authors of the chapters are well established and distinguished authorities in their respective areas of research.

It was a great pleasure to edit this volume and communicate with all the contributors and the editor-in-chief Prof. John Walker. I would like to thank all the contributors for their excellent work.

*Coventry, UK* *Ioannis P. Nezis*

# Contents

*Preface* . . . . . *v*
*Contributors* . . . . . *ix*

1 Fluorescence Microscopy and Immunoblotting for Mitophagy in Budding Yeast . . . . . 1
***Yuki Nakayama and Koji Okamoto***

2 Methods to Monitor Nucleophagy in Yeast . . . . . 15
***Ziyang Li and Hitoshi Nakatogawa***

3 Synthetic Dimerization Approaches for In Vivo Studies in *Saccharomyces cerevisiae* . . . . . 27
***Jana Elsässer and Claudine Kraft***

4 Assays to Study Mitophagy in Plants . . . . . 39
***Chengyang Li, Juncai Ma, Tian Zhen, Byung-Ho Kang, and Pengwei Wang***

5 Assessing Neuronal Mitophagy During Development in the Nematode *Caenorhabditis elegans* . . . . . 55
***Christina Ploumi, Antonis Roussos, and Konstantinos Palikaras***

6 In Vivo Monitoring of Nucleophagy in *Caenorhabditis elegans* . . . . . 67
***Georgios Konstantinidis and Nektarios Tavernarakis***

7 Analysis of Mitophagy Reporters in *Drosophila* . . . . . 79
***Alvaro Sanchez-Martinez, Aitor Martinez, and Alexander J. Whitworth***

8 Studying Selective Autophagy of Protein Aggregates Using Particles Induced by Multimerization (PIMs) . . . . . 95
***Giel Korsten and Lukas C. Kapitein***

9 Methods for Monitoring ER-Phagy . . . . . 109
***Marisa Di Monaco, Marian Peteri, and Jin Rui Liang***

10 Quantitative Super-Resolution Imaging of ER-Phagy Initiation in Cells . . . . . 127
***Ashwin Balakrishnan, Marius Glogger, and Mike Heilemann***

11 Fluorescence Methods to Measure Pexophagy . . . . . 141
***Francesco G. Barone, Sylvie Urbé, and Michael J. Clague***

12 Methods for Monitoring Mitophagy Using mt-Keima . . . . . 151
***Alexandra J. Gilsrud and Derek P. Narendra***

13 Monitoring Mitophagy Dynamics in Live Cells . . . . . 161
***Christina Karantanou and Sofia-Iris Bibli***

14 Methods for Monitoring NCOA4-Mediated Ferritinophagy . . . . . 177
***Thuy T. P. Nguyen and Joseph D. Mancias***

15 Method to Purify Phase-Separated p62 Bodies Using Fluorescence-Activated Particle Sorting . . . . . 191
***Shuhei Takada, Masaaki Komatsu, and Hideaki Morishita***

16 In Silico Assessment of LIR Motif-Containing Proteins Binding to ATG8-Family Proteins ........ 197
***Mamas Louca, Bridget Penman, Alexander Cameron, and Ioannis P. Nezis***

17 High-Throughput Screening for LC3/GABARAP Binders Utilizing the Fluorescence Polarization Assay ........ 203
***Martin P. Schwalm, Johannes Dopfer, Stefan Knapp, and Vladimir V. Rogov***

18 Thermodynamic Characterization of LC3/GABARAP:Ligand Interactions by Isothermal Titration Calorimetry ........ 219
***Johannes Dopfer, Martin P. Schwalm, Stefan Knapp, and Vladimir V. Rogov***

19 Determination of hATG8 Binding Selectivity of AIM (Autophagy-Interacting Motif) Peptides Using Native Electrospray Ionization Mass Spectrometry ........ 237
***Andrew Brennan, Neil Oldham, Robert Layfield, and Daniel Scott***

*Index* ........ *247*

## Contributors


ASHWIN BALAKRISHNAN • *Institute of Physical and Theoretical Chemistry, Goethe-University Frankfurt, Frankfurt, Germany*

FRANCESCO G. BARONE • *Biochemistry, Cell and Systems Biology, Institute of Systems, Molecular and Integrative Biology, University of Liverpool, Liverpool, UK; Biochemistry Section, Surgical Neurology Branch, National Institute of Neurological Disorders and Stroke, National Institutes of Health, Bethesda, MD, USA*

SOFIA-IRIS BIBLI • *Department of Vascular Dysfunction, European Center for Angioscience (ECAS), Medical Faculty Mannheim, Heidelberg University, Mannheim, Germany; Institute for Vascular Signalling, Centre for Molecular Medicine, Goethe University, Frankfurt am Main, Germany; German Centre for Cardiovascular Research (DZHK), Heidelberg-Mannheim, Germany*

ANDREW BRENNAN • *Department of Life Sciences, University of Bath, Bath, UK*

ALEXANDER CAMERON • *School of Life Sciences, University of Warwick, Coventry, UK*

MICHAEL J. CLAGUE • *Biochemistry, Cell and Systems Biology, Institute of Systems, Molecular and Integrative Biology, University of Liverpool, Liverpool, UK*

MARISA DI MONACO • *MRC Protein Phosphorylation & Ubiquitylation Unit, University of Dundee, Dundee, UK*

JOHANNES DOPFER • *Institute of Pharmaceutical Chemistry, Goethe University Frankfurt, Frankfurt, Germany; Structural Genomics Consortium, BMLS, Goethe University Frankfurt, Frankfurt, Germany*

JANA ELSÄSSER • *Institute of Biochemistry and Molecular Biology, ZBMZ, Faculty of Medicine, University of Freiburg, Freiburg, Germany; Faculty of Biology, University of Freiburg, Freiburg, Germany*

ALEXANDRA J. GILSRUD • *Inherited Movement Disorders Unit, Neurogenetics Branch, National Institute of Neurological Disorders and Stroke, Division of Intramural Research, National Institutes of Health, Bethesda, MD, USA*

MARIUS GLOGGER • *Optical Imaging Competence Centre, Friedrich-Alexander-Universität Erlangen-Nürnberg, Erlangen, Germany*

MIKE HEILEMANN • *Institute of Physical and Theoretical Chemistry, Goethe-University Frankfurt, Frankfurt, Germany*

BYUNG-HO KANG • *School of Life Sciences, Centre for Cell & Developmental Biology and State Key Laboratory of Agrobiotechnology, The Chinese University of Hong Kong, Hong Kong, China*

LUKAS C. KAPITEIN • *Cell Biology, Neurobiology and Biophysics, Department of Biology, Faculty of Science, Utrecht University, Utrecht, The Netherlands; Centre for Living Technologies, Alliance TU/e, WUR, UU, UMC Utrecht, Utrecht, The Netherlands*

CHRISTINA KARANTANOU • *Department of Vascular Dysfunction, European Center for Angioscience (ECAS), Medical Faculty Mannheim, Heidelberg University, Mannheim, Germany; Institute for Vascular Signalling, Centre for Molecular Medicine, Goethe University, Frankfurt am Main, Germany*

STEFAN KNAPP • *Institute of Pharmaceutical Chemistry, Goethe University Frankfurt, Frankfurt, Germany; Structural Genomics Consortium, BMLS, Goethe University Frankfurt, Frankfurt, Germany*

MASAAKI KOMATSU • *Department of Physiology, Juntendo University Graduate School of Medicine, Bunkyo-ku, Tokyo, Japan*

GEORGIOS KONSTANTINIDIS • *Institute of Molecular Biology and Biotechnology, Foundation for Research and Technology-Hellas, Heraklion, Crete, Greece*

GIEL KORSTEN • *Cell Biology, Neurobiology and Biophysics, Department of Biology, Faculty of Science, Utrecht University, Utrecht, The Netherlands*

CLAUDINE KRAFT • *Institute of Biochemistry and Molecular Biology, ZBMZ, Faculty of Medicine, University of Freiburg, Freiburg, Germany; CIBSS – Centre for Integrative Biological Signalling Studies, University of Freiburg, Freiburg, Germany*

ROBERT LAYFIELD • *School of Life Sciences, University of Nottingham, Nottingham, UK*

CHENGYANG LI • *National Key Laboratory for Germplasm Innovation & Utilization of Horticultural Crops, College of Horticulture and Forestry Sciences, Huazhong Agricultural University, Wuhan, China; Hubei Hongshan Laboratory, Wuhan, China*

ZIYANG LI • *Cell Biology Center, Institute of Innovative Research, Tokyo Institute of Technology, Yokohama, Japan; School of Life Science and Technology, Tokyo Institute of Technology, Yokohama, Japan*

JIN RUI LIANG • *MRC Protein Phosphorylation & Ubiquitylation Unit, University of Dundee, Dundee, UK*

MAMAS LOUCA • *School of Life Sciences, University of Warwick, Coventry, UK*

JUNCAI MA • *School of Life Sciences, Centre for Cell & Developmental Biology and State Key Laboratory of Agrobiotechnology, The Chinese University of Hong Kong, Hong Kong, China*

JOSEPH D. MANCIAS • *Division of Radiation and Genome Stability, Department of Radiation Oncology, Dana-Farber Cancer Institute, Harvard Medical School, Boston, MA, USA; Department of Radiation Oncology, Brigham and Women's Hospital, Harvard Medical School, Boston, MA, USA*

AITOR MARTINEZ • *MRC Mitochondrial Biology Unit, University of Cambridge, Cambridge, UK*

HIDEAKI MORISHITA • *Department of Physiology, Juntendo University Graduate School of Medicine, Bunkyo-ku, Tokyo, Japan; Department of Molecular Cell Biology, Graduate School of Medical Sciences, Kyushu University, Higashi-ku, Fukuoka, Japan*

HITOSHI NAKATOGAWA • *Cell Biology Center, Institute of Innovative Research, Tokyo Institute of Technology, Yokohama, Japan; School of Life Science and Technology, Tokyo Institute of Technology, Yokohama, Japan*

YUKI NAKAYAMA • *Graduate School of Frontier Biosciences, Osaka University, Osaka, Japan*

DEREK P. NARENDRA • *Inherited Movement Disorders Unit, Neurogenetics Branch, National Institute of Neurological Disorders and Stroke, Division of Intramural Research, National Institutes of Health, Bethesda, MD, USA*

IOANNIS P. NEZIS • *School of Life Sciences, University of Warwick, Coventry, UK*

THUY T. P. NGUYEN • *Division of Radiation and Genome Stability, Department of Radiation Oncology, Dana-Farber Cancer Institute, Harvard Medical School, Boston, MA, USA*

KOJI OKAMOTO • *Graduate School of Frontier Biosciences, Osaka University, Osaka, Japan*

NEIL OLDHAM • *School of Chemistry, University of Nottingham, Nottingham, UK*

KONSTANTINOS PALIKARAS • *Department of Physiology, Medical School, National and Kapodistrian University of Athens, Athens, Greece*

BRIDGET PENMAN • *School of Life Sciences, University of Warwick, Coventry, UK; Zeeman Institute, University of Warwick, Coventry, UK*

Marian Peteri • *MRC Protein Phosphorylation & Ubiquitylation Unit, University of Dundee, Dundee, UK*
Christina Ploumi • *Department of Physiology, Medical School, National and Kapodistrian University of Athens, Athens, Greece*
Vladimir V. Rogov • *Institute of Pharmaceutical Chemistry, Goethe University Frankfurt, Frankfurt, Germany; Structural Genomics Consortium, BMLS, Goethe University Frankfurt, Frankfurt, Germany*
Antonis Roussos • *Department of Physiology, Medical School, National and Kapodistrian University of Athens, Athens, Greece*
Alvaro Sanchez-Martinez • *MRC Mitochondrial Biology Unit, University of Cambridge, Cambridge, UK*
Martin P. Schwalm • *Institute of Pharmaceutical Chemistry, Goethe University Frankfurt, Frankfurt, Germany; Structural Genomics Consortium, BMLS, Goethe University Frankfurt, Frankfurt, Germany*
Daniel Scott • *School of Life Sciences, University of Nottingham, Nottingham, UK*
Shuhei Takada • *Department of Physiology, Juntendo University Graduate School of Medicine, Bunkyo-ku, Tokyo, Japan*
Nektarios Tavernarakis • *Institute of Molecular Biology and Biotechnology, Foundation for Research and Technology-Hellas, Heraklion, Crete, Greece; Department of Basic Sciences, Faculty of Medicine, University of Crete, Heraklion, Crete, Greece*
Sylvie Urbé • *Biochemistry, Cell and Systems Biology, Institute of Systems, Molecular and Integrative Biology, University of Liverpool, Liverpool, UK*
Pengwei Wang • *National Key Laboratory for Germplasm Innovation & Utilization of Horticultural Crops, College of Horticulture and Forestry Sciences, Huazhong Agricultural University, Wuhan, China; Hubei Hongshan Laboratory, Wuhan, China*
Alexander J. Whitworth • *MRC Mitochondrial Biology Unit, University of Cambridge, Cambridge, UK*
Tian Zhen • *National Key Laboratory for Germplasm Innovation & Utilization of Horticultural Crops, College of Horticulture and Forestry Sciences, Huazhong Agricultural University, Wuhan, China; Hubei Hongshan Laboratory, Wuhan, China*

# Chapter 1

# Fluorescence Microscopy and Immunoblotting for Mitophagy in Budding Yeast

**Yuki Nakayama and Koji Okamoto**

## Abstract

Selective removal of excess or damaged mitochondria is an evolutionarily conserved process that contributes to mitochondrial quality and quantity control. This catabolic event relies on autophagy, a membrane trafficking system that sequesters cytoplasmic constituents into double membrane-bound autophagosomes and delivers them to lysosomes (vacuoles in yeast) for hydrolytic degradation and is thus termed mitophagy. Dysregulation of mitophagy is associated with various diseases, highlighting its physiological relevance. In budding yeast, the pro-mitophagic single-pass membrane protein Atg32 is upregulated under prolonged respiration or nutrient starvation, anchored on the surface of mitochondria, and activated to recruit the autophagy machinery for the formation of autophagosomes surrounding mitochondria. In this chapter, we provide protocols to assess Atg32-mediated mitophagy using fluorescence microscopy and immunoblotting.

**Key words** Mitochondria, Atg32, Budding yeast, Fluorescence microscopy, Immunoblotting

## 1 Introduction

Mitochondria are smart energy converters that produce most of the cellular ATP through oxidative phosphorylation, dynamically acting as the metabolic center in the cell [1]. During energy conversion, they also generate reactive oxygen species whose accumulation causes oxidative stress, ultimately leading to mitochondrial dysfunction [2]. To maintain mitochondrial quality for proper cytoplasmic ATP levels, damaged mitochondria must be eliminated from the healthy population. Cells must also constantly adjust mitochondrial quantity to changes in cellular energy demand and respiratory conditions [3]. When cellular ATP consumption decreases, surplus mitochondria are subjected to be excluded from the cytoplasm, thereby maintaining a balance between energy demand and supply [4]. One of the cellular pathways that contribute to mitochondrial quality and quantity control is mitophagy, the

Ioannis P. Nezis (ed.), *Selective Autophagy: Methods and Protocols*, Methods in Molecular Biology, vol. 2845,
https://doi.org/10.1007/978-1-0716-4067-8_1,

selective degradation of mitochondria as a whole organelle via autophagy [5]. The basic principles of mitophagy are highly conserved from yeast to humans, and defects in mitophagy lead to mitochondrial impairment and subsequent pathogenesis in diverse tissues and organs, implying mitophagy as a physiologically relevant process in almost all eukaryotes [6, 7].

In the budding yeast *Saccharomyces cerevisiae*, mitophagy is strongly induced under prolonged nonfermentable growth or nitrogen starvation in a manner dependent on Autophagy-related 32 (Atg32), a transmembrane protein of 59 kDa that acts as a receptor for the autophagy machinery [8, 9]. Cells under these conditions exhibit a fraction of mitochondria transported to the lumen of vacuoles, lytic organelles containing a variety of hydrolytic enzymes for degradation like lysosomes in mammals [8, 9]. In contrast to most proteins rapidly degraded by vacuolar hydrolases, β-barrel-shaped fluorescent proteins such as GFP and mCherry are quite resistant to vacuolar degradation and even stay fluorescent and accumulate in the vacuole [8, 9]. Based on these characteristics, we have established mito-DHFR-mCherry, a mitophagy probe that is localized to the mitochondrial matrix [10]. Respiration- and starvation-induced mitophagic processes share common mechanisms, but they differ in the list of core Atg proteins required for autophagosome formation [11]. It should also be noted that the latter event can occur in respiratory-deficient cells or in the presence of respiratory inhibitors [11], which allows us to investigate most of the nonessential deletion mutants.

This chapter describes two conventional methods to analyze Atg32-mediated mitophagy in *S. cerevisiae* under prolonged nonfermentable growth or nitrogen starvation. Using mito-DHFR-mCherry, fluorescence microscopy can visualize mitochondria that are transported to the vacuole, and immunoblotting can semi-quantify the amount of free mCherry, indicating degradation of mitochondria in the vacuole.

## 2 Materials

### *2.1 Generation of DNA Cassette for Yeast Transformation*

1. Plasmid: pBSII-*TEF^p-mito-DHFR-mCherry-CgHIS3* (*see* **Note 1**). This plasmid can be requested and sent from our laboratory.
2. PCR primers for amplification of the TEF^p-mito-DHFR-mCherry-CgHIS3 cassette that can be integrated into the *his3* locus: forward primer, 5′- AATGTGATTTCTTCGAA GAATATACTAAAAAATGAG
 CAGGCAAGATAAACGAAGGCAAAGCGACGGTATC-GATAAGCTTG-3′; reverse primer, 5′-GGTATACA

TATATACACATGTATATATATCGTATGCTG-CAGCTTTAAATAATCGGTGTCACAAGCGCGCAATTAACCCTC-3′. These long primers should be purified by an oligonucleotide purification cartridge. The concentration of these primers is 0.1 mM.

### 2.2 Construction of Yeast Strains for Mitophagy Assays

1. Yeast strain: *S. cerevisiae* BY4741, *MATa his3Δ1 leu2Δ0 met15Δ ura3Δ* (*see* **Note 2**).
2. Rich liquid medium supplemented with dextrose (YPD): 1% (w/v) yeast extract, 2% (w/v) peptone, 2% (w/v) dextrose. YP medium (yeast extract, peptone) and dextrose are prepared separately in water, sterilized by autoclaving, and mixed together. Store at room temperature.
3. Synthetic solid medium supplemented with dextrose and histidine-dropout (SD-His): 0.17% (w/v) yeast nitrogen base without amino acids and ammonium sulfate, 0.5% (w/v) ammonium sulfate, 2% (w/v) dextrose, amino acid/nucleobase solutions without histidine, 2% (w/v) agar. S medium (yeast nitrogen base, ammonium sulfate), agar, and dextrose are prepared separately in water, sterilized by autoclaving, and mixed together. Supplement them with filter-sterilized amino acid/nucleobase solutions without histidine. Store at 4 °C.
4. 1 M lithium acetate (LiAc). Filter-sterilize and store at room temperature.
5. 0.1 M LiAc. Filter-sterilize and store at room temperature.
6. 60% (w/v) polyethylene glycol (PEG) (average molecular weight: 3350). Filter-sterilize and store at room temperature.
7. 10 mg/mL single-stranded salmon sperm DNA (ssDNA). Boil for 5 min and store at 4 °C.
8. TEF$^{P}$-mito-DHFR-mCherry-CgHIS3 DNA cassette: *see* Subheading 3.1. Store at −20 °C.
9. Sterile water.

### 2.3 Fluorescence Microscopy for Live Cell Imaging

1. BY4741 *his3Δ1::TEF$^{P}$-mito-DHFR-mCherry::CgHIS3* (positive control strain) [11].
2. BY4741 *his3Δ1::TEF$^{P}$-mito-DHFR-mCherry::CgHIS3 atg32Δ* (negative control strain) [11].
3. BY4741 *his3Δ1::TEF$^{P}$-mito-DHFR-mCherry::CgHIS3 atg15Δ* (negative control strain) [12].
4. Synthetic liquid medium supplemented with dextrose and casamino acids (SDCA): 0.17% (w/v) yeast nitrogen base without amino acids and ammonium sulfate, 0.5% (w/v) ammonium sulfate, 0.5% (w/v) casamino acids, 2% (w/v) dextrose, 20 μg/mL adenine sulfate, 20 μg/mL L-tryptophan, 20 μg/mL

uracil. SCA medium (yeast nitrogen base, ammonium sulfate, casamino acids) and dextrose are prepared separately in water, sterilized by autoclaving, and mixed together. Supplement them with filter-sterilized amino acid/nucleobase solutions and store at room temperature.

5. Synthetic liquid medium supplemented with dextrose, glycerol, and casamino acids (SDGlyCA): 0.17% (w/v) yeast nitrogen base without amino acids and ammonium sulfate, 0.5% (w/v) ammonium sulfate, 0.5% (w/v) casamino acids, 0.1% (w/v) dextrose, 3% (v/v) glycerol, 20 μg/mL adenine sulfate, 20 μg/mL L-tryptophan, 20 μg/mL uracil. SCA medium (yeast nitrogen base, ammonium sulfate, casamino acids), dextrose, and glycerol are prepared separately in water, sterilized by autoclaving, and mixed together. Supplement them with filter-sterilized amino acid/nucleobase solutions and store at room temperature.
6. YPD: *see* Subheading 2.2, **item 2**.
7. Nitrogen-depleted synthetic liquid medium supplemented with dextrose (SD-N): 0.17% (w/v) yeast nitrogen base without amino acids and ammonium sulfate, 2% (w/v) dextrose. Yeast nitrogen base and dextrose are prepared separately in water, sterilized by autoclaving, and mixed together. Store at room temperature.

### 2.4 Immunoblotting for Quantification of Mitophagy

1. BY4741 *his3Δ1::TEF$^{P}$-mito-DHFR-mCherry::CgHIS3*: *see* Subheading 2.3, **item 1**.
2. BY4741 *his3Δ1::TEF$^{P}$-mito-DHFR-mCherry::CgHIS3 atg32Δ*: *see* Subheading 2.3, **item 2**.
3. BY4741 *his3Δ1::TEF$^{P}$-mito-DHFR-mCherry::CgHIS3 atg15Δ*: *see* Subheading 2.3, **item 3**.
4. SDCA: *see* Subheading 2.3, **item 4**.
5. SDGlyCA: *see* Subheading 2.3, **item 5**.
6. YPD: *see* Subheading 2.2, **item 2**.
7. SD-N: *see* Subheading 2.3, **item 7**.
8. 0.1 M NaOH. Store at room temperature.
9. Sample buffer: 60 mM Tris–HCl, pH 6.8, 5% (v/v) glycerol, 2% (w/v) SDS, 20 mM DTT, 0.00025% (w/v) bromophenol blue. Store at 4 °C.
10. Polyacrylamide gel containing 10% (w/v) acrylamide/bisacrylamide (ratio 37.5:1).
11. Polyvinylidene fluoride (PVDF) membrane (pore size: 0.45 μm): Immobilon-P Membrane (IPVH00010, EMD Millipore). Activate in methanol before use.

12. Blocking buffer: TBS-T (*see* Subheading 2.4, **item 15**), 5% (w/v) skim milk.
13. Incubation buffer A for primary antibody: TBS-T, 0.1% (w/v) skim milk.
14. Incubation buffer B for secondary antibody: TBS-T, 1% (w/v) skim milk.
15. Washing buffer (Tris Buffered Saline with Tween 20, TBS-T): 25 mM Tris–HCl, pH 7.4, 138 mM NaCl, 2.7 mM KCl, 0.1% Tween-20.
16. Primary antibody: mouse monoclonal anti-mCherry antibody ([1C51] ab125096, Abcam), mouse monoclonal anti-Por1 antibody ([16G9E6BC4] ab110326, Abcam), and mouse monoclonal anti-Pgk1 antibody ([22C5D8] ab113687, Abcam). Store at −20 °C.
17. Secondary antibody: horseradish peroxidase (HRP)-conjugated rabbit polyclonal anti-mouse IgG (H + L) (315–035-003, Jackson ImmunoResearch). Store at −20 °C.
18. Enhanced chemiluminescence (ECL) detection reagents: Western Lightning Plus-ECL (NEL103001EA, PerkinElmer). Store at 4 °C.

## 3 Methods

Perform all procedures at room temperature unless otherwise indicated.

### 3.1 Generation of DNA Kock-in Cassette for Yeast Transformation

The TEF$^{p}$-mito-DHFR-mCherry-CgHIS3 DNA cassette (3.2 kb) can be amplified by a PCR reaction (*see* a standard reaction below) with the template plasmid and primers (*see* Subheading 2.1, **items 1 and 2**). The resulting DNA solution is typically at a concentration of 0.5 mg/mL.

1. Prepare a PCR reaction mixture containing PCR buffer for KOD-Plus-Neo (KOD-401, TOYOBO), 0.2 mM dNTPs, 1.5 mM $MgSO_4$, 0.3 μM forward/reverse primers, 200 ng/mL template plasmid, and 20 U/mL KOD-Plus-Neo. A typical reaction volume is 0.2 mL (four aliquots of 50 μL in 0.2-mL PCR tubes).
2. Perform the reaction using a thermal cycler with the following program: 1 cycle of (1) 94 °C, 2 min; 30 cycles of (2) 98 °C, 10 s, (3) 55 °C, 30 s, (4) 68 °C, 3 min; 1 cycle of (5) 68 °C, 10 min; followed by (6) 4 °C, ever.
3. Purify the reaction mixture using a spin column and elute DNA into sterile water.

4. Measure the DNA concentration using a spectrophotometer and confirm the lengths of DNA fragments by agarose gel electrophoresis.
5. Store the DNA solution at −80 °C before use.

### 3.2 Construction of Yeast Strains for Analyzing Mitophagy

The PCR-amplified TEF$^{P}$-mito-DHFR-mCherry-CgHIS3 DNA cassette, containing terminal 60 nucleotides derived from the upstream and downstream of the *HIS3* gene, can be introduced into yeast cells using a standard method. The introduced cassette is then inserted into the *HIS3* locus via homologous recombination, and the transformants can be selected based on histidine prototrophy.

1. Grow cells to mid-log phase ($OD_{600}$ = 0.8–1.2) in 3 mL YPD at 30 °C (*see* **Note 3**).
2. Transfer 0.5 $OD_{600}$ units of cells to a 1.5-mL microcentrifuge tube, harvest them by centrifugation (16,000× *g*, 10 s), and discard the supernatant using a micropipette.
3. Wash the cells once with 0.5 mL sterile water, harvest them by centrifugation (16,000× *g*, 10 s), and discard the supernatant using a micropipette.
4. Resuspend the cells in 0.5 mL 0.1 M LiAc and incubate at 30 ° C for 10 min.
5. Harvest the cells by centrifugation (16,000× *g*, 10 s) and thoroughly discard the supernatant using a micropipette.
6. Add 60.5 μL 60% (w/v) PEG, 10.2 μL 1 M LiAc, 1 μL 10 mg/mL ssDNA, 10 μL 0.4 mg/mL DNA cassette, and 10 μL sterile water to the cell pellet in the indicated order and vortex vigorously.
7. Incubate the cells at 30 °C for 30 min, followed by a heat shock at 42 °C for 20 min.
8. Harvest the cells by centrifugation (16,000× *g*, 10 s) and discard the supernatant using a micropipette.
9. Add 50 μL sterile water to the cell pellet and vortex vigorously.
10. Spread the cell suspension on an SD-His agar plate and incubate at 30 °C for 3–4 days.
11. Restreak the colonies on fresh SD-His plates and incubate at 30 °C for 1–2 days.
12. Grow the transformed cells to mid-log phase in 3 mL SDCA and confirm the expression of mito-DHFR-mCherry using fluorescence microscopy and immunoblotting (*see* **Note 4**).
13. Resuspend the cells in YPD containing 15% glycerol and store at −80 °C before use.

### 3.3 Fluorescence Microscopy for Mitophagy in Living Cells

Transport of mitochondria to the vacuole in cells expressing the mitophagy probe mito-DHFR-mCherry can be examined under a fluorescence microscope (*see* Figs. 1 and 2) [10, 11]. When cells under respiratory conditions are grown to the late-log phase (Gly 24 h), mito-DHFR-mCherry indicates mitochondrial short tubules and fragments that are hardly colocalized with the vacuolar marker Vph1-GFP. Strikingly, cells at the stationary phase (Gly 72 h) exhibit Atg32-dependent formation of spherical mCherry patterns typical of vacuolar localizations overlapped with Vph1-GFP (*see* **Note 5**). Under the same conditions, cells lacking Atg15, a vacuolar phospholipase [13, 14], accumulate small mCherry dots within the Vph1-GFP-positive vacuolar membranes, indicating that mitochondria-containing vesicles are indeed incorporated into the vacuolar lumen. Similarly, mitophagy is induced by prolonged nitrogen starvation in a manner dependent on Atg32. When cells undergo drastic bulk autophagy (-N 6 h), mitochondria keep forming elongated tubules that are merely colocalized with Vph1-GFP. Subsequently, cells under long-term nitrogen depletion display spherical mCherry patterns overlapped with Vph1-GFP in an Atg32- or Atg15-dependent fashion (-N 24 h).

#### 3.3.1 Mitophagy Under Long-Term Respiration

1. Take a small pellet of cells expressing mito-DHFR-mCherry from the agar plate and grow them to the stationary phase ($OD_{600}$ > 3) in 3 mL SDCA at 30 °C under shaking (180 rpm).
2. Inoculate the culture media containing 0.006–0.012 $OD_{600}$ units of cells to 3 mL SDCA and grow them to the mid-log phase ($OD_{600}$ = 0.8–1.2, typically for 16–18 h) at 30 °C under shaking (180 rpm).
3. Inoculate the culture media containing 0.07 $OD_{600}$ units of cells to 7 mL SDGlyCA in a 50-mL Erlenmeyer flask and incubate them at 30 °C under shaking (180 rpm).
4. Transfer the culture media at the 24, 48, and 72 h time points containing 0.5 $OD_{600}$ units of cells to a 1.5-mL microcentrifuge tube, harvest them by centrifugation (16,000× *g*, 10 s), and discard a portion of the supernatant using a micropipette to leave 5 μL of cell suspension (100 $OD_{600}$/mL).
5. Pipette 1.6–2.0 μL of the cell suspension onto a microscope slide and cover it with an 18-$mm^2$ square cover slip without sealing.
6. Observe the cells using a fluorescence microscope equipped with filter sets for mCherry and GFP.

#### 3.3.2 Mitophagy Under Prolonged Starvation

1. Take a small pellet of cells expressing mito-DHFR-mCherry from an agar plate and grow them to the stationary phase ($OD_{600}$ > 4) in 3 mL YPD at 30 °C under shaking (180 rpm).

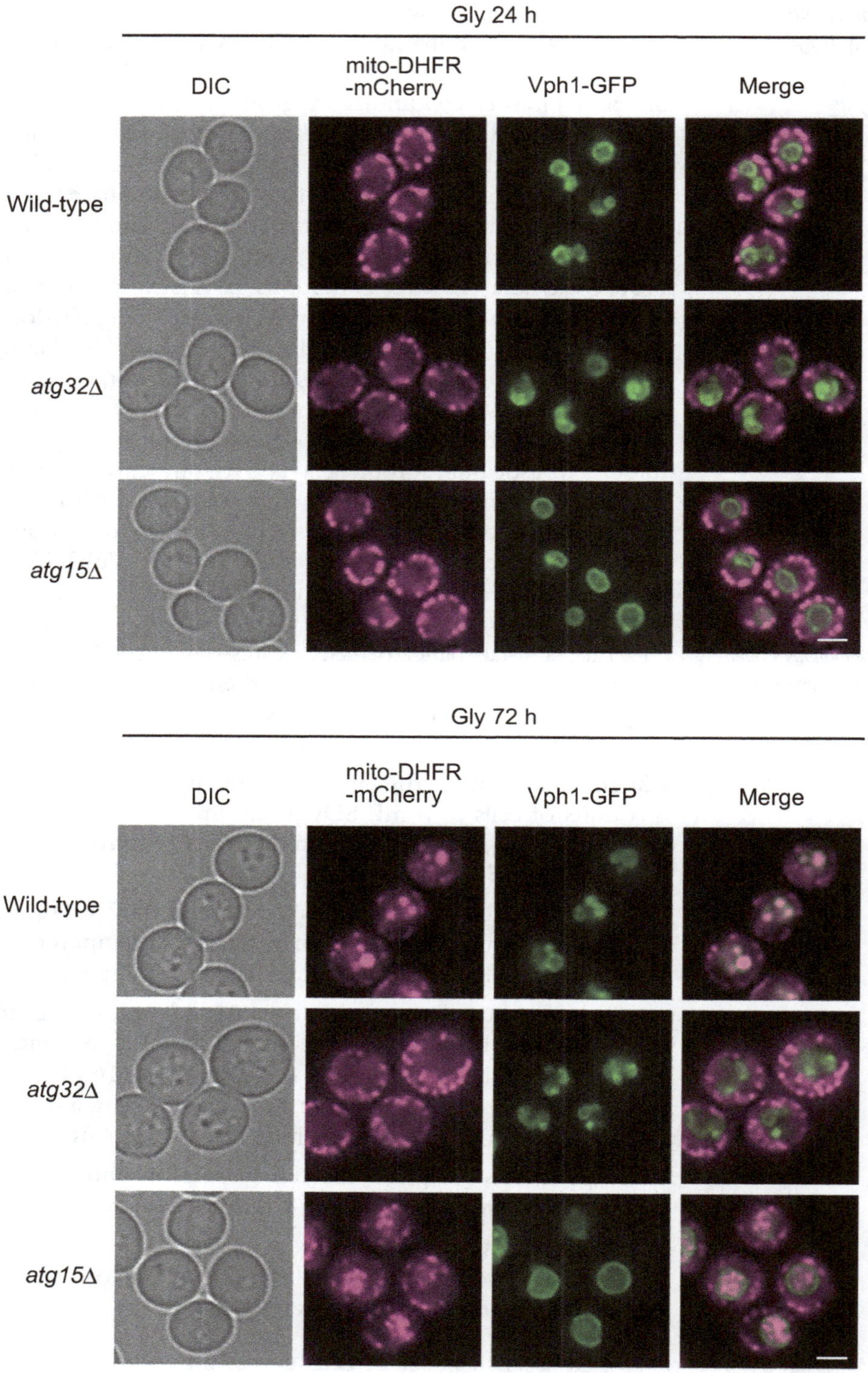


**Fig. 1** Fluorescence imaging for mitophagy under long-term respiration. Wild-type, *atg32*-null, and *atg15*-null cells expressing mito-DHFR-mCherry and Vph1-GFP were grown in SDGlyCA (Gly), collected at the 24 and 72 h time points, and observed using fluorescence microscopy. *DIC* differential interference contrast. Scale bar, 2 μm

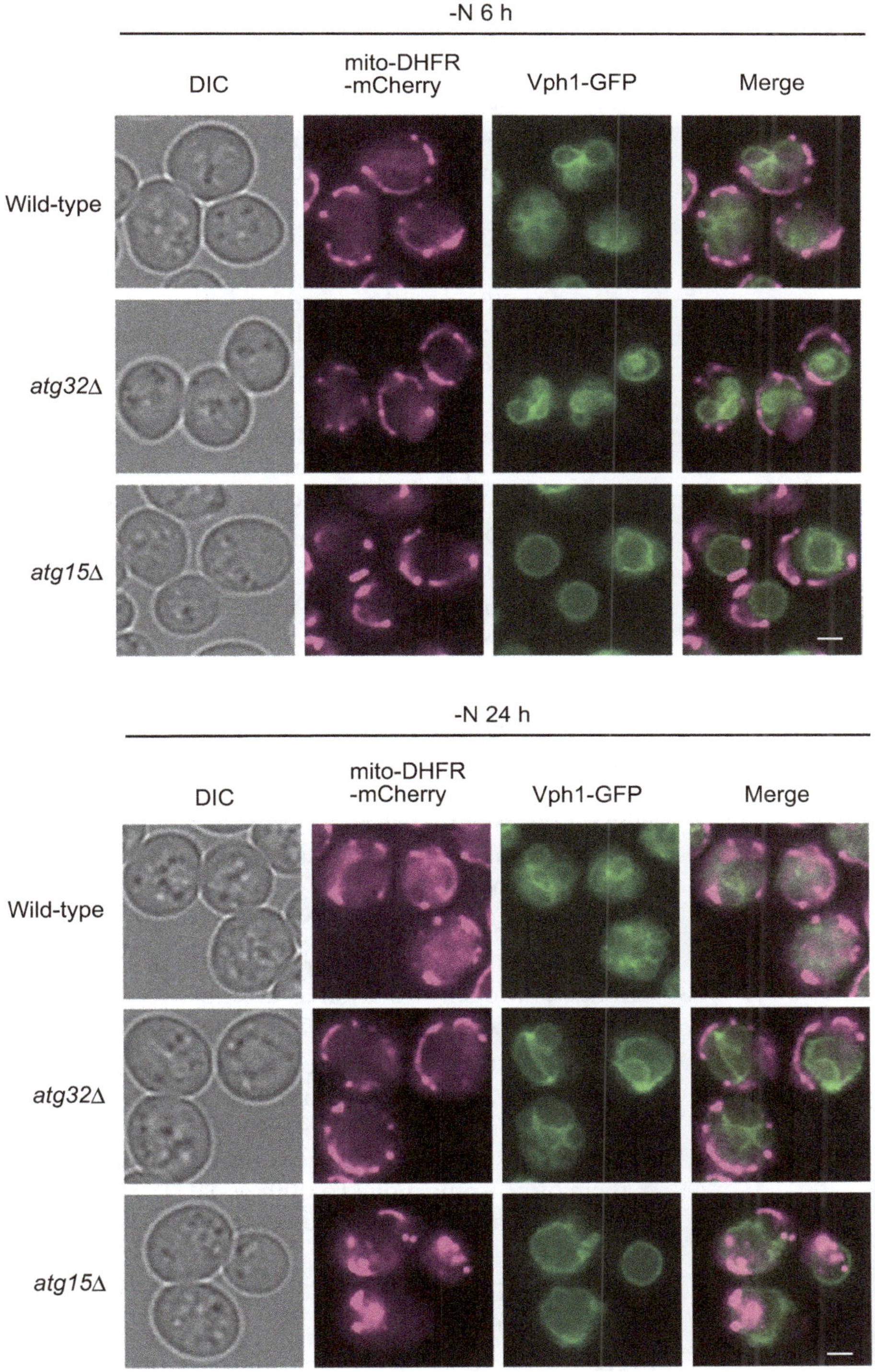


**Fig. 2** Fluorescence imaging for mitophagy under prolonged starvation. Wild-type, *atg32*-null, and *atg15*-null cells expressing mito-DHFR-mCherry and Vph1-GFP were grown in SD-N (-N), collected at the 6 and 24 h time points, and observed using fluorescence microscopy. *DIC* differential interference contrast. Scale bar, 2 μm

2. Inoculate the culture media containing 0.007–0.028 $OD_{600}$ units of cells to 7 mL YPD using a 50-mL Erlenmeyer flask and grow them to mid-log phase ($OD_{600} = 0.8$–1.2, typically for 12–15 h) at 30 °C under shaking (180 rpm) (*see* **Note 6**).
3. Transfer the culture media containing 7 $OD_{600}$ units of cells to a 15-mL conical centrifuge tube, harvest them by centrifugation (1600× *g*, 3 min), and completely discard the supernatant using a micropipette.
4. Resuspend the cells in 1 mL SD-N, transfer them to a 1.5-mL microcentrifuge tube, harvest them by centrifugation (16,000× *g*, 10 s), and discard the supernatant using a micropipette.
5. Wash the cells once more in 1 mL SD-N, harvest them by centrifugation (16,000× *g*, 10 s), and discard the supernatant using a micropipette (*see* **Note 7**).
6. Resuspend the cells in 1 mL SD-N, transfer the cell suspension to 6 mL SD-N (1 $OD_{600}$/mL) in a 50-mL Erlenmeyer flask, and incubate them at 30 °C under shaking (180 rpm).
7. Prepare a sample at the 6, 12, and 24 h time points for observation as described in Subheading 3.3.1, **steps 4–6**.

### *3.3.3 Immunoblotting for Quantification of Mitophagy*

To assess mitophagy in a more quantitative way, the mitophagy probe, mito-DHFR-mCherry, allows us to analyze the degradation of mitochondria by immunoblotting (*see* Figs. 3 and 4) [10, 11]. Upon mitophagy, this mitochondrial matrix-localized probe is processed to generate free mCherry, which is highly resistant to vacuolar proteases. Thus, Atg32- and Atg15-dependent accumulation of free mCherry semi-quantitatively indicates the relative level of mitophagy (*see* **Note 8**) [15, 16].

1. Prepare cell cultures under long-term respiration and prolonged nitrogen starvation as described in Subheading 3.3.1, **steps 1–3**, and Subheading 3.3.2, **steps 1–6**, respectively.
2. Transfer 1 $OD_{600}$ unit of cells to a 1.5-mL microcentrifuge tube at the 24, 48, and 72 h time points for mitophagy under long-term respiration, or 6, 12, and 24 h time points for mitophagy under prolonged nitrogen starvation, harvest them by centrifugation (16,000× *g*, 10 s), discard the supernatant using a micropipette, and store the cell pellet at −80 °C before use.
3. Add 0.1 mL 0.1 M NaOH to the cell pellet, vortex vigorously, and keep them for 5 min.
4. Harvest the cells by centrifugation (16,000× *g*, 1 min), and discard the supernatant using a micropipette.
5. Add 50 μL sample buffer to the cell pellet, and vortex vigorously.

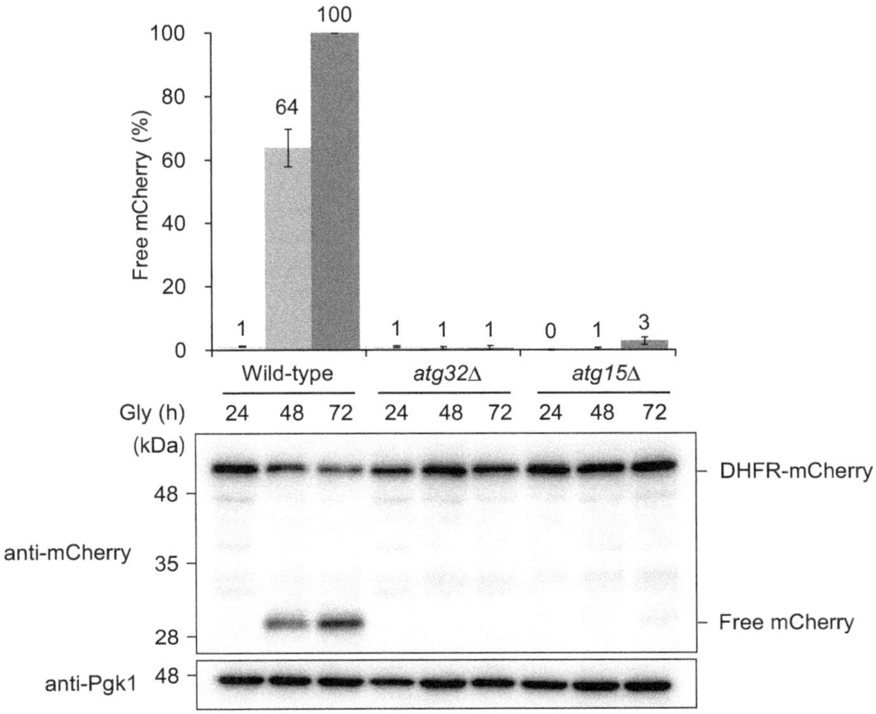


**Fig. 3** Immunoblot detection for mitophagy under long-term respiration. Wild-type, *atg32*-null, and *atg15*-null cells expressing mito-DHFR-mCherry and Vph1-GFP were grown in SDGlyCA (Gly), collected at the 24, 48, and 72 h time points, and subjected to SDS-PAGE and immunoblotting. The signal intensity value of free mCherry in wild-type cells at the 72-h time point was set to 100%. Data represent the averages of three experiments, with error bars indicating standard deviations. Pgk1 was monitored as a loading control

6. Boil the cell suspension for 3 min and keep them on ice for 1 min.
7. Separate the supernatant by centrifugation (16,000× *g*, 1 min) and transfer it to a 1.5 mL microcentrifuge tube.
8. Analyze 5–10 μL samples (corresponding to 0.1–0.2 $OD_{600}$ units of cells) by 10% SDS-PAGE and blot them to a PVDF membrane.
9. Block the membrane with 5% skim milk in TBS-T for 30 min.
10. Incubate the membrane with 0.1% skim milk in TBS-T containing the anti-mCherry (1:2000 dilution) or anti-Pgk1 (1: 10,000 dilution) antibody for 1 h (*see* **Note 9**).
11. Wash the membrane three times with TBS-T for 5 min each.
12. Incubate the membrane with 1% skim milk in TBS-T containing HRP-conjugated anti-mouse IgG (1:10,000 dilution) for 1 h.
13. Wash the membrane three times with TBS-T for 5 min each.

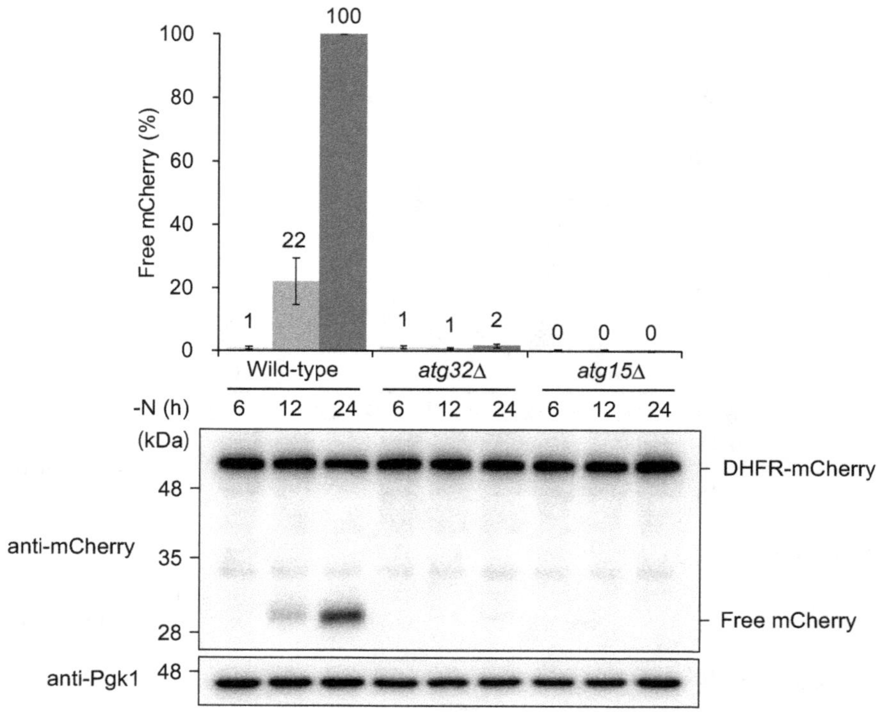


**Fig. 4** Immunoblot detection for mitophagy under prolonged starvation. Wild-type, *atg32*-null, and *atg15*-null cells expressing mito-DHFR-mCherry and Vph1-GFP were grown in SD-N (-N), collected at the 6, 12, and 24 h time points, and subjected to SDS-PAGE and immunoblotting. The signal intensity value of free mCherry in wild-type cells at the 24-h time point was set to 100%. Data represent the averages of three experiments, with error bars indicating standard deviations. Pgk1 was monitored as a loading control

14. Incubate the membrane with ECL detection reagents and detect protein signals using a luminescent image analyzer (*see* **Note 10**).
15. For free mCherry bands, prepare three western blots, perform immunodecoration and ECL detection, and quantify protein signals using a luminescent image analyzer.

## 4 Notes

1. The plasmid pBSII-*TEF$^p$-mito-DHFR-mCherry-CgHIS3* contains the pBluescript II backbone (a high-copy *E. coli* vector with an ampicillin-resistance marker) and the yeast expression cassette encoding *mito-DHFR-mCherry* and *CgHIS3* genes. The mitochondrial matrix-localized fusion protein mito-DHFR-mCherry consists of dihydrofolate reductase (DHFR)-mCherry fused at the C-terminus of the mitochondrial targeting sequence derived from the filamentous fungus *Neurospora crassa* ATP synthase subunit 9 (amino acids 1–69).

CgHis3 is an imidazoleglycerol-phosphate dehydratase derived from the yeast pathogen *Candida glabrata* and can complement histidine auxotrophy of the *S. cerevisiae* BY4741 strain. The expression of *mito-DHFR-mCherry* and *CgHIS3* is under the control of the strong constitutive *TEF2* (*TEF*$^{p}$) and endogenous *CgHIS3* promoters, respectively.

2. Other conventional *S. cerevisiae* strains can also be used to generate derivatives for mitophagy assays using the protocols described here. It should, however, be noted that BY4741 grows faster in nonfermentable medium and undergoes drastic mitophagy.
3. To perform high-efficiency transformation followed by homologous recombination to integrate a knock-in cassette into the chromosome, it is important to use cells at mid-log phase.
4. When it reaches the mitochondrial matrix, mito-DHFR-mCherry is processed by the mitochondrial processing peptidase to lose the mitochondrial targeting sequence and become a mature form (DHFR-mCherry). Live cell imaging with fluorescence microscopy can visualize DHFR-mCherry as elongated tubules at the cell periphery (*see* Figs. 1 and 2). In addition, immunoblotting with the anti-mCherry antibody can detect DHFR-mCherry as a 52 kDa band (*see* Figs. 3 and 4).
5. The monomeric fluorescent protein mCherry, which is a DsRed derivative from *Discosoma* sea anemones, forms a β-barrel fold that is highly resistant to vacuolar proteases and remains fluorescent inside the vacuolar lumen.
6. To fully activate autophagy under nitrogen deprivation, cells should be pre-grown in nutrient-rich YPD medium to mid-log phase.
7. If the cell pellet contains a residual amount of YPD, autophagy cannot be fully activated. Thus, washing steps with SD-N should be performed carefully to remove it as thoroughly as possible.
8. Accumulation of free mCherry, which can be detected as a 28 kDa band on the immunoblot membrane, depends on its Atg32-mediated delivery to the vacuole and Atg15-mediated turnover in the vacuole (*see* Figs. 3 and 4). It should be noted that these events may vary by growth phase and genetic background.
9. The mouse monoclonal mCherry antibody ([1C51] ab125096, Abcam) is quite specific and does not provide any strong cross-reacting bands for yeast whole-cell extracts.
10. The protein expression profiles of Pgk1, a cytosolic 3-phosphoglycerate kinase for glycolysis and gluconeogenesis, should also be examined as a loading control.

## Acknowledgments

We thank Noriko Kondo-Okamoto for constructing yeast strains and establishing the original methods. This work was supported in part by the Osaka University International Joint Research Promotion Programs (Type A+) (to K.O.) and the JST SPRING (Grant Number JPMJSP2138) (to Y.N.).

### References

1. Chandel NS (2021) Mitochondria. Cold Spring Harb Perspect Biol 13:a040543
2. Scheibye-Knudsen M, Fang EF, Croteau DL, Wilson DM 3rd, Bohr VA (2015) Protecting the mitochondrial powerhouse. Trends Cell Biol 25:158–170
3. Picard M, Shirihai OS (2022) Mitochondrial signal transduction. Cell Metab 34:1620–1653
4. Chan DC (2020) Mitochondrial dynamics and its involvement in disease. Annu Rev Pathol 15: 235–259
5. Onishi M, Yamano K, Sato M, Matsuda N, Okamoto K (2021) Molecular mechanisms and physiological functions of mitophagy. EMBO J 40:e104705
6. Uoselis L, Nguyen TN, Lazarou M (2023) Mitochondrial degradation: mitophagy and beyond. Mol Cell 83:3404–3420
7. Picca A, Faitg J, Auwerx J, Ferrucci L, D'Amico D (2023) Mitophagy in human health, ageing and disease. Nat Metab 5: 2047–2061
8. Okamoto K, Kondo-Okamoto N, Ohsumi Y (2009) Mitochondria-anchored receptor Atg32 mediates degradation of mitochondria via selective autophagy. Dev Cell 17:87–97
9. Kanki T, Wang K, Cao Y, Baba M, Klionsky DJ (2009) Atg32 is a mitochondrial protein that confers selectivity during mitophagy. Dev Cell 17:98–109
10. Kondo-Okamoto N, Noda NN, Suzuki SW, Nakatogawa H, Takahashi I, Matsunami M, Hashimoto A, Inagaki F, Ohsumi Y, Okamoto K (2012) Autophagy-related protein 32 acts as autophagic degron and directly initiates mitophagy. J Biol Chem 287:10631–10638
11. Eiyama A, Kondo-Okamoto N, Okamoto K (2013) Mitochondrial degradation during starvation is selective and temporally distinct from bulk autophagy in yeast. FEBS Lett 587:1787–1792
12. Calvelli H, Krigman J, Onishi M, Narendra DP, Sun N, Okamoto K (2020) Detection of mitophagy in mammalian cells, mice, and yeast. Methods Cell Biol 155:557–579
13. Teter SA, Eggerton KP, Scott SV, Kim J, Fischer AM, Klionsky DJ (2001) Degradation of lipid vesicles in the yeast vacuole requires function of Cvt17, a putative lipase. J Biol Chem 276:2083–2087
14. Epple UD, Suriapranata I, Eskelinen EL, Thumm M (2001) Aut5/Cvt17p, a putative lipase essential for disintegration of autophagic bodies inside the vacuole. J Bacteriol 183: 5942–5955
15. Eiyama A, Okamoto K (2015) Protein N-terminal acetylation by the NatA complex is critical for selective mitochondrial degradation. J Biol Chem 290:25034–25044
16. Sakakibara K, Eiyama A, Suzuki SW, Sakoh-Nakatogawa M, Okumura N, Tani M, Hashimoto A, Nagumo S, Kondo-Okamoto N, Kondo-Kakuta C, Asai E, Kirisako H, Nakatogawa H, Kuge O, Takao T, Ohsumi Y, Okamoto K (2015) Phospholipid methylation controls Atg32-mediated mitophagy and Atg8 recycling. EMBO J 34:2703–2719

# Chapter 2

# Methods to Monitor Nucleophagy in Yeast

Ziyang Li and Hitoshi Nakatogawa

## Abstract

The selective degradation of nuclear components via autophagy, termed nucleophagy, is an essential process observed from yeasts to mammals and crucial for maintaining nucleus homeostasis and regulating nucleus functions. In the budding yeast *Saccharomyces cerevisiae*, nucleophagy occurs in two different manners: one involves autophagosome formation for the sequestration and vacuolar transport of nucleus-derived vesicles (NDVs), and the other proceeds with the invagination of the vacuolar membrane for the uptake of NDVs into the vacuole, termed macronucleophagy and micronucleophagy, respectively. This chapter describes methods to analyze and quantify activities of these nucleophagy pathways in yeast.

**Key words** Nucleophagy, Nucleus, Vacuole, Yeast, Immunoblotting, Fluorescence microscopy

## 1 Introduction

The cell nucleus is a pivotal eukaryotic organelle, which is bound by the nuclear envelope and houses genetic information in the form of chromosomes. The nucleus also contains different intranuclear structures, including the nucleolus responsible for ribosome biogenesis. In the nuclear envelope, nuclear pore complexes (NPCs) mediate molecule exchange between the nucleus and cytoplasm, a process essential for gene expression and the regulation of various nucleus functions. Thus, the nucleus is central to overall cellular function and stability [1, 2]. Recently, the role of cellular degradation mechanisms in maintaining and renewing nuclear components has garnered increasing attention. Autophagy-dependent degradation, specifically targeting nuclear components and termed nucleophagy, has been suggested to serve as a nuclear quality control mechanism. Deficiencies in nucleophagy are implicated in various disorders, including tumorigenesis, cell senescence, and cell death, emphasizing the biological significance of this catabolic pathway [3].

Autophagy is a collective term for pathways for the degradation of cells' own components in lysosomes or vacuoles, including

Ioannis P. Nezis (ed.), *Selective Autophagy: Methods and Protocols*, Methods in Molecular Biology, vol. 2845,
https://doi.org/10.1007/978-1-0716-4067-8_2,

macroautophagy and microautophagy [4–6]. While macroautophagy sequesters degradation targets within double-membrane vesicles called autophagosomes, which transport them into lysosomes/vacuoles, microautophagy invaginates the lysosomal/vacuolar membrane for the direct uptake of degradation targets into the organelle. Recent studies highlight the involvement of both macroautophagy and microautophagy in degrading diverse nuclear components—ranging from nucleoplasmic and nucleolar proteins to NPCs, nucleus-derived vesicles (NDVs), and DNA—across a spectrum of physiological and pathological situations [7]. In macroautophagy of nuclear components, termed macronucleophagy, nucleophagy receptor proteins localize to the nucleus, recruit core autophagy-related (Atg) proteins responsible for autophagosome biogenesis, and bind to Atg8-family proteins on forming autophagosomal membranes through the Atg8-family interacting motif. In the budding yeast *Saccharomyces cerevisiae*, Atg39 is transcriptionally upregulated upon the deprivation of nitrogen sources such as amino acids or cell exposure to endoplasmic reticulum (ER) stress, localizes to the outer nuclear membrane (ONM), and acts as a receptor for macronucleophagy (Fig. 1a) [8–10]. The intriguing feature of Atg39 is its ability to link the ONM and inner nuclear membrane (INM). Its association with the INM via its amphipathic helices prevents Atg39's leakage into the ER and also contributes to nuclear envelope deformation to generate NDVs in conjunction with autophagosome formation [11, 12]. Although specificity in the degradation of nuclear proteins via Atg39-dependent macronucleophagy remains unclear, this pathway does not efficiently degrade NPCs [13, 14]. Instead, NPCs embedded in NDVs are enclosed within autophagosomes via the interactions of Atg8 with the nucleoporin Nup159 and an unknown receptor [13, 14]. Macronucleophagy in mammalian cells has been reported to degrade nuclear lamina proteins, such as lamin B1 and lamin A/C, which interact with the Atg8-family protein LC3, contributing to oncogene-induced senescence, tumorigenesis suppression, and DNA damage response [15, 16]. LC3 also interacts with the protein deacetylase SIRT1 for its degradation via macronucleophagy, and this degradation was suggested to have protective roles in aging and disease [17].

In the process of microautophagic degradation of nuclear components in *S. cerevisiae*, termed piecemeal microautophagy of the nucleus or micronucleophagy, the nuclear envelope forms a bleb along with the invagination of the vacuolar membrane at the nucleus-vacuole junction (NVJ) (Fig. 1b). Subsequently, both the nuclear envelope and the vacuolar membrane pinch off to release an NDV enclosed by the vacuolar membrane into the vacuolar lumen for degradation. The interaction between the ONM protein Nvj1 and the vacuolar membrane protein Vac8 is crucial for establishing the NVJ and thus for micronucleophagy [18]. In mammalian cells,

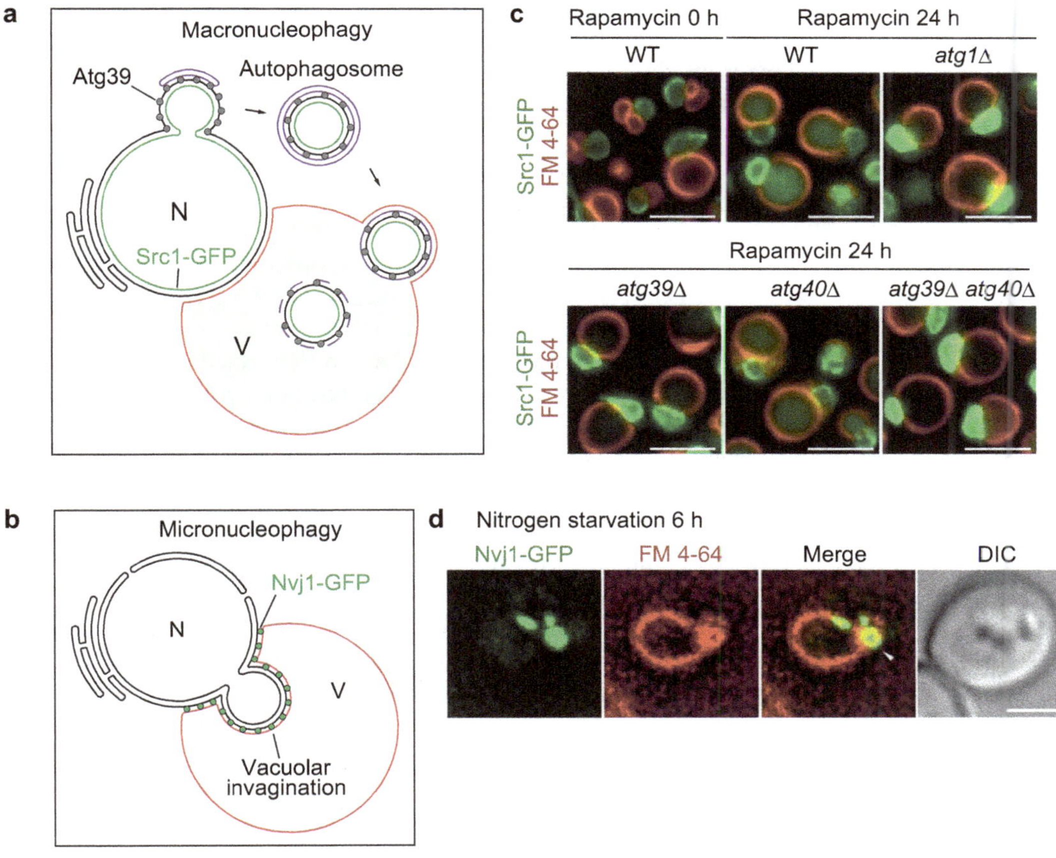


**Fig. 1** Live cell imaging of macro- and micronucleophagy in yeast. (**a**, **b**) Schematic diagrams of macronucleophagy (**a**) and micronucleophagy (**b**) in yeast. (**c**) Fluorescence microscopy of macronucleophagy in yeast. Yeast cells expressing Src1-GFP were grown in SD + CA + ATU medium and the vacuolar membranes were stained with FM4–64. These cells were treated with rapamycin for 24 h and examined by fluorescence microscopy. Scale bars, 5 μm. (Reproduced from Ref. [8]) (**d**) Fluorescence microscopy of micronucleophagy in yeast. Wild-type cells expressing Nvj1-GFP were incubated in SD-N medium for 6 h and observed under a fluorescence microscope. Arrowhead, micronucleophagic vesicle. Scale bar, 2 μm

a recent study reported a novel micronucleophagy pathway, in which the ONM swells during the unfolded protein response and is selectively captured by endolysosomes for degradation after the alleviation of ER stress [19].

This chapter presents two established protocols for studying nucleophagy in *S. cerevisiae* following nitrogen starvation or rapamycin treatment. First, fluorescence microscopy allows the visualization of the transport of nuclear proteins into the vacuole in living cells. Secondly, quantitative measurement of nuclear protein degradation can be achieved through immunoblotting. By combining robust yeast genetics with high-throughput techniques, these approaches are also applicable for screening genes or chemicals that selectively promote or inhibit nucleophagy-related pathways. Similar methods have also been shown to be available for studying macronucleophagy in the fission yeast *Schizosaccharomyces pombe* [20].

## 2 Materials

### 2.1 Yeast Cell Culture

1. Nutrient-rich liquid medium supplemented with dextrose (YPD): 1% (w/v) yeast extract, 2% (w/v) peptone, and 2% (w/v) dextrose. YP medium (yeast extract and peptone) and dextrose are separately dissolved in deionized water and mixed after autoclaving. Store at room temperature.
2. Synthetic liquid medium supplemented with dextrose and casamino acids (SD + CA + ATU): 0.17% (w/v) yeast nitrogen base without amino acids and ammonium sulfate, 0.5% (w/v) ammonium sulfate, 0.5% (w/v) casamino acids, 2% (w/v) dextrose, 20 μg/mL adenine sulfate, 20 μg/mL L-tryptophan, and 20 μg/mL uracil. S + CA medium (yeast nitrogen base, ammonium sulfate, and casamino acids) and dextrose are separately dissolved in deionized water and mixed together after autoclaving. Supplement them with filter-sterilized amino acid and nucleobase solutions. Store at room temperature.
3. Nitrogen-deprived liquid medium supplemented with dextrose (SD-N): 0.17% (w/v) yeast nitrogen base without amino acids and ammonium sulfate and 2% (w/v) dextrose. Yeast nitrogen base and dextrose are separately dissolved in deionized water and mixed together after autoclaving. Store at room temperature.
4. 1 mg/mL Rapamycin dissolved in 90% (v/v) ethanol and 10% (v/v) Tween-20. Store at −30 °C.
5. Autoclaved deionized water. Store at room temperature.

### 2.2 Yeast Strains

*S. cerevisiae* BY4741 strains used in this method can be found in previous studies [8]. Gene disruption and tagging were performed using a standard PCR-based method [21].

BY4741 MATa *his3Δ1 leu2Δ0 met15Δ0 ura3Δ0* (wild-type cells and its derivatives).

### 2.3 Observation of Macro- and Micronucleophagy by Fluorescence Microscopy

1. 1.6 mM FM4–64 dissolved in deionized water (Invitrogen). Store at 4 °C.

### 2.4 Detection of Macronucleophagy by Immunoblotting

1. 20% (v/v) Trichloroacetic acid (TCA) solution. Store at 4 °C.
2. Acetone. Store at −30 °C.
3. Sodium dodecyl sulfate (SDS) sample buffer: 60 mM Tris–HCl, pH 7.5, 5% (v/v) glycerol, 2% (w/v) SDS, 20 mM dithiothreitol (DTT), and 0.005% (w/v) bromophenol blue. Store at −30 °C (*see* **Note 1**).
4. 0.5 mm YZB zirconia beads.
5. 10% Acrylamide gel containing 10% acrylamide/bisacrylamide (37.5:1).
6. Polyvinylidene fluoride (PVDF) membrane (Millipore, IPVH00010).
7. Washing buffer (Tris-buffered saline with Tween 20, TBS-T) pH 7.4: 25 mM Tris–HCl, 137 mM NaCl, 2.68 mM KCl, and 0.1% (v/v) Tween-20.
8. Blocking buffer: TBS-T (see Subheading 2.4, **item 7**) with 2% (w/v) skim milk.
9. Primary antibody: mouse monoclonal anti-GFP antibody (Clontech, 632381). Store at 4 °C.
10. Secondary antibody: antibody against mouse IgG (Jackson Immuno Research Laboratories, 315-035-003) conjugated with horseradish peroxidase (HRP). Store at 4 °C.
11. Enhanced chemiluminescence detection reagent: FEMTO-GLOW Plus HRP Substrate (Michigan Diagnostics, LLC, SHRPE21008).

## 3 Methods

### 3.1 Fluorescence Microscopy of Macro- and Micronucleophagy

As green fluorescent protein (GFP) is resistant to proteases and acidic environments in the yeast vacuolar lumen, the vacuolar transport of a nuclear protein of interest (nPOI) fused with GFP can be examined under a fluorescence microscope (*see* **Note 2**). In cells grown to the logarithmic phase, the GFP-fused nPOI highlights the nuclear envelope, while the lipophilic dye FM4-64 stains the vacuolar membrane in red, and no GFP signal is observable within the vacuole (*see* **Note 3**). Upon cell treatment with rapamycin or incubation in SD-N medium, nucleophagy is induced, which transports NDVs containing nPOI-GFP to the vacuole for degradation. Due to GFP's stability in the vacuolar lumen, a GFP signal becomes evident in the vacuolar lumen, indicating the vacuolar transport and degradation of nPOI-GFP. Nuclear proteins such as the ONM protein Hmg1, the INM protein Heh1 (also called Src1), the nucleoplasm protein Tal1, and the nucleolar protein Nop1 were previously shown to be degraded in a macronucleophagy-dependent manner by means of this method

[8, 11], thus they could serve as controls. It is also important to examine whether the deletion of *ATG39* or core *ATG* genes such as *ATG1* abolishes the increase in vacuolar GFP fluorescence derived from nPOI-GFP in order to conclude that nPOI-GFP is transported to the vacuole via macronucleophagy (Fig. 1c).

In micronucleophagy, the vacuolar membrane invaginates at NVJs to take up NDVs into the vacuolar lumen after cell treatment with rapamycin or nitrogen starvation. Therefore, micronucleophagy results in the formation of Nvj1-positive NDVs surrounded by the vacuolar membrane within the vacuole, which can be observed under a fluorescence microscope (Fig. 1d). Unlike macronucleophagy, probably due to much slower degradation of the vacuolar membrane surrounding NDVs compared to that of the inner autophagosomal membrane (autophagic body membrane), Nvj1-GFP or nPOI-GFP is retained in the NDVs, and therefore, a GFP signal is not dispersed within the vacuole. While we describe here the visualization of Nvj1 and the vacuolar membrane to detect micronucleophagic vesicles, fluorescence tagging of nuclear proteins can allow for observing their sequestration within micronucleophagic vesicles.

#### 3.1.1 Observation of Macronucleophagy upon Rapamycin Treatment

1. Grow yeast cells expressing nPOI-GFP to log phase ($OD_{600}$ 0.6–0.8) in 5 mL of SD + CA + ATU medium using a 14 mL Falcon round-bottom tube at 30 °C.
2. Add 2.5 μL of 1.6 mM FM4-64 stock solution to the culture.
3. Incubate the cells at 30 °C for 120 min, during which they grow to late log phase ($OD_{600}$ 1.5–2.0) and FM4-64 enters the cells and enriches in the vacuolar membrane via endocytosis (*see* **Note 4**).
4. Spin down the cells by centrifugation (3000 rpm, 2 min), and discard the supernatant completely using a micropipette.
5. Resuspend the cells in 5 mL of fresh SD + CA + ATU medium, and add 1 μL of 1 mg/mL rapamycin stock solution to the culture.
6. Incubate the cells at 30 °C for 6 ~ 24 h to induce macronucleophagy.
7. Transfer 200 μL of the culture to a 1.5 mL microcentrifuge tube, spin down the cells by centrifugation (3000 rpm, 15 s), and discard the supernatant to leave about 5 μL for cell suspension.
8. Pipette 1.5 μL of the cell suspension onto a microscope slide, and cover with a cover slip.
9. Observe the cells using a fluorescence microscope with devices for the detection of GFP and RFP fluorescence signals.

#### 3.1.2 Observation of Micronucleophagy During Nitrogen Starvation

1. Grow yeast cells expressing Nvj1-GFP to early mid-log phase ($OD_{600}$ 0.6–0.8) in 5 mL of SD + CA + ATU medium using a 14 mL Falcon round-bottom tube at 30 °C.
2. Add 2.5 μL of 1.6 mM FM4–64 stock solution to the culture.
3. Incubate the cells at 30 °C for 120 min, during which they grow to late log phase ($OD_{600}$ 1.5–2.0) and FM4-64 enters the cells and enriches in the vacuolar membrane via endocytosis.
4. Spin down the cells by centrifugation (3000 rpm, 2 min), and discard the supernatant completely using a micropipette.
5. Suspend the cells in 5 mL of autoclaved deionized water.
6. Repeat **steps 4** and **5**, spin down the cells by centrifugation, and discard the supernatant completely using a micropipette (*see* **Note 5**).
7. Resuspend the cells in 5 mL of SD-N and incubate them at 30 ° C for 6 h to induce micronucleophagy (*see* **Note 6**).
8. Prepare samples for fluorescence microscopy observation as described in Subheading 3.1.1, **steps 7–9**.

### 3.2 Detection of Macronucleophagy by Immunoblotting

In addition to its application in fluorescence imaging, GFP tagging of nuclear proteins can also allow for monitoring their degradation by immunoblotting. Vacuolar transport and degradation of nPOI-GFP via macronucleophagy result in vacuolar protease-resistant GFP fragments, and therefore the magnitude of macronucleophagy can be estimated by quantifying these GFP fragments (Fig. 2a). However, since GFP fragments can also be generated in a macronucleophagy-independent manner, the dependence on *ATG39* must be examined to conclude whether and how much macronucleophagy contributes to the degradation of the nuclear protein. The generation of GFP fragments from the GFP-fused nuclear proteins listed in the fluorescence microscopy-based method (Hmg1, Heh1/Src1, Tal1, and Nop1) is almost completely abolished by disruption of *ATG39* (Fig. 2b and c) [8, 11]. In addition, it should be noted here that this GFP cleavage assay tends to be unsuitable for detection of nuclear protein degradation via micronucleophagy. Probably because degradation of nuclear proteins via micronucleophagy is much slower than that via macronucleophagy, as mentioned in the section for fluorescence microscopy of macro- and micronucleophagy, disruption of *NVJ1* has almost no or only slight impact on degradation of nuclear proteins when evaluated by this assay (Fig. 2c) [8, 22].

1. Grow yeast cells expressing nPOI-GFP to late-log phase ($OD_{600}$ 1.5–2.0) in 5 mL of SD + CA + ATU medium using a 14 mL Falcon round bottom tube at 30 °C.

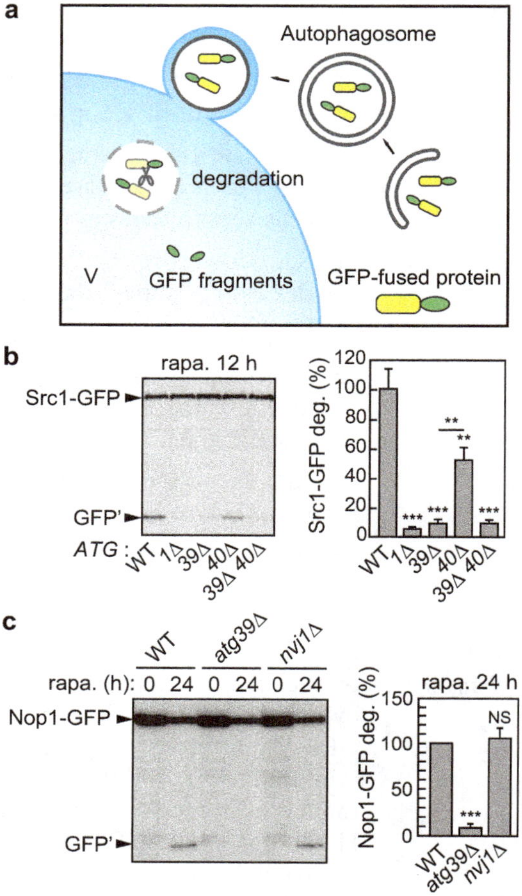


**Fig. 2** Immunoblotting analysis of macronucleophagy in yeast. (**a**) A schematic diagram of GFP cleavage assay. (**b**, **c**) Yeast cells expressing Src1-GFP (**b**) or Nop1-GFP (**c**) were treated with rapamycin for the indicated time periods and examined by immunoblotting using an anti-GFP antibody. To quantify the degradation of nPOI-GFP, the band intensity of GFP fragments (GFP') was divided by the sum of the intensities of nPOI-GFP and GFP', and the values were normalized with those of wild-type (WT) cells. (Reproduced from Ref. [8])

2. Add 1 μL of 1 mg/mL rapamycin stock solution to the culture, and incubate at 30 °C for 6 ~ 24 h to induce macronucleophagy.
3. Transfer 1 mL of the culture to a 1.5 mL microcentrifuge tube, harvest cells by centrifugation (15,000 g at 4 °C, 1 min), and discard the supernatant using a micropipette. Store the cell pellets at −80 °C until the next step.
4. Suspend the cells in 200 μL of 20% TCA solution, and keep on ice for 15 min.

5. Spin down the cells by centrifugation (15,000 g at 4 °C, 5 min), and discard the supernatant using a micropipette.
6. Suspend the pellets in 500 μL of cold acetone using a bath sonicator containing iced water.
7. Spin down the cells by centrifugation (15,000 g at 4 °C, 5 min), and discard the supernatant using a micropipette.
8. Dry the pellets at room temperature for 10 min with the lids of microcentrifuge tubes opened.
9. Suspend the pellets in (the value of $OD_{600}$ units × 50) μL of SDS sample buffer by mixing at 65 °C for 10 min.
10. Transfer the samples to 1.5 mL microcentrifuge tubes with screw caps containing 0.5 g of 0.5 mm YZB zirconia beads.
11. Disrupt the cells using a FastPrep-24 (MP Biomedicals) (6.5 m/s for 60 s) (*see* **Note 7**).
12. Puncture the bottom of the tubes with a needle and set the tubes in new 1.5 mL microcentrifuge tubes. Centrifuge at 2500 rpm for 2 min to recover the samples in the new tubes, leaving the beads in the old ones (*see* **Note 8**).
13. Subject 10 μL of the samples to electrophoresis using a 10% SDS-PAGE gel, and transfer proteins in the gel to a PVDF membrane.
14. Block the membrane with 2% skim milk in TBS-T at room temperature for 30 min.
15. Incubate the membrane with 2% skim milk in TBS-T containing the anti-GFP antibody (1:2500 dilution) at room temperature for 90 min.
16. Wash the membrane three times with TBS-T for 5 min each.
17. Incubate the membrane with 2% skim milk in TBS-T containing HRP-conjugated anti-mouse IgG (1:5000 dilution) at room temperature for 30 min.
18. Wash the membrane three times with TBS-T for 5 min each.
19. Incubate the membrane with chemiluminescence detection reagents and detect protein signals using a luminescent image analyzer.

## 4 Notes

1. Urea-containing SDS sample buffer (37.5 mM MOPS-NaOH (pH 6.8), 2% SDS, 100 mM DTT, and 4 M urea) is better for analyzing some nuclear membrane proteins including Atg39.
2. Instead of GFP, the red fluorescent protein mCherry, which is resistant to vacuolar degradation and acidic environments, can

also be used for monitoring nucleophagy with the methods provided here.

3. While the vacuole undergoes fragmentation when yeast cells are transferred to nitrogen starvation medium, rapamycin treatment generally causes vacuole enlargement.
4. When yeast cells are continuously incubated with FM4–64, the dye typically stains membrane structures throughout the endocytic pathway. However, once FM4-64 is removed, it takes approximately 90–120 min for the staining to become specific to vacuolar membranes (and endosomes).
5. If a residual amount of SD + CA medium remains, nucleophagy cannot be fully activated. Thus, these washing steps should be performed carefully.
6. Micronucleophagy takes place at a very low rate after rapamycin treatment or nitrogen starvation.
7. Although cell disruption is required to completely extract proteins from the cells, some proteins can be efficiently extracted without these steps.
8. The remaining samples can be stored at −30 °C. When analyzing the frozen samples, incubate them at 65 °C for 10 min before SDS-PAGE.

## Acknowledgments

Research in the authors' group is supported in part by KAKENHI Grants-in-Aid for Scientific Research JP19H05708 and JP17H01430 from the Ministry of Education, Culture, Sports, Science and Technology of Japan (to H.N.) and AMED Grant Number JP21gm1410004 (to H.N.).

### References

1. Tiku V, Antebi A (2018) Nucleolar function in lifespan regulation. Trends Cell Biol 28:662–672
2. Dundr M, Misteli T (2001) Functional architecture in the cell nucleus. Biochem J 356:297
3. Papandreou ME, Tavernarakis N (2019) Nucleophagy: from homeostasis to disease. Cell Death Differ 26:630–639
4. Yang Z, Klionsky DJ (2010) Eaten alive: a history of macroautophagy. Nat Cell Biol 12: 814–822
5. Schuck S (2020) Microautophagy – distinct molecular mechanisms handle cargoes of many sizes. J Cell Sci 133:jcs246322
6. Ohsumi Y (2014) Historical landmarks of autophagy research. Cell Res 24:9–23
7. Li Z, Nakatogawa H (2022) Degradation of nuclear components via different autophagy pathways. Trends Cell Biol 32:574–584
8. Mochida K, Oikawa Y, Kimura Y, Kirisako H, Hirano H, Ohsumi Y, Nakatogawa H (2015) Receptor-mediated selective autophagy degrades the endoplasmic reticulum and the nucleus. Nature 522:359–362
9. Mizuno T, Muroi K, Irie K (2020) Snf1 AMPK positively regulates ER-phagy via expression control of Atg39 autophagy receptor in yeast ER stress response. PLoS Genet 16:e1009053

10. Mizuno T, Irie K (2021) Msn2/4 transcription factors positively regulate expression of Atg39 ER-phagy receptor. Sci Rep 11:11919
11. Mochida K, Otani T, Katsumata Y, Kirisako H, Kakuta C, Kotani T, Nakatogawa H (2022) Atg39 links and deforms the outer and inner nuclear membranes in selective autophagy of the nucleus. J Cell Biol 221:e202103178
12. Chandra S, Mannino PJ, Thaller DJ, Ader NR, King MC, Melia TJ, Lusk CP (2021) Atg39 selectively captures inner nuclear membrane into lumenal vesicles for delivery to the autophagosome. J Cell Biol 220:e202103030
13. Tomioka Y, Kotani T, Kirisako H, Oikawa Y, Kimura Y, Hirano H, Ohsumi Y, Nakatogawa H (2020) TORC1 inactivation stimulates autophagy of nucleoporin and nuclear pore complexes. J Cell Biol 219:e201910063
14. Lee CW, Wilfling F, Ronchi P, Allegretti M, Mosalaganti S, Jentsch S, Beck M, Pfander B (2020) Selective autophagy degrades nuclear pore complexes. Nat Cell Biol 22:159–166
15. Dou Z, Xu C, Donahue G, Shimi T, Pan JA, Zhu J, Ivanov A, Capell BC, Drake AM, Shah PP et al (2015) Autophagy mediates degradation of nuclear lamina. Nature 527:105–109
16. Li Y, Jiang X, Zhang Y, Gao Z, Liu Y, Hu J, Hu X, Li L, Shi J, Gao N (2019) Nuclear accumulation of UBC9 contributes to SUMOylation of lamin A/C and nucleophagy in response to DNA damage. J Exp Clin Cancer Res 38:67
17. Xu C, Wang L, Fozouni P, Evjen G, Chandra V, Jiang J, Lu C, Nicastri M, Bretz C, Winkler JD et al (2020) SIRT1 is downregulated by autophagy in senescence and ageing. Nat Cell Biol 22:1170–1179
18. Kvam E, Goldfarb DS (2007) Nucleus-vacuole junctions and piecemeal microautophagy of the nucleus in S. cerevisiae. Autophagy 3:85–92
19. Kucińska MK, Fedry J, Galli C, Morone D, Raimondi A, Soldà T, Förster F, Molinari M (2023) TMX4-driven LINC complex disassembly and asymmetric autophagy of the nuclear envelope upon acute ER stress. Nat Commun 14:3497
20. Zou CX, Ma ZH, Di Jiang Z, Pan ZQ, Xu DD, Suo F, Shao GC, Dong MQ, Du LL (2023) The ortholog of human REEP1-4 is required for autophagosomal enclosure of ER-phagy/nucleophagy cargos in fission yeast. PLoS Biol 21:e3002372
21. Knop M, Siegers K, Pereira G, Zachariae W, Winsor B, Nasmyth K, Schiebel E (1999) Epitope tagging of yeast genes using a PCR-based strategy: more tags and improved practical routines. Yeast 15:963–972
22. Golam Mostofa M, Rahman MA, Koike N, Yeasmin AMST, Islam N, Waliullah TM, Hosoyamada S, Shimobayashi M, Kobayashi T, Hall MN et al (2018) CLIP and cohibin separate rDNA from nucleolar proteins destined for degradation by nucleophagy. J Cell Biol 217:2675–2690

# Chapter 3

# Synthetic Dimerization Approaches for In Vivo Studies in *Saccharomyces cerevisiae*

**Jana Elsässer and Claudine Kraft**


## Abstract

Synthetic tethering approaches induced by chemical means offer precise control over protein interactions in cells. They enable the manipulation of when, where, and how proteins interact, making it possible to study their functions, dynamics, and cellular consequences at a molecular level. These methods are versatile, reversible, and adaptable, allowing the dissection of complex cellular processes and the engineering of cellular functions. Here, we describe two chemically induced dimerization systems in the model organism *Saccharomyces cerevisiae*. Using the autophagy pathway as an example, we show how these approaches can be used to dissect molecular events in cells.


**Key words** Synthetic tethering, Dimerization, FKBP, Cnb1, FRB, Autophagy, Rapamycin, FK506

## 1 Introduction

Chemically induced tethering approaches are a valuable tool in cellular research, allowing precise control over protein interactions. They allow tight temporal control, which is essential for the study of dynamic cellular processes. These methods also enable spatial control, redirecting proteins to specific cellular locations to investigate the consequences of altering subcellular localization. Importantly, chemical tethering can bypass the need for endogenous cellular factors or protein–protein interactions, making it possible to study the isolated functions of individual proteins within complex networks without disturbing other cellular processes (Fig. 1a). Here, we describe different chemical tethering approaches suitable for studies in *Saccharomyces cerevisiae* and highlight important considerations to bear in mind when employing them. Furthermore, we demonstrate their use in addressing molecular events in the autophagy pathway.

Autophagy is a cellular degradation pathway in eukaryotic cells that recycles cell material by packaging it into a double-membrane vesicle, the autophagosome, and delivering it to the lytic

Ioannis P. Nezis (ed.), *Selective Autophagy: Methods and Protocols*, Methods in Molecular Biology, vol. 2845,
https://doi.org/10.1007/978-1-0716-4067-8_3,

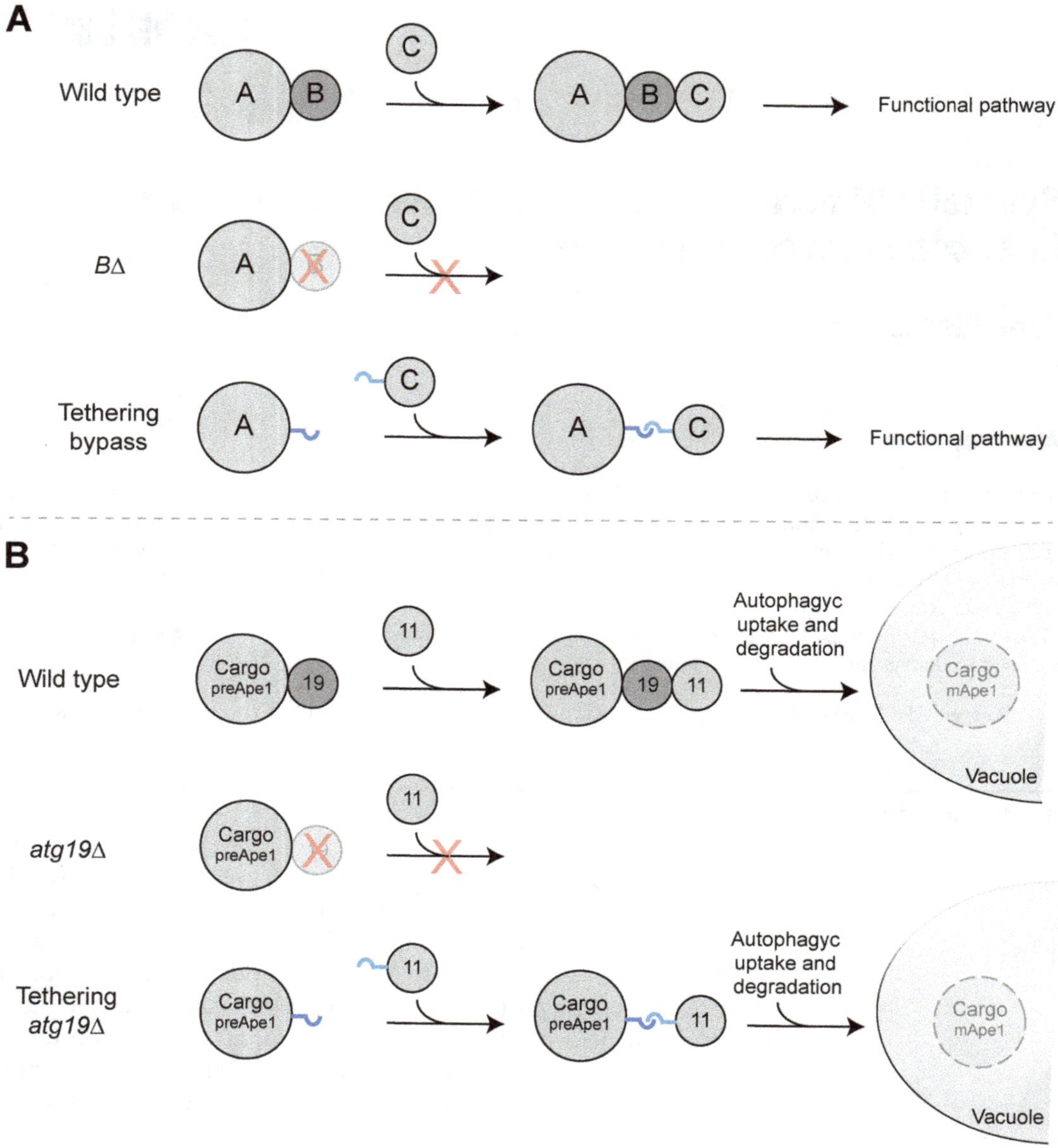


**Fig. 1** Chemically induced synthetic tethering. (**a**) Schematic representation of the principle of synthetic tethering systems to bypass functions. A component B, which is required to bridge two other factors, A and C, within a complex ABC, is deleted so that A and C fail to interact. A and C are then tethered (A–C) in the absence of B. If B only serves to bridge the binding of A with C, then synthetic tethering should restore function. (**b**) Schematic representation of the preApe1-Atg19-Atg11 interaction. In wild-type cells, the functional Cvt pathway ensures that the preform of Ape1 (preApe1) is delivered to the vacuole, where it undergoes proteolytic processing to become mature Ape1 (mApe1). However, in the absence of the cargo receptor Atg19, vacuolar delivery is compromised, and preApe1 remains in its cytosolic, unprocessed state. Tethering preApe1 synthetically to Atg11 in the absence of Atg19 restores the delivery of preApe1 to the vacuole and hence bypasses the receptor function

compartment for degradation. In the yeast *S. cerevisiae*, the Cytoplasm-to-vacuole targeting (Cvt) pathway is a selective autophagic pathway that transports the hydrolase aminopeptidase 1 (Ape1) to the vacuole, where the precursor form matures into an active enzyme [1, 2]. The cytosolic preform of Ape1 (preApe1) is recognized by its cargo receptor Atg19. The preApe1-Atg19 cargo-receptor complex is then escorted by the scaffold protein Atg11 to the vacuole, where further autophagy factors catalyze the formation of autophagosomal membranes around the cargo and guide its subsequent uptake by the vacuole (Fig. 1b, [3]). We previously applied synthetic tethering approaches to investigate whether the function of the cargo receptor Atg19 can be bypassed when synthetically tethering the cargo directly to Atg11.

First, we used an inducible synthetic tethering approach (*see* **Notes 1** and **2**) based on the interaction of rapamycin with the human protein FKBP (12-kDa FK506 binding protein) and FRB, a 93 amino acid FKBP-rapamycin binding domain of the mammalian target of rapamycin (mTOR, amino acids 2021-2113 [4]) [5, 6]. FKBP binds rapamycin to form an immunosuppressive complex in humans [7]. The FKBP-rapamycin complex binds to the FRB domain of the mTOR complex, forming a stable trimeric assembly. This mode of interaction can be utilized by synthetic engineering: if two proteins of interest, protein A and protein B, are tagged with FKBP and FRB, respectively, then the addition of rapamycin induces the formation of the trimeric complex and thereby the association of the two proteins (Fig. 2a). As rapamycin is a small molecule capable of entering the cell, it can simply be added to the growth medium. To probe the function of Atg19, we created FKBP-Ape1 and Atg11-FRB fusion proteins and expressed these in *atg19Δ* cells. In the absence of rapamycin, *atg19Δ* deletion cells showed an impaired Cvt pathway as expected (Fig. 2a(i), [2]). The addition of rapamycin, however, induced the recruitment of Atg11 to Ape1 and restored the Cvt pathway, suggesting that the function of Atg19 can be bypassed by synthetic tethering (Fig. 2a(i), [8]).

A disadvantage of this approach is that yeast possesses homologs of both FKBP and FRB. Therefore, a mutant background needs to be used to prevent competing interactions with rapamycin. Fpr1 is the yeast homolog of human FKBP and can be removed without major cellular phenotypes. FRB domains are found in both Tor1 and Tor2. In a wild-type cell, therefore, rapamycin addition results in Tor1 and Tor2 inhibition, leading to enhanced autophagy and Cvt pathway progression (Fig. 2b). Tor2 is crucial for cell viability, and the lack of Tor1 results in various severe cellular phenotypes. To preserve their normal function while conferring resistance to rapamycin, the *tor1-1* and *tor2-2* were introduced instead of eliminating these proteins [9, 10]. The *fpr1Δ* and *fpr1Δ tor1-1 tor2-2* mutant backgrounds do not affect autophagy function, and upon rapamycin treatment do not lead to enhanced

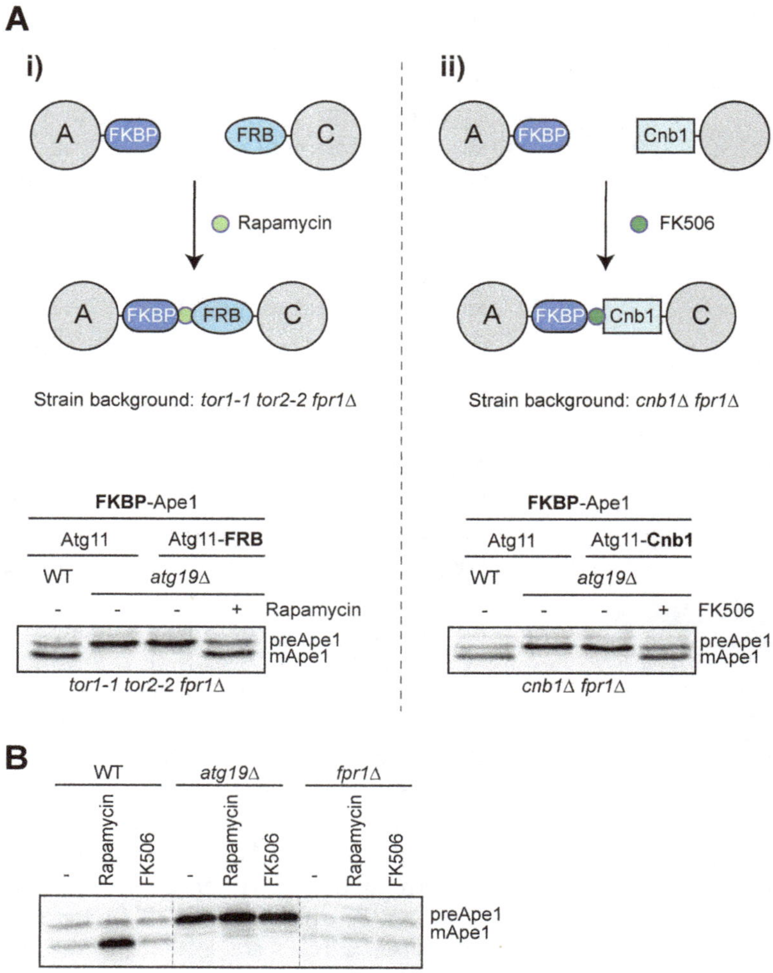


**Fig. 2** Induction of heterodimerization through the use of rapamycin or FK506. (**a**) (i) The FKBP-FRB-rapamycin system: Two proteins of interest are tagged with FKBP and FRB, respectively. These fusion proteins are co-expressed in cells, and upon the introduction of rapamycin to the growth medium, they become physically connected. This cellular setup requires the specific genetic mutations *fpr1Δ*, *tor1-1*, and *tor2-2* to prevent the binding of rapamycin to endogenous proteins, thus ensuring the selective interaction of the tagged proteins of interest. (ii) The FKBP-Cnb1-FK506 system: This tethering setup is similar to the FKBP-FRB system shown in (i), but instead of FRB employs Cnb1, and FK506 as the tethering molecule. This setup requires the genetic mutations *fpr1Δ* and *cnb1Δ*. (**b**) Effect of rapamycin and FK506 on Cvt pathway progression: Whereas FK506 does not accelerate the Cvt pathway, rapamycin addition enhances Ape1 maturation in wild-type cells due to the inhibition of TORC1

autophagy and Cvt pathway progression (Fig. 2b, [8]). Nevertheless, it is essential to verify this for the specific pathway under investigation in each case.

As the introduction of these mutations is rather tedious, and mutating Tor proteins might not be ideal when studying autophagy, we also used another tethering system independent of the interaction with FRB. FKBP also binds another small molecule, FK506, which induces the formation of a trimeric complex with Cnb1 (Calcineurin subunit B) [11]. Similar to the FKBP-FRB system, two proteins of interest are tagged with FKBP and Cnb1 (Fig. 2a(ii)). For this approach, only two mutations are required, *fpr1Δ cnb1Δ*. These deletions exhibit no severe phenotypic effects and do not affect the autophagy pathway but restore functionality upon FK506 treatment, mirroring observations in the FKBP-FRB system (Fig. 2, [8, 12]).

Several subsequent investigations using these systems have also provided molecular insight into other steps in the autophagy pathway [8], including the study of assembly processes by fluorescence microscopy [13], as well as the redirection of autophagosome formation to an ectopic site [14]. Taken together, chemically induced tethering approaches provide a powerful tool to dissect and manipulate molecular events within cells with a high degree of precision, offering insights into complex cellular processes and pathways.

## 2 Materials

### 2.1 Yeast Growth Media

1. 2×S solution (1 L total volume): Weigh 3.4 g of Yeast nitrogen base w/o Amino acids and Ammonium sulfate, and 10 g of Ammonium sulfate. Dissolve in ultra-purified water using an ion exchange cartridge (mQ water) until the final volume is reached. Autoclave the mixture and store it at Room temperature(RT).
2. 10× Complete supplement, drop-out: -aa (10× CSM-aa): Weigh 690 mg CSM-aa powder (-aa indicates the specific drop-out amino acid(s) required for the selection of transformed plasmids) and add 1 L of mQ water. Add a magnetic stirrer and let it stir at 60 °C until it dissolves (it takes up to 3 h). Filter it sterile and store the medium at 4 °C.
3. 50% Glucose stock solution: Weigh 2750 g glucose-monohydrate and add 2 L of mQ water to a glass beaker. Place it on a magnetic stirrer at 160–180 °C, 150 rpm. Add glucose stepwise in 15-min intervals to the mQ water and dissolve the whole 2750 g glucose in 2 L mQ water. It takes 2 to 3 h to dissolve. Fill up the solution until 5 L with MQ water and let it stir for 15–20 min. Distribute the 5 L of 50%

glucose to smaller flasks and autoclave as soon as possible to avoid contamination. Store the stock solution at RT.

4. 1 L Synthetic Defined Medium, drop-out: -aa (SD-aa): Mix 500 mL of 2×S solution, 100 mL of 10x CSM-aa, 40 mL of 50% glucose solution (final concentration 2%) and 360 mL of sterile mQ water. Store the medium at RT.
5. 1 L Synthetic Defined Medium plates, drop-out: aa (SD-aa agar plates): Weigh 1.7 g of Yeast Nitrogen base w/o Amino acids and Ammonium sulfate, 5 g of Ammonium sulfate, and 28 g of Agar. Add 800 mL of mQ water and let it mix on a magnetic stirrer for a few minutes before autoclaving. Take the media out of the autoclave and mix it on a magnetic stirrer to cool down. When the media has cooled down to at least 60 °C, the preparation can be continued. Pay attention that the media doesn't cool down too much, as the agar will solidify again. Add sterile 100 mL of 10x CSM-aa solution and sterile 40 mL of 50% glucose solution (final concentration 2%). Pour the plates, let them dry, and store them at RT.
6. One-step transformation buffer: Weigh 23.5 g of PEG 3550 (or PEG 4000) (final concentration is 47%). Add 12 mL of 1 M LiAc mQ water up to 50 mL. Filter sterilize, don't autoclave, and store at RT.
7. 3× Urea loading buffer (ULB+): Weigh 14.4 g of urea (final concentration 8 M). Add 3.5 mL of 1 M Tris–HCl (pH 6.8) (final concentration 116 mM), 2 mL of 70% glycerol (final concentration 4.9%), and 300 μL of 14.33 M β-mercaptoethanol (final concentration 143 mM). Fill up to 30 mL with 10% Sodium-Dodecyl-Sulfate (SDS) and add 8% bromphenol blue. Heat slightly to dissolve and prepare 1 mL aliquots. Store them at −20 °C.

### 2.2 Chemicals and Antibodies

1. 1 M Dithiothreitol (DTT): Add 1.55 g of DTT powder to 10 mL of mQ water (final concentration 1 M). Prepare aliquots of 1 mL each and store them at −20 °C.
2. 10 mg/mL Salmon sperm DNA (ssDNA): Add 500 mg of ssDNA to 50 mL of 1×TAE. Let it dissolve at 4 °C for 2 to 3 days. Boil the solution for 20 min under the sterile hood. Prepare 1 mL aliquots and store them at −20 °C.
3. 10 mM Tacrolimus (FK506): Add 8.039 mg of FK506 to 1 mL of DMSO/EtOH (final concentration 10 mM) to prepare the stock solution. Use 3 μM to induce tethering by directly pipetting into the culture medium. Store the stock solution at −20 °C. FK506 is ordered as a powder (LC laboratories), which is soluble in DMSO or ethanol (*see* **Note 3**).
4. 10 mM Rapamycin: Add 9.15 mg to 1 mL of DMSO/EtOH (final concentration 10 mM) to prepare the stock solution.

Use 0.75 μM to induce tethering by directly pipetting into the culture medium. Store the stock solution at −20 °C. Rapamycin is ordered as a powder (LC laboratories) which is soluble in DMSO or Ethanol (*see* **Note 3**).

5. 100% Trichloroacetic acid (TCA): Add 1101.32 g of TCA to 800 mL of distilled water. Fill up with distilled water until the volume reaches 1 L to prepare the 100% (W/V) TCA stock solution. Add a final concentration of 7% to your sample. Store the 100% (w/v) TCA solution at RT, protected from sunlight.
6. Anti-Ape1 antibody: The polyclonal anti-Ape1 antibody was generated by immunizing rabbits with a synthetic peptide corresponding to amino acids 168–182 [8].
7. Anti-Atg11 antibody: The polyclonal anti-Atg11 antibody was generated by immunizing rabbits with recombinant Atg11 corresponding to amino acids 685–1178 [8].

### 2.3 Yeast Background Mutations

Generate gene deletions, tagging, and mutations by homologous recombination, mating, and dissection [15].

1. For the generation of the *tor1-1* mutation, mutate serine 1972 to arginine [10].
2. For the generation of the *tor2-2* mutation, mutate serine 1975 to arginine [10, 16].

### 2.4 Tethering Tags

The following tags need to be added to the proteins of interest. They function both as N-terminal and C-terminal fusions. We usually also add a linker region (e.g., multiple GGS) between the tag and the protein of interest. Tagging of proteins can be done either at their genomic locus or on plasmids.

FKBP:

MGVQVETISPGDGRTFPKRGQTCVVHYTGMLEDGKKF
DSSRDRNKPFKFMLGKQEVIRGWEEGVAQMSVGQ
RAKLTISPDYAYGATGHPGIIPPHATLVFDVELLKLE

FRB:

MILWHEMWHEGLEEASRLYFGERNVKGMFEVLEPLHA
MMERGPQTLKETSFNQAYGRDLMEAQEWCRKYMK
SGNVKDLL$^{2098}$QAWDLYYHVFRRISK

Cnb1:

MGAAPSKIVDGLLEDTNFDRDEIERLRKRFMKLDRDSS
GSIDKNEFMSIPGVSSNPLAGRIMEVFDADNSGDVD
FQEFITGLSIFSGRGSKDEKLRFAFKIYDIDKDGFISN
GELFIVLKIMVGSNLDDEQLQQIVDRTIVENDSDG
DGRLSFEEFKNAIETTEVAKSLTLQYDV

## 3 Methods

### 3.1 Yeast Transformation

This step is only required when adding the proteins of interest on plasmids; when tagging is done genomically, this step can be omitted.

1. Mix 85 μL of One-Step buffer with 10 μL of 1 M DTT and 5 μL of ssDNA (0.05 mg final DNA concentration).
2. Add 1 μg of each plasmid.
3. Add freshly streaked yeast cells from an agar plate that has been grown overnight (approximately the amount of a pea).
4. Vortex the mixture and incubate it at 45 °C for 30 min.
5. Plate the mixture on an SD-aa agar plate with the appropriate drop-out to select the plasmid(s) added, using glass beads to evenly distribute the mixture. Incubate the plate at 30 °C. It takes around 2 to 3 days for single colonies to appear.

### 3.2 Yeast Growth and Induction of Tethering

1. Inoculate 5 mL of synthetic drop-out medium with a few single colonies from the yeast transformation plate. Start this pre-culture the day before the experiment. Let the yeast grow for at least two doubling cycles (2 h is the normal doubling time in SD medium) throughout the day.
2. In the evening, measure the $OD_{600}$. According to the measured $OD_{600}$, dilute the culture for overnight growth so that it reaches the logarithmic growth phase (0.5–1.5 OD units/mL) the next morning. To induce tethering, add 0.75 μM rapamycin or 3 μM FK506 directly to the culture. For the mock treatment, add the respective volume of solvent to the culture. Incubate the cultures for the appropriate time (*see* **Notes 4** and **5**). Here, we incubated for 4 h.

### 3.3 TCA Precipitation and Western Blot Analysis

1. Take samples immediately upon tether addition and at each time point of interest. In parallel, take control samples from a culture that has been mock-treated. Take 1.4 mL of the culture and add 98 μL of 100% TCA (7% final concentration). TCA precipitates the proteins in the sample. When working with TCA wear gloves since it irritates the skin. Incubate TCA samples on ice for at least 10 min before proceeding with the experiment or store them at −20 °C.
2. Precool a centrifuge to 4 °C and spin samples at 15,500 x g for 10 min.
3. Keep samples on ice and remove the supernatant with a pipette. Add 1 mL of acetone to the sample and vortex. Centrifuge at 15,500 x g for 2 min.
4. Remove the supernatant and let the pellet dry under the flow hood with an open lid (it takes around 10 min).

5. When the pellet is dry, resuspend cell pellets in 3x urea loading buffer. To calculate the μL of urea loading buffer for resuspension, multiply the measured $OD_{600}$ by the mL of harvested culture and the number 25 (e.g., the pellet of 1 mL culture of $OD_{600}$ 0.5 is resuspended in 12.5 μL urea loading buffer). Boil the sample at 95 °C for 5 min. Vortex every min to dissolve the pellet.
6. Briefly spin and load 5 μL for each sample, which corresponds to approx. 20 μg of protein. Analyze by SDS-PAGE and Western blotting.

## 4 Notes

1. Optogenetics: This chapter describes two inducible synthetic tethering systems that allow the tethering of two proteins of interest upon the addition of a tethering molecule to the growth medium. Alternative tethering approaches exist, such as optogenetic approaches that use light to induce the binding of two proteins of interest. In optogenetic tethering approaches, two dimerization domains, attached to the proteins of interest, are unable to interact; however, upon exposure to a specific wavelength, these domains change their conformation, resulting in their interaction. Using light as the "tethering agent" does not require the deletion of competing factors in yeast, facilitating strain generation. A further advantage is that, in contrast to a chemical molecule, a light beam can be applied to a specific region of a cell, thereby allowing the analysis of certain regions rather than the entire cell. Some wavelengths used, however, are rather damaging to cells. Also, a suitable microscope with specific lasers and filters needs to be available.
2. Constitutive tethering: The tethering approaches presented here are all inducible systems. Tethering only occurs upon the addition of a tethering molecule (or light in the case of optogenetics). This allows the comparison of different conditions, before and after tethering. There are further tethering approaches that are noninducible, such as the binding of GFP (Green Fluorescent Protein) to GBP (GFP binding protein). GFP and GBP are fused to the proteins of interest. Upon coexpression of these fusion constructs in cells, the GFP-tagged protein will bind to the GBP-tagged protein; however, this occurs constitutively. Hence, proteins of interest colocalize permanently.
3. Costs: The costs for rapamycin and FK506 treatments are around 0.06 cents and 0.3 cents per mL induced yeast culture, respectively.

4. 24-h tethering assay: In rich media, cells keep dividing during tethering experiments. Make sure cells do not exceed the logarithmic growth phase during longer incubations. Dilute cultures during the experiment, if necessary, with media containing the tethering molecule. Overgrown cells starve because of nutrient depletion, which can alter the experimental readout.
5. Tethering time points: Tethering approaches are also useful in visualizing protein localizations and interactions within cells [13]. Since tethering induced by the addition of rapamycin or FK506 occurs rapidly, it is possible to observe these localization changes using fluorescence microscopy within minutes after the addition of the tethering molecule. However, when tethering is employed for endpoint assays, such as tracking the accumulation of a degradation product (as shown in Fig. 2), the incubation time with the tethering agent typically extends over several hours to allow the readout signal to accumulate. It is therefore essential to adjust the incubation time according to the specific requirements of the experiment.

## Acknowledgments

The Kraft laboratory has received funding from the Deutsche Forschungsgemeinschaft (DFG, German Research Foundation) SFB 1381 (Project ID 403222702), GRK 2606 (Project ID 423813989), under Germany's Excellence Strategy (CIBSS-EXC-2189-Project ID 390939984), and from the European Research Council (ERC) under the European Union's Horizon 2020 research and innovation programme under grant agreement No 769065. This work reflects only the authors' view, and the European Union's Horizon 2020 research and innovation programme is not responsible for any use that may be made of the information it contains. The authors declare no competing financial interests.

### References

1. Harding TM, Morano KA, Scott SV, Klionsky DJ (1995) Isolation and characterization of yeast mutants in the cytoplasm to vacuole protein targeting pathway. J Cell Biol 131:591–602
2. Scott SV, Guan J, Hutchins MU et al (2001) Cvt19 is a receptor for the cytoplasm-to-vacuole targeting pathway. Mol Cell 7:1131–1141
3. Yorimitsu T, Klionsky DJ (2005) Atg11 links cargo to the vesicle-forming machinery in the cytoplasm to vacuole targeting pathway. Mol Biol Cell 16:1593–1605
4. Chen J, Zheng XF, Brown EJ, Schreiber SL (1995) Identification of an 11-kDa FKBP12-rapamycin-binding domain within the 289-kDa FKBP12-rapamycin-associated protein and characterization of a critical serine residue. Proc Natl Acad Sci 92:4947–4951
5. Choi J, Chen J, Schreiber SL, Clardy J (1996) Structure of the FKBP12-rapamycin complex

interacting with binding domain of human FRAP. Science 273:239–242

6. Kolos JM, Voll AM, Bauder M, Hausch F (2018) FKBP ligands—where we are and where to go? Front Pharmacol 9:1425
7. Harding MW, Galat A, Uehlingt DE, Schreibert SL (1989) A receptor for the immunosuppressant FK506 is a cis-trans peptidyl-prolyl isomerase. Nature 341:758–760
8. Torggler R, Papinski D, Brach T et al (2016) Two independent pathways within selective autophagy converge to activate Atg1 kinase at the vacuole. Mol Cell 64:221–235
9. Lorenz MC, Heitman J (1995) TOR mutations confer rapamycin resistance by preventing interaction with FKBP12-rapamycin. J Biol Chem 270:27531–27537
10. Heitman J, Movva NR, Hall MN (1991) Targets for cell cycle arrest by the immunosuppressant rapamycin in yeast. Science 253:905–909
11. Sio SO, Suehiro T, Sugiura R et al (2005) The role of the regulatory subunit of fission yeast Calcineurin for in vivo activity and its relevance to FK506 sensitivity. J Biol Chem 280:12231–12238
12. Brezovich A, Schuschnig M, Ammerer G, Kraft C (2015) An *in vivo* detection system for transient and low-abundant protein interactions and their kinetics in budding yeast. Yeast 32: 355–365
13. Hollenstein DM, Gómez-Sánchez R, Ciftci A et al (2019) Vac8 spatially confines autophagosome formation at the vacuole. J Cell Sci 132: jcs.235002
14. Hollenstein DM, Licheva M, Konradi N et al (2021) Spatial control of avidity regulates initiation and progression of selective autophagy. Nat Commun 12:7194
15. Longtine MS, Mckenzie A III, Demarini DJ et al (1998) Additional modules for versatile and economical PCR-based gene deletion and modification in Saccharomyces cerevisiae. Yeast 14:953–961
16. Cardenas ME, Heitman J (1995) FKBP12-rapamycin target TOR2 is a vacuolar protein with an associated phosphatidylinositol-4 kinase activity. EMBO J 14:5892–5907

# Chapter 4

## Assays to Study Mitophagy in Plants

**Chengyang Li, Juncai Ma, Tian Zhen, Byung-Ho Kang, and Pengwei Wang**

### Abstract

Like most eukaryotic cells, mitophagy is essential in plant development and stress response. Several recent studies have revealed proteins that regulate this process, such as Friendly (FMT) and TraB family proteins (TRB), which are plant-unique mitophagy regulators so far. Here, we describe methods for studying mitophagy activity in plants through conventional microscopy and the use of loss-of-function mutants, such as using transgenic mitochondrial marker lines followed by image analysis, chemical inhibitor treatment, and plant phenotype studies. These methods can be used in combination to identify the putative mitophagy regulators and understand their functions in mitochondrial-related activities in plants.

**Key words** *Nicotiana benthamiana*, Arabidopsis thaliana, Mitochondria, Mitophagy, De-etiolation

## 1 Introduction

To maintain cellular homeostasis, damaged mitochondria need to be recognized and engulfed by the autophagosome membrane for degradation [1]. This process is known as mitophagy, which is involved in various biological processes, such as plant development, de-etiolation, stress response, and programmed cell death [2]. Mitophagy is initiated by the recognition of unwanted mitochondria, followed by their engulfment by autophagosome membranes. This process is regulated by core autophagy proteins and mitophagy receptors [2, 3]. Most of the receptor proteins localize/recruit to the outer membrane of mitochondria and interact with ATG8 through a conserved sequence (ATG8 interaction motif, AIM) [4, 5]. Although our understanding of mitophagy has increased dramatically over the last few years [6–12], the knowledge of mitophagy receptors is still limited in plants, possibly due to the lack of molecular markers and the difficulty in quantifying mitophagy levels in plant cells.

The authors Chengyang Li and Juncai Ma have equally contributed to this chapter.

Ioannis P. Nezis (ed.), *Selective Autophagy: Methods and Protocols*, Methods in Molecular Biology, vol. 2845,
https://doi.org/10.1007/978-1-0716-4067-8_4,

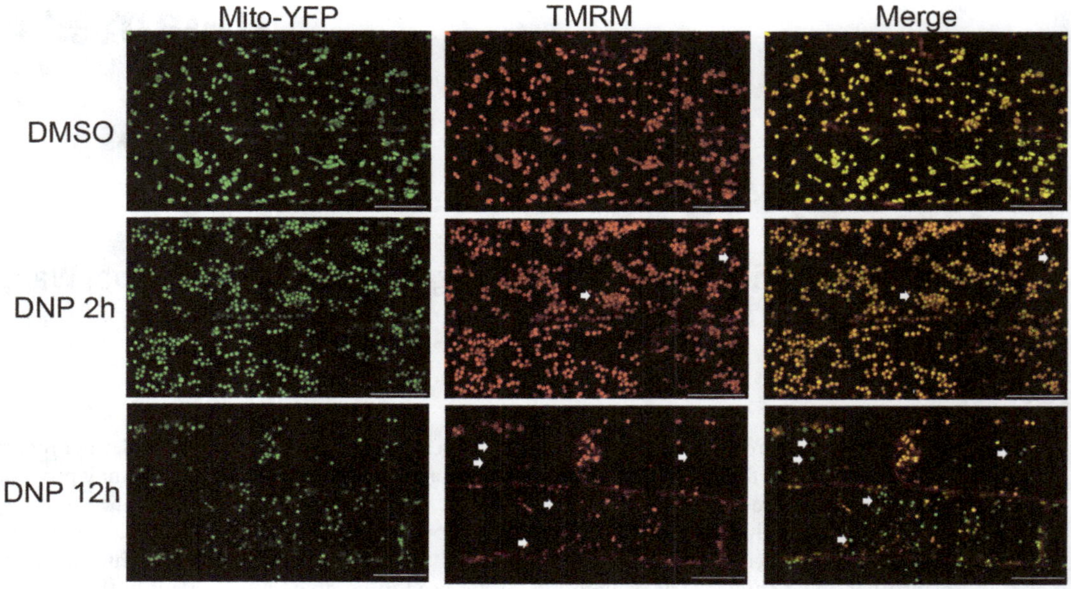


**Fig. 1** Transgenic Arabidopsis expressing Mito-YFP treated with DNP to induce mitochondrial depolarization. After 2 h of treatment, a few mitochondria are weakly labeled by TMRM (arrow); after prolonged treatment (12 h), the majority of mitochondria have lost their membrane potential and cannot be stained by TMRM (arrows pointing to the mitochondria that are Mito-YFP positive but TMRM negative). (The image is from a previous publication [9])

Cells from the root elongation zone or leaf epidermal tissue are commonly used for studying organelle dynamics and autophagy in plants [13, 14]. However, mitophagosomes are normally difficult to capture in these systems under normal growth conditions. Therefore, various drug treatments are needed to facilitate the observation and quantification of mitophagy activity. One example is concanamycin A (Conc A), which is a V-ATPase inhibitor that inhibits lytic vacuole activities [9, 13]. Mature mitophagosomes are transported to vacuoles for degradation, and vacuole-accumulated mitochondria enclosed by membranes are a clear indication of mitophagy at the ultrastructural level [8]. These structures were much more abundant when cells were treated with Conc A, and the activity of mitophagy can be estimated based on their numbers in the vacuole. Some chemical inhibitors are also widely used to induce mitophagy activity, such as 2,4-dinitrophenol (DNP), a mild mitochondrial uncoupler that can be used to induce mitochondrial depolarization and mitophagy in Arabidopsis (Fig. 1). Moreover, tetramethylrhodamine ethyl ester perchlorate (TMRE) and tetramethylrhodamine (TMRM) are fluorescent dyes that are sensitive to mitochondrial membrane potential (MMP). Both of them can be used to study mitochondrial activity and mitophagy at the light microscopy level [8, 9] as well.

On the other hand, transmission electron microscopy (TEM) is also a useful method to study the process of mitophagy [8]. Recently, we performed TEM analysis, TMRE staining, and biochemical assays to study the contribution of mitochondria in the de-etiolation of Arabidopsis (a process that a plant shoot undergoes in response to light, also known as greening) and demonstrated that this process also leads to the accumulation of compromised mitochondria and promotes their mitophagy degradation [8]. In this chapter, we present methods that we developed recently for studying mitophagy in plants, at both cell and whole-organism levels.

## 2 Materials

### 2.1 Plant Materials and Medium

1. Both wild-type and transgenic *Nicotiana benthamiana* (stably expressing a mitochondrial marker) are grown on soil in a plant growth room (24 °C, 70% relative humidity, 14 h light/10 h dark cycle) (*see* **Note 1**).
2. Arabidopsis (Col-0, mutants, or genetically transformed lines) are grown on MS medium in a growth chamber (22 °C, 70% relative humidity, 14 h light/10 h dark cycle).
3. Murashige and Skoog (MS) medium: one liter of solid medium contains 4.6 g MS basal salt mixture, 10 g sucrose, and 8 g plant agar.
4. Liquid MS medium: prepare as above without plant agar.

### 2.2 Microscopy--Related Materials

1. TMRE or TMRM fluorescent dye (Thermo Scientific).
2. Glass slides and coverslips.
3. Razor blade.
4. Leica Microsystem Immersion Oil Type F.

### 2.3 Plasmids and Agrobacterium-Mediated Transformations

1. Mitochondria marker: Mito-mCherry [15].
2. Mitophagy receptor: GFP-TRB1 [9].
3. Suppressor of posttranscriptional gene silencing (PTGS): pCB301-p19 [16].
4. Agrobacterium strain: GV3101.

### 2.4 Chemical Inhibitor Treatment

1. 2,4-dinitrophenol (DNP, Sigma) stock solution: 50 mM DNP in DMSO and store at −20 °C before use.
2. Concanamycin A (Conc A, APExBio) stock solution: 1 mM Conc A in DMSO and store at −20 °C before use.
3. Dimethyl sulfoxide (DMSO).
4. Tissue culture plate (12 or 24 wells).

### 2.5 Immunoblot

1. 1× PBST buffer: 0.137 M sodium chloride, 0.0027 M potassium chloride, 0.008 M sodium phosphate dibasic, 0.002 M potassium phosphate monobasic, and 0.05% Tween-20.
2. Lysis buffer: 50 mM Tris-HCl (pH 8.0), 150 mM NaCl, 0.5 mM ethylenediaminetetraacetic acid (EDTA), 0.5% (v/v) Triton X-100, 1× Complete Protease Inhibitor Cocktail (Roche).
3. 5× Laemmli sample buffer: 10% SDS, 500 mM DTT, 50% glycerol, 500 mM Tris-HCl (pH 6.8), 0.05% bromophenol blue dye.
4. 10× SDS-PAGE running buffer: weigh 25 g Tris base, 175 g glycine, and 10 g SDS, dissolve in 0.8 L distilled water, and adjust the volume to 1 L.
5. 1× SDS-PAGE running buffer: dilute 100 mL of 10× SDS-PAGE running buffer in 900 mL of distilled water.
6. 10× transfer buffer: weigh 33 g Tris base and 144 g glycine, dissolve in 0.8 L distilled water, and adjust the volume to 1 L.
7. 1× transfer buffer: dilute 100 mL of 10× transfer buffer in 700 mL distilled water and add 200 mL methanol.
8. Primary antibody: ATG8 (Agrisera), Actin (ACT; Agrisera), voltage-dependent anion channel 1 (VDAC1; Agrisera), L-galactono-1,4-lactone dehydrogenase (GLDH; Agrisera), and isocitrate dehydrogenase (IDH; Agrisera).
9. Secondary antibody: Goat anti-rabbit IgG.

### 2.6 Electron Microscopy

1. Cryoprotectant for HPF: 0.15 M sucrose solution.
2. Cryosubstitution medium: 2% OsO4 in acetone.
3. Embed-812 resin (Electron Microscopy Sciences).
4. Uranyl acetate solution: 0.75 g uranyl acetate dissolved in 25 mL 70% MeOH.
5. Lead citrate solution: 0.665 g lead (II) nitrate and 0.880 g sodium citrate ($Na_3(C_6H_5O_7)\cdot 2H_2O$) dissolved in 24 mL degassed distilled water, and add 1 mL $CO_2$-free 4 M NaOH to facilitate precipitate dissolution.

### 2.7 Equipment

1. Microcentrifuge.
2. Western blotting apparatus.
3. ChemiDoc MP imaging system.
4. Confocal microscope.
5. High-pressure freezer, type B aluminum planchettes for HPF, and freeze-substitution device.
6. Transmission electron microscope.

7. Ultramicrotome and diamond knife.
8. Formvar film-coated slot grids (2 mm × 1 mm slot copper grids).
9. Digital camera.

#### 2.8 Computer Programs

1. Adobe Photoshop.
2. ImageJ.
3. GraphPad Prism.

## 3 Methods

Please note that the methods described in Subheadings 3.1, 3.2, 3.3, 3.4 and 3.5 are based on microscopical observations at different scales; some biochemical assays are not discussed, such as ATG8-PE and free GFP assays [13, 17] that determine the autophagic flux. We strongly recommend using multiple approaches for mitophagy studies in plants.

### 3.1 Quick Identification of Mitophagy Regulators Using the N. benthamiana Transient Expression System

The method described here can be used for a quick determination of putative mitophagy regulators using agrobacteria-mediated transient expression and Conc A treatment.

1. Use 5-week-old *N. benthamiana* transgenic plants (mito-mCherry) and infiltrate Agrobacterium carrying the candidate constructs (e.g., GFP-TRB1) into the lower leaf epidermal cells (*see* **Note 1**).
2. After 24 h of recovery, infiltrate 1 μM Conc A (or DMSO as a control) into the same leaf segments that have been pre-infiltrated.
3. Normally 24–36 h after Conc A treatment, dissect a leaf section (25–50 $mm^2$) and mount it on a slide with a coverslip.
4. Using a laser scanning confocal microscope, capture images from the middle section of cells, where the central vacuole and accumulated cargoes are visible (*see* **Note 2**). Normally, 15–20 images are collected for statistical analysis.
5. Using ImageJ, adjust the image background to an appropriate level (Image—Adjust—Threshold, select Dark background—Apply). Then, use the Analyze Particles Plugin to count the number of mitochondria in the vacuole (Analyze—Analyze Particles—Set the Size to 0.2) (*see* **Note 3**).
6. To export the result, select Display Results, Clear Results, and Summarize. Then, use GraphPad Prism to generate a bar chart and perform statistical analysis.

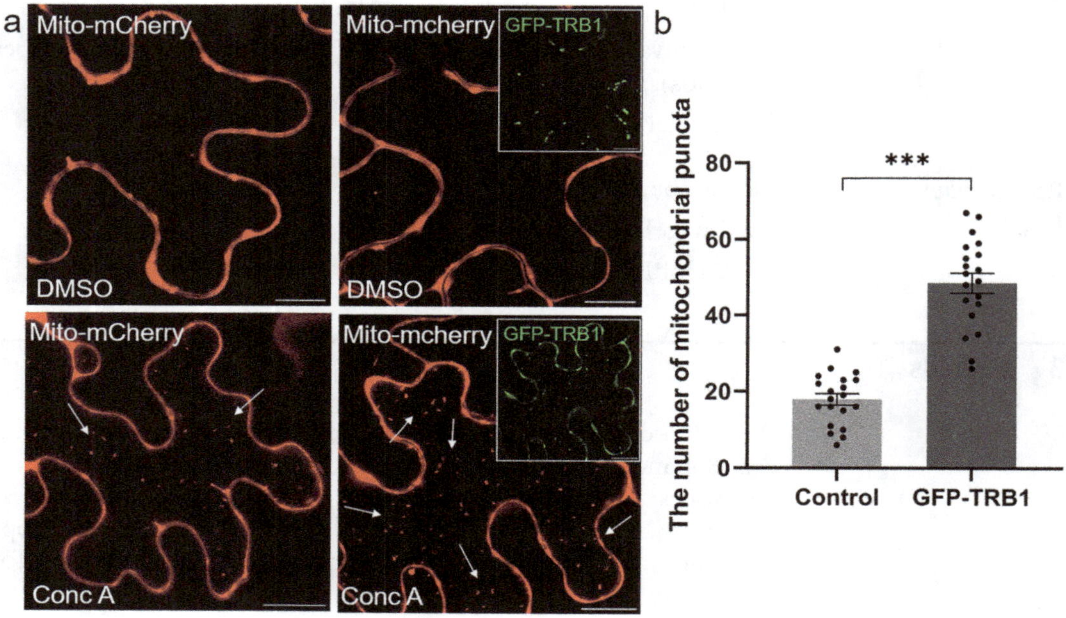


**Fig. 2** DNP and Conc A treatment of transgenic *N. benthamiana* stably expressing Mito-mCherry. (**a**) In the leaf epidermal cells, *GFP-TRB1* overexpression induced more vacuole-accumulated mitochondria after Conc A treatment. (**b**) Statistical analysis of (**a**) suggests an increased level of mitophagy. The white arrows indicate the signals of mitochondria in the vacuole. $n = 20$ for mitochondrial puncta analysis, $^{***}P < 0.001$ in two-tailed unpaired $t$ tests (Scale bar = 10 μm)

7. As shown in Fig. 2a, the images show the cross-section of the vacuole of leaf epidermal cells, with accumulated mitochondria/mitophagy-related structures visible after Conc A treatment. In contrast, when DMSO treatment is applied, no signal can be identified in the vacuole. According to the quantification results (Fig. 2b), we suggest that the transient expression of GFP-TRB1 promotes mitophagy and mitochondria-to-vacuole transport.
8. In addition, co-expression studies with autophagy markers (e.g., ATG8a, ATG8e) can also be used to examine if such vacuole-accumulated mitochondria are mitophagosome-related [9].

### 3.2 Mitochondrial Membrane Potential (MMP) Test After Uncoupler Treatment

TMRM staining identifies mitochondria with reduced MMP after DNP treatment [9], which is a sign of damaged mitochondria. In cells with dysfunctional mitophagy (e.g., *trb1 trb2* mutant), more damaged mitochondria are expected to accumulate in the cytoplasm during mitochondrial stress.

1. Grow Arabidopsis on solid MS medium, and use the seedlings 5–7 days after germination.

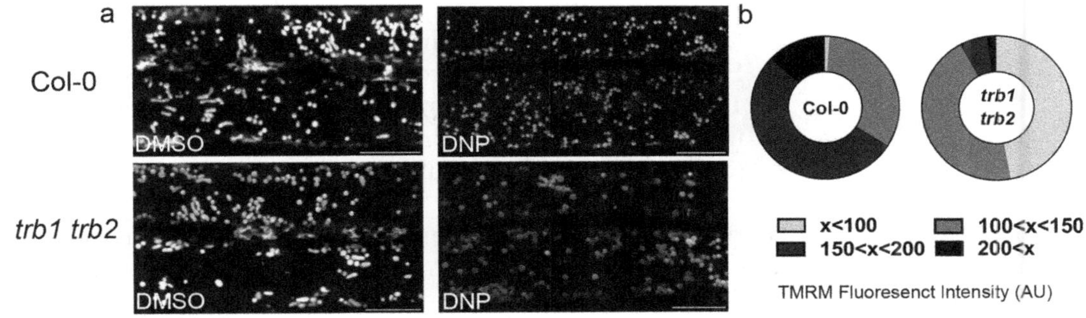


**Fig. 3** DNP treatment and TMRM labeling of Arabidopsis. (**a**) Arabidopsis (Col-0 and *trb1trb2*) were treated with DNP for 2 h, and their mitochondrial membrane potential was labeled by TMRM. (**b**) In the *trb1trb2* mutant, the proportion of mitochondria with reduced TMRM labeling is significantly increased, suggesting that the large population of mitochondria accumulated in the cytoplasm is dysfunctional in the mutants. $n = 100$ mitochondrial numbers examined over 20 cells for mitochondrial membrane potential analysis (Scale bar = 10 μm). (The image is from a previous publication with some modifications [9])

2. Prepare a fresh DNP working solution (50 μM in liquid MS medium) and transfer Arabidopsis seedlings to a 12-well cell culture plate containing the DNP working solution, and incubate them for 2 h (*see* **Note 4**).
3. Prepare a fresh TMRM working solution (500 nM in liquid MS medium), transfer the DNP-treated or control Arabidopsis seedlings onto a glass slide, and add an appropriate amount of TMRM staining solution. After staining for 10 min, wash out the staining solution and replace it with liquid MS medium, then put a coverslip on.
4. Using a laser scanning confocal microscope, image the cortical section of the cells from the root elongation zone (*see* **Note 5**). The excitation and detection setting for TMRM are 552 nm and 590–640 nm, respectively. Use a 63X or 100X oil objective, and collect 15–20 images for statistical analysis.
5. Use ImageJ, adjust all images to 8-bit. Then, use the 'oval' selection to measure the mean grey value of each mitochondrion (Analyze—Set measurements, select mean grey value—Oval selection—Measure). Normally, 10 mitochondria are selected randomly for each cell, and at least 100 mitochondria are used in total.
6. According to their fluorescence intensity, the mitochondria are classified into different categories. For example, fluorescence intensity smaller than 100 AU likely represents damaged mitochondria with reduced MMP.
7. Classify and count the number of mitochondria in each category (Fig. 3), and generate a chart using GraphPad Prism.

8. As shown in Fig. 3a, an overall decrease in TMRM signal is found after DNP treatment in different genetic backgrounds. However, the ratio of mitochondria with a signal lower than 150 AU increases in the *trb1trb2* mutant (Fig. 3b), indicating more damaged mitochondria are accumulated in the cytoplasm (as a result of dysfunctional mitophagy).

### 3.3 DNP and Conc A Treatment of Transgenic Arabidopsis Expressing Mitochondrial Markers

The method described here uses wild-type Arabidopsis and the respective loss-of-function mutants for comparative studies, providing direct functional evidence of mitophagy regulators.

1. Grow Arabidopsis on solid MS medium, and use the seedlings 5–7 days after germination.
2. Transfer Arabidopsis seedlings to a 12-well cell culture plate containing fresh Conc A working solution (1 μM in liquid MS), and incubate them overnight for about 12 h.
3. The next day, add DNP stock to the Conc A liquid MS to a final concentration of 50 μM (*see* **Note 6**), and further incubate the seedlings for 2 h.
4. Mount the Arabidopsis seedlings on a glass slide with a coverslip. Use a laser scanning confocal microscope to image the middle section of the cells from the root elongation zone. Normally, 15–20 images are collected for statistical analysis.
5. Perform image adjustment and statistical analysis as described in Subheading 3.1, **steps 5** and **6**.
6. As shown in Fig. 4, the vacuole-accumulated mitochondrial signals are reduced in the *trb1trb2* mutant after chemical treatment, suggesting impaired activity.

Mitochondrial damage releases reactive oxygen species and induces cell stress [18]. Thus, unsuccessful removal of damaged/

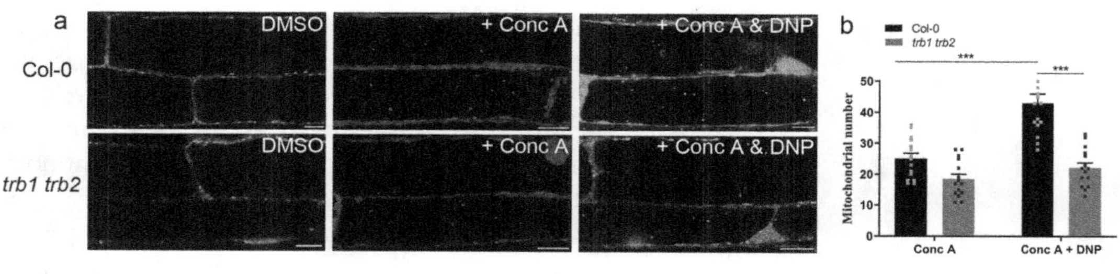


**Fig. 4** DNP and Conc A treatment of transgenic Arabidopsis expressing Mito-mCherry. (**a**–**b**) Under non-stressed conditions, the mitochondrial turnover rate is similar in both plants. However, the number of vacuole-accumulated mitochondria is significantly reduced in the *trb1trb2* mutants treated with DNP, suggesting that the turnover of damaged mitochondria is blocked. $n = 20$ cells examined over three independent experiments for confocal quantification, error bars are SEM, *$0.01 < P < 0.05$, ***$P < 0.001$ in two-way ANOVA, Sidak's multiple comparisons test (Scale bar = 10 μm). (The image is from a previous publication with some modifications [9])

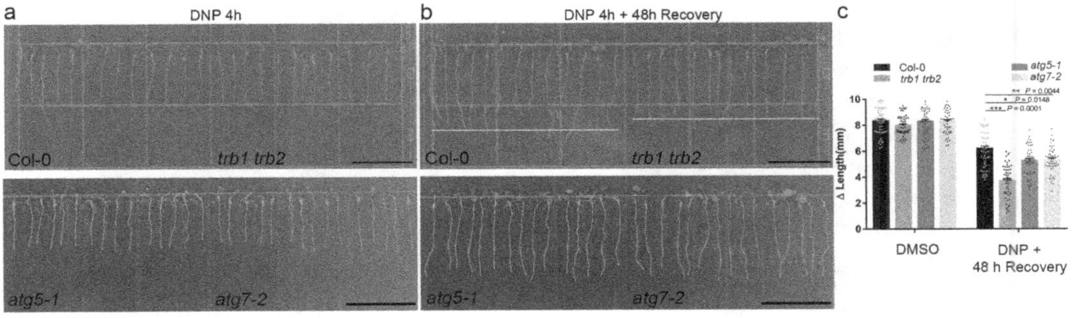


**Fig. 5** The recovery of Arabidopsis seedlings for 48 h after DNP treatment of Arabidopsis seedlings. (**a**) DNP treatment for 4 h; (**b**) recovery after 48 h of DNP treatment; (**c**) statistical analysis. The root growth after recovery (Δlength) of the mutant plants is significantly reduced compared to that in the wild type, suggesting that the DNP-induced mitochondrial dysfunction has more impact on plant development when the function of TRB1 and TRB2 is impaired. $n = 90$ independent plants for root analysis, error bars are SEM, *$0.01 < P < 0.05$, **$0.001 < P < 0.01$, ***$P < 0.001$ in two-way ANOVA, Dunnett's multiple comparisons test (Scale bar = 15 mm). (The image is from a previous publication with some modifications [9])

### 3.4 Phenotype Analysis of Arabidopsis Recovery After DNP Treatment

depolarized mitochondria may have negative impacts on plant growth [8–10]. This method aims to study the involvement of mitophagy under stress at the whole organism level.

1. Transfer 5-day-old Arabidopsis seedlings onto MS solid medium containing 50 μM DNP or DMSO, and leave them for 4 h (*see* **Note** 7).
2. Transfer seedlings onto solid MS medium without drugs to recover and take images using a digital camera (Fig. 6a).
3. Let the plant recover for 2 days and take the second image with the same camera settings (Fig. 6b).
4. With ImageJ, select the entire roots and measure the length using the 'Segmented Line' tool.
5. The root length measured from the first and second images is used to calculate the value of Δlength, which is used to generate a bar chart with GraphPad Prism.
6. As shown in Fig. 5, both mitophagy mutant (*trb1trb2*) and autophagy mutant (*atg5-1, ag7-2*) plants exhibit relatively slow recovery after DNP treatment compared to the wild-type plants (*see* **Note 8**).

### 3.5 TEM Analysis of Mitophagy During De-etiolation of Arabidopsis Seedlings

The ultrastructural study of mitophagosomes is crucial for mitophagy analysis. Here, we applied TEM to study mitophagy in Arabidopsis cotyledon cells during de-etiolation as an example (Fig. 6).

1. Grow seedlings on solid half MS medium horizontally in the dark at 22 °C for 5 days, and start illumination for 0–12 h (*see* **Note 9**).

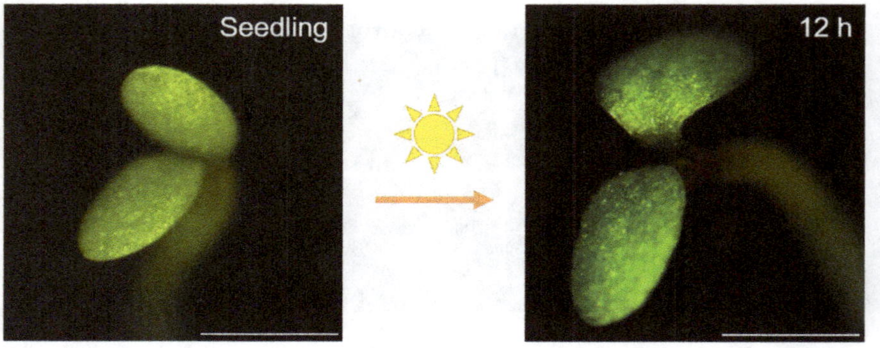


**Fig. 6** De-etiolation of Arabidopsis seedlings after light illumination. Illumination for 12 h enables Col-0 seedlings to complete the de-etiolation process (Scale bar = 1 mm)

2. Dissect cotyledons from 5-day-old Arabidopsis seedlings quickly and transfer them to planchettes with 0.15 M sucrose solution (*see* **Note 10**). Freeze the cotyledon samples with a Leica ICE HPF machine, followed by freeze-substitution at −80 °C for 24 h in anhydrous acetone containing 2% OsO4 in acetone using an AFS machine.
3. Warm up the samples slowly to room temperature over 48 h and remove excess OsO4 by rinsing with anhydrous acetone three times. Separate cotyledon samples from planchettes and embed them in Embed-812 resin.
4. After resin polymerization, cut out cotyledon samples and mount them on plastic stubs with super glue.
5. Trim the block (trapezoidal block face) and prepare 100 nm thin sections using an ultramicrotome. Collect section ribbons on slot grids and air-dry (*see* **Note 11**).
6. Post-stain sections with uranyl acetate for 2 min and lead citrate for 4 min. Wash the sections with distilled water and air-dry.
7. Image the sections by TEM (e.g., Hitachi 7400 TEM, Hitachi-High Technologies).
8. As shown in Fig. 7, mitophagosomes of cotyledon cells can be observed when dark-grown seedlings are exposed to light for 6–8 h.

### 3.6 Immunoblot Analysis of Mitophagy During De-etiolation of Arabidopsis Seedlings

In addition, western blot-based flux assays with mitochondrial-specific antibodies revealed changes in mitochondrial protein levels in Arabidopsis seedlings.

1. Grow seedlings and start illumination as in Subheading 3.5, **step 1**.
2. Freeze the seedlings in a mortar using liquid nitrogen and quickly grind the frozen plants.

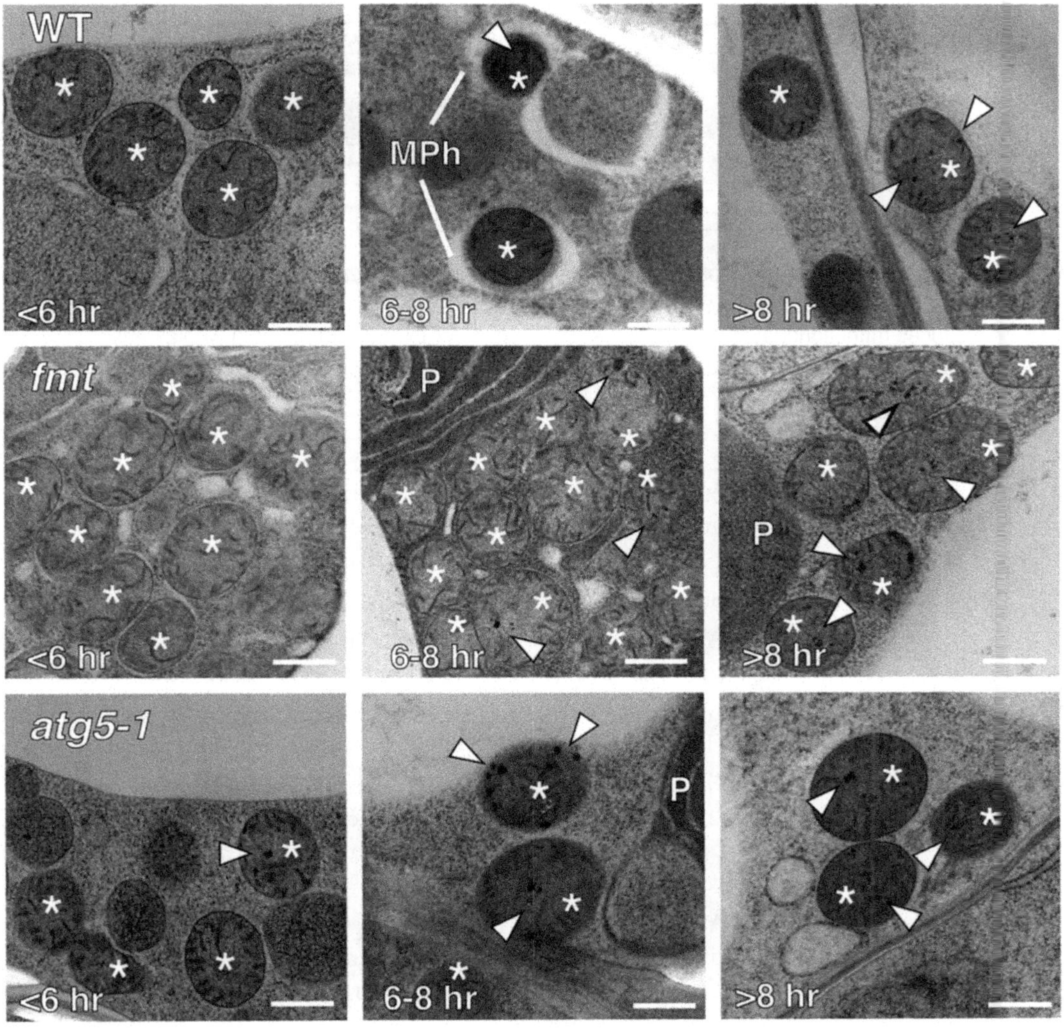


**Fig. 7** TEM micrographs of mitochondria (asterisks) in WT, *fmt*, and *atg5-1* cotyledon cells after light illumination (<6 h, 6–8 h, and > 8 h). Densely aggregated mitochondria in WT, *fmt*, and *atg5-1* are marked with arrowheads. P indicates plastid (Scale bar = 500 nm). (The image is from a previous publication with some modifications [8])

3. Add lysis buffer containing PIC without detergent (Triton X-100) into the mortar on ice.
4. After slow defrosting on ice, transfer the liquid samples to 1.5 mL tubes and centrifuge at 10,000× *g* for 10 min at 4 °C.
5. Gently transfer the supernatant (soluble fraction) into new 1.5 mL tubes and add the 5× sample loading dye to the soluble fraction samples.
6. Boil the samples at 95 °C for 10 min.
7. Prepare the 15% SDS-PAGE gel.

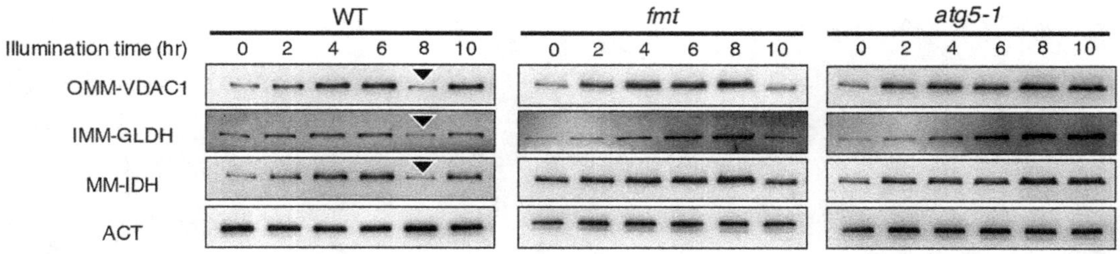


**Fig. 8** Immunoblot analyses of mitochondrial proteins in WT, *fmt*, and *atg5-1* seedlings during de-etiolation. VDAC1, GLDH, IDH, and actin (ACT) were examined for an OMM, an IMM, a matrix (MM), and a loading control protein. Reduced mitochondrial proteins at 8 h samples from WT extracts are marked with asterisks in WT. No such changes were observed in the *atg5-1* and *fmt* mutant lines. (The image is from a previous publication with some modifications [8])

8. Load the samples and run at 80 V for 30 min with 1× running buffer. Resolve the samples for an additional 60 min at 120 V until the blue front band reaches the bottom of the gel.
9. Transfer resolved polypeptides from the gel to a PVDF membrane at 80 V for 90 min with 1× transfer buffer.
10. After blotting, soak the blotted membrane in 1× PBST with 5% milk powder for 1 h. Wash with PBST three times.
11. Incubate the membrane with primary antibodies against mitochondrial proteins (5 μg/μL in PBST) for 2 h at room temperature and then wash with PBST three times.
12. Incubate the membrane with corresponding secondary antibodies (1:8000 in PBST) for 1 h at room temperature and then wash with PBST three times.
13. Add Western blotting detecting reagents (ECL Western Blotting Substrates) onto the membrane and expose it to the ChemiDoc MP imaging system (Fig. 8).

## 4 Notes

1. The transgenic *N. benthamiana* plants expressing a mitochondrial marker (Mito-mCherry) are described in a previous study [19]. For these experiments, it is essential to use stable plants, as the expression of Mito-Cherry is consistent from cell to cell.
2. The time for image acquisition should be reduced to a minimum, as the vacuole-accumulated fluorescence signals move very fast.
3. The SIZE should be adjusted based on the background of each image. If the background is very clean, it can be set to 0.1–0.15; whereas, if the background signal is noisy, it could be set to 0.2–0.25. Generally, the initial setting of SIZE needs

to be compared with the visual counting results to obtain an optimized setup.

4. For all drug treatments performed in this study, the liquid MS medium supplemented with an equal amount of DMSO is used as the control.
5. For comparison, images need to be taken from the same focal plane of a cell and the same section of the root tissue. Furthermore, microscopical settings (e.g., laser intensity, gain value) should remain consistent for different materials. For example, use TMRM-labeled mitochondria in the wild-type as a reference and optimize the settings to avoid overexposure.
6. Gently mix the solution with a dropper. Avoid using a pipette with fine tips, which may accidentally cause damage.
7. Let the MS medium cool down to below 50 °C before adding DNP. The transfer of Arabidopsis seedlings needs to be as gentle as possible; do not try to position and drag the seedlings on the solid medium, as this can cause damage and prevent them from growing.
8. Here, the classical autophagic mutants *atg5-1* and *atg7-2* were used as positive controls.
9. Consistent illumination intensity is crucial for reproducible induction of mitophagy during the de-etiolation process. We recommend using a light intensity of 130–150 μmol/($m^2 \cdot$ s) to illuminate etiolated Arabidopsis seedlings, and 5-day-old seedlings with adequate nutrients are optimal for inducing mitophagy.
10. The basal area of cotyledons connected to the apical hook area is more suitable for examining mitophagy and mitochondrial depolarization.
11. We recommend slot grids instead of mesh or finder grids for serial section analysis because areas of interest could be blocked from imaging by the metallic grid bars. Protocols for serial sectioning, immunogold labeling, and electron microscopy analysis of Arabidopsis tissues including root and cotyledon samples are described in Kang [20], Liang et al. [21], and Liang et al. [22].

## Acknowledgments

The project was supported by the National Natural Science Foundation of China (NSFC) grants (nos. 32261160371 and 92254307), the Fundamental Research Funds for the Central Universities (2662023PY011), and the Young Scientist Fostering Funds for the National Key Laboratory for Germplasm Innovation and Utilization of Horticultural Crops (Horti-PY-2023-001) for P.W. This work was also supported by the Hong Kong Research Grant Council (GRF 14110823, GRF 14109222, GRF 14113921, N_CUHK462/22, and C4002-20 W) and the Chinese University of Hong Kong for B-H.K.

## References

1. Vargas JNS, Hamasaki M, Kawabata T, Youle RJ, Yoshimori T (2023) The mechanisms and roles of selective autophagy in mammals. Nat Rev Mol Cell Biol 24:167–185
2. Broda M, Millar AH, Van Aken O (2018) Mitophagy: a mechanism for plant growth and survival. Trends Plant Sci 23:434–450
3. Palikaras K, Lionaki E, Tavernarakis N (2018) Mechanisms of mitophagy in cellular homeostasis, physiology and pathology. Nat Cell Biol 20:1013–1022
4. Noda NN, Ohsumi Y, Inagaki F (2010) Atg8-family interacting motif crucial for selective autophagy. FEBS Lett 584(7):1379–1385
5. Jacomin AC, Samavedam S, Promponas V, Nezis IP (2016) iLIR database: a web resource for LIR motif-containing proteins in eukaryotes. Autophagy 12(10):1945–1953
6. Ren K, Feng L, Sun S, Zhuang X (2021) Plant mitophagy in comparison to mammals: what is still missing? Int J Mol Sci 22(3):1236
7. Li F, Chung T, Vierstra RD (2014) Autophagy-related11 plays a critical role in general autophagy- and senescence-induced mitophagy in Arabidopsis. Plant Cell 26(2): 788–807
8. Ma J, Liang Z, Zhao J et al (2021) Friendly mediates membrane depolarization-induced mitophagy in Arabidopsis. Curr Biol 31(9): 1931–1944
9. Li C, Duckney P, Zhang T et al (2022) TraB family proteins are components of ER-mitochondrial contact sites and regulate ER-mitochondrial interactions and mitophagy. Nat Commun 13(1):5658
10. Yan H, Zhuang M, Xu X et al (2023) Autophagy and its mediated mitochondrial quality control maintain pollen tube growth and male fertility in Arabidopsis. Autophagy 19(3): 768–783
11. Duckney P, Li C, Hussey PJ, Wang P (2023) TraB, a novel plant ER-mitochondrial contact site protein functions as a mitophagy receptor in plants. Autophagy 19(7):2148–2150
12. Kacprzak SM, Van Aken O (2023) FRIENDLY is required for efficient dark-induced mitophagy and controlled senescence in Arabidopsis. Free Radic Biol Med 204:1–7
13. Kacprzak SM, Van Aken O (2022) Carbon starvation, senescence and specific mitochondrial stresses, but not nitrogen starvation and general stresses, are major triggers for mitophagy in Arabidopsis. Autophagy 18(12): 2894–2912
14. Sparkes I, Runions J, Kearns A et al (2006) Rapid, transient expression of fluorescent fusion proteins in tobacco plants and generation of stably transformed plants. Nat Protoc 1: 2019–2025
15. Nelson BK, Cai X, Nebenführ A (2007) A multicolored set of in vivo organelle markers for co-localization studies in Arabidopsis and other plants. Plant J 51(6):1126–1136
16. Voinnet O, Rivas S, Mestre P et al (2003) An enhanced transient expression system in plants based on suppression of gene silencing by the p19 protein of tomato bushy stunt virus. Plant J 33:949–956
17. Luo M, Zhuang X (2018) Analysis of autophagic activity using ATG8 lipidation assay in *Arabidopsis thaliana*. Bio Protoc 8(12):e2880
18. Huang S, Van Aken O, Schwarzländer M, Belt K, Millar AH (2016) The roles of mitochondrial reactive oxygen species in cellular signaling and stress response in plants. Plant Physiol 171(3):1551–1559
19. Guo Y, Bao Z, Deng Y, Li Y, Wang P (2022) Protein subcellular localization and functional studies in horticultural research: problems, solutions, and new approaches. Hortic Res 10(2):uhac271

20. Kang BH (2010) Electron microscopy and high-pressure freezing of Arabidopsis. Methods Cell Biol 96:259–283
21. Liang Z, Zhu N, Mai KKK, Liu Z, Tzeng D, Osteryoung KW, Zhong S, Staehelin LA, Kang BH (2018) Thylakoid-bound polysomes and a dynamin-related protein, FZL, mediate critical stages of the linear chloroplast biogenesis program in greening arabidopsis cotyledons. Plant Cell 30(7):1476–1495
22. Liang Z, Yeung WT, Ma J, Mai KKK, Liu Z, Chong YF, Cai X, Kang BH (2022) Electron tomography of prolamellar bodies and their transformation into grana thylakoids in cryo-fixed Arabidopsis cotyledons. Plant Cell 34(10):3830–3843

# Chapter 5

# Assessing Neuronal Mitophagy During Development in the Nematode *Caenorhabditis elegans*

Christina Ploumi, Antonis Roussos, and Konstantinos Palikaras

## Abstract

Preserving mitochondrial homeostasis is vital, particularly for the energetically demanding and metabolically active nerve cells. Mitophagy, the selective autophagic removal of mitochondria, stands out as a prominent mechanism for efficient mitochondrial turnover, which is crucial for proper neuronal development and function. Dysfunctional mitochondria and disrupted mitophagy pathways have been linked to a diverse array of neurological disorders. The nematode *Caenorhabditis elegans*, with its well-defined nervous system, serves as an excellent model to unravel the intricate involvement of mitophagy in developing neurons. This chapter describes the use of Rosella biosensor in *C. elegans* to monitor neuronal mitophagy, providing a user-friendly platform for screening genes and drugs affecting mitophagic pathways under physiological conditions or in the context of neurodevelopmental pathologies.

**Key words** Mitophagy, Mitochondria, Neurons, Development, *Caenorhabditis elegans*, Neurodevelopmental diseases, Rosella biosensor

## 1 Introduction

Mitochondrial function is pivotal for upholding cellular and organismal homeostasis. Beyond serving as primary energy-producing hubs, mitochondria actively coordinate a diverse array of vital processes that regulate metabolism, cellular fate, immunity, and adaptation to several stress responses [1, 2]. Disruptions in mitochondrial abundance, structure, and function can often be detrimental to organismal physiology. To counteract such interventions, cells have evolved intricate quality control mechanisms, including fusion and fission events (widely known as mitochondrial dynamics) and the mitochondrial unfolded protein response ($UPR^{mt}$), to prevent or promptly repair mitochondrial damage. In parallel, or as a downstream response to failures in the aforementioned mechanisms, mitochondrial biogenesis and clearance pathways are coordinated to preserve a healthy pool of mitochondria,

Ioannis P. Nezis (ed.), *Selective Autophagy: Methods and Protocols*, Methods in Molecular Biology, vol. 2845, https://doi.org/10.1007/978-1-0716-4067-8_5,

indispensable not only for energy production but also for several other processes [3, 4].

Mitophagy is the predominant catabolic pathway for mitochondrial turnover. It entails the selective sequestration of superfluous or damaged mitochondria into autophagosomes, followed by their subsequent fusion with lysosomes, where they are finally degraded. Mitophagic pathways can occur through ubiquitin-dependent or independent processes in the presence of highly specialized receptor and adaptor proteins that respond to various intracellular and environmental cues [5, 6]. The PINK1/Parkin pathway, named after the PTEN-induced putative kinase 1 (PINK1) and the E3 ubiquitin ligase Parkin (PARK2/PRKN), is one of the most well-studied mitophagy mechanisms [7, 8]. Under normal conditions, newly synthesized PINK1 is rapidly transferred to the mitochondrial intermembrane space, where it undergoes proteolytic cleavage by the residual presenilin-associated rhomboid-like protease (PARL). Truncated PINK1 is then released to the cytosol, where it is subjected to ubiquitination and degradation through the ubiquitin-proteasome system (UPS) [9–11]. In depolarized mitochondria, PINK1 processing and degradation are blocked and thus it is stabilized in the outer mitochondrial membrane (OMM), where it recruits Parkin and activates its E3 ubiquitin ligase activity [12]. In turn, activated Parkin ubiquitinates OMM-localized proteins, which attract ubiquitin-binding adaptors to mediate the recruitment of the autophagic machinery and the execution of mitochondrial degradation [6, 13]. There have also been characterized OMM receptors of protein or lipid nature that mediate the engulfment of damaged mitochondria into autophagosomes independently of ubiquitination. Notable examples include BCL2/adenovirus E1B 19-kDa-interacting protein 3 (BNIP3), BNIP3-Like/NIP-3-Like Protein X (BNIP3L/NIX), FUN14 domain-containing protein 1 (FUNDC1), and cardiolipin [14].

Maintaining mitochondrial health is of paramount importance, especially for the nervous system, to support its vital functions throughout development and aging. Mitochondria-dependent metabolic rewiring is a critical step for the activation of quiescent neural stem cells and for efficient dendritogenesis and synaptogenesis at later developmental stages [15, 16]. Proper mitochondrial activity is also crucial to ensure the survival and function of mature postmitotic neurons. Neuronal cells are capable of transmitting information by propagating electrical signals over extremely long distances, a complex task that requires huge amounts of energy supply. As such, neurons are highly dependent on mitochondria for adenosine triphosphate (ATP) production and for calcium buffering to support synaptic transmission and plasticity. It is not surprising, then, that failure in mitochondrial quality control mechanisms is among the major contributing factors for the onset and progression of neurodegeneration [17]. Indeed, an ever-expanding amount of studies during the last decades has provided evidence for impaired mitophagic pathways in prevalent neurodegenerative diseases [18, 19]. Although there are still limited reports, deregulation of mitophagy has been correlated with the incidence of neurodevelopmental disorders, including autism spectrum disorders (ASD) and epileptic seizures [20–22]. Therefore, there is an urgent need to elucidate the precise role of mitophagy during neuronal development in order to understand its potential implications in neurodevelopmental diseases.

The nematode *Caenorhabditis elegans* emerges as a promising model system for unraveling the intricate involvement of mitophagy during neuronal development and its potential implications in neurodevelopmental disorders. Despite its simplicity, this tiny hermaphrodite nematode possesses a well-defined nervous system comprised of precisely 302 neuronal cells, the lineage of which has been fully characterized. This allows researchers to investigate the cellular and molecular mechanisms underlying various aspects of neuronal function, from neurodevelopment to synaptic transmission [23–25]. In this chapter, we are going to describe how to use the Rosella biosensor to monitor neuronal mitophagy during development in *C. elegans* [4, 26]. This approach can be used as an easy tool for screening genes and/or drugs that interfere with mitophagic pathways during development under physiological conditions or within the specific context of neurodevelopmental diseases, utilizing relevant nematode models.

## 2 Materials

### 2.1 Disposables

1. Microscope slides (25.4 × 76.2 × 1 mm).
2. Microscope coverslips (18 × 18 mm).
3. Fingernail polish for sealing the coverslips.
4. Eyelash hair attached to a toothpick with wood glue: for transferring the worms into the tetramisole droplet for microscopic observation.
5. 2% agarose pads: Add 1 g agarose powder to 50 mL distilled water ($ddH_2O$) and boil it in a microwave until it is completely dissolved. While the solution is still hot, leave a droplet (approximately 100 μL) on top of a microscope slide. Put a second slide on top and press until a large gel pad is formed. Wait for about 2 min and carefully remove the second slide, trying not to disturb the agarose droplet. Agarose pads need to be prepared prior to microscopic observation and should not be stored for long-term use.
6. Platinum wire pick for transferring the worms to NGM plates.

7. Petri plates (60 × 15 mm) for the maintenance of *C. elegans* nematodes.
8. Petri plates (92 × 16 mm) for bacterial growth.
9. Micropipettes (1–20, 20–200, 100–1000 μL) and tips.
10. 1.5 mL microcentrifuge tubes.
11. Conical glass flasks for bacterial cultures.

### 2.2 Equipment

1. Stereomicroscope (e.g., KERN).
2. Fluorescent microscope (e.g., Invitrogen by Thermo Fisher Scientific EVOS M7000).
3. Confocal microscope (Zeiss LSM 900).
4. Standard tabletop centrifuge.
5. Incubators for the maintenance of *C. elegans* nematodes.
6. Laboratory stirrer.
7. Analytical balance and weight balancer.
8. Magnets.
9. Autoclave machine.
10. −80 °C refrigerators for the maintenance of bacterial stabs and nematode strains.
11. Laboratory shaker/vortex.

### 2.3 Nematode Strains

- N2 (Bristol): *C. elegans* wild isolate.
- IR1864: N2;*Ex*[$p_{unc-119}$::TOMM-20::Rosella + *rol-6 (su1006)*] [26].
- IR2284*: pink-1(tm1779);Ex*[$p_{unc-119}$::TOMM-20::Rosella + *rol-6(su1006)*] [26].
- IR2285*: pdr-1(gk448);Ex*[$p_{unc-119}$::TOMM-20::Rosella + *rol-6 (su1006)*] [26].
- *drp-1(tm1108);Ex*[$p_{unc-119}$::TOMM-20::Rosella + *rol-6 (su1006)*] (generated in this study).

### 2.4 Bacterial Food Source

OP50: Uracil auxotroph *Escherichia coli* bacteria. For liquid OP50 bacterial culture, inoculate a single colony [grown overnight (16-18 h) on an LB agar plate at a 37 °C incubator] using a sterilized tip or inoculation needle into a conical flask containing 50 mL of LB medium. Incubate the culture for 5–6 h in a shaking incubator at 37 °C. Seed 200 μL of the culture in the middle of freshly prepared NGM plates and let them grow overnight (16–18 h) at room temperature (RT).

### 2.5 Reagents

#### 2.5.1 Reagents for Bacterial Cultures

1. *Luria Bertani (LB) liquid medium (1 L)*: For the preparation of 1 L liquid LB medium, dissolve 10 g sodium chloride (NaCl), 10 g tryptone, and 5 g yeast extract in 800 mL distilled $H_2O$ ($ddH_2O$). Stir the mix until it is completely dissolved. Adjust the pH to 7.0 with sodium hydroxide (NaOH) and bring the volume to 1 L with $ddH_2O$. Dispense the solution into flasks and sterilize by autoclaving.
2. *LB solid medium (1 L)*: For the preparation of 1 L liquid LB medium, mix the ingredients of the aforementioned recipe for liquid LB with 15 g agar in 800 mL $ddH_2O$ (the powder will not be completely dissolved, as agar undergoes complete dissolution after autoclaving). Adjust the pH to 7.0 with NaOH and sterilize by autoclaving. Bring the volume to 1 L with sterile $ddH_2O$ and stir the medium until it is cooled enough (55–60 °C). Dispense LB medium into 92 × 16 mm petri dishes (20 mL per plate).

#### 2.5.2 Reagents for Nematode Growth and Maintenance

1. *Nematode Growth Medium (NGM)*: For the preparation of 1 L NGM, dissolve 3 g NaCl, 2.5 g bacto-peptone, 0.2 g streptomycin, and 17 g agar in 700 mL $ddH_2O$ and sterilize by autoclaving. Stir the medium at RT until it cools down to 55–60 °C. Add 1 mL of 1 M calcium chloride ($CaCl_2$), 1 mL of 1 M magnesium sulfate ($MgSO_4$), 25 mL of 1 M potassium phosphate ($KPO_4$), 1 mL of cholesterol (5 mg/mL in 100% ethanol), and 1 mL of nystatin (10 mg/mL in 70% ethanol). All solutions, except for cholesterol and nystatin, need to be sterilized by autoclaving. Bring the volume to 1 L with sterile $ddH_2O$ and stir for 5 min. By using a peristaltic pump, dispense the NGM medium into 60 × 15 petri dishes (10 mL per plate). Leave the NGM plates to dry for at least 16 h at RT. For long-term maintenance (up to 3 weeks), you can store the plates at 4 °C.
2. *M9 minimal medium buffer*: For the preparation of 1 L M9 buffer, dissolve 3 g monopotassium phosphate ($KH_2PO_4$), 6 g disodium phosphate ($Na_2HPO_4$), and 5 g NaCl in 800 mL $ddH_2O$. Sterilize by autoclaving. When the medium is cooled down to 55–60 °C, add 1 mL of sterile 1 M $MgSO_4$ and bring the volume to 1 L with sterile $ddH_2O$.
3. *Bleaching solution (for synchronization of nematode populations and removal of unwanted contaminating bacteria)*: For the preparation of 10 mL bleaching solution, add 1 mL of 5 N NaOH and 2 mL of 5% sodium hypochlorite (NaOCl) to 7 mL of sterile $ddH_2O$. This solution can be used for the synchronization of nematode populations and for removing unwanted contamination in the NGM-seeded plates.

4. *Tetramisole (for anesthetizing the nematodes)*: For the preparation of 100 mM stock solution, dissolve 0.24 g of tetramisole hydrochloride powder in 10 mL of sterile M9 buffer and store it at 4 °C. For the working solution, dilute the stock in M9 to a final concentration of 40 mM and keep it at RT.
5. *Urolithin A (UroA)*: For the preparation of 110 mM stock solution, weigh 0.025 g of UroA powder (SantaCruz/Sc-475514) using an analytical weigh balancer and dissolve it in 1 mL of Dimethyl sulfoxide (DMSO). To prepare UroA-supplemented NGM plates, add 5 μL of 110 mM UroA to 195 μL of sterile $ddH_2O$. When UroA is added, the solution becomes cloudy. Vortex for 10 s and dispense the solution onto the bacterial lawn of a freshly seeded NGM plate (final concentration per plate: 50 μM). Leave the plates to dry for at least 30 min at RT, protected from light.

#### 2.5.3 Software

1. FIJI/ImageJ.
2. GraphPad Prism software package (GraphPad Software Inc., San Diego, USA).

## 3 Methods

### 3.1 Preparing Nematode Populations of Specific Developmental Stages

The life cycle of *C. elegans* encompasses four distinct larval stages (L1 to L4) before transitioning to adulthood. The duration of each larval stage is finely orchestrated, providing a glimpse into the temporal dynamics of its development [27, 28]. In order to acquire a sufficient quantity of nematodes at specific developmental stages, you should follow the procedure outlined below:

1. From the IR1864 strain, transfer 20 L4-staged animals with the rolling phenotype into 4 freshly seeded OP50 plates (20 animals per plate). Incubate the plates at 20 °C (timepoint 0) [*see* **Note 1**].
2. After 16 h, in all 4 plates, there should be plenty of eggs and L1 larvae. Remove all the adult animals (parental generation) from all the 4 plates in order to prevent egg-laying production.
3. Immediately after the removal of the parental generation, select one of the 4 plates (plate#1) and gather the L1 larvae by adding 200 μL of M9 buffer onto the bacterial lawn of the plate and collect the liquid into a 1.5 mL microcentrifuge tube.
4. Centrifuge the tube at 3000 rpm for 10 s in a tabletop centrifuge. Carefully discard 175 μL of the supernatant without disturbing the L1-containing pellet.
5. Add 25 μL of concentrated tetramisole solution (80 mM) and mix by pipetting. The final tetramisole concentration in the

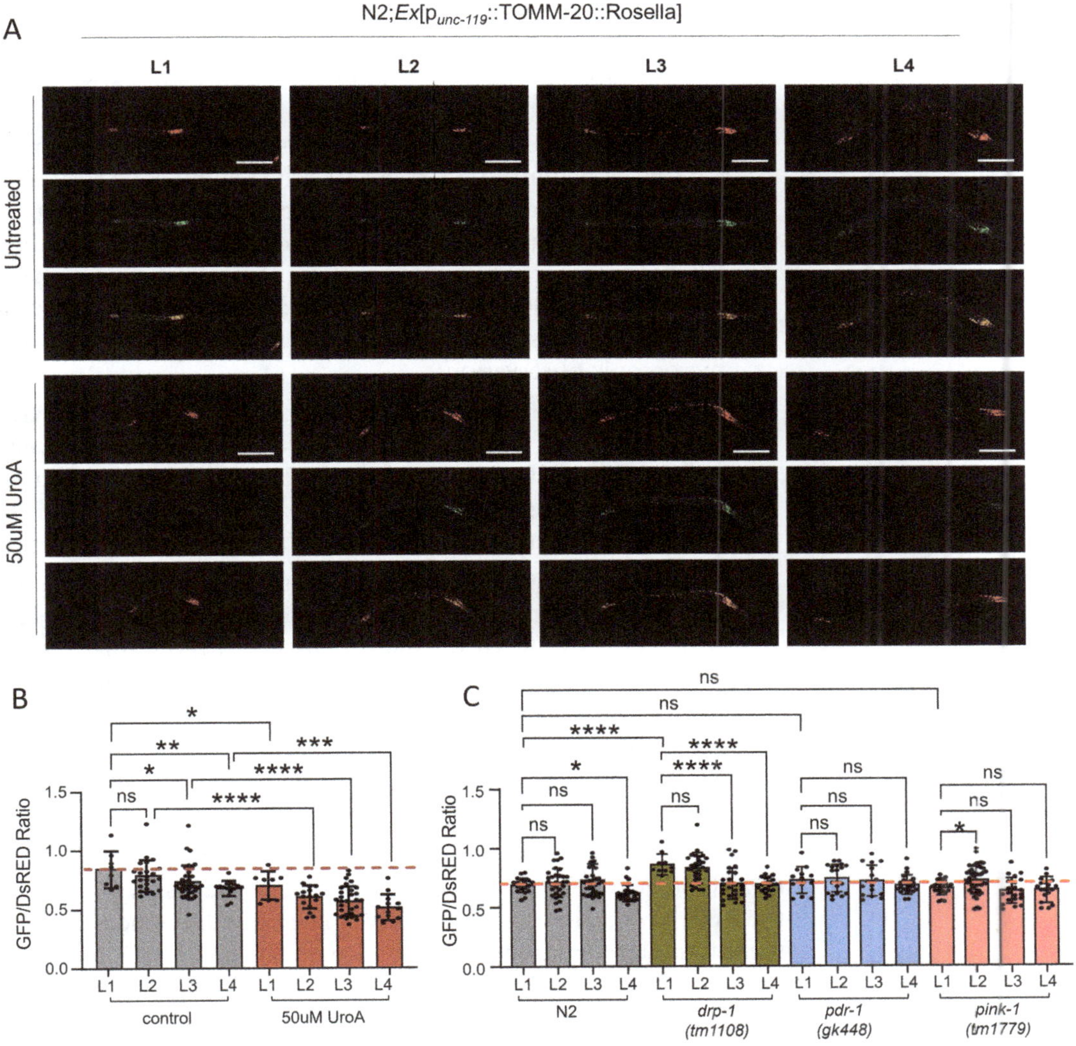


**Fig. 1** Microscopic analysis of neuronal mitophagy during development in *C. elegans*. (**a**) Representative confocal images of control and UroA-treated transgenic nematodes expressing the Rosella biosensor in neuronal mitochondria (Strain IR1864) at L1, L2, L3, and L4 larval stages. Scale bar 100 μm. (**b**) Quantification analysis of neuronal mitophagy in control and UroA-treated transgenic nematodes by determining the GFP to DsRED ratio. A decreased GFP/DsRed ratio of the Rosella biosensor indicates neuronal mitophagy stimulation. Values represent mean ± SD ($n = 25$–30; ns $P > 0.05$, *$P < 0.05$, **$P < 0.01$, ***$P < 0.001$, ****$P < 0.0001$, one-way ANOVA/ Tukey's multiple comparison test). (**c**) Quantification analysis of neuronal mitophagy in N2 (wild-type), *drp-1(tm1108)*, *pdr-1(gk448)*, and *pink-1(tm1779)* mutant nematodes at L1, L2, L3, and L4 larval stages. Values represent mean ± SD ($n = 25$–30; ns $P > 0.05$, *$P < 0.05$, ****$P < 0.0001$, one-way ANOVA/ Tukey's multiple comparison test)

solution should be 40 mM. Transfer 15 μL from the solution and directly place it onto an agarose pad for microscopic observation. This sample corresponds to the L1 larval-staged animals.

6. At 24 h after timepoint 0 (**step 1**), most of the L1 larvae should have developed into L2 larvae. Take the second plate (plate#2) and follow the same procedure described in **steps 3–5** in order to acquire L2-staged animals for microscopic observation. It is important to note that distinguishing the rolling phenotype is challenging for both L1 and L2 larvae; for that reason, washing the whole plate is preferred over selecting them individually for microscopic observation.
7. At 36 h after timepoint 0 (**step 1**), the progeny should have developed into L3. From the third plate (plate#3), pick 30–40 L3-staged animals with the rolling phenotype using an eyelash-hair-attached toothpick and place them directly in a 15 μL drop of tetramisole 40 mM on an agarose pad.
8. At approximately 44 h after timepoint 0 (**step 1**), the progeny should have developed into L4. From the fourth plate (plate#4), pick 30–40 L3-staged animals with the rolling phenotype using an eyelash-hair-attached toothpick and place them directly in a 15 μL drop of tetramisole 40 mM on an agarose pad.

To standardize our method, we employed Urolithin A(UroA)-treated animals as a positive-control condition for induced mitophagy [29] (Fig. 1a, b) (*see* **Note 2**). In order to acquire UroA-treated nematodes at specific developmental stages, we followed the exact procedure described above. The only difference was that the parental generation was treated with 50 μM UroA from the point of hatching. Afterwards, 20 L4-staged animals from the UroA-treated parental generation were transferred to 4 UroA-seeded plates (20 animals per plate).

### 3.2 Sample Preparation for Imaging

As mentioned above, for L1 and L2 larval stages, place 15 μL of 40 mM tetramisole solution (including the larvae) on a 2% agarose pad. Cover gently with a coverslip and seal with fingernail polish to prevent the animals from drying out during the imaging process. For the L3 and L4 larvae, transfer 30–40 L3/L4-staged animals with the rolling phenotype using an eyelash-hair-attached toothpick directly into a 15 μL drop of tetramisole 40 mM on an agarose pad, and follow the same procedure with the coverslip and sealing.

### 3.3 Microscopic Analysis of Neuronal Mitophagy Using Rosella Biosensor

To monitor neuronal mitophagy, we used the IR1864 strain, which expresses pan-neuronally (*unc-119* promoter) the Rosella biosensor fused to a mitochondrial-targeting sequence (amino-terminal fragment of TOMM-20). Rosella serves as a dual fluorescent reporter, concurrently expressing a pH-sensitive GFP and a pH-insensitive DsRED. Therefore, in principle, under normal conditions, neuronal mitochondria are labeled with both fluorescent proteins (green and red). However, when mitochondria are subjected to mitophagy, the GFP signal is quenched due to the acidic lysosomal environment [26].

For the analysis of neuronal mitophagy in larvae, follow the procedure described below:

1. Observe the animals under a fluorescent microscope at 10× magnification and acquire images of whole animal bodies at 488 nm (green channel) and 615 nm (red channel). Ensure that all images are acquired using the same settings (exposure time, brightness, contrast). Ignore animals that seem to be damaged (mishandling during preparation).
2. Use the ImageJ/FIJI image processing package to quantify separately the mean intensities of the red and green channels for each animal body.
3. Divide the individual values of green versus red fluorescence for each animal and plot the ratios on a graph using GraphPad Prism or any related software.

Under mitophagy-inducing conditions, we expect a notable decrease in the green versus red ratio. This shift occurs as an increased number of mitochondria are sequestered to the acidic lysosomes for degradation. As shown in Fig. 1a, b, mitophagy levels do not change significantly during the L1 to L2 transition in wild-type animals. However, L3- and L4-staged animals display a slight induction in mitophagy (reduced GFP/DsRED ratio). As expected, UroA-treated animals demonstrate a significantly induced mitophagic rate compared to their control counterparts. Remarkably, similarly to the control group, the UroA-treated animals exhibit elevated mitophagy during development.

We also monitored neuronal mitophagy upon depletion of genes implicated in the regulation of mitophagy: *drp-1(tm1108)*, *pink-1(tm1779)*, and *pdr-1(gk448)*. DRP-1 is the nematode ortholog of dynamin-related protein 1 (DRP1), a GTPase involved in mitochondrial fission and has been reported to contribute to the segregation of damaged mitochondrial fragments triggering their clearance through mitophagy [30, 31]. PINK-1 and PDR-1 are the nematode orthologs of the well-known mitophagy inducers PINK1 and Parkin, respectively [7]. As shown in Fig. 1c, *pink-1(tm1779)* and *pdr-1(gk448)* mutants fail to induce mitophagy at the L4 larval stage, as opposed to wild-type animals. However, *drp-1(tm1108)* mutants exhibit a mitophagy induction pattern comparable to that of wild-type animals, suggesting that mitophagy induction at the L4 stage acts independently of DRP-1.

## 4 Notes

1. Nematodes should be cultivated under well-fed conditions for at least three generations prior to the experiment. In general, starvation has a direct impact on autophagic pathways and developmental processes and thus may lead to misleading conclusions.
2. Specific genetic interventions or drugs may interfere with the developmental process. Therefore, it is essential to acknowledge these alterations and standardize accordingly the time points for each developmental stage to efficiently synchronize nematode populations. Noteworthy, the literature provides information regarding the area/size for each larval stage, which could be helpful for distinguishing the different larval stages [32].

## 5 Concluding Remarks

Despite considerable progress in monitoring mitophagy, it is crucial to acknowledge that the precise impact of mitochondrial dysfunction and impaired mitophagy on neuronal development is not well-defined. By employing Rosella biosensor in *C. elegans*, our methodological approach serves as a practical and efficient tool for screening genes and drugs that may modulate mitophagic activity in developing neurons. In essence, unraveling the molecular pathways governing mitochondrial function and mitophagy at early life stages will contribute to the development of effective therapeutic strategies for delaying or even preventing neurological defects in the context of neurodevelopmental diseases.

## Acknowledgments

This project was funded by the Fondation Santé (19656) and the European Union (European Research Council; ERC), under grant agreement "ERC-GA101077374-SynaptoMitophagy" to KP. Views and opinions expressed are, however, those of the author(s) only and do not necessarily reflect those of the European Union or the European Research Council. Neither the European Union nor the granting authority can be held responsible for them.

**Author Contributions** C.P. and A.R. performed the experiments and prepared the original draft of the manuscript. C.P., A.R., and K.P. revised and edited the manuscript.

## References


1. Spinelli JB, Haigis MC (2018) The multifaceted contributions of mitochondria to cellular metabolism. Nat Cell Biol 20(7):745–754
2. Shen K, Pender CL, Bar-Ziv R, Zhang H, Wickham K, Willey E, Durieux J, Ahmad Q, Dillin A (2022) Mitochondria as cellular and organismal signaling hubs. Annu Rev Cell Dev Biol 38(1):179–218
3. Ploumi C, Daskalaki I, Tavernarakis N (2017) Mitochondrial biogenesis and clearance: a balancing act. FEBS J 284(2):183–195
4. Palikaras K, Lionaki E, Tavernarakis N (2015) Coordination of mitophagy and mitochondrial biogenesis during ageing in *C. elegans*. Nature 521(7553):525–528
5. Wang S, Long H, Hou L, Feng B, Ma Z, Wu Y, Zeng Y, Cai J, Zhang D-w, Zhao G (2023) The mitophagy pathway and its implications in human diseases. Signal Transduct Target Ther 8(1):304
6. Onishi M, Yamano K, Sato M, Matsuda N, Okamoto K (2021) Molecular mechanisms and physiological functions of mitophagy. EMBO J 40(3):e104705
7. McWilliams TG, Muqit MMK (2017) PINK1 and Parkin: emerging themes in mitochondrial homeostasis. Curr Opin Cell Biol 45:83–91
8. Harper JW, Ordureau A, Heo JM (2018) Building and decoding ubiquitin chains for mitophagy. Nat Rev Mol Cell Biol 19(2): 93–108
9. Jin SM, Lazarou M, Wang C, Kane LA, Narendra DP, Youle RJ (2010) Mitochondrial membrane potential regulates PINK1 import and proteolytic destabilization by PARL. J Cell Biol 191(5):933–942
10. Yamano K, Youle RJ (2013) PINK1 is degraded through the N-end rule pathway. Autophagy 9(11):1758–1769
11. Sekine S, Youle RJ (2018) PINK1 import regulation; a fine system to convey mitochondrial stress to the cytosol. BMC Biol 16(1):2
12. Koyano F, Okatsu K, Kosako H, Tamura Y, Go E, Kimura M, Kimura Y, Tsuchiya H, Yoshihara H, Hirokawa T, Endo T, Fon EA, Trempe JF, Saeki Y, Tanaka K, Matsuda N (2014) Ubiquitin is phosphorylated by PINK1 to activate parkin. Nature 510(7503): 162–166
13. Lazarou M, Sliter DA, Kane LA, Sarraf SA, Wang C, Burman JL, Sideris DP, Fogel AI, Youle RJ (2015) The ubiquitin kinase PINK1 recruits autophagy receptors to induce mitophagy. Nature 524(7565):309–314
14. Fritsch LE, Moore ME, Sarraf SA, Pickrell AM (2020) Ubiquitin and receptor-dependent Mitophagy pathways and their implication in neurodegeneration. J Mol Biol 432(8): 2510–2524
15. Son G, Han J (2018) Roles of mitochondria in neuronal development. BMB Rep 51(11): 549–556
16. Rangaraju V, Lewis TL Jr, Hirabayashi Y, Bergami M, Motori E, Cartoni R, Kwon SK, Courchet J (2019) Pleiotropic mitochondria: the influence of mitochondria on neuronal development and disease. J Neurosci 39(42): 8200–8208
17. Evans CS, Holzbaur ELF (2020) Quality control in neurons: mitophagy and other selective autophagy mechanisms. J Mol Biol 432(1): 240–260
18. Fivenson EM, Lautrup S, Sun N, Scheibye-Knudsen M, Stevnsner T, Nilsen H, Bohr VA, Fang EF (2017) Mitophagy in neurodegeneration and aging. Neurochem Int 109:202–209
19. Palikaras K, Lionaki E, Tavernarakis N (2018) Mechanisms of mitophagy in cellular homeostasis, physiology and pathology. Nat Cell Biol 20(9):1013–1022
20. Wang YM, Qiu MY, Liu Q, Tang H, Gu HF (2021) Critical role of dysfunctional mitochondria and defective mitophagy in autism spectrum disorders. Brain Res Bull 168:138–145
21. Napoli E, Song G, Panoutsopoulos A, Riyadh MA, Kaushik G, Halmai J, Levenson R, Zarbalis KS, Giulivi C (2018) Beyond autophagy: a novel role for autism-linked Wdfy3 in brain mitophagy. Sci Rep 8(1):11348
22. Panda SP, Dhurandhar Y, Agrawal M (2022) The interplay of epilepsy with impaired mitophagy and autophagy linked dementia (MAD): a review of therapeutic approaches. Mitochondrion 66:27–37
23. Markaki M, Tavernarakis N (2020) *Caenorhabditis elegans* as a model system for human diseases. Curr Opin Biotechnol 63:118–125
24. Bessa C, Maciel P, Rodrigues AJ (2013) Using *C. elegans* to decipher the cellular and molecular mechanisms underlying neurodevelopmental disorders. Mol Neurobiol 48(3):465–489
25. Roussos A, Kitopoulou K, Borbolis F, Palikaras K (2023) *Caenorhabditis elegans* as a model system to study human neurodegenerative disorders. Biomol Ther 13(3):478
26. Fang EF, Hou Y, Palikaras K, Adriaanse BA, Kerr JS, Yang B, Lautrup S, Hasan-Olive MM, Caponio D, Dan X, Rocktäschel P, Croteau DL, Akbari M, Greig NH, Fladby T,

Nilsen H, Cader MZ, Mattson MP, Tavernarakis N, Bohr VA (2019) Mitophagy inhibits amyloid-β and tau pathology and reverses cognitive deficits in models of Alzheimer's disease. Nat Neurosci 22(3):401–412

27. Corsi AK, Wightman B, Chalfie M (2015) A transparent window into biology: a primer on *Caenorhabditis elegans*. Genetics 200(2): 387–407
28. Altun F, Hall DH (2009) Introduction to *C. elegans* anatomy. In WormAtlas
29. Ryu D, Mouchiroud L, Andreux PA, Katsyuba E, Moullan N, Nicolet-Dit-Félix AA, Williams EG, Jha P, Lo Sasso G, Huzard D, Aebischer P, Sandi C, Rinsch C, Auwerx J (2016) Urolithin A induces mitophagy and prolongs lifespan in *C. elegans* and increases muscle function in rodents. Nat Med 22(8):879–888
30. Twig G, Elorza A, Molina AJ, Mohamed H, Wikstrom JD, Walzer G, Stiles L, Haigh SE, Katz S, Las G, Alroy J, Wu M, Py BF, Yuan J, Deeney JT, Corkey BE, Shirihai OS (2008) Fission and selective fusion govern mitochondrial segregation and elimination by autophagy. EMBO J 27(2):433–446
31. Tanaka A, Cleland MM, Xu S, Narendra DP, Suen DF, Karbowski M, Youle RJ (2010) Proteasome and p97 mediate mitophagy and degradation of mitofusins induced by Parkin. J Cell Biol 191(7):1367–1380
32. Uppaluri S, Brangwynne CP (2015) A size threshold governs *Caenorhabditis elegans* developmental progression. Proc Biol Sci 282(1813):20151283

# Chapter 6

# In Vivo Monitoring of Nucleophagy in *Caenorhabditis elegans*

**Georgios Konstantinidis and Nektarios Tavernarakis**

## Abstract

The autophagy-lysosomal pathway enables the controlled degradation of cellular contents. Nucleophagy is the selective autophagic recycling of nuclear components upon delivery to the lysosome. Although methods to monitor and quantify autophagy as well as selective types of autophagy have been developed and implemented in cells and in vivo, methods monitoring nucleophagy remain scarce. Here, we describe a procedure to monitor the autophagic engagement of an endogenous nuclear envelope component, i.e., ANC-1, the nematode homologue of the mammalian Nesprins in vivo, utilizing super-resolution microscopy.

**Key words** ANC-1, Autophagy, *Caenorhabditis elegans*, LGG-1, Nesprin, Nucleophagy, Nucleus, Super-resolution microscopy

## 1 Introduction

The linker of nucleoskeleton and cytoskeleton (LINC) complexes comprise a conserved network of proteins supporting nucleus and nuclear envelope architecture [1]. They mediate fundamental cellular functions, including nuclear positioning, nucleoskeleton-cytoskeleton communication, mechanotransduction, homolog pairing in meiosis, DNA damage repair, and others [2, 3]. LINC complexes span the nuclear envelope and consist of Sad1 and UNC-84 (SUN) proteins at the inner nuclear membrane and Klarsicht, ANC-1, and Syne homology (KASH) proteins in the outer nuclear membrane [1]. Defects in nuclear positioning and mutations in LINC components are associated with a wide variety of human diseases [2, 3].

Previous studies in *Caenorhabditis elegans* (*C. elegans*) have contributed to our current understanding regarding nuclear positioning during development [4–7]. ANC-1 (orthologous to mammalian SYNE1/Nesprin-1 and SYNE2/Nesprin-2), is one of the

Ioannis P. Nezis (ed.), *Selective Autophagy: Methods and Protocols*, Methods in Molecular Biology, vol. 2845,
https://doi.org/10.1007/978-1-0716-4067-8_6,

four KASH proteins in *C. elegans*, which are localized to the outer nuclear membrane where they connect the nucleus to the cytoskeleton [8–10]. ANC-1 mediates nuclear anchorage through its interaction with the SUN protein UNC-84 at the inner nuclear membrane and actin cytoskeleton in the cytoplasm [10]. In addition, ANC-1 localizes to the endoplasmic reticulum (ER) membrane to position nuclei as well as other organelles in place [11]. Nesprin mutations perturbing Nesprin/Emerin/Lamin interactions can cause uncoupling of the nucleoskeleton from the cytoskeleton. Such mutations lead to two distinct human neurological or myopathic disease phenotypes, a feature similar to that found in laminopathies. Specifically, Nesprin dysregulated levels or Nesprin mutations are associated with spinocerebellar ataxias [12–16], autism [17, 18], bipolar disorders [19, 20], major depression [20], Emery-Dreifuss Muscular Dystrophy [21, 22], dilated cardiomyopathy [23], arthrogryposis multiplex congenita [24], cancer [25–32], and hearing loss [33].

The dynamics of the nucleus and the nuclear envelope confer plasticity and support nuclear functions. Quality control mechanisms ensure the maintenance of nuclear architecture and integrity. For example, the endosomal sorting complex required for transport (ESCRT) machinery functions to surveil nuclear pore complex (NPC) assembly, remodel the nuclear envelope in response to mechanical cues or rupture, and accomplish sealing of the postmitotic nuclear envelope [34]. Nucleophagy serves as the selective subtype of autophagy targeting nuclear components for degradation in lysosomes [35]. Autophagic engagement of nuclear content, such as DNA fragments, histones, and lamins, has been shown in different physiological or pathological settings in cells as well as in vivo, using predominantly fluorescence microscopy [36–46]. On the one hand, immuno- or direct staining enables the identification of endogenous cargo into autophagosomes in fixed samples by colocalization with autophagic proteins such as LC3, since a mammalian receptor of nucleophagy has not been identified yet. On the other hand, a fluorescently tagged protein of interest can be monitored in living cells by exogenous expression concomitantly with an autophagy reporter protein. Here, we describe a procedure to monitor nucleophagy by capturing the autophagic engagement of an endogenous nuclear component in vivo, using super-resolution microscopy.

Autophagy is a dynamic process, and visualizing subtle or rare autophagosome formation events can be challenging. Furthermore, autophagic engagement of nuclear components is rarely captured, especially under physiological conditions. In order to increase the probability of monitoring such events, two approaches are followed. First, in order to induce increased autophagic or selective nucleophagic activity, *daf-2(RNAi)* (Fig. 1) or nuclear insults such as DNA damage (etoposide or mitomycin C treatment)

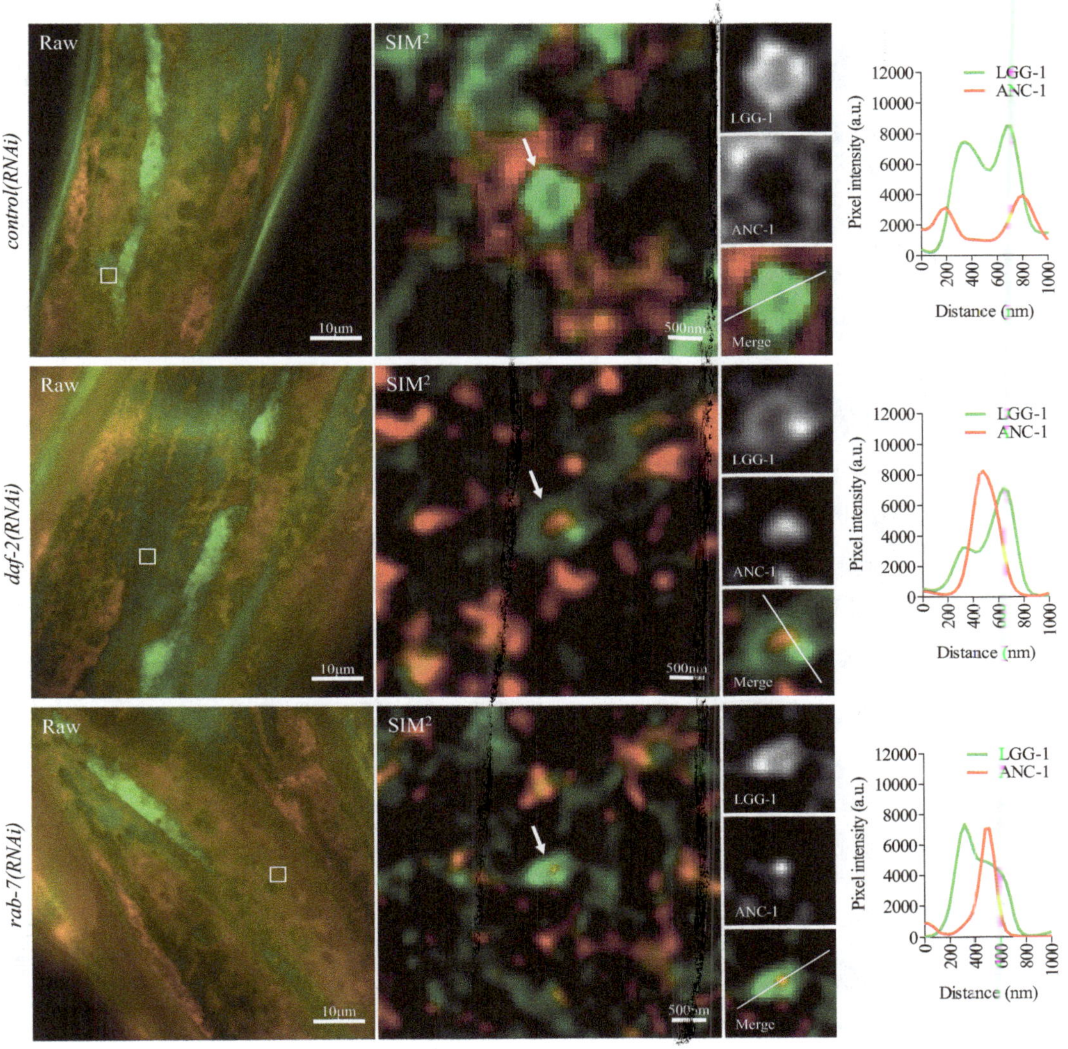


**Fig. 1** Monitoring nucleophagy upon genetic manipulation of autophagy, in vivo. The first of the 15 lattice SIM phase images is presented as a raw image with merged 488 nm (GFP::LGG-1) and 561 nm (mKate2::ANC-1) channels. SIM² images show areas of SIM processing corresponding to the boxes in raw images. White arrows point to autophagosomes, which are presented in both separate channels (LGG-1 and ANC-1) and merged images. Pixel intensity measurements correspond to the respective white line of 1 μm length. Note ANC-1 engulfment into LGG-1 positive structures upon autophagy induction [*daf-2(RNAi)*] or inhibition of autophagosome-lysosome fusion [*rab-7(RNAi)*]

(Fig. 2) are applied, respectively. The latter could serve as a specific nucleophagy induction condition. Second, by blocking the autophagosome-lysosome fusion step, it is possible to accumulate autophagic structures in order to quantify the autophagosome formation events as well as monitor their contents [47]. Lysosomotropic agents, such as chloroquine, or *daf-2(RNAi)* (Fig. 1) serve as autophagic flux inhibitors. The second perturbation serves as a

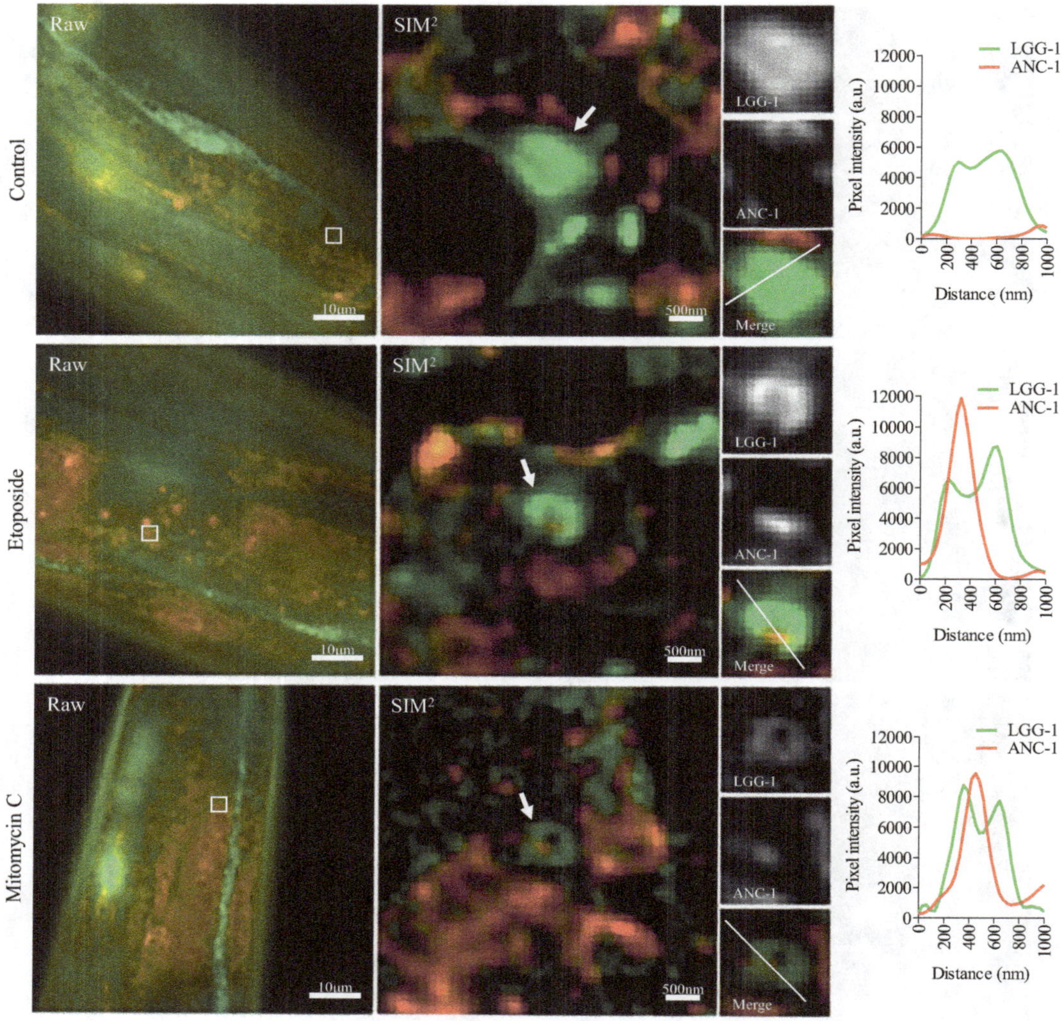


**Fig. 2** Monitoring nucleophagy upon DNA damage, in vivo. The first of the 15 lattice SIM phase images is presented as a raw image with merged 488 nm (GFP::LGG-1) and 561 nm (mKate2::ANC-1) channels. SIM$^2$ images show areas of SIM processing corresponding to the boxes in raw images. White arrows point to autophagosomes, which are presented in both separate channels (LGG-1 and ANC-1) and merged images. Pixel intensity measurements correspond to the respective white line of 1 μm length. Note ANC-1 engulfment into LGG-1 positive structures upon DNA damaging conditions

general approach to accumulate autophagic structures. Autophagy/nucleophagy induction and autophagic flux inhibition can be combined in order to increase the levels of nucleophagic events monitored.

Given that the nuclear envelope and ER form a contiguous network, ER-phagy and nucleophagy may occur concurrently [48, 49]. For example, the yeast protein Yep1, the ortholog of human receptor accessory proteins (REEP1-4), is essential for the autophagosomal enclosure of cargo membrane structures derived from both the ER and nuclear envelope [50]. Yep1 function is specific to these two types of selective autophagy but not to bulk autophagy. Therefore, the above approach enables the identification of autophagic cargo composition and the monitoring of nuclear/ER membrane-specific capture. The application described here can shed light on numerous other unexplored segments of nucleophagy in vivo, such as nucleophagic signaling, regulation, and molecular mechanisms, in health and disease.

## 2 Materials

Prepare all solutions using analytical grade reagents.

### 2.1 Caenorhabditis elegans *Strains*

All experiments are carried out using hermaphrodite worms. The strain DA2123, *adIs2122* [*lgg-1p::GFP::lgg-1+rol-6(su1006)*], is provided by the Caenorhabditis Genetics Center (https://cgc.umn.edu/), which is funded by NIH Office of Research Infrastructure Programs (P40 OD010440). The strain UD626: *yc72* [*mKate2::anc-1b*] is a kind gift from Daniel A. Starr. The strain IR2969: *yc72* [*mKate2::anc-1b*]; *adIs2122* [*lgg-1p::GFP::lgg-1+rol-6(su1006)*] was generated in a previous study [46].

### 2.2 *Bacteria Strains*

OP50 uracil auxotroph *Escherichia coli*: Streak OP50 bacteria on a 100 mm plate from a glycerol stab and grow overnight at 37 °C. Inoculate a single OP50 colony into 50 mL of sterile liquid LB medium. Incubate for 5 h in a shaking incubator at 37 °C. Seed 200 μL of the OP50 culture in the middle of freshly prepared NGM plates, and let bacteria grow overnight at room temperature. Make use of the plates within the next 2 days. Store liquid OP50 culture at 4 °C. HT115 (DE3) *Escherichia coli* are used for *RNAi* experiments.

### 2.3 *Stock Solutions*

1. 1 M $CaCl_2$ (100 mL): Dissolve 14.7 g $CaCl_2$ in 100 mL ultrapure water and autoclave. Store at room temperature.
2. 1 M $MgSO_4$ (100 mL): Dissolve 24.65 g $MgSO_4$ in 100 mL ultrapure water and autoclave. Store at 4 °C.
3. 5 mg/mL cholesterol (200 mL): Dissolve 1 g cholesterol in 200 mL ethanol. Do not autoclave. Store at 4 °C.
4. 10 mg/mL nystatin (200 mL): Dissolve 2 g nystatin in 200 mL 70% ethanol. Do not autoclave. Mix thoroughly before use. Store at 4 °C.
5. 1 M $KPO_4$ (100 mL): Dissolve 102.2 g $KH_2PO_4$ and 57.06 g $K_2HPO_4$ in 100 mL ultrapure water, adjust pH to 6.0, and autoclave. Store at 4 °C.

6. 5 N NaOH (50 mL): Dissolve 10 g NaOH in 50 mL ultrapure water. Store at room temperature.
7. 2% $NaN_3$ (50 mL): Dissolve 1 g $NaN_3$ in 50 mL M9 solution. Store at room temperature.
8. 100 mg/mL ampicillin (10 mL): Dissolve 1 g ampicillin in 10 mL ultrapure water. Aliquot and store at −20 °C.
9. 5 mg/mL tetracycline (10 mL): Dissolve 50 mg tetracycline in 10 mL ethanol. Aliquot and store at −20 °C.
10. IPTG master stock (1 M): Dissolve 2.38 g isopropyl β- d-1-thiogalactopyranoside in 10 mL ultrapure water. Aliquot and store at −20 °C.
11. IPTG stock (20 mM): Mix 20 μL IPTG (1 M) with 980 μL ultrapure water. Aliquot and store at −20 °C.
12. Etoposide (50 mM): Dissolve 25 mg etoposide in 850 μL DMSO. Aliquot and store at −20 °C.
13. Mitomycin C (1 mg/mL): Dissolve 2 mg mitomycin C in 2 mL ultrapure water. Aliquot and store at −20 °C.

### 2.4 Working Solutions

1. M9 minimal medium solution (1 L): Dissolve 3 g $KH_2PO_4$, 6 g $Na_2HPO_4$ and 5 g NaCl in 800 mL ultrapure water and sterilize by autoclaving. Cool down at room temperature to approximately 50–55 °C. Add 1 mL $MgSO_4$ (1 M) post sterilization. Top up to 1 L with sterile ultrapure water. Store at room temperature.
2. Bleaching solution (10 mL): Mix 7 mL ultrapure water, 2 mL common NaOCl (5%), and 1 mL NaOH (5 N). Store at room temperature.
3. 0.1% $NaN_3$ (1 mL): Mix 50 μL $NaN_3$ (2%) with 950 μL ultrapure water. Store at room temperature.

### 2.5 Growth Media

1. Nematode growth medium (NGM) (1 L): Dissolve 3 g NaCl, 2.5 g bactopeptone, 0.2 g streptomycin, and 17 g agar in 800 mL ultrapure water and sterilize by autoclaving. Cool down at room temperature to approximately 50–55 °C. Add 1 mL $CaCl_2$ (1 M), 1 mL $MgSO_4$ (1 M), 1 mL cholesterol (5 mg/mL), 1 mL nystatin (10 mg/mL), and 25 mL $KPO_4$ (1 M). Top up to 1 L with sterile ultrapure water. Dispense 7 mL to 60 mm dishes and let solidify at room temperature. Store at 4 °C the next day.
2. Luria–Bertani (LB) liquid medium (1 L): Dissolve 10 g NaCl, 10 g tryptone, and 5 g yeast extract in 1 L ultrapure water. Stir until the mixture is completely dissolved. Sterilize by autoclaving.

3. Luria–Bertani (LB) solid medium (1 L): Dissolve 10 g NaCl, 10 g tryptone, 5 g yeast extract, and 15 g agar in 1 L ultrapure water. The powder will not be dissolved completely; the agar will be dissolved after autoclaving. Sterilize by autoclaving and cool down at room temperature to approximately 50–55 °C. Dispense 17 mL medium to 100 mm petri plates and let solidify at room temperature. Store at 4 °C the next day.
4. *RNAi* growth media: Follow the procedure for NGM without the addition of streptomycin. Add 500 μL ampicillin (100 mg/mL) post sterilization. Add 60 μL tetracycline (5 mg/mL) right before use and spread evenly. Let plates dry for 1 h.

## 3 Methods

### 3.1 Maintenance

The strain IR2969 is maintained well-fed for at least three generations at 20 °C. Do not let the worms undergo starvation to any degree.

### 3.2 Bleaching

Harvest gravid adult worms from one 60 mm plate with 1.6 mL of M9 solution. Transfer the solution to a 1.5 mL tube. Spin down and discard the M9 solution. Wash with 1 mL of M9 solution. Spin down and discard the M9 solution. Add 500 μL of bleaching solution. Rotate at room temperature for 2–3 min or until worms are dissolved and only eggs are left. Centrifuge for 30 s at maximum speed. Discard the bleaching solution. Wash with 1 mL of M9 solution, centrifuge for 30 s at maximum speed, and discard the M9 solution. Repeat the wash. In the last wash, keep approximately 100 μL of M9 solution. Resuspend the eggs and place them on fresh OB50 60 mm plates.

### 3.3 RNAi Silencing

1. Day 1: Streak *control(RNAi)*, *rab-7(RNAi)*, and *daf-2 (RNAi)* HT115 onto LB/ampicillin/tetracycline plates. Grow overnight at 37 °C.
2. Day 2: Grow overnight starter cultures from single colonies in 5 mL LB/ampicillin/tetracycline at 37 °C (shaking).
3. Day 3: Mix 70 μL of the overnight culture with 1 mL LB/ampicillin and incubate for 4 h at 37 °C (shaking). Mix 900 μL of the culture with 100 μL IPTG (20 mM) and seed 200 μL from each strain in the middle of respective *RNAi* plates. Let bacteria grow overnight at room temperature. Make use of them the next day (*see* **Note 1**).

### 3.4 Chemical Treatments

1. DMSO control: Mix 7 μL of DMSO with 133 μL of ultrapure water.

2. 50 μM etoposide: Mix 7 μL of etoposide (50 mM) with 133 μL of ultrapure water.
3. 20 μg/mL mitomycin C: Use 140 μL of mitomycin C (1 mg/mL).

   Inactivate OP50 loan on 60 mm NGM plates using ultraviolet crosslinking. Apply each of the above mixtures to one UV-inactivated OP50 plate for each condition and spread evenly. Let the plates dry for 1 h at room temperature and place eggs after bleaching (*see* **Note 2**).

### 3.5 Sample Preparation for Imaging

To prepare 2% agarose pads, add 0.5 g of agarose into 25 mL of M9 solution. Heat in the microwave (do not boil) until the agarose is completely dissolved. While the agarose is still hot, add a droplet of approximately 1 cm in diameter to the middle of a microscope glass slide using a 1 mL pipette (*see* **Note 3**). Immediately place another glass slide vertically on top to spread the agarose evenly and flatten it to a larger area. Let the agarose solidify for 1 min, then carefully remove the second glass slide without disturbing the agarose layer. Keep the agarose pad in the center of the glass slide. Store at room temperature (*see* **Note 4**). Add 3 μL of $NaN_3$ to the center of the agarose pad slide (*see* **Note 5**). Transfer worms from the plate to the agarose pad using an eyelash hair pick. Cover with an 18 mM × 18 mM coverslip. Add more $NaN_3$ at the edges of the coverslip until all coverslip area is covered to keep the worms wet. Seal and fix by applying nail polish at the edge of the coverslip. Proceed to imaging.

### 3.6 Super Resolution Structured Illumination Microscopy (SIM) and Analysis

Acquire fluorescent images using a ZEISS Elyra 7 microscope equipped with a Plan apochromat 63×/1.4 oil DIC M27 objective. Record mKate2 fluorescence using a 561 nm track with 30% intensity and 50 ms exposure. Record GFP fluorescence using a 488 nm track with 20% intensity and 50 ms exposure. Use the Carl Zeiss Zen 3.0 SR FP2 software for acquisition and $SIM^2$ analysis with default settings in weak live mode. Use ImageJ to crop and overlay images as well as measure pixel intensity. Create intensity plots using GraphPad PRISM software.

## 4 Notes

1. In case worms have to be moved to new *RNAi* plates in order to reach day 1 without being starved, prepare fresh overnight cultures every second day, follow the rest of the procedure, and transfer. Maintain streaking plates for up to 1 week.
2. In case worms have to be moved to new treatment plates in order to reach day 1 without being starved, prepare fresh etoposide or mitomycin C plates the same day and transfer.

3. If not all the quantity of the agarose solution is going to be used at once, aliquot the agarose solution in 1.5 mL tubes and store them at 4 °C for a week. Resolve at 100 °C in a water bath when needed and follow the protocol. Do not vortex or mix vigorously. Bubbles remain in the agarose solution for long and form gaps in the agarose pad.
4. In order to avoid dehydration of the agarose layer, which could interfere with the physiology of the worms before microscopy, we always use the agarose pads 10 min to 8 h after preparation.
5. Since in this protocol GFP::LGG-1 imaging is taking place, use only $NaN_3$ as an anesthetic. Levamisole may cause nonphysiological punctate structures [51].

## Acknowledgments

We thank the *Caenorhabditis* Genetics Center (https://cgc.umn.edu/) and Daniel A. Starr for providing the DA2123 and UD626 strains, respectively. We thank Margarita-Elena Papandreou for generating the IR2969 strain. Work in the authors' laboratory is funded by the European Research Council under grant agreement "ERC-GA695190-MANNA," the European Union Horizon 2020 FETOPEN project "Dynamic," under the grant agreement "GA-863203," the Hellenic Foundation for Research and Innovation (H.F.R.I.) under the "1st Call for H.F.R.I. Research Projects to support Faculty members and Researchers and the procurement of high-cost research equipment" (Project Number: HFRI—FM17C3-0869, NeuroMitophagy), and the General Secretariat for Research and Innovation of the Greek Ministry of Development and Investments.

### References

1. Jahed Z et al (2021) Molecular models of LINC complex assembly at the nuclear envelope. J Cell Sci 134(12)
2. Turkmen AM, Saik NO, Ullman KS (2023) The dynamic nuclear envelope: resilience in health and dysfunction in disease. Curr Opin Cell Biol 85:102230
3. Bone CR, Starr DA (2016) Nuclear migration events throughout development. J Cell Sci 129(10):1951–1961
4. Horvitz HR, Sulston JE (1980) Isolation and genetic characterization of cell-lineage mutants of the nematode Caenorhabditis elegans. Genetics 96(2):435–454
5. Sulston JE, Horvitz HR (1977) Post-embryonic cell lineages of the nematode, Caenorhabditis elegans. Dev Biol 56(1):110–156
6. Sulston JE, Horvitz HR (1981) Abnormal cell lineages in mutants of the nematode Caenorhabditis elegans. Dev Biol 82(1):41–55
7. Sulston JE et al (1983) The embryonic cell lineage of the nematode Caenorhabditis elegans. Dev Biol 100(1):64–119
8. Hedgecock EM, Thomson JN (1982) A gene required for nuclear and mitochondrial attachment in the nematode Caenorhabditis elegans. Cell 30(1):321–330
9. Malone CJ et al (1999) UNC-84 localizes to the nuclear envelope and is required for nuclear migration and anchoring during C. Elegans

development. Development 126(14): 3171–3181

10. Starr DA, Han M (2002) Role of ANC-1 in tethering nuclei to the actin cytoskeleton. Science 298(5592):406–409
11. Hao H et al (2021) The Nesprin-1/−2 ortholog ANC-1 regulates organelle positioning in C. Elegans independently from its KASH or actin-binding domains. elife 10
12. Gros-Louis F et al (2007) Mutations in SYNE1 lead to a newly discovered form of autosomal recessive cerebellar ataxia. Nat Genet 39(1): 80–85
13. Izumi Y et al (2013) Cerebellar ataxia with SYNE1 mutation accompanying motor neuron disease. Neurology 80(6):600–601
14. Dupre N et al (2007) Clinical and genetic study of autosomal recessive cerebellar ataxia type 1. Ann Neurol 62(1):93–98
15. Noreau A et al (2013) SYNE1 mutations in autosomal recessive cerebellar ataxia. JAMA Neurol 70(10):1296–1231
16. Razafsky D, Hodzic D (2015) A variant of Nesprin1 giant devoid of KASH domain underlies the molecular etiology of autosomal recessive cerebellar ataxia type I. Neurobiol Dis 78:57–67
17. O'Roak BJ et al (2011) Exome sequencing in sporadic autism spectrum disorders identifies severe de novo mutations. Nat Genet 43(6): 585–589
18. Yu TW et al (2013) Using whole-exome sequencing to identify inherited causes of autism. Neuron 77(2):259–273
19. Psychiatric GWAS Consortium Bipolar Disorder Working Group (2011) Large-scale genome-wide association analysis of bipolar disorder identifies a new susceptibility locus near ODZ4. Nat Genet 43(10):977–983
20. Green EK et al (2013) Association at SYNE1 in both bipolar disorder and recurrent major depression. Mol Psychiatry 18(5):614–617
21. Zhang Q et al (2007) Nesprin-1 and -2 are involved in the pathogenesis of Emery Dreifuss muscular dystrophy and are critical for nuclear envelope integrity. Hum Mol Genet 16(23): 2816–2833
22. Puckelwartz MJ et al (2009) Disruption of nesprin-1 produces an Emery Dreifuss muscular dystrophy-like phenotype in mice. Hum Mol Genet 18(4):607–620
23. Puckelwartz MJ et al (2010) Nesprin-1 mutations in human and murine cardiomyopathy. J Mol Cell Cardiol 48(4):600–608
24. Attali R et al (2009) Mutation of SYNE-1, encoding an essential component of the nuclear lamina, is responsible for autosomal recessive arthrogryposis. Hum Mol Genet 18(18):3462–3469
25. Sjoblom T et al (2006) The consensus coding sequences of human breast and colorectal cancers. Science 314(5797):268–274
26. Marme A et al (2008) Loss of Drop1 expression already at early tumor stages in a wide range of human carcinomas. Int J Cancer 123(9):2048–2056
27. Tessema M et al (2008) Promoter methylation of genes in and around the candidate lung cancer susceptibility locus 6q23-25. Cancer Res 68(6):1707–1714
28. Doherty JA et al (2010) ESR1/SYNE1 polymorphism and invasive epithelial ovarian cancer risk: an ovarian cancer association consortium study. Cancer Epidemiol Biomarkers Prev 19(1):245–250
29. Stransky N et al (2011) The mutational landscape of head and neck squamous cell carcinoma. Science 333(6046):1157–1160
30. Shah SP et al (2012) The clonal and mutational evolution spectrum of primary triple-negative breast cancers. Nature 486(7403):395–399
31. Schoppmann SF et al (2013) Novel clinically relevant genes in gastrointestinal stromal tumors identified by exome sequencing. Clin Cancer Res 19(19):5329–5339
32. Matsumoto A et al (2015) Global loss of a nuclear lamina component, lamin A/C, and LINC complex components SUN1, SUN2, and nesprin-2 in breast cancer. Cancer Med 4(10):1547–1557
33. Horn HF et al (2013) The LINC complex is essential for hearing. J Clin Invest 123(2): 740–750
34. Ungricht R, Kutay U (2017) Mechanisms and functions of nuclear envelope remodelling. Nat Rev Mol Cell Biol 18(4):229–245
35. Konstantinidis G, Tavernarakis N (2021) Autophagy of the nucleus in health and disease. Front Cell Dev Biol 9:814955
36. Park YE et al (2009) Autophagic degradation of nuclear components in mammalian cells. Autophagy 5(6):795–804
37. Rello-Varona S et al (2012) Autophagic removal of micronuclei. Cell Cycle 11(1): 170–176
38. Ivanov A et al (2013) Lysosome-mediated processing of chromatin in senescence. J Cell Biol 202(1):129–143
39. Changou CA et al (2014) Arginine starvation-associated atypical cellular death involves mitochondrial dysfunction, nuclear DNA leakage,

and chromatin autophagy. Proc Natl Acad Sci USA 111(39):14147–14152

40. Lan YY et al (2014) Dnase2a deficiency uncovers lysosomal clearance of damaged nuclear DNA via autophagy. Cell Rep 9(1): 180–192
41. Dou Z et al (2015) Autophagy mediates degradation of nuclear lamina. Nature 527(7576): 105–109
42. Akinduro O et al (2016) Constitutive autophagy and Nucleophagy during epidermal differentiation. J Invest Dermatol 136(7): 1460–1470
43. Baron O et al (2017) Stall in canonical autophagy-lysosome pathways prompts Nucleophagy-based nuclear breakdown in neurodegeneration. Curr Biol 27(23): 3626–3642 e6
44. Lu X, Djabali K (2018) Autophagic removal of farnesylated carboxy-terminal lamin peptides. Cells 7(4)
45. Li Y et al (2019) Nuclear accumulation of UBC9 contributes to SUMOylation of lamin A/C and nucleophagy in response to DNA damage. J Exp Clin Cancer Res 38(1):67
46. Papandreou M-E, Konstantinidis G, Tavernarakis N (2023) Nucleophagy delays aging and preserves germline immortality. Nat Aging 3(1):34–46
47. Klionsky DJ et al (2021) Guidelines for the use and interpretation of assays for monitoring autophagy (4th edition)(1). Autophagy 17(1): 1–382
48. Mochida K et al (2015) Receptor-mediated selective autophagy degrades the endoplasmic reticulum and the nucleus. Nature 522(7556): 359–362
49. Mochida K et al (2022) Atg39 links and deforms the outer and inner nuclear membranes in selective autophagy of the nucleus. J Cell Biol 221(2)
50. Zou CX et al (2023) The ortholog of human REEP1-4 is required for autophagosomal enclosure of ER-phagy/nucleophagy cargos in fission yeast. PLoS Biol 21(11):e3002372
51. Zhang H et al (2015) Guidelines for monitoring autophagy in Caenorhabditis elegans. Autophagy 11(1):9–27

# Chapter 7

# Analysis of Mitophagy Reporters in *Drosophila*

**Alvaro Sanchez-Martinez, Aitor Martinez, and Alexander J. Whitworth**


## Abstract

Mitophagy is the degradation of mitochondria via the autophagy-lysosome system, disruption of which has been linked to multiple neurodegenerative diseases. As a flux process involving the identification, tagging, and degradation of subcellular components, the analysis of mitophagy benefits from the microscopy analysis of fluorescent reporters. Studying the pathogenic mechanisms of disease also benefits from analysis in animal models in order to capture the complex interplay of molecular and cell biological phenomena. Here, we describe protocols to analyze mitophagy reporters in *Drosophila* by light microscopy.


**Key words** Mitophagy, *Drosophila*, Mitochondria, Reporter, Light microscopy, mito-QC, mtx-QC, Brain, Muscle, Neurodegeneration

## 1 Introduction

Mitophagy is the degradation of mitochondria by the autophagy-lysosome system, whereby mitochondria (usually "damaged," dysfunctional, or excess mitochondria) are identified, labeled, engulfed by a phagophore membrane, and trafficked to the lysosome. Fusion of the autophagosome with the lysosome results in the degradation of its contents, i.e., the mitochondria. This is generally viewed as a mechanism to remove mitochondria that are no longer functional or that are sufficiently damaged that they risk rupturing and releasing cell-death-inducing factors. Since the mitochondrial content of cells can also change due to changes in cellular circumstances, such as cell cycle, cell type, and bioenergetic demand, we can view mitophagy broadly as a homeostatic quality control process. However, since mitochondria are essential organelles required for generating the majority of cellular energy (in the form of ATP), as well as many other crucial cellular functions, this destructive process must be tightly regulated, the mechanisms of which have been reviewed elsewhere [1–3].

Mitophagy as a term was coined in 2005 [4], but the phenomenon has been studied for considerably longer. The molecular

Ioannis P. Nezis (ed.), *Selective Autophagy: Methods and Protocols*, Methods in Molecular Biology, vol. 2845,
https://doi.org/10.1007/978-1-0716-4067-8_7,

players have been extensively studied in yeast, some of which are not obviously conserved in higher species, such as *Atg32*. Intense interest was garnered following the initial observations that the products of genes linked to Parkinson's disease, namely PINK1 and Parkin, and later multiple other Parkinson's genes, are linked to damage-induced mitophagy. Hence, a central focus of studying mitophagy in vivo has been in the nervous system, but other highly energetically demanding tissues, such as muscle, are also intensively studied. Thus, the methods described in this chapter will focus on imaging mitophagy in neuromuscular tissue.

Mitophagy is inherently a flux process, involving the generation of molecular signals triggering their intended cascade of cellular events, followed by their loss by degradation. Thus, methods to monitor this process must carefully consider the identity of the marker being analyzed and its place in the flux process (see Klionsky et al. [5] for a detailed discussion). As an example, a common marker of general autophagy is the level of lipidated LC3, or LC3-II, readily detectable by immunoblotting or imaging. LC3 is lipidated during the formation of the autophagophore membrane, which is subsequently degraded upon fusion with the lysosome. Hence, an increase in LC3-II could indicate either an induction of autophagy activation (increasing the production of LC3-II) or a reduction in autophagic degradation/flux (reduced turnover of LC3-II). The interpretation of this has profoundly different implications for the mechanistic action of the events being experimentally manipulated. The safest approach to analyze the impact of experimental manipulations on mitophagy/autophagy is to take into account the flux process. In the context of LC3-II analysis, the best practice is to analyze the impact of manipulations under steady-state conditions alongside treatment with agents that block autophagic flux, typically chemical inhibitors of lysosome fusion/function.

A widely used alternative to this immunoblotting approach in the presence and absence of flux inhibitors is microscopy-based methods that take advantage of the differentially labile nature of different fluorophores. Green fluorescent protein (GFP) is readily quenched by low pH, such as that occurring inside lysosomes, while other fluorophores such as mCherry are not. This establishes a simple system by which autophagy cargoes, or the autophagosome membrane itself, can be dual-labeled with mCherry and GFP. In the cytosol (neutral pH), both GFP and mCherry fluoresce, typically identified as a yellow-green signal when the channels are merged. However, once inside the lysosome (acidic pH), GFP is quenched to leave a "red-only" signal from the mCherry. Quantification of the "red-only" signal (often visualized as puncta) gives a direct read-out of the end stage of autophagic flux. This was successfully used to tag LC3, and the elegant simplicity of this approach has now been adopted as an autophagy flux reporter for many other cargoes.

Mitophagy has been analyzed across the gamut of experimental model systems. In the context of understanding neurodegeneration, such as occurs in Parkinson's disease, animal models must be employed at some stage to fully appreciate the in vivo complexities. *Drosophila* offers a powerful genetic model organism that has provided tremendous insights into fundamental cell biological processes as well as informing pathological processes relevant to human disease. Due to the burgeoning interest in mitophagy in the context of Parkinson's and other neurodegenerative conditions, we and others have adopted these pH-sensitive fluorescent reporters to detect and monitor mitophagy in *Drosophila* tissues. Here we describe protocols that we have developed and optimized over the past few years in our lab to analyze mitophagy reporters by conventional light microscopy. Initially, we adopted the first such reporter, termed "mito-QC," a dual fluorophore (mCherry+GFP) targeted to the outer mitochondrial membrane (OMM) [6, 7]. We and others have also used a mitochondria matrix-targeted variant of this method, termed "mtx-QC" [8], which can be used and analyzed in the same way as mito-QC and is discussed further below (*see* **Note 10**). We have also generated and characterized similar mitophagy reporters, namely mt-Keima [9] and mito-SRAI [10], but in our hands, these did not present any distinct advantages over mito-/mtx-QC. In fact, we found that mt-Keima to be more limiting than mito-QC since it must be imaged live as the differential fluorescence is not fixable, but the signal was rather weak and bleached very quickly under standard image acquisition conditions [7]. Analyzing mito-SRAI, which was reported to give superior sensitivity than mito-QC, we found that this was not noticeably brighter or more sensitive than mito-QC, but concerningly caused modest behavioral defects indicating that its expression was not entirely benign (data not shown). For these reasons, we have favored the use of mito-QC and mtx-QC.

Below, we provide detailed protocols for the analysis of mitophagy by mito-QC/mtx-QC. However, it is worth reflecting that other experimental approaches are possible, each with different pros and cons regarding throughput and/or spatiotemporal resolution. These include what could be considered the "gold standard" for autophagy-related analysis: electron microscopy (EM) or correlative light and electron microscopy (CLEM) [5]. Of course, these techniques provide unparalleled spatial resolution but are limited in throughput and don't readily allow the interrogation of dynamic processes. Alternatively, mass spectrometry analysis of pulse-chase isotope-labeled proteins can give a molecular-level insight into the turnover rates of individual proteins, detailing changes in half-life decay times. While submitochondrial compartments can be biochemically fractionated for analysis, further spatial resolution is limited. Since this approach relies on degradation rates of proteins or complex macromolecular structures at a population scale, the time-scale of degradation can range from hours to days, making temporal resolution challenging. With these considerations, analysis of mitophagy reporters by advanced light microscopy provides an excellent "middle ground" to follow the spatial and temporal dynamics of the lysosomal degradation of mitochondrial components. Here, we describe protocols, and the principles behind them, that we have developed and optimized, which can be readily adapted to researchers' local setups.

## 2 Special Materials

### 2.1 Fly Lines

1. UAS-mito-QC (BDSC_91640 and BDSC_91641).
2. UAS-mtx-QC (BDSC_602462, BDSC_602463, BDSC_602464).
3. nSyb-GAL4 (BDSC_51635): pan-neuronal driver used for larval and adult brain expression.
4. Mef2-GAL4 (BDSC_27390): muscle driver used for adult thorax expression.

### 2.2 Equipment

1. Dissection forceps (e.g., Dumont Tweezer 0303-5-PO).
2. Dissection scissors (e.g., FST 15000-00).
3. Sylgard plate (90 mM petri dish, bottom plate filled with Sylgard (SLS)).
4. Dissecting pins (e.g., FST 26002-20).
5. Microscope slides.
6. Square coverslips thickness #0 and #1.5.
7. Round coverslips thickness #1.5.
8. Hole reinforcement ring label stickers.

### 2.3 Reagents

1. PBS 1×.
2. Fixative (formaldehyde 4%-PBS 1×, pH 7.0).
3. Mounting media (e.g., ProLong and Vectashield).

## 3 Methods

Researchers will need to perform standard genetic crosses to combine the UAS-mito-QC or UAS-mtx-QC reporter lines with a desired GAL4 driver line. It is important to select an appropriate GAL4 line to drive expression in the required target tissue and at a suitable level. Too low, and the signal will be very weak to detect; too high, and the expression of the reporter can affect organelle dynamics and function, often fragmenting mitochondria. Here, we detail the analysis of brain and muscle where pan-neuronal (nSyb-GAL4) or pan-muscle (Mef2-GAL4) drivers have been used. Carry out all procedures at room temperature unless otherwise specified.

### 3.1 Third Instar Larval Brain Dissection

1. Pick "wandering" third instar larvae (5 days old). On a Sylgard plate, pin down the larvae with the dorsal side up just below the mouth hooks (anterior) and between the spiracles at the posterior end without stretching the larvae too much (Fig. 1a). Place a drop of PBS 1× on top, sufficient to cover the larva.
2. With dissecting scissors, make a small cut across the posterior end and use that opening to cut along the top of the cuticle towards the anterior end, stopping before the mouth hooks (Fig. 1b).
3. Remove some intestines with dissecting forceps by pulling out from the posterior end.
4. Pin out the larvae fillet in four "corners" (Fig. 1c) and remove any remaining unwanted internal tissues, leaving behind the tissue(s) of interest, e.g., the larval brain. If needed, readjust the anterior pin to expose the larval brain.
5. Remove the PBS 1× solution and add a drop of fixative (formaldehyde 4%-PBS 1×, pH 7.0; *see* **Note 1**). Incubate for 20–25 min.
6. Remove the fixative and wash twice for 5 min with PBS 1×.
7. Cut off the spiracles at the posterior end and remove the optic lobes (these are thick and/or protruding tissues) so the ventral nerve cord (VNC) can be in contact with the coverslip (Fig. 2a).

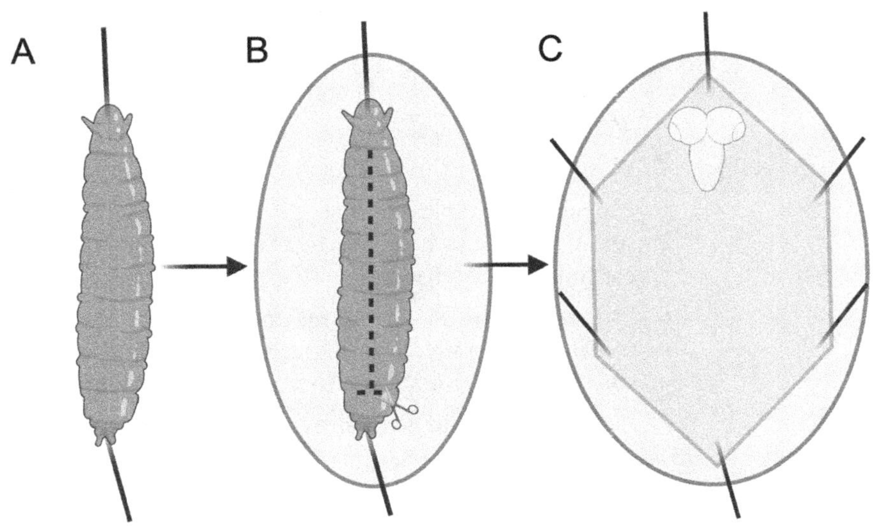


**Fig. 1** Dissection pattern followed to obtain the larval fillet

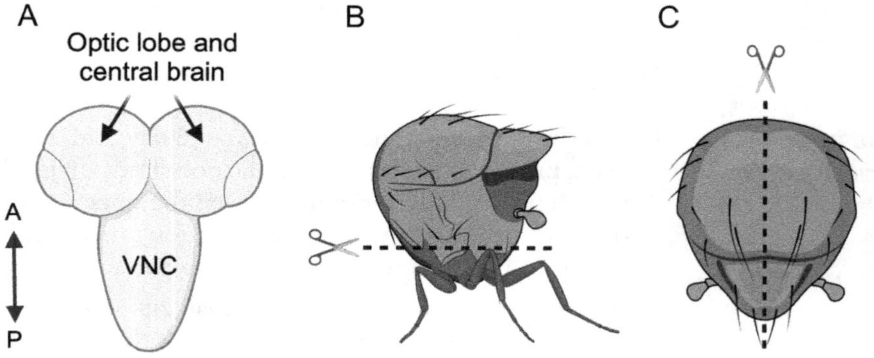


**Fig. 2** Schematics of (**a**) third instar larval brain and (**b**, **c**) adult thorax

8. Remove all the pins and transfer the fillet with the brain onto a slide with 100 μL of Prolong mounting media. Align the fillets so the brain is exposed and put a coverslip #1.5 on top ("spacers" are not usually necessary here, see below). The samples are ready to be imaged once the mounting media has been set (*see* **Note 2**). Keep samples in the dark until ready for microscopy.

#### *3.2* Drosophila melanogaster *Adult Brain Dissection*

1. Put the fly on a Sylgard plate in a drop of PBS 1× to perform the dissection. Pull the head off with dissecting forceps from the thorax and discard the body.
2. With two sets of fine forceps, peel off the cuticle from the proboscis towards one eye carefully, separating the cuticle and connective tissue from the soft neuronal tissue. Proceed to the other side until the brain is loose (*see* **Note 3**). While the brain is still attached to the exoskeleton, remove as much of the bright white connective tissue as possible.
3. Transfer the dissected brain to a new drop of fixative (formaldehyde 4%-PBS 1×, pH 7.0; *see* **Note 1**) for 30 min (*see* **Note 4**).
4. Wash the brain three times for 10 min in PBS 1×.
5. Prepare a suitable microscope slide with a "spacer" to avoid squashing the sample. We use "hole reinforcement ring" stickers (e.g., Avery, Ryman).
6. Put 10 μL of ProLong mounting media in each hole and transfer the brains with forceps (maximum 3 per well). Make sure the posterior side of the brain is facing up.
7. Put a small round coverslip on top (*see* **Note 5**).

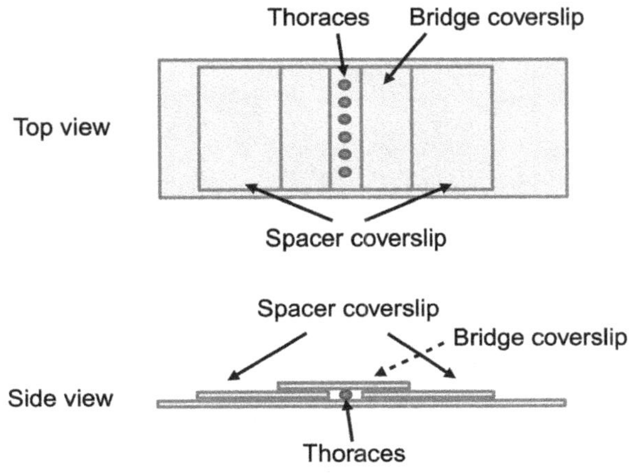


**Fig. 3** Diagram of the mounting method for adult thoraces

### *3.3* Drosophila melanogaster *Adult Thorax Dissection*

1. Put the fly on a Sylgard plate in a drop of fixative (formaldehyde 4%-PBS 1×, pH 7.0; *see* **Note 1**) to perform the dissection. Remove the wings, head, abdomen, and gut with forceps and discard.
2. Remove the legs and ventral section of the thorax with scissors (Fig. 2b).
3. Cut longitudinally the cuticle on the dorsal side using the neck hole as a starting point towards the scutellum (Fig. 2c). By doing so, both sides of the thorax should split apart but, if needed, make a clean cut through the middle section of the thorax to split the hemi-thoraces.
4. Incubate on the Sylgard plate in fixative for 25 min (formaldehyde 4%-PBS 1×, pH 7.0; *see* **Note 1**).
5. Wash the thoraces with PBS 1× twice for 10 min.
6. Mount the thoraces on a slide using coverslips with thickness #0 as spacers (thicker than . . . described above to accommodate the thicker sample) onto which the final coverslip thickness #1.5 will form a "bridge" (*see* **Notes 6** and 7). Stick the spacer coverslips with nail polish (Fig. 3).
7. Add Vectashield mounting media (*see* **Note 8**) in between the spacer coverslips and put thoraces facing upward so the indirect flight muscles are visible.
8. Put the coverslip thickness #1.5 (this will depend on the objective used in the microscope) on top of the thoraces, bridging the spacer coverslips.
9. Let the samples rest for at least 1 h, and they will be ready for image acquisition. Until then, keep them in dark conditions.

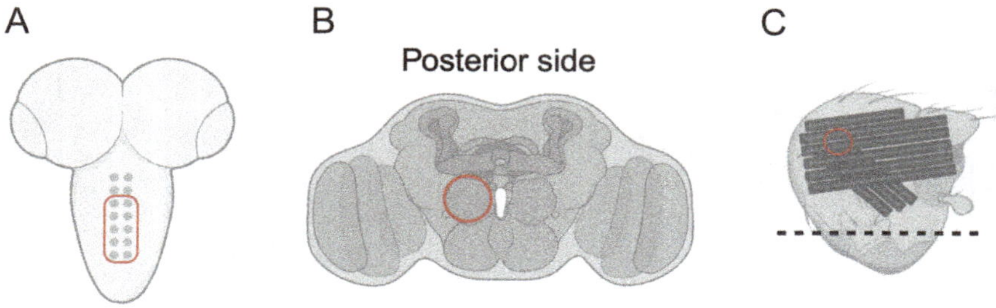


**Fig. 4** Schematic of the approximate regions where microscopy images are taken. (**a**) Larval brain, (**b**) adult brain and (**c**) adult thorax

### *3.4 Image Acquisition of Larval and Adult Brains Using a Spinning Disk Confocal*

For larval and adult brains, due to the thickness and complexity of the tissue, imaging with a spinning disk confocal microscope gives the necessary resolution without bleaching the sample. We use the Andor Dragonfly spinning disk microscope (Oxford Instruments Group), equipped with a Nikon Plan-Apochromat 100×/1.45 NA oil immersion objective and iXon camera.

1. Prepare the microscope as per local procedures. Place larval or adult brain-containing slides on the microscope and locate the sample/tissue/region of interest (ROI) through the eyepieces using brightfield and green fluorescence.
2. The degree of mitophagy can vary within tissue, e.g., central versus more peripheral areas. For all tissue types, choose an area within the tissue that will be a consistent ROI for imaging between samples. This will greatly aid reproducible quantification of the degree of mitophagy within the target tissue. For larval brains, we target the ROI as a parallel cluster of ≈100 neurons located in the medial region of the dorsal part of the ventral nerve cord (Fig. 4a). For adult brains, we select a region of cell bodies that is just lateral to the esophagus (see Fig. 4b). When locating the ROI, be fast and use low laser intensity to avoid bleaching.
3. Adjust optimum exposure time, gain, and laser intensity levels of red (594 nM) and green (488 nM) channels, and image all the samples within the same batch of experiments with the same settings.
4. Visualize the neurons on the computer screen and define the z-stack, starting from the dorsal towards the ventral. Typically, an approximately 10–13 μm depth stack is taken with a step size of 0.2 μm. Set the imaging mode so that red and green channels are imaged sequentially for each plane to avoid bleed-through between channels.
5. Repeat **steps 3–5** if the ROI does not fit in one plane.
6. Take one or two images per animal, and a target n of at least 5–6 brains per genotype should be analyzed.

### 3.5 Image Acquisition of Adult Thoraces Using a LSM Confocal

For adult thoraces, we use a Zeiss LSM 880 confocal microscope (Carl Zeiss MicroImaging) equipped with Nikon Plan-Apochromat 63×/1.4 NA oil immersion objectives, which allows better resolution for muscle tissue than the spinning disk while maintaining acceptably short acquisition times.

1. Prepare the microscope as per local procedures. Place adult thorax-containing slides on the microscope and locate the sample/tissue/region of interest (ROI) through the eyepieces using brightfield and green fluorescence.
2. For thoraces, we target the ROI as an area of mitochondria and muscle fibers located towards the dorsal-anterior side of the tissue (Fig. 4c). When locating the ROI, be fast and use low-laser intensity to avoid bleaching.
3. Adjust the optimum exposure time, gain, and laser intensity levels of red (594 nM) and green (488 nM) channels, and image all the samples within the same batch of experiments with the same settings.
4. Visualize the mitochondrial network on the computer screen and define the z-stack, starting from more superficial tissue towards the deeper layer of the tissue. Typically, an approximately 9–10 μm depth stack is taken with a step size of 1 μm. Set the imaging mode so that red and green channels are imaged sequentially for each plane to avoid bleed-through between channels.
5. Take one or two images per animal and a target "n" of around 8–12 thoraces per genotype should be analyzed.

Other tissues can also be analyzed in a similar fashion. For example, we have also analyzed third instar larvae epidermal cells [7]. The sample preparation method can be followed as described above in Subheading 3.1 up to **step 4**, carefully removing the brain and the muscles with dissecting forceps to expose the underlying epidermal cells. The anterior and posterior parts of the fillet can be cut away to keep a flat square surface, retaining segments A2 to A6 for imaging. Imaging and quantification can be performed as described.

#### 3.5.1 Quantification

The overarching principle for quantifying mitophagy using fluorescent reporters is that we want to quantify the relative amount of lysosome-acidified reporter (red signal with green quenched) versus reporter at neutral pH (unquenched red and green signals). For mito-QC, we count the number of "red-only" puncta, so we need to subtract the green channel (GFP) from the red channel (mCherry). Other mitophagy reporters, such as mt-Keima, report a more graded shift from green to red fluorescence, requiring the relative quantification of red versus green levels. Since we currently only use the mito-QC/mtx-QC reporter in our lab, the method for quantification described here focuses on the analysis of "red-only" puncta. This can be done in a number of ways depending on the available software. We describe the use of Imaris (purchased software) or FIJI (free software), which have pros and cons. Imaris provides high-accuracy segmentation of signals in 3D but requires a good, strong signal since the mitolysosomes are considered as the red dots that are not present in the green channel. In contrast, while FIJI is free, current methods cannot be used on Z-stacks (3D volumes); either representative images need to be selected and analyzed individually, or only very short Z-stacks can be compressed into a maximum projection image.

### 3.6 Quantification Using Imaris Software for Larval Brains

This method is based on the Imaris analysis software version 9.0.2 but can be adapted to other versions.

1. Open the image. Adjust the levels of red and green in the control panel so all the images within the same batch of experiments have the same starting point.
2. Discard all images where the mitochondrial network is not well defined. This is key when generating the "mito mask" (green channel) that needs to be subtracted from the red to get "red only puncta."
3. Create a surface corresponding to one individual cell or group of cells. Select "skip automatic creation, edit manually."
4. In "Contour" and "Mode," select "Distance" as Drawing Mode.
5. In "Board," click on "Draw" and draw the outline of the cell/cells. It can be applied to the whole section, and count the number of cells before creating the surface.
6. Click on "Copy" and "Paste" to apply the surface to all the slices of the stack. Then click on "Create surface."
7. Click on "Mask All" to only keep the channels corresponding to the surface newly created.
8. Create a mask for each channel one by one. The masked channels will appear in the list of channels (display adjustment window).
9. Create a rendering 3D surface. Leave automatic creation.
10. In the same sidebar, select the channel that the 3D rendered surface will be based on (my mask GFP mito channel), then select "Surface detail": 0.2 μm (*see* **Note 9**).
11. Adjust the threshold so that the 3D-rendered surface matches the masked channel. Maintain consistency across the different samples being analyzed.

12. Mask the signal from the red channel that overlaps the green channel. Mask the red channel from the 3D rendered surface, this creates a new channel "masked masked." Tick the box "Set the voxels inside surface to."
13. Create the spots by clicking on the "Spots" symbol and select "different spot sizes."
14. Select the source channel (masked masked RFP) and enter the estimated diameter of the spots (0.5).
15. Select "intensity center" of the "masked masked" channel (channel where the dots are) and adjust the threshold of the number of dots to match what is seen on the screen.
16. Select "local contrast" and adjust the threshold so that the signal from the red channel (not the artificial spherical dot) matches the white surface. Finish the analysis by clicking on the double arrow.
17. The total number of spots is indicated in the "stats" section of the "spots." At this point, other parameters such as puncta size can also be obtained here.

### 3.7 Quantification Using FIJI Plug-in "mito-QC Counter" Software for Adult Brains

1. Download and install the "mito-QC Counter" macro for FIJI as described in Montava-Garriga et al. [11].
2. Open the image with the following settings in the *Bio-Formats Import Options* window: *View stack with: Hyperstack* and *Color mode: Composite.*
3. Adjust the levels of red and green channels in the control panel (*Image>Adjust> Brightness/Contrast...*) so all the images within the same batch of experiments have the same levels. Adjust as necessary.
4. Discard images/slices where neurons cannot be clearly differentiated and the mitochondrial network is too clumped and/or debris is present.
5. Select the "*Rectangle*" tool in the FIJI tool selection panel (typically top left-hand side of the floating window) to draw a square around the neurons of interest and define the ROI.
6. Open the ROI Manager window (*Analyze>Tools>ROI Manager...*) and add the drawn square/ROI by clicking *Add [t].* In this way, the size of the ROI will be saved for future images.
7. Crop the image (*Image>Crop)* and condense all the z-stacks into a maximum projection image by clicking *Image>Stacks>Z Project....*
8. Overlay the ROI to the image (*Ctrl + B*) and save the maximum projection image in a separate folder (*File>Save As*) for subsequent analysis.

9. Once all the maximum projection images with selected ROIs for each genotype are saved in a folder, mitophagy can be analyzed in batches using the "mito-QC Counter" macro batch mode.
10. Open the "mito-QC Counter" macro (*Plugins> mQC Counter*) and select batch mode. Select the green and red channels for the images (Green channel = 1, Red channel = 2; or the other way around). The settings need to be optimized for each tissue/experimental setting as recommended by the developers [11]. For adult brains, we found that the best settings were: *Radius for smoothing images = 1, Ratio threshold = 0.5*, and *Red channel thresh: stdDev above mean = 1.*
11. Click *OK* and select the folder with images to be analyzed. The plug-in will run and provide:
    (a) A csv file with all results.
    (b) A mCherry/GFP ratio image.
    (c) An image with an overlay of cell ROIs and detected peaks.
    (d) A "mitophagy mask" image, in which the original image is masked by dots allowing visualization of how mitophagy was quantified. Verify that the plug-in is picking the red dots correctly.

    These will automatically open and be saved within the selected folder.
12. The total number of mitolysosomes is obtained, but the best representation of mitophagy is obtained from the number of mitolysosomes per area (MLs/$\mu m^2$). Total mitolysosome area per area (*MLtotalArea/*$\mu m^2$) and mean mitolysosome area per area are also given (*MLmeanArea/*$\mu m^2$).
13. Generate a single value for each animal, and a target n ≥ 6–8 animals per genotype should be analyzed.

### 3.8 Quantification Using FIJI Plug-in mito-QC Counter Software for Adult Thoraces

1. Download and install the "mito-QC Counter" macro for FIJI as described in Montava-Garriga et al. [11].
2. Open the image with the following settings in the *Bio-Formats Import Options* window: *View stack with: Hyperstack* and *Color mode: Composite.*
3. Adjust the levels of red and green channels in the control panel (*Image>Adjust> Brightness/Contrast...*) so all the images within the same batch of experiments have the same levels. Adjust as necessary.
4. Choose a single plane in the z-stack where muscle bundles are not damaged, and the tissue looks representative. Ideally, analyze all planes and finally average the values per individual.

5. Select the "*Rectangle*" tool in the FIJI tool selection panel (typically, the top left-hand side of the floating window) to draw a square around the mitochondrial network and muscle fibers of interest and define the ROI. Alternatively, the ROI may be better fitted using the Polygon selection tool.
6. Open the ROI manager window (*Analyze>Tools>ROI Manager...*) and add the drawn ROI by clicking *Add [t]*.
7. Crop the image (*Image>Crop*).
8. Overlay the ROI to the image (*Ctrl+B*) and save the image in a separate folder (*File>Save As*) for subsequent analysis.
9. Once all the images with selected ROIs for each genotype are saved in a folder, mitophagy can be analyzed in batches using the "mito-QC Counter" macro batch mode.
10. Open the "mito-QC Counter" macro (*Plugins> mQC Counter*) and select batch mode and repeat the analysis as in 3.7.10. For adult thoraces, we found that the best settings were: *Radius for smoothing images* = *1*, *Ratio threshold* = *0.8–2*, and *Red channel thresh: stdDev above mean* = *1*.
11. Generate a single value for each animal, and a target n ≥ 8–10 animals per genotype should be analyzed.

## 4 Notes

1. The fixative for mitophagy analysis must be at pH 7.0. Because of the pH sensitivity of the reporter, any deviation from pH 7.0 will impact the detection of red-only puncta.
2. Prolong mounting media needs at least 8–16 h to set before it can be imaged, and for this tissue, it works very well. Alternatively, other suitable mounting media (e.g., Vectashield, *see* **Note 8**) can be used following the manufacturer's instructions.
3. The brain should be clean from any connective tissue and trachea that might interfere with image acquisition.
4. Thorax and brain incubation times in fixative should be between 25 and 30 min, giving some flexibility while dissecting.
5. When placing the small round coverslip, be extremely careful so the brain stays visibly within the well.
6. Coverslip #0 spacers are needed to avoid the thoraces being squashed by the coverslip #1.5 used on top while allowing the tissue to be in contact with it.
7. The slides used to mount the thoraces should be made hours or a day in advance to allow the nail polish to dry.

8. The use of Vectashield allows the samples to be imaged on the same day, which is critical to visualize all the red-only puncta in this tissue. Other suitable mounting media that allows this can also be used.
9. The surface detail number should be based on the resolution capability of the microscope used.
10. Comparison between mito-QC and mtx-QC: As mentioned above, we have optimized these methods for two similar mitophagy reporters, mito-QC and mtx-QC, which differ only in their subcellular localization. The mito-QC is targeted to the OMM, while the mtx-QC is localized to the mitochondrial matrix. We have performed a comparative analysis of the reporters side-by-side in larval neurons. In general, the reporter constructs look highly similar with essentially identical labeling of the mitochondrial network. We have noticed, however, a consistently lower number of mitolysosomes (red-only puncta) reported by mtx-QC compared to mito-QC. A direct comparison of these two reporters can be found in our previous work [8].

    We believe that this difference is a genuine reflection of the turnover of different mitochondrial compartments and could be due to a number of reasons that are not yet fully explored. One reason could be the relative localization of the reporter proteins. For instance, the mitochondrial targeting sequence used in mito-QC is taken from Fis1, which serves to tether the reporter to the OMM. In comparison, mtx-QC uses a mitochondrial targeting sequence from human COX8A, which robustly delivers reporter proteins selectively to the mitochondrial matrix. Therefore, the differential mitochondrial localization—OMM versus matrix—also has the potential to report on the selective degradation of mitochondrial subcompartments. While most research analyzing mitophagy will envisage wholesale degradation of entire mitochondrial units, it is also appreciated that mitochondria can be degraded in a piecemeal fashion, with subcompartments being selectively isolated for degradation. Further work is required to better understand the different dynamics, and the application of super-resolution microscopy may be essential.

## Acknowledgments

We thank the many members of the Whitworth group who have contributed to the optimization of these methods over the years and for discussions around mitophagy in vivo. We also thank colleagues in the field for helpful technical and conceptual discussions. Ongoing work in the Whitworth group is funded by the Medical

Research Council (MC_UU_00028/6). AM was supported by the Basque Government Postdoctoral Fellowship (POS_2022_2_0045). Stocks made by the Whitworth group and discussed here are deposited at the Bloomington *Drosophila* Stock Center (BDSC), which is supported by grant NIH P40OD018537. Figures were generated using BioRender.

## References

1. Palikaras K, Lionaki E, Tavernarakis N (2018) Mechanisms of mitophagy in cellular homeostasis, physiology and pathology. Nat Cell Biol 20(9):1013–1022
2. Wang S, Long H, Hou L, Feng B, Ma Z, Wu Y et al (2023) The mitophagy pathway and its implications in human diseases. Signal Transduct Target Ther 8(1):304
3. Onishi M, Yamano K, Sato M, Matsuda N, Okamoto K (2021) Molecular mechanisms and physiological functions of mitophagy. EMBO J 40(3):e104705
4. Lemasters JJ (2005) Selective mitochondrial autophagy, or mitophagy, as a targeted defense against oxidative stress, mitochondrial dysfunction, and aging. Rejuvenation Res 8(1):3–5
5. Klionsky DJ, Abdel-Aziz AK, Abdelfatah S, Abdellatif M, Abdoli A, Abel S et al (2021) Guidelines for the use and interpretation of assays for monitoring autophagy (4th edition) (1). Autophagy 17(1):1–382
6. Allen GF, Toth R, James J, Ganley IG (2013) Loss of iron triggers PINK1/parkin-independent mitophagy. EMBO Rep 14(12): 1127–1135
7. Lee JJ, Sanchez-Martinez A, Zarate AM, Beninca C, Mayor U, Clague MJ et al (2018) Basal mitophagy is widespread in drosophila but minimally affected by loss of Pink1 or parkin. J Cell Biol 217(5):1613–1622
8. Sanchez-Martinez A, Martinez A, Whitworth AJ (2023) FBXO7/ntc and USP30 antagonistically set the ubiquitination threshold for basal mitophagy and provide a target for Pink1 phosphorylation in vivo. PLoS Biol 21(8): e3002244
9. Sun N, Yun J, Liu J, Malide D, Liu C, Rovira II et al (2015) Measuring in vivo mitophagy. Mol Cell 60(4):685–696
10. Katayama H, Hama H, Nagasawa K, Kurokawa H, Sugiyama M, Ando R et al (2020) Visualizing and modulating mitophagy for therapeutic studies of neurodegeneration. Cell 181(5):1176–87 e16
11. Montava-Garriga L, Singh F, Ball G, Ganley IG (2020) Semi-automated quantitation of mitophagy in cells and tissues. Mech Ageing Dev 185:111196

# Chapter 8

# Studying Selective Autophagy of Protein Aggregates Using Particles Induced by Multimerization (PIMs)

**Giel Korsten and Lukas C. Kapitein**

## Abstract

Selective autophagy of protein aggregates, called aggrephagy, is vital for maintaining cellular homeostasis. Classically, studying aggrephagy has been challenging due to the infrequent occurrence of autophagic events and the lack of control over the specificity and timing of protein aggregation. We previously reported two variants of a PIM (particles induced by multimerization) assay that enable the formation of chemically induced, fluorescently labeled protein aggregates in cells. PIMs are recognized by the selective autophagy machinery and are subsequently degraded in the lysosome. By making use of pH-sensitive fluorescent proteins, such as GFP or mKeima, the PIM assay allows for direct visualization of aggregate clearance in cells. Here, we describe a protocol for the use of the PIM assay to study aggrephagy in live and fixed cells.

**Key words** Aggrephagy, Live-cell imaging, Inducible dimerization

## 1 Introduction

The formation of potentially toxic protein aggregates containing damaged and mutated proteins is a hallmark of neurodegenerative diseases and aging [1]. In order to combat the presence of protein aggregates, cells employ a selective autophagy process called aggrephagy. Aggrephagy is initiated by ubiquitination of the aggregated protein and subsequent accumulation of selective autophagy receptors (SARs). Then, an isolation membrane, or omegasome, forms around the aggregate. This membrane expands and ultimately seals, forming an autophagosome around the protein aggregate. The autophagosome then fuses with late endosomes (LE) and lysosomes (LY), resulting in the acidification and clearance of the protein aggregate [2].

Studying selective autophagy is challenging since the process occurs infrequently under basal conditions. To study the aggrephagy pathway, previous research has therefore focused on the induction of protein aggregates using either chemical induction

Ioannis P. Nezis (ed.), *Selective Autophagy: Methods and Protocols*, Methods in Molecular Biology, vol. 2845,
https://doi.org/10.1007/978-1-0716-4067-8_8,

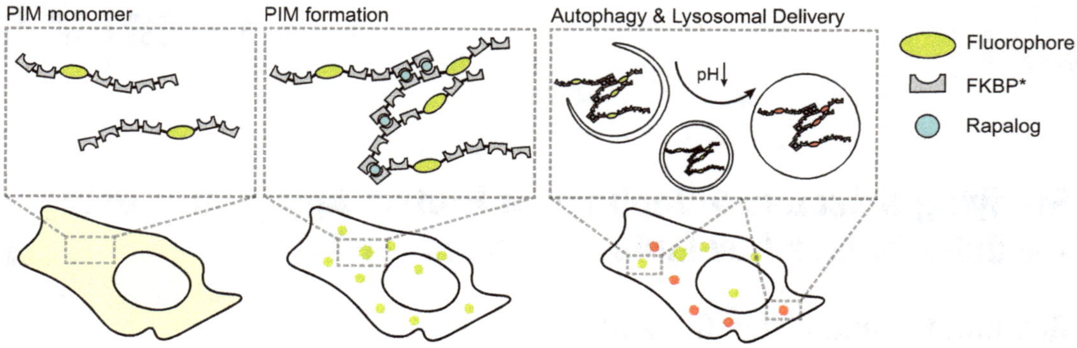


**Fig. 1** Schematic representation of the PIM assay. Dual-PIM or mKeima-PIM constructs are soluble and homogeneously distributed throughout the cytosol prior to rapalog addition. The addition of rapalog results in rapid multimerization and aggregate formation of PIM monomers. PIM aggregates are recognized by the aggrephagy machinery and undergo selective lysosomal degradation. The presence of mCherry/EGFP or mKeima fluorophores allows monitoring of acidification, indicating entry into lysosomes

of protein misfolding or the expression of aggregation-prone proteins [3, 4]. These approaches have provided valuable insight into the aggrephagy process, but they suffer from several drawbacks. While the expression of aggregation-prone proteins often yields aggregation in cells, this approach does not provide temporal control over aggregation [3–5]. Furthermore, the expression of mutant proteins, such as mutant huntingtin, can impair autophagy and preclude analysis of the physiological aggrephagy pathway [5–7]. Conversely, the formation of aggregates by inducing proteotoxic stress can be precisely timed [3], but it affects a large part of the proteome and likely also impairs the autophagy pathway.

In order to mitigate these issues, we previously introduced the PIM (particles induced by multimerization) assay as a tool to study aggrephagy in live and fixed cells [8]. The PIM construct is comprised of multiple multimerization domains, including four FKBP* homodimerizing domains. The addition of the rapamycin analog AP20187 (hereafter rapalog) enables the rapid formation of protein aggregates (Fig. 1). To monitor the autophagic clearance of these inducible aggregates, an mCherry-EGFP tag was added to the PIM construct (termed dual-PIM). This dual-tag approach was previously used to study the acidification of p62 inclusions and autophagosomes [3, 9]. In short, monitoring dual-PIMs relies on the difference in resistance to lysosomal acidification and breakdown between GFP and mCherry fluorophores. Upon lysosomal acidification, GFP rapidly and irreversibly quenches while mCherry fluorescence remains functional (Fig. 2a). Therefore, ratiometric imaging of the GFP/mCherry signal serves as a direct readout for lysosomal delivery of aggregates. This approach yields a robust tool to visualize aggrephagy of individual aggregates in cells. Notably, because mCherry, unlike GFP, resists lysosomal degradation, dual-PIMs can be used in both live- and fixed-cell assays [9, 10].

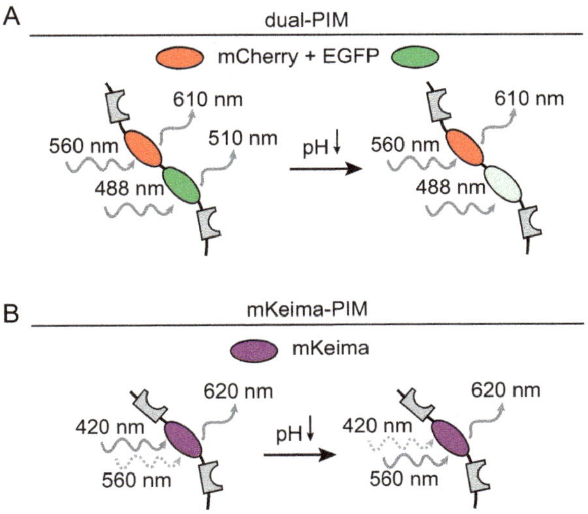


C

| | dual-PIM | mKeima-PIM |
|---|---|---|
| Green emitting fluorophore compatibility | - | + |
| Live-cell imaging | + | + |
| Fixed-cell imaging | + | - |
| Ease of use; e.g. visibility by eye | High | Medium (at neutral pH) |

**Fig. 2** Characteristics of the dual- and mKeima-PIM tools. (**a**) Schematic representation of the dual-PIM tool. At neutral pH, dual-PIM fluorophores mCherry and EGFP are both functional and show emission at 610 nm and 510 nm, respectively. Acidification results in the loss of EGFP emission at 510 nm. (**b**) Representation of mKeima-PIM construct containing the mKeima fluorophore. Emission of mKeima occurs consistently at 620 nm, while efficient excitation of mKeima depends on pH. At neutral pH, the mKeima excitation spectrum peaks at 420 nm, while excitation using 560 nm light is most efficient at low pH. (**c**) Schematic depicting the experimental applicability of dual-PIM and mKeima-PIM

However, since dual-PIMs rely on continuous imaging of GFP to determine the acidification status of PIMs, they cannot be used in combination with other green-emitting fluorophores. This limits the applicability of dual-PIMs for multicolor live-cell imaging. Therefore, we developed mKeima-tagged PIM (mKeima-PIM) to facilitate live-cell imaging of aggrephagy in combination with (endogenous) GFP-tagged proteins [11]. In the mKeima-PIM constructs, the mCherry-EGFP tag is exchanged for an mKeima fluorophore [12]. mKeima has a pH-sensitive bimodal excitation spectrum, peaking at 440 nm in neutral pH environments and at 568 nm at low pH. Moreover, like mCherry, mKeima is resistant to lysosomal degradation and has an emission spectrum peaking around 620 nm (Fig. 2b). These characteristics allow for simultaneous imaging of green-emitting fluorophores. We showed that mKeima-PIMs enable the visualization of aggregate clearance in combination with markers for various stages of the aggrephagy pathway, such as endogenously tagged LE/LY marker Rab7 [12]. Together, our previous reports established PIMs as a novel inducible aggrephagy probe and provided valuable insight into the spatiotemporal regulation of aggregate clearance.

Here, we describe an in-depth protocol for probing the selective autophagy of protein aggregates using dual-PIM and mKeima-PIM constructs, highlighting the relative strengths of each tool.

This protocol will enable researchers to study aggrephagy using fixed- and live-cell approaches, providing an easy-to-use template for developing (high-throughput) light microscopy analysis of aggrephagy without the side effects of expressing toxic proteins or inducing general protein stress.

## 2 Materials

### 2.1 Constructs

1. The constructs used in this protocol are all available on Addgene.
   (a) Dual-PIM: Addgene plasmid #111758; http://n2t.net/addgene:111758; RRID:Addgene_111758.
   (b) mKeima-PIM: Addgene plasmid #174650; http://n2t.net/addgene:174650; RRID:Addgene_174650.
2. The constructs were amplified using regular DH5alpha bacteria and prepared using the Qiagen Plasmid Plus Midi Sample Kit (Qiagen, cat. 12941). DNA was eluted in Milli-Q and stored at −20 °C.
3. *See* **Note 4.1**. for recommendations on cloning strategies.

### 2.2 Cell Culture

1. In this protocol, we used HeLa-WT (ATCC) and previously described HeLa-Rab7 endogenous knock-in cells [12]. *See* **Note 4.2.1**. for additional information on cell line compatibility.
2. Dulbecco's Modified Eagle Medium (DMEM) supplemented with 10% fetal calf serum, penicillin, and streptomycin.

### 2.3 Sample Preparation: Transfection and PIM Formation

Cell Seeding

1. Cells.
2. Round micro cover glasses, No. 1, 18 or 25 mm (VWR, cat. 48380-046 and cat. 48380-080).
3. Tweezers.
4. 12-well or 6-well plastic cell culture plates.

Transfection

1. Optimem (Gibco, Thermo Fisher Scientific, cat. 15392402).
2. Fugene 6 (Promega, cat. E2692).
3. DNA constructs, as described above.

PIM Formation

1. Rapalog (Takara, B/B homodimerizer, cat. 635059). Rapalog is aliquoted upon arrival in order to avoid multiple freeze-thaw cycles. The stock concentration is 0.5 mM.

Live Cell: Sample Mounting

1. Attofluor Cell Chamber (Thermo Fisher Scientific, cat. A7816) or Ludin chamber, depending on the coverslip size.
2. Imaging medium (regular culturing medium or DMEM without phenol red; Thermo Fisher Scientific, cat. 21063029).

Fixed Cell: Sample Fixation

1. 4% prewarmed PFA.
2. PBS 1×.
3. Prolong Diamond Antifade Mountant (Fisher Scientific, cat. P36970).
4. Microscope slides.

### 2.4 Microscopy

Microscopes

1. For the mKeima-PIM assay, we used a Nikon Eclipse Ti Microscope equipped with:
   (a) ASI motorized stage (MS-2000-XYZ) with Piezo top plate and on-stage incubation chamber (Tokai Hit, INUBG2E-ZILCS).
   (b) Perfect Focus System (Nikon).
   (c) CSU-X1-A1 confocal head (Yokagawa).
   (d) Plan Apo VC 60×, N.A. 1.4, oil immersion objective (Nikon).
   (e) Evolve 512 EMCCD (Photometrics).
   (f) Excitation lasers used for excitation; Vortran Stradus 445 nm (mKeima-PIM neutral pH), Cobolt Jive 561 nm (mKeima-PIM low pH), and Cobolt Calypso 491 nm (GFP).
   (g) Filter cubes; ET-mCherry (49008, Chroma) and ET-GFP (49002, Chroma).
   (h) MetaMorph 7.7 software (Molecular Devices).
2. For imaging of the dual-PIM assay, we used a Nikon Eclipse TE2000E equipped with:
   (a) Motorized stage (Prior) equipped with a Tokai Hit incubation chamber (INUBG2E-ZILCS).
   (b) Perfect Focus System (Nikon).
   (c) Plan Fluor 40×, NA 1.3, or Plan APO 60×, NA 1.4 oil immersion objectives (both Nikon).
   (d) pco.edge 4.2 CLHS sCMOS camera (Excelitas Technologies).
   (e) CoolLED pE4000 (CoolLED) setup for illumination.

(f) Filter cubes; ET-EGFP (49002, Chroma), ET-mCherry (49008, Chroma).

(g) μManager software [13].

### 2.5 Data Analysis

1. Fiji (Fiji is just ImageJ, v1.53t).
2. Comdet (v. 0.5.5.) plugin for Fiji.
   (a) For a guide on installation, see (https://github.com/UU-cellbiology/ComDet). There, you will also find extensive information on the Comdet particle-detection pipeline.
3. Microsoft Excel.
4. GraphPad Prism v. 9.4.0.

## 3 Methods

### 3.1 . Constructs

1. In Fig. 2, we provide a brief overview of the fluorophore characteristics of dual-PIM (Fig. 2a) and mKeima-PIM (Fig. 2b), as well as the applicability and general strengths of each tool (Fig. 2c).

### 3.2 Cell Culture

1. Maintain cells in full DMEM prior to splitting and seeding.
2. Seed cells on glass coverslips 1 day prior to transfection.
   (a) Note that the optimal density and amount of cells per coverslip might depend on the specific experimental setup. In general, we recommend carefully titrating densities to prevent overcrowding, since this will hamper both sample acquisition and downstream analysis.

### 3.3 Sample Preparation: Transfection and PIM Formation

1. Transfect seeded cells ±24 h prior to the start of the experiment. For this, we transfect 1 μg of plasmid DNA per 18 mm coverslip or 2 μg of plasmid DNA for a coverslip of 23 mm in diameter. For additional information on expression using transient transfection, (*see* **Notes 4.2.2**. and **4.2.3**).
   (a) First transfer the appropriate amount of plasmid DNA to a sterile Eppendorf tube. Add Optimem (100 μL per well) and resuspend well by pipetting.
   (b) Add Fusegene6 to the mixture (3 μL Fusegene6 per 1 μg of DNA).
   (c) Mix well by resuspending vigorously or by briefly vortexing (3–5 s). Briefly spin down using a tabletop centrifuge. Let the mixture incubate for 10–15 min.
   (d) Add the transfection mix drop by drop to the well.

2. After ±24 h of transfection, samples can be used for protein aggregation induction. For an illustrated overview of the PIM formation protocol used here, see Fig. 1. Rapalog addition and washout can be performed in culturing plates (for fixed-cell imaging or prior to mounting and transfer to the microscope) but can also be done successfully on stage.
   (a) Replace the culturing medium with medium (≤ 250 μL for 12-well plate, ≤ 500 μL for 6-well plate) containing rapalog (500 nM).
   (b) After 45 min, remove unbound rapalog by washing once and then adding culturing medium.
   (c) Depending on the experimental timeline, cells can now be mounted or fixed at the desired time points. *See* **Note 4.3**. for further recommendations on fixations.

### 3.4 Microscopy

In Fig. 3, we provide some example images of dual-PIM fixed-cell (Fig. 3a, b) and mKeima-PIM live-cell (Fig. 3c, e) experiments.

Fixed-Cell Microscopy

1. Remove the culturing medium and wash cells with 1× PBS.
2. Add prewarmed 4% PFA and incubate for 10 min.
3. Wash three times using 1× PBS.
   (a) If no further immunofluorescent labeling is required, proceed to **step 4**. Alternatively, samples can be permeabilized and probed with antibodies.
4. Stain nuclei with DAPI (1:1000 diluted in 1× PBS) for 2 min.
5. Wash three times using 1× PBS.
6. Dip the coverslip in Milli-Q and let dry prior to mounting.
7. Cells can now be mounted using Prolong Diamond. Apply a small drop of Prolong Diamond mounting medium on a microscopy slide and place the coverslip face-down on the drop. Press gently and dry overnight at RT.

Live-Cell Microscopy for Dual-PIM and mKeima-PIM

1. Mount cells in the appropriate imaging ring using sterile tweezers.
2. Add imaging medium. Use a minimum of 500 μL for long-term imaging (> 6 h).
   (a) Alternatively, to avoid evaporation and gas exchange, imaging rings can be sealed by placing a coverslip on top and removing excess medium. When sealing imaging rings, make sure to use imaging medium that contains

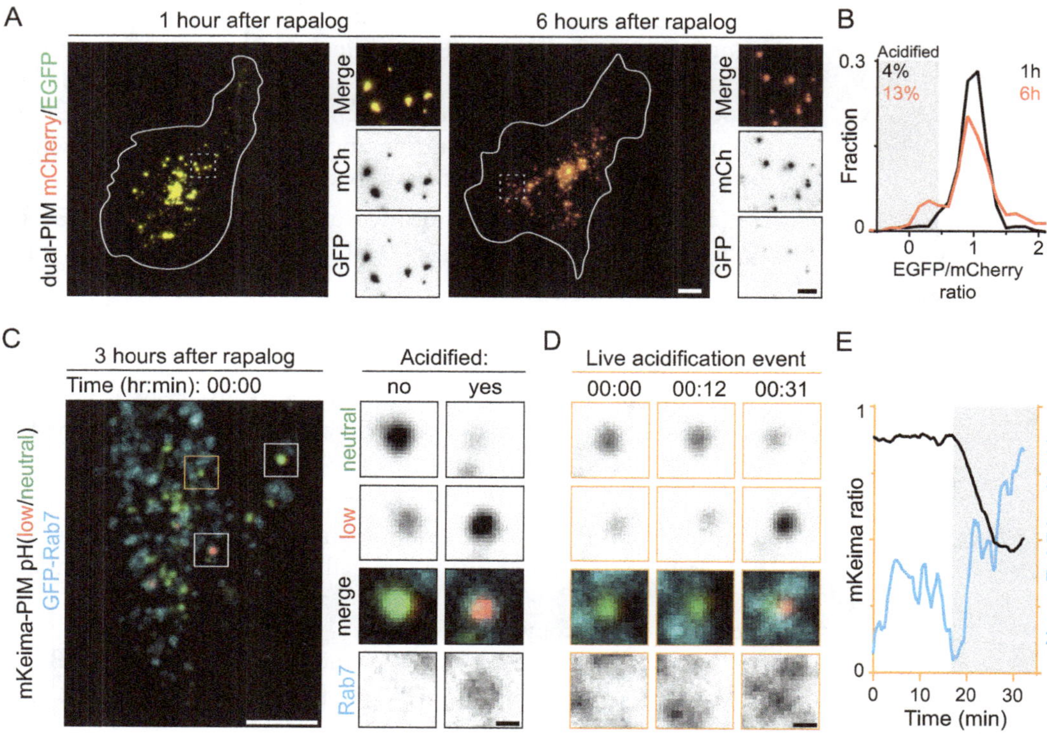


**Fig. 3** Application of dual-PIM and mKeima PIM assays to monitor aggrephagy. (**a**) Representative images of HeLa-WT cells expressing dual-PIM after 1 and 6 h of rapalog addition. Zooms (dashed white boxes) show a number of dual-PIM aggregates in neutral (dual positive, yellow) or acidic compartments (mCherry positive, red). (**b**) Frequency distribution of EGFP/mCherry signal intensity of individual aggregates from cells shown in (**a**). The emergence of a peak at low EGFP/mCherry ratios indicates the increased clearance of dual-PIM aggregates (13% vs. 4%). (**c**) Representative image of HeLa-KI cells expressing endogenously tagged GFP-Rab7 and mKeima-PIM 3 h after rapalog addition. Zooms (solid white boxes) show examples of mKeima-PIMs at neutral- and low pH (Acidified no/yes). Only mKeima-PIMs in low-pH environments colocalize with high levels of Rab7. (**d**, **e**) Zooms (orange boxes) show time-lapse imaging of the acidification of a single mKeima-PIM aggregate coinciding with the accumulation of GFP-Rab7. (**e**) Graph showing the quantification of the acidification event shown in D. Scale bars represent 10 μm (**a**, **c**), 2 μm (zooms of **a**), and 1 μm (zooms of **c** and **d**)

$CO_2$. For this, place imaging medium in an incubator supplemented with 5% $CO_2$ overnight (e.g., a 15 mL tube with the cap unscrewed).

3. Locate PIM-expressing cells.

   (a) When using dual-PIM, we generally locate cells expressing PIMs using the mCherry signal. Be aware that dual-PIM-expressing cells can be difficult to locate prior to rapalog addition due to low expression levels. Aggregated dual-PIMs should be readily visible in the mCherry channel at all stages post rapalog addition (1–24 h).

(b) When using mKeima-PIM, we generally locate expressing cells by excitation with a 561 nm laser due to a better signal-to-noise ratio. At the very early stage of aggregation, visualization using a 445 nm laser might be favored due to the lack of cleared PIMs.

4. For long-term imaging of aggregate turnover in whole cells, use 5–10 min intervals. To limit phototoxicity, this interval can be increased.

5. Imaging settings for fast, high-resolution live-cell imaging of single acidification events and aggregate dynamics:

   (a) Time intervals of ≤20 s between frames.

   (b) Z-range of ±5–8 μm depending on cell type with z-steps of ≤0.5 μm.

   (c) Exposure time at approximately 100 ms per frame, but this can be changed depending on laser power and the required speed of acquisition.

### 3.5 Data Analysis

For an overview of the data analysis pipeline and example images, see Figs. 3 and 4.

1. For the analysis of whole cell aggregate clearance, use the Comdet plugin to perform automated detection and measurements of single particles and colocalization analysis.

   (a) First, ensure the image contains only the channels that you want to be included into the Comdet detection.

   (b) Draw an ROI around the object of interest (e.g., delineating one cell). Note that the plugin can also run without a prespecified ROI. It will then do particle detection on the entire image. *See* **Note 4.4.1**. for additional information on ROI selection.

   (c) Run Comdet v. 0.5.5. (Plugins→Comdet v.0.5.5.). See Fig. 4 for a summary of the Comdet pipeline and detection/colocalization parameters. A link to a comprehensive explanation of the plugin can be found in (*see* **Note 4.4.2**).

   (d) Ensure the "Calculate colocalization?" box is marked and set the threshold for colocalization, called "Max Distance," at ±3 pixels. For visualization and further analysis, ROIs detected can be visualized and added to the "ROI manager" tab. Comdet also produces a "Summary" table. If you want to use this table for data collection of multiple ROIs, select "Append" from the dropdown menu.

   (e) Per channel, specify the detection parameters. For the data presented in this protocol, we selected "Include larger particles?" and "Segment larger particles?," and

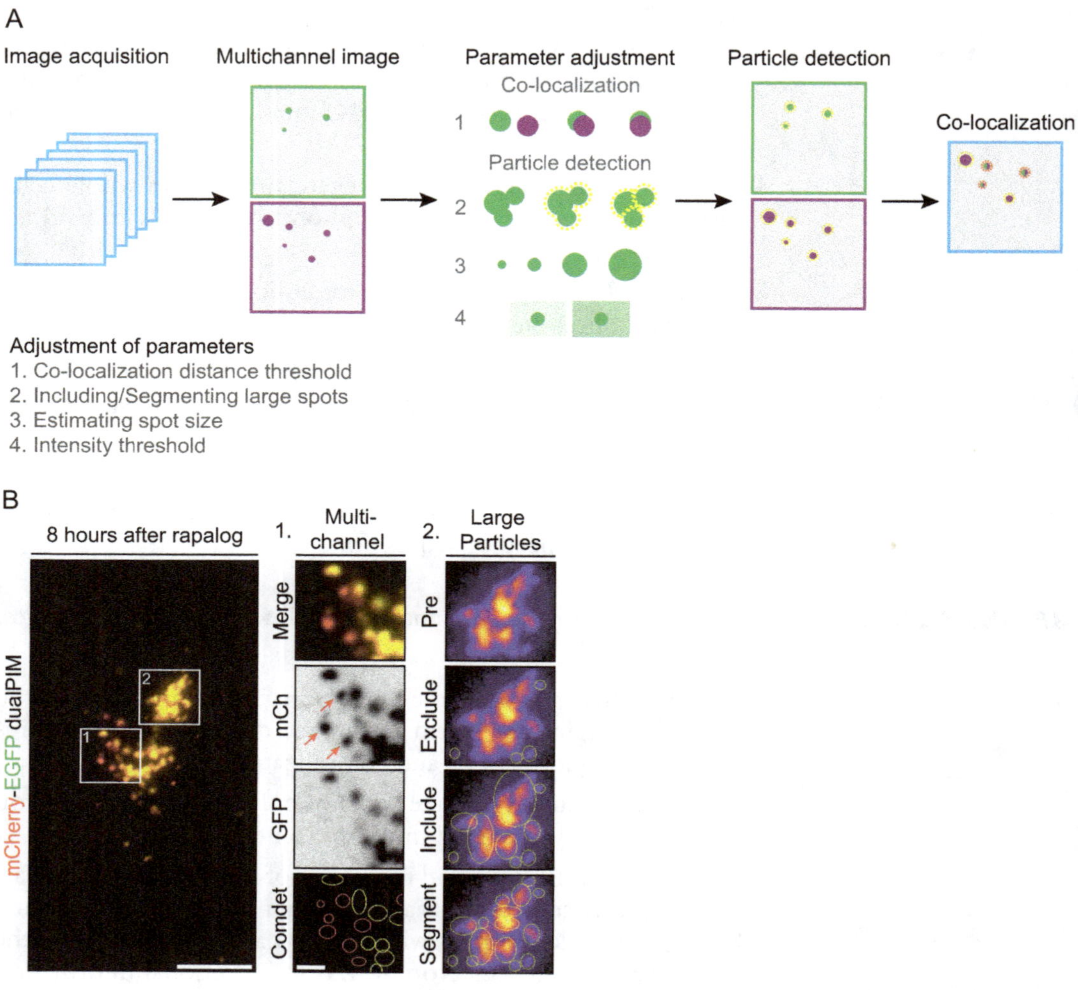


**Fig. 4** Automated particle detection and colocalization analysis using Comdet. (**a**) Schematic representation of Comdet detection and colocalization pipeline. For colocalization of detections in multichannel images, first select the colocalization distance threshold (1). Next, per channel, select particle detection parameters: (2) Determine whether Comdet should include or exclude larger particles from the detection and whether larger particles should be segmented. (3) Set the estimated particle size in pixels. (4) Finally, adjust the intensity threshold value. These values can be checked using the "preview" function and adjusted to facilitate detection of particles in all channels of the image. (**b**) Representative image of HeLa-WT cell expressing the dualPIM construct 8 h after rapalog addition. Zooms on the left show mCherry-EGFP positive aggregates at neutral pH and acidified aggregates that are only positive for mCherry (red arrows). Annotation of the Comdet detection shows particles assigned as mCherry positive (red) or dual positive (yellow) puncta. Zooms on the right show Comdet detection of larger particles. Scale bar represents 10 μm and 2 μm (zooms)

set "Approximate particle size" at 3–4 pixels and the intensity threshold at 8–10 depending on the channel.

(f) Comdet now generates "Summary" and "Results" tables containing all detections, intensities, and colocalization analysis. Save the "Log" window with the parameters used during detection as a reference. *See* **Note 4.4.3**. for notes on Comdet application.

(g) Next, transfer the data to Excel and calculate acidification ratios. Since Comdet measurements are already background corrected, the intensity measurement (Integrated density) can be used directly to calculate the ratios.

- To determine the mKeima ratio $r_{\mathrm{mKeima}}$, divide intensity at neutral pH by the total intensity of the mKeima signal in both channels.

$$r_{\mathrm{mKeima}} = \frac{I_{\mathrm{neutral\ pH}}}{I_{\mathrm{neutral\ pH}} + I_{\mathrm{low\ pH}}}$$

- To determine the dual-PIM ratio $r_{\mathrm{dual\text{-}PIM}}$, simply divide the intensity of the EGFP signal by the mCherry signal intensity.

$$r_{\mathrm{dual-PIM}} = \frac{I_{\mathrm{EGFP}}}{I_{\mathrm{mCherry}}}$$

- This provides an acidification ratio between 0 and 1, indicating cleared and non-cleared aggregates, respectively.

(h) It is possible that the value of mKeima-PIM intensities is slightly negative, due to the Comdet background subtraction of aggregates in highly acidic (low $I_{\mathrm{neutral\ pH}}$) or neutral (low $I_{\mathrm{low\ pH}}$) environments. We set these values to 0. The resulting ratios still accurately reflect the acidification status of the aggregates and are either 0 or 1.

(i) After calculating ratios of dual-PIM intensities, we perform normalization by fitting a single (cells at early time points) or double Gaussian (cells at later time points) distribution in GraphPad PRISM. We then use the peak of this Gaussian to normalize individual ratios. In the case of a double Gaussian fit, we use the peak associated with the higher EGFP/mCherry ratio, since this peak represents the aggregates in neutral environments.

(j) The (normalized) dual- and mKeima-PIM ratios can then be visualized using a frequency distribution plot (Fig. 3b). Another convenient estimation of aggrephagy activity can be made by calculating the amount of aggregates with low dual-PIM ratios (e.g., <0.4).

2. For analysis of single events over time (example shown in Fig. 3f), an aggregate can be manually traced to measure acidification (for example data, see Fig. 3c, e).

(a) For each frame, using Fiji, draw a small ROI around a single aggregate and measure all channels (mCherry/EGFP or mKeima-neutral/mKeima-low including third channel, e.g., Rab7-GFP).

(b) Obtain background measurements by performing multiple measurements per channel for each frame around the aggregate, where background signal in the cytosol is homogeneously distributed.

(c) Subtract the average background intensity from each channel to calculate the background-corrected intensity values.

(d) The acidification ratio can then be calculated as described above.

(e) Normalize the acidification ratio by dividing the EGFP/mCherry or mKeima ratios by the average ratio of the first ±5 frames (Fig. 3e).

(f) The acidification of single aggregates can then be plotted as a single curve. Overlaying the intensities of GFP-tagged proteins in the aggregate ROI can then be used to visualize the dynamic association and dissociation of these factors during the aggrephagy process.

(g) Multiple acidification events can then be averaged by synchronization based on the onset of acidification (ratio falling <0.9).

## 4 Notes

### 4.1 Cloning and Viral Production of PIM Constructs

1. Naturally, the PIM module can be appended with other fluorophores or other sequences. Be aware that cloning using the PIM module can be challenging. The construct contains repetitive sequences due to the multiple FKBP domains. Especially, PCR amplification can yield unexpected, unwanted results. We recommend avoiding PCR amplification of the construct if possible. If alternatives are unavailable, purification and excision of single bands of PCR product on a DNA gel prior to ligation, together with whole plasmid sequencing of the finalized construct, are recommended.
2. Note that the production of lentiviral particles containing a correct copy of the original PIM construct and its successful integration in target cells is likely challenging. We speculate this is a consequence of the highly repetitive sequences found in the PIMs.

### 4.2 Expression of PIM Constructs

1. In this protocol, we used WT and Rab7-KI HeLa cells, but we previously successfully transfected PIMs into multiple other cell lines (including HEK293, RPE-1, and U2OS).
2. For the data provided in this protocol we used PIM constructs that used a crippled β-actin promoter (Addgene plasmid # 111758 or Addgene plasmid # 174650) to ensure moderate expression levels. In our hands, the level of protein expression driven by this promoter is sufficient to form PIMs in most transfected cells without creating overexpression artifacts, such as aggregation prior to rapalog addition. In some instances, higher expression levels might be desirable. For this purpose, we also deposited a pGW1-dual-PIM variant under a strong CMV promoter. Alternatively, one could suggest varying the amount of DNA transfected per well.
3. In our hands, transfection of PIM constructs using Fusegene6 is well tolerated by cultured cells. To optimize cell viability; (1) make sure to use DNA preps that are kept no longer than 1 year at 4 °C, (2) consider replacing the medium with full DMEM after 6–8 h of transfection.

### 4.3 Fixation and Antibody Labeling

1. Fluorescence of non-degraded mCherry and EGFP is maintained after fixation, and the remaining fluorescent signal allows for robust imaging and quantification of mCherry/EGFP ratios. Therefore, we opt not to probe the PIMs with secondary antibodies. We recommend keeping fixed samples at 4 °C for short-term storage and suggest imaging within 3 days after fixation.

### 4.4 Comdet: Aggregate Detection and Analysis

1. When automating larger batches of particle detection, make sure to perform the Comdet detection on the entire image. Since the threshold for detection of spots relies on the background values, it is critical that these are included. Therefore, do not crop images to precise cellular or subcellular scales. Instead, a restrictive ROI can be placed. Comdet will then only detect particles inside the ROI but will determine the threshold based on the entire image.
2. An elaborate explanation of the installation, particle detection, and use of the plugin can be found here; https://github.com/UU-cellbiology/ComDet/wiki.
3. The optimal settings for Comdet detection of aggregates will depend on the imaging setup, sample, and research question. As a convenient starting point for optimizing detection, we recommend the following: a particle size of >3 pixels is sufficient to detect PIMs without introducing noise. The desired signal-to-noise ratio (SNR) depends heavily on the imaging

setup and will vary between 4 and 20. Comdet offers a convenient "preview" function that allows for easy optimization.

## References


1. Hipp MS, Kasturi P, Hartl FU (2019) The proteostasis network and its decline in ageing. Nat Rev Mol Cell Biol 20:421–435
2. Jahreiss L, Menzies FM, Rubinsztein DC (2008) The itinerary of autophagosomes: from peripheral formation to kiss-and-run fusion with lysosomes. Traffic 9:574–587
3. Pankiv S, Clausen TH, Lamark T et al (2007) p62/SQSTM1 binds directly to Atg8/LC3 to facilitate degradation of ubiquitinated protein aggregates by autophagy. J Biol Chem 282: 24131–24145
4. Bjørkøy G, Lamark T, Brech A et al (2005) p62/SQSTM1 forms protein aggregates degraded by autophagy and has a protective effect on huntingtin-induced cell death. J Cell Biol 171:603–614
5. Ashkenazi A, Bento CF, Ricketts T et al (2017) Polyglutamine tracts regulate beclin 1-dependent autophagy. Nature 545:108–111
6. Martin DDO, Ladha S, Ehrnhoefer DE, Hayden MR (2015) Autophagy in Huntington disease and huntingtin in autophagy. Trends Neurosci 38:26–35
7. Wong YC, Holzbaur ELF (2014) The regulation of autophagosome dynamics by huntingtin and HAP1 is disrupted by expression of mutant huntingtin, leading to defective cargo degradation. J Neurosci 34:1293–1305
8. Janssen AFJ, Katrukha EA, van Straaten W et al (2018) Probing aggrephagy using chemically-induced protein aggregates. Nat Commun 9: 4245
9. Kimura S, Noda T, Yoshimori T (2007) Dissection of the autophagosome maturation process by a novel reporter protein, tandem fluorescent-tagged LC3. Autophagy 3:452–460
10. Katayama H, Yamamoto A, Mizushima N et al (2008) GFP-like proteins stably accumulate in lysosomes. Cell Struct Funct 33:1–12
11. Katayama H, Kogure T, Mizushima N et al (2011) A sensitive and quantitative technique for detecting autophagic events based on lysosomal delivery. Chem Biol 18:1042–1052
12. Janssen AFJ, Korsten G, Nijenhuis W et al (2021) Direct observation of aggregate-triggered selective autophagy in human cells. J Cell Sci 134:jcs258824
13. Edelstein AD, Tsuchida MA, Amodaj N et al (2014) Advanced methods of microscope control using μManager software. J Biol Methods 1:e10

# Chapter 9

# Methods for Monitoring ER-Phagy

**Marisa Di Monaco, Marian Peteri, and Jin Rui Liang**

## Abstract

The endoplasmic reticulum (ER) serves as a central hub for protein synthesis, folding, and lipid biosynthesis in eukaryotic cells. Maintaining ER homeostasis is essential for optimal cellular function, and one mechanism that has garnered attention is endoplasmic reticulum-specific autophagy, or ER-phagy. ER-phagy selectively removes specific ER portions, playing a pivotal role in cellular health and adaptation to environmental stressors. ER-phagy can be induced by diverse cellular conditions such as amino acid starvation, disruption of ER quality control mechanisms, and accumulation of misfolded ER protein, highlighting cellular adaptability and the significance of ER-phagy in stress responses. Clinically relevant mutations in ER-phagy receptors are implicated in various diseases, underlining the fundamental importance of ER-phagy in ER homeostasis. Here, we provide comprehensive protocols and general considerations while investigating ER-phagy using three fundamental techniques—Western blotting, immunofluorescence, and flow cytometry—commonly used in ER-phagy detection and quantitation.

**Key words** Autophagy, ER-phagy, Endoplasmic reticulum, Selective autophagy, Western blotting, Immunofluorescence, FACS, Fluorescent reporters

## 1 Introduction

The intricate endoplasmic reticulum (ER) network in eukaryotic cells is a central hub for protein synthesis, folding, and lipid biosynthesis. To maintain ER homeostasis, cells have evolved sophisticated mechanisms to ensure optimal ER functionality. One such mechanism that has garnered significant attention in recent years is endoplasmic reticulum-specific autophagy, commonly referred to as ER-phagy.

ER-phagy is a selective form of autophagy. The selective removal of specific portions of the endoplasmic reticulum, either superfluous or aggregate-concentrated ER fractions, is pivotal for cellular health and adaptation to environmental stressors. In recent years, an increasing number of ER-specific autophagy receptors (also known as ER-phagy receptors) such as FAM134A/B/C, RTN3L, TEX264, CCPG1, SEC62, ATL3, CDK5RAP3, and

Ioannis P. Nezis (ed.), *Selective Autophagy: Methods and Protocols*, Methods in Molecular Biology, vol. 2845,
https://doi.org/10.1007/978-1-0716-4067-8_9,

CALCOCO1 have been uncovered and demonstrated to confer specificity to the degradation of the ER [1–9]. These ER-phagy receptors are mostly ER-transmembrane or ER-interacting proteins that invariably contain at least one LC3/GABARAP-interacting motif (LIR/GIM) and/or additional domains that recruit autophagy components to the ER surface.

ER-phagy can be induced by various cellular conditions, including amino acid starvation, endoplasmic reticulum stress, and the accumulation of misfolded ER proteins. These stimuli not only highlight the adaptability of cells in response to changing environments but also underscore the importance of ER-phagy in cellular stress responses. Indeed, clinically relevant mutations of ER-phagy receptors have been implicated in various diseases, including neurodevelopmental disorders, pancreatic and hepatic inflammations, and cancer [10, 11].

As most endogenous ER protein degradation during ER-phagy is subtle, fluorescently tagged ER-phagy reporter systems serve as a more quantitative approach to measure ER-phagy. The most commonly used reporter systems involve the use of tandem-fluorescent RFP-GFP (TF) protein tag fused to an ER-localized protein of interest (Table 1) [12]. Our lab established two fluorescence-based ER-phagy reporter systems which allowed high-throughput and quantitative measurement of ER-phagy. Using these reporter systems, we not only identified Atlastins as key ER membrane remodeling factors preceding ER-phagy [13] but also performed a genome-wide CRISPR-interference screen that uncovered many other novel ER-phagy regulators, including a role of UFMylation in ER stress and ER-phagy regulation [14].

One such example is the mCherry-eGFP-RAMP4 reporter, also known as the ER-autophagy tandem reporter (EATR; mCherry is a variant of RFP protein), which fuses the TF reporter to the ER transmembrane protein RAMP4 [13]. In the neutral cytosolic environment, both mCherry and eGFP are fluorescent. When ER-phagy is activated, a portion of the ER is localized within the lysosomes. The low lysosomal pH quenches the eGFP signal, whereas mCherry is resistant to protonation and thus remains fluorescent. Hence, a shift towards red mCherry signal or the appearance of red puncta lacking GFP signal indicates lysosomal degradation of the ER.

An alternative to the TF-tag approach is the mKeima reporter. mKeima is a monomeric fluorescent protein with different excitation wavelengths at neutral (440 nm) and acidic (561 nm) pH [15]. Thus, a comparison of the emitted signal after excitation with the two wavelengths allows quantification of the lysosomal portion of mKeima. Both the TF- and mKeima-reporter systems can be used to detect and quantify ER-phagy using flow cytometry and immunofluorescence.

**Table 1**
**Overview of ER-phagy reporter systems**

| Reporter plasmid | Description | Applications | Addgene number |
|---|---|---|---|
| mCherry-eGFP-RAMP4 ("EATR") | RAMP4 is a component of ER translocon complex but is ubiquitously expressed in both ER sheets and ER tubules upon overexpression | IF, WB, FACS | 109014 |
| mCherry-RAMP4 ("CCER") | | WB, IF | 118391 |
| ssCALR-RFP-GFP-KDEL | ER-targeting signal peptide from Calreticulin (ssCALR) fused with RFP and GFP at N-terminus and ER retention signal (KDEL) fused to the C-terminus. Ubiquitously expressed and retained in the ER lumen | IF, WB, FACS | 128257 |
| ssCALR-mCherry-KDEL | | WB, IF | Available upon request |
| FLAG-GFP-TMEM192 | TMEM192 is a lysosomal transmembrane protein used as a lysosomal marker | IF, to use with mCherry-RAMP4/ ssCALR-mCherry-KDEL for visualization of ER within the lysosomal lumen | 134630 |
| CLIMP63-mCherry-eGFP | CLIMP63 is an ER sheet-specific protein. Use for the examination of sheet-specific ER-phagy | IF, WB, FACS | 139871 |
| mCherry-eGFP-REEP5 | REEP5 is an ER tubule-specific protein. Use for examination of tubule-specific ER-phagy | IF, WB, FACS | 139870 |

Note that all plasmids suggested here are lentiviral-based (*IF* Immunofluorescence, *WB* Western blotting, *FACS* fluorescence-activated cell sorting)

In addition, the strategic addition of a cleavable linker between the ER protein and mCherry protein allows the cleavage of mCherry from the ER protein by lysosomal proteases. The exceptional stability of the cleaved mCherry within the lysosome can be detected by Western blot via an approach termed "Cleavage of mCherry from the ER" (CCER, described in Subheading 4) [13]. CCER can also be assessed using fluorescence microscopy where ER degradation can be confirmed as an overlapping signal of mCherry and fluorescently labeled/tagged lysosomal markers such as LAMP1 and TMEM192.

Overall, this chapter aims to provide in-depth instructions on using these ER-phagy reporter systems (complemented with orthogonal validation via endogenous ER protein measurement) to detect ER-phagy. We focus on three fundamental protocols—Western blotting, immunofluorescence, and flow cytometry—

which are routinely used in the detection, quantitation, and characterization of ER-phagy. We also provide advice on the consideration of the appropriate ER-phagy reporter system and a lentiviral protocol for the generation of reporter cell lines.

## 2 Materials

### 2.1 General Cell Culture Materials

1. Dulbecco's Modified Eagle's Medium supplemented with 1% L-Glutamine solution, 1% Sodium Pyruvate, 1% Penicillin and Streptomycin, 10% Fetal bovine serum, 1% Nonessential Amino Acids.
2. Trypsin-EDTA (0.05%), phenol red.
3. Phosphate Buffered saline (PBS)-sterile filtered, pH 7.4.
4. OptiMEM.
5. Cell culture incubator set at 37 °C, 5% $CO_2$.
6. Laminar flow hood suitable for viral work.
7. Cell line of choice for ER-phagy work.
8. Earle's Balanced Salt Solution (EBSS).
9. Relevant treatment media/compounds (Table 2).

**Table 2**
**Chemical compounds commonly used to induce/study ER stress and ER-phagy**

| Chemical compound | Mechanism of action | Recommended concentration and treatment duration |
|---|---|---|
| Tunicamycin | Inhibits the first step in the biosynthesis of N-linked glycans resulting in protein misfolding | 2.5–5 μg/mL for 2–24 h |
| Thapsigargin | Inhibitor of the sarco/ER calcium ATPase. High concentration of thapsigargin also results in lysosomal inhibition and general autophagy blockage | 3 μM for 2 h |
| Dithiothreitol | Blocks disulfide-bond formation of newly synthesized proteins in the cytosol | 2–3 nM for 8 h |
| Cyclopiazonic acid | A reversible inhibitor of the sarco/ER calcium ATPase | 10 μM for 12 h |
| Bafilomycin A1 or Concanamycin A/Folimycin | These are selective inhibitors of the vacuolar type H +-ATPase (V-ATPase), which prevents lysosomal acidification. Prolonged neutralization of the lysosomes will cause secondary inhibition of autophagosome fusion with the lysosomes | 50–100 nM for 4–24 h |

All recommended concentrations and treatment durations should be optimized for the cell line of interest and the assay used. In addition, certain ER stress-inducing compounds could be cell-type dependent ER-phagy inducers, which have not been conclusively shown as universal ER-phagy inducers

10. Doxycycline (if using Tet-inducible ER-phagy reporter system).
11. Tissue culture plates/dishes.

#### 2.2 Materials for Lentiviral Work

1. Polyethylenimine Max-transfection grade (PEI MAX).
2. Viral packaging plasmids:
   (a) pCMV-dR8.2 dVPR (Addgene #8455).
   (b) pCMV-VSV-G (Addgene #8454).
3. Plasmid for the reporter system of choice (see Table 1).
4. HEK293T/FT cells.
5. Tabletop centrifuge.

#### 2.3 Materials for Western Blotting

1. RIPA buffer (50 mM Tris–HCl, pH 8.0, with 150 mM sodium chloride, 1.0% Igepal CA-630 (NP-40), 0.5% sodium deoxycholate, and 0.1% sodium dodecyl sulfate).
2. PBS, pH 7.4.
3. Halt™ Protease inhibitor cocktail (ThermoFisher Scientific).
4. Bradford protein assay reagent.
5. Benchtop centrifuge (chilled at 4 °C).
6. Bovine serum albumin (2 mg/ml in ddH2O).
7. Spectral plate reader.
8. 96-well flat bottom plate.
9. 4× LDS sample buffer (ThermoFisher Scientific).
10. Reducing agent (e.g., β-mercaptoethanol).
11. 4–12% NuPAGE gradient gels (ThermoFisher Scientific).
12. 1× MES buffer (ThermoFisher Scientific).
13. Nitrocellulose membrane.
14. Filter paper.
15. Wet transfer buffer.
16. Ponceau S staining solution.
17. 5% milk in TBS with 0.1% Tween-20 (TBST).
18. Western blot rocker.

#### 2.4 Materials for Immunofluorescence

1. Sterile glass coverslips.
2. PBS, pH 7.4.
3. 4% paraformaldehyde (PFA) in PBS. (*Recommendation*: Dilute one vial of 10 mL 16% PFA in 30 mL of 1× PBS, and aliquot 10 mL each into 15 mL falcons. Store unused aliquots at −20 ° C. As PFA is volatile and toxic, perform all PFA-related steps in the fume hood.)

4. 5% Bovine Serum Albumin (BSA) in PBS (*Recommendation*: Fully dissolve 2.5 g BSA into 50 mL PBS and store at 4 °C).
5. 0.1% digitonin in 5% BSA. (*Recommendation*: First dissolve 0.5 g of digitonin in 10 mL of water to generate 5% digitonin. Add 200ul of 5% digitonin into 10 mL of 5% BSA/PBS.)
6. Relevant primary antibody suitable for IF.
7. Relevant fluorescent-tagged secondary antibody.
8. DAPI (4′,6-diamidino-2-phenylindole) fluorescent dye.
9. Suitable mounting media.

### 2.5 Materials for Flow Cytometry

1. FACS tubes (check tube compatibility with your local FACS machine).

## 3 Methods

### 3.1 Generating ER-Phagy Reporter Cell Lines by Lentiviral Transduction

ER-phagy reporter systems are modular, and depending on individual context, different ER proteins can be fused with the reporter by molecular cloning (see overview in Table 1). For example, instead of fusion with RAMP4, which localizes the reporter to the ER membrane surface, both EATR and CCER reporters can be reengineered to target the ER lumen. This can be achieved by fusing the TF-reporter with an ER targeting signal from Calreticulin at the N-terminus (ssCALR) and a tetrapeptide ER retention signal ("KDEL") at the C-terminus, which targets the fluorescent proteins to the ER lumen (see Table 1). Since most ER proteins have similar ER-targeting signals, retention motifs, or transmembrane domains either at the amino- or carboxyl-terminus, it is critical to consider the fusion position of the TF-reporter relative to these regions to avoid disrupting their ER localization. It is also worth considering whether a stable or inducible (e.g., doxycycline (Dox) inducible) expression system works for your cell type of choice. When using a stable expression system, we have found that certain cell lines with high basal ER-phagy have proportionately higher basal levels of mCherry over eGFP signal that could augment the interpretation of stress-induced ER-phagy. In this case, a Dox-inducible reporter system pulse-labels a fraction of ER, thus allowing tracking of those ER fractions instead of the total ER population.

*WARNING: Please consult your local health and safety guidelines for lentiviral work.*

1. Seed HEK293T/FT cells into a 6-well plate at $2.5 \times 10^6$/well.
2. Perform reverse transfection (i.e., same time as cell seeding):
   (a) For a 6-well plate transfection, prepare the transfection mixture in Table 3.

**Table 3**
**Transfection reaction setup for the generation of lentiviral particles**

| | 1× reaction |
|---|---|
| OptiMEM | 100 μL |
| pCMV-dR8.2 dVPR | 800 ng |
| pCMV-VSV-G | 200 ng |
| ER-phagy reporter plasmid | 1000 ng |
| PEI MAX (3ul per 1ug of total plasmid) | 6 μL |

3. Mix by gentle vortexing and then briefly spin down the mixture. Incubate the mixture at room temperature for 15 min.
4. Add the transfection mixture dropwise to the HEK293FT cells. Gently swirl the plate to ensure even distribution.
5. Incubate the cells at 37 °C with 5% $CO_2$ for 48 hrs.
6. After 48 hrs, harvest the virus by collecting the conditioned media from the transfected cells.
7. To prevent accidental carryover of HEK293T/FT cells, perform centrifugation using a tabletop centrifuge at maximum speed for 5 min. Transfer the media to a fresh tube, excluding the last 50 μL of media. Alternatively, filter the viral media through a 0.4 μm filter. We recommend the former approach for small-scale virus production in 6-well plates to reduce media loss.
8. The lentiviral media can be used immediately for transduction or stored at −80 °C for long-term storage. If required, the transfected cells can be topped up with fresh medium to harvest more virus the next day.
9. Lentiviral transduction of target cells: Seed your cell line of choice at the desired density into a 6-well or 12-well plate. We recommend ±50% confluency at the time of transduction. Depending on the cell type, lentiviral transduction can be performed on the day of seeding or the day after. Same-day transduction is described below.
10. Add the lentiviral media at a 1:5–1:2 ratio (lentiviral to fresh media ratio—optimize according to your cell type) to the seeded cells and gently swirl to mix the media.
11. Incubate the cells with the virus for 48 h.
12. After the incubation period, start selection for transduced cells by replacing the old medium with media containing the appropriate antibiotic. It is recommended to include one well/plate of nontransduced cells in the selection process as a control,

thus allowing for an indication of a selection endpoint by assessing the complete cell death of the nontransduced cells.

(a) The choice of antibiotic will depend on the plasmid backbone used.

(b) The concentration of antibiotic used will need to be optimized based on the cell type. We recommend the following ranges as starting points: Puromycin (0.5–4 μg/mL), G418/Neomycin (400–2000 μg/mL), Blasticidin (2–10 μg/mL), Hygromycin (100–500 μg/mL).

(c) Multiple rounds of selection might be required, with the media being changed every 48 h and cells being split if necessary.

(d) Once selection is over, we recommend single-cell seeding of the ER-phagy reporter cells to further isolate clones with the most robust stress-induced ER-phagy. From experience, we have found that clonal reporter cells also provide more uniform ER-phagy responses compared to a "pool" of cells with a variable number of lentiviral integrations.

### 3.2 Inducing ER-Phagy

Lysosomal ER turnover can be elicited by a variety of physiological and pathological signals. Most commonly, these are pleiotropic signals that activate a plethora of catabolic pathways, including but not limited to ER-phagy. Pleiotropic signals such as starvation and chemical compounds mean ER-phagy responses form an epiphenomenon of a much larger response, such as macroautophagy, which exceeds and includes the Unfolded Protein Response (UPR). Thus far, amino acid starvation is considered the most robust inducer of ER-phagy [1, 3, 13, 14] and is also commonly used as a positive control for ER-phagy activation while exploring other ER-phagy inducers. ER stress can be induced using a variety of chemical compounds that either perturb ER homeostasis, trigger the UPR, increase protein misfolding, or disrupt proteasomal degradation [5, 6]. It is worth noting that while ER stress has been reported to induce ER-phagy in some cases, this could vary depending on the cell type and choice of ER stress treatment (with optimization of the stressor concentration and duration).

For all ER-phagy inducers, appropriate control conditions should be set up using co-treatment of inducer with Bafilomycin A1 or Concanamycin A (lysosomal proton pump inhibitors) to block lysosomal degradation and, by extension, autophagy (see Table 2). It is also recommended that at least one well/plate of cells be incubated in fully supplemented media (unstressed condition) alongside any treatment condition as a negative control.

*3.2.1 Amino Acid Starvation*

1. Cells are typically seeded on Day 0 in fully supplemented media so that they reach 80–90% confluency on the day of treatment.
2. After 24 h, amino acid starvation can be induced using EBSS. It is noteworthy that prolonged starvation will cause a certain amount of cell death. We recommend 8–24 h of EBSS treatment.

*3.2.2 Chemical Compound Treatment*

1. Cells are typically seeded on Day 0 in fully supplemented media so that they reach 80–90% confluency on the day of treatment.
2. The compound can either be directly spiked into the media of each plate condition or can be added to a media master mix before being divided amongst conditions. The latter is recommended when smaller dosages are being administered to account for pipetting error.

### 3.3 ER-Phagy Measurement by Western Blotting

ER-phagy can be measured by Western blotting in two ways. The first approach relies on the cleavage of the flexible linker between the pH-resistant mCherry and the ER protein (e.g., the CCER assay in Table 1). This results in the formation and accumulation of a smaller molecular weight mCherry-only species (Fig. 1). Since mCherry is protease-resistant and accumulates within the lysosome for an extended period before being degraded, the extent of mCherry-only fragment accumulation is directly proportional to the extent of ER-phagy. The CCER approach is a good starting point to gain evidence of ER-phagy occurrence prior to in-depth characterization of the identity of the endogenous ER protein being degraded.

The second approach is via direct measurement of endogenous ER-resident proteins. However, it is worth noting that most ER-resident proteins only demonstrate a subtle decrease in total protein levels even during active ER-phagy, with the exception of certain ER-phagy receptor proteins such as FAM134B and TEX264, which are more pronouncedly decreased when measured

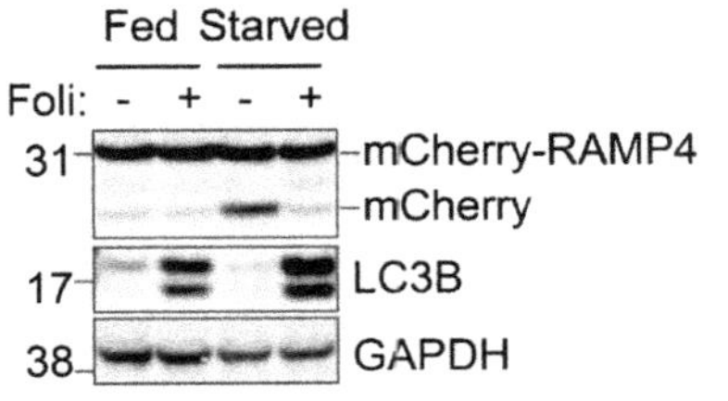


**Fig. 1** Using the CCER reporter system to measure ER-phagy. mCherry fused to ER-resident RAMP4 (full-length CCER) gets cleaved in lysosomes upon amino acid starvation-induced ER-phagy, generating the smaller cleaved mCherry fragment. Co-incubation with folimycin (Foli) inhibits lysosomal proteases, thus preventing mCherry cleavage, confirming the mCherry cleavage to be a lysosomal process. (Figure taken with permission from Liang et al. [13])

by Western blotting. In addition, different subcompartments of the ER (e.g., ER tubules versus ER sheets) could be selectively targeted for ER-phagy by different ER-phagy receptors and/or under different ER-phagy stimuli. Therefore, when using the second approach, one must carefully consider the choice of representative ER protein markers as a readout for their observed ER-phagy. A list of commonly used ER protein markers and validated antibodies is suggested in Table 4.

**Table 4**
**Some suggested ER-phagy markers and their respective validated antibodies for Western blot analyses**

| Antibody target | Notes | Recommended antibody (Cat. no. & company) |
|---|---|---|
| mCherry | Used for CCER assay. Detects both full-length mCherry-KDEL and cleaved mCherry | Ab183628 (Abcam) |
| FAM134A/ RETREG2 | ER-phagy receptor | NBP2-76380 (Novus Biologicals) |
| FAM134B/ RETREG1 | ER-phagy receptor | 21537-1-AP (ProteinTech) |
| FAM134C/ RETREG3 | ER-phagy receptor | PA5-53433 (Invitrogen) |
| TEX264 | ER-phagy receptor | HPA017739 (Sigma) |
| SEC62 | ER-phagy receptor | Ab140644 (Abcam) |
| CCPG1 | ER-phagy receptor. ER stress induces transcriptional upregulation of CCPG1 | 13861-1-AP (ProteinTech) |
| LC3B | Conversion of LC3I to LC3II is a marker of general autophagy. This antibody detects both forms | NB100-2220 (Novus Biologicals) |
| REEP5 | ER tubule marker | 14643-1-AP (ProteinTech) |
| CLIMP63 | ER sheet marker | A302-257A (Bethyl) |
| RAMP4 | ER sheet marker (Note that endogenous RAMP4 associates with the ER translocon and is enriched at the ER sheets. However, overexpressed RAMP4 are ubiquitously present in all ER compartments) | Ab184571 (Abcam) |
| ATL1 | Localized to ER three-way junctions | #12728 (Cell Signaling) |
| ATL2 | Localized to ER three-way junctions | NBP1-78733 (Novus Biologicals) |
| ATL3 | Localized to ER three-way junctions, ER-phagy receptor | A303-313A (Bethyl) |

#### 3.3.1 Cell Harvesting and Lysate Preparation for Western Blotting

1. After treatment for the desired amount of time, aspirate the medium and carefully wash the cells twice with PBS to remove residual media.
2. Incubate cells on ice and add 1X RIPA lysis buffer supplemented with Halt™ protease inhibitor cocktail to lyse the cells. For a 6-well plate, add 100–150 μL of RIPA per well. Gently tap the plate to spread the lysis buffer across the well. Incubate the plate on a rocker in the cold room for 10 min. For the CCER assay, a concentrated lysate is desirable as the cleaved mCherry band is generally faint, and you will need to load a high amount of proteins (e.g., 30–40 μg per well in a 10-well gel) for robust visualization.
3. Collect the lysates into prechilled microcentrifuge tubes and centrifuge at top speed (>10,000 g) for 5 min at 4 °C to remove the insoluble fraction and genomic DNA.
4. Perform a standard Bradford protein assay and normalize cell lysate to the same concentration.
5. Prepare samples for Western blotting by adding 4X SDS sample buffer at a 1:4 ratio to the final sample volume and a reducing agent (e.g., β-mercaptoethanol) at a 1:10 ratio to the final sample volume.
6. Briefly vortex and heat the samples at 85 °C for 5 min. Next, centrifuge the tubes briefly to collect condensation formed on the caps of the tubes.
7. The samples are now ready to be resolved by SDS gel electrophoresis. Alternatively, samples can now be stored at −20 °C until further use—in this case, ensure that the samples are reheated before loading onto SDS gels.

#### 3.3.2 SDS-Polyacrylamide Gel Electrophoresis and Western Blotting

1. We recommend using NuPAGE 4–12% gradient gels with MES buffer to achieve good separation of both small and large proteins. For the CCER assay, we recommend loading 30–40 μg of proteins in a 10-well gel (adjust accordingly if you use other gels such as 12-well or 15-well gels).
2. Run the gel at 180 V for 50 min or until the blue dye-front reaches the bottom.
3. Next, transfer the proteins from the gel to the nitrocellulose membrane using the standard wet transfer method with wet transfer buffer (25 mM Tris, 192 mM glycine, 20% methanol).
4. After the transfer, incubate the membrane with Ponceau S stain for 1–2 min with gentle rocking on a Western blot rocker. Next, remove background staining by rinsing the membrane with water to visualize the protein bands.
5. Wash off the Ponceau S stain by washing the membrane in TBS-T for 5 min, and then block the membrane using 5%

skimmed powdered milk in TBS-T for at least 15 min. Then, wash the membrane twice with TBS/PBS for 3 min each time.

6. Incubate the membrane with primary antibodies overnight at 4 °C or for 1–4 h at room temperature. A nonexhaustive list of commonly used ER protein markers and their validated antibodies is suggested in Table 4. When preparing the primary antibody dilutions, start by following the recommendations of the manufacturer for the specific antibody and optimize according to your cell type.
7. After incubating the membrane with primary antibodies, wash the membrane three times with TBST for 3 min each time.
8. Next, incubate the membrane with the respective secondary antibodies. Follow the manufacturer's recommendations for the dilution of secondary antibodies and incubation conditions. When using secondary antibodies conjugated with light-sensitive fluorescent dyes (such as the LI-COR secondary antibodies), cover the incubation container with tin foil to protect the membrane from light.
9. After secondary incubation, wash the membrane twice with TBST for 3 min each time, followed by a final wash with TBS/PBS for 5 min. The final wash without Tween-20 is especially critical when using the LI-COR imaging system, as Tween-20 causes high background autofluorescence.
10. Visualize the protein bands using the system of choice.

#### 3.3.3 Analysis of the mCherry Cleavage

1. To quantify ER-phagy using the CCER approach, measure the band intensity of full-length mCherry-RAMP4 and cleaved mCherry (*see* **Note 1**).
2. Perform the first normalization by dividing the cleaved mCherry band intensity by that of the full-length mCherry-RAMP4. This step will account for technical variation from any unequal loading.
3. Perform the second normalization by dividing the cleaved/full-length ratio of each sample by that of your untreated control sample. This step will account for any basal ER-phagy activity, and the end result will show only changes in ER-phagy due to experimental treatment.

### 3.4 ER-Phagy Measurement by Immunofluorescence

Immunofluorescence allows visual confirmation and quantification of ER co-localization with autophagy components (e.g., LC3 family proteins, WIPI-II, or ULK1) and its eventual localization within the lysosomes. ER-phagy detection via immunofluorescence can be performed either via antibody-based staining, plasmid expression of organelle-specific fluorescent proteins (e.g., mCherry-tagged ER with GFP-tagged lysosome, see Table 1), or a combination of antibody staining and plasmid expressions. As most ER proteins are transmembrane or membrane-associated proteins, the milder digitonin-based permeabilization is recommended over Triton-X to preserve ER membrane integrity.

#### 3.4.1 Cell Seeding and Fixation

1. Place sterile glass coverslips into a 12-well plate (*see* **Note 2**).
2. When seeding cells for the immunofluorescence experiment, optimize the cell number so that they reach no more than 70–80% confluence on the last day of your experiment before fixation. When imaging the ER, lower cell confluence is generally desired to allow space for cells to spread out. This allows better visualization of the ER network.
3. The next day, perform ER-phagy induction as appropriate. Since ER within the lysosomes will be degraded and thus not detectable by immunofluorescence, it is necessary to include Bafilomycin A1 (50–100 nM) either at the same time or in the last 4–6 h of ER-phagy treatment (see the section on ER-phagy/ER stress inducers). However, if you are using stable expression of mCherry/RFP-tagged ER and GFP-tagged lysosome, Bafilomycin A1 treatment is generally not required as RFP-variants are stable in the lysosome and are observable even without neutralizing the lysosome.
4. Fixation: At the end of ER-phagy treatment, wash cells twice with 1 mL of PBS, and then add 1 mL of 4% PFA to each well. Incubate at room temperature for 15 min. You can then either wrap the plate with Parafilm and store the fixed samples in the fridge (stable up to 1 month) or proceed directly to the next step.
5. PFA Quenching: Remove the 4% PFA from the wells. Add 500 μL of 5% BSA in PBS, quench for 5 min. *Warning*: PFA is toxic by skin contact and inhalation. Always follow your local health and safety procedure for disposal of PFA waste.
6. If you are using cell lines with fluorescently tagged ER and fluorescently tagged lysosome/autophagy, you can proceed directly to DAPI nuclear staining and mounting onto glass slides (see **step 5–8** of the next section). Otherwise, continue with the next section (immunostaining of ER and lysosome/autophagosome).

#### 3.4.2 Immunostaining of ER and Lysosome/Autophagosomes

1. Permeabilization and blocking step: Remove the quenching buffer, replace it with 500 μL of 0.1% digitonin in 5% BSA, and incubate for 15 min (*see* **Note 3**). *Warning*: Digitonin is a hazardous toxic substance, so it must be handled with care.
2. Preparation of the primary antibody: Prepare the primary antibody dilution in 0.1% digitonin-5%BSA buffer following the respective antibody datasheet. Only 50–100 μL of diluted

antibody is required per coverslip, so it is recommended to make a master mix depending on how many coverslips you are processing.

3. Primary antibody incubation:
   - (a) Using a container with a lid (we recommend using a 15 cm plate), create a wet chamber by putting wetted filter paper inside the container. Overlay a piece of Parafilm on the filter paper. Using a marker pen, map the Parafilm with the same grid as your 12-well plate.
   - (b) Then, add 50–100 μL of your diluted antibody onto the Parafilm.
   - (c) Using tweezers, carefully remove the coverslip from the 12-well plate, and gently dab off excess liquid from the coverslip onto a piece of tissue paper. Place the coverslip *cell-side down* so that the cells are directly in contact with the antibody.
   - (d) Incubate for 1 hr. at room temperature or overnight in the cold room.
   - (e) Using tweezers, transfer the coverslips back into the 12-well plate, this time *cell-side up*. Gently wash the coverslips three times with PBS.
4. Secondary antibody incubation (similar steps to primary antibody incubation):
   - (a) Prepare the secondary antibody dilution in 0.1% digitonin to 5% BSA buffer following the manufacturer's instructions. You can make a master mix based on the number of coverslips you need to process.
   - (b) Put the secondary antibody onto fresh Parafilm inside the wet chamber, and incubate the coverslips cell-side *down* onto the secondary antibody.
   - (c) Incubate for 30 min to 1 h at room temperature. *Important*: Secondary antibodies for immunofluorescence are conjugated with light-sensitive fluorescent dyes, so incubation must be shielded from light (either in a drawer or covered with tin foil).
   - (d) Transfer the coverslips cell-side *up* into the 12-well plate. Gently wash the coverslips three times with PBS.
5. Nuclear staining: Prepare a 1:1000 dilution of DAPI in PBS (make sure this is constantly shielded from light), add 50–100 μL to each coverslip, and incubate at room temperature for 10 min. Perform a 1× PBS wash. Coverslips are now ready to be mounted onto glass slides (*see* **Note 4**).
6. Coverslip mounting: Add a small drop of mounting media to the glass slide and gently remove the coverslip from the plate

using tweezers. Dip the coverslip into ddH2O once and dab the edges on clean tissue paper to remove excess water.

7. With an angle, gently place the coverslip cell-side *down* on the mounting media. Allow the media to slowly spread to cover the surface of the coverslip. Ensure that there are no air bubbles trapped between the coverslip and the glass slide.
8. Store mounted slides shielded from light until ready for microscopy imaging; room temperature is usually appropriate.

### 3.5 ER-Phagy Measurement by Flow Cytometry

Compared to Western blotting and immunofluorescence, the automated and rapid nature of flow cytometry allows high-throughput measurement, for example, assessing the effect of a list of gene knockdowns or overexpressions on ER-phagy. In addition, flow cytometry requires far fewer cells compared to Western blotting for representative measurement and is thus the desired method for experiments where cell number is a limitation (e.g., when using primary cell types).

ER-phagy measurement by flow cytometry is commonly achieved using the EATR or mKeima system, which relies on subtle changes in fluorescence signals of the reporter based on the surrounding environmental pH (see Subheading 2 and Table 1). Since the mKeima system requires specific laser filter modification that might not be accessible depending on local FACS setup and availability, we recommend using the EATR system, which simply measures GFP and RFP signals. Henceforth, we will describe FACS measurement of ER-phagy using the EATR system.

#### 3.5.1 Harvesting Cells for FACS ER-Phagy Analysis

1. Seed cells with either stable expression or Tet-inducible expression of the EATR reporters into a 24- or 12-well plate such that the cells are 80% confluent the next day for ER-phagy treatment. If using the Tet-inducible system, doxycycline can be added at the same time as seeding.
2. The next day, perform ER-phagy induction treatment as appropriate. Ensure that you have the appropriate negative control (untreated cells in full media) and positive control (amino acid starvation is the most robust ER-phagy inducer) to help establish the flow cytometry fluorescence gates.
3. On the day of harvest, wash cells once with PBS and detach cells using 20–50 μL of trypsin per well. Replace cells in the incubator for optimal trypsin activity.
4. After 3–5 min, retrieve the culture plate from the incubator, gently tap the plate, and inspect under the microscope to ensure that cells are fully detached.
5. Add 250–500 μL of full media to each well to stop the trypsin activity. Pipette the media 3–5 times to fully detach cells and transfer cells to a FACS tube.

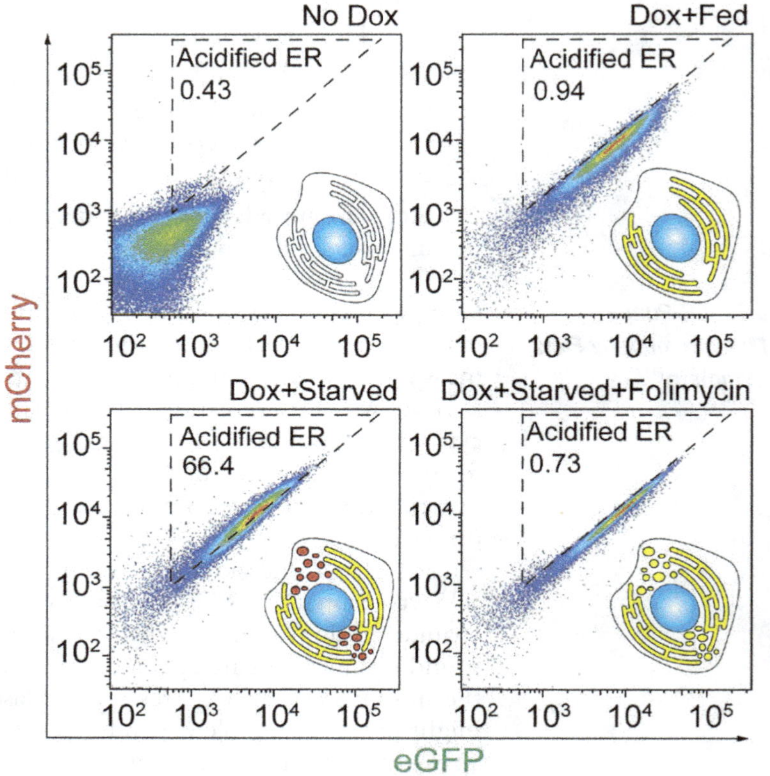


**Fig. 2** Strategy for the setup and measurement of ER-phagy by flow cytometry. The FACS plot describes the use of a Tet-inducible EATR system. Upon the addition of doxycycline (Dox), the cell population shifts to a diagonal distribution that reflects the expression of both eGFP and mCherry signals (top right plot; Dox + Fed). Using this fed/full media condition as a negative control, we can outline the "Acidified ER/ER-phagy" gate, which excludes the bulk of the population as there is only minimal basal ER-phagy (0.94%) happening under fed conditions. Upon starvation, lysosome-localized ER lose their eGFP signal while retaining the mCherry signal. This translates to a shift in the cell population to the left, entering the "Acidified ER" gate (bottom left plot). Treatment with Folimycin, which neutralizes the lysosome, restores the eGFP signal even under starvation conditions, thus resulting in the reverse shift of the population back to the right (bottom right plot). This validates that the EATR system measures lysosomal ER degradation. (Figure taken with permission from Liang et al. [13])

6. Proceed directly to flow cytometry analysis with the harvested live cells (*see* **Note 5**).

*3.5.2 Setup of FACS ER-Phagy Analysis*

1. Using your negative control sample, first establish the voltage/intensity of all relevant lasers.
   (a) Begin by setting up the Forward Scatter-Area and Side Scatter-Area measurements to gate for "live cells."

(b) Based on your "live cells," further gate for your "single cells" based on Forward Scatter-Area versus Forward Scatter-Height.

(c) Based on your "single cells," set up the eGFP and mCherry signals on the X- and Y-axis, respectively (*see* Fig. 2).

(d) Since we assume little to no ER-phagy happening at the fed/basal state, use your negative control sample as a baseline to draw out a triangular gate in your eGFP vs. mCherry fluorescence plot that *excludes* the bulk of the cell population—this will be your "Acidified ER" gate or "ER-phagy" (*see* Fig. 2).

2. To ensure that your "Acidified ER" gate is set up correctly, analyze your positive control (starved) sample next. Most of your positive control cells should now shift into the "Acidified ER" gate, i.e., you now have an increase in the percentage of cells undergoing ER-phagy during amino acid starvation.
3. Once you have optimized your sorting layout, you can now analyze your remaining samples. Ideally, you should aim to sort at least 10,000 effective cells per sample (*see* **Note 6**).

## 4 Notes

1. The cleaved mCherry band will always be around 30 kDa, no matter which target ER protein it is fused with.
2. To prevent contamination, always use sterile coverslips and tweezers. We recommend ethanol washing the coverslips (5 min on a rocker is sufficient) and allowing them to air dry in the tissue hood before use, or placing the plate with coverslips and tweezers in the tissue culture hood under UV irradiation for 15 min.
3. Digitonin is a reversible membrane permeabilizing agent, so it needs to be present in every subsequent step to ensure that your primary and secondary antibodies are delivered into the cells.
4. DAPI staining occupies the 405 nm range (blue channel) and is optional. If you are doing triple staining or a combination of fluorescent plasmid expressions where the blue channel/405 nm range is occupied, DAPI staining should be omitted.
5. Since PFA fixation has been observed to neutralize the lysosome, resulting in the partial reactivation of GFP fluorescence, we always perform live-cell FACS analysis when using the EATR and mKeima systems. If immediate analysis is not

possible, cells can be kept on ice for up to 4 h without significant changes in fluorescence signals.

6. These 10,000 effective cells refer to single cells that express the reporter system for meaningful measurement of ER-phagy rather than the total cell number.

## References


1. Khaminets A, Heinrich T, Mari M, Grumati P, Huebner AK, Akutsu M, Liebmann L, Stolz A, Nietzsche S, Koch N et al (2015) Regulation of endoplasmic reticulum turnover by selective autophagy. Nature *522*:354–358. https://doi.org/10.1038/nature14498
2. Grumati P, Morozzi G, Hölper S, Mari M, Harwardt M-LI, Yan R, Müller S, Reggiori F, Heilemann M, Dikic I (2017) Full length RTN3 regulates turnover of tubular endoplasmic reticulum via selective autophagy. eLife *6*: e25555. https://doi.org/10.7554/eLife.25555
3. Chino H, Hatta T, Natsume T, Mizushima N (2019) Intrinsically disordered protein TEX264 mediates ER-phagy. Mol Cell *74*:909–921.e6. https://doi.org/10.1016/j.molcel.2019.03.033
4. An H, Ordureau A, Paulo JA, Shoemaker CJ, Denic V, Harper JW (2019) TEX264 is an endoplasmic reticulum-resident ATG8-interacting protein critical for ER remodeling during nutrient stress. Mol Cell *74*:891–908.e10. https://doi.org/10.1016/j.molcel.2019.03.034
5. Smith MD, Harley ME, Kemp AJ, Wills J, Lee M, Arends M, von Kriegsheim A, Behrends C, Wilkinson S (2018) CCPG1 Is a non-canonical autophagy cargo receptor essential for ER-phagy and pancreatic ER proteostasis. Dev Cell *44*:217–232.e11. https://doi.org/10.1016/j.devcel.2017.11.024
6. Fumagalli F, Noack J, Bergmann TJ, Cebollero E, Pisoni GB, Fasana E, Fregno I, Galli C, Loi M, Soldà T et al (2016) Translocon component Sec62 acts in endoplasmic reticulum turnover during stress recovery. Nat Cell Biol *18*:1173–1184. https://doi.org/10.1038/ncb3423
7. Chen Q, Xiao Y, Chai P, Zheng P, Teng J, Chen J (2019) ATL3 Is a tubular ER-phagy receptor for GABARAP-mediated selective autophagy. Curr Biol 29:846–855.e6. https://doi.org/10.1016/j.cub.2019.01.041
8. Stephani M, Picchianti L, Gajic A, Beveridge R, Skarwan E, Sanchez de Medina Hernandez V, Mohseni A, Clavel M, Zeng Y, Naumann C et al (2020) A cross-kingdom conserved ER-phagy receptor maintains endoplasmic reticulum homeostasis during stress. eLife 9: e58396. https://doi.org/10.7554/eLife.58396
9. Nthiga TM, Kumar Shrestha B, Sjøttem E, Bruun J-A, Bowitz Larsen K, Bhujabal Z, Lamark T, Johansen T (2020) CALCOCO1 acts with VAMP-associated proteins to mediate ER-phagy. EMBO J *39*:e103649. 10.15252/embj.2019103649
10. Hübner CA, Dikic I (2020) ER-phagy and human diseases. Cell Death Differ 27:833–842. https://doi.org/10.1038/s41418-019-0444-0
11. Wilkinson S (2019) ER-phagy: shaping up and destressing the endoplasmic reticulum. FEBS J *286*:2645–2663. https://doi.org/10.1111/febs.14932
12. Liang JR, Corn JE (2022) A CRISPR view on autophagy. Trends Cell Biol *32*:1008–1022. https://doi.org/10.1016/j.tcb.2022.04.006
13. Liang JR, Lingeman E, Ahmed S, Corn JE (2018) Atlastins remodel the endoplasmic reticulum for selective autophagy. J Cell Biol 217:3354–3367. https://doi.org/10.1083/jcb.201804185
14. Liang JR, Lingeman E, Luong T, Ahmed S, Muhar M, Nguyen T, Olzmann JA, Corn JE (2020) A genome-wide ER-phagy screen highlights key roles of mitochondrial metabolism and ER-resident UFMylation. Cell 180:1160–1177.e20. https://doi.org/10.1016/j.cell.2020.02.017
15. Katayama H, Kogure T, Mizushima N, Yoshimori T, Miyawaki A (2011) A sensitive and quantitative technique for detecting autophagic events based on lysosomal delivery. Chem Biol 18:1042–1052. https://doi.org/10.1016/j.chembiol.2011.05.013

# Chapter 10

# Quantitative Super-Resolution Imaging of ER-Phagy Initiation in Cells

**Ashwin Balakrishnan, Marius Glogger, and Mike Heilemann**

## Abstract

Selective autophagy of the endoplasmic reticulum (ER-phagy) is a mechanism that is necessary for degrading damaged ER components and preventing cells from experiencing ER stress. Various ER-phagy receptors orchestrate this process by building protein assemblies with dedicated functions. In order to understand the molecular building principles of ER-phagy, it is important to reveal the assembly of ER-phagy receptors in a temporal and functional context. However, direct visualization is hampered by the diffraction limit in light microscopy. Super-resolution microscopy (SRM) can bypass this limitation and resolve single proteins and nanoscale protein clusters in cells. In particular, DNA points accumulation for imaging in nanoscale topography (DNA-PAINT) is a powerful technology that can resolve individual protein clusters in cells and provide information on their molecular composition. Here, we report a step-by-step protocol on how to utilize DNA-PAINT to perform super-resolution imaging of ER-phagy receptors in fixed cells. In addition, we provide a detailed explanation of image generation, cluster analysis, and molecular quantification.

**Key words** ER-phagy, Super-resolution microscopy, DNA-PAINT, Exchange-PAINT, Quantitative microscopy

## 1 Introduction

The endoplasmic reticulum (ER) is a large cellular organelle important for lipid synthesis, posttranslational modification of proteins, protein secretion to different parts of the cell, and provides residence for ribosomes [1]. A continuous turnover of the ER is necessary for maintaining cellular homeostasis, and its quality control is mediated by different pathways. One important and crucial pathway involves the selective autophagy of the ER, termed ER-phagy [1, 2]. ER-phagy is involved in selectively degrading portions of the ER that harbor unfolded or aggregated proteins or cytotoxic proteins, thereby maintaining ER homeostasis in normal and stress-induced conditions ranging from starvation to microbial infection. One mechanism through which the ER-phagy

Ioannis P. Nezis (ed.), *Selective Autophagy: Methods and Protocols*, Methods in Molecular Biology, vol. 2845,
https://doi.org/10.1007/978-1-0716-4067-8_10,

machinery achieves selective degradation is through cargo selection mediated by ER-phagy receptors [3]. Recent studies have shown ER-phagy receptors are restructured through clustering to form functional complexes [4, 5].

Understanding how proteins organize into functional complexes requires their characterization in situ. Optical super-resolution microscopy facilitates the visualization of nanoscale protein clusters and today represents a powerful tool that has revolutionized our understanding of cell biology [6]. A prominent technology is single-molecule super-resolution microscopy (single-molecule localization microscopy, SMLM), which, next to achieving near-molecular spatial resolution [7], provides access to quantitative information, such as protein nano-cluster size [8] and composition [9, 10]. The core principle of all SMLM methods is the spatial and temporal separation of the fluorescence signal of single emitters, which enables their individual localization with high spatial precision and the reconstruction of super-resolved fluorescence images [7]. This separation can be achieved by exploiting the photophysics of fluorophores [11] or by employing weak-affinity labels that transiently bind to and unbind from a target molecule [12]. The latter approach brings the advantage of continuous imaging that is independent of photobleaching, sequential multi-target imaging through probe exchange, and robust quantification from fluorophore label binding kinetics. The original conceptualization of SMLM with weak-affinity fluorophores (points accumulation for imaging in nanoscale topography, PAINT) used the lipid bilayer-binding fluorophore Nile Red [13]. The integration of weak DNA oligonucleotide pairs flexibilized this approach by introducing programmability (DNA-PAINT) [14] and allowed visualizing [15, 16] and quantifying protein networks in cells [17, 18]. In addition, multi-target imaging is easily possible through buffer exchange (exchange-PAINT) [19].

One class of ER-phagy receptors is the FAM134 family. FAM134 receptors provide critical functions in maintaining ER homeostasis, and their absence results in deformation and swelling of the ER network [20]. FAM134 receptors carry a conserved LC3 interaction region (LIR) at their C-terminus, and a membrane domain that consists of a reticulon homology domain (RHD) with two hydrophobic regions embedded in the membrane bridged by a flexible cytoplasmic linker [21]. Recently, a combination of biochemical methods, molecular dynamics (MD), and DNA-PAINT super-resolution microscopy showed that ubiquitination of FAM134B is the driving force for the initiation of ER-phagy, mediating receptor–receptor interactions and nano-cluster formations in cells [4]. In another study, it was found that phosphorylation precedes ubiquitination of FAM134B and FAM134C, leading to the subsequent activation of ER-phagy flux [5]. Both studies employed DNA-PAINT super-resolution microscopy to visualize single FAM134 protein clusters and analyze their composition, highlighting the opportunities of this technology in autophagy research. The underlying concepts of super-resolution imaging, cluster analysis, and protein number quantification are discussed in this article.

## 2 Materials

### 2.1 Sample Preparation

1. Primary antibodies against ER-phagy receptors (for example, rabbit anti-FAM134B (Genscript)) and associated proteins (e.g., mouse anti-REEP5 (Genscript)).
2. DNA (docking strand)-conjugated secondary antibodies (commercially available (Nanotag or Massive Photonics) or custom-labeled [16]) against the primary antibodies (*see* **Note 1**).
3. U2-OS cells seeded on fibronectin-coated Ibidi μ-slide VI chambers (Ibidi GmbH) at 70% confluency.
4. Methanol-free formaldehyde (FA).
5. Phosphate-buffered saline (PBS).
6. Immunofluorescence (IF) buffer: PBS, 10% (v/v) fetal bovine serum (FBS), 0.1% (w/v) saponin (Sigma-Aldrich).
7. 125 nM gold bead solution (250 μM) as a fiducial marker (A11-125-NPC-DIH-1-25; Nanopartz).
8. ATTO 655-labeled DNA (imager strands) (Eurofins) freshly diluted to 50 pM in PBS with 0.5 M NaCl, pH 8.3, in DNA LoBind tubes (Eppendorf).

### 2.2 DNA-PAINT Imaging

A widefield microscope capable of imaging in highly inclined laminated optical sheet (HILO) mode [23] and suitable for SMLM is required. The data shown in this chapter were recorded with a commercial SMLM microscope (N-STORM, Nikon) equipped with the following main components:

1. 100× 1.49 NA oil immersion objective (Nikon).
2. Autofocus system (perfect focus system, Nikon).
3. EMCCD camera (iXon Ultra X8884, Andor) and acquisition software (NIS Elements, Nikon; custom-written auxiliary software, μManager [24]).
4. 647 nM laser (Agilent Technologies) operated at 1.1 $kWcm^{-2}$.
5. Dichroic mirror: HC Quad 410/504/582/669 (AHF).
6. Optional: A microfluidic device (the data shown in this chapter were recorded with a PlexFlo (Bruker) controlled through Vutara SRX 7.0.00rc07 software with an Elbow Luer connector (Ibidi) that can fit the Ibidi μ-slide chamber and connect to the tubings of the microfluidics device) (*see* **Note 2**).

### 2.3 Image Reconstruction and Analysis

Any SMLM software for reconstruction and cluster analysis and any statistical graphing software for postprocessing, we used the following:

1. Picasso (for image reconstruction and cluster analysis based on Density-Based Spatial Clustering of Applications with Noise (DBSCAN) [25] of protein nano-clusters) [16].
2. SR-Tesseler (Voronoi tessellation-based cluster analysis for micro-clusters) [26].
3. OriginPro (OriginLabs).

## 3 Methods

### 3.1 Sample Preparation

The sample preparation involves the fixation of cells with minimal cell perturbation, hence using a mild detergent like saponin. The following steps were followed for the data shown in this chapter:

1. The seeded cells were fixed with a preheated solution of 4% (v/v) FA at 37 °C for 30 min (*see* **Note 3**).
2. The cells were washed thrice with PBS and blocked with immunofluorescence (IF) buffer for 10 min at room temperature (RT).
3. The IF buffer was then replaced with primary antibodies diluted in IF buffer at a suitable dilution (here, a dilution of 1:200) and incubated at RT for 1.5 h. Alternatively, the sample is incubated overnight at 4 °C.
4. The sample was washed thrice to remove unbound antibodies with incubation times >5 min per washing step.
5. The solution was then replaced with secondary antibodies diluted in IF buffer (here, a dilution of 1:100) and incubated at RT for 1.5 h.
6. Right before the end of the incubation, the gold bead stock solution was sonicated for 10 min and diluted 1:5 (v/v) in PBS.
7. After incubation, the sample was washed thrice with PBS (>5 min) and the diluted gold bead solution was added and incubated for 10 min.
8. The sample was washed thrice with PBS and then postfixed with 4% (v/v) FA at RT for 10 min and washed again thrice with PBS.
9. The sample is now ready for imaging and can be stored for approximately 2 weeks at 4 °C in PBS with 0.01% (w/v) $NaN_3$.

### 3.2 Exchange DNA-PAINT Imaging

1. Switch on the necessary components of the microscope and start the image acquisition software. Choose appropriate settings for the camera (EMCCD or sCMOS); for the data shown in this chapter, we used an EMCCD camera (iXon Ultra X8884, Andor) and the following parameters:
   (a) EM gain: 200 (adapt for other cameras to ensure optimal detection of photon signal over readout noise).
   (b) Preamp gain: 5.
   (c) Frame transfer: on.
   (d) Readout rate: 5 MHz.
   (e) Exposure time: 100 ms (this value was used to reflect the lifetime of the DNA hybrid; adapt accordingly if other DNA sequences are used).
   (f) Frames: 20,000 (adapt to achieve at least 10–20 binding events per docking strand, which can be calculated from Eq. (1):

$$B = n \cdot C \cdot t \cdot k_{on} \tag{1}$$

   where $B$ is the number of binding events per docking strand, $n$ is the number of binding sites per docking strand (*see* **Note 1**), $C$ is the concentration of imager strands, $t$ is the imaging time and $k_{on}$ is the association rate constant.
   (g) For example, the R1 strand with 100 ms exposure time and a 20,000-frame measurement would give 5 binding sites × 50 pM × 2000 s × (37 × 10^6) $M^{-1}$ $s^{-1}$ = 18.5 binding events per docking strand on average.
2. Add the first diluted imager strand (50 pM) buffer to the sample (always prepare fresh imager strand solutions in DNA-LoBind tubes before imaging).
3. Set a drop of oil on the objective, place the sample onto the microscope stage, and bring the cells into focus using bright-field illumination. Switch to fluorescence, illuminate the sample with low laser power, and change the TIRF mirror such that the signal is uniform and the illumination is HILO (Fig. 1a) (this considers that the ER is a highly convoluted organelle spanning the entire cell (Fig. 1b)) (*see* **Note 4**).
4. Set appropriate laser power (we operated at 1.1 kWcm$^{-2}$, measured with widefield illumination after the objective) and start an image series acquisition.
5. Once the acquisition is finished, wash the sample thrice with PBS (incubate 5 min per wash) using a microfluidics device to keep drift as low as possible. Follow it up by adding the next imager strand diluted in buffer and start another acquisition (Fig. 1c).

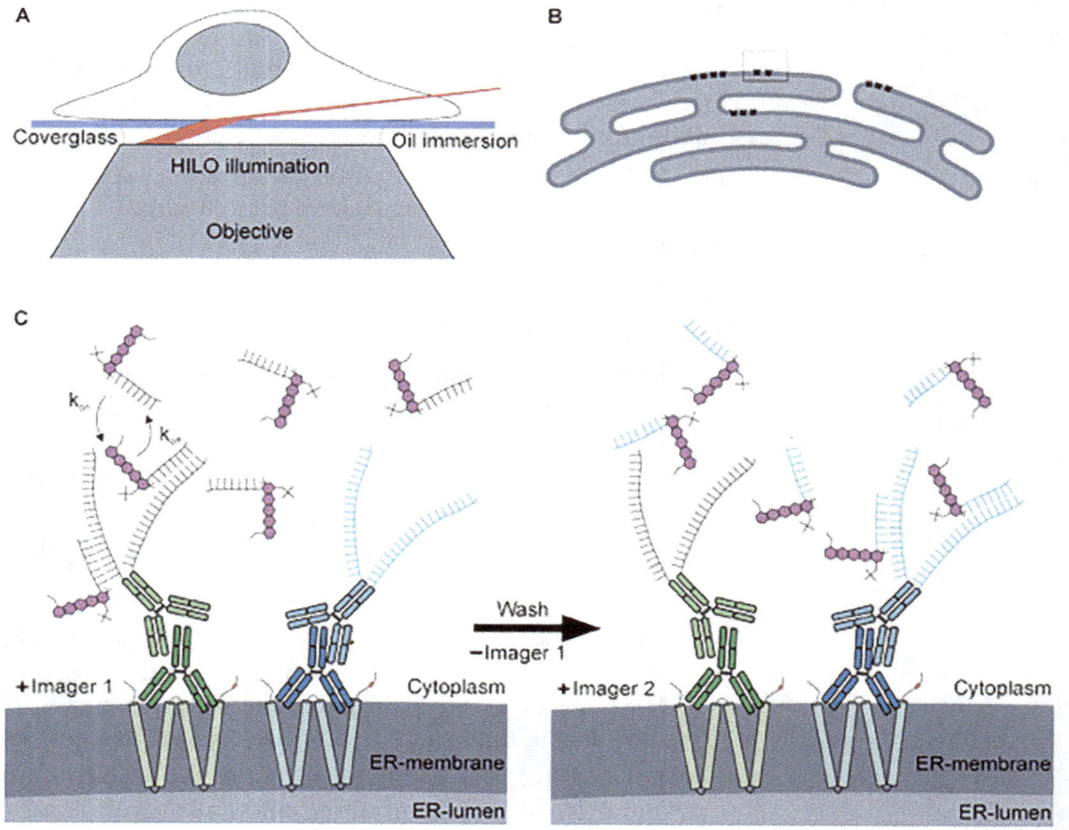


**Fig. 1** Principle of exchange DNA-PAINT for ER-phagy receptors. (**a**) Schematic of imaging a cell with HILO illumination, where the laser hits the cover glass at an oblique angle (red: laser beam). (**b**) Schematic of the ER as a convoluted structure with ER-phagy associated receptors on its membrane. (**c**) Zoomed-in view of **b**, visualizing exchange DNA-PAINT imaging for two different targets. Note: The schematic is not to scale

6. Repeat **steps 4** and **5** by introducing new imager strands depending on the number of targets.
7. Acquire a phase contrast widefield image at the end which can later be used to define the border of the acquired cell.

### *3.3 Image Reconstruction and Analysis*

The raw imaging data (movie of 20,000 frames) contains single-molecule fluorescence events arising from the binding and unbinding of imager strands to the binding sites offered by docking strands. Each single-molecule binding event shows up as a symmetric "spot" (point spread function (PSF)) in the images. The first step in image processing is to determine the center of mass of each spot (PSF) and reconstruct a super-resolution image from all single-molecule coordinates. Here we use *Picasso v0.3.3* [16] to perform centroid localization and image reconstruction. *Picasso* is provided as a set of different modules, each for a different function.

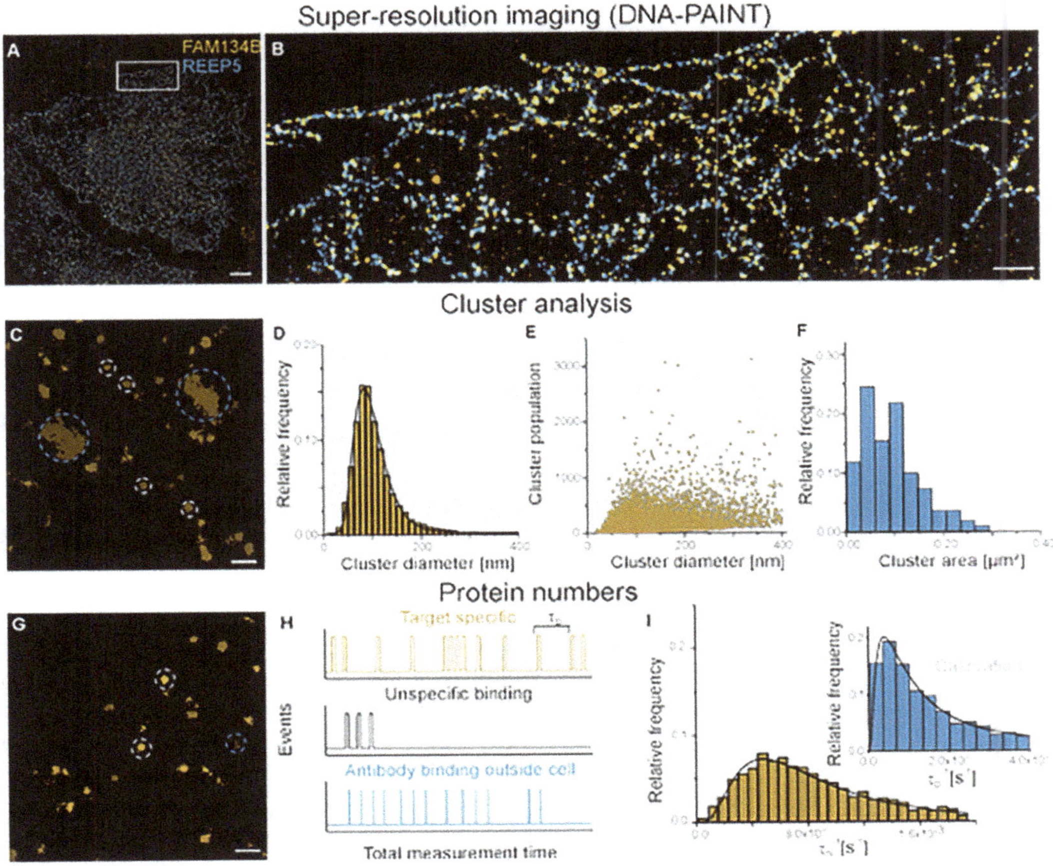


**Fig. 2** DNA-PAINT imaging and cluster analysis of ER-phagy receptors. (**a**) Reconstructed exchange DNA-PAINT image of FAM134B (yellow) and REEP5 (blue) and a zoom-in of the image (**b**). (**c**) Zoomed-in image of FAM134B clusters derived from DBSCAN analysis. Both nanoclusters (dotted white circles) and microclusters (dotted blue circles) are visible. (**d**) Frequency histogram of FAM134B nanocluster diameters. (**e**) Correlation between the number of localizations per cluster (Cluster population) and nanocluster diameter. (**f**) Frequency histogram of FAM134B microcluster areas. (**g**) Zoomed-in image of FAM134B clusters after DBSCAN analysis, visualizing target-specific and unspecific localizations. (**h**) Schematic of fluorescence events spread over the measurement time. The average difference between events gives the dark time ($\tau_D$). (**i**) Frequency histogram of dark times from FAM134B clusters (yellow) and calibration clusters (blue). Scale bar: (**a**) 10 μm; (**b**) 1 μm; (**c**, **d**) 500 nM

1. Open the raw imaging data (image series) in the *Picasso Localize* module and set appropriate localization parameters by opening *Analyze→Parameters*. In our case, we used the following parameters:
   (a) Box side length: 7 pixels (adjust for single-molecule PSF size).
   (b) Net gradient: 40,000 (gradient-based identification of boxes that contain single-molecule spots, adjust

depending on signal-to-noise ratio (SNR) and observe multiple frames by checking *Preview* to visualize if only the signal is selected; increase the net gradient if noise is selected).

(c) EM gain: 200 (set the value used in measurement).

(d) Baseline: 205 (average signal noise level of the camera without light exposure; determine the background per pixel in images using image processing tools such as Fiji [27]).

(e) Sensitivity: 4.78 (dependent on the camera, usually available from the camera datasheet).

(f) Quantum efficiency: 0.95 (depends on the spectral range and camera sensitivity; available from the camera datasheet).

(g) Pixel size: 158 nm (camera pixel size/lens magnification).

(h) Fitting method: maximum likelihood estimation (*MLE, integrated Gaussian*) (*see* **Note 5**).

2. Start the centroid localization using *Analyze→Localize*. The centroid of all single-molecule spots is now calculated, and the lateral coordinates of the centroids are stored in a file which can then be rendered as the super-resolved image. Two files will be generated, a .hdf5 (localization data) and a .yaml (settings) file.
3. Repeat **steps 1** and **2** for the other image series acquired.
4. Open the .hdf5 file in the *Picasso Render* module. This visualizes the reconstructed super-resolution image from the image series (Fig. 2a, b) (*see* **Note 6**). The next step is to perform a drift correction by picking localizations arising from gold beads. For this, select at least 5 beads in the image area using the pick tool (*Tools→Pick*); pick beads outside the cell area which are easier to identify (*see* **Note 7**). Visualize their fluorescence intensity over time using *Tools→select picks (trace)*, and accept or decline the spots corresponding to fiducial markers. Drift correction using fiducial markers can then be performed by selecting *Postprocess→ Undrift from picked*. Save the drift-corrected image as a .hdf5 file. Repeat step 4 for other image series acquired (*see* **Note 8**).
5. Open the drift-corrected .hdf5 file in the *Picasso Filter* module (*see* **Note 9**). This file summarizes various parameters for each single-molecule detection event, which can be further filtered by plotting them each as histograms using *Plot→Histogram* or together using *Plot→2D Histogram* and selecting a range by drawing a selection pane (*see* **Note 10**). Typical parameters and ranges are the following (adapt if necessary):

(a) Point spread function (PSF) widths $s_x$,$s_y$: 0.5–1.5 (pixel).

(b) Localization precisions $lp_x$, $lp_y$: 0–0.2 (pixel).

(c) Further properties such as photons, ellipticity, and background can be filtered additionally to remove outliers if necessary.

Save the filtered data as a .hdf5 file.

6. Open the filtered .hdf5 file in the *Picasso Render* module. Calculate the average localization precision through the nearest neighbor-based analysis (NeNA) [28] by *View→ Show info→ Calculate*. Single-molecule binding events that are detected in multiple frames are linked to one localization using *Postprocess→ Link localizations* and the following two parameters:

   (a) Max distance: 3 x NeNA value (could be 1–6 × NeNA, depending on the signal density; use lower values for high-density data, *see* **Note 11**).

   (b) Max transient dark frames: At least 5 frames (depends on the fluorophore; 5 to 15 frames (100 ms exposure time) is a common range for ATTO 655 and reflects the photophysical properties of the fluorophore under the imaging conditions).

   Save the linked image as a .hdf5 file.

   After the above postprocessing, a super-resolved image can be generated for each target. Images can be superimposed using the position of the fiducial markers.

   The next steps explain quantitative cluster analysis (*see* **Note 12**).

7. For DBSCAN analysis, open the linked .hdf5 file in *Picasso Render* and choose *Postprocess→ Clustering→ DBSCAN*. DBSCAN requires the following two parameters:

   (a) Radius: 3 × NeNA value.

   (b) Density: 10 localizations.

   Both radius and density depend on the density of labels in a nano-cluster. Explore the parameter space of radius vs density and evaluate results for the optimal parameter settings. The optimum value can be found by opening the resultant *dbscan* .hdf5 file using *Picasso Render* and picking clusters to observe if their traces are indeed true signals. For the optimum radius and density, most clusters picked should exhibit true signals (*see* **Note 13**). Optionally, use available tools to evaluate parameter choice [8, 29].

8. After the analysis, three files are created automatically (*Picasso* v0.3.3): one .hdf5 file with the suffix *dbscan*, one .hdf5 file with the suffix *dbclusters*, and one .yaml file (*see* **Note 14**). The

*dbscan* file contains the localizations attributed to clusters and can be visualized using *Picasso Render* (Fig. 2c); the *dbclusters* file contains properties including cluster area in pixels and the number of localizations per cluster. Cluster width can then be calculated from cluster area by assuming that the clusters are circular, then visualized as a frequency histogram (Fig. 2d) or in relation to the number of localizations per cluster (using statistical graphics software) (Fig. 2e).

9. It is recommended to perform kinetic filtering of DBSCAN data to remove apparent clusters that arise from nonspecific binding of imager strands to cellular structures. For kinetic filtering, the localizations in the *dbscan* .hdf5 file are loaded into the *Picasso Filter* module. Plot the column *mean frame* as a histogram (*Plot→Histogram*) and select the central peak that occurs at half the number of frames of the imaging series. If needed, additional filtering using the standard deviation of the mean frame, *std_frames*, can be performed.
10. For Voronoi tessellation, open the linked localizations file in *Picasso Render* and export it as a .csv file using *File→Export localizations* and choosing *.csv for Thunderstorm*. Open the .csv file in *SR-Tesseler*. Use the ROI tool (penultimate icon from the right) to select an ROI. In the control window, choose the *Voronoi diagram* tab and *Voronoi construction* sub-tab, and click *Create polygons* to perform tessellation. Move to the *Objects tab* and set the *Density factor* to the local density calculated from the DBSCAN analysis, uncheck the other parameters, and click *Set density factor*. Then click *Create objects*, which creates a table with the area of the objects detected within the ROI. The results can then be exported using *Export stats* and plotted as a frequency histogram (using statistical graphics software) (Fig. 2f).
11. For cluster number quantification, open the *dbscan* .hdf5 file in *Picasso Render* (Fig. 2g). Pick a few hundred clusters in different regions in the cell using *Tools→Pick* and visualize them using *Tools→select picks (trace)*. Accept only clusters that exhibit localization events uniformly over the measurement time (Fig. 2h top) and reject the rest (Fig. 2h middle). Use *Tools→Pick similar* to pick clusters similar to the ones that were accepted. Save the properties of the picked clusters using *File→Save pick properties* (*see* **Note 15**).
12. The *pickproperties* file contains the column *dark_mean*, which gives the mean dark times of the clusters/regions selected, related to the numbers of binding sites in a cluster. A frequency histogram can be calculated from these values and plotted with a log-normal curve to provide the mean dark time of all

selected clusters (using statistical graphics software) (Fig. 2i) (*see* **Note 16**).

13. Repeat **step 12** for clusters that are both outside the cell and show repetitive and transient binding (Fig. 2h bottom). These can be attributed to antibodies unspecifically binding in the sample and can be used as dark time calibration. Repeat **step 12** on the selected clusters (Fig. 2i inlay). The ratio between the mode of the dark time distribution of target receptors (**step 12**) and calibration clusters (**step 13**) allows for relative quantification (*see* **Note 17**).

## 4 Notes

1. For the data shown in this chapter, we used custom-labeled secondary antibodies (donkey anti-rabbit (AffiniPure) and goat anti-mouse (AffiniPure)) conjugated to speed-optimized R1 or R2 docking strands R1 and R2 (for nomenclature and DNA sequences, see Table 1) [22]. In its original implementation, a docking strand binds to one imager strand, whereas the speed-optimized docking strands, owing to their sequence design, allow multiple overlapping binding sites for an imager strand. In our case, we used R1 and R2 docking strands with five overlapping binding sites.
2. As an alternative to a pressure-controlled microfluidics system, syringe plungers could be used.
3. When using Ibidi µ-slide, use a maximum of 180 µL of working volume and add any new solution always on the same end and remove it always from the opposite end for the whole experiment. When removing any solution, make sure there are a few left in the channel, as removing everything can lead to bubbles in the channel. Avoid getting bubbles into the channel.
4. Since the ER is a highly convoluted structure (Fig. 1b) stretching hundreds of nm at the perimeter of a cell and getting thicker closer to the nucleus, we focus on a plane that covers

**Table 1**
**Sequences of DNA strands used in this chapter**

| DNA strands used in this chapter | Sequence |
|---|---|
| R1 imager strand | 5′-AGG AGG A-3′-ATTO 655 |
| R1 (5×) docking strand | Antibody-5′-TCC TCC TCC TCC TCC TCC T-3′ |
| R2 imager strand | 5′-GGT GGT G-3′-ATTO 655 |
| R2 (5×) docking strand | Antibody-5′-ACC ACC ACC ACC ACC ACC A-3′ |

the maximum surface area of the ER, which is close to the surface. Hence, it is recommended after finding a cell, focus on the glass surface and then change it going up to hit the right plane.

5. When optimizing parameters for image analysis, users might choose between least square (*LQ Gaussian*) or MLE-based fitting options for fast or precise localization of emitters, respectively.
6. In the *Picasso Render* module, the contrast of the image can be adjusted for better visualization using *View→Display Settings* and adjusting the *Min.Density and Max. Density values* nested under *Contrast.*
7. Make sure sufficient fiducial markers are visible in the ROI (>5) to guarantee precise drift correction of the image. Spots arising from fiducial markers display constant fluorescence over all or most frames throughout the measurement, whereas signals arising from binding-unbinding events are intermittent.
8. In the case that there is no fiducial marker available in the measurement, a drift correction with redundant cross-correlation (RCC) [30] can be performed.
9. After every image processing step, the dataset can be further filtered for new properties using the *Picasso filter* module.
10. In the *Picasso Filter* module, when visualizing histograms, the selection pane selects the range and can't be undone, and the drift-corrected .hdf5 file needs to be reopened in case the wrong range is selected.
11. When performing NeNA analysis for the average localization precision, it is recommended to check the visual quality of the fit using *Show plot.* If the fit is not good, repeat Subheading 3.3 **step 6**.
12. ER-phagy receptors form functional conjugates by clustering together; however, their nature of cluster formation or size has been eluded until recently. Gonzalez et al. [4] and Berkane et al. [5] have shown the existence of different cluster sizes, termed nano- and micro-clusters as they occupy areas of a few $nm^2$ vs a few $\mu m^2$. For nano-cluster-based analysis, the DBSCAN algorithm fares better, while for micro-cluster analysis, Voronoi tessellation fares better. Note that it is recommended to first perform DBSCAN to understand the spread of different clusters and, depending on the result, continue with Voronoi tessellation.
13. Steps could be performed in addition to filtering based on intensity traces over time to find the optimum radius and density parameter for DBSCAN. One step would be to visually check the resultant clusters to observe whether a defined

cluster is, in actuality, smaller clusters merged together or one big cluster defined as several small clusters. Another would be to use the *Picasso Filter* module to filter out any outliers in the *dbcluster* .hdf5 file.

14. Performing DBSCAN with the most recent version of Picasso, v0.6.5 (at the time of writing), requires the user to select *Save cluster_centers* in addition to *Postprocess→Clustering→DBSCAN*. This would yield four files, two .yaml files and two .hdf5 files (*clustered and cluster_centers respectively*). The clustered file contains the localizations attributed to clusters, and *cluster_centers* contains the cluster area in $nm^2$ and the number of localizations per cluster.
15. Recheck the picked clusters if they are indeed inside the area of the cell. If not, use *Tools→Clear picks* and redo Subheading 3.3 **step 11** manually without using the *Pick similar* tool.
16. As an alternative, the mean dark times/cluster could be plotted as histograms and fitted with an exponential cumulative distribution function (Eq. 2) to get the cumulative dark time of all clusters [18].

$$y = 1 - \exp\left(\frac{-t}{\tau_D}\right) \tag{2}$$

where $t$ is the measurement time and $\tau_D$is the cumulative dark time.

17. Relative quantification means that the number of antibodies binding to a target protein nano-cluster is determined by relating the number of single-molecule detection events in the nano-cluster to those recorded for a reference protein (single antibody aside a cell). For absolute quantification, we recommend using a primary monoclonal antibody or nanobody conjugated to a single docking strand (degree of labeling = 1).

## Acknowledgments

We thank Anshu Khatri and Mahil Lambert for proofreading. This work was supported by the Deutsche Forschungsgemeinschaft (DFG, German Research Foundation), SFB 1177 (project-id: 259130777), and INST 161/778-1 FUGG.

### References

1. Chino H, Mizushima N (2020) ER-Phagy: quality control and turnover of endoplasmic reticulum. Trends Cell Biol 30:384–398
2. Reggiori F, Molinari M (2022) ER-phagy: mechanisms, regulation, and diseases connected to the lysosomal clearance of the endoplasmic reticulum. Physiol Rev 102:1393–1448
3. Dikic I (2018) Open questions: why should we care about ER-phagy and ER remodelling? BMC Biol 16:1–5

4. González A, Covarrubias-Pinto A, Bhaskara RM et al (2023) Ubiquitination regulates ER-phagy and remodelling of endoplasmic reticulum. Nature 618:394–401
5. Berkane R, Ho-Xuan H, Glogger M et al (2023) The function of ER-phagy receptors is regulated through phosphorylation-dependent ubiquitination pathways. Nat Commun 14: 8364
6. Valli J, Garcia-Burgos A, Rooney LM et al (2021) Seeing beyond the limit: a guide to choosing the right super-resolution microscopy technique. J Biol Chem 297:100791
7. Sauer M, Heilemann M (2017) Single-molecule localization microscopy in eukaryotes. Chem Rev 117:7478–7509
8. Nieves DJ, Pike JA, Levet F et al (2023) A framework for evaluating the performance of SMLM cluster analysis algorithms. Nat Methods 20:259–267
9. Dietz MS, Heilemann M (2019) Optical super-resolution microscopy unravels the molecular composition of functional protein complexes. Nanoscale 11:17981–17991
10. Fricke F, Dietz MS, Heilemann M (2015) Single-molecule methods to study membrane receptor oligomerization. ChemPhysChem 16: 713–721
11. Fürstenberg A, Heilemann M (2013) Single-molecule localization microscopy-near-molecular spatial resolution in light microscopy with photoswitchable fluorophores. Phys Chem Chem Phys 15:14919–14930
12. Albertazzi L, Heilemann M (2023) When weak is strong: a plea for low-affinity binders for optical microscopy. Angew Chem Int Ed Engl 62:e202303390
13. Sharonov A, Hochstrasser RM (2006) Wide-field subdiffraction imaging by accumulated binding of diffusing probes. Proc Natl Acad Sci USA 103:18911–18916
14. Jungmann R, Steinhauer C, Scheible M et al (2010) Single-molecule kinetics and super-resolution microscopy by fluorescence imaging of transient binding on DNA origami. Nano Lett 10:4756–4761
15. Schröder MS, Harwardt M-LIE, Rahm JV et al (2021) Imaging the fibroblast growth factor receptor network on the plasma membrane with DNA-assisted single-molecule super-resolution microscopy. Methods 193:38–45
16. Schnitzbauer J, Strauss MT, Schlichthaerle T et al (2017) Super-resolution microscopy with DNA-PAINT. Nat Protoc 12:1198–1228
17. Böger C, Hafner A-S, Schlichthärle T et al (2019) Super-resolution imaging and estimation of protein copy numbers at single synapses with DNA-point accumulation for imaging in nanoscale topography. Neurophotonics 6: 035008
18. Jungmann R, Avendaño MS, Dai M et al (2016) Quantitative super-resolution imaging with qPAINT. Nat Methods 13:439–442
19. Jungmann R, Avendaño MS, Woehrstein JB et al (2014) Multiplexed 3D cellular super-resolution imaging with DNA-PAINT and exchange-PAINT. Nat Methods 11:313–318
20. Reggio A, Buonomo V, Berkane R et al (2021) Role of FAM134 paralogues in endoplasmic reticulum remodeling, ER-phagy, and collagen quality control. EMBO Rep 22:e52289
21. Bhaskara RM, Grumati P, Garcia-Pardo J et al (2019) Curvature induction and membrane remodeling by FAM134B reticulon homology domain assist selective ER-phagy. Nat Commun 10:2370
22. Strauss S, Jungmann R (2020) Up to 100-fold speed-up and multiplexing in optimized DNA-PAINT. Nat Methods 17:789–791
23. Tokunaga M, Imamoto N, Sakata-Sogawa K (2008) Highly inclined thin illumination enables clear single-molecule imaging in cells. Nat Methods 5:159–161
24. Edelstein AD, Tsuchida MA, Amodaj N et al (2014) Advanced methods of microscope control using μManager software. J Biol Methods 1. https://doi.org/10.14440/jbm.2014.36
25. Ester M, Kriegel H-P, Sander J, Xu X A density-based algorithm for discovering clusters in large spatial databases with noise https://cdn.aaai.org/KDD/1996/KDD96-037.pdf. Accessed 8 Jan 2024
26. Levet F, Hosy E, Kechkar A et al (2015) SR-Tesseler: a method to segment and quantify localization-based super-resolution microscopy data. Nat Methods 12:1065–1071
27. Schindelin J, Arganda-Carreras I, Frise E et al (2012) Fiji: an open-source platform for biological-image analysis. Nat Methods 9: 676–682
28. Endesfelder U, Malkusch S, Fricke F, Heilemann M (2014) A simple method to estimate the average localization precision of a single-molecule localization microscopy experiment. Histochem Cell Biol 141:629–638
29. Verzelli P, Nold A, Sun C et al (2022) Unbiased choice of global clustering parameters for single-molecule localization microscopy. Sci Rep 12:22561
30. Wang Y, Schnitzbauer J, Hu Z et al (2014) Localization events-based sample drift correction for localization microscopy with redundant cross-correlation algorithm. Opt Express 22:15982–15991

# Chapter 11

# Fluorescence Methods to Measure Pexophagy

**Francesco G. Barone, Sylvie Urbé, and Michael J. Clague**

## Abstract

We outline our approach for studying the selective autophagy of peroxisomes (pexophagy), using fluorescence microscopy in tissue cell culture models. Ratiometric reporters, which specifically localize to peroxisomes, allow a quantitative assessment of pexophagy in fixed and live cells, as well as whole organisms. We discuss chemical and physiological inducers of pexophagy and any overlap with the induction of mitophagy.

**Key words** Pexophagy, Peroxisomes, Autophagy, Fluorescence microscopy, Pexo-QC, Keima-SKL

## 1 Introduction

The peroxisome is an intriguing organelle with well-established roles in lipid metabolism and in the handling of reactive oxygen species (ROS) [1, 2]. Furthermore, peroxisomes have recently been linked to ferroptosis, an important cell death pathway [3, 4]. Their cellular homeostasis involves a delicate balance between peroxisome biogenesis, fission, and selective autophagy, hereafter referred to as pexophagy. The study of peroxisomal dynamics originally relied on electron microscopical observation or the use of GFP-tagged peroxisomal markers, which give indications of size and number [5]. The parameter of peroxisomal turnover, via autophagic degradation, can now be derived from specific ratiometric fluorescent reporter molecules [6, 7]. This is important, as it has been proposed that defects in pexophagy contribute to the majority of peroxisome biogenesis disorders [8, 9].

The two most popular such reporters are a tandem construct of Green (GFP) and Red (Cherry) fluorescent proteins and Keima. Here we refer to the combination of EGFP and mCherry as -QC, for quality control, adopting this nomenclature from McWilliams et al. description of a mitophagy reporter [10]. Both Keima and QC respond to the low pH environment in lysosomes and thereby measure the final stage of the autophagy process. GFP is

Ioannis P. Nezis (ed.), *Selective Autophagy: Methods and Protocols*, Methods in Molecular Biology, vol. 2845,
https://doi.org/10.1007/978-1-0716-4067-8_11,

conformationally unstable under these acidic conditions, whilst mCherry is relatively stable leading to solely red fluorescent autolysosomes. Keima, on the other hand, undergoes a pH-dependent shift in its excitation spectrum. QC probes can be used on fixed cells, whilst Keima probes (e.g., mKeima; monomeric Keima) are only suited for live cell imaging [11]. Both reporters lend themselves to expression in whole organism models, such as mice, flies, and worms [10, 12–14].

How is peroxisomal targeting of such reporters achieved? The peroxisome possesses a unique matrix protein import system that recognizes a tripeptide motif (SKL) and does not require protein unfolding [15]. Hence, this can be used to direct reporters to the peroxisomal matrix. Here we detail the design and benchmarking of pexophagy reporters for quantitative assessment in mammalian cell culture models.

## 2 Materials

### *2.1 Reporter Choice and Organelle Targeting*

The first class of organelle-specific autophagy reporters are essentially adaptations of reporters, which were designed to assess non-selective autophagy through appendage to LC3 as an autophagosome targeting signal [16]. A variety of these have been described in the literature, but the simplest version, mRFP-GFP-LC3, is available from Addgene (plasmid # 21074; ptfLC3)). The Ganley laboratory has notably substituted the targeting sequence with a mitochondrial protein, Fis1 (mito-QC), through which they created a mouse mitophagy reporter model, later also adopted in fly models [10, 13]. It should be noted that Fis1 also localizes to peroxisomes [17]. So, although the bulk of mito-QC clearly localizes to mitochondria, this is technically a dual organelle autophagy reporter, and newer variants now target other mitochondrial outer membrane proteins (*see* **Note 1**). A key advantage of the -QC probes is that they are fixable and hence suitable for postmortem analysis. They may also be most suitable if a heated microscope stage in a controlled atmosphere is not readily available for live cell imaging. Typically, images of cells expressing the EGFP-mCherry reporters are acquired sequentially with a 488 nM laser for EGFP (525 nm emission filter) and a 561 nM laser for mCherry (617 nM emission filter).

The acid protease-resistant coral protein mKeima emits at 620 nM, but has a bimodal excitation spectrum peaking at 440 and 586 nM, corresponding to neutral and ionized states of its phenolic hydroxyl moiety (pKa ~6.5) [11]. One advantage of mKeima is that because of this large Stoke's shift, it can be used in conjunction with other green-emitting fluorophores. A cloning plasmid for mKeima is available via Addgene (plasmid # 54546). Time resolution is limited by switching between excitation wavelengths. For ease of representation, images are pseudo-colored red or green depending on excitation wavelength with lysosomes represented as red, mirroring the colors of -QC reporters. Magenta may be substituted for red to ensure compatibility with color blindness.

By substituting existing targeting information with the tripeptide peroxisomal matrix targeting signal (SKL), we and others have made both types of pexophagy reporters, which we refer to as pexo-QC and pexo-Keima respectively [5, 6, 14, 18–21]. A construct targeting mCherry-EGFP to peroxisomes via the peroxisomal integral membrane protein PMP34 is available from Addgene (plasmid number 109017) [22]. For cell-based studies, we would recommend generating a stable cell line, such as hTERT-RPE1-Keima-SKL cells, which we have used, to provide a uniformity of expression [23]. We have also generated pexo-Keima and pexo-QC *Drosophila* strains, which can be used for gaining a whole-body overview of pexophagy, or for the specific expression in tissues of choice (*see* **Note 2**) [14].

## 3 Methods

### 3.1 Validating Reporter Localization and Specificity

For pexo-QC: PMP70 (also known as ABCD3) and Catalase (CAT) provide good markers for the peroxisomal membrane and matrix respectively (*see* **Note 3**). A third emission channel in the blue or far-red (e.g., AlexaFluor405 or 647 labeled secondary antibody) will be necessary.

For pexo-Keima, co-expression of a GFP-labeled peroxisomal protein (e.g., PMP70) will be necessary.

In most cells under basal conditions, one is presented with a mix of red/green versus red puncta. Red puncta can be colocalized with a lysosomal marker, which can be achieved for pexo-QC by staining fixed cells with anti-LAMP1 or anti-LAMP2 antibodies. Alternatively, for pexo-Keima (which loses green-red discrimination upon fixing), CFP-tagged LAMP1 or LAMP2 can be transfected into the cells.

### 3.2 Validating Selective Autophagy

To check that red puncta are the result of an autophagy pathway, we would recommend depleting a core autophagy gene such as ATG7 using siRNA as a control and monitoring its depletion by Western blotting. A number of chemicals have recently been introduced that elevate pexophagy and mitophagy (see Table 1, Fig. 1, and *see* **Note 1**).

1. Day 1: Seed cells in a 6-well plate. For hTERT-RPE1 cells, we typically seed $1.2 \times 10^5$ cells per well. Perform this experiment in duplicate, one for imaging assessment of pexophagy and another one for Western blot validation of knockdown

**Table 1**
**Summary of compounds known to induce pexophagy (and mitophagy)**

| Drug | Source | Mechanism |
|---|---|---|
| MLN4924 | C1231; Chemgood | BNIP3/BNIP3L elevation via inhibition of CRL$^{VHL}$ and CRL$^{FBXL4}$ |
| Deferiprone (DFP) | Sigma | BNIP3/BNIP3L elevation by iron chelation |
| Compound 39 (CMPD-39) | HY-141659MedChemExpress | USP30 inhibition and promotion of NBR1 dependent pathway |

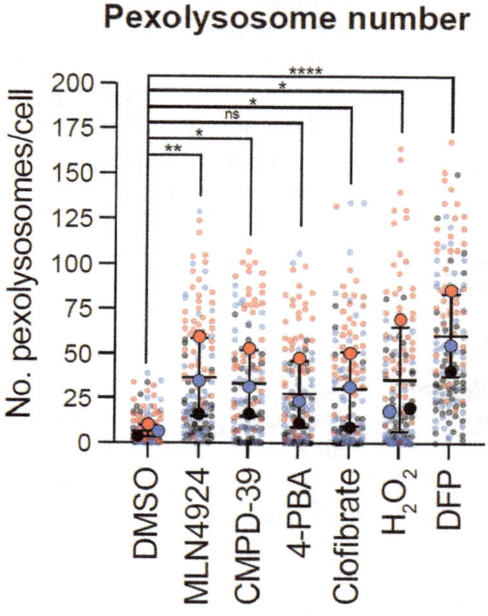


**Fig. 1** Graph illustrates the number of pexolysosomes per cell in hTERT-RPE1 cells stably expressing the Keima-SKL pexophagy reporter. CMPD-39 (1 μM) was administered for 96 h before imaging. For all other conditions, cells were treated for 24 h with MLN4924 (1 μM), 4-PBA (1 mM), clofibrate (20 μM), hydrogen peroxide ($H_2O_2$, 100 μM), and deferiprone (DFP, 1 mM). Quantification of the data from three independent color-coded experiments is shown. Mean and SD are indicated; >40 cells were quantified per condition in each replicate experiment. One-way ANOVA and Bonferroni's multiple comparison test were conducted. $^{*}P < 0.05$, $^{**}P < 0.01$, $^{***}P < 0.001$, $^{****}P < 0.0001$. (Image reused from Barone et al. [23] under a Creative Commons License)

efficiency and impact on peroxisome marker levels used above for IF (PMP70, Catalase).

2. Day 2: Begin a 72-h depletion of ATG7. Treat cells with 40 nM of non-targeting (NT1) or ATG7 siRNA oligonucleotides, using Lipofectamine RNAi-MAX (Invitrogen) according to the manufacturer's instructions.

3. Day 3: Replate cells into imaging dishes ($2 \times 10^5$). For live cell imaging, we typically use IBIDI μ-Dish (81156; IBIDI). This allows cells to recover from trypsinization, minimizing potential variation due to the trypsinization step.
4. Day 5: Process cells for imaging and in parallel lyse cells for Western blot analysis.

### 3.3 Quantitation of Pexophagy

1. Various parameters are of interest, the number and size of pexolysosomes per cell, as well as the number of acidic pexolysosomes versus total peroxisomes, which we term the pexophagy index. One may effectively combine on-microscope observation and manual counting of pexolysosomes in fixed pexo-QC expressing cells, but ideally, one seeks to remove any subjectivity and have an automated analysis workflow [24]. Different microscopes may come with relevant proprietary software, often at a significant cost. Here we highlight a free-to-use analysis software developed by Ganley and colleagues, originally for quantification of mitophagy, which we have adapted for pexophagy. mito-QC counter is offered as a macro within the popular FIJI image analysis suite [25, 26].
2. For live cell imaging, we routinely use a 3i Marianas spinning disk microscope, with 40× or 63× oil objective, NA1.4 equipped with a heated stage enclosed within a controlled atmosphere chamber (5% $CO_2$) and an Evolve EMCCD camera.
3. For pexo-Keima, excitation wavelengths of 445 and 561 nM are used sequentially both in combination with a 617 nM bandpass filter. We denote these "green" and "red/magenta" respectively, which reflects a pseudo-coloring applied for visualization purposes.
4. Pexo-QC reporters are acquired sequentially with a 488 nM laser for EGFP (525 nm emission filter) and a 561 nM laser for mCherry (617 nM emission filter).

### 3.4 Processing Images

1. Open images using publicly available FIJI software (e.g., version 2.1) with mito-QC counter installed (available from https://github.com/graemeball/mQC_counter).
2. Make sure to import them using the Bio-Formats import options. Choose settings for batch analysis using the single image mode. First, one needs to define the settings for batch analysis. Open a picture with high pexophagy, e.g., after treatment with MLN4924 [23].
3. Adjust the "green" channel to visualize cell boundaries (Image - > Adjust - > Brightness/Contrast).
4. Using the "*Freehand Selection*" tool, manually define cell or tissue regions for quantification as regions of interest (ROIs).

This can be achieved by using the "*ROI manager*" function, from either single Z-plane (for flat cells) or maximal intensity projection images (*see* **Note 4**). Press "t" after each cell has been delineated.

5. OPTIONAL: Each time a ROI is drawn for an individual cell, press "CTRL + B" to create an "*image overlay*" and save these images in a new folder. This crucial step is needed for batch processing analysis.
6. Reset the Brightness/Contrast (B&C window).
7. Save as a .tiff image ("CTRL" + "S") in a new folder, preferably one folder for each experimental condition.
8. Repeat this process for all the images per condition.
9. Start the macro.
10. Channel Assignment and Threshold selection for mCherry/EGFP intensity Ratio analysis.
11. Upon starting "mitoQC counter" for pexo-QC analysis, assign GFP and mCherry (or Keima445 and Keima561 in the case of Keima-based reporter) channels to their corresponding numbers in the source image (e.g., Channel 1 or Channel 2).
12. Select the "Batch Box" if you want to perform batch analysis from previously saved "*overlay images*," streamlining analysis efficiency.
13. Define the mCherry/EGFP intensity ratio threshold, commencing with the default settings. In case the outcomes prove unsatisfactory with the default threshold parameters, iterate this process and adjust the thresholds to eliminate false negatives and false positives in the analysis.

### 3.5 Automated File Generation for Subsequent Analysis

The plugin automatically generates .csv files, facilitating subsequent analysis, whether applied to individual images or processed in batch. In addition, the plugin will automatically generate and save the mCherry/EGFP ratio image, overlay image with ROIs with detected pexolysosomes, and a "pexophagy mask" image which allows the visualization of pexophagy.

### 3.6 Data Representation

1. The number of pexolysosomes per cell or pexophagy index (number of pexolysosomes (red)/pexolysosomes (red) + peroxisomes (red = green)) are often plotted.

A minimum of three biological repeats is recommended with >30 cells per condition in each individual experiment. Plot the spread of values for each point as shown in the example (Fig. 1).

## 4 Notes

1. It is important to dissociate selective autophagy from simple bulk or bystander autophagy [27]. Earlier studies have used oxidative stress to induce pexophagy, but this induces widespread autophagy [28]. There are often multiple pathways that may elicit selective removal of an organelle. For example, loss of mitochondrial membrane potential leads to the removal of mitochondria via the PINK1/Parkin pathway [29]. However, this pathway does not constitute the major mitophagy pathway under basal conditions, which is likely mediated by two mitophagy adaptors, BNIP3 and BNIP3L (otherwise known as NIX) [13, 30–33]. BNIP3 and BNIP3L are transcriptionally regulated by HIF1α and thus can be elevated by lowering the oxygen tension. This can be effectively mimicked by iron chelation, which inhibits prolyl hydroxylase or by MLN4924, a specific UBA7 inhibitor of neddylation. MLN4924 inhibits $CRL^{VHL}$ which controls HIF1α stability and $CRL^{FBXL4}$ which directly controls BNIP3/BNIP3L stability [33–35]. Interestingly, BNIP3 and BNIP3L can also directly insert into peroxisomal membranes to induce pexophagy [21, 23]. This offers a mechanism to coordinate the removal of these two organelles, which are known to cooperate in various aspects of metabolism. There is a significant overlap in the membrane composition of these two organelles [36]. They share a transmembrane-anchored deubiquitylase, USP30, which regulates both pexophagy and mitophagy [6, 37–39]. At mitochondria, specific USP30 inhibition leads to elevation of PINK1-dependent pathways leading to mitophagy [6, 39, 40]. We have also shown that USP30 inhibitors such as compound 39 (CMPD-39) also induce pexophagy in a PINK1-independent manner. This pathway is distinct from the BNIP3/BNIP3L pathway and requires the ubiquitin-binding selective autophagy adaptor NBR1 [18, 23, 41]. As USP30 is a high-value target in multiple disease settings (e.g., Parkinson's Disease), there are multiple small molecule inhibitors that have been developed belonging to the cyanopyrrolidine or benzosulphonamide classes [39, 40, 42–44].

   We recently surveyed the magnitude of pexophagy induction in pexo-Keima expressing RPE1 cells following the application of a number of chemicals for 24 or 96 h (CMPD-39), summarized in Table 1 and Fig. 1. Amongst these, the high specificity of CMPD-39 towards a nonessential protein (USP30) is very well tolerated for long-term treatments. DFP gives a response beyond that achieved with MLN4924, indicating that other effects beyond the elevation of HIF1α may be contributing.

2. We expect interest in pexophagy research to grow as its role in metabolism and disease is now clearly established [8, 9]. The corresponding mitochondrial fluorescent reporters have been used to great effect in FACS-based sorting for CRISPR/Cas9 screens [33, 35]. The reporters described above offer the same opportunity for unbiased discovery of pexophagy regulators. The pexophagy probes have also recently been introduced into whole organism models, *Caenorhabditis elegans* and *Drosophila melanogaster*, opening up the field for analysis of candidate genes linked to disease [14].
3. Catalase requires a fully functioning peroxisomal import pathway so it is not a good marker for immature peroxisomes, but will overlap extensively with pexo-Keima and pexo-QC, whereas PMP70 will also detect immature peroxisomes which cannot yet import the probe.
4. The mito-QC counter plug-in deals with a single image. Many tissue culture cells are relatively flat, e.g., RPE1, U2OS cells. In these cases, we choose a single focal plane representing the largest cross-sectional area of the nucleus. For thicker samples, we use a maximal image projection, which provides the highest data value encountered for each pixel of volumetric data.

## Acknowledgments

FB has received PhD funding from the Wellcome Trust (102172/Z/13Z). MC is a Royal Society Industry Fellow INF\R2\212031.

### References

1. Islinger M, Grille S, Fahimi HD, Schrader M (2012) The peroxisome: an update on mysteries. Histochem Cell Biol 137(5):547–574. https://doi.org/10.1007/s00418-012-0941-4
2. Wanders RJA et al (2023) The physiological functions of human peroxisomes. Physiol Rev 103(1):957–1024. https://doi.org/10.1152/physrev.00051.2021
3. Tang D, Kroemer G (2020) Peroxisome: the new player in ferroptosis. Signal Transduct Target Ther 5(1):273. https://doi.org/10.1038/s41392-020-00404-3
4. Zou Y et al (2020) Plasticity of ether lipids promotes ferroptosis susceptibility and evasion. Nature 585(7826):603–608. https://doi.org/10.1038/s41586-020-2732-8
5. Jo DS et al (2020) Loss of HSPA9 induces peroxisomal degradation by increasing pexophagy. Autophagy 16(11):1989–2003. https://doi.org/10.1080/15548627.2020.1712812
6. Marcassa E et al (2018) Dual role of USP30 in controlling basal pexophagy and mitophagy. EMBO Rep 19:e45595. https://doi.org/10.15252/embr.201745595
7. Riccio V et al (2019) Deubiquitinating enzyme USP30 maintains basal peroxisome abundance by regulating pexophagy. J Cell Biol 218(3):798–807. https://doi.org/10.1083/jcb.201804172
8. Waterham HR, Ferdinandusse S, Wanders RJ (2016) Human disorders of peroxisome metabolism and biogenesis. Biochim Biophys Acta 1863(5):922–933. https://doi.org/10.1016/j.bbamcr.2015.11.015
9. Nazarko TY (2017) Pexophagy is responsible for 65% of cases of peroxisome biogenesis disorders. Autophagy 13(5):991–994. https://doi.org/10.1080/15548627.2017.1291480

10. McWilliams TG et al (2016) Mito-QC illuminates mitophagy and mitochondrial architecture in vivo. J Cell Biol 214(3):333–345. https://doi.org/10.1083/jcb.201603039
11. Katayama H et al (2011) A sensitive and quantitative technique for detecting autophagic events based on lysosomal delivery. Chem Biol 18(8):1042–1052. https://doi.org/10.1016/j.chembiol.2011.05.013
12. Sun N et al (2015) Measuring in vivo mitophagy. Mol Cell 60(4):685–696. https://doi.org/10.1016/j.molcel.2015.10.009
13. Lee JJ et al (2018) Basal mitophagy is widespread in Drosophila but minimally affected by loss of Pink1 or parkin. J Cell Biol 217(5):1613–1622. https://doi.org/10.1083/jcb.201801044
14. Barone FG et al. (2023) Whole organism and tissue specific analysis of pexophagy in Drosophila. bioRxiv:2023.2011.2017.567516. https://doi.org/10.1101/2023.11.17.567516
15. Skowyra ML, Feng P, Rapoport TA (2023) Towards solving the mystery of peroxisomal matrix protein import. Trends Cell Biol. https://doi.org/10.1016/j.tcb.2023.08.005
16. Kimura S, Noda T, Yoshimori T (2007) Dissection of the autophagosome maturation process by a novel reporter protein, tandem fluorescent-tagged LC3. Autophagy 3(5):452–460. https://doi.org/10.4161/auto.4451
17. Koch A et al (2005) A role for Fis1 in both mitochondrial and peroxisomal fission in mammalian cells. Mol Biol Cell 16(11):5077–5086. https://doi.org/10.1091/mbc.e05-02-0159
18. Deosaran E et al (2013) NBR1 acts as an autophagy receptor for peroxisomes. J Cell Sci 126(Pt 4):939–952. https://doi.org/10.1242/jcs.114819
19. Nazarko TY et al (2014) Peroxisomal Atg37 binds Atg30 or palmitoyl-CoA to regulate phagophore formation during pexophagy. J Cell Biol 204(4):541–557. https://doi.org/10.1083/jcb.201307050
20. Dolese DA et al (2022) Degradative tubular lysosomes link pexophagy to starvation and early aging in C. elegans. Autophagy 18(7):1522–1533. https://doi.org/10.1080/15548627.2021.1990647
21. Wilhelm LP et al (2022) BNIP3L/NIX regulates both mitophagy and pexophagy. EMBO J:e111115. https://doi.org/10.15252/embj.2022111115
22. Liang JR, Lingeman E, Ahmed S, Corn JE (2018) Atlastins remodel the endoplasmic reticulum for selective autophagy. J Cell Biol 217(10):3354–3367. https://doi.org/10.1083/jcb.201804185
23. Barone FG, Urbe S, Clague MJ (2023) Segregation of pathways leading to pexophagy. Life Sci Alliance 6(5). https://doi.org/10.26508/lsa.202201825
24. Lee RM et al (2024) Believing is seeing—the deceptive influence of bias in quantitative microscopy. J Cell Sci 137(1). https://doi.org/10.1242/jcs.261567
25. Schindelin J et al (2012) Fiji: an open-source platform for biological-image analysis. Nat Methods 9(7):676–682. https://doi.org/10.1038/nmeth.2019
26. Montava-Garriga L, Singh F, Ball G, Ganley IG (2020) Semi-automated quantitation of mitophagy in cells and tissues. Mech Ageing Dev 185:111196. https://doi.org/10.1016/j.mad.2019.111196
27. An H, Harper JW (2018) Systematic analysis of ribophagy in human cells reveals bystander flux during selective autophagy. Nat Cell Biol 20(2):135–143. https://doi.org/10.1038/s41556-017-0007-x
28. Li H et al (2021) The peroxisome-autophagy redox connection: a double-edged sword? Front Cell Dev Biol 9:814047. https://doi.org/10.3389/fcell.2021.814047
29. Narendra D, Tanaka A, Suen DF, Youle RJ (2008) Parkin is recruited selectively to impaired mitochondria and promotes their autophagy. J Cell Biol 183(5):795–803. https://doi.org/10.1083/jcb.200809125
30. Novak I et al (2010) Nix is a selective autophagy receptor for mitochondrial clearance. EMBO Rep 11(1):45–51. https://doi.org/10.1038/embor.2009.256
31. Ney PA (2015) Mitochondrial autophagy: origins, significance, and role of BNIP3 and NIX. Biochim Biophys Acta 1853(10 Pt B):2775–2783. https://doi.org/10.1016/j.bbamcr.2015.02.022
32. McWilliams TG et al (2018) Basal mitophagy occurs independently of PINK1 in mouse tissues of high metabolic demand. Cell Metab 27(2):439–449. e435. https://doi.org/10.1016/j.cmet.2017.12.008
33. Elcocks H et al (2023) FBXL4 ubiquitin ligase deficiency promotes mitophagy by elevating NIX levels. EMBO J:e112799. https://doi.org/10.15252/embj.2022112799
34. Nguyen-Dien GT et al (2022) FBXL4 suppresses mitophagy by restricting the accumulation of NIX and BNIP3 mitophagy receptors. bioRxiv:2022.2010.2012.511867 https://doi.org/10.1101/2022.10.12.511867

35. Cao Y et al (2023) A mitochondrial SCF-FBXL4 ubiquitin E3 ligase complex degrades BNIP3 and NIX to restrain mitophagy and prevent mitochondrial disease. EMBO J:e113033. https://doi.org/10.15252/embj.2022113033
36. Costello JL, Passmore JB, Islinger M, Schrader M (2018) Multi-localized proteins: the peroxisome-mitochondria connection. Subcell Biochem 89:383–415. https://doi.org/10.1007/978-981-13-2233-4_17
37. Liang JR et al (2015) USP30 deubiquitylates mitochondrial Parkin substrates and restricts apoptotic cell death. EMBO Rep 16(5):618–627. https://doi.org/10.15252/embr.201439820
38. Bingol B, Sheng M (2016) Mechanisms of mitophagy: PINK1, parkin, USP30 and beyond. Free Radic Biol Med 100:210–222. https://doi.org/10.1016/j.freeradbiomed.2016.04.015
39. Rusilowicz-Jones EV et al (2022) Benchmarking a highly selective USP30 inhibitor for enhancement of mitophagy and pexophagy. Life Sci Alliance 5(2). https://doi.org/10.26508/lsa.202101287
40. Rusilowicz-Jones EV et al (2020) USP30 sets a trigger threshold for PINK1-PARKIN amplification of mitochondrial ubiquitylation. Life Sci Alliance 3(8). https://doi.org/10.26508/lsa.202000768
41. Rasmussen NL, Kournoutis A, Lamark T, Johansen T (2022) NBR1: the archetypal selective autophagy receptor. J Cell Biol 221(11). https://doi.org/10.1083/jcb.202208092
42. Kluge AF et al (2018) Novel highly selective inhibitors of ubiquitin specific protease 30 (USP30) accelerate mitophagy. Bioorg Med Chem Lett 28(15):2655–2659. https://doi.org/10.1016/j.bmcl.2018.05.013
43. Phu L et al (2020) Dynamic regulation of mitochondrial import by the ubiquitin system. Mol Cell 77(5):1107–1123 e1110. https://doi.org/10.1016/j.molcel.2020.02.012
44. Fang TZ et al (2023) Knockout or inhibition of USP30 protects dopaminergic neurons in a Parkinson's disease mouse model. Nat Commun 14(1):7295. https://doi.org/10.1038/s41467-023-42876-1

# Chapter 12

# Methods for Monitoring Mitophagy Using mt-Keima

**Alexandra J. Gilsrud and Derek P. Narendra**


## Abstract

Mitochondria-targeted Keima (mt-Keima) is a pH-sensitive, acid-stable fluorescent protein used for the quantification of mitophagy. Mt-Keima contains a mitochondrial matrix targeting sequence and has bimodal excitation with peaks at 440 nM in neutral environments and 586 nM in acidic environments. From this bimodal excitation, a ratiometric signal may be calculated to quantify mitophagy in live cells. This chapter describes procedures for measuring mitophagy by flow cytometry and live cell confocal microscopy with mt-Keima.


**Key words** Mitophagy, Mitochondria, Selective autophagy, Parkin, PINK1

## 1 Introduction

Mitophagy is the selective autophagic degradation of mitochondria (reviewed in [1]). It regulates mitochondrial number and quality by capturing mitochondria in the cytosol and delivering them to the acidic lumen of a lysosome for degradation by hydrolases. Loss of mitophagy has been implicated in a number of neurodegenerative disorders, including Parkinson's Disease and Alzheimer's Disease. Indeed, mutations in the E3-ubiquitin ligase Parkin and the kinase PINK1, which form a quality control mitophagy pathway, are the leading cause of recessive Parkinson's disease.

To better study mitophagy pathways and subsequent dysfunction, many fluorescent protein-based strategies have been developed, most of which utilize the differential pH of the mitochondrial matrix (when free in the cytosol) and the lysosomal lumen. Among these tools is the fluorescent mitophagy reporter mitochondria-targeted Keima (mt-Keima), in which the coral-derived fluorescent protein mKeima is fused at its N-terminus to the mitochondrial matrix targeting signal from human COX8A (1–25) [2–5].

mKeima has distinct excitation peaks in its neutral form (with a maximum of 440 nM) and its ionized form (with a maximum of

Ioannis P. Nezis (ed.), *Selective Autophagy: Methods and Protocols*, Methods in Molecular Biology, vol. 2845, https://doi.org/10.1007/978-1-0716-4067-8_12,

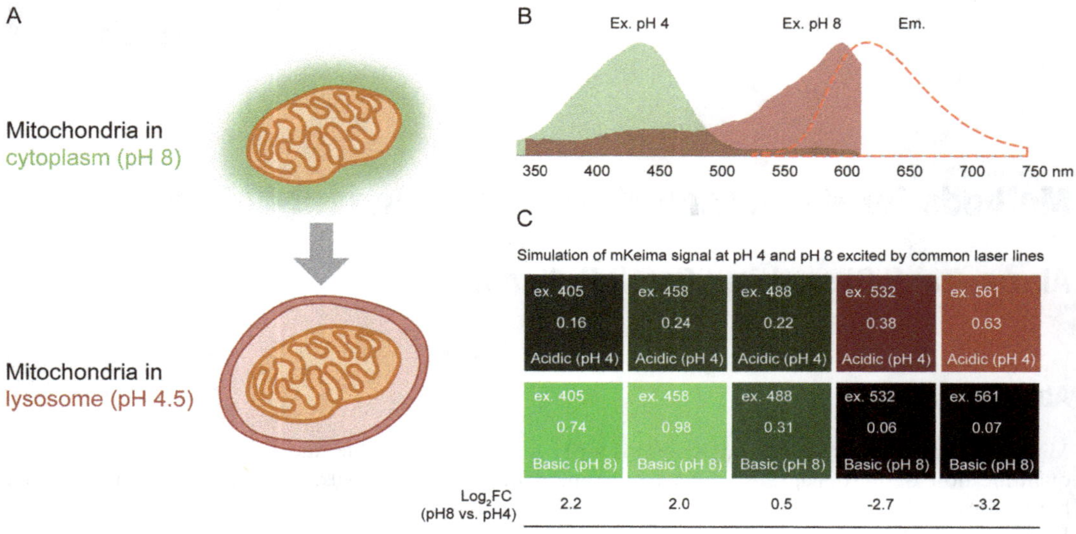


**Fig. 1** Localization of mt-Keima to the matrix of free mitochondria vs. mito-lysosomes is reflected in the ratio of fluorescence following excitation with blue vs. yellow light. (**a**) Cartoon represents the differential excitation spectrum of mt-Keima in the alkaline environment of the mitochondrial matrix (pH 8) vs. the acidic environment of a mito-lysosome (pH 4.5). (**b**) Spectra for mKeima excited at pH 4 and pH 8, generated from spectral data in FPbase [4, 7]. (**c**) Simulation of mKeima fluorescence at pH 4 and pH 8, using common laser lines. The combinations of ex. 458 and ex. 561 and ex. 405 and ex. 561 give the highest fold change in fluorescence between mKeima at pH 4 and pH 8. Fluorescent signal is more intense with ex. 458 compared to ex. 405. Images were generated using FIJI/Image lookup tables, green (for blue lasers) and red (for yellow lasers), based on spectral values in (**b**)

590 nM), both of which have an emission peak at 620 nM [4] (Fig. 1a, b). These different excitation spectra are caused by a reversible *trans*-to-*cis* conformational change in its chromophore. The chromophore is stabilized in *cis* by a hydrogen bond with Ser142 at low pH. At high pH, Ser142 stabilizes deprotonated Asp157 (p*K*a of ~5.7) instead, freeing the chromophore to adopt the *trans* conformation. Additionally, as with many coral fluorescent proteins, mKeima is acid-stable and resistant to lysosomal hydrolases, giving it a long half-life within the lysosome.

The ratio of mt-Keima fluorescence sequentially excited with blue light and yellow light distinguishes mt-Keima in free mitochondria (with a pH of 8) from mitochondria in the lysosome (also called mito-lysosomes) (with a pH of 4.5–5) [3]. Both the matrix targeting of mKeima and the high ratiometric change in fluorescence contribute to mt-Keima's high sensitivity, but its reversibility limits its use to live cell applications [6]. Practically, differential excitation of the two mKeima populations can be achieved with many commonly used laser lines, including 405 nM, 458 nM, or 488 nM to excite neutral mKeima and 531 nM or 561 nM to excite ionized mKeima (Fig. 1c). Of these, the combination of 405 nM or 458 nM with 561 nM theoretically yields the greatest ratiometric difference between the two mKeima populations, with the greatest quantal yields for 458 nM and 561 nM (calculated from data deposited in FPbase [7]). Because mKeima emits at 620 nM, it can be used in combination with fluorescent proteins or other fluorophores that emit green, yellow, or near-infrared light, including BFP, GFP, and YFP.

Here, we demonstrate the use of mt-Keima in flow cytometry and confocal microscopy to quantify mitophagy in live mammalian cells, using Parkin-induced mitophagy as an example. In this example, we demonstrate how mt-Keima distinguishes the activity of the wildtype Parkin from a disease-causing variant, p.C431F, harboring a mutation in its active site.

## 2 Materials

Flow Cytometry

1. Cell line expressing mt-Keima. Stable cell line generated using pHAGE-mt-mKeima plasmid (addgene #131626).
2. 24-well cell culture plate.
3. 96-well u-bottom plate.
4. 10 mM Carbonyl cyanide 3-chlorophenylhydrazone (CCCP), catalog #C2759, prepared in DMSO.
5. Pure Dimethyl Sulfoxide (DMSO).
6. Trypsin-EDTA 0.05%.
7. Cell culture media: 89% Dulbecco's Modified Eagle Medium (DMEM), 10% cell culture-grade fetal bovine serum (FBS), 1% penicillin–streptomycin–neomycin, filter sterilized.
8. Fugene HD, catalog #E2311.
9. Opti-MEM medium.
10. Plasmid containing the protein of interest.
11. Centrifuge with 96-well plate rotor attachment.
12. Cell sorting media, in DI $H_2O$: 145 mM NaCl, 5 mM KCl, 1.8 mM $CaCl_2$, 0.8 mM $MgCl_2$, 10 mM Hepes, 10 mM glucose. Check and adjust pH to 7.3 and add 1 mg/mL immunoglobulin-free albumin, filter sterilized.
13. Flow cytometer (Luminex Cellstream is used in the example).
14. Software for analysis (i.e., Python).
15. Micropipettes 1 mL, 200 μL, 10 μL, 2 μL.
16. Multichannel micropipette 200 μL.
17. 5 mL serological pipette.

Microscopy

1. Cell line expressing mt-Keima.
2. Ibidi 8-well slides.
3. Cell culture media.
4. 10 mM CCCP stock prepared in DMSO.
5. Pure DMSO.
6. Fugene 6 HD, catalog# E2691.
7. Opti-MEM medium.
8. Plasmid containing the gene of interest.
9. Confocal microscope with 445 nM (or 405 nM or 488 nM) and 561 nM (or other orange) laser lines (Olympus FV3000 is used in the example).
10. Microscope incubation chamber.
11. 60× oil immersion lens.
12. Immersion oil.
13. Software for analysis (e.g., FIJI/ImageJ).
14. Micropipettes: 1 mL, 200 μL, 10 μL, 2 μL.
15. Kim Wipes.

## 3 Methods

### 3.1 Easuring Mitophagy with mt-Keima by Flow Cytometry

Day 1

1. Warm cell culture media in a 37 °C bead bath for 1 h.
2. Thaw or split mt-Keima-expressing cells following standard cell culture procedures.
3. In a 24-well cell culture plate, seed 50,000 cells per well in 500 μL of cell culture media. For triplicate, seed seven times as many wells as desired conditions, three for CCCP and three for the vehicle. The additional control wells will be left untransfected and will be used to set the YFP gate on the flow cytometer.
4. Incubate HeLa cells overnight at 37 °C.

Day 2

1. Bring Fugene HD and Opti-MEM to room temperature (RT).
2. For each well, combine 7.5 μL Opti-MEM, 0.75 μL Fugene HD, and 125 ng of DNA (in 1 μL) in a microcentrifuge tube.
3. Vortex DNA:Fugene HD mixture.
4. Incubate at RT for 20 min.

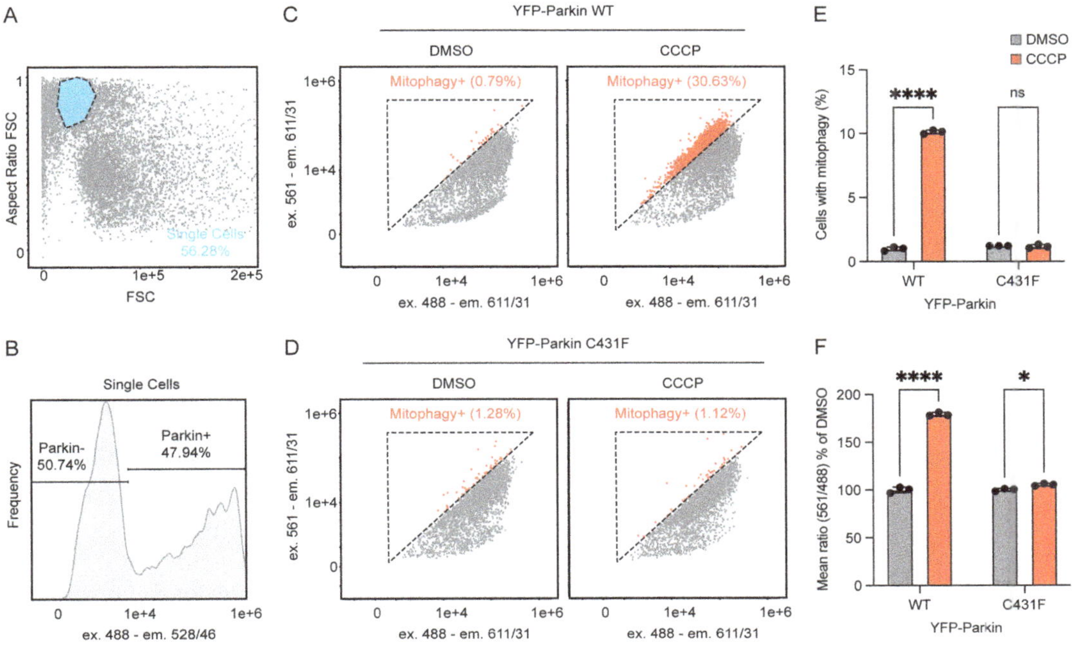


**Fig. 2** Measuring mitophagy with mt-Keima by flow cytometry. (**a**) Scatterplot depicting the gate for single cells. (**b**) Scatterplot depicting the gate for YFP-Parkin cells in HeLa cells stably expressing mt-Keima and transiently transfected with YFP-Parkin. (**c**, **d**) Cells with mitophagy have a higher whole-cell ratio of ex. 488 nM (x-axis) vs. ex. 561 nM (y-axis) (with emission measured by a bandpass filter centered on 611 nM for both), as reflected by the percentage of events in the triangular gates (dotted lines). Activation of Parkin-dependent mitophagy with the mitochondrial uncoupler CCCP results in a greater percentage of cells with mitophagy in cells expressing wildtype YFP-Parkin. By contrast, cells expressing the pathogenic mutant of YFP-Parkin p.C431F do not show a substantial increase in mitophagy with CCCP treatment. (**e**) Data from (**c**, **d**) can be represented as the number of cells in the "mitophagy+" gate (dotted lines in **c**, **d**). (**f**) Alternatively, data from (**c, d**) can be represented as the ratio of fluorescence emitting at 611 nM following excitation at 561 nM vs. 488 nM. Data have been scaled as a percent of the vehicle (DMSO) treated condition

5. Add 9.25 μL to each well dropwise.
6. Incubate cells overnight at 37 °C.

Day 3

1. Warm cell culture media and Trypsin-EDTA in a 37 °C bead bath for 1 h.
2. Initialize and calibrate the flow cytometer.
3. Dilute 10 mM CCCP 1:1000 in cell culture media for the treatment condition; dilute DMSO 1:1000 in cell culture media for the control condition.
4. Aspirate old cell culture media from each well and immediately add media containing CCCP or DMSO to the treatment and control wells, respectively.

5. Incubate cells with CCCP for 3 h at 37 °C.
6. Aspirate the media.
7. Rinse each well with 500 μL of PBS and aspirate.
8. Add 150 μL of Trypsin-EDTA 0.05% to each well and incubate at 37 °C for 3—5 min until cells lift off the plate.
9. Add 150 μL of cell culture media to deactivate Trypsin (by divalent ions $Ca^{2+}$ and $Mg^{2+}$ in media).
10. Mix cells by pipetting up and down 3–5 times to achieve a single-cell suspension. (Note: filtering through a cell strainer may be necessary to prevent instrument clogging if using high-speed stream-in-air flow cytometers. We have not found this to be necessary for benchtop microcapillary flow cytometers.)
11. Transfer 300 μL of the cell mixture to the corresponding well in a 96-well u-bottom plate.
12. Pellet cells using a swinging-bucket centrifuge spun at 400 g for 3 min at RT. (This step can also be performed at 4 °C, and samples can be kept on ice for the remainder of the protocol if needed to slow mitophagic flux.)
13. Aspirate (or tap out) excess media, being very careful not to disturb the cell pellet.
14. Using a multichannel pipette, resuspend cell pellets in 50 μL of Cell Sorting Media.
15. Turn on the 488 nM and 561 nM lasers. (Note: 405 nM or 458 nM may be used in place of 488 nM if available; 532 nM may be used in place of 561 nM if 561 nM is not available.)
16. Gate for single-cell events by plotting the aspect ratio of FSC-Height vs. FSC-Area on a standard flow cytometer. In the example, we use FSC-Height vs. the FSC aspect ratio, an imaging parameter that is available with the CellStream flow cytometer (Fig. 2a).
17. Using a sample expressing all fluorescent proteins, adjust laser power (or photomultiplier tube (PMT) voltage) for each fluorescent protein so that the fluorescent intensity of each channel is approximately in the middle of the dynamic range. In our example, this includes ex. 488/em. 528/46 (for GFP), ex. 488/em. 611/31 (for neutral mKeima), and ex. 561/em. 611/31 (for ionized mKeima).
18. In this example, the cells have been transiently transfected with YFP-Parkin to compare the activities of WT and mutant Parkin. To establish gates for cells specifically expressing YFP, load the control sample that was not transfected with YFP. Using the 488 nM laser and the appropriate YFP filter (528/46 bandpass filter, in our example), create a gate that excludes the background autofluorescence in the control sample (Fig. 2b).

19. Switching to a vehicle-treated sample (DMSO in this case) transiently transfected with YFP-Parkin, next, establish a gate for mitophagy-positive cells. First, create a dot plot (of single-cell and YFP-positive gated events) for basic mKeima vs. acidic mKeima fluorescence. In this example, ex. 488/em. 611/31 (x-axis) vs. ex. 561/em. 611/31 (y-axis) is plotted (Fig. 2c, d). In untreated cells with low basal mitophagy, events should fall along a diagonal in the dot plot. Next, draw a right triangular gate so that its hypotenuse is parallel to this diagonal line. We position the triangle close enough so that a few events (<5%) are positive in the control condition.
20. Set the flow cytometer to record at least 5000 events per condition. In this example, we specify these events to be single cells and YFP+.
21. Acquire samples for all conditions.
22. An increase in mitophagy can be represented by the percent of mitophagy-positive cells within the triangle gate established above (Fig. 2e). Alternatively, it can be represented as the mean of the ratio of mKeima (ionized) to mKeima (neutral) for events in each sample (Fig. 2f). This can be calculated and represented in real-time on some flow cytometers. In our example, we exported the data for ex. 488/em. 611/31 (x-axis) vs. ex. 561/em. 611/31 (y-axis) fluorescence of each event and calculated offline using a Python script, but this can also be calculated in Excel.

### 3.2 Measuring Mitophagy with mt-Keima by Confocal Microscopy

Day 1

1. Warm cell culture media in a 37 °C bead bath for one hour.
2. Thaw or split cells stably expressing mt-Keima using standard cell culture techniques.
3. Seed 8-well ibidi slides (or similar covered chamber glass slides for live cell imaging) with 25,000 cells per well in 250 μL cell culture media. Seed six wells per condition, and seed one additional well for setting microscope parameters.

Day 2

1. Bring Fugene HD and Opti-MEM to RT (RT).
2. For each well, combine 3.75 μL opti-MEM, 0.38 μL Fugene HD, and 62.5 ng DNA (in 1 μL) in a microcentrifuge tube.
3. Vortex the Fugene HD mixture.
4. Incubate at RT for 20 min.
5. Add 5.1 μL to each well dropwise.
6. Incubate cells overnight at 37 °C.

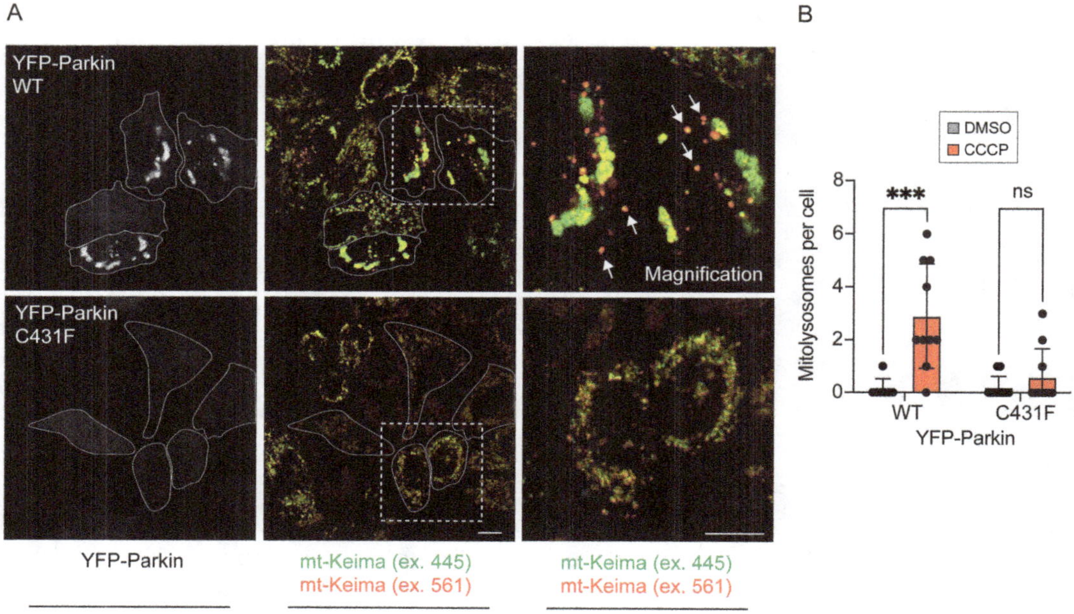


**Fig. 3** Measuring mitophagy with mt-Keima by confocal microscopy. (**a**) Confocal images of HeLa cells expressing mt-Keima and transiently transfected with wildtype (top) or mutant (p.C431F, bottom) YFP-Parkin. Images in left panels show YFP-Parkin. Images in middle and right panels show mt-Keima acidic (red) and neutral (green) channels merged. Red dots in the merge are mito-lysosomes, whereas mitochondria free in the cytosol appear yellow. (**b**) Graph depicts mitolysosomes per cell counted from images, such as that represented in (**a**). Similar to measurements by flow cytometry (in Fig. 2), measurements by microscopy show induction of mitophagy by wildtype but not mutant YFP-Parkin following depolarization of mitochondria with CCCP

Day 3

1. Warm cell culture media and Trypsin-EDTA in a 37 °C bead bath for one hour.

1. Heat microscope incubation chamber to 37 °C.
2. Dilute 10 mM CCCP (1:1000) in cell culture media and dilute DMSO (1:1000) in cell culture media.
3. Aspirate media and immediately add media containing CCCP or DMSO gently to cells.
4. Place cells back in the $CO_2$ incubator for 3 h at 37 °C.
5. After incubation, place ibidi slide on microscope. In this example, we used a 60× oil objective.
6. Set microscope to acquire images in two stages: first exciting with the 561 nM laser and recording 570–670 nM emission, and then exciting with the 445 nM laser and recording 570-670 nM emission.

7. Using the additional well, adjust laser power and PMT voltage for each channel, maximizing signal intensity without saturating.
8. When adjusting laser power or acquiring images, avoid imaging the same field more than once. This will lead to photoconversion (*see* **Notes 1**). When adjusting focus or previewing an image, only use the 561 nM laser. When searching or adjusting focus in widefield, likewise use a filter set for red fluorescence like TRITC or Texas Red to minimize blue light exposure.
9. Acquire images for all conditions.
10. Images can be analyzed by counting the number of acidic puncta, puncta with high 561/445 Keima signal, in the cells (Fig. 3a, b). These are mito-lysosomes, mt-Keima inside of the lysosomal lumen.

## 4 Notes

1. When imaging cells expressing mt-Keima, we have found that the 561 nM laser should be used before the 445 nM laser, as the blue laser can photo-convert mKeima at a neutral pH, shifting its excitation maximum toward 570 nM (Fig. 4a, b). This reduces the 561/458 ratio of neutral vs. ionized mKeima and decreases the signal-to-noise for discriminating mito-lysosomes from mitochondria (Fig. 4b, c). We hypothesize that blue light switches the chromophore of mKeima from *trans* to *cis*, similar to the *trans*-to-*cis* switch reported for other widely used photoactivatable and photoswitchable fluorescent proteins [8]. For this reason, caution should be used if imaging a single field multiple times. We also recommend searching for fields of interest and focusing using yellow light to minimize exposure to blue light before imaging. The photoconversion of mKeima may be stronger with line scanning confocal microscopy than with imaging methods that use less intense light, such as widefield, spinning disc, or resonance scanning.

## Acknowledgments

This work was supported in part by the Intramural Research Program of the NINDS, National Institutes of Health: project number 1ZIA-NS003169.

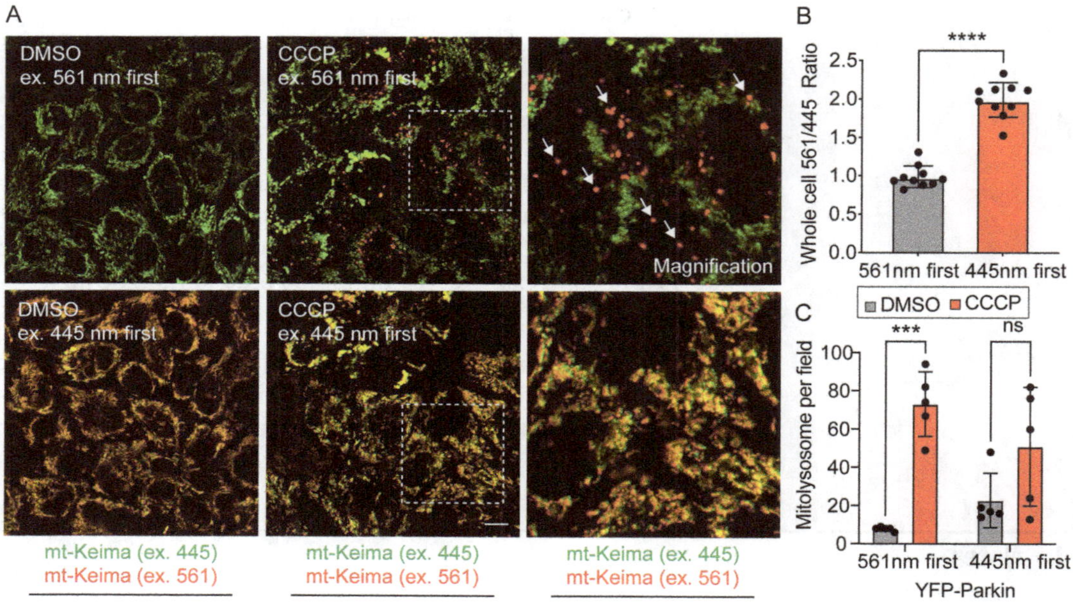


**Fig. 4** Imaging acidic mt-Keima first avoids photoswitching of mKeima with blue light. (**a**) Representative images showing that the order of sequential acquisition of images with the 445 nM or the 561 nM lasers yields different results, likely due to photoswitching of mKeima on exposure to blue light. (**b**) Graph demonstrates that the whole cell 561/458 ratio changes depending on the order of channel acquisition. (**c**) Graph demonstrates how incorrect order of channel acquisition reduces sensitivity for detecting mitolysosomes, due to a decrease in the ratiometric difference in signal between mito-lysosomes and mitochondria in cytoplasm

## References


1. Narendra DP (2021) Managing risky assets—mitophagy in vivo. J Cell Sci 134:jcs240465. https://doi.org/10.1242/jcs.240465
2. Kogure T, Karasawa S, Araki T, Saito K, Kinjo M, Miyawaki A (2006) A fluorescent variant of a protein from the stony coral Montipora facilitates dual-color single-laser fluorescence cross-correlation spectroscopy. Nat Biotechnol 24:577–581. https://doi.org/10.1038/nbt1207
3. Katayama H, Kogure T, Mizushima N, Yoshimori T, Miyawaki A (2011) A sensitive and quantitative technique for detecting autophagic events based on lysosomal delivery. Chem Biol 18:1042–1052. https://doi.org/10.1016/j.chembiol.2011.05.013
4. Violot S, Carpentier P, Blanchoin L, Bourgeois D (2009) Reverse pH-dependence of chromophore protonation explains the large stokes shift of the red fluorescent protein mKeima. J Am Chem Soc 131:10356–10357. https://doi.org/10.1021/ja903695n
5. Henderson JN, Osborn MF, Koon N, Gepshtein R, Huppert D, Remington SJ (2009) Excited state proton transfer in the red fluorescent protein mKeima. J Am Chem Soc 131:13212–13213. https://doi.org/10.1021/ja904665x
6. Liu Y-T, Sliter DA, Shammas MK, Huang X, Wang C, Calvelli H, Maric S, Narendra D (2021) Mt-Keima detects PINK1-PRKN mitophagy in vivo with greater sensitivity than mito-QC. Autophagy:1–10. https://doi.org/10.1080/15548627.2021.1896924
7. Lambert TJ (2019) FPbase: a community-editable fluorescent protein database. Nat Methods 16:277–278
8. Groenhof G, Bouxin-Cademartory M, Hess B, De Visser SP, Berendsen HJC, Olivucci M, Mark AE, Robb MA (2004) Photoactivation of the photoactive yellow protein: why photon absorption triggers a trans-to-cis isomerization of the chromophore in the protein. J Am Chem Soc 126:4228–4233. https://doi.org/10.1021/ja039557f

# Chapter 13

# Monitoring Mitophagy Dynamics in Live Cells

**Christina Karantanou and Sofia-Iris Bibli**


## Abstract

The purpose of this protocol is to provide a comprehensive, stepwise guide for assessing mitophagy flux utilizing a live-cell mt-KEIMA approach. The proposed protocol is sensitive, reproducible, quantitative, and easy to perform. While mitophagy has been extensively studied, current methodologies primarily focus on terminal measurements, neglecting the dynamic aspect of this process. Hence, the introduction of this straightforward live-cell mitophagy tracing protocol enables real-time monitoring of the dynamics of mitochondrial selective autophagy, thereby enhancing the ability to draw conclusions regarding key regulators and the reversibility of the process. The assay employs a lentiviral approach to induce mt-KEIMA expression in primary or immortalized cell lines. Subsequently, the respective mitophagy reporter cells are observed using a live-cell imaging system at specific time intervals, and further quantification allows the detection of mitophagy flux. This protocol has proven efficacious in investigating mitophagy flux, including responses to chemical inducers or genetically modified cells over time. Notably, this approach is well-suited for large throughput screening of chemicals or appropriate gene-editing libraries that may influence mitophagy responses in cells.


**Key words** Mitophagy, mt-Keima, Flux, Live-cell imaging, Screening

## 1 Introduction

Mitophagy constitutes a selective autophagy process crucial for maintaining cellular mitochondrial homeostasis. Defective mitophagy has been implicated in various pathological conditions, such as neurodegenerative diseases, cancer, and aging [1, 2]. Therefore, investigating mitophagy is of paramount significance, addressing an imperative need.

The dynamic nature of mitophagy poses challenges in achieving reliable and quantitative measurements, particularly in the context of in vivo studies, prompting the necessity for more advanced tools. Electron microscopy (EM) has traditionally served as the gold

**Supplementary Information** The online version contains supplementary material available at https://doi.org/10.1007/978-1-0716-4067-8_13.

Ioannis P. Nezis (ed.), *Selective Autophagy: Methods and Protocols*, Methods in Molecular Biology, vol. 2845, https://doi.org/10.1007/978-1-0716-4067-8_13,

standard method for the examination of mitophagy over the years. Notably, EM holds the distinction of being the pioneering technique through which mitochondria were initially visualized within autophagosomes [3, 4]. Nonetheless, EM falls short in capturing the time-dependent dynamics of mitochondrial turnover, and its reliance on terminal evaluations hinders the real-time evaluation of the reversibility of this phenomenon.

Recently, more streamlined approaches to study mitophagy were introduced [5]. In particular, initial methodologies involved colocalization fluorescent analysis of mitochondria and autophagosomal markers [6, 7] or genetically modified cells for autophagy receptors, e.g., light chain 3 (LC3), were employed, wherein mitochondrial degradation was assessed by flow cytometry or immunofluorescence of labeled mitochondria using mitotracker dyes [8]. However, in most of these studies, the mitochondrial protein turnover rates do not allow the distinction between mitochondrial degradation mediated by mitochondrial proteases or through lysosomal delivery. Although, these approaches provided valuable insights into the process, yet the transient nature of LC3-based autophagy probes, which prevents the monitoring of dynamic mitophagy over extended time intervals in a sensitive and quantitative manner, coupled with the high false-positive rates due to LC3 aggregation, prompted the development of probes that selectively target mitochondrial autophagy. Two clusters of probes have been described, including specific bioprobes for the detection of autolysosome-enclosed mitochondria (cyanine dye, HQO [9]; Mtphagy Dye [10]; MitoBlue [11]; Atto 647 N dye [12]; MitoCD [13]) or fluorescent reporter systems (mKEIMA [14]; mito-QC [15]; mito-SRAI [16]; mCherry-eGFP-LC3B [17]; Cox8-EGFP-mCherry [18]; mito-timer [19]) for the detection of mitochondria localization. Among these, the mKeima probe exhibits pH-dependent characteristics and is composed of the coral-derived acid-stable fluorescent protein Keima, featuring an emission spectrum peaking at 620 nm and a bimodal excitation spectrum at 440 and 586 nm [20]. The neutral state of the chromophore (corresponding to 440 nm) prevails in high pH, while the ionized state (corresponding to 586 nm) predominates in low pH conditions. Notably, mKeima is resistant to lysosomal degradation, overcoming the stability and LC3 degradation limitations observed in LC3 reporter systems. Interestingly, the mitochondrial variant of mKeima (mt-Keima) was engineered by fusing a tandem repeat of the cytochrome C oxidase subunit VIII (COX VIII) pre-sequence to mKeima, leading to its localization in the matrix of mitochondria [20]. An alternative pH-sensitive probe, mito-QC (mCherry-GFP-FIS1), relies on the quenching of the green fluorescent protein (GFP) signal, but not the mCherry signal, in a tandem mCherry-GFP tag on the mitochondrial protein fission 1 (FIS1) [21], occurring when mitochondria enter the acidic lysosomal environment.

On the other hand, mito-timer, a pH-independent probe, is based on a color shift of the mitochondrial matrix-targeted dsRed1-E5 during maturation, offering temporal and spatial information on protein turnover [19]. Both the mito-QC and mt-Keima reporter systems have been successfully implemented in vivo and integrated into transgenic mouse models to identify the dynamic nature of mitophagy in tissues, providing insights into its spatially differential regulation within a given tissue [15, 22].

Since its discovery, the mt-Keima probe has been utilized in numerous studies for the quantification of mitophagy, relying on fluorescent microscopy or flow cytometric analysis at single time points [23–28]. However, in contrast to static imaging techniques or single time point measurements, live-cell imaging approaches to study mitophagy would offer the advantage of monitoring the process and its reversibility under stable environmental conditions (e.g., temperature, humidity, and $CO_2$ levels). This approach involves automated microscopes for high-quality time-lapse imaging and incorporates appropriate software for fast and straightforward analysis, akin to physiological states. Overall, live-cell imaging is a potent technique for monitoring cellular processes, including cell division and migration, as well as studying the dynamics of processes such as vesicular trafficking and autophagy.

In the current protocol, we have combined the benefits of the live-cell imaging approach with the sensitivity of the mt-Keima fluorescent probe to dynamically monitor mitophagy in cells in real-time. However, the approach can be transferred to any mitophagy reporter probe. Lentiviruses expressing mt-Keima were generated from HEK 293 T cells, titrated using a qualitative strip-based approach, and efficiently incorporated into HeLa cells. Subsequent culture of these cells in the IncuCyte S3 live-cell imaging system allowed different treatment approaches with pharmacological mitophagy inducers, quantitative measurement of mitophagy, and validation of the reversibility of the process. A combination of antimycin A and oligomycin A was used to induce and monitor mitophagy in real-time within a 72 h window. The system provides a spatial tool for quantifying mitophagy flux based on the shift in excitation wavelength and subsequent fluorescent detection. Results are presented as the ratio of non-mitophagic to mitophagic cells within the population.

## 2 Materials

### 2.1 Production of Mt-Keima Lentivirus

1. HEK 293T cells (ATCC).
2. HEK 293T growth medium: Dulbecco's Modified Eagle's Medium/Nutrient Mixture F-12, DMEM/F-12, without phenol red, containing L-Glutamine, supplemented with 10% fetal

calf serum (FCS), 100 U/mL penicillin, and 100 U/mL streptomycin solution.

3. Transfection growth medium: HEK 293 T growth medium supplemented with a fresh 10 mM (4-(2-hydroxyethyl)-1-piperazineethanesulfonic acid) HEPES solution (*see* **Note 1**).
4. 0.05% Trypsin-EDTA with phenol red.
5. Opti-MEM™/Reduced Serum Medium.
6. Plasmids: 0.41 μg/cm$^2$ pHAGE-mt-mKeima, 0.38 μg/cm$^2$ Pax2, 0.21 μg/cm$^2$ pMD2.G (Addgene).
7. 0.2 M polyethylenimine (PEI) stock solution.
8. 100 mm cell culture-compatible dishes.
9. 10 mL polypropylene syringe.
10. 0.45 μm syringe filter.
11. Pipettes and tips.
12. Polypropylene 1.5 mL tubes.
13. Ultrapure water.
14. 37 °C, 5% $CO_2$ incubator.
15. Laminar flow hood.
16. Water bath at 37 °C.
17. Cell counter.
18. Lenti-X GoStix Plus Takara kit: GoStix cassettes, CHASE buffer, p24 positive control.
19. IncuCyte S3 (Sartorius Lab Instruments GmbH & Co. KG).
20. −80 °C freezer.

### 2.2 Mt-Keima/HeLa Cell Line Generation

1. HeLa cells (ATCC).
2. HeLa growth medium: DMEM supplemented with 0.11 g/L sodium pyruvate, 2 mM L-Glutamine, 4.5 g/liter glucose, 10% FCS, 100 U/mL penicillin, and 100 U/mL streptomycin solution.
3. 0.05% trypsin-EDTA with phenol red.
4. 10 mg/mL polybrene stock solution.
5. 6-well cell culture-compatible plate.
6. IncuCyte S3.
7. −150 °C freezer.

### 2.3 Experimental Example

1. Dimethyl sulfoxide (DMSO).
2. 10 mmol/L Antimycin A.
3. 10 mmol/L Oligomycin A.

4. 6-well cell culture-compatible plate.
5. IncuCyte S3.

### 2.4 Data Analysis

1. IncuCyte S3.
2. IncuCyte S3 software.
3. Microsoft Excel software.
4. GraphPad Prism.

## 3 Methods

Carry out all procedures at room temperature (RT) and in sterile conditions unless otherwise specified. Keep the buffers at RT prior to use.

### 3.1 Production of Mt-Keima Lentivirus from HEK 293T Cells

Lentiviral vectors need special handling to ensure safety (*see* **Note 2**).

1. Culture HEK 293T cells at 37 °C, 5% $CO_2$ in HEK 293T growth medium.
2. Plate 2.5 million HEK 293 T cells in one 100 mm dish with 10 mL of HEK 293T growth medium (*see* **Note 3**).
3. 24 h after plating and when cells are at 60% confluence, replace the medium with 9 mL of transfection growth medium.
4. Prepare two different 1.5 mL polypropylene tubes named A and B. In tube A, add 700 μL of Opti-MEM™ and the respective volumes of 0.41 μg/cm$^2$ pHAGE-mt-mKeima, 0.38 μg/cm$^2$ Pax2, 0.21 μg/cm$^2$ pMD2.G (*see* **Note 4**). Mix well by pipetting. In tube B, add 650 μL of Opti-MEM™ and 50 μL of 0.04 mol/L PEI. Mix well by pipetting. Incubate both solutions (tube A and tube B) for 5 min at RT. Add solution A to solution B, mix well, and incubate at RT for 15–20 min. Add the mixed solutions on top of the cells dropwise and incubate them in a 37 °C, 5% $CO_2$ incubator (Fig. 1a). At this stage, the cells can already be placed in the IncuCyte S3 to monitor the development of fluorescence (Fig. 1b, Video 1).
5. 24 h after transfection, replace the medium with 9 mL of fresh transfection growth medium.
6. 72 h later, collect the cell supernatant with a 10 mL polypropylene syringe. Filter the collected supernatant through a 0.45 μm syringe filter and make aliquots of 2 mL.
7. Discard the plate in special S2 biosafety waste.
8. All viral preparations can be used directly to infect recipient cells or can be stored at −80 °C until use (*see* **Note 5**).

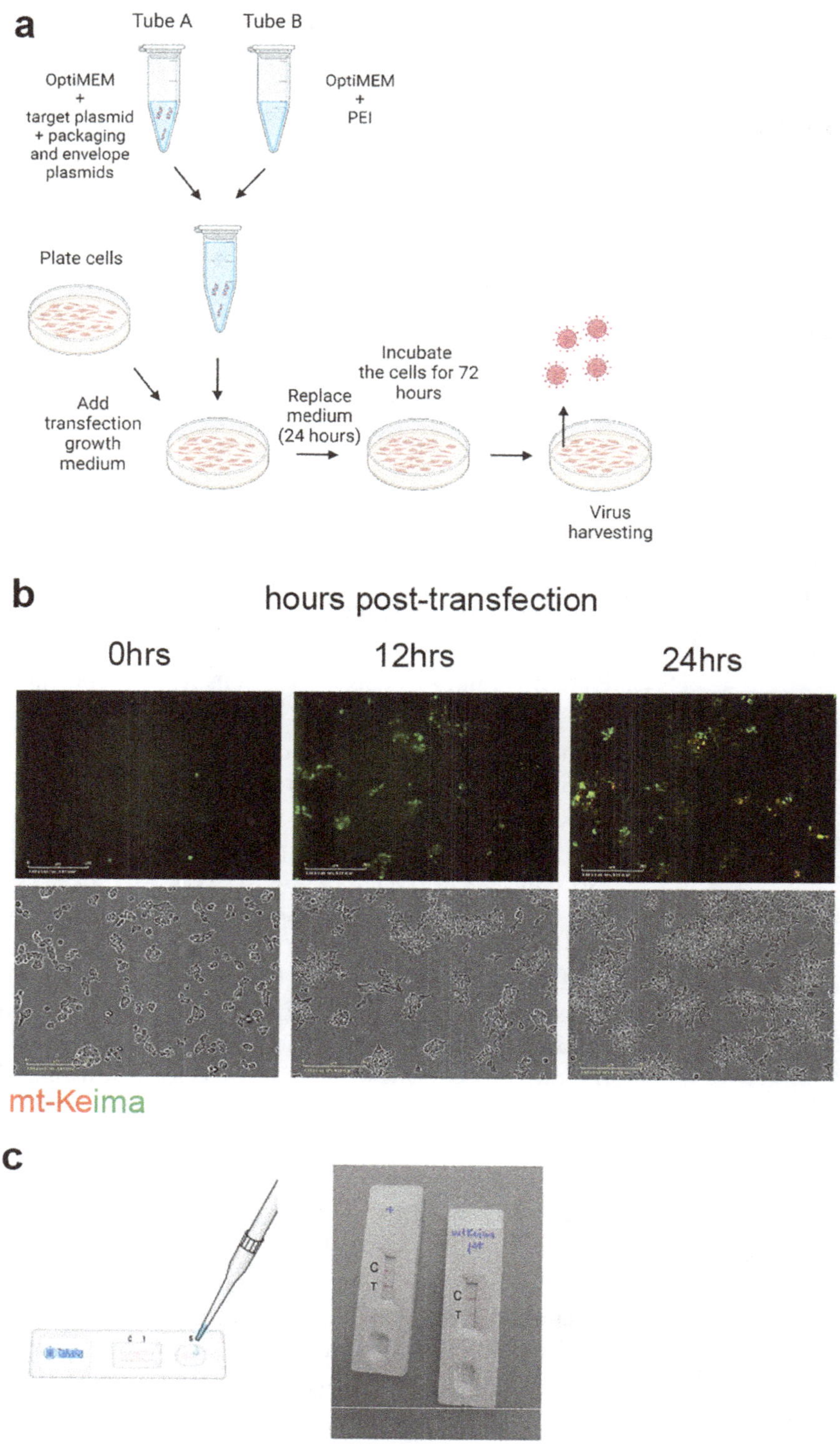


**Fig. 1** Production of mt-Keima lentivirus from HEK 293 T cells. (**a**) Schematic representation of mtkeima lentivirus production from HEK 293 T cells. (**b**) Representative brightfield and fluorescent images of transfected mt-Keima HEK 293 T cells 0, 12, and 24 h post-transfection using the IncuCyte S3 system. (**c**) Qualitative lentiviral titer test using the Lenit-X GoStix Plus

9. To ensure successful viral generation, label 2 cassettes from the Lenti-X GoStix Plus kit: one with "+" as the positive control and one with the name of the virus, i.e., "mt-Keima."
10. For the mt-Keima virus, transfer 20 μL of the cell supernatant to a 1.5 mL Eppendorf and add 60 μL of the assay buffer (CHASE buffer). Take all the 80 μL and add them dropwise to the cassette according to the manufacturer's instructions. For the positive control (included in the kit), mix 20 μL with assay buffer following the manufacturer's instructions and add it dropwise to the respective cassette (Fig. 1c).
11. Leave them for 10 min at RT, until the band is showing. The Lenti-X GoStix Plus Takara kit compares the intensity of the bands to the internal p24 standard control (positive ctrl) (Fig. 1c).

### 3.2 Stable Expression of Mt-Keima in HeLa Cells

1. Culture HeLa cells at 37 °C, 5% $CO_2$ in HeLa growth medium.
2. The day prior to infection, plate 0.3 million HeLa cells in a well of a 6-well plate with 2 mL of HeLa growth medium.
3. On the day of infection (24 h later), prepare the viral medium: prepare dilution of mt-Keima lentivirus with HeLa growth medium at a 1:1 ratio (for a well of a 6-well plate, add 1 mL of lentivirus and 1 mL of HeLa growth medium). Add 8 μg/mL of polybrene. To achieve this, add 1.6 μL of polybrene stock solution to each well (*see* **Note 6**). Remove the medium from the cells and substitute it with the viral medium.
4. 24 h postinfection, replace the viral medium with HeLa growth medium.

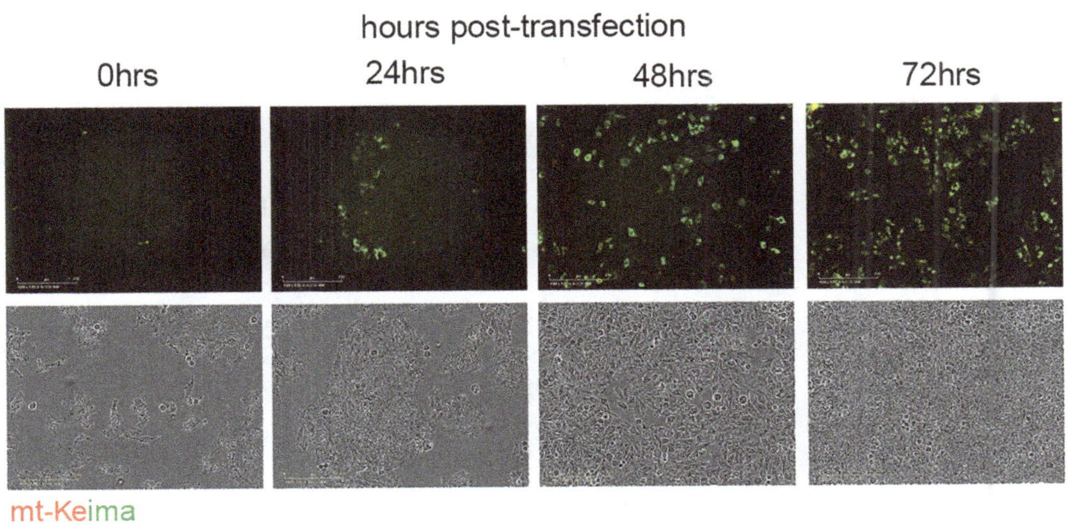


**Fig. 2** Stable expression of mt-Keima in HeLa cells. Representative brightfield and fluorescent images of infected HeLa cells with the mtkeima lentivirus at 0, 24, 48, and 72 h postinfection

5. The mt-Keima-infected HeLa cells can be transferred to the IncuCyte S3 right after infection. This allows monitoring of the development of fluorescence over time (Fig. 2).
6. After infecting HeLa cells with the mt-Keima virus and confirming the expression of the construct with imaging in the IncuCyte S3, stable cell lines are produced. These can be subcultured for further treatments and downstream applications or can be cryopreserved at −150 °C and thawed for subsequent experiments.

### 3.3 Experimental Example: Live-Cell Imaging of Mt-Keima HeLa Cells Treated with Antimycin a and Oligomycin a to Monitor the Mitophagic Flux Using IncuCyte S3

#### 3.3.1 Experimental Design

1. Culture the mt-Keima/HeLa stable cell line at 37 °C, 5% $CO_2$ in HeLa growth medium.
2. Plate 0.3 million mt-Keima-positive HeLa cells per well in 2 wells of a 6-well plate in 2 mL of HeLa growth medium.
3. 4 h later, when cells are attached, transfer the plate into the incubator of IncuCyte S3 and monitor for 12 h (or the desired time window).
4. 12 h after plating, replace the medium as follows:
   (a) Well 1: Add 2 mL of HeLa growth medium and 2 μL of DMSO.
   (b) Well 2: Add 2 mL of HeLa growth medium with final concentrations of 10 μM Antimycin A and 10 μM Oligomycin A.
5. Monitor for an additional 24 h (or the desired time window).
6. Return cells to normal culture media for another 16 h (or the desired time window).

#### 3.3.2 Experimental Setup in IncuCyte S3

1. Open the IncuCyte main window and select “Schedule to Acquire” to initiate a new scanning experiment (Fig. 3a).
2. On the pop-up window, click on the + symbol to add a new vessel (Fig. 3b).
3. On the pop-up window, select “Scan on Schedule” (Fig. 3c) and click Next.
4. On the pop-up window, select “New” (Fig. 3d) and click Next.
5. On the pop-up window, select “Standard” (Fig. 3e) and click Next.
6. On the pop-up window, select Cell-by-cell option: “None,” Image Channels: “Phase,” “Green,” “Red,” Objective: “10×” or “20×” (Fig. 3f), and click Next.
7. On the pop-up window, select the appropriate plate type (Fig. 3g) and click Next.
8. On the pop-up window, select the appropriate location for your current experiment (Fig. 3h) and click Next.

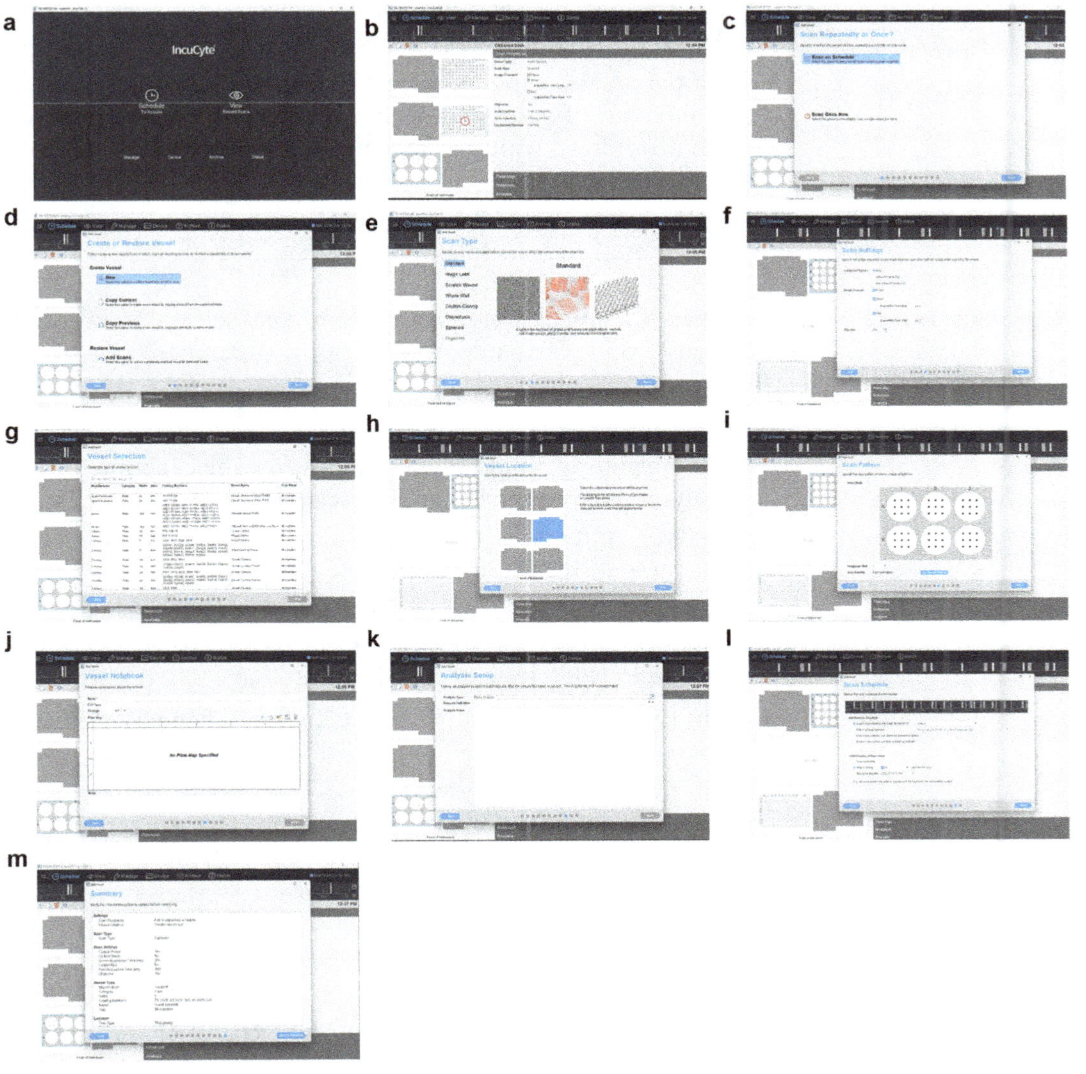

**Fig. 3** Experimental setup in IncuCyte S3. (**a**) Select "Schedule To Acquire." (**b**) Select add new vessel ("**+**" symbol). (**c**) Select "Scan on Schedule." (**d**) Select "New." (**e**) Select Scan Type: "Standard." (**f**) Select Cell-by-cell option: "None," Image Channels: "Phase," "Green," "Red," Objective: "10×" or "20×." (**g**) Select the appropriate plate type. (**h**) Select the Vessel Location. (**i**) Select Scan Pattern (recommended "9" images per well for a 6-well plate). (**j**) Name your vessel. (**k**) Select Analysis Type: "Basic Analyzer." (**l**) Select "Create new schedule with scans at intervals of" and choose the appropriate time points from the drop-down window (recommended every "4 h"). (**m**) Review the summary and select "Add to Schedule"

9. On the pop-up window, select the appropriate Scan Pattern, recommended "9" images per well for a 6-well plate (Fig. 3i), and click Next.
10. On the pop-up window, name your experiment appropriately (Fig. 3j) and click Next.

11. On the pop-up window, select Analysis Type: select "Basic Analyzer" (Fig. 3k) and click Next.
12. On the pop-up window, select "Create new schedule with scans at intervals of" and choose the appropriate time points from the drop-down window (recommended every "4 h") (Fig. 3l) and click Next.
13. On the pop-up window, review the summary (Fig. 3m) and select "Add to Schedule."

#### 3.3.3 Image Analysis

1. Open the IncuCyte main window and select "View recent Scans" to initiate a new analysis of the experiment (Fig. 4a).
2. On the pop-up window, choose the experiment to be analyzed (Fig. 4b).
3. On the pop-up window, you can now see the experiment to be analyzed, and in the left corner, select the button for "Launch Analysis" (Fig. 4c).
4. On the pop-up window, select "Create New Analysis Definition" (Fig. 4d) and click Next.
5. On the pop-up window, select Analysis Type: select "Basic Analyzer" (Fig. 4e) and click Next.
6. On the pop-up window, select Image Channels: "Phase," "Green," "Red" (Fig. 4f), and click Next.
7. On the pop-up window, choose appropriate representative images to define the analysis parameters. Here, we recommend choosing at least two different time points from your positive controls (Fig. 4g, h) and click Next.
8. On the pop-up window, adjust Segmentation: "Top-Hat," Radius: "100,00," Threshold: "20,000" (Fig. 4i), and click Next.
9. On the pop-up window, select Channel Masks: "Green object" (Fig. 4j) to view that your criteria from the previous item suffice to include all positive Green cells. Repeat for the Red Channel, then click Next.
10. On the pop-up window, select the appropriate desired time points and wells to apply the analysis (Fig. 4k) and click Next.
11. On the pop-up window, name your analysis under the "Definition name" tab (Fig. 4l) and click Next.
12. On the pop-up window, click "OK" and then "Finish" (Fig. 4m).

#### 3.3.4 Data Export and Results

1. Open the IncuCyte main window and select "View recent Scans" to view your analysis performed in Subheading 3.3.3 (Fig. 5a).

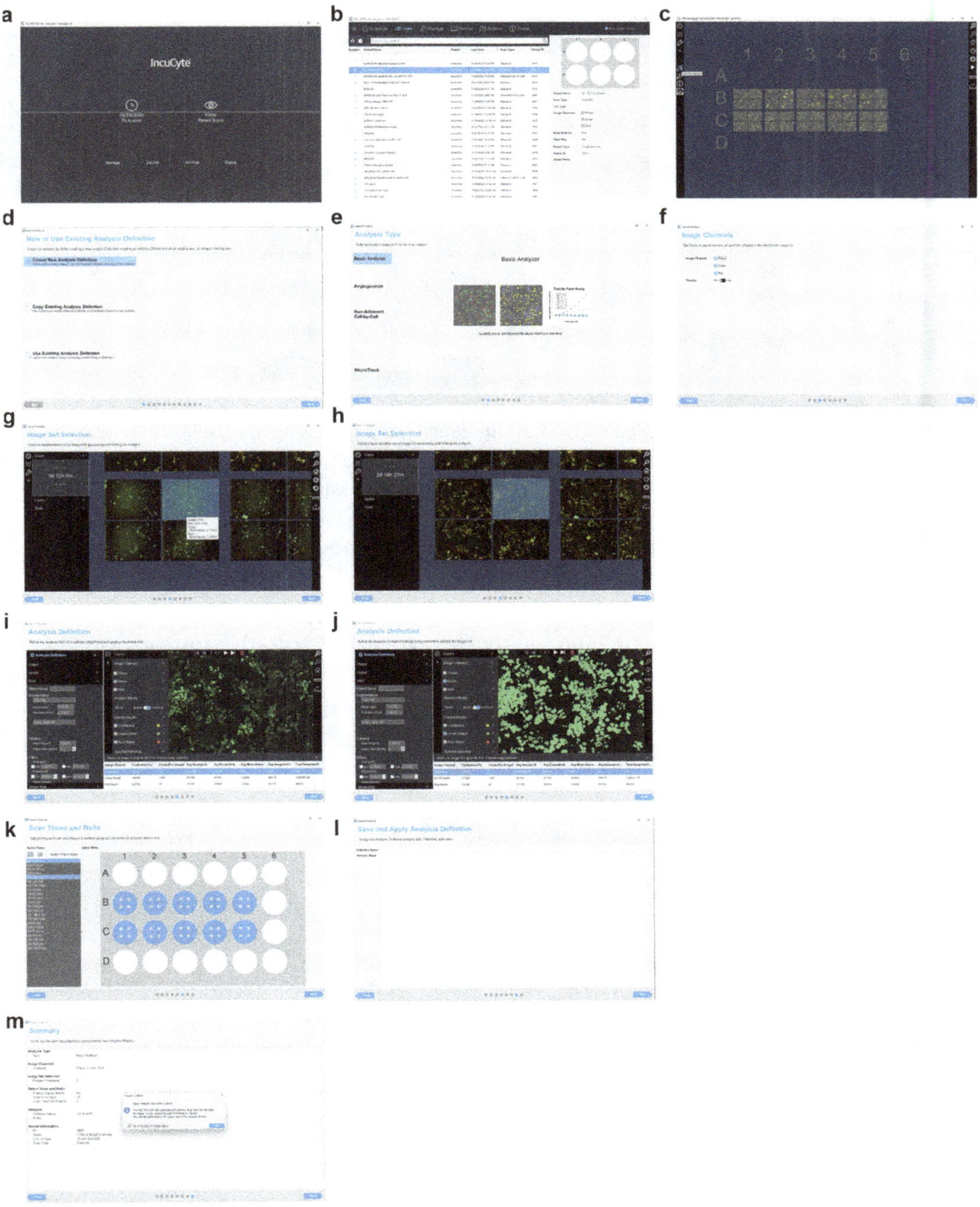

**Fig. 4** Image viewing and analysis setup. (**a**) Select "View Recent Scans." (**b**) Select the appropriate experiment for analysis. (**c**) In the pop-up window, you see the experiment to be analyzed and in the left corner, select the button for "Launch Analysis." (**d**) Select "Create New Analysis Definition." (**e**) Select Analysis Type "Basic Analyzer." (**f**) Select Image Channels "Phase," "Green," "Red." (**g-h**) Choose appropriate representative images for at least two desired time points. (**i**) Adjust Segmentation: "Top-Hat," Radius: "100,00," Threshold "20,000." (**j**) Select from the channel mask "Green Object." (**k**) Select the appropriate desired time points and wells to apply the analysis. (**l**) Give an appropriate "Definition name." (**m**) Save and apply the analysis definition

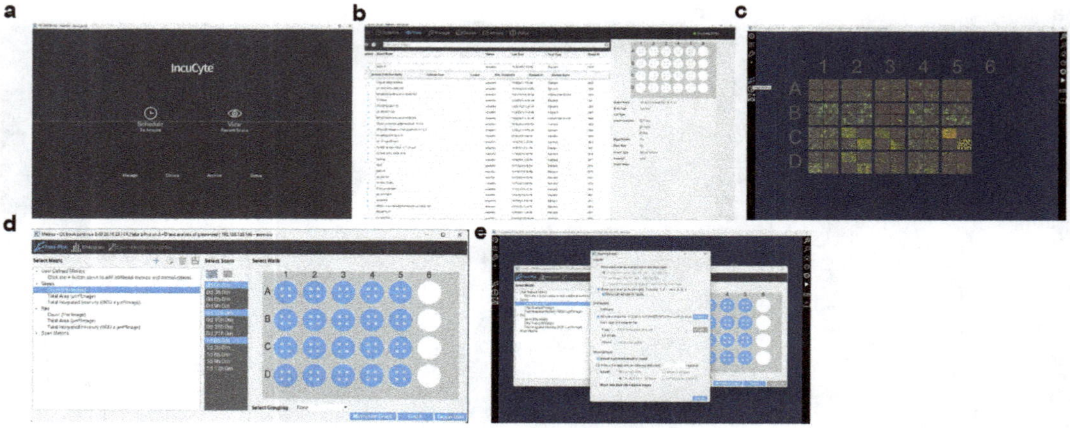

**Fig. 5** Data export. (**a**) Select "View Recent Scans." (**b**) Select the appropriate analyzed experiment. (**c**) Your analyzed experiment is shown; choose "Graph Metrics" in the left corner. (**d**) Select the appropriate times and wells. (**e**) Select the appropriate position for storing the analysis and export style

2. On the pop-up window, choose the analyzed experiment (Fig. 5b).
3. On the pop-up window, you can now see the analyzed experiment, and in the left corner, select the button for "Graph Metrics" (Fig. 5c).
4. On the pop-up window, select the appropriate time points and wells that you wish to visualize the results (Fig. 5d) and click "Export Data."
5. On the pop-up window, select the folder where your analyzed data will be transferred and click "Export" (Fig. 5e).
6. Representative data are shown in Fig. 6 and Video 2 (Solvent) and 3 (Antimycin A and Oligomycin A treatment). The addition of Antimycin A and Oligomycin A 12 h after monitoring the cells in normal growth media, increased mitophagic flux within the first 24 h. Subsequent removal of the trigger reduced mitophagic responses within 2 h in HeLa cells (*see* **Note** 7).

## 4 Notes

1. HEPES is used here to increase the titer of the virus produced, as it prevents rapid shifts in the pH that can occur due to the addition of transfection reagents.
2. The work needs to be done in a biosafety level 2 (BSL-2) laboratory, and proper training of the investigator is required in compliance with institutional regulations and guidelines.

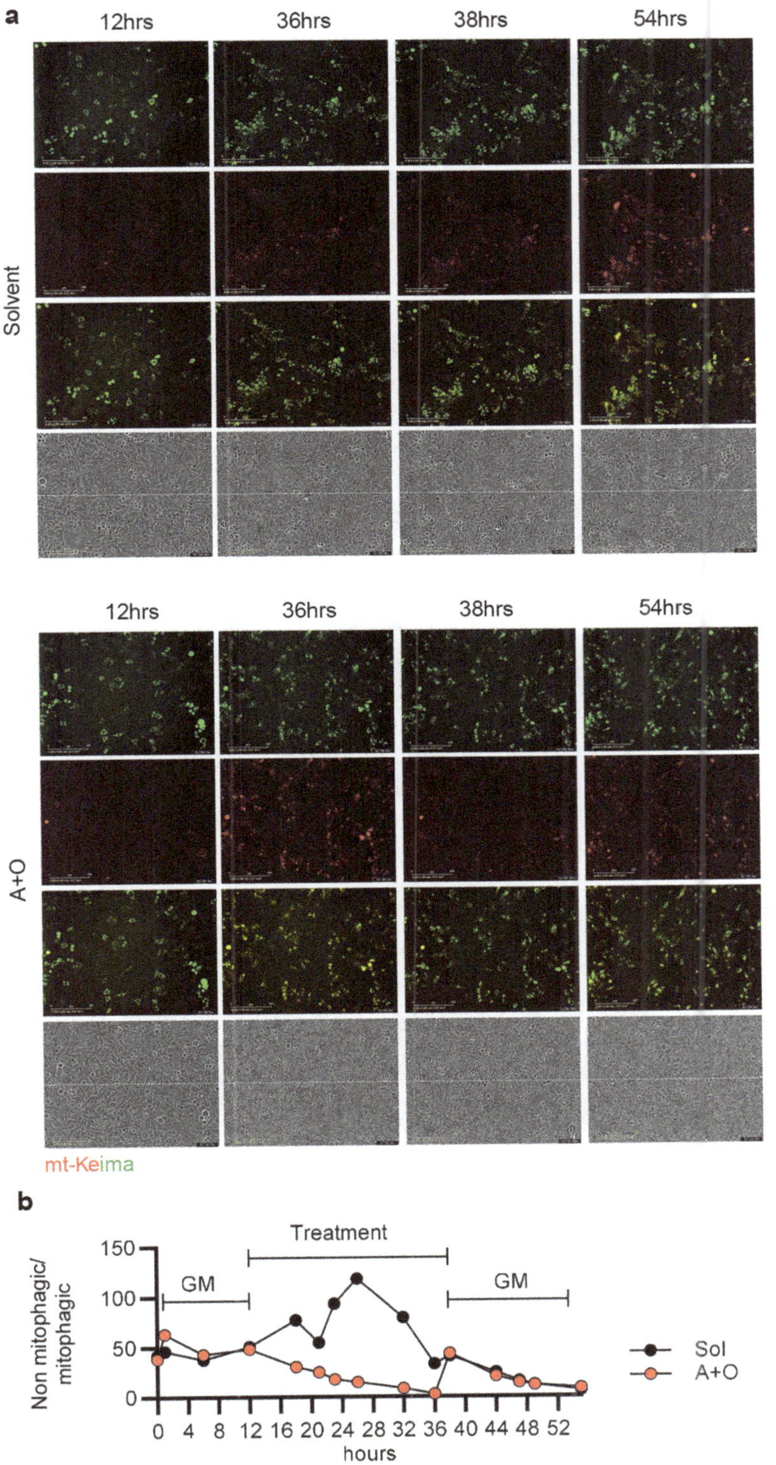


**Fig. 6** Representative data. (**a**) Representative brightfield and fluorescent images of HeLa cells maintained in growth media (GM) for 12 h after seeding and subsequently treated with solvent for an additional 40 h, or treated with Antimycin A and Oligomycin A (A+O) for 24 h and then returned for an additional 16 h to normal growth media. (**b**) Line graph presenting the change in mitophagy flux over time of cells treated as in panel **a**

3. The counting and plating of the cells depend on the investigator's handling; therefore, 2.5 million cells is an approximate number. On the day of transfection, the cells need to be attached to the bottom of the dish at around 60% confluence and have the characteristic spindle-shaped morphology of HEK 293 T cells.
4. The volume depends on the concentration of the produced plasmid. The plasmid production involves a series of steps, including bacterial transformation, selective growth, plasmid extraction, and plasmid storage, which can be performed in the investigator's laboratory.
5. Multiple freezing-thawing cycles may influence the structure of the virus or lead to viral protein denaturation, thus reducing its infectivity. Making aliquots of the viruses can prevent these effects.
6. The use of polybrene enhances the ability of the viral particles to attach to the recipient cells.
7. The effect of Antimycin A and Oligomycin A on mitophagy induction, cell proliferation, or cell death can vary depending on the cell type. It is recommended to image the cells every 4 h, including the selection of the "phase" channel, so the investigator can evaluate the effect of the drugs on these parameters independently of mitophagy.

## Acknowledgments

This work was supported by the Deutsche Forschungsgemeinschaft SFB1177 Project F08; Project ID 259130777 to S.-I.B and the Emmy Noether Programme BI 2163/1-1 to S.-I.B.

### References

1. Palikaras K, Lionaki E, Tavernarakis N (2018) Mechanisms of mitophagy in cellular homeostasis, physiology and pathology. Nat Cell Biol 20(9):1013–1022. https://doi.org/10.1038/s41556-018-0176-2
2. Picca A, Faitg J, Auwerx J et al (2023) Mitophagy in human health, ageing and disease. Nat Metab 5:2047. https://doi.org/10.1038/s42255-023-00930-8
3. Mizushima N, Yoshimori T, Levine B (2010) Methods in mammalian autophagy research. Cell 140(3):313–326. https://doi.org/10.1016/j.cell.2010.01.028
4. Novikoff AB, Essner E (1962) Cytolysomes and mitochondrial degeneration. J Cell Biol 15(1):140–146. https://doi.org/10.1083/jcb.15.1.140
5. Montava-Garriga L, Ganley IG (2020) Outstanding questions in mitophagy: what we do and do not know. J Mol Biol 432(1):206–230. https://doi.org/10.1016/j.jmb.2019.06.032
6. Dolman NJ, Chambers KM, Mandavilli B et al (2013) Tools and techniques to measure mitophagy using fluorescence microscopy. Autophagy 9(11):1653–1662. https://doi.org/10.4161/auto.24001
7. Matsuda N, Sato S, Shiba K et al (2010) PINK1 stabilized by mitochondrial depolarization recruits Parkin to damaged mitochondria and activates latent Parkin for mitophagy. J Cell

Biol 189(2):211–221. https://doi.org/10.1083/jcb.200910140

8. Schweers RL, Zhang J, Randall MS et al (2007) NIX is required for programmed mitochondrial clearance during reticulocyte maturation. Proc Natl Acad Sci USA 104(49): 19500–19505. https://doi.org/10.1073/pnas.0708818104
9. Liu Y, Zhou J, Wang L et al (2016) A cyanine dye to probe mitophagy: simultaneous detection of mitochondria and autolysosomes in live cells. J Am Chem Soc 138(38):12368–12374. https://doi.org/10.1021/jacs.6b04048
10. Iwashita H, Torii S, Nagahora N et al (2017) Live cell imaging of mitochondrial autophagy with a novel fluorescent small molecule. ACS Chem Biol 12(10):2546–2551. https://doi.org/10.1021/acschembio.7b00647
11. Sánchez MI, Vida Y, Pérez-Inestrosa E et al (2020) MitoBlue as a tool to analyze the mitochondria-lysosome communication. Sci Rep 10(1):3528. https://doi.org/10.1038/s41598-020-60573-7
12. Han Y, Li M, Qiu F et al (2017) Cell-permeable organic fluorescent probes for live-cell long-term super-resolution imaging reveal lysosome-mitochondrion interactions. Nat Commun 8(1):1307. https://doi.org/10.1038/s41467-017-01503-6
13. Ye Z, Wei L, Geng X et al (2019) Mitochondrion-specific blinking fluorescent bioprobe for nanoscopic monitoring of mitophagy. ACS Nano 13(10):11593–11602. https://doi.org/10.1021/acsnano.9b05354
14. Wang C (2020) A sensitive and quantitative mKeima assay for mitophagy via FACS. Curr Protoc Cell Biol 86(1):e99. https://doi.org/10.1002/cpcb.99
15. McWilliams TG, Prescott AR, Allen GFG et al (2016) mito-QC illuminates mitophagy and mitochondrial architecture in vivo. J Cell Biol 214(3):333–345. https://doi.org/10.1083/jcb.201603039
16. Katayama H, Hama H, Nagasawa K et al (2020) Visualizing and modulating mitophagy for therapeutic studies of neurodegeneration. Cell 181(5):1176–1187.e16. https://doi.org/10.1016/j.cell.2020.04.025
17. Moore AS, Holzbaur ELF (2019) Imaging the dynamics of mitophagy in live cells. Methods Mol Biol 1880:601–610. https://doi.org/10.1007/978-1-4939-8873-0_39
18. Ma X, Ding W-X (2021) A fluorescence imaging based-assay to monitor mitophagy in cultured hepatocytes and mouse liver. Liver Res 5(1):16–20. https://doi.org/10.1016/j.livres.2020.12.002
19. Hernandez G, Thornton C, Stotland A et al (2013) MitoTimer: a novel tool for monitoring mitochondrial turnover. Autophagy 9(11): 1852–1861. https://doi.org/10.4161/auto.26501
20. Katayama H, Kogure T, Mizushima N et al (2011) A sensitive and quantitative technique for detecting autophagic events based on lysosomal delivery. Chem Biol 18(8):1042–1052. https://doi.org/10.1016/j.chembiol.2011.05.013
21. Allen GFG, Toth R, James J et al (2013) Loss of iron triggers PINK1/Parkin-independent mitophagy. EMBO Rep 14(12):1127–1135. https://doi.org/10.1038/embor.2013.168
22. Sun N, Yun J, Liu J et al (2015) Measuring in vivo mitophagy. Mol Cell 60(4):685–696. https://doi.org/10.1016/j.molcel.2015.10.009
23. Bingol B, Tea JS, Phu L et al (2014) The mitochondrial deubiquitinase USP30 opposes parkin-mediated mitophagy. Nature 510(7505):370–375. https://doi.org/10.1038/nature13418
24. Hoshino A, Wang W-J, Wada S et al (2019) The ADP/ATP translocase drives mitophagy independent of nucleotide exchange. Nature 575(7782):375–379. https://doi.org/10.1038/s41586-019-1667-4
25. Kageyama Y, Hoshijima M, Seo K et al (2014) Parkin-independent mitophagy requires Drp1 and maintains the integrity of mammalian heart and brain. EMBO J 33(23):2798–2813. https://doi.org/10.15252/embj.201488658
26. Lazarou M, Sliter DA, Kane LA et al (2015) The ubiquitin kinase PINK1 recruits autophagy receptors to induce mitophagy. Nature 524(7565):309–314. https://doi.org/10.1038/nature14893
27. Liu K, Li X, Li Z et al (2022) Evaluating mitophagy in embryonic stem cells by using fluorescence-based imaging. Front Cell Dev Biol 10:910464. https://doi.org/10.3389/fcell.2022.910464
28. Yamano K, Wang C, Sarraf SA et al (2018) Endosomal Rab cycles regulate Parkin-mediated mitophagy. elife 7:e31326. https://doi.org/10.7554/eLife.31326

# Chapter 14

# Methods for Monitoring NCOA4-Mediated Ferritinophagy

Thuy T. P. Nguyen and Joseph D. Mancias


## Abstract

Ferritinophagy is a selective form of autophagy in which ferritin, the primary intracellular iron storage protein complex, is targeted by NCOA4 (Nuclear receptor coactivator 4) to the lysosome for degradation. NCOA4-mediated ferritinophagy plays a crucial role in cellular iron metabolism, influencing iron homeostasis, heme synthesis, mitochondrial respiratory function, and ferroptosis, an iron-dependent form of cell death. Targeting ferritinophagy has emerged as a potential anticancer therapeutic strategy. In this context, we provide a flowchart of the procedures and accompanying protocols for monitoring ferritinophagic flux.


**Key words** Ferritinophagy, NCOA4, Ferritin, FTH1, Iron

## 1 Introduction

Iron is an essential element for life as it is a cofactor for biochemical processes such as oxygen transport, DNA synthesis, and ATP production. However, an excess of cellular iron can increase redox-active iron, which triggers DNA damage, lipid peroxidation, and contributes to various forms of cell death, most notably ferroptosis [1, 2]. Precise control of cellular iron levels is achieved through a network of proteins involved in iron import, export, and storage. In the cytosol, excess iron is stored in ferritin nanocages, which are composed of 24 subunits of ferritin heavy chain (FTH1) and ferritin light chain (FTL) [3]. Iron stored in ferritin is released under conditions of iron depletion or high iron demand via NCOA4 (Nuclear receptor coactivator 4)-mediated ferritinophagy, a selective autophagy process whereby NCOA4 recognizes and recruits ferritin nanocages to the autophagosome-lysosome system for degradation, thereby maintaining the cytosolic labile iron pool (LIP) [4, 5]. Dysregulation of ferritinophagy impacts human health, contributing to conditions like anemia and other iron-related disorders [6–9]. Here we provide a framework for the evaluation of flux through the ferritinophagy pathway.

Ioannis P. Nezis (ed.), *Selective Autophagy: Methods and Protocols*, Methods in Molecular Biology, vol. 2845,
https://doi.org/10.1007/978-1-0716-4067-8_14,

While ferritinophagic flux is detectable in almost all cells basally, it is maximally upregulated under iron-depleted conditions to maintain the cytosolic labile iron pool [6, 10, 11]. Emerging data also demonstrates that ferritinophagy is active in iron overload to traffic ferritin to the lysosome for storage as hemosiderin and for degradation [12, 13]. Furthermore, many other physiological and pathophysiological conditions have been proposed to activate or involve NCOA4-mediated ferritinophagy, including a role in regulating sensitivity to ferroptosis, an iron-dependent form of cell death [14]. Mechanistically, NCOA4 and FTH1 (the ferritin subunit that NCOA4 recognizes) are required for ferritinophagy. NCOA4-FTH1 complexes can be trafficked to the lysosome via ATG8-dependent and independent pathways [15, 16]. However, there are additional NCOA4-mediated trafficking destinations for ferritin beyond the lysosome, including an NCOA4-mediated exocytosis pathway [17]. NCOA4-mediated ferritinophagy activity is regulated by NCOA4 protein levels and localization. In acute iron loading, HERC2, a ubiquitin E3 ligase, recognizes an iron-sulfur cluster on NCOA4 and targets NCOA4 for proteasomal degradation [6, 18]. In addition, during the early phase of iron loading, iron binding to NCOA4 induces a liquid-phase separation such that NCOA4 is physically sequestered from FTH1 in cytoplasmic condensates [13, 15]. Both processes allow for ferritin accumulation and iron storage during iron loading. Numerous additional mechanisms for controlling NCOA4 levels and ultimately ferritinophagy have been reported [19]. Ultimately, NCOA4-ferritin condensates are targeted for degradation during iron depletion or for degradation or lysosomal storage as hemosiderin during prolonged periods of iron loading [13]. Given the complexities and interconnected nature of NCOA4 and cellular iron metabolism, measuring NCOA4-mediated ferritinophagy flux requires an understanding of not just the levels of NCOA4 and ferritin but also NCOA4 and ferritin localization. Furthermore, given the dependency on NCOA4, genetic ablation of NCOA4 can be critical to understanding the reliance on ferritinophagic flux for any condition being evaluated. Here, we present a workflow for measuring NCOA4-mediated ferritinophagy flux in cell culture systems (Fig. 1).

## 2 Materials

### 2.1 For Western Blot Analysis

1. RIPA lysis buffer: RIPA buffer (50 mM Tris–HCl pH 7.4, 150 mM NaCl, 2 mM EDTA, 1% NP-40, 0.1% SDS) stored at 4 °C. Add fresh 1 mM DTT and 1× Halt™ protease and phosphatase inhibitor cocktail to RIPA buffer just before harvesting the cell pellets for lysis.

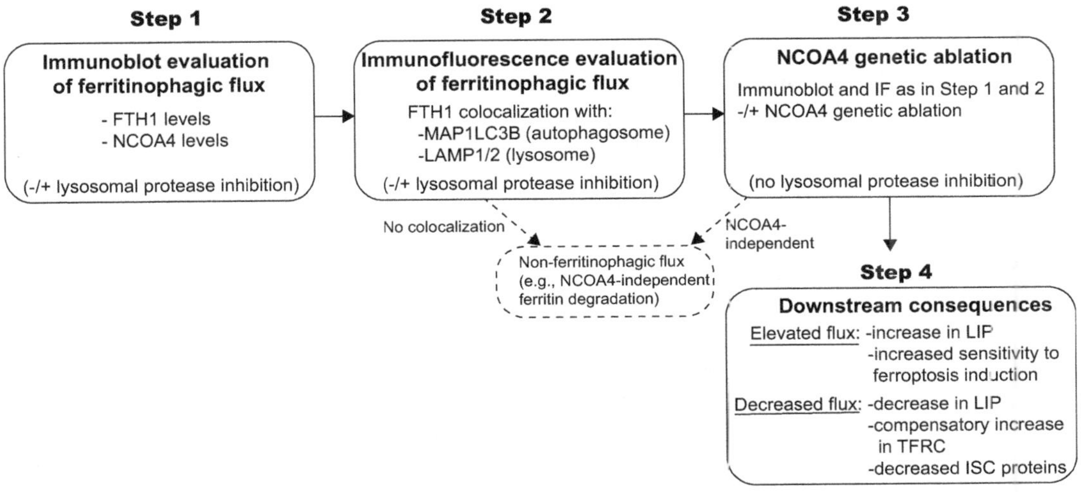


**Fig. 1** Flowchart of the procedures for measuring ferritinophagic flux. Step 1: Measure NCOA4 and FTH1 levels via immunoblotting. Step 2: Measure FTH1 and MAP1LC3B (autophagy) or LAMP1/2 (lysosome) colocalization via immunofluorescence. Step 3: NCOA4 genetic ablation can be used to specifically implicate NCOA4 in FTH1 turnover. Step 4: Downstream consequences can be evaluated: Elevated flux: Measure the cellular labile iron pool level to understand whether ferritinophagy activation altered the labile iron pool (LIP). Evaluating sensitivity to ferroptosis as a downstream consequence of increased ferritinophagic flux can also be performed. Downstream readouts of decreased ferritinophagy are related to the general effects that may be seen with a decrease in the labile iron pool, such as a compensatory increase in iron import (e.g., increased TFRC) and decreased iron-sulfur cluster (ISC) protein levels. Finally, if the change in FTH1/NCOA4 levels cannot be rescued by lysosomal inhibitors (or via NCOA4 genetic ablation) and there is poor colocalization between FTH1/NCOA4 and lysosome, an NCOA4-independent ferritin degradation process may be active

2. Urea lysis buffer: 7 M urea, 1× Halt™ protease and phosphatase inhibitor cocktail in RIPA buffer. Prepare fresh and keep at RT for use.
3. Benchtop centrifuge.
4. Pierce™ BCA Protein Assay Kits.
5. 2× Laemmli sample buffer.
6. SDS-PAGE gels, ideally TGX™ Precast 4–20% gradient gels (Bio-Rad, Cat#5671094) run in Tris–Glycine buffer.
7. Heat block.
8. **Kaleidoscope** pre-stained protein molecular weight marker.
9. SDS-PAGE running buffer: 25 mM Tris base (3.03 g/L), 192 mM glycine (14.4 g/L), and 1% SDS (1 g/L) dissolved in distilled water. Make fresh for each gel run. Do not reuse the running buffer. Reusing the buffer can affect reproducibility since the ionic strength and pH of the buffer change during the run.
10. Western Blot transfer equipment (Bio-Rad).

11. Transfer buffer: 25 mM Tris base (3.03 g/L), 192 mM glycine (14.4 g/L), and 20% methanol dissolved in distilled water.
12. PVDF or nitrocellulose membrane.
13. TBST (pH 7.4): 20 mM Tris–HCl, 150 mM NaCl, and 0.1% Tween 20 dissolved in distilled water.
14. Blocking solution (pH 7.5): 5% nonfat dry milk in TBST.
15. Methanol.
16. Orbital shaker.
17. Primary antibodies: Anti-NCOA4 (or anti-ARA70) antibody (1:100, Santa Cruz, Cat# sc-373739 (works for human and murine cell lines) or 1:1000, Cell Signaling, Cat# 66849S (only for human cell lines)), anti-FTH1 antibody (1:1000, Cell Signaling, Cat #4393), anti-β-actin antibody (1:10,000, Sigma, Cat# A5441), and anti-IREB2 antibody (1:1000, Cell Signaling, Cat# 37135) in 5% BSA or blocking buffer.
18. Secondary antibodies: Goat anti-Rabbit IgG (H + L) Secondary Antibody, HRP (Thermo Scientific, Cat #31460), anti-mouse IgG, HRP-linked Antibody (Cell Signaling, Cat #7076S).
19. Pierce™ ECL Western Blotting Substrate (Thermo Scientific, Cat# PI32209).
20. SuperSignal™ West Pico PLUS Chemiluminescent Substrate (Thermo Scientific, Cat# 34580).

### 2.2 For Immunofluorescence Assays

1. 12-well plates.
2. 1.5 mm glass coverslips of appropriate size.
3. Phosphate-buffered saline (PBS).
4. 4% paraformaldehyde (PFA): Prepared by diluting 16% PFA solution (Electron Microscopy Science, Cat #15710) in PBS.
5. ProLong™ Gold Antifade Mountant with DNA stain DAPI (Invitrogen, Cat# P36931).
6. 0.5% Triton X-100 in PBS.
7. Chilled methanol at −20 °C for at least 4 h before fixing cells (for MAP1LC3B immunostaining only).
8. Normal goat serum (Vector Laboratories, Cat# S-1000-20).
9. Blocking solution: 0.2% (w/v) powdered nonfat milk, 2% normal goat serum, 0.1 M glycine, 1% BSA, and 0.01% Triton X-100 in PBS. Prepare fresh and maintain at 4 °C until use.
10. Primary antibody: Anti-LAMP1 antibody (1:100, Cell Signaling, Cat #9091), anti-LAMP2 antibody (1:50, Abcam, Cat# ab25631), anti-FTH1 antibody (1:100, Cell Signaling, Cat# 4393) or anti-ferritin polyclonal antibody (1:400, Rockland,

Cat# 200-401-090-0100), anti-NCOA4 (1:50, Invitrogen, Cat# PA5-96398), and anti-MAP1LC3B antibody (1:100, Cell Signaling, Cat# 3868).

11. Secondary antibody: Goat anti-rabbit IgG(H + L) highly cross-adsorbed antibody Alexa Fluor 488 (1:500, Invitrogen, Cat# A-11034), goat anti-mouse IgG(H + L) highly cross-adsorbed antibody Alexa Fluor 594 (1:500, Invitrogen, Cat# A-11005).
12. Forceps.
13. Microscopic glass slides.
14. Software for image analysis such as ImageJ.
15. Software for statistical analysis such as GraphPad Prism.

### 2.3 Ferritinophagy Modulators

1. Lysosomal protease inhibitors including E64-d (MedChemExpress, Cat# HY-100229, 10 μg/mL) & pepstatin A (EMD Millipore, Cat# 561481, 10 μg/mL) or cathepsin B-specific inhibitor CA-074 [20] (Cat# A1926, ApexBio, 15 μM). Lysosomal acidification inhibitor such as Bafilomycin A1 (Sigma, 50 nM), Iron chelators including deferoxamine (DFO, Sigma, 100 μM), and deferasirox (DFX, Seleckchem, 30 μM).
2. shNCOA4, or siNCOA4: Lentiviral shRNA plasmid clones (pLKO.1) from the RNAi Consortium collection against NCOA4 (Gene ID: 8031, NM_001145263.1) shNCOA4-1: 5′- CCCAGGAAGTATTACTTAATT -3′ (TRCN0000019724); shNCOA4-2: 5′- GGCCCAGGAAG TATTACTTAA-3′ (TRCN0000236185); and shGFP: 5′-GC AAGCTGACCCTGAAGTTCAT -3′ (Addgene [Cambridge, MA, United States] plasmid #30323); or siNCOA4-1: 5′ ACAAAGAUCUAGCCAAUCA 3′ and siNCOA4-2: 5′ GACCUUAUUUAUCAGCUUA 3′ (Invitrogen) [4, 6].

## 3 Methods

### 3.1 Monitoring Ferritinophagy Using Immunoblotting

#### 3.1.1 Overview

To monitor ferritinophagic flux using immunoblotting, NCOA4 and FTH1 protein levels are measured in cell lysates from cells treated in the absence and presence of lysosomal protease inhibitors under both control and "treatment" conditions of interest, e.g., compound treatment, RNAi-mediated knockdown, or knockout samples. The hallmark of activated ferritinophagy flux is a decrease in the protein level of FTH1 that is rescued by inhibition of lysosome-mediated degradation (Fig. 2). NCOA4 levels may be decreased or increased upon activation of ferritinophagy; however, lysosomal protease inhibitor treatment would rescue or further increase, respectively, NCOA4 levels.

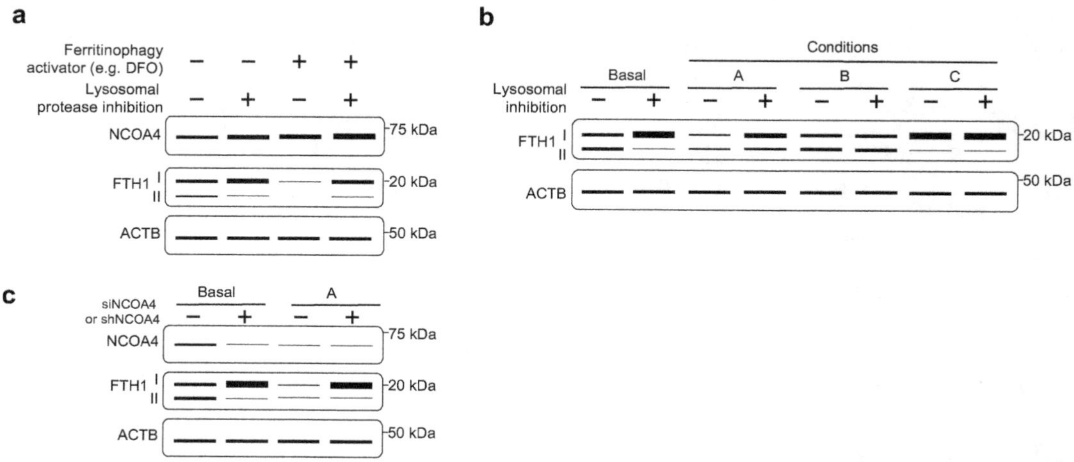


**Fig. 2** Summary of representative Western blot data. (**a**) A schematic illustration of the detection of ferritinophagic flux by immunoblotting. FTH1 and NCOA4 are degraded by ferritinophagy; therefore, the amounts of FTH1 and NCOA4 degraded by ferritinophagy provide an estimate of ferritinophagic flux. Iron depletion, for instance by DFO treatment, enhances NCOA4 and FTH1 lysosomal degradation. Inhibition of lysosomal degradation by lysosomal protease inhibitors (e.g., E64d and Pepstatin A or Cathepsin B inhibitors) or with Bafilomycin $A_1$, causes accumulation of FTH1, and this increment reflects the amount of FTH1 and NCOA4 that would have been degraded by autophagy during treatment. In this example, there is basal ferritinophagic flux in the cell that is further activated in response to DFO treatment. Note that NCOA4 levels may increase or decrease with ferritinophagic activation but regardless will increase with lysosomal inhibition reflecting the pool of NCOA4 degraded in the lysosome during flux. Note that FTH1 can demonstrate two bands (here labeled I and II). The faster-migrating band (II) is indicative of a lysosomal processed form of FTH1. (**b**) A schematic illustration of the interpretation of FTH1 immunoblotting results. Ferritinophagy flux in untreated (basal) cells is estimated by an increase in FTH1 amount under lysosomal inhibitor treatment. Condition "A" has a decreased amount of FTH1 compared to basal conditions, which is rescued in the presence of lysosomal inhibitors, suggesting increased ferritinophagic flux in these cells. Condition "B" has a level of FTH1 that does not respond to the inhibitor, indicating that ferritinophagy is not activated in these cells. Condition "C" has an increased amount of FTH1 compared to basal conditions which does not increase with lysosomal inhibition, suggesting decreased ferritinophagic flux in these cells. (**c**) A schematic illustration of the determination of NCOA4-dependent ferritinophagy. Knockdown of NCOA4 blocks the degradation of FTH1, therefore causing the accumulation of FTH1 in the cells. Condition "A" has a decrease in FTH1 levels, which is rescued upon genetic suppression of NCOA4, indicating Condition "A" induces ferritinophagic flux in an NCOA4-dependent manner

#### 3.1.2 Protocols

All steps are performed at room temperature (RT), and working solutions are kept at RT unless otherwise indicated.

1. To ensure robust protein detection for quantification, we recommend plating enough cells in poly-D lysine-coated 6-well plates to ensure at least 50 μg protein lysate (e.g., $5 \times 10^5$ HeLa cells) (*see* **Note 1**).
2. On the day of inducing/inhibiting ferritinophagy, cells should be at an appropriate density for experimental analysis, e.g., 80% confluency.

3. Perform experimental manipulation of cultured cells, e.g., compound treatment, knockdown, as appropriate (*see* **Note 2**). Ensure there are two wells for each condition (one that will be treated with lysosomal protease inhibitors and the other as a control).
4. Add lysosomal protease inhibitors, 2 μg/mL E64d and Pepstatin A or 10 μM CA-074 and incubate between 6 and 16 h. Alternatively, a short treatment of lysosomal acidification inhibitor 100 nM BafA1 for 2 h can be used (*see* **Note 3**). For each condition tested, include an untreated sample (DMSO) as a negative control. Concentrations, timing of lysosomal inhibitors addition and length of treatment may need to be optimized depending on the experimental "treatment" being evaluated and the cell line used.
5. After treatments are completed, place plates on ice, aspirate media and wash cells in each well with 1 mL ice-cold 1× PBS (*see* **Note 4**).
6. Remove excess liquid, then add 1 mL ice-cold 1× PBS to each well.
7. Scrape and collect the cells using a cell scraper and pipet cell suspension into a chilled, prelabeled 1.5 mL Eppendorf tube. Centrifuge the tubes at 500× *g* for 5 min at 4 °C. Discard the supernatant and add RIPA or Urea lysis buffer (*see* **Note 5**) to lyse cells. For a well of a 6-well plate, use ~40 μL lysis buffer. Add Benzonase (Millipore, Cat #70664) if needed to shear DNA and reduce sample viscosity.
8. Incubate the mixtures on ice for 30 min, mixing 2–3 times during incubation time.
9. If necessary, sonicate for 20–30 s to complete cell lysis and shear DNA to reduce sample viscosity.
10. Centrifuge at maximum speed (~21,000× *g*) for 15 min at 4 °C. Transfer the supernatant to a new pre-chilled tube and keep on ice. Discard the pellet.
11. Measure protein concentration using BCA kit per the manufacturer's protocol.
12. Prepare samples with 2× Laemmli sample buffer and boil samples for 5 min at 95 °C. Vortex to mix well and spin at 15000× *g* for 5 min at room temperature.
13. Load 50 μg protein/lane onto a 4–20% SDS-PAGE gel along with a molecular weight marker.
14. Run the gel at 100 V for 5 min, then increase the voltage to 150 V and complete the run in about 40–60 min (stop the power supply when the dye front reaches the end of the gel).

15. Transfer the protein from the gel to nitrocellulose or PVDF membrane at a constant current of 0.2 A for 1 h.
16. After transfer, wash the membrane with PBS or TBST for 5 min at RT.
17. Incubate the membrane in 10 mL of blocking solution for 1 h at RT. Wash the membrane with TBST for 5 min at RT.
18. To evaluate the levels of NCOA4, FTH1, and ACTB (β-actin) loading control, cut the membrane at 37 and 50 kDa. The strip below the 37 kDa marker is probed with FTH1, the 37–50 kDa strip is probed with ACTB, and the strip above 50 kDa is probed with NCOA4. If desired, MAP1LC3B and SQSTM1 can be probed separately to assay global autophagic flux (*see* **Notes 6–8**).
19. Incubate the membranes with the primary antibody overnight at 4 °C.
20. Wash the membrane with TBST for 3 × 5 min at RT.
21. During the washing step, prepare secondary antibodies and incubate the membrane for 1 h at RT.
22. Wash the membrane again with TBST for 3 × 5 min at RT.
23. Develop the blots with the ECL Western Blotting Substrate Kit.
24. Image the blots using standard X-ray film and developer or a CCD camera imaging device. Save the images and quantify the intensity of the immunoblot with ImageJ, as necessary.
25. Analyze the results for changes in ferritinophagic flux (*see* **Note 9**, text and Fig. 2).

### 3.2 Monitoring Ferritinophagy Using Fluorescence Microscopy

#### 3.2.1 Overview

In tandem with immunoblotting of NCOA4 and FTH1 levels, colocalization of NCOA4 and FTH1 in autophagic and lysosome structures in the presence of lysosomal protease inhibitors is required to evaluate ferritinophagic flux (Fig. 1). Here, we describe how to measure the colocalization of FTH1 and autophagosome markers such as MAP1LC3B [10] and lysosomal markers such as LAMP1 [21] or LAMP2 [4, 10, 22] to determine autophagic/lysosomal localization and turnover (*see* **Note 10**). The increase in the number of colocalized FTH1 puncta in the presence of a lysosome inhibitor, compared to that in the absence of the inhibitor, represents the amount of ferritin that would have been degraded during the treatment period. Autophagic/lysosomal localization is particularly important as NCOA4 (and FTH1) can form cytoplasmic condensates that can be mistaken for autophagic/lysosomal structures if no colocalization marker is included.

*3.2.2 Protocol*

All steps are performed at room temperature, and solutions are kept at room temperature unless otherwise indicated.

1. Prepare a 12-well tissue culture plate by adding the methanol-immersed 15 mm glass coverslips to each well at a slanted angle to allow the methanol to evaporate (~30 min).
2. After the methanol has evaporated, gently tap the plate to slide the coverslip flat onto the bottom of the well. The desired number of cells can now be plated into the plate.
   (a) For cells that are not adherent, we recommend coating the coverslips with poly-D-lysine as some cell lines tend to detach during treatments such as EBSS starvation. To do so, add 1 mL poly-D-lysine solution to each well and incubate for 10 min, wash twice with 1 mL autoclaved water, then once with 1 mL PBS.
3. Seed cells into the plate (usually 2–5 × $10^5$ cells per well). The cell number is adjusted based on cell size and the period of drug treatment.
4. After cells attach to the plate (usually overnight), treat cells with drugs and/or the ferritinophagy modulators, as above (*see also* **Note 11**).
5. At the end of the treatments, aspirate media and wash 2X with 1 mL PBS.
6. Fix cells with 1 mL 4% PFA in PBS at RT. Alternatively, for MAP1LC3B immunofluorescence, fix cells in chilled methanol for 15 min at −20 °C.
7. Wash 3X with RT PBS to remove the fixation solution. Pipette onto a corner of the slide and aspirate the opposite corner. If not continuing immediately, leave 1 mL PBS in each well and cover the plates with parafilm to prevent evaporation. Store at 4 °C until ready to stain.
8. If continuing immediately, wash an additional 2X with PBS and then permeabilize cells with 0.25% Triton X-100 in PBS for 5 min. Skip this step if fixation is done via methanol.
9. Wash 3X in PBS.
10. Block the coverslips for 30 min at RT using blocking solution, then wash 3X with PBS.
11. Prepare primary antibodies as indicated in PBS (*see* **Note 12**).
12. Prepare a humidified chamber by placing wet filter paper inside the stain tray. Add parafilm on top of the 12-well plate cover.
13. Add 30–50 μL antibody mix in a drop on the parafilm. Use forceps to take out the coverslip, tap it on the filter paper to remove PBS completely, and then place the coverslip with the cells facing down on the antibody drop. In this way, cells are

covered by the antibody solution. A piece of labeling tape with the treatment condition and antibody used noted can be placed next to the coverslip.

14. Repeat these steps for all the samples needed to stain. After finishing, put the plate in the humidified chamber and incubate overnight at 4 °C.
15. The next day, place the coverslips back in the 12-well plate containing PBS (cells facing up). Wash 3X in PBS and add both secondary antibodies for 1 h at RT in the dark.
16. Wash off secondary antibodies with PBS and mount for imaging to visualize in a fluorescence microscope.
17. Typically, imaging is performed at 63× or 100× magnification, and at least 20 cells per condition are imaged.
18. To determine colocalization, calculate the Pearson coefficient and count the number of colocalized puncta (FTH1/MAP1LC3B, FTH1/LAMP1, or FTH1/LAMP2) using ImageJ software. A further method for comparison between treatments is to calculate the relative fraction of intensity of ferritin staining in colocalized puncta versus the total intensity in the cell.

## 4 Notes

1. The expression level of NCOA4 and FTH1 varies among cell lines. To obtain an optimal NCOA4 immunoblot, we recommend loading ~50 μg protein per lane and developing the membrane with SuperSignal™ West Pico PLUS Chemiluminescent Substrate.
2. Positive control: To maximize ferritinophagic flux, cells can be loaded with ferric ammonium citrate (FAC, Fisher Scientific, 36–180 μM added to cells for 24 h) to induce the level of FTH1 in the cells. Then, the cells are removed from FAC completely, and an iron chelator (e.g., deferoxamine (DFO, 50–100 μM) or deferasirox (DFX, 50–100 μM) can be added for 6 to 12 h. This leads to robust NCOA4-dependent ferritinophagy [4, 13].
3. Lysosomal protease inhibitors are the preferred method for inhibition of lysosome-mediated degradation as opposed to inhibitors of lysosomal acidification (e.g., Bafilomycin $A_1$) as inhibition of lysosomal acidification blocks endolysosomal trafficking of Transferrin-iron complexes, thereby perturbing cellular iron homeostasis beyond ferritinophagy.
4. MAP1LC3B-I and MAP1LC3B-II (lipidated form) are sensitive to degradation, although MAP1LC3B-I is more labile.

These proteins are sensitive to freeze-thaw cycles and SDS sample buffers. Fresh samples should be analyzed rapidly to prevent protein degradation.

5. NCOA4 can form detergent-insoluble condensates in the cytoplasm (with or without FTH1) under basal conditions with a further increase under iron-replete conditions. To collect and evaluate both the soluble and insoluble fractions of NCOA4 and FTH1 by SDS-PAGE, urea lysis buffer should be used [13, 23].
6. Anti-NCOA4 (or anti-ARA70) antibody (Santa Cruz, Cat# sc-373739) reacts with both human and mouse NCOA4, while the anti-NCOA4 antibody (Cell Signaling, Cat#66849S) only reacts with human NCOA4.
7. Anti-NCOA4 antibody (Santa Cruz, Cat# sc-373739) has significant background, including at bands just above and below where NCOA4 migrates (69 kDa). NCOA4 siRNAs or shRNAs can be used to ensure the correct band for NCOA4. To minimize off-target antibody binding, the NCOA4 blot can also be cut just above 80 kDa.
8. Note that FTH1 can demonstrate two bands (labeled I and II in Fig. 2). The faster-migrating band (II) is indicative of a lysosomal processed form of FTH1 [10, 24].
9. Genetic NCOA4 ablation (typically RNAi-mediated knockdown) is frequently employed to demonstrate the specificity of FTH1 turnover to the NCOA4-mediated ferritinophagy pathway. Instead of lysosomal protease inhibition, NCOA4 knockdown would be employed following standard procedures for either siRNA or shRNA-mediated knockdown [4, 6]. NCOA4 knockdown would rescue FTH1 turnover under the "treatment" conditions (see also Fig. 2).
10. We do not use fluorescently tagged (tandem RGP-GFP, Keima, etc.) FTH1 or NCOA4 to determine ferritinophagic flux as the exogenous expression of FTH1 or NCOA4 (especially at supraphysiologic levels) can alter cellular iron metabolism (overexpression of FTH1 leads to a decrease in the labile iron pool and overexpression of NCOA4 can itself induce ferritinophagy), thereby interfering with any effect of the experimental conditions being evaluated.
11. Knockdown of NCOA4 can similarly be used to demonstrate that loss of NCOA4 abrogates FTH1/ferritin localization to autophagic/lysosomal structures under the treatment conditions being evaluated.
12. While NCOA4 colocalization with FTH1 and MAP1LC3B/LAMP1/LAMP2 would further demonstrate evidence of NCOA4-mediated ferritinophagy, NCOA4 immunostaining

is challenging given the lack of suitable antibodies. Immunofluorescence of endogenous NCOA4 has been reported (using Invitrogen, Cat #PA5-96398, and LAMP1/2); however, in our experience, the background of this antibody is high. We recommend that investigators validate any NCOA4 antibody used for immunofluorescence for the specificity of the signal using NCOA4 genetic ablation.

## 5 Concluding Remarks

Here we have described existing techniques and methods to monitor ferritinophagic flux in mammalian cells. In vitro, Western blotting and immunofluorescence-based colocalization are the simplest and most effective methods for quantifying flux. Measuring ferritinophagic flux in vivo in animal models remains challenging and requires the development of additional tools and model systems. In general, we have applied Western blotting techniques from whole tissues in tandem with conditional knockout mouse models to confirm ferritinophagy flux in a particular cell type (a *Ncoa4*$^{tm1.1Jdom}$/J conditional knockout strain is available at The Jackson Laboratory https://www.jax.org/strain/033295) [8]. NCOA4 immunohistochemistry from mouse and human tissues has been reported, but we have been unable to replicate NCOA4-specific staining (as determined using wild-type and Ncoa4 knockout animals) in our laboratory.

The downstream consequences of altered ferritinophagic flux in cell culture systems can also be measured to further understand the importance of ferritinophagy under the experimental conditions being evaluated (Fig. 1; protocols for this are beyond the scope of this methods paper). For instance, one consequence of increased ferritinophagic flux may be an increase in the labile iron pool. The labile iron pool ($Fe^{2+}$) can be measured using FerroOrange or by monitoring the levels of IREB2, which is sensitive to labile iron pool levels (see these Refs. [4, 7] for protocols). An increase in the labile iron pool can increase sensitivity to ferroptosis, an iron-dependent form of cell death [14]. Of note, an increased ferritinophagic flux on its own does not induce ferroptosis but rather increases sensitivity to ferroptotic inducers. Downstream readouts of decreased ferritinophagy are related to the general effects that are demonstrated with a decrease in the labile iron pool, such as a compensatory increase in iron import (e.g., increased TFRC), decreased iron-sulfur cluster protein levels, and a decrease in mitochondrial oxidative phosphorylation (due to decreases in ISC-containing components) [8].

## References


1. Galy B, Conrad M et al (2023) Mechanisms controlling cellular and systemic iron homeostasis. Nat Rev Mol Cell Biol 25:133
2. Dixon SJ, Stockwell BR (2014) The role of iron and reactive oxygen species in cell death. Nat Chem Biol 10(1):9–17
3. Koorts AM, Viljoen M (2007) Ferritin and ferritin isoforms I: structure–function relationships, synthesis, degradation and secretion. Arch Physiol Biochem 113(1):30–54
4. Mancias JD, Wang X et al (2014) Quantitative proteomics identifies NCOA4 as the cargo receptor mediating ferritinophagy. Nature 509(7498):105–109
5. Kidane TZ, Sauble E et al (2006) Release of iron from ferritin requires lysosomal activity. Am J Physiol Cell Physiol 291(3):C445–C455
6. Mancias JD, Pontano Vaites L et al (2015) Ferritinophagy via NCOA4 is required for erythropoiesis and is regulated by iron dependent HERC2-mediated proteolysis. elife 4: e10308
7. Santana-Codina N, Gableske S et al (2019) NCOA4 maintains murine erythropoiesis via cell autonomous and non-autonomous mechanisms. Haematologica 104(7): 1342–1354
8. Santana-Codina N, Del Rey MQ et al (2022) NCOA4-mediated ferritinophagy is a pancreatic cancer dependency via maintenance of iron bioavailability for iron–sulfur cluster proteins. Cancer Discov 12(9):2180–2197
9. Jahng JWS, Alsaadi RM et al (2019) Iron overload inhibits late stage autophagic flux leading to insulin resistance. EMBO Rep 20(10): e47911
10. Dowdle WE, Nyfeler B et al (2014) Selective VPS34 inhibitor blocks autophagy and uncovers a role for NCOA4 in ferritin degradation and iron homeostasis in vivo. Nat Cell Biol 16(11):1069–1079
11. Li X, Lozovatsky L et al (2020) NCOA4 is regulated by HIF and mediates mobilization of murine hepatic iron stores after blood loss. Blood 136(23):2691–2702
12. Asano T, Komatsu M et al (2011) Distinct mechanisms of ferritin delivery to lysosomes in iron-depleted and iron-replete cells. Mol Cell Biol 31(10):2040–2052
13. Kuno S, Fujita H et al (2022) Iron-induced NCOA4 condensation regulates ferritin fate and iron homeostasis. EMBO Rep 23(5): e54278
14. Santana-Codina N, Gikandi A et al (2021) The role of NCOA4-mediated ferritinophagy ferritinophagy in ferroptosis. In: Florez AF, Alborzinia H (eds) Ferroptosis: mechanism and diseases, Advances in experimental medicine and biology. Springer, Cham, pp 41–57
15. Ohshima T, Yamamoto H et al (2022) NCOA4 drives ferritin phase separation to facilitate macroferritinophagy and microferritinophagy. J Cell Biol 221(10):e202203102
16. Goodwin JM, Dowdle WE et al (2017) Autophagy-independent lysosomal targeting regulated by ULK1/2-FIP200 and ATG9. Cell Rep 20(10):2341–2356
17. Yanatori I, Richardson DR et al (2021) CD63 is regulated by iron via the IRE-IRP system and is important for ferritin secretion by extracellular vesicles. Blood 138(16):1490–1503. https://doi.org/10.1182/blood.2021010995
18. Kuno S, Iwai K (2023) Oxygen modulates iron homeostasis by switching iron sensing of NCOA4. J Biol Chem 299(5):104701
19. Dai Y, Zhu C et al (2023) Mycobacterium tuberculosis hijacks host TRIM21-and NCOA4-dependent ferritinophagy to enhance intracellular growth. J Clin Invest 133(8): e159941
20. Mai TT, Hamaï A et al (2017) Salinomycin kills cancer stem cells by sequestering iron in lysosomes. Nat Chem 9(10):1025–1033
21. Li J, Huang Q et al (2023) Role of liensinine in sensitivity of activated macrophages to ferroptosis and in acute liver injury. Cell Death Discov 9:189. https://doi.org/10.1038/s41420-023-01481-3
22. Ravichandran M, Hu J et al (2022) Coordinated transcriptional and catabolic programs support iron-dependent adaptation to RAS–MAPK pathway inhibition in pancreatic cancer. Cancer Discov 12(9):2198–2219
23. Rai M, Curley M et al (2021) Analysis of proteostasis during aging with western blot of detergent-soluble and insoluble protein fractions. STAR Protoc 2(3):100628
24. Cohen LA, Gutierrez L et al (2010) Serum ferritin is derived primarily from macrophages through a nonclassical secretory pathway. Blood 116(9):1574–1584

# Chapter 15

# Method to Purify Phase-Separated p62 Bodies Using Fluorescence-Activated Particle Sorting

**Shuhei Takada, Masaaki Komatsu, and Hideaki Morishita**


## Abstract

p62 bodies are ubiquitin-positive cytoplasmic condensates formed by liquid–liquid phase separation. They are targeted by selective autophagy and play important roles in intracellular quality control and stress responses. However, little is known about their constituents. In this chapter, we describe a method for purifying p62 bodies using fluorescence-activated particle sorting. This method contributes to the identification of novel components of p62 bodies under various physiological and stress conditions.


**Key words** p62/SQSTM1 body, Ubiquitinated proteins, Liquid–liquid phase separation, Autophagy, Selective autophagy, Cell sorter, Fluorescence-activated particle sorting

## 1 Introduction

Under physiological and stress conditions, various biomolecular condensates are formed by liquid–liquid phase separation (LLPS). Among them, p62 bodies are cytoplasmic ubiquitin-positive condensates whose formation requires the self-oligomerization of p62 and its interaction with polyubiquitinated proteins [1–3]. p62 bodies are targeted by selective autophagy through the recruitment of autophagy receptors (such as Next to BRCA1, gene 1 [NBR1] and tax1-binding protein 1 [TAX1BP1]) and autophagy-related (Atg) proteins, such as FAK family kinase-interacting protein of 200 kDa (FIP200) and Atg8 family proteins on autophagosomes [4]. Kelch-like ECH-associated protein 1 (KEAP1) is another protein recruited to p62 bodies to activate NF-E2-related factor 2 (NRF2), a transcription factor critical for antioxidative stress responses [5]. The accumulation of p62 bodies due to impaired autophagic clearance results in persistent activation of the antioxidant stress signaling pathway, followed by life-threatening diseases, including hepatocellular carcinoma [3]. Thus, the autophagy-mediated homeostasis of p62 bodies is indispensable for human

Ioannis P. Nezis (ed.), *Selective Autophagy: Methods and Protocols*, Methods in Molecular Biology, vol. 2845, https://doi.org/10.1007/978-1-0716-4067-8_15,

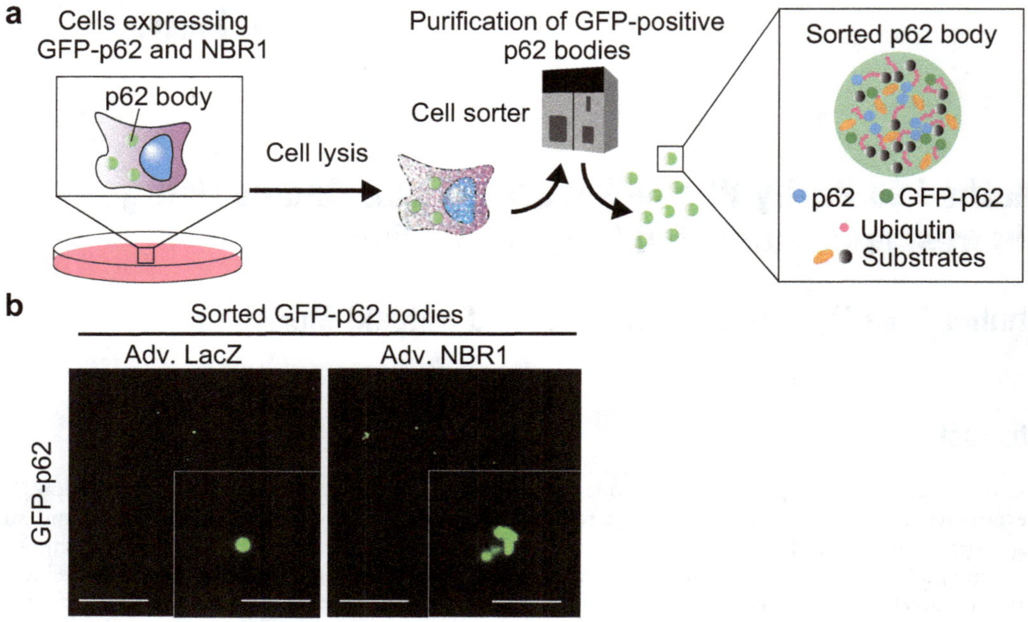


**Fig. 1** (**a**) Schematic representation of p62 body purification method. (**b**) Fluorescence micrographs of GFP-positive p62 bodies purified from GFP-p62-expressing Huh-1 cells infected with LacZ or NBR1 adenoviruses (Adv.) for 24 h. Scale bars: 50 μm (main panels) and 2.5 μm (inset panels)

health. Despite recent progress in elucidating the degradation mechanisms and pathophysiological functions of p62 bodies, little is known about their components.

Recently, we developed a method to purify p62 bodies using fluorescence-activated particle sorting (FAPS). This method purifies GFP-labeled p62 bodies from lysates of GFP-p62-expressing cells using a cell sorter [6] (Fig. 1a). We modified a previously described method for purifying P-bodies [7]. To efficiently purify p62 bodies, we overexpressed NBR1, a binding partner of p62 [6]. NBR1 decreases the liquidity of p62 bodies and increases their abundance in cells, probably by slowing their autophagic turnover [8]. Purified p62 bodies can be used in various assays, including fluorescence microscopy, western blotting, and mass spectrometry. Mass spectrometry of the isolated p62 bodies revealed that they contained not only known components such as p62, NBR1, and TAX1BP1 but also novel components such as the supramolecular protein complex vault [6]. Thus, the FAPS-based method of purification is a powerful tool for identifying novel p62 body components, particularly under various stress conditions. In this chapter, we present an overview of the FAPS-based p62 body purification method.

## 2 Materials

All solutions are prepared using ultrapure and deionized water and analytical-grade reagents. Unless otherwise noted, all reagents and solutions should be stored at 4 °C.

### 2.1 Fluorescent Labeling of p62 Bodies and Their Stabilization in Cells

1. Cells: Huh-1 (human hepatocellular carcinoma cell line, Riken, JCRB0199) stably expressing GFP-p62 at low levels [6] (*see* **Note 1**).
2. Culture medium: Dulbecco's modified Eagle's medium/high glucose supplemented with 10% fetal bovine serum (FBS), 2 mM L-glutamine, 5 U/mL penicillin, and 50 μg/mL streptomycin.
3. Culture dish: 10-cm culture dish.
4. NBR1 and LacZ adenoviruses (control) [6] (*see* **Note 2**).

### 2.2 Preparation of Cell Lysates

1. Phosphate-buffered saline (PBS).
2. 10 mL syringe with a 27G 1/2 needle.
3. Lysis buffer #1: 50 mM Tris [pH 7.4], 1 mM EDTA, 150 mM NaCl, 0.2% Triton X-100, cOmplete EDTA-free protease inhibitor cocktail, 32.5 U/mL RNasin Ribonuclease Inhibitor.
4. 500 mM $MgSO_4$.
5. 1 M $CaCl_2$.
6. 5 U/μL DNase I.
7. Lysis buffer #2: 50 mM Tris [pH 7.4], 1 mM EDTA, 150 mM NaCl, 0.2% Triton X-100, cOmplete EDTA-free protease inhibitor cocktail, 160 U/mL RNasin Ribonuclease Inhibitor.
8. Cell Strainer (352235; Falcon).
9. 15 mL Falcon tubes.

### 2.3 Purification of p62 Bodies Using a Cell Sorter

1. Cell sorter: Cell Sorter SH800S (SONY) fitted with a disposable 70-μm nozzle.
2. Sheath liquid: IsoFlow (8599600; BECKMAN COULTER).

## 3 Methods

### 3.1 Fluorescent Labeling of p62 Bodies and Their Stabilization in Cells

1. Thaw and culture GFP-p62-expressing cells using a 10-cm culture dish at 37 °C in a 5% $CO_2$ incubator. Split cells every 2–3 days.
2. Twenty-four hours prior to infection with NBR1 adenovirus, seed the appropriate number of GFP-p62-expressing cells (e.g., $1 \times 10^6$ cells/dish) in at least two 10 cm culture dishes.

3. Infect the GFP-p62-expressing cells with NBR1 adenovirus and incubate the cells at 37 °C for 24 h (*see* **Note 3**).

### 3.2 Preparation of Cell Lysates

1. Twenty-four hours after infection with the NBR1 adenovirus, discard the culture medium from 10-cm culture dishes and wash cells twice with ice-cold PBS.
2. Add 1 mL of lysis buffer #1 to the dish from which PBS was removed and collect cells in a 2-mL tube using a cell scraper.
3. Add 1 mL of lysis buffer #1 to the dish and collect the remaining cells in the same 2-mL tube using a cell scraper.
4. Slowly homogenize the cells in lysis buffer #1 using a 10 mL syringe with a 27G 1/2 needle, passing the solution through the syringe 20 times while ice cold.
5. Centrifuge the sample at 200 × *g* for 5 min at 4 °C to remove the nuclei.
6. Collect the supernatant in a new 2-mL tube and add an appropriate volume of each stock solution (500 mM $MgSO_4$, 1 M $CaCl_2$, 5 U/μL DNase I) to digest DNA contaminants and make a final concentration of 10 mM $MgSO_4$, 1 mM $CaCl_2$, and 4 U/mL DNase I.
7. Mix the sample by inversion five times, wrap it in aluminum foil, shield it from light, and incubate it at room temperature for 30 min.
8. Centrifuge the sample at 10,000 × *g* for 7 min at 4 °C.
9. Discard the supernatant and gently wash the pellet at the bottom of the tube with 500 μL of lysis buffer #1.
10. Centrifuge the sample at 10,000 × *g* for 7 min at 4 °C.
11. Discard the supernatant and resuspend the pellet at the bottom of the tube in lysis buffer #2 (2 mL/10-cm culture dish).
12. Pass the sample through a Cell Strainer and store it in the 15-mL tube at 4 °C.

### 3.3 Purification of p62 Bodies Using a Cell Sorter

1. Place the above sample in a cell sorter and gate the GFP-positive fraction detected by back-scattered light and green fluorescence (*see* **Note 4**).
2. Collect the GFP-positive fraction (*see* **Note 5**).
3. During sample exchange, flush the sample line with lysis buffer #2 for 1 min.
4. Transfer the sorted sample to a 1.5-mL tube and centrifuge the sample at 20,000 × *g* for 7 min at 4 °C. Carefully discard the supernatant, leaving approximately 100 μL of sorting fluid. Repeat the same procedure as many times as necessary using the same tube until all sorted samples can be pelleted and concentrated in a single tube (*see* **Note 6**).

5. Use the resuspended samples (100 μL in sheath liquid) for each assay (*see* **Note 7**). GFP-positive p62 bodies are visible by fluorescence microscopy (Fig. 1b).

## 4 Notes

1. To efficiently purify p62 bodies, we recommend using cell lines expressing high levels of endogenous p62 and containing some spontaneously formed p62 bodies (e.g., Huh-1). To avoid the overproduction of p62 bodies, the expression level of GFP-p62 should not be higher than the endogenous level. To this end, we used a retrovirus and a cell sorter to establish Huh-1 cells that stably express low levels of GFP-p62. To ensure experimental stability, a single cloning of cells with appropriate expression is recommended.
2. The multiplicity of infection (MOI) of the NBR1 adenovirus for Huh-1 cells was 7.5, but optimization is required for each cell type. Instead of an adenovirus, transient or stable expression of NBR1 can be achieved by plasmid transfection or retroviral infection. pMRX-IRES-puro-3xFLAG-NBR1 is available from Addgene (#204546).
3. Overexpression of NBR1 causes p62 bodies to become gel-like structures from liquid [6]. To confirm the morphology of NBR1-positive p62 bodies, it is recommended to carry out immunofluorescence analysis using anti-NBR1 (Santa Cruz Biotechnology, sc-130380) and anti-p62 antibodies (Proteintech, 18420-1-AP).
4. Forward scattering (FSC), backscattering (BSC), and GFP channels should all be set to logarithmic. When sorting with Sony SH800S, the quality setting of the cell sorter is set to "Semi-Purity," and the sample pressure is adjusted so that 10,000 events/s can be measured. Flow rates exceeding 13,000 events/s increase the risk of the sample line or nozzle clogging. If sorting with BD FACS Aria2, set the flow rate to 3000 events/s or fewer.
5. To adjust the sensor gain for each channel, including the FSC, BSC, and GFP channels, it is recommended to check for the presence or absence of a GFP-positive population on the display scale using appropriate cells such as (1) cells not expressing GFP-p62 or NBR1, (2) cells expressing GFP-p62 and LacZ, and (3) cells expressing GFP-p62 and NBR1. The threshold and sensor gain must be set for each experiment. If using the SH800S system, setting FSC: sensor gain 10 (1–10 steps), threshold 0.08%, BSC: sensor gain 38%, and GFP: sensor

gain 43% should be sufficient to observe GFP-positive fractions.

6. The amount of protein in purified p62 bodies that can be collected in 1 day is approximately 1500–3000 ng. If the number of p62 bodies is insufficient, the number of cells should be increased.
7. The pellets can be lyophilized using a Flexi-Dry MP freeze-dryer (FTS Systems) and stored at −80 °C.

## Acknowledgments

M.K. was supported by a Grant-in-Aid for Scientific Research on Innovative Areas (19H05706), a Grant-in-Aid for Scientific Research (S) (24H000060), and a Grant-in-Aid for International Leading Research (23K20044). H.M. was supported by a Grant-in-Aid for Scientific Research (B) (23H02441), a Grant-in-Aid for Transformative Research Areas (24H01901), SGH Cancer Research Grant (A2023_18), and Advanced Research and Development Programs for Medical Innovation (AMED-PRIME, 23gm6410019h0003). S. T. received support from a Grant-in-Aid for JSPS Research Fellows (23KJ1927).

### References

1. Sun D, Wu R, Zheng J, Li P, Yu L (2018) Polyubiquitin chain-induced p62 phase separation drives autophagic cargo segregation. Cell Res 28(4):405–415. https://doi.org/10.1038/s41422-018-0017-7
2. Zaffagnini G, Savova A, Danieli A et al (2018) p62 filaments capture and present ubiquitinated cargos for autophagy. EMBO J 37(5):e98308. https://doi.org/10.15252/embj.201798308
3. Kurusu R, Morishita H, Komatsu M (2023) p62 bodies: cytosolic zoning by phase separation. J Biochem. https://doi.org/10.1093/jb/mvad089
4. Turco E, Savova A, Gere F et al (2021) Reconstitution defines the roles of p62, NBR1 and TAX1BP1 in ubiquitin condensate formation and autophagy initiation. Nat Commun 12(1):5212. https://doi.org/10.1038/s41467-021-25572-w
5. Komatsu M, Kurokawa H, Waguri S et al (2010) The selective autophagy substrate p62 activates the stress responsive transcription factor Nrf2 through inactivation of Keap1. Nat Cell Biol 12(3):213–223. https://doi.org/10.1038/ncb2021
6. Kurusu R, Fujimoto Y, Morishita H et al (2023) Integrated proteomics identifies p62-dependent selective autophagy of the supramolecular vault complex. Dev Cell 58(13):1189–1205.e11. https://doi.org/10.1016/j.devcel.2023.04.015
7. Hubstenberger A, Courel M, Benard M et al (2017) P-Body purification reveals the condensation of repressed mRNA regulons. Mol Cell 68(1):144–157e5. https://doi.org/10.1016/j.molcel.2017.09.003
8. Sanchez-Martin P, Sou YS, Kageyama S, Koike M, Waguri S, Komatsu M (2020) NBR1-mediated p62-liquid droplets enhance the Keap1-Nrf2 system. EMBO Rep 21(3):e48902. https://doi.org/10.15252/embr.201948902

# Chapter 16

# In Silico Assessment of LIR Motif-Containing Proteins Binding to ATG8-Family Proteins

**Mamas Louca, Bridget Penman, Alexander Cameron, and Ioannis P. Nezis**

## Abstract

Selective autophagic degradation of cellular components has been shown to be mediated by the interaction of LIR motif-containing proteins with ATG8-family proteins. Here, we present a detailed methodology for the in silico evaluation of potential binding between LIR motif-containing proteins and ATG8-family proteins. We visualize AlphaFold-predicted protein complexes using PyMOL to assess potential interactions, providing an effective computational tool for this purpose.

**Key words** LIR motif, Atg8-family proteins, AlphaFold, p62

## 1 Introduction

Understanding protein–protein interactions is essential for deciphering cellular processes and signaling pathways. Of particular interest are interactions between proteins containing a LIR motif and ATG8 family proteins, which are crucial for selective autophagy [1].

Our proposed in silico approach involves generating AlphaFold-predicted protein complexes [2, 3]. As we show below, PyMOL visualization of these complexes enables differentiation between the case where a wild-type protein binds to an ATG8-family protein through its LIR motif and the case where a mutant protein (with experimentally verified disruption in LIR binding) fails to interact in the same way.

## 2 Methods

### 2.1 ColabFold2: Protein Structure Prediction

Protein structure prediction was performed using ColabFold2, an adaptation of the AlphaFold2 algorithm that runs on the Google Collaboratory platform. ColabFold2 provides a user-friendly

Ioannis P. Nezis (ed.), *Selective Autophagy: Methods and Protocols*, Methods in Molecular Biology, vol. 2845,
https://doi.org/10.1007/978-1-0716-4067-8_16,

interface and computational resources for accurate and efficient protein structure prediction [4].

### 2.2 ColabFold2 Prediction Workflow

1. *Access ColabFold2 Website*: To initiate the ColabFold2 prediction workflow, navigate to the ColabFold2 website, which hosts an adapted version of the AlphaFold2 algorithm designed to run on the Google Collaboratory platform. The website link is: https://colab.research.google.com/github/sokrypton/ColabFold/blob/main/AlphaFold2.ipynb [Accessed 20 June 2023]. The version utilized here is ColabFold v1.5.2 (*see* **Note 1**).
2. *Input Preparation*: Prepare the protein sequences of your ATG8 family protein (bait protein) and the LIR motif-containing ligand (prey protein) in FASTA format. You can retrieve these sequences from databases such as UniProt or PubMed. Concatenate the two sequences together, separating them with a colon (:) to represent the interaction between the bait and prey proteins. For instance, if exploring the interaction between LC3B (bait) and p62 (prey) [5], combine their sequences as follows:

Bait sequence (LC3B): *MPSEKTFKQRRTFEQRVEDVRLIREQHPTKIPVIIERYKGEKQLPVLDKTKFLVPDHVNM*SE
LIKIIRRRLQLNANQAFFLLVNGHSMVSVSTPISEVYESEKDEDGFLYMVYASQETFGMKLSV

Prey sequence (p62): *MASLTVKAYLLGKEDAAREIRRFSFCCSPEPEAEAEAAAGPGPCERLLSRVAALFPALRP* GGF
QAHYRDEDGDLVAFSSDEELTMAMSYVKDDIFRIYIKEKKECRRDHRPPCAQEAPRN MVHPNVICDGCNGPV
VGTRYKCSVCPDYDLCSVCEGKGLHRGHTKLAFPSPFGHLSEGFS H
SRWLRKVKHGHFGWPGWEMGPPGNWSPRPPRAGEARPGPTAESASGPSEDPSVNFLKNV GESVAAALSPLGIEV
DIDVEHGGKRSRLTPVSPESSSTEEKSSSQPSSCCSDPSKPGGNV
EGATQSLAEQMRKIALESEGRPEEQMESDNCSGGDDDWTHLSSKEVDPSTGELQSLQMPE SEGPSSLDPS
QEGPTGLKEAALYPHLPPEADPRLIESLSQMLSMGFSDEGGWLTRLLQTK NYDIGAALDTIQYSKHPPPL

Input sequence (combined): *MPSEKTFKQRRTFEQRVEDVRLIREQHPTKIPVIIERYKGEKQLPVLDKT*KFLVPDHVN
MSELIKIIRRRLQLNANQAFFLLVNGHSMVSVSTPISEVYESEKDEDGFLYMVYASQETFGMKLSV:MASLTV
KAYLLGKEDAAREIRRFSFCCSPEPEAEAEAAAGPGPCERLLSRVAALFPALRPGGFQAHYRDEDGDL
VAFSSDEELTMAMSYVKDDIFRIYIKEKKECRRDHRPPCAQEAPRNMVHPNVICDGCNGPVVGTRYKCSVCP
DYDLCSVCEGKGLHRGHTKLAFPSPFGHLSEGFSHSRWLRKVKHGHFGWPGWEMGPPGNWSPRPPRAGEARP
GPTAESASGPSEDPSVNFLKNVGESVAAALSPLGIEVDIDVEHGGKRSRLTPVSPESS
STEEKSSSQPSSCCSDPSKPGGNVEGATQSLAEQMRKIALESEGRPEEQMESDNCSGGDDDWTHLSSKEVDPST
GELQSLQMPESEGPS
SLDPSQEGPTGLKEAALYPHLPPEADPRLIESLSQMLSMGFSDEGGWLTRLLQTKNYDIGAALDTIQYSKHPPPL

3. *Assign Job Name*: Specify a job name for the ColabFold2 prediction, which will also serve as the name of the output folder containing the generated files.
4. *Default Parameter Settings*: Utilize the default settings provided on the ColabFold2 website for the prediction process. These default configurations encompass various parameters, including model selection, minimum coverage, maximum templates, maximum MSA size, and Amber relaxation.

5. *Execute the Job*: Submit the prediction job on the ColabFold2 platform, which will utilize the available computational resources and infrastructure to perform the protein structure prediction.
   (a) Click on "Runtime" in the menu bar of the ColabFold2 website, and from the dropdown, select "Run All" to initiate the execution of the ColabFold2 prediction workflow.
   (b) Monitor Progress: During the execution, you can monitor the progress of the job as it proceeds through the different steps of the prediction workflow.
   (c) Download the Output: Once the prediction is completed, the resulting files, including the PDB file of the predicted protein complex, will be automatically generated and downloaded. The PDB file labeled *"rank 1"* and *"relaxed"* corresponds to the highest confidence predicted protein complex *("rank 1")* and represents the structure after Amber relaxation *("relaxed")*. Use this *"rank 1"* and *"relaxed"* version for further analysis.
6. *Result Analysis*: Utilize molecular visualization tools, such as PyMOL, to analyze and visualize the predicted protein structures obtained from the ColabFold2 prediction. These structures will provide insights into the potential interactions between the bait (ATG8-family protein) and the prey (LIR motif-containing ligand), facilitating the study of protein–protein interactions.

## 3 Molecular Visualization Workflow Using PyMOL

1. *Retrieve the Predicted Protein Complex*: Locate the predicted protein complex file in PDB format that was generated and downloaded from the ColabFold2 output folder. This file contains the 3D coordinates of the predicted protein complex, representing the interaction between the ATG8-family protein (bait) and the LIR motif-containing protein (prey).
2. *Open the File in PyMOL*: Open PyMOL, or your molecular visualization tool of choice. Then, load the PDB file of the predicted protein complex into PyMOL by clicking on "File" in the menu bar and selecting "Open" from the dropdown menu. Navigate to the location of the PDB file and open it in PyMOL.
3. *Label the Molecular Chains*: In the PyMOL graphical interface, identify and label the molecular chains corresponding to the

ATG8 family protein (bait) and the LIR motif-containing protein (prey). This can be done by selecting the respective chains and assigning labels or color-coding them for clarity. For example, you can label the ATG8 family protein as "Bait" and the LIR motif-containing protein as "Prey."

4. *Label the LIR Docking Site on the Bait Protein*: Identify the residues that make up the LIR docking site on the ATG8-family protein (bait) and label them within PyMOL. You can select the specific residues using the PyMOL selection tools and then apply labels or change their colors to highlight the LIR docking site.
5. *Label the LIR Motif on the Prey Protein*: Similarly, identify the residues that constitute the LIR motif on the LIR motif-containing protein (prey) and label them within PyMOL. Use the selection tools to choose the relevant residues and apply labels or change their colors to highlight the LIR motif.
6. *Visual Analysis of Protein–Protein Interaction*: With the protein chains, LIR docking site, and LIR motif labeled in PyMOL, visually analyze the predicted protein complex to determine if the two proteins are indeed interacting through the LIR docking site and LIR motif. Observe the spatial arrangement and potential interactions between the labeled residues to assess the plausibility of the predicted interaction.
7. *Interpretation of Results*: If the labeled residues of the LIR docking site on the bait protein and the LIR motif on the prey protein are closely positioned and exhibit favorable interactions, such as hydrogen bonding or electrostatic attractions, it may suggest a potential interaction between the two proteins through the LIR motif and LIR docking site. The alignment of these interactions with the predicted LIR motif and LIR docking site enhances confidence in the accuracy of ColabFold2's computational prediction (see Fig. 1). Conversely, the absence of favorable interactions or significant spatial distance between residues may indicate a lower likelihood of interaction through the LIR motif and LIR docking site or a less accurate prediction (*see* Fig. 1).

## 4 Notes

1. It is anticipated that new versions of ColabFold will be released in the future. There may be some variations in the results depending on the version of software used.

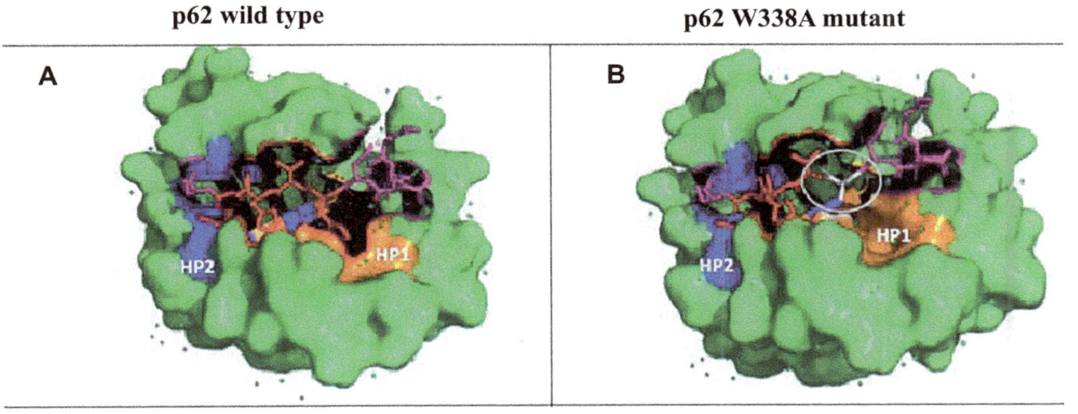


**Fig. 1** AlphaFold-predicted protein complexes of various ligands binding to LC3B protein. Comparing predicted interaction complexes with wild-type ligands against ligands with mutations that have experimentally been shown to disrupt LC3b binding. (**a**) Wild-type p62 binding to LC3B, (**b**) p62 sequestosome (W338A) mutation. LC3B is represented by a translucent green surface with the hydrophobic pockets (HP1 & HP2) that form the binding site for the LIR motif colored, orange for HP1 and blue for HP2. The ligand p62 is represented by a cyan-colored peptide backbone structure. The residues that form the extended LIR motif are colored in magenta, and the residues that comprise the core-LIR are colored in red. The residues that have been experimentally shown to disrupt LC3B binding are colored in white

## References

1. Rogov VV, Nezis IP, Tsapras P et al (2023) Atg8 family proteins, LIR/AIM motifs and other interaction modes. Autophagy Rep 2(1): 27694127.2023.2188523
2. Bryant P, Pozzati G, Elofsson A (2022) Improved prediction of protein-protein interactions using AlphaFold2. Nat Commun 13
3. Zhu W, Shenoy A, Kundrotas P, Elofsson A (2023) Evaluation of AlphaFold-multimer prediction on multi-chain protein complexes. Bioinformatics 39
4. Mirdita M, Schütze K, Moriwaki Y et al (2022) ColabFold: making protein folding accessible to all. Nat Methods 19:679–682
5. Pankiv S et al (2007) p62/SQSTM1 binds directly to Atg8/LC3 to facilitate degradation of ubiquitinated protein aggregates by autophagy. J Biol Chem 282:24131–24145

# Chapter 17

# High-Throughput Screening for LC3/GABARAP Binders Utilizing the Fluorescence Polarization Assay

**Martin P. Schwalm, Johannes Dopfer, Stefan Knapp, and Vladimir V. Rogov**


## Abstract

The characterization of interactions between autophagy modifiers (Atg8-family proteins) and their natural ligands (peptides and proteins) or small molecules is important for a detailed understanding of selective autophagy mechanisms and for the design of potential Atg8 inhibitors that affect the autophagy processes in cells. The fluorescence polarization (FP) assay is a rapid, cost-effective, and robust method that provides affinity and selectivity information for small molecules and peptide ligands targeting human Atg8 proteins.

This chapter introduces the basic principles of FP assays. In addition, a case study on peptide interaction with human Atg8 proteins (LC3/GABARAPs) is described. Finally, data analysis and quality control of FP assays are discussed for the proper calculation of $K_i$ values for the measured compounds.


**Key words** Fluorescence polarization, $K_i$ determination, LC3, GABARAP, LIR motif, High-throughput screening

## 1 Introduction

### 1.1 Role of LC3/GABARAPs in Autophagy-Related Protein–Protein Interactions

Autophagy is a conserved cellular pathway that aims to degrade and recycle a variety of cellular components (protein aggregates, cellular organelles, invading bacteria, etc.) and thus maintain cellular homeostasis (reviewed in [1, 2]). Autophagy is manifested by the double-membrane vesicle called autophagosome, whose initiation, maturation, and fusion with lysosomes/vacuoles are controlled by ~40 autophagy-related (Atg) proteins [3, 4]. Among these, ubiquitin-like Atg8 family proteins, including human LC3A, LC3B, LC3C, GABARAP, GABARAPL1, and GABARAPL2 proteins (LC3/GABARAP hereafter), play an essential role. They are the key regulators of autophagosome biogenesis and also mediate a plethora of protein–protein interactions—with core Atg's,

Authors Martin P. Schwalm and Johannes Dopfer have equally contributed to this chapter.

Ioannis P. Nezis (ed.), *Selective Autophagy: Methods and Protocols*, Methods in Molecular Biology, vol. 2845,
https://doi.org/10.1007/978-1-0716-4067-8_17,

autophagy receptors, and autophagy cargo [5–7]. Recent investigations have revealed that LC3/GABARAPs interact with their partner proteins via relatively short LC3-interaction region (LIR) motifs, typically located in unstructured regions of autophagy receptors and other LC3/GABARAP-interacting proteins [8, 9]. LC3/GABARAPs provide a distinct interaction surface called the LIR docking site (LDS), which serves for the specific recognition and binding of most functional LIR motifs. The LDS consists of two hydrophobic pockets (HP1 and HP2), formed by residues of a central β-strand and the adjacent α-helices within LC3/GABARAPs. The LIR motif from the most investigated autophagy receptor p62/SQSTM (p62 thereafter) has a low micromolar affinity to the LC3/GABARAP proteins and displays no significant selectivity to any of the human Atg8 orthologs [10]. Another interacting surface (UDS, UIM docking site, [11]) is located on the opposite side of LC3/GABARAP, but UDS-mediated interactions with endogenous interaction partners have not been characterized yet.

The interactions between LC3/GABARAP and their interaction partners are still in the focus of modern research. A variety of complex biophysical methods have been implemented to investigate the affinity and selectivity of these interactions (reviewed in [9]). However, each method has its own experimental limitations, and taken alone, is not able to provide complete information on interactions involving LC3/GABARAP with their binding partners. Reliable methods are especially important to verify recently developed small molecules binding to LC3/GABARAPs that may serve for the development of autophagy-related protein degraders, as to date, targeted protein degradation (TPD) approaches are mainly based on small molecules that recruit E3 ligases [12]. Due to the key role of LC3/GABARAP proteins in the recruitment of cargo for lysosomal degradation, a similar mechanism has been proposed exploiting LC3/GABARAP proteins for TPD [13]. Despite the interest in LC3/GABARAP-interacting small molecules either as autophagy modulators/inhibitors or ligands for the development of chimeric molecules for TPD, screening campaigns have only yielded a few low-affinity binders [14]. Hence, more screening campaigns are required for the identification of LC3/GABARAP binders, together with a high-throughput method for affinity and selectivity determination. This protocol describes the use of a tracer-based fluorescence technique for the screening and characterization of LC3/GABARAP inhibitors.

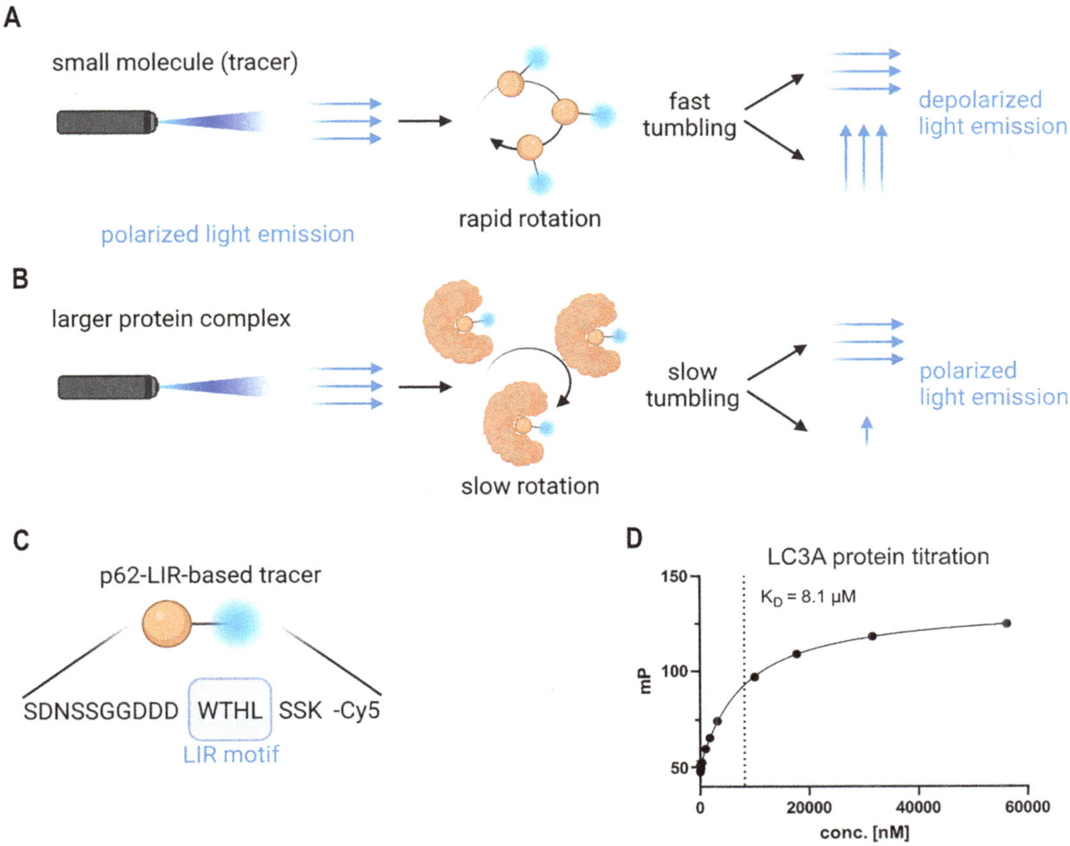


**Fig. 1** Principle of the fluorescence polarization assay. (**a**) and (**b**) the different mechanisms of free (**a**) and protein-bound (**b**) tracer. (**a**) The unbound tracer is excited by polarized light, and fast tumbling leads to the emission of depolarized light. (**b**) Binding to a protein partner increases the overall mass of the fluorophore-protein complex, leading to slow tumbling and therefore lower emission of depolarized light. (**c**) Schematic overview of the p62 LIR-derived tracer used for fluorescence polarization assays for LC3/GABARAPs. (d) An exemplary protein titration curve obtained from LC3A titration against the p62 LIR tracer, with the $K_D$ value determined according to 3.2, is given at the top (created with biorender.com)

### *1.2 Fluorescence Polarization to Study LC3/GABARAP:Ligand Interactions*

Fluorescence polarization is based on the relation between the hydrodynamic radius of a molecule in solution and its Brownian motion. Light emitted by a fluorophore (e.g., Cy5) after excitation with polarized light, will be depolarized to a certain extent. Since the excited state of fluorophores possesses a limited lifetime, the rotational speed of this fluorophore influences the level of depolarization (Fig. 1a). Therefore, increasing the mass of the fluorophore complex will decrease the rotational speed together with the amount of depolarization, which is measurable (Fig. 1b). This principle can be utilized by fluorescently labeling a peptide (here, we used a 17-meric peptide containing the prototypical p62 LIR motif) which is known to bind to the protein of interest (in our case, all 6 human LC3/GABARAP proteins) in the low micromolar $K_D$ region as shown in Fig. 1c. The small Cy5-labeled p62 LIR peptide has a much faster rotational speed (tumbling) as compared to the LC3/GABARAP-bound state, where its mass is drastically increased. The mass increase upon binding slows down the tumbling of the attached fluorophore and allows measurements of peptide binding to LC3/GABARAP proteins by tracking the amount of depolarized light emission. To implement this principle for ligand identification, displacement experiments reveal inhibitory constant values ($K_i$) for each tested ligand, allowing affinity determination of ligands binding to the LDS of LC3/GABARAP (as the p62 LIR interacts with LC3/GABARAPs exclusively by LDS). Subsequent comparison of the $K_i$ values for all six human LC3/GABARAPs will provide the selectivity profile of the measured ligand. Despite its robustness, the fluorescence polarization assay has some limitations, including problems with very high and very low ligand affinity, ligand-induced polarization quenching and interference, nonspecific ligand binding, etc. Therefore, the FP method should be implemented carefully with a number of positive and negative controls.

This protocol describes how to purify all 6 human Atg8 homologues from *Escherichia coli* cells to subsequently use these proteins for the establishment of fluorescence polarization binding assays for high-throughput ligand screening and binder characterization.

## 2 Materials

All compounds were dissolved in 100% DMSO to a final concentration of 10 mM. Expression plasmids for LC3A, LC3B, LC3C, GABARAP, GABARAPL1, and GABARAPL2 are available upon request.

### 2.1 Protein Expression and Purification

1. *E. coli* Rosetta (DE3) chemically competent cells.
2. SOC Medium.
3. LB Medium: 25 g LB (Luria/Miller), 1 L water. Autoclave before bacterial transformation.
4. LB-agar plates: 35 g LB Agar (Lennox), 1 L water, Kanamycin stock solution, Chloramphenicol stock solution.
5. TB Medium: 50.8 g, 1 L water. Autoclave before growing bacteria.
6. IPTG stock solution: 0.5 M Isopropyl-ß-D-thiogalactopyranoside, water.
7. Master buffer: 500 mM NaCl, 0.5 mM Tris-(2-carboxyethyl)-phosphine Hydrochloride/TCEP, 30 mM 4-(2-Hydroxyethyl)

piperazine-1-ethanesulfonic acid/HEPES at pH 7.5, 5% glycerol, water.

8. Lysis buffer: master buffer, Complete protease inhibitor (Sigma Aldrich), lyophilized DNase I (Sigma Aldrich).
9. Sterile filter (0.45 μm).
10. Washing buffer: master buffer, 20 mM Imidazole.
11. Elution buffer: master buffer, 300 mM Imidazole.
12. NiNTA affinity chromatography column (gravity flow or pump system).
13. Gel casting buffer (pH 8.8): 1.5 M Tris–HCl, water.
14. Gel casting buffer (pH 6.8): 0.5 M Tris–HCl, water.
15. Running buffer: 25 mM Tris–HCl, 200 mM Glycine, 0.1% (w/v) SDS, water.
16. SDS page resolving gel (20%): 2.1 mL water, 26.7 mL 30% acrylamide, 10.4 mL casting buffer pH 8.8, 400 μL 10% (w/v) SDS, 400 μL 10% Ammonium persulfate/APS solution, 40 μL N,N,N′,N′-Tetramethylethylenediamine/TEMED.
17. SDS page stacking gel (5%): 11.7 mL water, 2.7 mL 30% acrylamide, 5.2 mL casting buffer pH 6.8, 200 μL 10% (w/v) SDS, 200 μl 10% Ammonium persulfate/APS solution, 20 μL N,N,N′,N′-Tetramethylethylenediamine/TEMED.
18. Staining solution: i.e., ROTI®Blue quick (Carlroth).
19. Protein ladder: i.e., CozyHi™ Prestained Protein Ladder (HighQu).
20. His-tagged purified TEV protease.
21. Spin concentrator columns (3 and 10 kDa cut-off).
22. SEC buffer: 250 mM NaCl, 0.5 mM TCEP, 30 mM HEPES at pH 7.5, and 5% glycerol.
23. Size exclusion chromatography system.

### 2.2 FP Measurements

1. Fluorescence polarization (FP) buffer: 0.05% Tween 20 in a buffer containing 30 mM HEPES at pH 7.5, 250 mM NaCl, 5% glycerol, and 0.5 mM TCEP.
2. p62 LIR tracer (tracerDB ID: T000015, sequence: SDNSSGGDDDWTHLSSK-Cy5, obtained from Genecust).
3. Black 348 well low volume plates (alternatively 1536 well plates).
4. Assay master mix: FP buffer containing determined concentrations of protein and tracer.
5. Test compounds.

## 3 Methods

### 3.1 Protein Expression and Purification

Since fluorescence polarization is a biophysical method for in vitro characterization of protein:ligand interactions, purified protein must be obtained first. For this, plasmids containing an open reading frame of individual human Atg8 homologues were used. These plasmids contain a T7 promoter for protein expression in *E. coli* Rosetta BL21 (DE3) cells and a TEV-protease cleavable $His_6$-tag. Protein expression and purification were carried out as described in the following:

#### 3.1.1 Bacterial Transformation

1. Thaw 50 μL of *E. coli* Rosetta (DE3) cells on ice and add 1–10 ng plasmid DNA. Incubate the cells for 20 min on ice, followed by a heat shock for 30 s at 42 °C. After the heat shock, incubate the cells on ice for an additional 2–3 min before adding SOC medium and incubating the cells for 1 h at 37 ° C while shaking (~300 rpm). Finally, recover the bacteria by centrifugation at 5000 × *g* for 5 min and discard the supernatant (the supernatant should be autoclaved or sterilized chemically). Add 50 μL of fresh LB medium and plate the cells on LB-agar plates containing 50 μg/mL Kanamycin and 34 μg/ mL Chloramphenicol followed by incubation at 37 °C overnight (*see* **Notes 1** and **2**).
2. On the following day, use a sterile pipet tip to pick 8–10 colonies and inoculate a 100 mL culture of TB medium containing 50 μg/mL Kanamycin and 34 μg/mL Chloramphenicol. Incubate the flask at 37 °C while shaking at 180 rpm overnight.

#### 3.1.2 Recombinant Protein Expression

1. Add 25 mL of the pre-culture to 2 L shaking flasks containing 1 L of TB medium containing 50 μg/mL Kanamycin and incubate at 37 °C while shaking at 120 rpm until the cultures reach an optical density at a wavelength of 600 nm ($OD_{600}$) of 0.8 using a 1 cm cuvette.
2. Cool down the culture flasks by incubation on ice or in a cold room while cooling down the shaker to 18 °C. After the culture has been cooled down, shake the cells for an additional 30 min at 18 °C and 120 rpm.
3. Add 1 mL of IPTG to each culture flask. Continue shaking at 18 °C at 120 rpm overnight.
4. Harvest the cells by centrifuging the cultures for 10 min at 5000 × *g* and discard the supernatant (the supernatant should be autoclaved or sterilized chemically). Collect the cell pellet in a 50 mL reaction tube which can be stored at −20 °C for future purification. For immediate purification, keep the cells on ice.
5. Resuspend the cells in lysis buffer (*see* **Note 3**).

6. Sonicate the cell suspension on ice for 15 min with alternating 5 s on- and 10 s off-pulse. Mix the lysate by stirring followed by another round of sonication (*see* **Note 4**).
7. Separate the lysate from cell debris by centrifugation at 4 °C and 50,000 × *g* for 1 h followed by lysate filtration using a 45 μm sterile filter.

### 3.1.3 Protein Purification

1. Load the filtered lysate either onto a NiNTA gravity flow column or use a pump system (e.g., Äkta Start, Cytiva) to load the lysate onto a NiNTA column. After loading the lysate, wash the column with 3 column volumes (CV) of washing buffer. For gravity flow elution, sequentially add 1 CV of each lysis buffer containing 100, 200, and last 300 mM Imidazole to obtain 3 elution fractions, each 1 CV. For elution using the pump system, change the buffer to elution buffer and monitor the A280 nm chromatogram while fractionating (*see* **Note 5**).
2. To evaluate the purity of recombinant proteins in each fraction, perform an SDS-PAGE (*see* **Note 6**) by loading samples (e.g., 10 μL) of the lysate before affinity chromatography, the wash, and either all 3 elution fractions (gravity flow) or all elution fractions (pump system) corresponding to the peak at A280 nm (include a protein marker for protein size estimation). Stain the gel using staining solution to visualize the bands corresponding to the purified proteins.
3. To cleave the $His_6$-tag, prepare 1 L of lysis buffer (store at 4 ° C). Pool fractions containing pure protein of the corresponding size (try to avoid impure fractions). Add approximately 500 μg purified His-tagged TEV protease per ~50 mg of the pure protein to the pooled protein fractions and transfer the mixture into a dialysis tube with <15 kDa cut-off pore size, which needs to be sealed carefully. Place the loaded tube in cooled lysis buffer and dialyze the protein while slowly stirring overnight (*see* **Note 7**).
4. The following day, carefully collect the liquid from the dialysis tube and load it onto a fresh NiNTA column while collecting the flow-through. Wash the column with 3 CV of washing buffer while collecting and elute with 3 CV of elution buffer while collecting the flow-through.
5. To assess whether the TEV cleavage was successful, run another SDS-PAGE gel by loading a marker, the "uncleaved" sample collected before dialysis, and the three fractions from the second round of affinity chromatography. Stain the gel to visualize the bands corresponding to the purified proteins without the tag by comparing the size difference between the "uncleaved" and "cleaved" fractions and isolate the collected fraction which contains most of the pure cleaved protein (pool, if possible).

6. Filter the (pooled) fraction utilizing a 0.22 μm syringe filter and perform a size exclusion chromatography (SEC) using SEC buffer. Collect all fractions showing a peak in the chromatogram at 280 nm and validate the protein size via a third SDS-PAGE gel.
7. Pool all pure fractions containing the protein of the expected size and concentrate the protein to approximately 2 mg/mL using a spin concentrator with a cut-off of 3 kDa by centrifugation at 4000 × *g* in a swinging bucket rotor at 4 °C.
8. The protein may be stored at −80 °C (up to 1 year) after snap freezing in liquid nitrogen in aliquots of 100–200 μL.

### 3.2 Tracer Affinity Determination

Before a reliable tracer displacement assay can be performed, the tracer $K_D$ needs to be determined for the successive calculation of inhibitory constant ($K_i$) values. Since in fluorescence polarization, the tracer is the signal-providing assay parameter, its concentration has to be constant, allowing only the variation of the protein concentration. Therefore, a protein titration will yield polarization values suitable for an exponential fit for the $K_D$ determination (Fig. 1d).

1. In fluorescence polarization assays (especially when using Cy5 as the dye), the use of a detergent to suppress nonspecific binding is often essential. Therefore, a Tween-20 containing buffer is used as described in the Materials chapter.
2. Determine the final tracer concentration by using different tracer concentration in FP buffer (*e.g.*, 30, 60, 90 nM) and evaluate which concentration yields a satisfactory signal for the unbound tracer, whilst keeping the tracer concentration as low as possible. The plate reader has to be equipped with a fluorescence polarization module for the desired dye (*e.g.*, Cy5). Alternatively, perform the protein titration experiment with different concentrations and identify the most satisfactory assay window (raw value difference between unbound and bound tracer should be >3) at the lowest tracer concentration that can still be used. Use the determined tracer concentration to add tracer to the assay buffer.
3. For protein titrations, perform a serial dilution of purified protein with assay buffer containing the tracer, starting from a concentration of approximately 60 μM down to 10 nM in 10–12 steps. Transfer 10 μL of the dilutions (in triplicate) to a 384-well plate (alternatively 5 μL in a 1536-well plate) and incubate the plate for 20 min at room temperature (RT).
4. Measure the plate in a microplate reader (equipped with an FP 590–50,675–50,675–50 filter module), yielding millipolarization values ($mP$) consisting of values for perpendicular fluorescence intensity ($I_{perp}$) and parallel fluorescence intensity ($I_{para}$).

The *G*-factor is an additional variable determining the experimental correction of the optics. *G* is either determined by the plate reader or by carrying out calibration experiments by calculating the ratio between vertically and horizontally polarized light [15]. Therefore, the *mP* value reported by the plate reader is calculated using the following formula [16] (*see* **Note 8**):

$$mP = 1000 * \frac{I_{para} - G * (I_{perp})}{I_{para} + G * (I_{perp})} \quad (1)$$

5. Plot the obtained *mP* values against the protein concentration on a linear X-axis. From this plot, the $K_D$ can be obtained by using an exponential fit. To obtain the most precise fit, the following formula can be used [17]:

$$\mathrm{LR} = \frac{(\mathrm{X} + \mathrm{L_{tot}} + K_D) - \sqrt{(\mathrm{X} + \mathrm{L_{tot}} + K_D)^2 - 4 * \mathrm{X} * \mathrm{L_{tot}}}}{2} \quad (2)$$

with

$$\mathrm{L} = \mathrm{L_{tot}} - \mathrm{LR} \quad (3)$$

and

$$\mathrm{Y} = \mathrm{BKG} + \mathrm{MF} * \mathrm{L} + \mathrm{FR} * \mathrm{MF} * \mathrm{LR} \quad (4)$$

Here, the variables describe the total tracer/ligand concentration (**$\mathbf{L_{tot}}$**), while **BKG** reflects the background fluorescence of the tracer without protein, and **MF** describes the molar fluorescence, which is determined by the division of Y units by X units. **LR** describes the concentration of the tracer/ligand bound to the protein and **L** the concentration of unbound tracer/ligand.

6. Fitting the curve with $K_D$ as a parameter yields a value for the $K_D$ describing the affinity of the tracer to the protein.

### 3.3 Dose-Dependent Peptide Displacement Assay for Inhibitor Affinity Determination

After determining the accurate $K_D$ value for the interaction between the target protein and tracer, displacement experiments can be performed to determine $K_i$ values for small molecules (Fig. 2). In dose-dependent experiments, the optimal tracer concentration remains the same as previously described, and the protein concentration is ideally equal to the determined $K_D$ if the $K_D$ is lower than 5 μM. If the $K_D$ is higher, use 5 μM as the protein concentration to avoid excessive amounts of protein within the assay system. To obtain positive control curves, non-labeled LIR-containing peptides (e.g., p62 LIR) may be used to achieve displacement of the Cy5-labeled p62 LIR tracer (Fig. 2b).

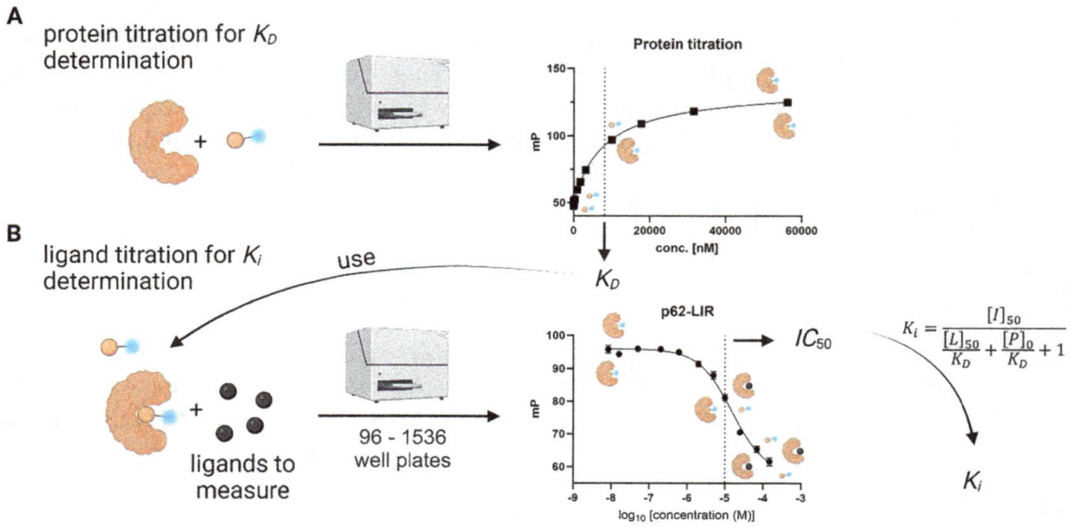


**Fig. 2** Scheme of the high-throughput fluorescence polarization assay. (**a**) Protein titration to determine the $K_D$ and optimal tracer concentration. (**b**) Determination of $K_i$ values in high-throughput mode (created with biorender.com)

1. Prepare an assay master mix by mixing the FP buffer with the tracer and adding the protein of interest at a concentration equal to the measured $K_D$ value.
2. For dose-response measurements, competing compounds can either be added by hand or by using small-volume liquid handlers. If no liquid handlers are available, doses need to be added in amounts that can be precisely pipetted by hand ($\geq 1$ μL), and therefore, compound addition is required prior to transferring to microplates. Small-volume liquid handlers can add nanoliters (nL) of compound solutions, and therefore, allowing compound addition after transferring the assay mix to the microplate. In order to obtain sufficient data points for reliable calculation of dose-dependent titration curves, measuring 11 different concentrations (+ DMSO as a negative control) is recommended. These concentrations should be equally distributed in $\log_{10}$ units in the range of approximately 100 μM to 10 nM (the highest concentration can be lowered for high-affinity binders). For a 384-well plate, transfer 10 μL per well, and 5 μL if 1536-well plates are used.
3. After either mixing the compounds with the FP assay buffer containing the tracer and protein and transferring them onto a microtiter plate (pipetting by hand) or adding the compounds to the mix distributed in a microplate (automated dispensing), incubate the plate for a minimum of 20 min at RT.

4. Read the plate using a microplate reader equipped with a polarization filter (e.g., FP 590-50 675-50 675-50 filter) (*see* **Note 9**).
5. Plot the obtained *mP* values from the microplate reader against the $\log_{10}$(compound concentration). If a positive control is used with known high affinity, 0% (DMSO negative control) and 100% displacement (positive control) can be used for data normalization.
6. Use a three-parameter fit (Hill Slope = −1.0) to fit the data, yielding an $IC_{50}$ value for each dose-dependent titration.
7. To calculate the inhibitory constant ($K_i$) for each compound, the following equation published by Nikolovska-Coleska et al. [18] is used:

$$K_{\mathrm{i}} = \frac{[I]_{50}}{\frac{[L]_{50}}{K_{\mathrm{D}}} + \frac{[P]_0}{K_{\mathrm{D}}} + 1} \tag{5}$$

Here, $[I]_{50}$ refers to the concentration of free inhibitor at 50% inhibition, $[L]_{50}$ is the concentration of free tracer at 50% inhibition, $[P]_0$ the concentration of free protein at 0%, and $K_D$ is the previously determined $K_D$ of the protein:tracer interaction.

For the calculation, the first parameter to be calculated is $[P]_0$, which requires the determined tracer $K_D$, the total concentration of tracer ($[L]_T$), and the total concentration of the protein used ($[P]_T$):

$$[P]_0 = \frac{-\left(K_{\mathrm{D}} + [L]_T - [P]_T\right) - \sqrt{\left(K_{\mathrm{D}} + [L]_T + [P]_T\right)^2 + 4 * [P]_T * K_{\mathrm{D}}}}{2} \tag{6}$$

The obtained value for the concentration of free protein at 0% inhibition can now be used for the calculation of the concentration of the free tracer at 0% inhibition ($[L]_0$):

$$[L]_0 = \frac{[L]_T}{1 + \frac{[P]_0}{K_{\mathrm{D}}}} \tag{7}$$

Next, the concentration of the protein:ligand complex at 0% inhibition ($[PL]_0$) needs to be calculated using the following equation:

$$[PL]_0 = \frac{[P]_T}{\frac{K_{\mathrm{D}}}{[L]_0} + 1} \tag{8}$$

Dividing the obtained concentration by 2, the concentration of the protein–ligand complex at 50% inhibition ($[PL]_{50}$) can be calculated:

$$[PL]_{50} = \frac{[PL]_0}{2} \tag{9}$$

The value for $[PL]_{50}$ can now be used for the calculation of the concentration of free tracer at 50% inhibition ($[L]_{50}$):

$$[L]_{50} = [L]_T - [PL]_{50} \tag{10}$$

Finally, the $IC_{50}$ obtained from the dose-dependent titration of the inhibitor in the form of the turning point of the sigmoidal fit needs to be converted into the concentration of free inhibitor at 50% inhibition ($[I]_{50}$) by using the following equation:

$$[I]_{50} = IC_{50} - [P]_T + K_D * \frac{[PL]_{50}}{[L]_{50}} + [PL]_{50} \tag{11}$$

Calculation of $[I]_{50}$ now allows for the calculation of the $K_i$ by using Eq. (5).

### 3.4 High-Throughput Screening of Compound Libraries Using Fluorescence Polarization

After measuring control compounds or control peptides such as unlabeled p62 LIR in a dose-dependent manner, single concentration (one-shot) measurements can be used for high-throughput hit identification campaigns. The obtained $mP$ value can also be used to estimate an affinity by comparison with the $mP$ value obtained from dose-dependent measurements. For high-throughput screening, 1536-well plates are ideally used to obtain up to 100,000 data points per day. If measurement in a 1536-well plate is not feasible, 384-well plates may be used.

1. Prepare an assay master mix by mixing the assay buffer with the tracer and adding the protein of interest at a concentration equal to the determined $K_D$ value.
2. Prepare compounds by diluting compound stocks to equal concentrations (10 mM recommended). Determine the compound concentration for screening according to the expected hit affinity. For low-affinity binders ($K_D$ range 10–1000 μM), screening at 100 μM is recommended, while screening at 10 μM neglects low-affinity binders, mostly focusing on low μM to nM binders.
3. If no liquid handling system is available, prepare a master plate (e.g., 96-well plate) containing 1 μL compound and add the master mix to obtain the desired concentration (e.g., add 100 μL master mix to 1 μL of 10 mM compound for screening 100 μM of the compound in the assay) before transferring the mixture in triplicates to the microplate. Using automation, the master mix can be distributed onto the microplate followed by the addition of the compounds in triplicates (e.g., add 50 nL of a 10 mM compound stock to 5 μL of master mix in a 1536-well plate).

4. Incubate the plate(s) at RT for a minimum of 20 min.
5. Read the plate in a microplate reader equipped with a polarization filter (e.g., FP 590-50 675-50 675-50 filter for Cy5) (*see* **Note 10**).
6. Using automation and stacker attachments on the microplate reader, one 1536-well plate can be measured in 20 min using dual channel detection (as available in the PHERAstar FSX platereader, BMG Labtech), allowing for 4600 data points in 1 h and 110,000 data points in 24 h.

## 4 Notes

1. Scale the amount of LB agar according to the number of plates needed. After melting the LB agar using a microwave let it cool down to 60 °C, a temperature at which most of the commonly used antibiotics remain stable. Add Kanamycin and Chloramphenicol to obtain final concentrations of 50 μg/mL and 34 μg/mL, respectively, and mix thoroughly. Next, pour the agar mixture into the plates and allow it to solidify. It is important to maintain a clean workspace throughout the process.
2. No clones after transformation: If no clones can be identified after transformation, either the DNA has insufficient purity/concentration or the cells were not handled correctly. To make sure that the plasmid is present in sufficient quality and amount, measure the concentration of the purified PCR product with a spectrophotometer. A purity check can be done by measurement of the A260/A280 ratio (expected between approximately 1.8 and 2) and the A260/A230 ratio (expected to be approximately 2.0 to 2.2). If the values are off or the concentration insufficient (e.g., < 20 ng/μL), repurify the plasmid DNA. If the DNA is in the expected range, thaw another aliquot of cells. For transformation, it is crucial to handle the cells carefully and always on ice. Do not vortex the cells and avoid large shear forces when pipetting. Make sure, that the recovery incubation for 1 h directly after transformation is carried out without the use of antibiotics and the plates contain the correct antibiotic in the correct concentration.
3. Add the protease inhibitor together with a few flakes of DNase I shortly before mixing the lysis buffer with the bacteria. Ensure that the protease inhibitor tablet has dissolved completely. If DNase I is not available, a polyethylenimine/PEI solution can be used to remove excess free DNA *after* lysis.
4. Insufficient cell lysis: Incomplete lysis of the cells can lead to a drastic loss in yield. Make sure that the cell lysate does not contain any visual particles or cell clumps and should be

coloured evenly throughout the whole beaker. Additionally, high viscosity indicates genomic DNA in the lysate which can be resolved through the addition of DNase I.

5. No protein after affinity chromatography: It can happen that no protein elutes after affinity chromatography. Since all human LC3/GABARAPs were already validated to be successfully expressed and purified in the described assay setup, no construct errors are expected. To ensure high amounts of soluble protein, keep the lysate constantly on ice and isolate the protein within 1 day. Do not store the lysate and ensure the correct pH value of the used buffers. Ensure that the wash buffer does not contain too high imidazole concentrations which lead to premature elution. Lastly, check for protein expression by comparison of all fractions on an SDS-PAGE gel. If no band is found, the protein expression might not have been induced by using IPTG.
6. After adding TEMED as the final ingredient, pour the gel solution quickly into the assembled plates. Smooth the surface of the gel and remove any bubbles by overlaying it with isopropanol. Polymerization takes up to 30 min and can be assessed by observing the consistency of the remaining gel solution. Once the resolving gel is fully cured, remove the isopropanol and prepare the stacking gel. Place the stacking gel on top of the resolving gel and insert the combs just before the gel overflows. Allow the gel to set for an additional 30 min.
7. Precipitated protein after dialysis: Some of the proteins tend to precipitate during dialysis. One reason is the instability of the constructs induced by the $His_6$-tag. The stability is therefore drastically increased after cleavage but not during cleavage. Additionally, the TEV protease can also precipitate during this procedure. Due to the high yield of the human LC3/GABARAPs expression constructs, a limited extent of precipitation is of no concern. In this case, centrifugation of the suspension, followed by filtration through 0.22 μm filters and subsequent size exclusion chromatography will recover all functional protein from the suspension and make them suitable for the biophysical assays.
8. No binding of the tracer observed: Since the interaction of the described p62 LIR-based peptide was already successfully shown [14], missing binding detection is most likely driven by the assay setup or the protein quality. Ensure protein identity through SDS-PAGE and/or mass spectrometry. Measure protein concentration in triplicates using a spectrophotometer or carry out a BCA assay [19]. Ensure that the correct filters are installed in the microplate reader and include positive (peptide)

and negative (DMSO) controls. Additionally, ensure that Tween-20 was added to the assay buffer.

9. No displacement of the tracer: Generally, no displacement of the tracer indicates that the compound used for displacement does not compete with the tracer, meaning undetectable binding at the identical binding site. However, fluorescent compounds might interfere with the tracer signal, leading to wrong readouts. Furthermore, insoluble compounds may also impair the readout of the assay and/or lead to protein precipitation.
10. Off-scale values during high-throughput screening: It was already observed that the one-shot experiments lead to values that are completely off-scale or show 100% displacement without interaction. In that case, check the compound colour, which mostly occurs to be colourful. Since the compound colour cannot be influenced, interfering compounds must be excluded or screened through a technique that is not based on optical detection.

## Acknowledgments

MPS, JD, SK, and VVR are grateful for support from the Structural Genomics Consortium (SGC), a registered charity (no: 1097737) that receives funds from Bayer AG, Boehringer Ingelheim, Bristol Myers Squibb, Genentech, Genome Canada through Ontario Genomics Institute, EU/EFPIA/OICR/McGill/KTH/Diamond Innovative Medicines Initiative 2 Joint Undertaking [EUbOPEN grant 875510], Janssen, Merck KGaA, Pfizer, and Takeda and by the German Cancer Research Center DKTK and the Frankfurt Cancer Institute (FCI). MPS is funded by the Deutsche Forschungsgemeinschaft (DFG, German Research Foundation), CRC1430 (Project-ID 424228829).

### References

1. Mizushima N, Yoshimori T, Ohsumi Y (2011) The role of Atg proteins in autophagosome formation. Annu Rev Cell Dev Biol 27:107–132. https://doi.org/10.1146/annurev-cellbio-092910-154005
2. Ohsumi Y (2014) Historical landmarks of autophagy research. Cell Res 24(1):9–23. https://doi.org/10.1038/cr.2013.169
3. Tooze SA, Yoshimori T (2010) The origin of the autophagosomal membrane. Nat Cell Biol 12(9):831–835. https://doi.org/10.1038/ncb0910-831
4. Dikic I, Elazar Z (2018) Mechanism and medical implications of mammalian autophagy. Nat Rev Mol Cell Biol 19(6):349–364. https://doi.org/10.1038/s41580-018-0003-4
5. Kirkin V, Rogov VV (2019) A diversity of selective autophagy receptors determines the specificity of the autophagy pathway. Mol Cell 76(2):268–285. https://doi.org/10.1016/j.molcel.2019.09.005
6. Johansen T, Lamark T (2020) Selective autophagy: ATG8 family proteins, LIR motifs and cargo receptors. J Mol Biol 432(1):80–103. https://doi.org/10.1016/j.jmb.2019.07.016
7. Rogov VV, Nezis IP, Tsapras P et al (2023) Atg8 family proteins, LIR/AIM motifs and

other interaction modes. Autophagy Rep 2(1): 2188523

8. Noda NN, Ohsumi Y, Inagaki F (2010) Atg8-family interacting motif crucial for selective autophagy. FEBS Lett 584(7):1379–1385. https://doi.org/10.1016/j.febslet.2010.01.018
9. Wesch N, Kirkin V, Rogov VV (2020) Atg8-family proteins-structural features and molecular interactions in autophagy and beyond. Cells 9(9). https://doi.org/10.3390/cells9092008
10. Wirth M, Zhang W, Razi M et al (2019) Molecular determinants regulating selective binding of autophagy adapters and receptors to ATG8 proteins. Nat Commun 10(1):2055. https://doi.org/10.1038/s41467-019-10059-6
11. Marshall RS, Hua Z, Mali S et al (2019) ATG8-binding UIM proteins define a new class of autophagy adaptors and receptors. Cell 177(3):766–781 e724. https://doi.org/10.1016/j.cell.2019.02.009
12. Nemec V, Schwalm MP, Muller S et al (2022) PROTAC degraders as chemical probes for studying target biology and target validation. Chem Soc Rev 51(18):7971–7993. https://doi.org/10.1039/d2cs00478j
13. Schwalm MP, Knapp S, Rogov VV (2023) Toward effective Atg8-based ATTECs: approaches and perspectives. J Cell Biochem. https://doi.org/10.1002/jcb.30380
14. Schwalm MP, Dopfer J, Kumar A et al (2023) Targeting LC3/GABARAP for degrader development and autophagy modulation. bioRxiv:2023.2010. 2005.560930
15. Hall MD, Yasgar A, Peryea T et al (2016) Fluorescence polarization assays in high-throughput screening and drug discovery: a review. Methods Appl Fluoresc 4(2):022001. https://doi.org/10.1088/2050-6120/4/2/022001
16. Piston DW, Rizzo MA (2008) FRET by fluorescence polarization microscopy. Methods Cell Biol 85:415–430. https://doi.org/10.1016/S0091-679X(08)85018-2
17. Motulsky HJ, Neubig RR (2010) Analyzing binding data. Curr Protoc Neurosci Chapter 7:Unit 75 https://doi.org/10.1002/0471142301.ns0705s52
18. Nikolovska-Coleska Z, Wang R, Fang X et al (2004) Development and optimization of a binding assay for the XIAP BIR3 domain using fluorescence polarization. Anal Biochem 332(2):261–273. https://doi.org/10.1016/j.ab.2004.05.055
19. Schwalm MP, Krämer A, Dölle A et al (2023) Tracking the PROTAC degradation pathway in living cells highlights the importance of ternary complex measurement for PROTAC optimization. Cell. Chem Biol 30(7):753–765 e758. https://doi.org/10.1016/j.chembiol.2023.06.002

# Chapter 18

# Thermodynamic Characterization of LC3/GABARAP:Ligand Interactions by Isothermal Titration Calorimetry

**Johannes Dopfer, Martin P. Schwalm, Stefan Knapp, and Vladimir V. Rogov**

## Abstract

Isothermal titration calorimetry (ITC) is a widely used technique for the characterization of protein–protein and protein–ligand interactions. It provides information on the stoichiometry, affinity, and thermodynamic driving forces of interactions. This chapter exemplifies the use of ITC to investigate interactions between human autophagy modifiers (LC3/GABARAP proteins) and their interaction partners, the LIR motif-containing sequences. The purpose of this report is to present a detailed protocol for the production of LC3/GABARAP-interacting LIR peptides using *E. coli* expression systems. In addition, we outline the design of ITC experiments using the LC3/GABARAP:peptide interactions as an example. Comprehensive troubleshooting notes are provided to facilitate the adaptation of these protocols to different ligand-receptor systems. The methodology outlined for studying protein–ligand interactions will help to avoid common errors and misinterpretations of experimental results.

**Key words** Isothermal titration calorimetry, *Kd* determination, Autophagy, LC3, GABARAP, LIR motifs

## 1 Introduction

### 1.1 Molecular Mechanisms of Interaction Between LC3/GABARAP Proteins and Their Interaction Partners

Autophagy is an essential cellular process for maintaining homeostasis by removing various unwanted cellular components (autophagy cargo) [1]. Different types of cargo, such as proteins and their aggregates, organelles, pathogens, and lipid droplets, are targeted for lysosomal degradation by autophagic pathways [2]. Macroautophagy, hereafter simply referred to as autophagy, employs double-membrane vesicles called autophagosomes as carriers to transport autophagosome-encapsulated cargo to lysosomes for degradation. A large number of autophagy-related (Atg) proteins coordinate this process [3]. The human homologs of the yeast Atg8 protein

Authors Johannes Dopfer and Martin P. Schwalm have equally contributed to this chapter.

Ioannis P. Nezis (ed.), *Selective Autophagy: Methods and Protocols*, Methods in Molecular Biology, vol. 2845,
https://doi.org/10.1007/978-1-0716-4067-8_18,

(termed LC3/GABARAP proteins) play a critical role in regulating selective autophagy machinery by coordinating the autophagosome biogenesis and tethering the autophagy receptors to the forming phagophore through multiple protein–protein interactions [4].

Recent studies have shown that the autophagy receptors (and other LC3/GABARAP-interacting proteins) contain a short sequence motif called the LC3 interacting region (LIR). This motif is the recognition sequence that mediates a variety of Atg8-related protein–protein interactions [5]. The LC3/GABARAP proteins have an LIR-docking site (LDS) that is recognized by canonical LIR motifs with moderate affinities. Variations and extensions of the canonical LIR modulate its affinity for LC3/GABARAP, resulting in binding affinities ranging from subnanomolar to millimolar [6]. An additional binding site on the surface of LC3/GABARAP proteins has been reported called the UIM-docking site [7]. This site has not been extensively characterized biochemically or in terms of sequence requirements present in interaction partners. Investigation of interactions between LC3/GABARAP proteins and LIR motifs on affinity and selectivity remains in the focus of the contemporary field of autophagy. Interactions mediated by the LIR binding site provide the basis for the design of autophagosome-tethering compounds (ATTECs [8, 9]), which may recruit proteins of therapeutic interest to the autophagy degradation machinery [10]. To elucidate the biophysical nature of the key LC3/GABARAP:ligand interactions, a comprehensive analysis of structural and thermodynamic factors driving these interactions is required.

### 1.2 ITC as a Tool for Detailed Investigation of Protein–Protein and Protein–Ligand Interactions

Calorimetric measurements determine the amount of heat released in an exothermic reaction or consumed in an endothermic reaction. The detection method used in isothermal titration calorimetry (ITC) is based on the power compensation required to maintain a constant temperature difference between two measurement cells. ITC is a general method for studying chemical reactions and molecular interactions based on a characteristic change in enthalpy. To study LC3/GABARAP:peptide or LC3/GABARAP:small molecule interactions, ITC marks a valuable, label-free method. A titration calorimeter consists of a sample cell and a reference cell enclosed in an adiabatic shield. To ensure isothermal conditions, the temperature difference between the reference cell is continuously monitored and adjusted by a heating module. When a ligand is titrated into a protein solution in the sample cell, either an endothermic or exothermic reaction will be initiated by the protein–ligand interaction. The resulting temperature difference between the sample and the reference cell is recorded using a Peltier device located between the two cells (Fig. 1). A feedback system regulates the power compensation required to maintain both cells at the defined temperature, which is recorded over time and

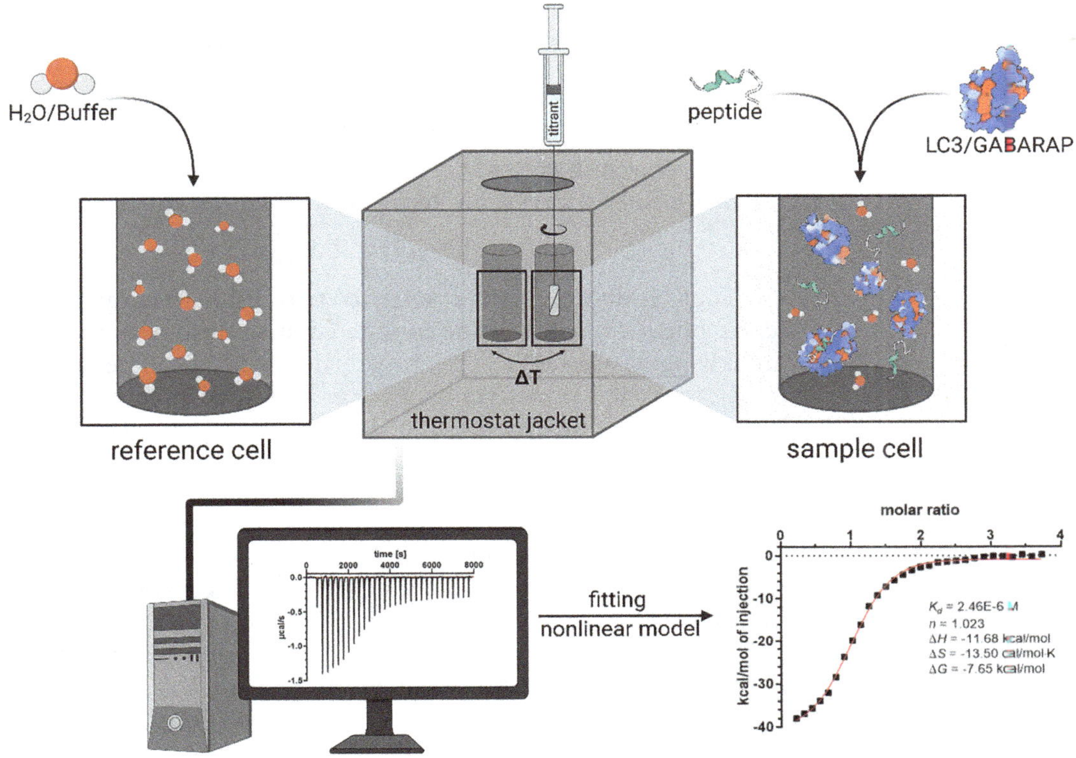


**Fig. 1** Schematic overview of a titration calorimeter consisting of a sample cell and a reference cell. The titrant is injected into the sample cell and mixed to ensure rapid diffusion. As a result of binding events, there is a difference in temperature, which is detected and transferred into a feedback system to maintain isothermal conditions. The recorded differential power signal is integrated for every injection peak and analyzed by fitting a nonlinear binding model

converted into a readout of µcal/s or µJ/s, often referred to as a thermogram [11, 12]. Integrating the injection peaks over time yields the energy required to return back to the baseline value. A binding isotherm is calculated by plotting the heat of each injection normalized to the amount of injected ligand against the molar ratio. By using a nonlinear fitting model, the association constant $K_a$, the total change in binding enthalpy $\Delta H$, and the stoichiometry ($n$) of the monitored interaction can be calculated. The simplest binding model assumes a single binding site describing the binding of a ligand (L) to a macromolecule (M). As described by the law of mass action, $K_a$ is related to the activity of the macromolecule–ligand complex and its constituents by:

$$K_a = \frac{[\mathrm{ML}]}{[\mathrm{M}][\mathrm{L}]} = \frac{1}{K_d}$$

The dissociation constant $K_d$ is inversely related to $K_a$. The Gibbs free energy of binding ΔG is linked to $K_a$ by the following equation:

$$\Delta G = -RT \ln K_a$$

with the universal gas constant R and the absolute temperature $T$. ΔG may also be expressed in terms of $\Delta H$ and the entropy $\Delta S$ as:

$$\Delta G = \Delta H - T\Delta S$$

Thus, a binding event consists of two distinct energy-related terms, both of which must be considered in order to understand its underlying intricacies [13]. A properly designed ITC experiment is therefore capable of determining the total change of enthalpy $\Delta H$, the association constant $K_a$, the stoichiometry coefficient $n$, and the Gibbs free energy ΔG of the process, allowing the calculation of the change in entropy $\Delta S$ at a certain temperature and buffer condition.

The shape of the binding isotherm depends on $K_a$, the total concentration of the macromolecule $[M]_t$ (if placed as a receptor in the sample cell), and the stoichiometry $n$ of the binding interaction [14]. The parameter describing this relationship is termed the $c$ value, and it can be expressed in terms of $K_a$ or $K_d$:

$$c = nK_a[M]_t = \frac{n[M]_t}{K_d}$$

Ideally, for accurate calculation of the thermodynamic parameters mentioned above, the value of $c$ should be between 10 and 500. A low $c$ value titration ($c < 5$) has to be performed if the solubility of M acts as a limiting factor for the adjustment of $[M]_t$ or low-affinity ligands (in the micromolar range) are expected (*see* **Note 1**). Experiments conducted under these conditions can still lead to an accurate determination of $K_a$ and consequently $\Delta G$ if the protein saturation is sufficient, the concentrations of both ligand and protein are well known, and the stoichiometry is fixed to the actual value. Caution should be taken by interpreting the value of $\Delta H$ in the case of low $c$ value titrations [15]. For $c$ values larger than 1000, the slope of the sigmoidal binding isotherm becomes exceedingly steeper, closely resembling a step function. Under high $c$ value conditions, the accurate determination of $K_a$ is not possible, while the measured enthalpy change of the binding event is still reliable. Lowering $[M]_t$ to reduce $c$ is limited by the sensitivity of the calorimeter. Therefore, one method for determining high-affinity binding equilibria relies on artificially reducing the binding affinity by altering the experimental conditions. For instance, temperature modification can be implemented in combination with the determination of thermodynamic linkage parameters. This approach permits the estimation of the Gibbs free energy change of the altered condition and its extrapolation back to the condition of interest. In the case of the change in temperature, this can be achieved by utilizing a modified van't Hoff equation and recorded values of the heat capacity changes for different temperature settings [16]. It is important to exercise caution, as extrapolation can introduce errors, and changes in experimental conditions may alter the stability of the system being studied, which could ultimately result in protein aggregation. Another approach for studying high-affinity ligands ($K_d$ < 10 nM) is a displacement titration. The method involves saturating the receptor with a weaker competitive ligand before titrating the high-affinity ligand, resulting in a reduced apparent affinity [17]. This is due to an energy penalty which is paid for displacing the weaker ligand from the binding site, increasing the Gibbs free energy change of the high-affinity binding event. For this experimental mode, at least two titrations are necessary: (I) the association constant of the lower affinity ligand $K_{low}$ is to be determined and (II) displacement of the weaker ligand by titration of the high-affinity ligand to yield its apparent association constant $K_{high}^{app}$. Assuming $n$ identical and independent binding sites, fitting a binding isotherm to the data given the additional information of $K_{low}$ and the concentration of the weaker ligand [X] is possible [18]. The following equations describe the relation between the association constants and the change in enthalpy:

$$K_{high}^{app} = \frac{K_{high}}{1 + K_{low}[\mathrm{X}]}$$

$$\Delta H_{high}^{app} = \Delta H_{\mathrm{high}} - \Delta H_{\mathrm{low}} \frac{K_{low}[\mathrm{X}]}{1 + K_{low}[X]}$$

This technique has been utilized by Li et al. to study the low nanomolar interactions between the LIR of ankyrins and different GABARAP isoforms [19], and this method has been described by another protocol [17]. The thermodynamic signatures of the binding of extremely weak ligands, such as fragments that may not have high enough affinity to reach saturation without encountering a solubility problem, can also be analyzed using the principle of displacement titrations. In this case, the macromolecule is incubated with the fragment of unknown affinity, which is then displaced by a molecule with known affinity that may contain the fragment as a substructure [20].

## 2 Materials

Peptides are either generated as described in this protocol or obtained from GeneCust. The expression plasmids for LC3A,

LC3B, LC3C, GABARAP, GABARAPL1, GABARAPL2, and human ANK3 LIR motif can be obtained upon request. All buffers and solutions are prepared with deionized water.

### 2.1 Peptide/Protein Expression and Purification

1. LB Medium: 25 g LB (Luria/Miller), 1 L water. Autoclave before bacterial transformation.
2. LB-agar plates: 35 g LB Agar (Lennox), 1 L water, Kanamycin stock solution, Chloramphenicol stock solution.
3. TB Medium: 50.8 g, 1 L water. Autoclave before growing bacteria.
4. IPTG stock solution: 0.5 M Isopropyl-ß-D-thiogalactopyranoside, water.
5. Master buffer: 500 mM NaCl, 0.5 mM Tris-(2-carboxyethyl)-phosphine Hydrochloride/TCEP, 30 mM 4-(2-Hydroxyethyl) piperazine-1-ethanesulfonic acid/HEPES at pH 7.5, 5% glycerol, water.
6. Lysis buffer: master buffer, Complete protease inhibitor (Sigma Aldrich), lyophilized DNase I (Sigma Aldrich).
7. Sterile filter (0.45 μm).
8. Washing buffer: master buffer, 20 mM Imidazole.
9. Elution buffer: master buffer, 300 mM Imidazole.
10. Gel casting buffer (pH 8.8): 1.5 M Tris–HCl, water. Dissolve Tris base in water and adjust pH to 8.8 using 6 N HCl.
11. Gel casting buffer (pH 6.8): 0.5 M Tris–HCl, water. Dissolve Tris base in water and adjust pH to 6.8 using 6 N HCl.
12. Running buffer: 25 mM Tris–HCl, 200 mM Glycine, 0.1% (w/v) SDS, water.
13. SDS page resolving gel (20%): 2.1 mL water, 26.7 mL 30% acrylamide, 10.4 mL casting buffer pH 8.8, 400 μL 10% (w/v) SDS, 400 μL 10% Ammonium persulfate/APS solution, 40 μL N,N,N′,N′-Tetramethylethylenediamine/TEMED.
14. SDS page stacking gel (5%): 11.7 mL water, 2.7 mL 30% acrylamide, 5.2 mL casting buffer pH 6.8, 200 μL 10% (w/v) SDS, 200 μL 10% Ammonium persulfate/APS solution, 20 μL N,N,N′, N′-Tetramethylethylenediamine/TEMED.
15. Staining solution: i.e., ROTI®Blue quick (Carlroth).
16. Protein ladder: i.e., CozyHi™ Prestained Protein Ladder (HighQu).
17. Spin concentrator columns (3 and 10 kDa cut-off).

### 2.2 ITC Measurements

1. ITC buffer (pH 7.0): 50 mM $Na_2HPO_4$, 100 mM NaCl, water.
2. Purified human LC3/GABARAP proteins in ITC buffer.

3. Synthetic p62 LIR peptide (sequence: SDNSSGGDDDWTH-LSS) 95% purity.
4. Synthetic AS_28 LIR peptide (sequence: RSRAQHQVENG-WWWMPHIDHNGGG) 95% purity.
5. Purified ANK3 LIR peptide (sequence: SMEDDWIEFSSE-EIREARQQAAASQS).

## 3 Methods

### 3.1 Peptide/Protein Expression and Purification

ITC measurements rely on highly purified protein in a well-controlled buffer system. The expression of the Atg8 homologs is described in the related Chap. 17 (Schwalm et al.). The plasmid designed for ANK3 peptide expression features a T7 promoter and it is suitable for expression in *Escherichia coli* Rosetta (DE3) cells, along with a ubiquitin-based tag that is cleavable by TEV protease. The solubility-enhancing Ub-leader consists of a modified ubiquitin and a $His_{10}$ sequence designed to increase the yield and allow for purification via His-tag-mediated affinity chromatography [21]. The subsequent steps for protein expression and purification are carried out as described:

#### 3.1.1 Bacterial Transformation

1. Thaw 50 μL of *E. coli* Rosetta (DE3) cells (Sigma-Aldrich) on ice and add approximately 1–10 ng of plasmid DNA. Incubate the cells on ice for 20 min, followed by a brief heat shock for 30 s at 42 °C.
2. After the heat shock, incubate the cells for 2–3 min on ice before adding SOC medium (provided in the Rosetta kit). Alternatively, LB-Medium can be used. Incubate the cells for 1 h at 37 °C while shaking at 300 rpm.
3. To concentrate the transformed cells, harvest the cells at 5000 × *g* for 5 min and remove the supernatant. Subsequently, 50 μL of fresh medium is added and the cells are plated on antibiotic-containing LB-agar plates (*see* **Note 2**). Incubate the plates at 37 °C overnight.
4. The following day, use a sterile pipette tip to inoculate 100 mL of TB culture medium containing 50 μg/mL Kanamycin and 34 μg/mL Chloramphenicol, followed by overnight incubation at 37 °C while shaking at 180 rpm.

#### 3.1.2 Recombinant Protein Expression

1. For each construct isolation, prepare a total of 4 L of sterile TB medium supplemented with 50 μg/ml Kanamycin distributed across four 2 L shaking flasks (1 L TB medium per flask).
2. Inoculate each of the expression cultures with 25 mL of the pre-culture, followed by incubation at 37 °C while shaking at 120 rpm to an optical density at 600 nm ($OD_{600}$) of 0.8.

3. Before inducing protein expression, cool the cultures to 18 °C and shake them for an additional 30 min at 120 rpm. Use 1 mL of a 0.5 M IPTG stock solution per flask and continue shaking overnight at 18 °C and 120 rpm.
4. Harvest the cells by centrifugation of the cultures for 10 min at 4000 × *g* and discard the supernatant. Transfer the cell pellet to a 50 mL tube, which can be stored at −20 °C for future purification. Keep the cells on ice if purification is continued.
5. For lysis, resuspend the cells in a lysis buffer to yield a total of approximately 100 mL (*see* **Note 3**).
6. Sonicate the cell suspension on ice for 15 min, using an alternating 5-s on and 10-s off cycle. After mixing the lysate, repeat sonication for another 15 min.
7. Separate the lysate from cellular debris by centrifugation at 4 °C and 50,000 *g* for 1 h followed by filtration using a 0.45 μm sterile filter.

#### 3.1.3 Protein Purification

1. The protein within the filtered lysate is subsequently isolated, utilizing a chromatography system (e.g., Äkta Start, Cytiva; alternatively, gravity flow columns can be used) to load the lysate onto a Ni-NTA affinity chromatography column (Cytiva).
2. After loading, wash the column with 3 column volumes (CV) of washing buffer.
3. For elution, change the buffer to the elution buffer in a stepwise manner and monitor the A280 nm chromatogram while fractionating.
4. Perform an SDS-PAGE (*see* **Note 4**) gel electrophoretic analysis of the cleared lysate (collected before affinity chromatography), the 20 mM imidazole wash fraction, and all elution fractions corresponding to a signal at A280 nm (include a Protein marker for protein size estimation). Add the protein ladder and run the SDS-PAGE at 200 V for 1 h.
5. Stain the gel using a staining solution to visualize the protein bands corresponding to the purified Ub-fused protein and assess the preliminary purity.
6. To cleave the $His_{10}$-tag from the purified protein, pool the fractions with corresponding protein size (while avoiding impure fractions) according to the SDS-PAGE gel and prepare a sample for the subsequent SDS-PAGE. Add approximately 500 μg purified His-tagged TEV protease to the pooled fractions and transfer the mixture into a 50 mL reaction tube. Incubate the tube on a tube roller mixer at 4 °C overnight.
7. Prepare another SDS-PAGE gel by loading the sample collected at **step 17** before cleavage and a sample collected from

the reaction tube after incubation with TEV protease to evaluate the cleavage efficiency. Note that the cleaved peptide is of ~3 kDa, hence use an SDS-PAGE gel of sufficient acrylamide concentration (~20%) to achieve appropriate resolution.

8. Stain the gel to visualize the bands corresponding to the purified peptide without the tag by comparing the size difference between the uncleaved and cleaved fractions.
9. Assuming that the molecular weight of TEV protease is 27 kDa, the uncleaved fusion protein is 15 kDa, and the Ub-tag is 12 kDa, the cleaved peptide can easily be separated by using a spin concentrator with a 10 kDa cut-off (Sigma Aldrich) and collection of the *flow-through*. Alternatively, size exclusion chromatography (SEC) on the prepacked Superdex 75 PG HighLoad 16/600 column can be used to purify the target peptide as well.
10. To yield adequately concentrated peptide for ITC experiments, use a spin concentrator with a 3 kDa pore size, which is sufficient given the larger hydrodynamic radius of the unstructured peptide. This also provides a convenient opportunity to exchange the buffer for the one used in the subsequent ITC experiments.
11. Ultimately, the identity and purity of the peptide can be confirmed through mass spectrometry.

### 3.2 ITC Measurement of Interactions Between LC3/GABARAP Proteins and LIR-Contained Peptides

#### 3.2.1 Experimental Design

For the measurements conducted in this work, we employed the standard volume Nano ITC calorimeter from TA Instruments, which is equipped with an active cell volume of 1.0 mL. The volume per injection and stir rate might vary depending on the system (refer to the manufacturer's manual if a comparable device is used). Throughout the experiment, microaliquots are injected into the sample cell, while the contents of the sample cell are kept in continuous motion by the rotation of the syringe and the paddle at its tip, as depicted in Fig. 1. The stirring rate should be sufficiently high to allow rapid diffusion of the macromolecule/ligand introduced during the injections. Excessive stirring, however, decreases the signal-to-noise ratio due to the formation of artificial heat signals caused by friction. The recommended stirring rate differs between instruments. Here, all experiments were conducted at 350 rpm (recommended 300–400 rpm) and yielded a stable baseline. The injection volume is dependent on the volume of the sample cell and the sensitivity of the calorimeter (5–10 μL for the Nano ITC). If the binding event causes large differences in heat, the volume of the aliquots may be reduced. It is common practice to use a smaller initial injection volume that is not taken into account when analyzing the data to exclude the diffusion of the cell contents into the syringe tip during equilibration from the

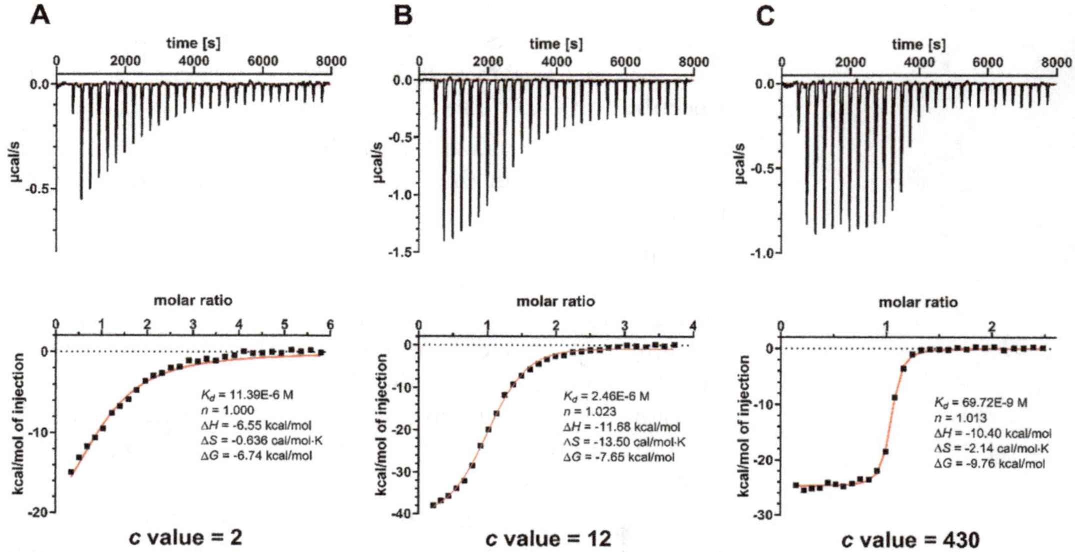


**Fig. 2** ITC experiments corresponding to the titration of p62 LIR (**a**), AS_28 LIR (**b**), and ANK3 LIR (**c**) against GABARAPL1. The upper panel depicts the thermograms of each binding event. Integrated heat data fitted using a one-site independent binding model along with the calculated thermodynamic parameters is shown

analysis. The spacing of the injections has to be large enough to avoid that binding heat peaks of each injection overlap. During our method optimization, the standard 300 s injection spacing was reduced to 250 s to shorten the total duration of the experiment. The number of injections is typically chosen to allow for the collection of enough data points to accurately fit the binding isotherm, while also generating an acceptable heat signal per injection (usually up to 30 injections).

The concentrations of macromolecule and ligand in each experiment depend on the extent of heat change upon binding and the anticipated binding constant. For a 1:1 binding reaction, the molar ratio (ligand to protein) at the final titration step must be at least 2. Using an injection volume of 8 μL and 30 injections, the neutralization point is reached after half of all injections (Fig. 2). This allows to collect enough data points at saturation and therefore enables a more reliable fit of the data. If the binding constant is already known, the concentration of the titrand (usually macromolecule in the sample cell) can be adjusted to set the $c$ value of the current titration inside the recommended experimental window of 5–500. If the binding constant is completely unknown and the stoichiometry $n$ equals one, use a molar ratio of 1:10 (titrand to titrant). If the $c$ value falls below 5, a ratio of 1:20 is recommended, and a sufficient saturation of at least 80% should be reached at the end of the titration (*see* **Note 5**).

The apparent heat recorded over time $Q_{app}$ is not only dependent on the heat of ligand binding to the macromolecule but also on the heats of dilution of the ligand $Q_{dil,\ L}$ and the macromolecule $Q_{dil,\ M}$. Additional nonspecific heat effects $Q_{ns}$ could contribute as well. Therefore, the corrected heat is given by:

$$Q_{corr} = Q_{app} - Q_{dil,L} - Q_{dil,M} - Q_{ns}$$

To account for these additive effects, blank titrations should be employed. $Q_{dil,\ L}$ can be assessed by titrating the ligand into buffer without the macromolecule, which usually also incorporates $Q_{ns}$. The same applies to $Q_{dil,\ M}$, whose values are generally too small to be considered. Due to this, at least one blank titration for the heat of dilution of the titrant is recommended [22]. In order to reduce the heat of dilution caused by mixing the buffers of the macromolecule and the ligand, both should be identical in composition (*see* **Note 6**). This may be achieved either by dialysis of both components in the same buffer (the dialysate may also be used to dissolve/dilute the ligand) or by multiple runs of buffer exchange with spin filters or buffer exchange columns.

*3.2.2 Running the Experiment*

The titrations of three peptides, p62 LIR, AS_28 LIR, and the LIR motif of ankyrin 3 (ANK3 LIR), will be discussed to provide an overview of how to conduct an incremental titration ITC experiment. AS_28 LIR is an artificial GABARAP-selective peptide discovered in a phage display against human Atg8 isoforms [23]. All peptides were titrated into buffer containing isolated GABARAPL1. For further details on the experimental parameters, please refer to Table 1. The herein presented method also applies to measurements of the other 5 homologs and their respective ligands.

All purified proteins should be equilibrated with ITC buffer. For proteins containing cysteine residues, such as LC3A or

**Table 1**
**Summary of the experimental parameters employed for the three measurements**

| | p62 LIR | AS_28 LIR | ANK3 LIR |
|---|---|---|---|
| Conc. peptide [μM] | 500 | 450 | 240 |
| Conc. protein [μM] | 25 | 30 | 30 |
| Ratio | 1:20 | 1:15 | 1:8 |
| Stir rate | 350 RPM | | |
| Temperature | 25 °C | | |
| # injections, spacing | 30, 250 s | | |
| Injection volume | 4 μL first, 8 μL others | | |

GABARAPL2, 0.5 mM of TCEP should be added. Buffer exchange was conducted via spin columns, with 5 consecutive runs of 10 min at 4 °C and 4000 *g*. For ligands dissolved in DMSO (here, p62 LIR and AS_28 LIR), the same relative amount of DMSO has to be added to the protein sample. Even small deviations in DMSO content between titrant and titrand cause large heats of dilution. Based on previous experiments, the human Atg8 homologs were shown to be stable at DMSO concentrations up to 5%.

1. To obtain high-quality data, samples should be degassed to avoid the formation of bubbles within the cell during the experiment, leading to unpredictable thermal signals.
2. Clean the cell with a suitable cleaning solution based on previous experiments (*see* **Note 7**). If a well-behaved system was studied (no apparent precipitation), rinsing with deionized water is sufficient. After aggregation or the use of immiscible organic solvents, it is necessary to employ detergents. The degassed water within the reference cell also has to be exchanged regularly.
3. Load the sample cell with the appropriate amount of overfill and load the syringe with the titrant as well. Overfill is an excess of sample volume over the volume of the sample cell. This improves the equilibration and the signal-to-noise ratio as much of the instrument signal is collected and regulated above the sample cell. To remove any remaining titrant from the syringe, rinse the needle with sample buffer and gently dry the tip using a lint-free tissue. Place the syringe in the sample cell and set the stirring speed to the desired level to allow the system to equilibrate. This is done by monitoring the differential power signal while waiting for the curve to reach a steady state (some devices support auto-equilibration).
4. Begin the experiment by collecting a baseline of at least 60 s. Observe the initial two injections to ensure sufficient spacing and modify the scheduled injections if necessary.
5. After completion of the titration, remove the syringe and clean the sample cell as mentioned above.

If the initial titration is successful, the blank titration of the ligand into the buffer must be performed identically.

#### 3.2.3 Data Analysis

Determination of $K_a$, $\Delta H$, and $n$ is achieved by nonlinear regression analysis of the experimentally determined data points. The algorithms in use and their implementation might differ, but in general, for a one-site binding model, the basis is to find the global minimum of an error surface of $K_a$ and $\Delta H$. To yield the apparent heat change for the $i$-th injection $\Delta Q_{i,\ app}$, the differential power signal for each peak is integrated over time. When the affinity of a ligand to the macromolecule is high, defined by a large $K_a$ or small $K_d$, nearly all the ligand molecules added in the beginning of the titration experiment bind to the macromolecule. This only results in minor changes of $\Delta Q_{i,\ app}$ for the first injections recorded. As the titration approaches the neutralization point, the absolute heat change decreases until it reaches saturation. At this point, the heat change per injection remains relatively constant. The one-site binding model assumes $n$ identical independent binding sites, and $K_a$ can be expressed in terms of Θ, the fraction of sites occupied, with [L] being the free concentration of the ligand:

$$K_a = \frac{\Theta}{(1 - \Theta)[\mathrm{L}]}$$

The total ligand concentration $[\mathrm{L}]_t$ can be expressed in terms of the known total concentration of the macromolecule $[\mathrm{M}]_t$:

$$[\mathrm{L}]_t = [L] + n\Theta[\mathrm{M}]_t$$

Combining the two prior equations leads to a quadratic equation shown below that can be solved for Θ, which is not directly accessible given the experimental setup:

$$\Theta^2 - \Theta\left[1 + \frac{[\mathrm{L}]_t}{n[\mathrm{M}]_t} + \frac{1}{n[\mathrm{M}]_t K_a}\right] + \frac{[\mathrm{L}]_t}{n[\mathrm{M}]_t} = 0$$

The total heat produced during the titration is given by:

$$Q = n\Theta \mathrm{M}_t \Delta H V_0$$

$V_0$ is the active volume of the sample cell that produces the heat signal. Solving the quadratic equation for Θ and substituting it into the previous equation leads to:

$$Q = \frac{n\mathrm{M}_t \Delta H V_0}{2}$$

$$\left[1 + \frac{[\mathrm{L}]_t}{n[\mathrm{M}]_t} + \frac{1}{n[\mathrm{M}]_t K_a} - \sqrt{\left(1 + \frac{[\mathrm{L}]_t}{n[\mathrm{M}]_t} + \frac{1}{n[\mathrm{M}]_t K_a}\right)^2 - \frac{4[\mathrm{L}]_t}{n[\mathrm{M}]_t}}\right]$$

Expressing $Q$ in terms of a change in heat per injection $\Delta Q_i$ is necessary to then apply the fitting procedure to the model.

$$\Delta Q_i = Q_i - Q_{i-1}$$

In order to account for the volume being displaced from the sample cell for each injection, corrections are employed for the total concentration of macromolecule and ligand. The model is then employed to fit the data points by initializing the parameters $K_a$, $\Delta H$, and $n$ (*see* **Note 8**), calculating $\Delta Q_i$ for each injection, and then refining them iteratively according to an error function, which describes the quality of the fit given the experimental data [22, 24, 25].

To conclude, after obtaining the experimental data, the analysis workflow consists of:

1. Adjusting the baseline to the peaks and defining appropriate windows for the numerical integration of the peaks.
2. The experimental parameters such as employed concentrations, injection volumes, sample cell volume, and temperature are either fetched from the files containing the raw data or entered manually (*see* **Note 9**).
3. Blank titrations are subtracted to correct the measured $Q_{app}$.
4. The correction for the dilution of the samples is usually done automatically within the available analysis software.
5. Select the appropriate model and fit the data in the case of LC3/GABARAP proteins, usually the one-site binding model. For further information on multiple binding sites, sequential binding, or displacement titrations, please refer to the literature [18, 26, 27].

## 4 Notes

1. In case of encountering protein aggregation during the experiment, this can be addressed by either altering the buffer system in use or reducing the stirring velocity, while guaranteeing the appropriate mixing of the sample. Crosslinking between proteins, due to disulfide bond formation, may also be a contributing factor which can be resolved through the addition of TCEP or 2-Mercaptoethanol. It is recommended to avoid the use of (2S,3S)-1,4-Bis(sulfanyl)butane-2,3-diol/DTT, because its ring-closing reaction generates significant heats. Another possible solution would be to conduct a reverse titration by adding the protein to the syringe and the low-solubility ligand to the sample cell. Generally speaking, the component of lower solubility should be placed into the sample cell.
2. Consider the amount of LB agar in relation to the number of plates required and the temperature tolerance of the antibiotics used (temperate agar at 60 °C before adding antibiotics). *See* **Note 2** in the corresponding Chap. 17 (Schwalm et al).
3. Add the protease inhibitor together with a few flakes of DNase shortly before mixing the buffer with the bacteria. Ensure that the protease inhibitor tablet has dissolved completely. If DNase is not available, a polyethylenimine/PEI solution can be used to remove excess free DNA *after* lysis.
4. After adding TEMED as the final ingredient, pour the gel solution quickly into the assembled plates. Smooth the surface of the gel and remove any bubbles by overlaying it with

isopropanol. Polymerization takes up to 30 min and can be assessed by observing the consistency of the remaining gel solution. Once the resolving gel is fully cured, remove the isopropanol and prepare the stacking gel. Place the stacking gel on top of the resolving gel and insert the combs just before the gel overflows. Allow the gel to set for an additional 30 min.

5. In case of missing saturation at the end of the titration, increase the ratio of the concentrations of ligand to macromolecule. If the rate of $\Delta Q_{i,\ app}$ decreases too rapidly, it may become necessary to reduce the molar ratio.
6. If no well-defined slope is apparent, even though the generated heat is sufficient, this might be due to a buffer mismatch between the sample cell and syringe material. Ensure that the sample has been prepared properly and every constituent apart from the analytes is equally abundant in both buffers. To further reduce nonspecific heats due to friction, keep additives with high viscosity such as glycerol at low concentrations. If the binding event is accompanied by an exchange of protons, the heat of protonation is recorded as well. The number of protons released per binding event is experimentally accessible [28]. 2-Amino-2-(hydroxymethyl)propane-1,3-diol/TRIS, for example, has a high enthalpy of protonation. By replacing it with another buffer system, heat effects that are due to protonation differences will be reduced.
7. The calorimeters and syringes used need to be cleaned thoroughly to reduce artifacts and noisy baselines (especially in experiments in which one of the binding partners precipitated). Refer to the manufacturer's manual for the compatibility of cleaning agents with the sample cells and the possible cleaning procedures.
8. Unexpected discrepancies in stoichiometry may result from imprecise concentration measurements provided as data inputs in the analysis software, solubility limits, or significant impurities in protein or ligand samples, partial unfolding of the macromolecule, or the selection of an inappropriate binding model.
9. Ensure that the cell volume, injection volume, and heat signal are properly calibrated; otherwise, the experimental data may be unreliable [29].

## Acknowledgments

MPS, JD, SK, and VVR are grateful for support by the Structural Genomics Consortium (SGC), a registered charity (no: 1097737) that receives funds from Bayer AG, Boehringer Ingelheim, Bristol

Myers Squibb, Genentech, Genome Canada through Ontario Genomics Institute, EU/EFPIA/OICR/McGill/KTH/Diamond Innovative Medicines Initiative 2 Joint Undertaking [EUbOPEN grant 875510], Janssen, Merck KGaA, Pfizer, and Takeda and by the German Cancer Research Center DKTK and the Frankfurt Cancer Institute (FCI). MPS is funded by the Deutsche Forschungsgemeinschaft (DFG, German Research Foundation), CRC1430 (Project-ID 424228829).

## References

1. Yang Z, Klionsky DJ (2010) Eaten alive: a history of macroautophagy. Nat Cell Biol 12(9):814–822. https://doi.org/10.1038/ncb0910-814
2. Kirkin V, Rogov VV (2019) A diversity of selective autophagy receptors determines the specificity of the autophagy pathway. Mol Cell 76(2):268–285. https://doi.org/10.1016/j.molcel.2019.09.005
3. Ohsumi Y (2014) Historical landmarks of autophagy research. Cell Res 24(1):9–23. https://doi.org/10.1038/cr.2013.169
4. Johansen T, Lamark T (2020) Selective autophagy: ATG8 family proteins, LIR motifs and cargo receptors. J Mol Biol 432(1):80–103. https://doi.org/10.1016/j.jmb.2019.07.016
5. Noda NN, Ohsumi Y, Inagaki F (2010) Atg8-family interacting motif crucial for selective autophagy. FEBS Lett 584(7):1379–1385. https://doi.org/10.1016/j.febslet.2010.01.018
6. Wesch N, Kirkin V, Rogov VV (2020) Atg8-family proteins-structural features and molecular interactions in autophagy and beyond. Cells 9(9). https://doi.org/10.3390/cells9092008
7. Martens S, Fracchiolla D (2020) Activation and targeting of ATG8 protein lipidation. Cell Discov 6:23. https://doi.org/10.1038/s41421-020-0155-1
8. Li Z, Wang C, Wang Z et al (2019) Allele-selective lowering of mutant HTT protein by HTT-LC3 linker compounds. Nature 575(7781):203–209. https://doi.org/10.1038/s41586-019-1722-1
9. Schwalm MP, Dopfer J, Kumar A et al (2023) Targeting LC3/GABARAP for degrader development and autophagy modulation. bioRxiv:2023.2010. 2005.560930
10. Paudel RR, Lu D, Roy Chowdhury S et al (2023) Targeted protein degradation via lysosomes. Biochemistry 62(3):564–579. https://doi.org/10.1021/acs.biochem.2c00310
11. Pierce MM, Raman CS, Nall BT (1999) Isothermal titration calorimetry of protein-protein interactions. Methods 19(2):213–221. https://doi.org/10.1006/meth.1999.0852
12. Ladbury JE, Chowdhry BZ (1996) Sensing the heat: the application of isothermal titration calorimetry to thermodynamic studies of biomolecular interactions. Chem Biol 3(10):791–801. https://doi.org/10.1016/s1074-5521(96)90063-0
13. Du X, Li Y, Xia YL et al (2016) Insights into protein-ligand interactions: mechanisms, models, and methods. Int J Mol Sci 17(2). https://doi.org/10.3390/ijms17020144
14. Wiseman T, Williston S, Brandts JF et al (1989) Rapid measurement of binding constants and heats of binding using a new titration calorimeter. Anal Biochem 179(1):131–137. https://doi.org/10.1016/0003-2697(89)90213-3
15. Turnbull WB, Daranas AH (2003) On the value of c: can low affinity systems be studied by isothermal titration calorimetry? J Am Chem Soc 125(48):14859–14866. https://doi.org/10.1021/ja036166s
16. Doyle ML, Hensley P (1998) Tight ligand binding affinities determined from thermodynamic linkage to temperature by titration calorimetry. Methods Enzymol 295:88–99. https://doi.org/10.1016/s0076-6879(98)95036-4
17. Velazquez-Campoy A, Freire E (2006) Isothermal titration calorimetry to determine association constants for high-affinity ligands. Nat Protoc 1(1):186–191. https://doi.org/10.1038/nprot.2006.28
18. Sigurskjold BW (2000) Exact analysis of competition ligand binding by displacement isothermal titration calorimetry. Anal Biochem 277(2):260–266. https://doi.org/10.1006/abio.1999.4402
19. Li J, Zhu R, Chen K et al (2018) Potent and specific Atg8-targeting autophagy inhibitory peptides from giant ankyrins. Nat Chem Biol 14(8):778–787. https://doi.org/10.1038/s41589-018-0082-8

20. Ruhmann E, Betz M, Fricke M et al (2015) Thermodynamic signatures of fragment binding: validation of direct versus displacement ITC titrations. Biochim Biophys Acta 1850(4):647–656. https://doi.org/10.1016/j.bbagen.2014.12.007
21. Rogov VV, Rozenknop A, Rogova NY et al (2012) A universal expression tag for structural and functional studies of proteins. Chembiochem 13(7):959–963. https://doi.org/10.1002/cbic.201200045
22. Freyer MW, Lewis EA (2008) Isothermal titration calorimetry: experimental design, data analysis, and probing macromolecule/ligand binding and kinetic interactions. Methods Cell Biol 84:79–113. https://doi.org/10.1016/S0091-679X(07)84004-0
23. Stolz A, Putyrski M, Kutle I et al (2017) Fluorescence-based ATG8 sensors monitor localization and function of LC3/GABARAP proteins. EMBO J 36(4):549–564. https://doi.org/10.15252/embj.201695063
24. Jelesarov I, Bosshard HR (1999) Isothermal titration calorimetry and differential scanning calorimetry as complementary tools to investigate the energetics of biomolecular recognition. J Mol Recognit 12(1):3–18. https://doi.org/10.1002/(SICI)1099-1352(199901/02)12:1<3::AID-JMR441>3.0.CO;2-6
25. Velazquez-Campoy A, Ohtaka H, Nezami A et al (2004) Isothermal titration calorimetry. Curr Protoc Cell Biol Chapter 17:Unit 17–18 https://doi.org/10.1002/0471143030.cb1708s23
26. Zhang YL, Zhang ZY (1998) Low-affinity binding determined by titration calorimetry using a high-affinity coupling ligand: a thermodynamic study of ligand binding to protein tyrosine phosphatase 1B. Anal Biochem 261(2):139–148. https://doi.org/10.1006/abio.1998.2738
27. Brautigam CA (2015) Fitting two- and three-site binding models to isothermal titration calorimetric data. Methods 76:124–136. https://doi.org/10.1016/j.ymeth.2014.11.018
28. Coussens NP, Schuck P, Zhao H (2012) Strategies for assessing proton linkage to bimolecular interactions by global analysis of isothermal titration calorimetry data. J Chem Thermodyn 52:95–107. https://doi.org/10.1016/j.jct.2012.02.008
29. Demarse NA, Quinn CF, Eggett DL et al (2011) Calibration of nanowatt isothermal titration calorimeters with overflow reaction vessels. Anal Biochem 417(2):247–255. https://doi.org/10.1016/j.ab.2011.06.014

# Chapter 19

# Determination of hATG8 Binding Selectivity of AIM (Autophagy-Interacting Motif) Peptides Using Native Electrospray Ionization Mass Spectrometry

**Andrew Brennan, Neil Oldham, Robert Layfield, and Daniel Scott**


## Abstract

Establishing the hATG8 binding selectivity of AIM (autophagy-interacting motif) sequences found within autophagy system proteins provides insights into their biological roles, and in the case of disease-associated AIM mutations, potential pathophysiological mechanisms. Given the sometimes small differences in affinity for an individual AIM amongst the six hATG8 proteins, establishing AIM preferences can be experimentally challenging. We describe a native mass spectrometry method that is suitable for detecting such differences, using synthetic AIM peptides and recombinant hATG8 proteins, to probe hATG8-AIM interactions in the gas phase. Binding preferences of a single AIM peptide against multiple hATG8s, or two AIM peptides against a single hATG8 (e.g., wild-type versus mutant AIM), may be determined.


**Key words** hATG8, Autophagy-interacting motif, Native mass spectrometry, Binding selectivity

## 1 Introduction

The six human autophagy-related 8 (hATG8) proteins can be divided into the LC3 (LC3A, LC3B, and LC3C) and GABARAP (GABARAP, GABARAPL1, and GABARAPL2) subfamilies. These hATG8s are posttranslationally lipidated and anchored in the autophagosomal membrane, where they recruit interacting partners such as autophagy receptors and associated cargoes to developing autophagosomes [1]. Specific biological roles for hATG8 family members are supported by gene knockdown experiments in cultured cells, indicating differences in their binding partners [2, 3].

hATG8 proteins recruit autophagy receptors and adapters via short consensus sequences known as the ATG8-interacting motifs (AIM) with the sequence [$\theta$]-$x_1$-$x_2$-[$\tau$], where $\theta$ is an aromatic amino acid and $\tau$ is hydrophobic. Although the hATG8 proteins exhibit a high degree of structural similarity, there is significant

Ioannis P. Nezis (ed.), *Selective Autophagy: Methods and Protocols*, Methods in Molecular Biology, vol. 2845,
https://doi.org/10.1007/978-1-0716-4067-8_19,

diversity in their sequences (as low as 31% similarity between GABARAP and LC3A) and subsequently they present differing binding surfaces for interaction with different AIM motifs. Indeed, improvements in understanding of AIM recognition of hATG8s now distinguish LC3-interaction and GABARAP-interaction motifs [4]. It is therefore important to study the binding of AIM sequences amongst the family of hATG8 proteins to understand mechanisms by which the processes of autophagy are mediated, with individual hATG8 proteins linked to specific functions by virtue of their varying interactions with a given AIM. These differences in binding affinity are often small (though amplified by avid binding in vivo) and hard to differentiate. We recently addressed this challenge utilizing recombinant hATG8 protein expression and synthesized AIM peptides in native mass spectrometry experiments, completing competition assays to determine binding preferences [5]. Specifically, we used the approach to demonstrate that an ALS-associated point mutation (L341V) mapping to the AIM (residues 332–343) of the SQSTM1/p62 autophagy receptor, although initially reported as a loss-of-function mutation with respect to LC3B binding [6], is in fact neomorphic, leading to incorrect recognition of other hATG8s [5]. We outline the methodology used to complete these experiments below, which should be broadly applicable to determining hATG8 binding preferences of other AIM sequences.

## 2 Materials

Standard molecular biology and protein production lab equipment are required, including incubators, autoclaves, centrifuges, spectrophotometers, gel electrophoresis apparatus, sonicators, liquid nitrogen, tube rollers, and a fast protein liquid chromatography (FPLC) system such as the AKTA Start (Cytiva). All buffers should be prepared in ultrapure deionized water.

### 2.1 Expression and Purification of hATG8 Proteins

1. hATG8 pGEX-4T-1 plasmids (available upon request).
2. *Escherichia coli* BL21 (DE3) (Novagen).
3. Sterile lysogeny broth (LB) containing ampicillin (100 μg/mL).
4. Sterile LB agar plates containing ampicillin (100 μg/mL).
5. 50 mL conical tubes.
6. Glycerol.
7. 1 M Isopropyl β-D-1-thiogalactopyranoside (IPTG).
8. Lysis buffer: 10 mM Tris–HCl, 150 mM NaCl, pH 7.5 (1 mM DTT for cysteine-containing proteins).

9. DNase I.
10. cOmplete™, EDTA-free Protease Inhibitor Cocktail.
11. Dithiothreitol (DTT).
12. Glutathione Sepharose 4B beads (Cytiva).
13. PD-10 column (Cytiva).
14. Cleavage buffer: 10 mM Tris–HCl, 150 mM NaCl, 25 mM $CaCl_2$, 1 mM DTT, pH 8.4.
15. Human α-thrombin (Invitrogen).
16. 5 mL HiTrap SP HP (Cytiva).
17. Ion Exchange buffer A: 5 mM potassium phosphate buffer, pH 7 (1 mM DTT for cysteine-containing proteins).
18. Ion Exchange buffer B: 5 mM potassium phosphate buffer, 1 M NaCl, pH 7 (1 mM DTT for cysteine-containing proteins).
19. 25 mM Ammonium acetate.
20. 5 × 5 mL HiTrap Desalt column (attaching 5 columns together improves separation).

### 2.2 Native Electrospray Ionization Mass Spectrometry (ESI-MS)

1. 25 mM Ammonium acetate.
2. Peptides can be purchased from a number of vendors or can be readily produced using standard solid-phase peptide synthesis methodologies. Twelve amino-acid peptides comprised of residues 332–343 of SQSTM1/p62 were used in our work with free N- and C-termini, supplied at >95% purity (Peptide synthetics).
3. Bruker Impact II mass spectrometer with an ESI source and quadrupole time-of-flight (qTOF) mass analyzer. Nano-ESI sources may also be used.
4. Compass acquisition and processing software.

## 3 Methods

### 3.1 Expression and Purification of hATG8 Proteins

#### 3.1.1 Expression of hATG8 Proteins

1. Transform ~100 ng pGEX-4T-1 plasmid containing the relevant hATG8 insert using heat shock into chemically competent *E. coli* BL21 (DE3) cells and plate on LB + ampicillin agar. Incubate overnight at 37 °C (*see* **Note 1**).
2. Inoculate a colony from the plate into 10 mL of LB + ampicillin and incubate overnight at 37 °C, 180 rpm shaking (*see* **Note 2**).
3. Add 500 μL of culture to sterile 50% glycerol, snap freeze in liquid nitrogen, and store at −80 °C for future use.

4. Inoculate the remaining overnight culture into 1 L of LB + ampicillin and incubate at 37 °C, 180 rpm shaking until the $OD_{600} = 0.7$.
5. Add IPTG to a final concentration of 1 mM and incubate overnight at 20 °C, 180 rpm shaking.
6. Harvest cells by centrifugation at 3000× *g* for 30 min.
7. Store the cell pellet at −80 °C.

*3.1.2 Purification of hATG8 Proteins (Fig. 1a)*

1. Thaw the cell pellet at room temperature before resuspending it in 4 mL of lysis buffer supplemented with 1 mg DNase I (100 μL of 10 mg/mL) and one protease inhibitor cocktail tablet in a 50 mL conical tube on ice.
2. To lyse the cells, place the tube containing the resuspended cells in an ice bath and clamp it in place with the sonicator probe submerged at the top of the sample.
3. Sonicate for 10 min with alternating 30-s sonication and 30-s rest intervals, avoiding excessive heating.
4. Clear the cell debris by centrifugation at 30,000× *g* for 30 min, then collect the supernatant and centrifuge again at 35,000× *g* for 30 min (*see* **Note 3**).
5. During centrifugation, add 1 mL of glutathione Sepharose 4B bead slurry to a PD-10 column and equilibrate it with 3 × 5 mL washes of lysis buffer by gravity elution.
6. Apply the clarified supernatant to the beads and incubate at 4 ° C on a tube roller at 30 rpm for 1 h.
7. Wash the beads with 5 mL of lysis buffer.
8. Equilibrate the protein-bound beads into cleavage buffer with 3 × 5 mL washes.
9. Add 15 units of human α-thrombin in 4 mL of cleavage buffer and incubate overnight at 4 °C on a tube roller at 30 rpm.
10. Elute the cleaved protein by gravity elution.
11. Equilibrate the HiTrap SP HP column with 25 mL of ion exchange buffer A at 3 mL/min on the FPLC system (*see* **Note 4**).
12. Dilute the 4 mL of cleaved protein to 40 mL in ion exchange buffer A (*see* **Note 5**).
13. Inject the protein sample onto the HiTrap SP HP column at 2 mL/min.
14. Wash with 25 mL of ion exchange buffer A at 3 mL/min.
15. Elute the protein using a linear gradient of 0–100% buffer B over 300 mL at 3 mL/min, collecting 10 mL fractions.

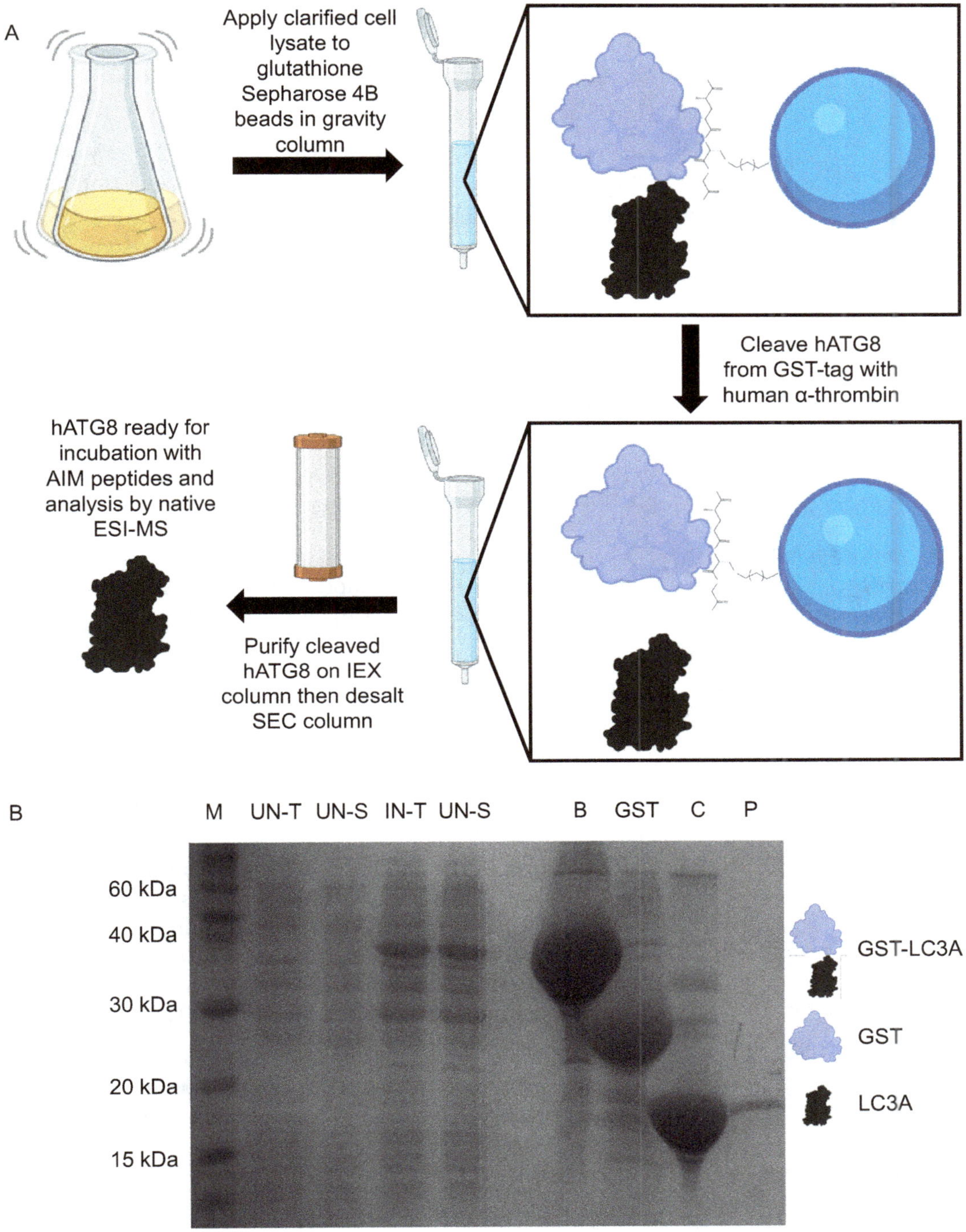


**Fig. 1** (**a**) Schematic of hATG8 protein expression and purification. Created with Biorender.com. (**b**) SDS-PAGE gel showing the purification of LC3A. No leaky expression of the protein is present in the uninduced total (UN-T) or soluble fractions (UN-S) but upon induction by IPTG, a band is observed corresponding to the GST-LC3A fusion. Equivalent band intensity in the induced total (IN-T) and soluble (IN-S) samples confirms soluble protein expression. After incubation with glutathione Sepharose 4B beads, a significant concentration of the desired fusion protein is observed (**b**). After thrombin cleavage, the GST tag is observed still bound to the

16. Analyze the fractions by SDS-PAGE to identify those containing the protein (Fig. 1b, *see* **Note 6**).
17. Confirm the hATG8 protein mass by MS (*see* **Note 7**).
18. Combine and freeze-dry the fractions.
19. Resuspend the freeze-dried protein in 2 mL of 25 mM ammonium acetate.
20. Equilibrate the HiTrap Desalt columns with 125 mL of 25 mM ammonium acetate at 3 mL/min on the FPLC system.
21. Inject the protein sample and flow 25 mM ammonium acetate at 2 mL/min.
22. Collect fractions manually by monitoring the 215 nm absorbance and conductivity as the protein will elute before the salt, resulting in a protein fraction, a salty protein fraction (where the peaks overlap), and a salt fraction.
23. Repeat lyophilization and resuspension in water of the protein fraction (3 cycles) to yield a pure desalted protein.

### 3.2 Native ESI-MS Selectivity Experiments

Prepare protein and peptide stocks at a concentration of 100 μM by resuspending the freeze-dried sample in 25 mM ammonium acetate (*see* **Note 8**). Two distinct experiment types are possible:

#### 3.2.1 Sample Preparation

1. Determination of the binding preference of a single hATG8 protein between two peptides, which allowed us to investigate the effect of an ALS mutant L341V peptide compared to the WT. In this case, prepare samples containing an equimolar mixture of the two AIM peptides (10 μM each, 20 μM total peptide concentration) and another set of samples containing the two peptides plus one hATG8 protein (20 μM) (*see* Fig. 2).
2. Determination of the binding preference of a single AIM peptide amongst a subfamily of hATG8 proteins (*see* **Note 9**). In this case, prepare samples containing an equimolar mixture of the LC3 or GABARAP subfamily proteins (10 μM each, 30 μM total protein concentration) and another set of samples containing the proteins plus one AIM peptide (30 μM) (*see* Fig. 3).

Incubate samples at room temperature for 30 min before analysis. It may also be useful to perform the measurements across a range of concentrations to observe dose-dependent effects.

#### 3.2.2 Data Collection

1. The instrument is controlled using Bruker Compass software.
2. Operate the mass spectrometer in positive ion mode under the following conditions: end plate offset, 500 V; capillary voltage,

**Fig. 1** (continued) beads along with significant impurities, and the cleaved protein is observed in the eluted fraction (**c**). After IEX chromatography, the pure protein is observed (P)

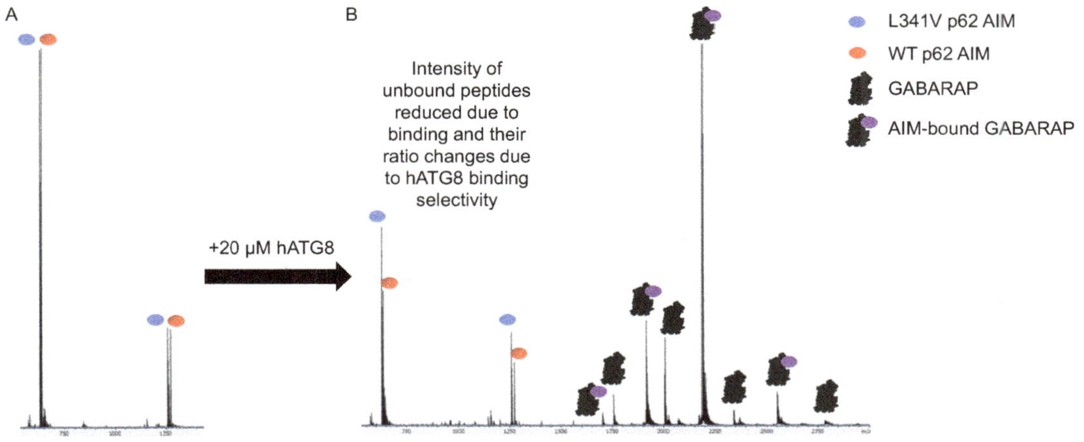


**Fig. 2** (**a**) Native ESI-MS spectrum showing a mixture of unbound WT p62 AIM and L341V p62 AIM peptides (10 μM each). (**b**) Native ESI-MS spectrum upon addition of GABARAP (20 μM) to the peptide mixture, which shows differential depletion of the unbound peptide signals due to binding selectivity. A mixture of bound and free hATG8 protein is also observed, but the two AIM-bound hATG8 signals could not be resolved. This shows a shift in the ratio of the unbound peptides upon addition of GABARAP, with the WT p62 AIM preferentially bound

3.5 kV; nebulizer pressure, 0.6 bar; dry gas pressure, 4 L/min; temperature, 150 °C.

3. Start data collection between 500 and 3000 m/z.
4. Inject the sample into the ESI-MS qTOF instrument sample chamber by direct infusion from a 500 μL Hamilton syringe, using a flow rate of 2 μL/min on a syringe pump.
5. Monitor data collection and ensure ion intensity is steady for at least 1 min.
6. Stop data collection, then stop the syringe pump.
7. To clean the ESI needle, directly infuse 25 mM ammonium acetate buffer until peptide/protein signals are no longer observed before injecting the next sample.
8. Collect each data point in triplicate.

#### 3.2.3 Data Analysis

1. Sum the intensities for the peaks corresponding to each protein, peptide, or bound species in a given spectrum.
2. Upon the addition of hATG8 protein to a mixture of two AIM peptides, the ratio of unbound peptide intensity should be determined, with the more depleted peptide being preferentially bound (*see* **Note 10**). For our work, we determined the ratio of unbound WT/L341V peptide to give a value between 0 and 1, with lower values indicating an increased preference for the WT and therefore a more significant effect of the mutation on binding to that hATG8 (*see* Fig. 2) [5].

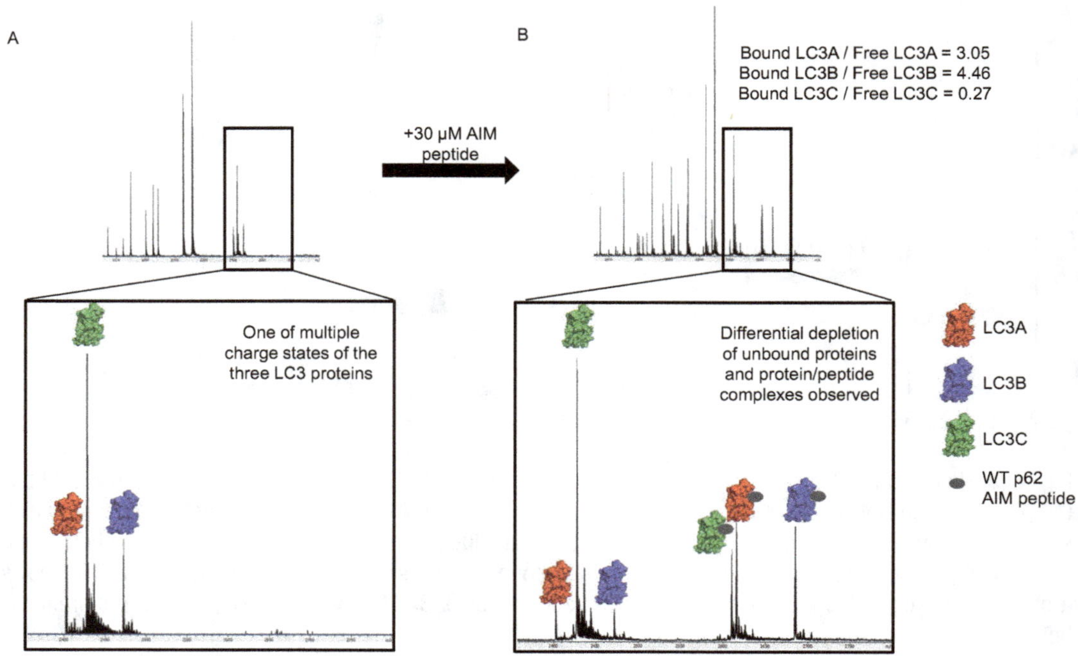


**Fig. 3** ESI-MS spectra showing the competition experiment for the LC3 subfamily binding to the WT p62 AIM peptide. (**a**) Sample containing the three LC3 subfamily proteins (10 μM each), which shows the proteins in their various charge states, with a highlighted section showing one charge state of each. (**b**) Sample containing the three LC3 subfamily proteins (10 μM each) and the WT p62 AIM peptide (30 μM), which shows the proteins and the protein–peptide complexes in their various charge states, with the same highlighted section showing the mixture of protein and protein/peptide complexes. The ratios for all the peak intensities corresponding to a bound protein can be summed and divided by the sum of the intensities of the free protein to give a measure of peptide binding in the mixture. This shows that LC3B is the preferred binding partner of the WT-AIM in this mixture and also indicates that most LC3C remains unbound

3. Upon the addition of an AIM peptide to a mixture of either of the two hATG8 subfamilies, the ratio of bound/free hATG8 protein intensities should be calculated for each protein in competition (*see* Fig. 3). The binding preference of the peptide can then be compared between peptides to determine changes in the binding equilibrium amongst the subfamily, such as the decrease in binding selectivity we observed due to the L341V mutation.
4. Use a *t*-test to determine *p*-values and therefore the significance of the shifts observed.

## 4 Notes

1. Cloning into pGEX-4T-1 results in the production of the desired protein with an N-terminal GST-tag, which we have

observed to increase protein yield and simplify the purification protocol compared to an untagged protein. The GST-tag cleavage with thrombin leaves an N-terminal Gly-Ser extension.

2. At this stage, it is advisable to pick multiple colonies and screen them for protein expression (typically 10 mL growths in conical tubes) by following the subsequent protocol on a small scale and analyzing by SDS-PAGE to identify the best colony to take forward, as variation is typically observed. Additionally, expression conditions may be optimized by further small-scale expression tests where the induction OD, temperature and time, and IPTG concentration may all be varied.

3. Samples of cell lysate before and after centrifugation should be collected for subsequent analysis by SDS-PAGE, which allows the amount of soluble and insoluble protein to be compared. If there is substantially more protein in the total sample (before centrifugation) than in the soluble sample (after centrifugation), then expression conditions should be optimized to minimize this (*see* Fig. 1).

4. The theoretically calculated isoelectric points of the hATG8 proteins are between 8 and 9.15, so they will bind to the sulfopropyl groups in the column in a buffer at pH 7. The isoelectric point may vary if a different protein construct or mutants are to be tested, so calculate this before proceeding as a different buffer pH (keep the buffer at least one pH unit away from the predicted isoelectric point) or potentially switching to a different column type may be required.

5. The conductivity of the sample can be measured as a proxy for salt concentration to confirm this is low enough to allow for protein binding to the column. The ion exchange buffer A has a conductivity of ~5 mS/cm, so ensure the sample is close to this.

6. Typically, this will yield a pure protein sample, but if there are impurities, a further size exclusion chromatography step can be used.

7. Typically, we will also analyze the protein samples using $^1$H NMR and circular dichroism spectroscopy to ensure folding and consistency across protein batches.

8. Protein concentration can be determined using the mass of desalted protein and subsequent calculation of the resuspension volume and confirmed using spectrophotometry at 280 nm.

9. Whilst it is possible to investigate all six hATG8 proteins in a single experiment, we observed peak overlap that prevented this analysis, making subfamily analysis preferable.

10. Depending on the resolution of the MS and the difference in mass between the two AIM peptides investigated it may also be possible to determine the ratio of bound:free peptide, but the ratio of unbound AIM peptide is sufficient to determine binding preference.

## References


1. Shpilka T, Weidberg H, Pietrokovski S, Elazar Z (2011) Atg8: an autophagy-related ubiquitin-like protein family. Genome Biol 12:226
2. Grunwald DS, Otto NM, Park JM et al (2020) GABARAPs and LC3s have opposite roles in regulating ULK1 for autophagy induction. Autophagy 16:600–614
3. Weidberg H, Shvets E, Shpilka T et al (2010) LC3 and GATE-16/GABARAP subfamilies are both essential yet act differently in autophagosome biogenesis. EMBO J 29:1792–1802
4. Rogov VV, Stolz A, Ravichandran AC et al (2017) Structural and functional analysis of the GABARAP interaction motif (GIM). EMBO Rep 18:1382–1396
5. Brennan A, Layfield R, Long J et al (2022) An ALS-associated variant of the autophagy receptor SQSTM1/p62 reprograms binding selectivity toward the autophagy-related hATG8 proteins. J Biol Chem 298:101514
6. Goode A, Butler K, Long J et al (2016) Defective recognition of LC3B by mutant SQSTM1/p62 implicates impairment of autophagy as a pathogenic mechanism in ALS-FTLD. Autophagy 12:1094–1104

# Index


## A

AlphaFold ............ 197, 198
ANC-1 ............ 67–70
*Arabidopsis thaliana* ............ 39
Atg8 family proteins ............ 16, 191, 197–200, 203
Autophagy ............ 2, 7, 13, 15, 27, 29, 31, 39, 40, 44, 47, 68, 69, 71, 79–81, 109, 110, 112, 115, 118, 120, 121, 141–143, 147, 162, 163, 191, 197, 204, 219, 220, 238
Autophagy interacting motif (AIM) ............ 39, 237–246
Autophagy-related 32 (Atg32) ............ 2, 3, 7, 80

## B

Binding selectivity ............ 237–246
Brains ............ 82–84, 86–91
Budding yeast ............ 1–13, 16

## C

*Caenorhabditis elegans* ............ 57, 58, 60, 61, 64, 67, 68, 71, 148
Cell sorter ............ 192–195
Cnb1 ............ 30, 31, 33

## D

Developments ............ 39, 47, 56, 57, 60, 61, 63, 64, 67, 75, 162, 165, 168, 188, 204
Dimerization ............ 27–36
DNA points accumulation for imaging in nanoscale topography (DNA-PAINT) ............ 128, 129, 131–133
*Drosophila* ............ 79–93, 143, 148

## E

ER-phagy ............ 70, 109–129, 132, 133, 138
Exchange-PAINT ............ 128

## F

Ferritin ............ 177–179, 183, 186, 187
Ferritinophagy ............ 177–188
FKBPs ............ 29–31, 33, 96, 104
FK506 ............ 29–32, 34–36
Fluorescence-activated particle sorting (FAPS) ............ 191–196
Fluorescence microscopy ............ 2, 3, 6–11, 13, 17–21, 31, 36, 68, 111, 183–186, 192, 195
Fluorescence polarization (FP) ............ 203–217
Flux ............ 43, 48, 69, 70, 80, 128, 156, 163, 168–173, 177–179, 181–184, 186–188
FTH1 ............ 177–179, 181–184, 186, 187

## G

GABARAPs ............ 203–217, 219–233, 237, 238, 242, 243

## H

hATG8 ............ 237–246
High-throughput screening ............ 212, 217

## I

Immunoblotting ............ 1–13, 17, 19, 21, 22, 80, 179, 181–183
Iron ............ 144, 147, 177–179, 181, 182, 186–188
Isothermal titration calorimetry (ITC) ............ 219–233

## K

$K_D$ determination ............ 210
$K_i$ determination ............ 206, 212
Keima-SKL ............ 143, 144

## L

LC3 ............ 16, 68, 80, 110, 120, 128, 142, 162, 203–217, 219–233, 237, 242, 244
LGG-1 ............ 69, 70, 75
Light microscopy ............ 40, 81, 82, 98
Liquid–liquid phase separation (LLPS) ............ 191
LIR motifs ............ 197, 200, 201, 204, 205, 220, 224, 228
Live cell imaging ............ 3, 13, 17, 97, 103, 142, 144, 145, 155, 163, 168–172



Ioannis P. Nezis (ed.), *Selective Autophagy: Methods and Protocols*, Methods in Molecular Biology, vol. 2845, https://doi.org/10.1007/978-1-0716-4067-8,

**M**

Mitochondria....1, 2, 7, 10, 39–41, 43–47, 49, 51, 55, 56, 61, 63, 79, 81, 82, 87, 92, 142, 147, 151, 152, 158, 160, 162
Mitophagy ....1–13, 39–41, 43–48, 51, 56, 57, 61–64, 79–93, 141–144, 147, 151, 153–155, 157, 158, 161–163, 173, 174
Mito-QC....81, 87–92, 142, 144, 162, 163
Mt-keima ....81, 87, 151–155, 158–160, 162–172
Muscles ....80, 82, 83, 85, 87, 90, 91

**N**

Native mass spectrometry....238
Nesprin ....68
Neurodegeneration....57, 81
Neurodevelopmental diseases....57, 64
Neurons ....56, 64, 86, 89, 92
*Nicotiana benthamiana*.... 41, 43, 44, 49
Nuclear receptor coactivator 4 (NCOA4)....177–179, 181–184, 186–188
Nucleophagy ....15–17, 19, 24, 68–71
Nucleus....15, 16, 67, 68, 133, 148

**P**

p62 ....96, 191–196, 205, 207, 212, 243
p62/SQSTM1 body ....238, 239
Parkin....56, 63, 80, 147, 151, 153, 156
Peroxisomes....141–144, 146, 148
Pexophagy ....141–144, 146–148
Pexo-QC....143, 144, 146, 148
PTEN-induced putative kinase 1 (PINK1) ....56, 63, 80, 147, 151

**Q**

Quantitative microscopy.... 10, 128

**R**

Rapamycin ....17–20, 22, 24, 29, 30, 32–36, 96
Reporters ....61, 68, 79–93, 110–117, 122, 126, 141–144, 146, 148, 151, 162, 163
Rosella biosensor....57, 61, 64

**S**

Screening ....17, 57, 64, 203–217
Selective autophagy....71, 95, 97, 127, 141, 143, 147, 161, 177, 191, 197, 220
Super resolution microscopy ....68, 92, 128
Synthetic tethering....28, 29, 35

**T**

Tris .... 178–180

**U**

Ubiquitinated proteins ....191

**V**

Vacuoles ....2, 7, 13, 15–17, 19, 20, 24, 28, 29, 40, 43, 44, 203

**Y**

Yeast ....2–6, 11, 13, 14, 17–22, 24, 29, 31–35, 59, 70, 72, 73, 80, 219

www.ingramcontent.com/pod-product-compliance
Lightning Source LLC
Chambersburg PA
CBHW080729130926
51338CB00015B/24

*9781071640661*